D1572987

Meteor Showers and Their Parent Comets

It is only in the past ten years that advanced computing techniques and painstaking observations have enabled the successful prediction and observation of meteor storms. Spectacular displays of "shooting stars" are created when the Earth crosses a meteoroid stream, causing the meteoroids to light up into meteors as they enter our atmosphere.

Meteor Showers and Their Parent Comets is a unique handbook for astronomers interested in observing meteor storms and outbursts. The author, a leading astronomer in the field and an active meteor storm chaser, explains how meteoroid streams originate from the decay of comets (and asteroids) and how they evolve into ever changing orbits by the gravitational pull of planets to cause meteor showers on Earth. He includes the findings of recent space missions that have visited comets and asteroids, the risk of meteoroid impacts on Earth, what showers to expect on other planets, and how meteor showers may have seeded the Earth with the ingredients that made life possible.

All known meteor showers are identified, accompanied by fascinating details on the most important showers and their parent comets. The book predicts when exceptional meteor showers will occur over the next 50 years, making it a valuable resource for both amateur and professional astronomers.

Astronomer PETER JENNISKENS completed his Ph.D. at Leiden University, the Netherlands, in 1992. He then worked as a National Research Council Associate at the Exobiology branch of the NASA Ames Research Center in Moffett Field, California, where he uncovered exotic properties of astrophysical ices, such as those in comets. Early in his studies, he became an amateur meteor astronomer with the Dutch Meteor Society. He has continued the study of meteor showers professionally at Ames and at the nearby SETI Institute, successfully predicting the α-Monocerotid meteor outburst in 1995. He went on to become the Principal Investigator of the NASA sponsored *Leonid Multi-Instrument Aircraft Campaign* that mobilized the scientific community to study 1998–2002 Leonid meteor storms. Amateurs continued to support his research. Dr Jenniskens is the chair of the Professional–Amateur Working Group of the IAU Commission 22 on meteoroids and interplanetary dust, and secretary of the IAU Commission 15 on the physical properties of minor bodies. In the course of writing this book, he identified the comet fragments remaining after the breakup that formed the meteoroid streams responsible for the Quadrantid and Phoenicid meteor showers, and in doing so he changed our ideas on how meteor showers predominantly originate.

METEOR SHOWERS AND THEIR PARENT COMETS

By Peter Jenniskens
The SETI Institute, California

CAMBRIDGE UNIVERSITY PRESS
Cambridge, New York, Melbourne, Madrid, Cape Town, Singapore, São Paulo

Cambridge University Press
The Edinburgh Building, Cambridge CB2 2RU, UK

Published in the United States of America by Cambridge University Press, New York

www.cambridge.org
Information on this title: www.cambridge.org/9780521853491

First published 2006

Printed in the United Kingdom at the University Press, Cambridge

A catalog record for this publication is available from the British Library

ISBN-13 978-0-521-85349-1 hardback
ISBN-10 0-521-85349-4 hardback

To my father,
Pierre Johan Jenniskens,

die altijd even bij me kwam staan
tijdens een waarneem aktie
totdat hij zelf ook een meteoor zag.

Contents

Preface

It was a warm summer evening in June in the light polluted Dutch city of Leiden in 1981 when I first sat down and gazed at the sky, waiting. A meteor appeared and I made a wish: "One more, please!" After 90 min, I had plotted four arrows on a chart of stars. That record still exists and has played a small role in the ongoing exploration of meteor showers. A very modest beginning to what has become a lifelong adventure.

In those days, we were resigned to the believe that our two most intense showers had no parent body, that meteor showers were as irregular as the weather (and more difficult to predict), that meteor storms came unannounced, and that this would always be so.

Today, we have reached an impressive milestone: about half of all large (>1 km sized) minor bodies approaching Earth's orbit have been discovered, two of which are the extinct comet nuclei that once produced the Geminid shower in December and the Quadrantid shower in January. The identification of the Quadrantid parent and several others were made in the course of writing this book.

Computers have revolutionized our insight into meteoroid stream dynamics. Meteor storm forecasting is now a reality. Over the years, amateur astronomers were witness to outbursts quite coincidentally. Now, storm chasing has become a popular pastime. In this book you will find much practical information about when to see meteor outbursts in the next 50 years and how they might manifest. We can look further into the future, but by 2050 the raw computing muscle of top-of-the-line computers is expected to have increased a million fold, at which time better predictions will surely be available than can be made now.

While writing this book, I found that many of our main meteor showers are the product of comet fragmentation. That new paradigm revives old ideas that had gone into submission after Fred Whipple proposed water vapor drag as the spring of meteoroid streams. If you are a professional astronomer, you will find in this book an overview of your work and that of colleagues who have helped illuminate the evolution of meteoroid streams, the physical properties of their parent bodies, their influx on Earth's atmosphere, their danger to satellites in orbit, and their role in the origin of life.

Acknowledgements

I have been fortunate to find a path in life that brings so much excitement from anticipation and surprise. While at NASA Ames and the SETI Institute, I have had the fortune of meeting many able researchers and program managers at NASA and the US Air Force, with open minds, who appreciate that meteors are a unique window on the universe around us and a door to both our past and our future.

Early on, *Hans Betlem* and *Rudolf Veltman* introduced me to the field. While observing, my father would stand by my side, and bear the cold just long enough to see at least one meteor. I found friends among members of the Dutch Meteor Society who were my teachers and guides, and who continued to support my work after I completed my studies at Leiden University and moved to NASA Ames Research Center and the SETI Institute. In the USA, I thank *Mike Koop* and members of the California Meteor Society for their unwavering support, and my partner in life *Charlie Hasselbach*, who smiled down on me and won my heart with meringue meteors of the sublime sort.

This book was written only because of the help of *Esko Lyytinen* and *Jérémie Vaubaillon*, who performed many numerical simulations. Others made contributions as well. *Bill Bottke* identified what might be extinct comet nuclei, *Peter Gural* calculated the visibility figures for future Moon impacts, *Giovanni Valsecchi* studied the link between 2003 EH_1 and comet C/1491 Y_1, *Emmanuel Jehin* observed 2003 EH_1, *Marco Fulle* studied the possibility of outbursts by ejection at aphelion, *Teemu Mäkinen* studied the water production rate of comet Tempel–Tuttle, and *Apostolos Cristou* calculated showers on other planets. *Brian Marsden* and *Daniel W. Green* of the Minor Planet Center provided comet light curve data and investigated several links between minor planets and meteoroid streams. *Joshua Kitchener* and copy editor *Louise Staples* assisted with the proof reading. Earlier versions of chapters were reviewed by *Sang-Hyeon Ahn*, *Josep Trigo*, *Iwan Williams*, and *Apostolos Cristou*. *Vladimir Porubčan*, head of the IAU Meteor Data Center, helped review the list of annual meteor showers.

For making other material available, I thank Shinsuke Abe, David Asher, Jack Baggaley, Hans Beltem, Nicholas Biver, Peter Brown, Donald Brownlee, Maurice Clark, Tony Cook, Gabriele Cremonese, Marco Fulle, Chet Gardner, Paul Gitto,

Peter Gural, Bo Gustafson, Robert Haas, Cathy Hall, David Harvey, Jane Houston-Jones, David Hughes, Vincent Icke, Eric James, Jim Jones, Hideki Kondo, Mike Koop, Marco Langbroek, Steven Lee, Marc de Lignie, Steve Lowry, Esko Lyytinen, Paul D. Maley, Keith Mason, Neil McBride, Dick McCrosky, Bob McMillan, Karen Meech, Wil Milan, Koen Miskotte, Rick Morales, Kenton Parker, Vladimir Porubčan, Tim Puckett, Jim Richardson, Galina Ryabova, Mikiya Sato, George Rossano, Ed Schilling, Seth Shostak, Chikara Shimoda, Pavel Spurný, Walter Steiger, Hans Stenbeak-Nielsen, Casper ter Kuile, Josep Trigo-Rodriguez, Jérémie Vaubaillon, Jun-Ichi Watanabe, Hal Weaver, Paul Wiegert, Iwan Williams, Jeff Wood, *NASA*, *ESO*, the astronomical journals of *MNRAS*, *A&A*, and many others. Ilkka Yrjölä and other participants in Global-MS-Net provided the radio-MS observations that are presented here. I thank *Harm Habing* of Leiden University, *Werner Pfau* of Jena University, and *Didier Despois* of the University of Bordeaux for their support and hospitality, and for creating the conditions needed to write this book. Over the years, I received support in my studies from the NASA Planetary Astronomy and Planetary Atmospheres programs and the NASA Astrobiology program, with local help from *David F. Blake*, *Mark Fonda*, *Greg Schmidt*, *Paul Wercinski*, and *Dave Jordan* at the NASA Ames Research Center, and especially from *Frank Drake*, *Chris Chyba*, *Tom Pierson*, *Debbie Kolyer*, *Sue Lehr*, *Hal Roey*, and *Brenda Simmons* of the dedicated staff from the SETI Institute. Thank You.

Part I
Introduction

1

How meteor showers were linked to comets

When we wish upon a falling star, we appeal to an ancient belief that the stars represent our souls and a meteor is one falling into the hereafter.[1] In Teutonic mythology, for example, your star was tied to heaven by a thread, spun by the hands of an old woman from the day of your birth, and when it snapped, the star fell and your life had ended.[2]

The Greek philosophers were the first to speculate on the nature of things without regard to ancient myths. Especially the world views of *Aristotle of Stagira* (384–322 BC) in his 350 BC book *Meteorology*[3] were widely quoted for over two thousand years, embraced by Christian religion, and passionately defended until into the eighteenth century. The Greeks held that all matter in the Universe is made of the elements "earth," "water," "air," and "fire." Aristotle was of the opinion that *shooting stars*, because of their rapid motion, occurred relatively nearby in the realm of the element "fire" above the layer of "air" that is now called our atmosphere. He believed that shooting stars were not caused by the falling of stars, but were caused by thin streams of a warm and dry "windy exhalation" (a mixture of the elements fire and air) that had risen from dry land warmed by the Sun. Those exhalations would rise above the moist parts of the atmosphere containing clouds (mixtures of "air" and "water"), into the realm of "fire." The more and the faster a thing moves, the more it is heated by friction and the more apt it is to catch fire. Hence, when the motion of the heavenly bodies stir the "fire," the exhalations can burst into flame at the point where they are most flammable. Once ignited, the flame would run along the path of the vapor and thus create a "torch" – what we now call either a *fireball* or a *bolide* (βολιδεσ) meaning "thrown spear."

Aristotle's peers and predecessors used the Greek adjective μετεωρον in its plural form to refer to all "atmospheric phenomena or anything in the heavens." It is the substantive use of the Greek μετεωροσ which means "raised," "lofty," or in a more

[1] E. Mozzani, *Le Livre des Superstitions – Mythes, Croyances et Légendes* (Paris: Bouquins, Robert Laffont, 1995), pp. 682–685.

[2] J. Grimm, *Deutsche Mythologie* (Berlin: Ferd. Duemmlers, 1876), p. 602.

[3] Aristotle (350 BC), *Meteorology*, book I, section 4, lines 32–34 (translation by E. W. Webster).

figurative sense, "sublime."[4] An eighteenth century meringue candy was called "meteors."

Meteor showers, Aristotle said, resulted from a very large exhalation that was scattered in small parts in many directions, when the hot "fire" element was squeezed from the cooling vapor like slippery fruit seeds pinched between one's fingers.

It is hard to picture Aristotle pinching his seeds and not knowing that meteor showers were radiating from a point in the sky (Fig. 1.1). But meteor showers were of no particular concern to Greek philosophers. Since Aristotle, meteor showers were considered part of our weather, a form of lightning. They were said to help sailors warn of upcoming storms.[5] For those less enlightened, meteor showers were either a good or a bad omen. The periodic meteor storm of April 3, 1095, for example, was mistaken by the Council at Clermont, France, *for a celestial monition that the Christians must precipitate themselves in like manner on the East*, when Pope Urban II called for the first crusades in November, 1095.[6]

The Leonid storm of 1833 changed all that and made meteor showers part of astronomy. It came at a time when *Isaac Newton*'s law of gravity had just been established. From that, it had been calculated how fast the Earth was moving around the Sun: with a speed of 30 kilometers per second (= km/s), or about 800 times the speed of a fast pitch in baseball. Even a small rock colliding with the Earth's atmosphere would find a violent end.

Meteor showers were now understood as being the result of *streams of meteoroids*, most no bigger than a grain of sand, approaching from one direction, before colliding with our atmosphere. Initially, this revelation created confidence that now all was understood, but predicting the return and activity of meteor showers proved to be as elusive as predicting the weather. In an age of rapidly expanding knowledge, many astronomers would start their career on a warm summer night during the Perseids, only soon to turn their attention to easier and more profitable problems such as Black Holes or the Age of the Universe.[7]

Only in the last ten years has the unyielding beast of a trillion particles finally been caged. We are not yet sure if all the bars will hold, but as in a zoo stocked for our pleasure, we now recognize a generous range of meteor shower manifestations, each providing clues about the minor planets at their source, which are nearly all comets.

4 J. A. Simpson and E. S. C. Weiner, *Oxford English Dictionary*, 2nd edn. 20 vols. (Oxford: Oxford University Press, 1989).

5 L. A. Seneca (AD 62), *Naturales Quaestiones*, book I, sections 1.1–12, 14.1–15.6, book 2, sections 55.2–3. Translated by Thomas H. Corcoran (Cambridge, MA: Harvard University Press and London: Heinemann, 1971).

6 J. W. Draper, *A History of the Intellectual Development of Europe* (New York: Harper Brothers, 1864); V. Clube and B. Napier, *The Cosmic Winter* (Oxford: Blackwell, 1990); A. McBeath, *WGN* **27** (1999), 318–326.

7 M. Beech, Meteor astronomy: a mature science? *Earth, Moon Planets* **43** (1988), 187–194; D. Hoffleit, From early sadness to happy old age. *Comments Astrophys.* **18** (1996), 207–221.

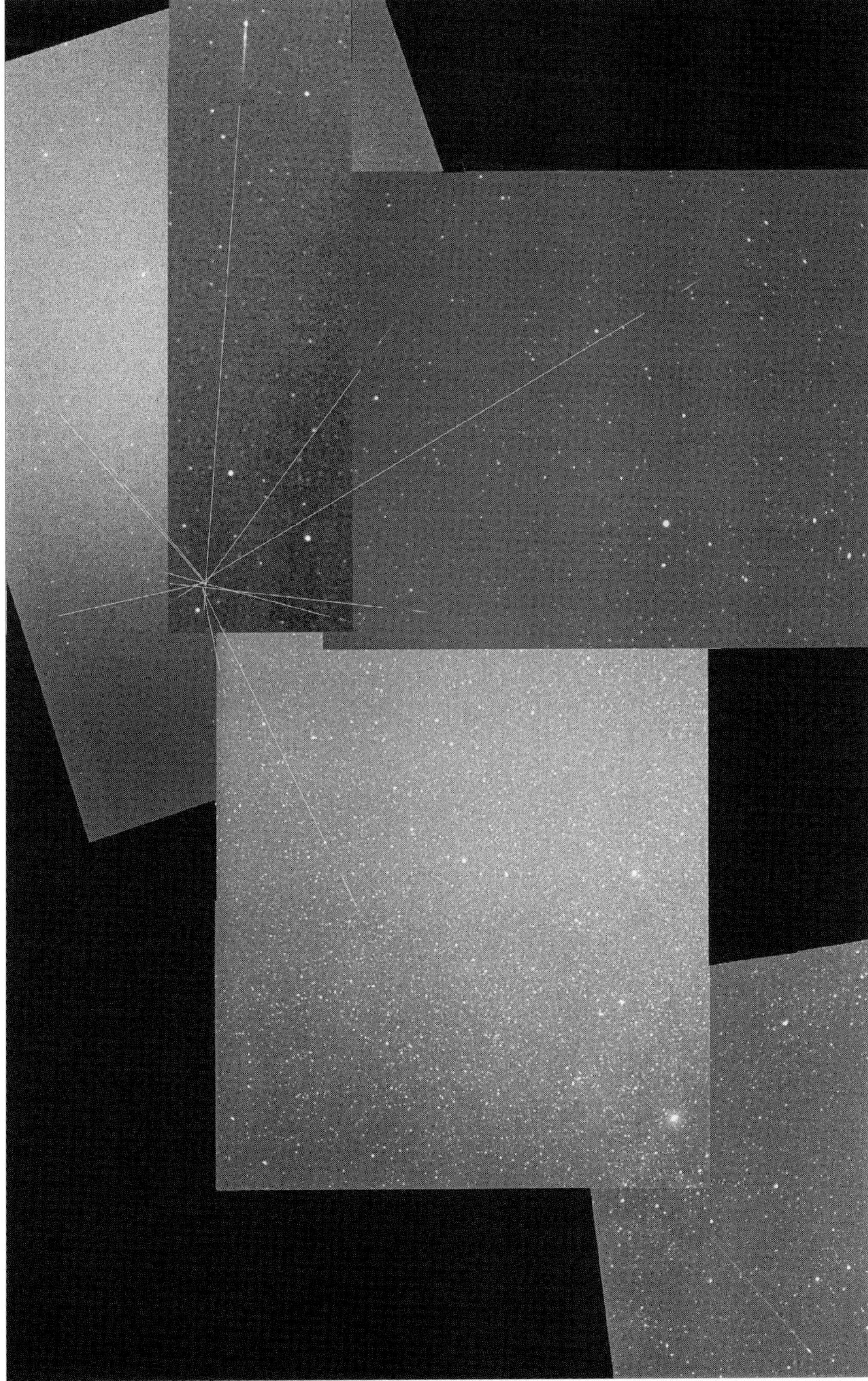

Fig. 1.1 The "radiant" of a meteor shower is the point in the sky from where the meteors appear to radiate, the head of Draco in this compilation of photographs of the 1985 Draconid outburst by members of the *Nippon Meteor Society*.

1.1 The quest to understand the nature of meteor showers

The first to keep careful records of meteor shower sightings were court-appointed astronomers in China, who were both time keepers and astrologers. Their motivation to do so was rooted in an eastern culture that considered its ruler "the emperor of all under heaven," the earthly counterpart of the heavenly god *Shang-ti*. The emperor maintained the harmony of Heaven and Earth by his actions in following the ritual and the prescripts of his forefathers precisely.[8] Any unrest in the sky was seen as a sign that something was amok with the emperor's rule. The astronomers at the royal court would gather such information from all over the empire. This included sightings of comets, fireballs, and meteor showers.

Meteor showers were known as periods of unusually high meteor rates. We now know that some repeat each year, called the *annual meteor showers*, and that there are also irregular showers called *meteor outbursts*. An example of meteor outbursts in recent years are those from the November Leonid showers. The rate in 1994, for example, was much higher than in 1995 ("Leo" in Fig. 1.2).

The oldest account linked to a modern shower is the exceptional *Lyrid* outburst of March 16, 687 BC (Julian calendar) during the Chou dynasty period, when it was written: "In the middle of the night, stars fell like rain." This account dates from more than two centuries before the philosopher *Confucius* (K'ung Fu-tze, 551–470 BC) and others like him transformed old ideas of knighthood into teachings of virtuous behavior as the basis of a good state.[9] We will explain later why this particular shower was seen so long ago.[10]

There are hundreds of such records in the Chinese, Japanese, and Korean literature. Table 1 gives a list of dated accounts prior to 1900, mostly compiled by Ishiro Hasegawa from Japan and Sang-Hyeon Ahn from Korea, building on work started in 1841 by *Edouard Biot*.[11] Table 1 also includes scattered references to clay tablets written in cuneiform script by the pre-Greek priest-astronomers of Mesopotamia from about 747 to 75 BC, who observed the Moon and planets for timekeeping and later also astrology, as well as references dating from the post-Greek Arabic Middle-East and from medieval Europe.

Most accounts are readily identified as the summer Perseids (Fig. 1.3), but many have no known present-day counterpart. Some are mere second-hand accounts of bright fireballs, or normal meteor activity seen in exceptionally clear nights (no Moon, no haze). The rest tell a story about meteor showers changing in time and place and about some very fortunate observers, now long forgotten.

[8] A. Pannekoek, *A History of Astronomy* (London: Allen and Unwin, 1961, New York: Dover, 1989).
[9] *Ibid.*
[10] C. P. Olivier, *Meteors*. (Baltimore, MD: Williams & Wilkins, 1925), p. 6. Olivier believes this account could have been a meteorite fall, from the alternative translation "there fell a star in the form of rain."
[11] M. Éd. Biot, *Catalogue Général des Étoiles Filantes et des Autres Météores Observés en Chine pendent 24 Siècles* (Paris: Imprimerie Royale, 1841).

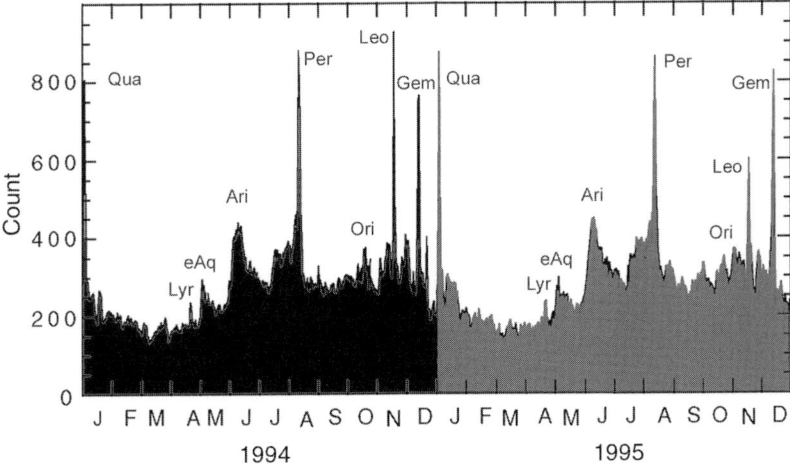

Fig. 1.2 Rate hikes in the daily count of meteors in the years 1994 and 1995, measured by Ilkka Yrjölä of Kuusankoski, Finland, by means of counting reflected radio signals from far away TV or radio stations. Note how the rates repeat year after year.[12]

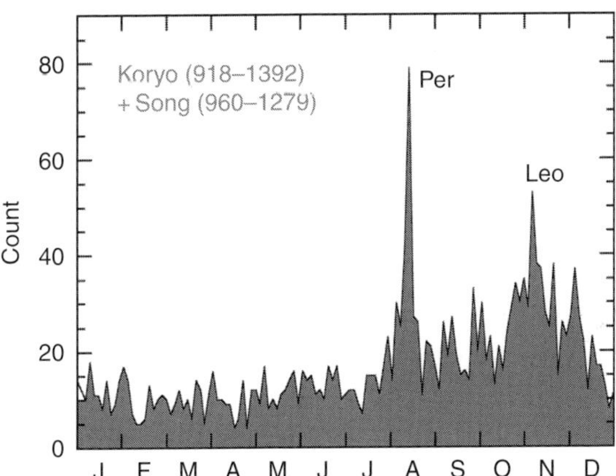

Fig. 1.3 Daily variations in meteor activity in the Middle Ages as reflected in the total daily number of shower reports from the Chinese Song and Korean Koryo dynasties, gathered by Sang-Hyeon Ahn.[13] Note the absence of the now prominent Quadrantid (Boo) and Geminid (Gem) showers.

Today, the most significant annual variations in meteor rates are due to the showers of *Quadrantids* (= Bootids) in early January, the *Lyrids* in April, the η-*Aquariids* in May (southern hemisphere), the δ-*Aquariids* in July, the *Perseids* in August, the *Orionids* in October, the *Taurids* and *Leonids* in November, and the *Geminids* in December. These

[12] I. Yrjölä and P. Jenniskens, Meteor stream activity VI. A survey of annual meteor activity by means of forward meteor scattering. *Astron. Astrophys.* **330** (1998), 739–752.

[13] S.-H. Ahn, Meteoric activities during the 11th century. *Mon. Not. R. Astron. Soc.* **358** (2004), 1105–1115; S.-H. Ahn, Meteors and showers a millennium ago. *Mon. Not. R. Astron. Soc.* **343** (2003), 1095–1100; S.-H. Ahn, Catalog of meteor showers and storms in Korean history. *J. Astron. Space Sci.* **21** (2004), 39–72.

showers are named after the constellation from where their meteoroids appear to approach us: Bootes, Lyra, Aquarius, Perseus, Orion, Taurus, Leo, and Gemini.

The discovery of the *radiant*, more than the periodic increase of rates, defines what is a *meteor shower*. That discovery was made in 1833, after elevated Leonid rates were first seen in 1831 and then a storm of Leonids was noticed by city guards in Europe on the night of November 12, 1832.[14] When the phenomenon repeated the next year, Professor *Denison Olmsted* (1791–1859)[15] at Yale College, "through the kindness of a friend, was awaked in season to witness the spectacle in much of its grandeur," the results of which were swiftly published in the *New Haven Daily Herald*. There were widespread reports of a radiant placed close to the star γ Leonis, stationary during the night.

Olmsted recognized that the radiant phenomenon was caused by bodies moving on parallel tracks entering Earth's atmosphere from the general direction of γ Leonis. Olmsted reached this conclusion based on the 1794 thesis by *Ernst Florens Friedrich Chladni* (1756–1827),[16] who had argued how meteors had to be caused by solid *meteoroids* entering Earth's atmosphere at high speed. Chladni wanted observers to measure the height of the meteors in the atmosphere by triangulation from simultaneous observations at two separated observing sites. In 1798, *Johann Benzenberg* (1777–1846) and *Heinrich Wilhelm Brande*s (1777–1834), students at the University of Göttingen, were encouraged by their professor (who collaborated with Chladni) to follow up, and they proved that meteors were higher than other weather phenomena and indeed had to move at astronomical speeds.

It was then remembered that, 33 years earlier, the famous German scientist and traveler *Alexander von Humboldt* on an expedition to south and middle America had seen, and described, a similar meteor storm in the early morning of November 12, 1799, while in Cumaná, Venezuela. We now know that the meteors peaked that year around 06:15 UT in a massive pile up of dust trails. Von Humboldt wrote that people old enough to remember recalled that the same phenomenon was seen about 30 years before. A pattern was recognized. During the research for this book, Jérémie Vaubaillon and the author set out to investigate this anecdote and we discovered that there was only one storm that season, and that storm happened to be visible from South America at 06:18 UT on November 9, 1771 under similar circumstances albeit not as intense as the later storm (Chapter 15).

The discovery of periodic Leonids and the phenomenon of the radiant quickly led to the discovery of other meteor showers. The January *Quadrantids* (1835) and the

[14] W. Olbers, Die Sternschnuppen. In *Jahrbuch für 1837*, ed. H. C. Schumacher. (Stuttgart: Cotta'schen Buchhandlung, 1837), pp. 36–64.

[15] D. Olmsted, Observations on the meteors of November 13th 1833. *Am. J. Sci. Arts* **25** (1834), 363–411; **26**, 132–174; A. C. Twining, *Am. J. Sci. Arts* **25** (1834), 320.

[16] E. F. F. Chladni, *Ueber Den Ursprung Der Von Pallas Gefundenen Und Anderen Aehnlichen Eisenmassen* (Riga: Hartknoch, 1794), 63 pp; E. F. F. Chladni, *Ueber Feuer Meteore Und Uber Die Mit Denselben Herabgefallenen Massen* (Wein: Heubner 1819), 424 pp.; M. Beech, The makings of meteor astronomy: part X. *WGN* **23** (1995), 135–140.

August *Perseids* (1835) were first made widely known by *Adolphe Quételet* in Brussels, founder of the Observatoire Royal de Bruxelles.[17] Quételet not only observed the Perseids, but found many earlier records, the oldest by the Dutch inventor of capacitance (the Leyden jar), the physicist Pieter (Petrus) van Musschenbroek (1692–1761),[18] who wrote in a publication that was printed in 1762: *Stellae (cadentae) potissimum mense Augusto post praegressum aestum trajici observantur, saltem ita in Belgio, Leydae et Ultrajecti.*[19]

In addition, a well-observed 1803 Lyrid outburst in the eastern United States led to the discovery of the weak annual April *Lyrid shower* in 1838 by *Edward Claudius Herrick* at Yale College,[20] to which, in October 1839, he added the discovery of the annual *Orionids*[21] (independently discovered also by Quételet[22] and Benzenberg). Johann Benzenberg[23] and *Eduard Heis* observed the Andromedids in 1838, following a 1798 sighting of an outburst by their colleague Brandes. Other major showers were not discovered until just after the next Leonid storm in 1866, which again raised interest in the topic of meteor showers.

For the next 150 years, visual meteor observations mostly concentrated on plotting meteors in search of new annual shower radiants. Best for that are *gnomonic star charts*, on which meteors move as straight lines and it is easily checked whether they radiate from a common circular area. British amateur astronomer *William Frederick Denning* of Bristol, witness of the 1866 Leonid storm at age 17, published thousands of such radiants at the turn of the century,[24] and several updates after that. He was so much respected as a meteor observer that the novelist H. G. Wells featured Denning as the "meteorite expert" (*sic*) in his 1898 *The War of the Worlds*. In 1935, the list was complimented with southern showers when New Zealander *Ronald Alexander McIntosh* published his *An Index to Southern Meteor Showers*.[25] Unfortunately, poorly drawn star charts and a common habit of accepting big circles for radiant association made many of these "showers" unreliable.

Better criteria were needed to recognize streams. This became possible in the mid-twentieth century when photographic and radar techniques first measured the atmospheric trajectory and speed of meteors and, from that, the orbit of the meteoroids in

[17] A. Quételet, *Correspond. Math. Phys. IX*, **184** (1837), 432–441; J. Sauval, Quételet and the discovery of the first meteor showers. *WGN* **25** (1997), 21–33.

[18] P. Van Musschenbroek, *Introductio ad Philosophiam Naturalem* (Lugdani Batavorum: Luchtman, 1762).

[19] Loosely translated: "Falling stars are observed in the middle of August more than at other times in the year given the rate of observed trails seen at least in such places as Belgium, Leyden, and Utrecht."

[20] E. Herrick, *Am. J. Sci. Arts* **34** (1838), 398; **35** (1839), 366; **36** (1840), 358.

[21] E. Herrick, *Am. J. Sci. Arts* **35** (1839), 366.

[22] A. Quételet, Catalogue des principales apparitions d'etoiles filantes. *Mém. l'Acad. Roy. Sci. Belles-Lett. Bruxelles* **12** (1839), 1–56.

[23] J. F. Benzenberg, *Die Sternschnuppen* (Hamburg: Perthes, 1839), 339 pp, p. 244 (Orionids), p. 331 (Andromedids).

[24] W. F. Denning, General catalogue of radiant points of showers and fireballs observed at more than one station. *Mem. R. Astron. Soc.* **53** (1899), 202–292; see also M. Beech, W. F. Denning – the doyen of amateur astronomers. *WGN* **26** (1998), 19–34.

[25] R. A. McIntosh, An index to southern meteor showers. *Mon. Not. R. Astron. Soc.* **95** (1935), 709–718; G. W. Wolf, Ronald Alexander McIntosh – not just a southern meteor pioneer. In *Proceedings IMC Belogradchik 1994* (Potsdam: International Meteor Organization, 1994), pp. 78–85.

space. New meteoroid streams were discovered now from their similar orbits. In one study, as much as 65% of all bright meteors were assigned to (mostly minor) meteor showers.[26]

Even with these tools, it continued to be a problem to recognize diffuse streams among the sporadics. This is especially the case for the imprecise orbits measured by radar in the past. Because different sets of sporadic meteoroid orbits were mixed in, and because showers were observed only intermittently, the same stream is often reported under a different name, creating much confusion about its identity. Many of the reported "streams" are groupings of meteoroids that do not originate from the same parent body.

1.2 Meteoroid streams as debris from comets

The association of meteor showers with comets was made only when it became clear how comets and meteoroids orbit the Sun. The first step was taken when observers of the 1833 Leonid storm, such as Olmsted, wanted to share their experiences and set out to predict the next Leonid storm. Olmsted recognized their periodic nature and suggested that clouds of meteoroids were moving in orbits around the Sun every six months, mistakenly attributing the 1803 *April Lyrid* outburst to the same repeating phenomenon responsible for the two spectacular Leonid storms of 1832 and 1833![27]

These ultra-short orbital periods tended to be believed, misled too by the discovery that some showers returned annually. From the now translated Chinese accounts, Herrick showed in 1837–38 that meteor showers were periodic on a sidereal rather than a tropical year.[28] When Quetelet raised once again the possibility of a link with the weather, mathematician *Hubert Anson Newton* of New Haven (in 1863) pointed out that the meteor showers did not come at the same time in the season. Unlike the weather, the Julian date of past Leonid storms had progressed by a month from October 13 in AD 902 to November 13 in 1833. During that time, the Earth's spin axis had gradually changed position. It completes a full circle every 25 792 years, a phenomenon called *precession*. As a result, the seasons fall progressively at a different position in Earth's orbit (the duration of a siderial and a tropical year differ by 1 day in 70.613 34 years). After taking this into account, Newton found that those Leonid storm dates nearly corresponded to the same position of Earth in its orbit.[29]

Not exactly to the same position, however. There was a remaining shift in the time of the peak, amounting to +29 min per orbit of 33.25 yr, which had to be on account of

[26] L. G. Jacchia and F. L. Whipple, Precise orbits of 413 photographic meteors. *Smithsonian Contrib. Astrophys.* **4** (1961), 97–129.

[27] D. Olmsted, Observations on the meteors of November 13th, 1833. *Am. J. Sci. Arts* **25** (1834), 363–411; **26**, 132–174; D. Olmsted, *Letters of Astronomy Addressed to a Lady* (New York: Harper & brothers, 1849), pp. 359–364.

[28] E. C. Herrick, *Am. J. Sci. Arts* **33** (1837), 176; **33** (1838), 354.

[29] H. A. Newton, Evidence of the cosmical origin of shooting stars derived from the dates of early star showers. *Am. J. Sci.* **36** (1863), 145–147; H. A. Newton, The original accounts of the displays in former times of the November Star-Shower. *Am. J. Sci.* **37** (1864), 377–389; **38**, 53–61; D. W. Hughes, The history of meteors and meteor showers. *Vistas Astron.* **26** (1982), 325–345.

other influences. Newton was also struggling with the periodicity of the returns. He favored periods of 354 d (1 − 1/33.25 yr); another suggestion was 375 d (1 + 1/33.25 yr), and another 33.25 yr. He predicted a return of the storms in 1866.

Astronomer John Couch Adams, better known for his role in the discovery of Neptune, later proved that only the last solution could be true. In April, 1867 Adams figured that the meteoroid orbits were also precessing and calculated that this +29 min/orbit was well matched by the expected combined effect in rotating the orbit from the gravitational pull by Jupiter (+20 min), Saturn (+7 min) and Uranus (+1 min), but only if the orbital period was the longer 33.25 yr. The proposed shorter orbits by Olmsted and Newton would not do.[30]

Before Adams made his arguments about the long orbital period of the Leonid shower, *Giovanni Virgínio Schiaparelli* (1835–1910) at Milan, of Mars *canali* fame, had found that most meteoroid orbits had to be very elongated. Mainly, because meteors were seen in the evening as well as morning hours in a numbers ratio of 1.4 ($=\sqrt{2}$), the ratio of speeds for meteoroids in circular and parabolic orbits. Shiaparelli concluded that meteoroids in general were moving on near-parabolic orbits. In a series of Italian papers that formed the basis of his 1866 book: *Note e riflessioni intorno alla teoria astronomica della stelle cadenti*,[31] he showed that the orbit of the Perseids, if nearly parabolic in shape, was very similar to *Theodor Ritter von Oppolzer*'s orbit for comet 1862 III (Swift–Tuttle).[32] Schiaparelli had discovered the source of the meteoroids.

Schiaparelli failed to find a comet for his Leonid orbit, because he used γ Leonis as the approximate position of the radiant, which was several degrees off. The first comet of 1866 (55P/Tempel–Tuttle) was recognized as the parent of the Leonid storms[33] shortly after *Urbain Jean Joseph Le Verrier* in France derived an orbit from a better radiant position in 1867.[34]

A third shower parent was identified in the metropolis of Vienna in 1867, when *Edmond Weiss*, looking for comets passing near Earth's orbit, found that the 1861 comet C/1861 G_1 (Thatcher) passed within 0.002 AU on April 20 and found evidence of an April Lyrid shower in the literature.[35] Later that year, *Johann Gottfried Galle* calculated the Lyrid orbit, assuming it was a parabola, and confirmed the association. He also first pointed to the Chinese account from 687 BC as a possible early Lyrid shower sighting.

It was now understood, given that a cloud of meteoroids from a distance would look like a comet, that comets and meteoroid streams, properly speaking, were identical.

[30] J. C. Adams, On the orbit of November meteors. *Mon. Not. R. Astron. Soc.* **27** (1867), 247–252.

[31] G. V. Schiaparelli, *Note e Riflessioni intorno Alla Teoria Astronomica delle Stelle Cadenti* (Firenze: Stamperia Reale, 1867), 132 pp. (Translated into German in 1871. *Entwurf einer astronomischen Theorie der Sternschnuppen*. Stettin: Th. V. d. Nahmer VIII, 268 pp, long the standard book on meteor astronomy.)

[32] M. J. V. Schiaparelli, Sur la relation qui existe entre les comètes et les étoiles filantes. *Astron. Nachr.* **68** (1967), 331.

[33] J. C. Adams, On the orbit of November meteors. *Mon. Not. R. Astron. Soc.* **27** (1867), 247–252.

[34] U. J. LeVerrier, *Comptes Rendus* **64** (1867), 94.

[35] E. Weiss, Bemerkungen über den Zusammenhang zwischen Cometen und Sternschnuppen. *Astron. Nachr.* **68** (1867), 381.

2

What is at the core of comets?

The most spectacular result of recent space missions to comets has been to show the spring of meteoroid streams, first when Giotto visited Halley in 1986. The return of comet Halley was highly anticipated. I was an undergraduate student of astronomy at *Leiden University* in the Netherlands and was invited to be a tour guide on a chartered DC-9 airplane to watch the comet above the usual deck of clouds. Two hundred people eager to see the scourge of legend sparkle in the sky paid $50 and were given six ten-minute laps over the North Sea, each time providing a new group a seat at the windows. I recall spending some extra time with an eyewitness of comet Halley's previous return in 1910. She had the gray hair and worn face of one outlasting Halley's 76 yr orbit. Sadly, her eyesight had suffered over the years and she never found the faint $+4^{m}$ fuzz of light in the constellation of Capricorn. She was thrilled nonetheless. This was her first time in a plane, and my first astronomical expedition.

The word *comet* comes from the Greek κομετεσ = "the hairy one." The Chinese astronomers called these objects *hui* or "broomstars" and tracked their position in the constellations, moving from one group of stars to the next over days or sometimes many weeks on account of their great distances. From a distance, these inferior planets of our solar system are fuzzy blobs, sometimes with a diffuse tail pointing away from the Sun. Prior to AD 1577, comets and shooting stars were all considered meteors (Fig. 2.1).[1] Even today, popular culture does not always make the correct distinction between comets, the minor planets in space, and the meteors caused by their debris impacting on Earth's atmosphere.

Danish astronomer *Tycho Brahe* (1546–1601) first proved that comets belong in the realm of astronomy by demonstrating from the lack of parallax between viewing the comet in the evening and the morning that the bright comet of 1577 was at least four times farther from Earth than the Moon. In 1610, amateur Sir William Lower proposed correctly that comets move in elongated ellipses, while Robert Hooke and Giovanni Borelli thought cometary orbits might be so elongated as to be barely open

[1] Illustration of Fig. 2.1 is from: A. M. Mallets, *Beschreibuing des ganzen Welt Kreisses* (Frankfurt am Main: Johann Adam Jung Verlag, 1719) (republished from: *Description de l'Universe* (Paris, 1684)). It illustrates the comment that comets, according to Apollonius, were considered part of the "wandering stars" by the Chaldeans (612–539 BC), who were the "New Babylonians" following the fall of the Assyrian empire.

Fig. 2.1 Comet types. A seventeenth century engraving by Mallets, depicting comets as if a shower of meteors.[2]

[2] *Ibid.*

ended, so-called *parabolic orbits*. *Isaac Newton* (1643–1727) in his 1687 book *Principia Mathematica* applied his new theory of gravitation, the core of which was that everything attracted everything else, to show that the comet of 1680 moved in an elliptical orbit, albeit nearly parabolic.

In the year 1705, British astronomer *Edmond Halley* (1656–1742) investigated the orbits of 24 comets and found that those of 1531, 1607, and 1682 were similar.[3] In light of Newton's new theory of gravity, Halley recognized that the slightly different episodes between returns came about on account of the gravitational attraction of the planets, called *planetary perturbations* of the orbit, and predicted the return of the comet in December of 1758. Halley died before the comet was seen again on Christmas day that year. This is now the first numbered and officially named comet, *1P/Halley*.

1P/Halley has been seen on each return since the earliest recorded sighting from China in 239 BC. Famous returns include that of AD 1066, the year which began the Norman conquest of England following the battle of Hastings, after which comet Halley was immortalized in fine needle work on the *Bayeux Tapestry*. The recovery in 1758 proved conclusively that Newton's law of gravity was valid as far out as comet Halley traveled from the Sun: a distance three times that of Saturn, the outermost planet known at the time.

The first scientific study of comets came with the next return of 1P/Halley in 1835, when more sophisticated instruments were available. Jets were observed for the first time in comet images, which led the German astronomer *Friedrich Wilhelm Bessel* (1784–1846) in 1836 to postulate, much ahead of his time, that dust particles were ejected in the direction of the Sun, which were then pushed back away from the Sun by an unknown repulsive force[4], now know to be radiation pressure from sunlight.[5]

2.1 The comet nucleus

The big riddle has always been what force could drive the ejection of meteoroids. Until into the twentieth century, many thought that comets were a *flying sand bank*,[6] a cloud of dust and pebbles, held together by their own mutual gravitational attraction. It was assumed that the Sun's tidal force (the difference in gravitational attraction between one side of the comet and the other) was enough to bring the meteoroids from moving around each other to moving in independent orbits. Indeed, seen from a great distance, meteoroid streams were expected to look like comets.

This impression was enforced when comet 3D/Biela broke apart in 1843 and was last seen one orbit later in 1852, and shortly thereafter spectacular storms of Andromedids were seen in 1872 and 1885. At the time, this took away much of the

[3] E. Halley (1705) *Astronomiae Cometicae Synopsis*. Philosophical Transactions.

[4] F.W. Bessel Beobachtungen ueber die physische Beschaffenheit des Halley'schn Kometen und dadurch veranlasste Bemerkungen. *Astron. Nachr.* **13** (1836), 185–232.

[5] S.A. Arrhenius, On the physical nature of the solar corona. *ApJ.* **20** (1904), 224–231.

[6] H. Schellen *Die Spektralanalyse*. (Brauschweig: Westermann, 1870), 452 pp; R.A. Lyttleton, On the origin of comets. *Mon. Not. R. Astron. Soc.* **108** (1948), 465–475.

early nineteenth century fears that comets might cause devastation when hitting Earth. Instead, comets were now expected to merely cause a brilliant meteor storm. In 1948, the British astronomer *Raymond A. Lyttleton* developed this idea into a comprehensive scenario where comets were formed as a loose swarm of ice and dust at the birth of the solar system directly from the condensation of a stream of interstellar dust and gas particles.

In reaction to that, meteor astronomer *Fred Lawrence Whipple* argued in 1950 that there had to be a solid core with his *icy conglomerate model* of a comet nucleus, nicknamed the *dirty snowball model*.[7] Astronomers talk of "models" when they discuss a simplified picture of something otherwise too complex to grasp. Many comets kept arriving back in the Earth's neighborhood a few days earlier or later than expected from Newton's laws of gravity alone. A force other than gravitation (a so-called *nongravitational force*) was needed to explain why. Whipple pictured a solid object at the center of the comet, the *comet nucleus*, consisting of a conglomeration of water ice with dust grains imbedded. He figured that the evaporation of the ice would cause a reaction force, a rocket effect. In a second paper,[8] Whipple calculated how the flowing water vapor would drag solid meteoroids into the vacuum of space against the gravitational field of the remaining mass, for the first time illuminating the birth of a meteoroid stream.

The ejection of matter will push the comet gradually into a different orbit, but only if the push is different before and after passing the Sun. The light curve of the comet will tend to be asymmetric.[9] This is possible when the comet spin axis (constant with respect to the stars) is tilted, in such a way that the comet presents a different side to the Sun. The change of the orbit is often expressed in terms of units "A_1," the radial *nongravitational acceleration* of the comet (radial as in acting along the line Sun–comet), and "A_2," the transverse nongravitational acceleration perpendicular to A_1 in the orbital plane. There is also a third term "A_3," the transverse nongravitational acceleration normal to the orbital plane responsible for changes of the orientation of the orbit.[10]

Always shrouded in a mist of dust particles, this "nucleus" was one of the solar system's best kept secrets. At the time, Whipple's postulation of a solid center was only that. That veil was lifted only by the spectacular images from the European *Giotto* spacecraft in 1986. Giotto traveled in the essential company of two Russian "VeGa" spacecraft (for Venus Galley – signifying an extended mission to 1P/Halley after a

[7] F. L. Whipple, A comet model. I. The acceleration of comet Encke. *Astrophys. J.* **111** (1950), 375–394.

[8] F. L. Whipple, A comet model. II. Physical relations for comets and meteors. *Astrophys. J.* **113** (1951), 464–474.

[9] M. Festou, H. Rickman and L. Kamél, The origin of nongravitational forces in comets. In *Asteroids, Comets, Meteors III. Proc. Meeting, Uppsala, Sweden, 12–16 June 1989*, ed. C.-I. Lagerkvist, H. Rickman, B. A. Lindblad, M. Lindgren, *Astron. Obs.* (1989), pp. 313–316.

[10] A_i is defined as $A_i = a_i/g(r)$, where the empirical function $g(r)$ is the ice sublimation rate, which changes with the heliocentric distance, and a_i is the orbital acceleration vector induced by the evaporation. Both are usually expressed in units of 10^{-8} AU/d^2. B. G. Marsden, Comets and nongravitational forces II. *Astron. J.* **74** (1969), 720–734.

flyby of Venus).[11] That year, I was glued to the television in anticipation. These remote travelers were penetrating the dense haze of the comet, their TV cameras ready, and just as in a dense fog we were expecting to see the actual nucleus emerge in crisp detail upon arrival. When the first pictures were finally broadcast, they were presented as contour images in false colors of pink, green, and yellow, leaving even the brightest commentators biting the dust. Nobody could tell what we were looking at, except that there was something bright in the center of the pictures at which the cameras were automatically pointing. I did not know it at the time, but that bright blob was the birth of a meteoroid stream!

Broadcasters were still clueless, talking about the 8 min it would take for radio signals to arrive on Earth and how wonderfully the mission was unfolding, when in the background mission control in Garching was clearly in disarray. Contact had been lost with the Giotto spacecraft just after the moment of closest approach. Later it was found that the spacecraft was knocked into a wobble by a collision with a large meteoroid and the transmission was interrupted. Responsible for that near knock-out was the sort of meteoroid that could have made a really nice Orionid or η-Aquariid meteor for some star gazer in the far future. Sadly enough, this particle did not survive the encounter.

The next day's six o'clock news finally showed the now familiar grayscale image of Halley's nucleus (Fig. 2.2). Whipple's solid core was a pitch-black rock covered with mountains and valleys, later measured at 15.3 by 7.2 by 7.2 km. Compare that to Mount Everest, which stands tall at only 8.85 km. This comet was shaped like a potato, covered by circular depressions that, in hindsight, were probably impact craters.[12] Only 4% of sunlight was reflected from the surface, a fraction called the *albedo*. That meant that the comet surface was as dark as charcoal and freshly paved asphalt road surfaces.

Since that time three other comets have been visited by NASA probes: 19P/Borrelly by Deep Space One, comet 81P/Wild 2 by Stardust, and comet 9P/Tempel 1 by Deep Impact. Comet 19P/Borrelly measured 8.8 by 3.6 by 0.8 km. Borrelly was even darker than Halley: only 2.9% of light on average reflected from the surface.[13] Looking at the image of comet Borrelly in Fig. 2.2, it is not hard to understand that after having lost kilometers of overlaying material this nucleus is only a small remaining core of what used to be a much larger comet. The result is a fairly smooth surface. In contrast, comet 81P/Wild 2 (pronounced "Vilt 2") may still have the impact craters from its formation time: a much more irregular surface with deep flat-bottomed pits. Its shape was almost spherical and the nucleus measured about

[11] The French space agency CNES cooperated in the VeGa program, which made it possible to deploy a copy of the German Giotto dust composition experiment (PIA) onboard the VeGa 1 (PUMA-1) and VeGa 2 (PUMA-2) spacecraft despite Cold War restrictions at the time.

[12] P. J. Stooke, A. Abergel, Morphology of the nucleus of comet P/Halley. *Astron. Astrophys.* **248** (1991), 656–668.

[13] D. T. Britt, D. C. Boice, B. J. Buratti *et al.*, The morphology and surface processes of comet 19P/Borrelly. *Icarus* **167** (2004), 45–53; B. Buratti, M. D. Hicks, L. A. Soderblom *et al.*, Deep Space 1 photometry of the nucleus of comet 19P/Borrelly. *Icarus* **167** (2004), 16–19.

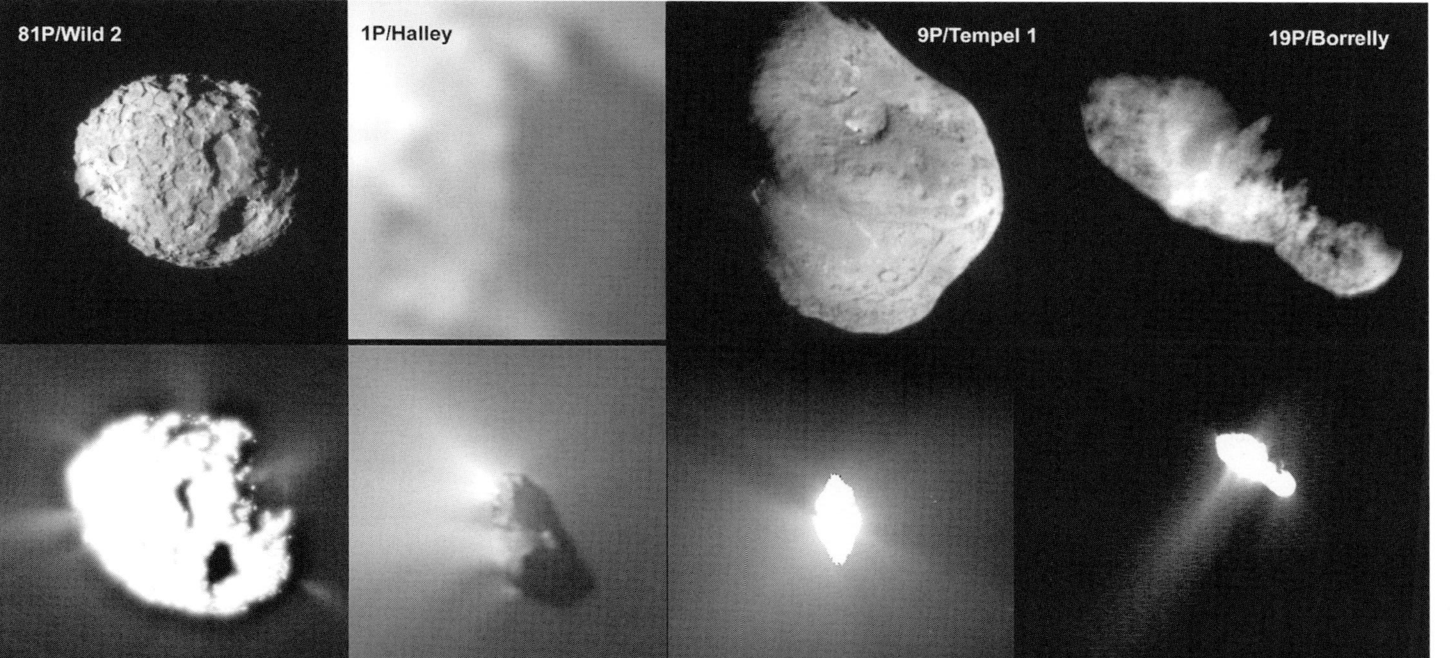

Fig. 2.2 Young, older, old: the nucleus of comets 81P/Wild 2, 1P/Halley, 9P/Tempel 1, and 19P/Borrelly. Stardust was about 500 km (311 miles) from the nucleus of 81P/Wild 2 when it took the picture on the left, showing flat bottomed craters and a dozen faint jets. Photo: NASA/JPL. Comet 1P/Halley, parent of the Orionid and η-Aquariid showers, was imaged by the Giotto spacecraft prior to its closest approach, which occurred on March 14, 1986 at 00:03:02 UT at a distance of 596 km. Photo: MPI für Aeronomie, Katlenburg-Lindau, ESA. Comet 9P/Tempel 1 approximately 5 min before Deep Impact's probe smashed into its surface on July 04, 2005, at 05:45 UT. This image was taken by the impacting targeting sensor on the probe. Photo: NASA/JPL-Caltech/UMD. 19P/Borrelly was observed by the Deep Space One probe on September 22, 2001. Photo: NASA/JPL.

5.5 by 4.0 by 3.0 km.[14] The 8 by 5 by 5 km sized comet 9P/Tempel 1 had terrain common to both comets, except for having rigid craters on the older terrain and for having a more facetted young terrain.

Indeed, most known comets have sizes of the order of 1–10 km. Very few comets are known to be smaller than 1 km across. Among the biggest comets, and the most spectacular in recent years, was comet *C/1995 O_1 (Hale–Bopp)*, which was brighter than the star Vega for 7 weeks in the spring of 1997. The nucleus of this comet has been estimated at 60 ± 20 km diameter.[15] And there are much larger comets out there!

2.2 The birth of a meteoroid stream

As the smoke of a big fire, the jets of comet 1P/Halley scattered the bright sunlight and put the dark nucleus in a glowing frame (Fig. 2.2). The jets of dust and molecules also made the Giotto pictures a poignant reminder that comets in the inner solar system are in the process of dying. Gradually losing material in fountains of meteoroids until, at last, the comet falls to dust. Multiple jets on Halley originated from two regions on the nucleus. Including seeps over the remainder, only 10–15% of the surface was active. On Borrelly, one strong jet and two fainter jets were observed near the northern pole of the spinning nucleus, while about 10% of the day-side surface was actively emitting water vapor. In contract, the fresher looking 81P/Wild 2 had a dozen jets (Fig. 2.1).

We do not know how the landscape will look on a human scale, except that it will be as dark as coal with a hazy sky overhead (Fig. 2.3). At the time of writing, we are still years away from landing on the surface of a comet for the first time. The satellite Rosetta has been launched to do so.

The surface of the nucleus is expected to be covered in a thick layer of meteoroids that have fallen back in the gravity field of the comet. Deep Impact showed that layer to be very fluffy and porous, with a bulk porosity of >60% and a bulk tensile strength of ~100 Pa. The crust acts as insulation against the heat of the Sun, protecting the ice below.

Jets might arise from landslides on steep slopes that expose the fresh ice underneath. NASA's Cassini mission observed such slides from the subsidence of the steep crater walls on the Saturn moon Phoebe (Fig. 2.4). The moon is pocketed with impact craters. Interestingly, bright spots were also observed on comet 82P/Wild 2 by Stardust (Fig. 2.4) and on comet Tempel 1 by Deep Impact, where the spots are only tens of meters in size and are not always found on steep slopes.

[14] Z. Sekanina, D. E. Brownlee, T. E. Economou, A. J. Tuzzolino and S. F. Green, Modeling the nucleus and jets of comet 81P/Wild 2 based on the stardust encounter data. *Science* **304** (2004), 1769–1774; A. J. Tuzzolino, T. E. Economou, B. C. Clark, P. Tsou, D. E. Brownlee, S. F. Green, J. A. M. McDonnell, N. McBride and M. T. S. H. Colwell, Dust measurements in the coma of comet 81P/Wild 2 by the dust flux monitor instrument. *Science* **304** (2004), 1776–1780; S. F. Green, J. A. M. McDonnell, N. McBride, M. T. S. H. Colwell, A. J. Tuzzolino, T. E. Economou, P. Tsou, B. C. Clark and D. E. Brownlee, The dust mass distribution of comet 81P/Wild 2. *J. Geophys. Res.* **109** (2004), E12S04.

[15] Y. R. Fernández, The nucleus of comet Hale–Bopp (C/1995 O1): size and activity. *Earth, Moon, Planets* **89** (2002), 3–25.

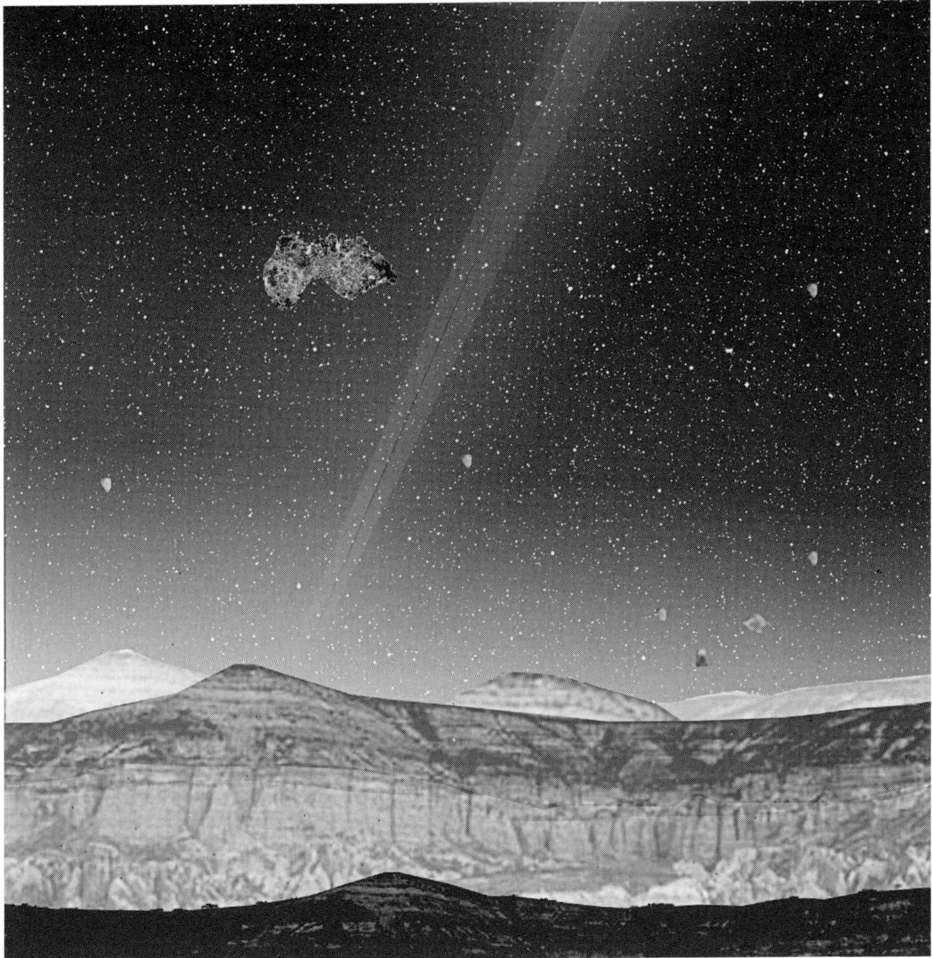

Fig. 2.3 Comet nucleus surface (author's artist impression, drawn prior to the 81P/Wild 2 encounter).

Several other scenarios have been proposed to explain the jets. Some suspect that the jets escape from narrow openings in the roof of still hidden subterranean caverns, in which case the bright spots could be areas of condensation of vapor around the cold vent. In such caves, water vapor pressure can build up, ejecting gas and dust at higher speeds. This would help to explain the very narrow width of the jets, only $\sim 5^\circ$, but leaves unanswered how the heat of sunlight can penetrate deep down into the caves.

Others suspect that the jets emanate from flat-bottomed sinkholes,[16] created when the roof of a subterranean cave collapsed. Vapor flowing from the walls will

[16] H. U. Keller, J. Knollenberg and W. J. Markiewicz, Collimation of cometary dust jets and filaments. *Planet. Space Sci.* **42** (1994), 367–382.

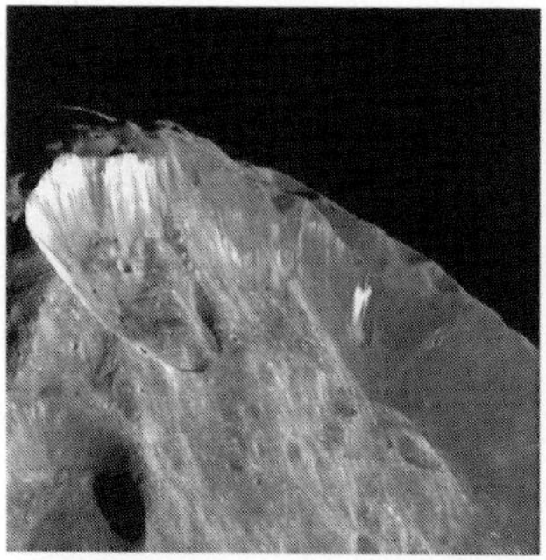
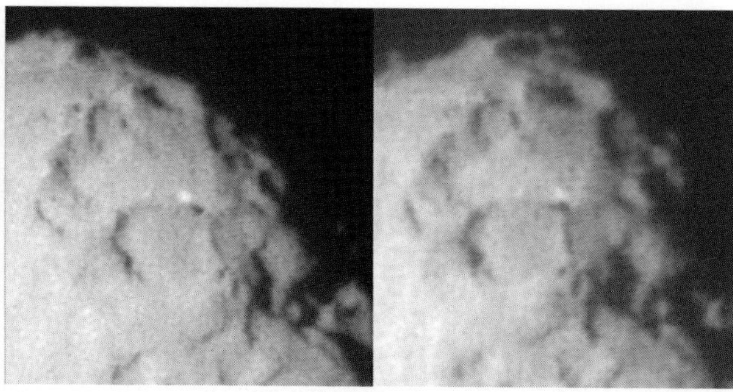

Fig. 2.4 Left: Saturn's moon Phoebe as seen by the Cassini–Huygens mission. The crater on the left is about 45 km (28 miles) in diameter. Photo: NASA/JPL. **Right: Stereo image of the 81P/Wild 2 surface**, showing a bright spot, flat-bottomed impact craters, and upturned ridges. Photo: NASA/JPL – Stardust.

concentrate in a narrow beam in the middle of the pit, creating a dust spike and a jet cone.[17] Comet 82P/Wild 2 had such flat-bottomed holes. However, these are now understood to be relatively old craters resulting from impacts in cohesive porous material.[18] If so, it remains a puzzle why some small fragments of the old surface still stand as tall and steep pinnacles.

One of the great discoveries of satellite missions is that caves and crevasses can also come from the internal structure of the comet. It turns out that the comets encountered so far have a low *bulk density*. When the volume of the nucleus can be measured from TV images, then the density equals the mass per volume. The mass of a comet nucleus is measured from the magnitude of the rocket effect on the orbit as envisioned by Whipple.

When something has a density of less than 1.04 g/cm^3, it will float in liquid water at room temperature. With solid rock at 3.5 g/cm^3 and ice at 0.96 g/cm^3, a comet nucleus was expected to measure somewhere in between. Instead, the density of comet 1P/Halley is only 0.55 ± 0.25 g/cm^3 and comet Borrelly has a density of 0.24 ± 0.06 g/cm^3. Less than pinewood at 0.8 g/cm^3. Halley's *Orionid shower* meteoroids have a similar density of 0.23 g/cm^3, or larger if fragmentation is considered,[19] but that is after the meteoroid has lost all ice and only a loose assembly of minerals and organic matter remains. Hence, some of the low density of comets has to be on account of the bulk morphology.

Internal caves and crevasses can result from a loose packing of km-sized cometesimals into a *rubble pile* (Fig. 2.5).[20] *Cometesimals* are the smaller units that once came together under mutual gravity to form a comet. In my opinion, this leads to a natural

[17] T. I. Gombosi, A heuristic model of the comet Halley dust size distribution. *ESA SP* **250** (1986), 167–171.
[18] D. E. Brownlee, F. Horz, R. L. Newburn *et al.*, Surface of young Jupiter Family Comet 81P/Wild 2: view from the Stardust spacecraft. *Science* **304** (2004), 1764–1769.
[19] F. Verniani, Meteor masses and luminosity. *Smithsonian Contrib. Astrophys.* **10** (1967), 181–195.
[20] P. R. Weissman, Are cometary nuclei primordial rubble piles? *Nature* **320** (1986), 242–244.

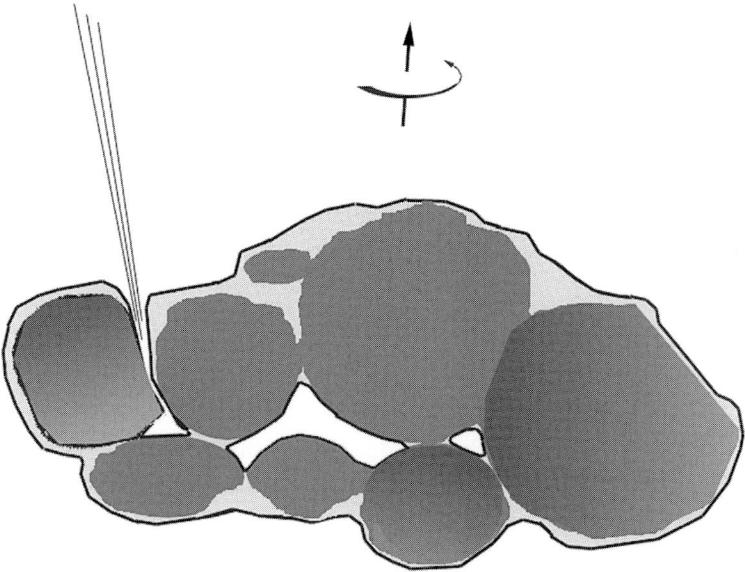

Fig. 2.5 Rubble pile nucleus with jet from crevasse between cometesimals, opened up by centrifugal forces.

formation mechanism for jets: outgassing will spin-up the nucleus leading to centrifugal forces on the cometesimals. These can open up deep crevasses between the cometesimals, resulting in jets being directed towards one of either of the spinning poles of the comet nucleus (Fig. 2.5).

Finally, dust grains can be dragged into space by escaping from millions of pores in the crust of the comet. It has even been suggested that such *seeps* can cause jets as a result of the larger scale topography of the comet.[21] Laboratory experiments show that even from a seep, the dust particles tend to leave the surface area perpendicularly, with directions spread by only *full-width-at-half-maximum* FWHM $= 19^\circ$ about nominal. Gas outflow, in contrast, is less confined, with a FWHM of $\sim 90^\circ$.[22]

2.3 The driving force: evaporation of ices

Whipple's old idea of a spherical snowball warmed by sunlight is pretty much redundant now that comets are found to be dark and covered by a crust. However, the main idea of dust particles being dragged along by water vapor is alive and kicking, because

[21] J.-F. Crifo, A. V. Rodionov, K. Szego and M. Fulle, Challenging a paradigm: do we need active and inactive areas to account for near-nuclear jet activity? *Earth, Moon, Planets* **90** (2002), 227–238.

[22] H. Kohl, K. Kölzer, E. Grün and K. Thiel, Dust-particle acceleration near simulated cometary surfaces: experimental results. In *Asteroids Comets Meteors III*, Uppsala, 1989 pp. 367–371.

it is a way to accelerate the grains. When ices evaporate, the outflowing vapor can push the grains.

Ices are small molecules that are solid at low temperature but evaporate at room temperature. The ice of comets is a mean cocktail of 79% water (H_2O), 13% carbon monoxide (CO), 2.8% dry ice (CO_2), 3.0% formaldehyde (H_2CO), 1.0% methanol (H_3COH), 1.2% ammonia (NH_3), and 0.08% hydrogen cyanide (HCN), amongst others. Of all these ices, water is the least volatile because it is most strongly bonded. All the other ices are trapped in a matrix of water ice. As a result, it is the evaporation of the water ice that drives much of the outgassing of comets.

The carbon monoxide molecule has such a low sublimation temperature that it is very unexpected to find CO in a comet. It should all have evaporated long ago. We now know that CO molecules are caged in the water ice and thus prevented by the strongly bonded structure of water molecules from evaporating until much higher temperatures. This is possible, because the water ice is not crystalline as in the snowflakes on Earth, but in a disordered amorphous form. Like window glass is an amorphous form of the mountain crystals of quartz (with impurities of soda and lime). When ice crystallizes into snowflakes, all the impurities are expelled. Amorphous water ice is a very interesting material and when I first came to the NASA Ames Research Center in California in 1993, I spent many hours probing its peculiar structure with a transmission electron microscope in experiments with David F. Blake (Fig. 2.6). We found that the amorphous water frost of interstellar grains, formed at temperatures $T < 15$ K, can rearrange into a more open structure when it is warmed a few tens of degrees. At even higher temperatures, this amorphous ice starts to become soft and turn into a glass, a viscous liquid, much like window glass when heated in an oven. This property of the ice was known before, but never thought to be important because the ice also quickly crystallizes into small solid cubic ice crystals. However, David and I found that most ice in the thin films we studied never crystallized completely and the ice continued to flow until all of it turned into the hexagonal ice crystals of snow at a much higher temperature! In the microscope, we saw that the ice retracted from the hydrophobic amorphous carbon substrate and formed little droplets as soon as its viscosity decreased. This exotic "restrained" amorphous form of (still very viscous) *liquid* water may occur naturally in comets.[23]

The heat of the Sun evaporates the water, increasingly as the comet approaches the Sun. The distance to the Sun is called "*r*" (from heliocentric *radius*) throughout this

[23] P. Jenniskens and D. F. Blake, Structural transitions in amorphous water ice and astrophysical implications. *Science* **265** (1994), 753–756; D. F. Blake and P. Jenniskens, The ice of life. *Sci. Am.* August (2001), 2–7; P. Jenniskens and D. F. Blake, Crystallization of amorphous water ice in the solar system. *Astrophys. J.* **473** (1996), 1104–1113.

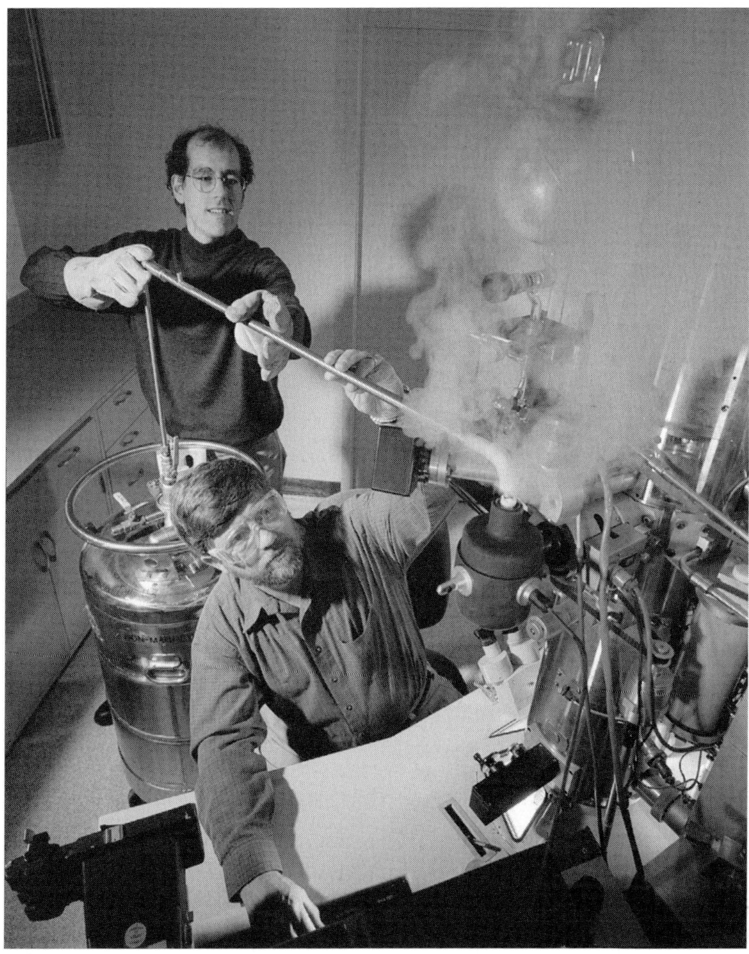

Fig. 2.6 Electron microscopy studies of the ice of comets. Author with Dr. David F. Blake (seated). Photo: NASA Ames Research Center.

book, usually measured to the center of mass of the solar system rather than the center of the Sun alone. That center of mass is inside the Sun (Chapter 13). This distance is expressed in terms of *Astronomical Units* = 149 597 870.691 km, approximately the distance between Earth and the Sun.[24] The Earth is always close to $r = 1$ AU, to within ± 0.02 AU on account of a slightly elliptical orbit.

[24] The formal definition of *Astronomical Unit* is the radius of an unperturbed circular orbit that a massless body would revolve about the Sun in $2(\pi)/k$ d (i.e., 365.256 89... d), where k is defined as the Gaussian constant exactly equal to 0.017 202 098 95. Since an AU is based on the radius of a circular orbit, 1 AU is slightly less than the average distance between the Earth and the Sun.

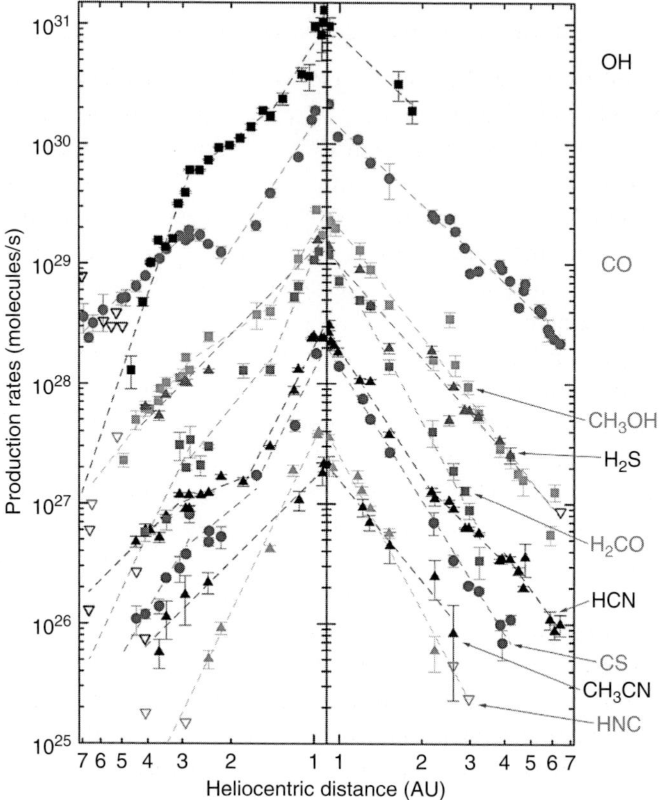

Fig. 2.7 The amount of ices lost from comet Hale–Bopp per second as a function of the distance from the Sun. "OH" marks the sublimation of water ice, CO the sublimation of carbon monoxide, and "CH_3OH" the sublimation of methanol (methyl alcohol). Image by Nicolas Biver, Meudon Observatory.

When a comet moves closer to the Sun, the comet nucleus surface will warm to a temperature of about: $T = 300/\sqrt{r}$ (with r in AU) degrees Kelvin. When water ice is exposed to the vacuum of space at temperatures above 175 K (at 3 AU), it will evaporate without melting, a process called *sublimation*, just as does the "dry ice" from a fire extinguisher. Before the water ice sublimates, less volatile molecules are lost from its matrix, fizzing out like the pop in soda. As expected, the first to go is carbon monoxide (CO, Fig. 2.7) and other weakly bonded compounds. For that reason, CO is often seen in comets as far out as 7 AU from the Sun ($T_s \sim 115$ K), beyond the orbit of Jupiter.

Only when the temperature of the ice rises above the sublimation temperature of water ice at distances less than 3 AU, in the case of Hale–Bopp when the comet crossed the asteroid belt, is water observed to leave the comet and to quickly fall apart into $H_2O \rightarrow H + OH$ (Fig. 2.7).

2.4 How to lift the grains from the comet surface

There are a number of ways for comets to lose solid particles. Whipple envisioned the dust particles to be imbedded in the flow of water vapor and calculated at what final speed those dust particles emerge from the water vapor jets (after taking into account the speed lost from escaping the gravity of the comet). That formula was based on the concept that a gas flow colliding with a particle will push it, in much the same way as a meteoroid entering the Earth's atmosphere is slowed down by collisions with air molecules. Indeed, Whipple's equation originated from his work on meteor trajectories in the atmosphere.

Whipple's equation for meteoroid ejection (Appendix A) is still used today, albeit with some modifications. There are a number of different formulas around, each taking into account certain aspects not considered by Whipple. For example, Whipple did not consider the presence of the jets that cause energy to be lost from a smaller surface area than where it is absorbed, the adiabatic expansion of the vapor when it flows into space, nor did he consider nonspherical grains that may sail more efficiently in the water vapor wind, or the delayed evaporation of ice that can propel the grains like a rocket. All these effects can potentially change the outcome of the ejection process dramatically.

One of the latest incarnations by Jean-François Crifo and Alex V. Rodionov[25] (the *Crifo ejection model*) was developed for the interpretation of comet images and is used in the meteor storm prediction software developed in the Ph.D. thesis work of Jerémié Vaubaillon, results of which are presented throughout this book. The software was developed in collaboration with thesis advisors François Colas and William Thuillot at the recently founded *Institut de Mécanique Céleste et de Calcul des Ephemerides* (IMCCE) in Paris, France, and was applied during Jerémié's postdoctoral stay at the SETI Institute in late 2004, where he worked with the author during the writing of this book.

Those modifications are important, but only insofar as they can be validated by observations. According to Whipple's formula, a typical +3 magnitude Leonid meteoroid ejected at perihelion would give a speed of $V_{ej} = 28.5$ m/s, which is 103 km/h.[26] Crifo's main modification is the inclusion that only a small fraction of the surface ejects meteoroids, resulting in an ejection speed of: $V_{ej} \leq 12.8$ m/s if 24% is active, or $V_{ej} \leq 14.9$ m/s if only 4% of the surface is active. This would mean a meteor shower at least half as wide.

It is important to realize that the coupling between gas and dust is poor and the ultimate speed of a meteoroid is only a small fraction of the outflow speed of the gas. The smaller particles are more efficiently accelerated than the large ones. In the case of comet C/Hale–Bopp, water vapor flowed out at 1200 m/s at $r = 0.9$ AU, decreased to

[25] J.-F. Crifo and A. V. Rodinov, The dependence of the circumnuclear coma structure on the properties of the nucleus. *Icarus* **127** (1997), 319–353.

[26] For a Leonid of $M = 0.008$ g and $\rho = 0.7$ g/cm^3, Tempel–Tuttle's diameter $D_c = 3.5$ km, $\rho_c = 0.5$ g/cm^3, $\Lambda = 1$, at a distance $r = 0.976$ AU from the Sun.

500 m/s at $r = 6$ AU, about a hundred times larger than the outflow speed of large meteoroids. Moreover, when the gas flows out more violently, it drags along more dust. While the gas production rate fell off with an r^{-2} power law away from the Sun (Fig. 2.5), the dust production rate of comet 1P/Halley fell off with a steeper $r^{-3.0 \pm 0.7}$ power law. Most dust was ejected when the comet was closest to the Sun.[27]

The ejection speed of these meteoroids is mostly determined by the initial acceleration in pits or just above the comet surface. Although the interaction with the gas continues as far out as $\sim$5 times the nucleus size, the water vapor loses most of its dragging force when it expands and becomes less dense only a few (tens of) meters above the surface of the comet.

In Whipple's picture, the grains are embedded in the ice and therefore, in a way, are already entrained in the gas flow when the water evaporates. We now know that comets are mostly dust with little ice (Chapter 15). In that case, the dust has to break from the rest of the comet before being dragged along in the vapor (the bonds between the grains have been eroded). More fragile material may be lost first, while more sturdy material may be lost in larger chunks that can fall back onto the surface of the comet.

Meteoroid fragmentation during and after ejection can also change the outcome. The main effect would be that small grains can have the lower ejection speed expected for the original larger grains. Indeed, fragmentation shortly after ejection is a common phenomenon.[28] When the Giotto satellite approached the nucleus of comet Halley, a large number of tiny attogram ($= 10^{-18}$ g) grains were discovered, thought to be the product of vigorous fragmentation of dust out to distances of 1 million km from the nucleus.[29] Larger grains often arrived in clusters. Also, the distribution of scattered sunlight from very small grains and the distribution of CO gas, presumably still evaporating from the fragmenting grains, were more persistent away from the comet nucleus than would be expected if there was no fragmentation.[30]

The reason for fragmentation is perhaps the continued evaporation of ice and the heating by sunlight. After ejection from the comet, the dust grains will first be under stress from the remaining ice turning into vapor, putting pressure on the walls of pores. Dark (absorbing) dust grains tend to warm to the point of sublimation in a few hours or less, evaporating any remaining ice before the particle has moved a few hundred kilometers from the surface. While dragged out by the vapor, the grains are repeatedly heated and cooled, while spinning in the bright sunlight. What remains after this process are the meteoroids that we see as meteors.

[27] P. D. Singh, W. F. Huebner, R. D. D. Costa, S. J. C. Landaberry and J. A. de Freitas Pacheco, Gas and dust release rates and color of dust in comets P/Halley (1986 III), P/Giacobini–Zinner (1985 XIII), and P/Hartley–Good (1985 XVII). *Planet. Space Sci.* **45** (1997), 455–467.

[28] H. U. Keller, M. L. Marconi and N. Thomas, Hydrodynamic implications of particle fragmentation near cometary nuclei. *Astron. Astrophys.* **227** (1990), L1–L4.

[29] N. G. Utterback and J. Kissel, Attogram dust cloud a million kilometers from comet Halley. *Astron. J.* **100** (1990), 1315–1322.

[30] Eberhardt P., Krankovwsky D., Schulte W. *et al.*, The CO and N_2 abundance in comet P/Halley. *Astron. Astrophys.* **187** (1987), 481–484.

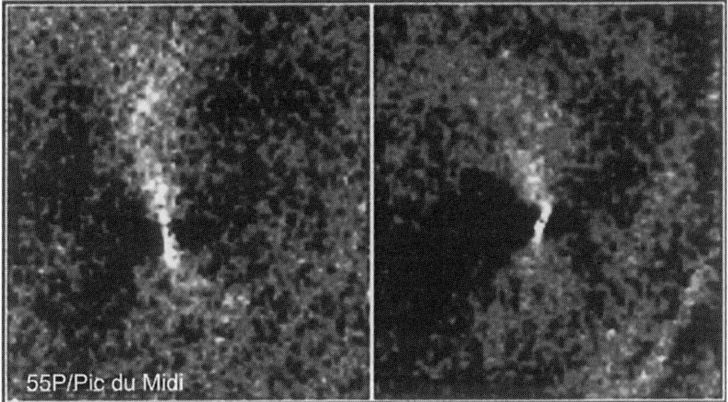

Fig. 2.8 The dust jets of comet 55P/Tempel–Tuttle the parent of the Leonid shower, by François Colas of IMCCE, taken at the Pic du Midi Observatory.[31]

After the vapor dissipates, dust continues to stream away from the nucleus into space, on its own independent orbit around the Sun. When the comet nucleus spins, a jet will point in different directions over one rotation, but the particles will continue to move outwards in nearly straight lines. This will cause the jet to have a corkscrew-shape (Fig. 2.8). The jet of Leonid parent 55P/Tempel–Tuttle in 1998 was located at a northern position on the nucleus at a small angle to the spin pole, judging from the small opening angle of the corkscrew motion. The comet is seen to spin with a period of 15.33 ± 0.02 h.[32] In comparison, 19P/Borrelly rotated with a period of 25.0 ± 0.5 h. The dust grains are ejected during daytime. When it is morning at the vent on the comet surface, the jet starts to sprout again and a new band of dust is deposited.

Water vapor can not drag along very large pieces. Whipple calculated that the *maximum sized* Leonid that can be lifted off the nucleus by water vapor drag (the particles having ejection speed of 0 m/s) would be about 19 cm in diameter, or some 2.7 kg in mass (Appendix A). Such a large fragment would cause a spectacular -11^{m} fireball, nearly as bright as a full the Moon and casting shadows. Instead Leonid fireballs as bright as -15^{m} have been reported during the recent Leonid storms.

[31] Measured on January 30, 1998, by Jean Lecacheux, Eric Frappa, and François Colas of Pic du Midi observatory.
[32] *Ibid.*

3

The formation of meteoroid streams

Meteoroid streams in space used to be invisible, their existence illuminated only by the meteor showers they caused on Earth. Then, in 1983, *dust trails* were discovered in the orbit of short-period comets. Dust grains absorb visible light, warm up, and re-emit that energy as thermal emission in the mid-infrared.

My Alma Mater at *Leiden Observatory* was deeply involved in the interpretation of data from the monumental 1983 all-sky survey of heat emissions at the mid-infrared wavelengths of 12, 25, 60, and 100 μm by the *InfraRed Astronomical Satellite* (*IRAS*), a joint project of the USA, UK, and the Netherlands. The observatory had a vested interest in the topic of interstellar dust, with my professor, Harm Habing, being one of the leading investigators of IRAS. As in so many astronomical institutes, meteor studies were delegated to amateurs. I was such an amateur, joining the ranks of the Dutch Meteor Society two years earlier.

When the news spread that the images from IRAS showed dust trails in the path of comets, I immediately suspected a link with meteor outbursts.[1] It was the excellent 1986 report by *Mark Sykes* and coworkers, with details of the width of the trails and estimates of the sizes of the dust grain,[2] that first alerted me to the trails, although the discovery was made by *John Davies* a few years earlier and published in a paper that discussed other things as well.[3]

John Davies, a scientist involved with the IRAS moving object project at the University of Hawai'i, recalls how he discovered the trails in the images of the IRAS satellite: "One day in August, 1983 the fast moving object detection software seemed to find a number of 'asteroids' all in the same patch of sky. None of these looked right and they could not have been a single object being detected several times as the motion would have been too erratic to be real, so I did not worry too much about them. However, to my surprise the next day several more

[1] P. Jenniskens, Stofsporen. *Radiant, J. DMS* **9** (1987), 73–74.

[2] M. V. Sykes, L. A. Lebofsky, D. M. Hunten and F. Low, The discovery of dust trails in the orbits of periodic comets. *Science* **232** (1986), 1115–1117.

[3] J. K. Davies, S. F. Green, A. J. Meadows, B. C. Stewart and H. H. Aumann, The IRAS fast-moving object search. *Nature* **309** (1984), 315–319.

appeared in a very similar region. This went on for several more days and eventually I tried plotting all the positions onto a map of the sky. The result was amazing, all the objects seemed to lie on a straight line! A closer look at the positions revealed that the structure pointed straight at the position of comet 10P/Tempel 2," from which the dust grains appeared to originate. John predicted that the trails would move along with the comet in projection on the sky, and saw that to be true when IRAS returned to the area a few days later.

Fig. 3.1 shows a compilation of IRAS images. Most of the emission from the zodiacal light in the center of the image has been removed to bring out the more subtle structures. The horizontal bands that remain after removing a smooth zodiacal light component are due to asteroidal dust grains in what are called the *zodiacal dust bands*. The irregular wisps above and below are the interstellar clouds of our galaxy seen from a great distance.

The comet dust trails are the thin lines stretching across the sky. A particularly bright one emanates from the position of comet 10P/Tempel 2, marked by an arrow. Another belongs to comet 2P/Encke, associated with the Taurids (just above the ecliptic plane in the center of the image). A third dust trail is from comet 7P/Pons–Winnecke, parent of the June Bootids.

What peaked my curiosity was that Davies and Sykes rejected the notion that dust trails could be responsible for meteor showers, because they could not identify any. Comet Encke was a known source of meteor showers, but the Encke trail was much more confined in space and distinct from the Taurid showers. Instead of days or months, it would take Earth only 1.4 h to cross the dust trail, measuring no more than 150 000 km (~0.001 AU) instead of $\sim 0.44 \times 0.05$ AU for the Orionids, for example. As Sykes wrote:[4] *Meteoroid streams are qualitatively very different from their trail counterparts in that they are far more dynamically evolved, are spread out over a vastly greater volume, and often have mean orbits whose nodes are significantly separated from their parents.* Sykes and Walker calculated that the dust trails were so dense that an observer would see more than 10 000 meteors per second (!) if Earth were to cross the dust trail near the position of comet 10P/Tempel 2. From the reported dust density of $\sim 3 \times 10^{-16}$ cm^{-3}, however, I calculated a zenith hourly rate of ZHR ~140 000, or at best ≤ 40 meteors per second were these to hit Earth at the speed of Draconid meteors.

I knew that there were meteor showers that lasted only ~1.4 h. In fact, they were more common than generally believed. In 1985, there were outbursts of the August *β-Hydrusids*, the October *Draconids*, and the November *α-Monocerotids*, while in 1986 there were *κ-Pavonids* in July, September *Aurigids*, and a burst of December *Ursids*.

As an undergraduate student, I set out to collect as much information about such meteor outbursts as I could find to establish the link between meteor outbursts and comet dust trails, in the process attempting to define what is the normal annual shower

[4] M. V. Sykes, D. J. Lien and R. G. Walker, The Tempel 2 dust trail. *Icarus* **86** (1990), 236–247; M. V. Sykes and R. G. Walker, Cometary dust trails. I Survey. *Icarus* **95** (1992), 180–210.

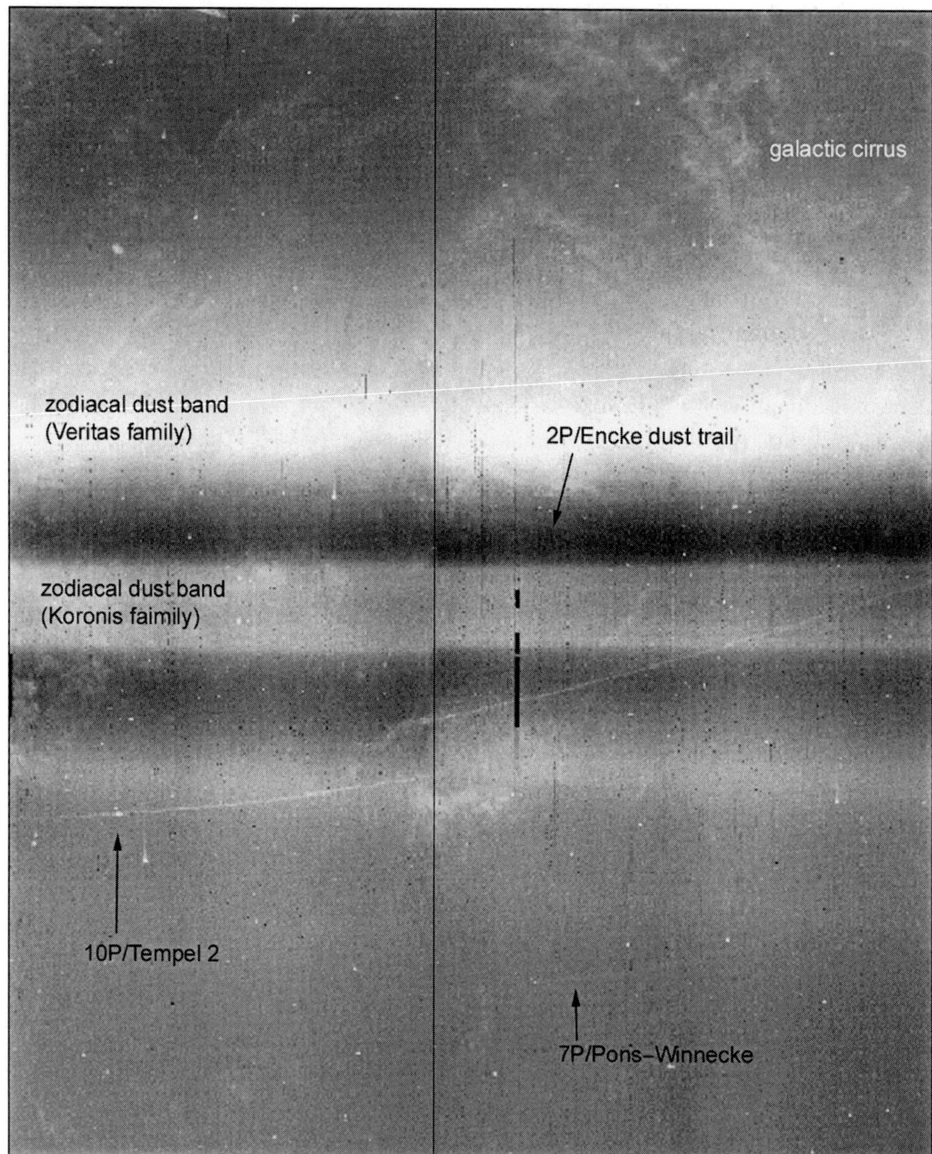

Fig. 3.1 Dust trails in IRAS images. A 40° × 50° area of the sky in a mosaic of 25 and 60 μm mid-infrared HCON-1 IRAS images, centered on the ecliptic plane. Dust trails of 2P/Encke, 10P/Tempel 2, and 7P/Pons–Winnecke are marked. Image courtesy of Mark Sykes, University of Arizona, and William Reach, IPAC.

activity.[5] At the same time, Slovak meteor astronomer *Lubor Kresák* (1927–1994) thought the same and first published papers discussing this connection in 1992 and 1993.[6]

While a student at Charles University in Prague, Kresák had observed the 1946 Draconid storm from Skalnaté Pleso Observatory in the northern mountains of Slovakia on the Polish border.[7] He continued to graduate in 1951 on a thesis "Structure, mass, and age of the comet Halley meteoroid stream." After his graduation, he worked at Skalnaté Pleso until 1955, where he discovered comets 41P/Tuttle–Giacobini–Kresák and C/1954 M_2 (Kresák–Peltier). He then worked at the Astronomical Institute of the Slovak Academy of Sciences in Bratislava on the dynamics of comets and asteroids and on meteor showers. He is remembered for insightful tools and diagrams to address the dynamical interpretation of observations.[8] His final work on attempting to link meteor showers with comet dust trails put a crown on a very rich and fruitful career covering both comets and meteor showers. Kresák died on January 20, 1994.

Later that year, meteor astronomer Duncan Steel wrote: *The relationship between these trails and the streams observed as meteor showers at the Earth is by no means clear at this stage.*[9] My inventory of annual shower activity based on visual observations of the Dutch Meteor Society (gathered by the Visual Section leader Rudolf Veltman) and the Western Australian Meteor Society (gathered by Jeff Wood) finally appeared in 1994,[10] and that of meteor outbursts and their relationship to IRAS dust trails in 1995.[11]

3.1 Comet coma and tail

Before discussing how comet dust trails are formed, let us first examine how the gas and smaller meteoroids of a comet move away from the nucleus. These make *tails* instead of trails (Fig. 3.2).

The *ion tail* of a comet is part of the remains of the evaporated ices. The vivid blue–green colour is a fluorescent molecular band emission from long-lived positively charged carbon-monoxide molecules (CO^+), after solar radiation knocked off an electron from CO molecules and thus created *ions*. The charged ions feel the Sun's magnetic field and are swept almost exactly in a direction away from the Sun along the magnetic field that emanates from Sun spots, distorted by the *solar wind* of charged

[5] P. Jenniskens, Meteor stream activity profiles from naked eye counts. In *Asteroids, Comets, Meteors III*, ed. C.-I. Lagerkvist, H. Rickman, B. A. Lindblad and M. Lindgren. (Uppsala: Uppsala University, 1989), pp. 535–538.
[6] L. Kresák, Cometary dust trails and meteor storms. *Astron. Astrophys.* **279** (1993), 646–660.
[7] I. P. Williams, Lubor Kresák (1927–1994). *Quart. J. R. Astron. Soc.* **35** (1994), 579.
[8] A. Carusi and G. Valsecchi, In memoriam – Lubor Kresák. In *Asteroids, Comets, Meteors 1993: Proc. 160th Int. Astronomical Union*, ed. A. Milani *et al.*, (Dordrecht: Kluwer, 1994) pp. 75–76.
[9] D. Steel, Meteoroid streams. In *Asteroids, Comets, Meteors 1993: Proc. 160th Int. Astronomical Union*, ed. A. Milani *et al.*, (Dordrecht: Kluwer, 1994) pp. 111–126.
[10] P. Jenniskens, Meteor stream activity. I. Annual streams. *Astron. Astrophys.* **287** (1994), 990–1013.
[11] P. Jenniskens, Meteor stream activity II. Meteor outbursts. *Astron. Astrophys.* **295** (1995), 206–235.

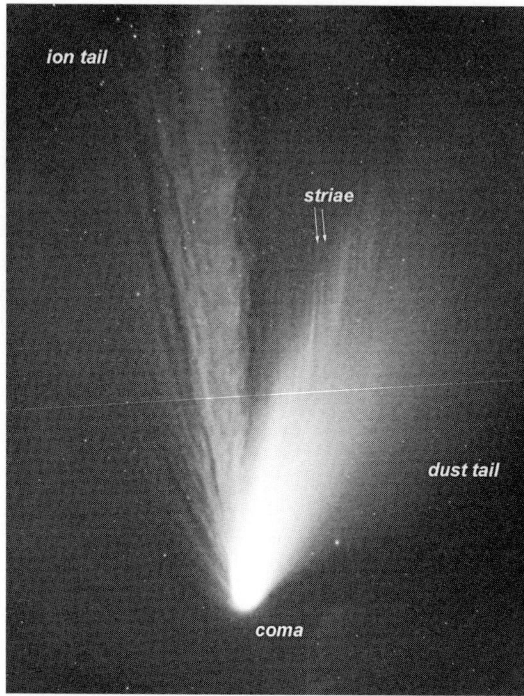

Fig. 3.2 Comet C/Hale–Bopp in a photo by Wei-Hao Wang (Institute for Astronomy at the University of Hawai'i). The photo was taken from Taiwan, at 20:38 UT on March 9, 1997, 22.3 d before perihelion.

particles that sweep past the comet at 500 km/s. The resulting filamentary structures can change in minutes. When the comet travels through magnetic polarization reversals, there are abrupt ion tail disconnections.

Neutral gas molecules (and large dust grains) hang around the nucleus to form a *coma*, rich in green CN and C_2 molecular band emissions. The latter do not originate from ices, but from the organic matter in the meteoroids.

The *dust tail* is the ensemble of all the small solid particles that are strongly affected by the solar radiation pressure. The tail is made visible mainly by scattered sun light, due to dust particles that are comparable to or smaller in size than the wavelength of light, that is up to about 10 μm. My other thesis advisor at Leiden University, Professor *J. Mayo Greenberg*, used to demonstrate how efficiently small particles scatter light by smoking a cigarette into the light of an overhead projector. A puff of smoke can scatter away so much light that it darkens the screen. In contrast, a piece of chalk leaves most of the light untouched despite having much more mass. Despite his love for light scattering, Mayo despised cigarettes and performed this trick only reluctantly, often to demonstrate how the comet surface can be dark because of the fine-grained and fluffy morphology of the surface materials, fallen back meteroids, by scattering light inwards.

3.2 Radiation pressure

Once ejected, the trajectories of these grains are determined by their ejection speed *and* by the momentum carried by the Sun's radiation. The force of impact by absorption or scattering of a light particle (a massless photon) is very small, but there are so many light particles that a small meteoroid can feel a strong push away from the Sun. This is called *radiation pressure*. Radiation pressure, being a factor β times as strong as gravity, lowers the pull from gravity by a factor $1-\beta$.

Heavy meteoroids are more difficult to blow off course. Hence meteoroid mass and the parameter β are often used interchangeably. However, it matters whether the grains are compact and spherical or fluffy, perhaps fractal, in shape. For spherical grains of diameter d (cm) and density ρ (in g/cm^3) and radiation pressure efficiency Q_{pr}, the relationship is (masses $> 10^{-12}$ g):[12]

$$\beta = 1.148 \times 10^{-4}\, Q_{pr}/\rho\, d \tag{3.1}$$

For even smaller grains (smaller than the wavelength of light), the absorption is not efficient and β drops off. The pressure efficiency includes the effect of *albedo*, which is the percentage of light absorbed. Better absorbing grains have a peak $\beta_{max} > 1$, for example $\beta_{max} = 1.8$ for iron grains, and $\beta_{max} > 5$ for graphite, while very transparent particles have $\beta_{max} \sim 0$. Because of all this, the ejection speeds of dust particles need to be calibrated by observations. It is found that the meteoroids in comet tails tend to have $\beta_{max} \sim 2.5$.

If the solar radiation pushes just as hard outward as the solar gravity pulls the particle inwards, then $\beta = 1.0$. At that moment, there is no net force and the particles continue to move on a straight line out of the solar system. Comet dust tail particles have β in the range 0.01–2.5, while dust trails and meteoroid streams typically have $\beta \sim 0.001$.

Those heavier particles follow curved elliptical orbits in the same plane as the comet, forming a thin sheet. When the Earth crosses the orbital plane of the comet it is possible for large dust grains, ejected at some prior time, to be in projection against the sky in front of the comet nucleus. The result is a spike pointing in the direction of the Sun. This is called the *antitail* of the comet (Fig. 3.3). Some amount of spreading of large grains is needed to get an antitail and that usually means that antitails are best seen after the comet has passed perihelion, permitting the dust moving on different orbits to separate far enough from the comet.

The apparent trajectories of the meteoroids after ejection from the comet can be calculated, assuming different levels of radiation pressure. Fig. 3.4 shows a *synchrone/syndyne diagram*. The curved dashed lines mark the position of a cloud of particles with

[12] From the ratio of the force of radiation pressure and the force of gravity: $\beta = (Q_{pr}\, L_{\odot} A/4\,\pi\, c)/G\, M_{\odot} M$, where $M_{\odot}$ and $L_{\odot}$ are the Sun's mass and total energy emitted per second, A is the projected cross-sectional area of the particles, G is the gravitational constant, c is the speed of light, and M is the mass of the metoroid. From: Z. Sekanina, M. S. Hanner, E. K. Jessbergr and M. N. Fomenkova, Cometary dust. In *Interplanetary Dust*, ed. E. Grün, B. Å. S. Gustafson, S. F. Dermott and H. Fechtig. (Berlin: Springer, 2001), pp. 95–161.

Fig. 3.3 The dust tail is a thin sheet as shown by Hale–Bopp when Earth crossed the plane of the comet's orbit on January 5, 1998 (Photo with the ESO 1.4 m Schmidt Telescope by Guido Pizarro). The inset shows the antitail of comet C/1995 Q_1 (Bradfield) as seen from slightly different perspectives while crossing the comet orbital plane, in images taken by Allessandro Dimai of the Associazione Astronomica Cortina, Obs. "Helmut Ullrich" at Col Drusciè – Italy, on September 30 (03:25 UT), October 04 (03:40 UT and 04:06 UT), and October 21 (04:06 UT).

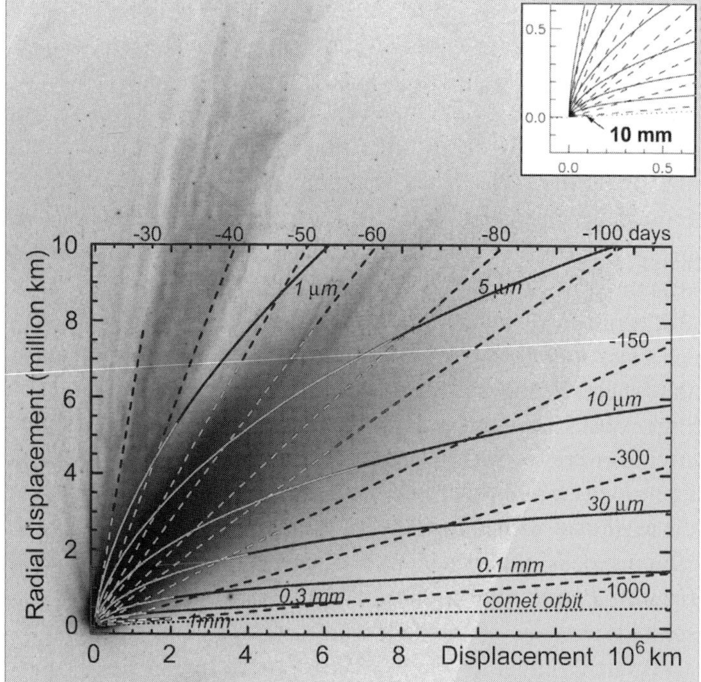

Fig. 3.4 Synchrones and syndynes overlaying the image of comet Hale–Bopp. The Sun is towards the bottom of the graph, the comet moves from right to left along the dotted line. Synchrone and syndyne lines show the position of dust grains ejected at different times for grains of different size, respectively, as calculated by Marco Fulle of Trieste Observatory, Italy. The time of ejection is in days prior to perihelion. The large mm–cm sized grains are very close to the comet nucleus and in the orbit of the comet (inset).

different β ejected at *the same time* and observed a while later. This line is called a *synchrone* (from the Greek word *chronos* = time, as in "chronological"). Synchrones become visible as dust streamers when there is a sudden brief outburst of comet activity. Comet Halley had at least six such streamers.[13] Banded rectilinear structures in the dust tail, separated from the nucleus, are called *striae*, and are caused by large dust particles that fell apart into innumerably more tiny dust particles a short time after ejection.[14] Those dust particles are then pushed outward, more so for finer grains. As a result, this swarm of particles spreads out into elongated stripes along synchrones.

The curved solid lines in Fig. 3.4 are the position of particles of *the same* β (or forces, proportional to mass) ejected at different times. Such a line is called a *syndyne* or more correctly *syndyname* (from the Greek word *dunamis* = power; the same root as for the English words dynamite and dynamo). The graph of syndynes and synchrones in the coordinate system of the comet show the age and mass of the particles at any position in the comet dust trail. *M. L. Finson* and *R. F. Probstein* developed a method using syndynes and synchrones to calculate the distribution of dust in the tail of a comet using the approach of adding the contributions from superimposed uniformly expanding shells, later used extensively and improved by Zdenek Sekanina, Marco Fulle, and others.[15]

3.3 The formation of dust trails

Comets do not leave the large meteoroids behind like a bar of soap in water. They initially move away from the comet only slowly (Fig. 3.3) and then spread quite dramatically in the form of a dust trail after one revolution. The formation of such structures was first described, in a manner, in the nineteenth century, notably as early as 1877 by the Russian astronomer *Theodor Brédikhine* of Moscow.[16] He was a very active and bold astronomer, responsible for introducing the names "synchrone" and "syndyne," but he also pursued many incorrect ideas. Based on the misconception that comet antitails were ejecta towards the Sun, Brédikhine correctly proposed that meteoroid streams were formed from nuclear ejections towards the Sun. To account for the antitails, he assumed that the grains were ejected with high enough speeds to populate a sheet of dust with elliptic, parabolic, and hyperbolic orbits.

Shortly after Fred Whipple calculated actual ejection speeds, which were much smaller, *Miroslav Plavec* (the Technical University, Prague) first described the

[13] P. Lamy, Ground-based observations of the dust emission from comet Halley. *Adv. Space Res.* **5** (1986), 317–323; K. Beisser and H. Boehnhardt, Evidence for the nucleus rotation in streamer patterns of comet Halley's dust trail. *Astron. Space Sci.* **139** (1987), 5–12.

[14] Z. Sekanina and J. A. Farell, Two dust populations of particle fragments in the striated tail of comet Mrkos 1957 V. *Astron. J.* **87** (1982), 1836–1853.

[15] M. L. Finson and R. F. Probstein, A theory of dust comets. I. Model and equations. *Astrophys. J.* **154** (1968), 327–352.

[16] From: C. P. Olivier, *Meteors* (Baltimore: Williams and Wilkins, 1925) pp. 207–211; Th. Brédikhine, Sur l'origine des étoiles filantes. *Bull. Soc. Imp. Nat. Moscou* 1888; Th. Brédikhine, *Bull. l'Acad. Imp. Sci. St. Petersbourg* **17** (1902), 181; R. Jägermann, *Professor Dr. Th. Bredichin's Mechanische Untersuchungen über Cometenformen.* (St. Petersburg: Voss, 1903).

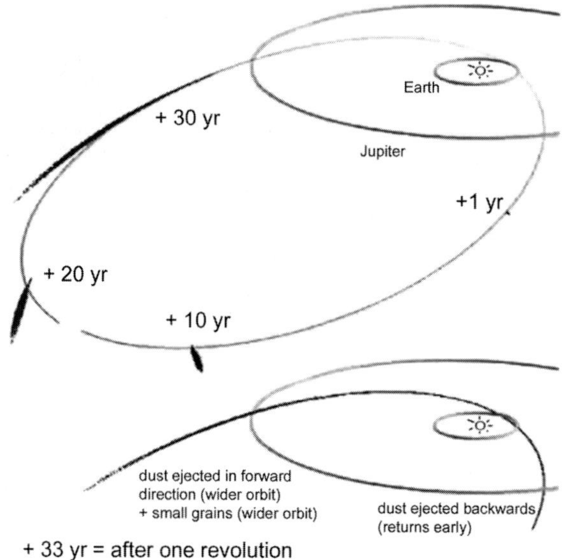

Fig. 3.5 Formation of a dust trail from dust ejected by comet 55P/Tempel–Tuttle, the parent of the Leonid shower, as calculated by Jérémie Vaubaillon, IMCCE.

formation of dust trails correctly as a result of slightly different orbital periods between the dust and the comet.[17] Hence, dust trails had been predicted to exist before they were first detected in space by IRAS.

Dust trails simply result from differences in the orbital period caused by ejection velocity and radiation pressure. The formation of a comet dust trail is illustrated in Fig. 3.5, which is a computation of the trajectory of 50 000 meteoroids ejected from comet 55P/Tempel–Tuttle in the year 1767. This figure from the Ph.D. thesis of Jérémie Vaubaillon shows nicely how the dust initially is a clump (a short dust *tail*) near the nucleus of the comet and only starts to spread out significantly when the dust arrives at the furthest point from the Sun. At this "aphelion," everything happens in slow-motion.

Most spreading along the comet orbit is established on the inward leg. That is also when planets influence meteoroids differently in different parts of the trail.

There is a 1:1 relationship between the conditions of ejection (and radiation pressure) and the final place of each particle in the dust trail (Fig. 3.5). The particles that were slowed down by ejection (or the larger particles that were pushed outward least by radiation pressure) will have the shortest orbit and return first, while the rest will follow later after completing a longer orbit. Under certain restrictions, the change in the meteoroid orbit and the subsequent dispersion of the dust can be expressed in analytical form (Appendix B).

[17] M. Plavec, A classification of the meteor streams. *Bull. Astron. Inst. Czechoslov.* **5** (1954), 15–21; M. Plavec, Ejection theory of the meteor shower formation I. Orbit of an ejected meteor. *Bull. Astron. Inst. Czechoslov.* **6** (1955), 20–26; M. Plavec, On the origin and early stages of meteor streams. *Ceskosl. Akad. ved. Astr. Ústav Publ.* **30** (1957), 93 (see *Nature* (1957) **179**, 1063).

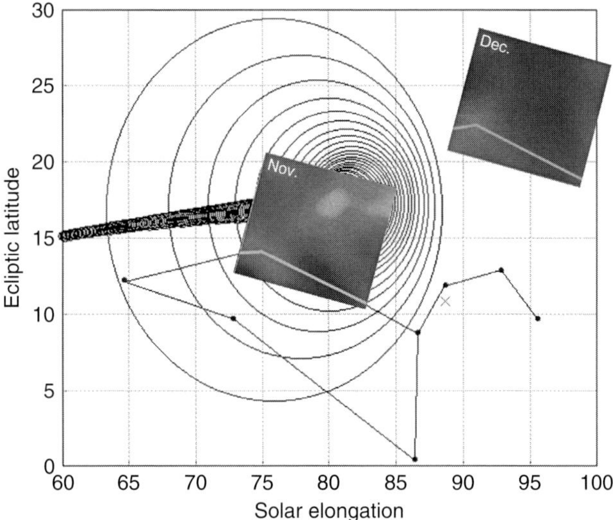

Fig. 3.6 The dust trail of comet 55P/Tempel–Tuttle in scattered sunlight in the direction of the approaching Leonid meteoroids. Results by Ryosuke Nakamura *et al.*[18]

Kresák[19] recognized that the smaller particles have the highest radiation pressure and will end up lagging the comet most. For that reason, meteor storms coming long after the passage of the comet are expected to consist of fainter meteors.

Indeed, Sykes and coworkers analyzed the eight dust trails detected by IRAS and found that they extended mostly behind the comet. The spreading along the comet orbit implied particle sizes of about 1 mm. Particles in front of the comet needed to be at least 6 mm in size,[20] because solar radiation pressure would delay the meteoroids to arrive after the comet if they were smaller. When hitting Earth's atmosphere slowly, 6 mm sized meteoroids cause a meteor of $+4.5^{\mathrm{m}}$ in a typical slow collision, while a fast collision would result in a bright $+0.2^{\mathrm{m}}$ meteor!

The grains were dark, their temperature implying that only 5% of light reflected back. The first optical detection of comet dust trails from scattered sunlight (the way we see the small particles in comet tails) was made by looking along the dust trail of comet 55P/Tempel–Tuttle at the time of the 1998 Leonid encounter. In setting up a coordinated observing campaign, I found that researchers at the University of Kobe in Japan had the expertise to do this experiment. Astronomer *Ryosuke Nakamura* and coworkers peeled away the light of stars, airglow, and zodiacal light from CCD frames taken at Hawai'i, to find a faint diffuse glow at the expected position of the

[18] R. Nakamura, Y. Fum, M. Ishiguro *et al.*, The discovery of a faint glow of scattered sunlight from the dust trail of the Leonid parent comet 55P/Tempel–Tuttle. *Astrophys. J.* **540** (2000), 1172–1176.

[19] L. Kresák, Orbital evolution of the dust streams released from comets. *Bull. Astron. Instit. Czechoslov.* **27** (1976), 35–46.

[20] M. V. Sykes, L. A. Lebofsky, D. M. Hunten and F. Low, The discovery of dust trails in the orbits of periodic comets. *Science* **232** (1986), 1115–1117; M. V. Sykes, D. J. Lien and R. G. Walker, The Tempel 2 dust trail. *Icarus* **86** (1990), 236–247; M. V. Sykes and R. G. Walker, Cometary dust trails. I – Survey. *Icarus* **95** (1992), 180–210.

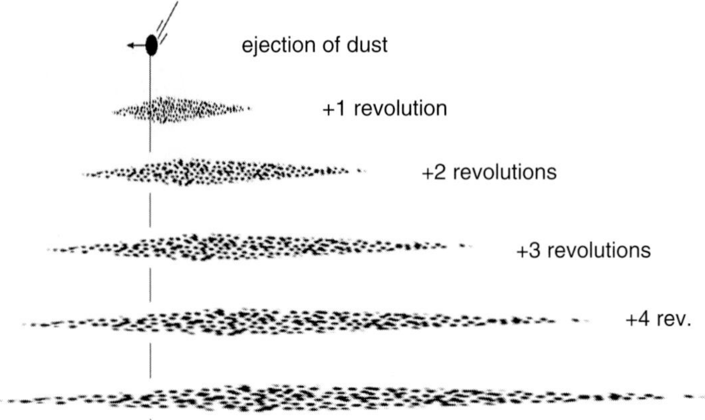

Fig. 3.7 Dispersion of dust in subsequent revolutions.

approaching meteoroids (Fig. 3.6). The November, 1998 image shows the scattered light from the approaching Leonid meteoroids (wire structure), while the December image (off-set) shows the absence of this emission a month later.

The first side view of a dust trail in scattered sunlight was obtained on February 14, 2002, by *Masateru Ishiguro* at Kiso Observatory.[21] It originated from the short-period comet 22P/Kopff, which was located at 3 AU from the Sun at that time, too far to detect an infrared signature.

If the orbital period of each meteoroid remains unchanged, the dust will continue to spread in the direction along the orbit of the comet (Fig. 3.7). The dust trail will increase in length proportionally to the number of revolutions, because the delays from each wider orbit add up. *Zdenek Sekanina* used this in an early attempt to trace meteor storms seen a certain time following the passage of a comet to the episode of ejection in the past.[22]

Finally, meteor storms do not get broader with age: the dust trails do not spread perpendicularly, because each particle ejected at a given point will return to that point if the orbit is not changed. Even if the orbit is changed by perturbations of the planets, those changes tend to be the same for all meteoroids in a cross section of the trail. As a result, the trail can get stretched, but will not broaden. Because of that, all meteor storms of a given stream tend to have much the same duration, even old dust trails.

[21] M. Ishiguro, J.-I. Watanabe, F. Usui *et al.*, First detection of an optical dust trail along the orbit of 22P/Kopff. *Astrophys. J. Lett.* **572** (2002), L117–L120.

[22] Z. Sekanina, Meteoric storms and formation of meteor streams. In *Asteroids, Comets, Meteoric Matter*, ed. C. Cristescu, W. J. Klepczynski and B. Millet (Bucharest: Ed. Acad. Republicii Soc., 1975), pp. 239–267.

4

Meteors from meteoroid impacts on Earth

In the fall of 1798, University of Göttingen students *Johann Friedrich Benzenberg* and *Heinrich Wilhelm Brandes* set out to prove new ideas about the nature of meteors. In 1714, Edmund Halley had challenged Aristotle by suggesting that fireballs are not slow burning terrestrial vapors but solid objects entering Earth's atmosphere at high speed, only later to rescind. Ernst Chladni first reasoned that opinion most convincingly in 1794.

Between September 11 and November 4, Benzenberg and Brandes observed 22 meteors simultaneously from two locations 15 km apart.[1] By teaming up in this manner, each meteor was seen from two different perspectives, against a different background of stars. Comparing star charts, they noticed to their surprise and frustration, that the parallax was much less than expected, which meant that the meteors had to be further away and above the lower layers of the atmosphere that cause the weather.[2] This was a spectacular result! Despite the short baseline, their measured end heights were in the correct range between 35 and 126 km altitude, and they found the meteors traveling at correct speeds of some hundred thousand kilometers per hour (~28 km/s). Unfortunately, due to measurement errors, some solutions gave upward going trajectories, and it took a re-analysis, more triangulations, and another forty years, before it was accepted that meteoroids are solid bodies that come with great speed from outside Earth's atmosphere.

4.1 How dust trails manifest at Earth

The meteoroids in a stream move on nearly parallel trajectories. Standing in the middle of it, warmly clothed in the scented night, with fog on your breath and staring at the sky, an observer on Earth sees all the meteoroids approach and, as soon as they hit the atmosphere, cause a shower of meteors to radiate from one point on the sky, called the *radiant*. TV addicts such as myself recognize the radiant as the direction

[1] H.W. Brandes and J.F. Benzenberg, Versuche, die Entfernungen, die Geschwindigkeit und die Bahnen der Sternschnuppen zu bestimmen. *Ann. Phys.* **6** (1800), 224.

[2] C. Hoffmeister, Hundertfuenfzig Jahre Meteorforschung. *Sterne* **24** (1948), 33–37.

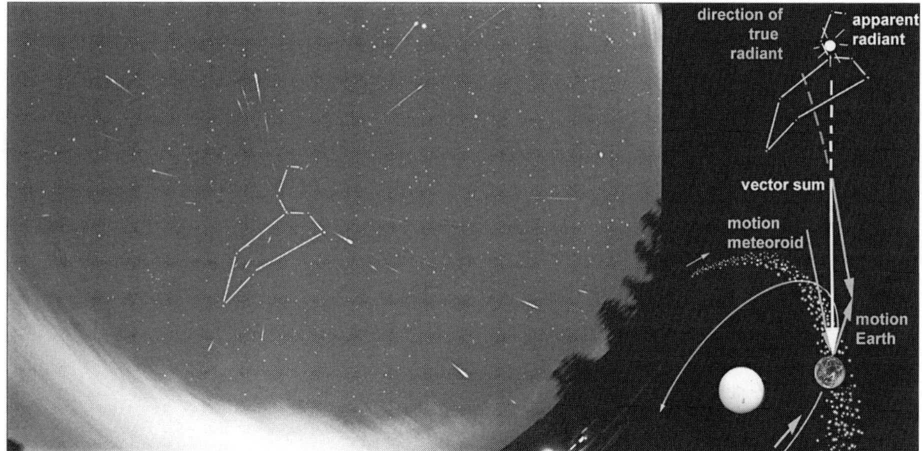

Fig. 4.1 The true and apparent radiant of the Leonid shower. The photo of the 2001 Leonid storm is by Ishiro Ohno of Kanazawa city, Japan. This image was made by combining photographs taken on ISO 800 film with a 15 mm F2.8 lens during 15:45–20:40 UT on November 18, 2001.

from which the stars are seen to approach in "10-forward." In that case, the radiant direction is purely determined by the direction of motion of the observer. For meteor showers, the radiant direction is a combination of the velocity of our spaceship (Earth) and that of the meteoroids.

We speak of the *velocity* of a meteor when both direction and magnitude matter, usually depicted by an arrow (vector) of given length and angle. *Speed* refers only to the magnitude of velocity, irrespective of direction. In order to find the radiant direction, one has to add the velocity arrows of both Earth and meteoroid. This is called a *vector sum*. In the case of the Leonids, Earth moves at about $V_E = 29.6$ km/s (ignoring the Earth's daily spin) in a direction slightly west from where the meteoroids are approaching at 41.1 km/s, both in a reference frame where the Sun is at rest. As a result, the apparent radiant in the head of Leo is slightly west from the "true radiant" (Fig. 4.1). The *true radiant* is the direction from where the meteoroids approach and the direction where Nakamura *et al.* discovered the diffuse glow of scattered sunlight (Chapter 3, Fig. 3.6).

When Leonid meteoroids fall in the gravity well of the Sun from far, they reach 41.1 km/s at Earth's orbit.[3] A collision with Earth creates a *geocentric velocity* (= from the perspective of the center of Earth) of: $V_g = 29.6 + 41.1 = 70.7$ km/s. This is strictly a vector sum, but here I ignore the small angle difference.

[3] The total energy of a comet of mass M in the solar system is the sum of the kinetic and the potential energy: $E = \frac{1}{2}MV^2 - GM_{\odot}M/r$, where V is the comet's velocity and r is its distance to the Sun, with $M_{\odot}$ denoting the mass of the Sun and G being the gravitational constant. This energy remains constant: while the potential energy decreases, the kinetic energy increases proportionally.

Depending on the time of day and the position of the observer on the globe (away from Earth's center), the speed is modified slightly (aberration, up to ~0.4 km/s) by the Earth's spin. The speed is further modified by falling into the gravity well of Earth, causing the meteoroid to speed up by another 11.2 km/s before reaching an altitude of 100 km above Earth's surface. This increases the geocentric velocity of the meteor to the observed *atmospheric velocity* (V_∞, just prior to being slowed down by air collisions). The gain in speed is the result of a transfer of potential energy into kinetic energy and therefore the gain is calculated as a sum of squares: $V_\infty = \sqrt{(70.7^2 + 11.2^2)} = 71.6\,\mathrm{km/s}$.

The direction of V_g is called the *geocentric radiant*, and is expressed in equatorial coordinates of *Right Ascension* (R.A.) and *Declination* (Decl.). The gravity of Earth also changes the direction of the meteoroid motion, moving the radiant to an apparently higher position on the sky. This phenomenon is called *zenith attraction* and is more pronounced for slow meteors. A method for calculating the change in radiant and speed is given in Appendix B. The gravitational attraction of Earth also increases the area over which Earth sweeps up dust (and hence the rate of meteors) by about a factor $1 + V_E^2/V_g^2$,[4] while the observed rate of meteors, the "zenith hourly rate", relates to the actual dust density in the stream approximately as $\sim\mathrm{ZHR}/V_g/V_\infty^{3.92}/(1 + V_E^2/V_g^2)$.[5] All radiants and speed mentioned in this book, unless specifically stated, are the geocentric radiant and speed, before the influence of zenith attraction.

Because Earth is always changing direction in its course around the Sun, the radiant moves from day to day. Despite William F. Denning's claims to the contrary in the early days of visual meteor shower observations, when he was the authority on the matter, radiants are not stationary. The change in direction of Earth's motion is 360° in 365.25 d, or about one degree each day. The position of the radiant changes accordingly. The Perseid shower radiant, as Denning himself showed for the first time, is in the constellation of Cassiopeia in early July and shifts towards Perseus in August, where the shower peaks, and then on into Camelopardalis (Fig. 4.2).[6]

The amount of radiant drift depends on the ecliptic latitude of the radiant and also on the distribution of meteoroid orbits. If on parallel orbits, this *daily drift* of the radiant will be along a small circle parallel to the ecliptic plane.

Finally, it is helpful to realize that the path of Earth through the meteoroid stream is also given by the sum of the velocity vectors (Fig. 4.3). The path makes a shallower angle if the geocentric velocity is larger. The measured duration of a meteor storm (W) is usually larger than the intrinsic width of the trail (W_t).

[4] E. J. Öpik, Collision probabilities with the planets and the distribution of interplanetary dust. *Proc. R. Irish Acad.* **54** (1951), 165–199.

[5] L. Kresák, Cometary dust trails and meteor storms. *Astron. Astrophys.* **279** (1993), 646–660.

[6] R. Arlt, Radiant ephemeris for the Perseid meteor shower. *WGN* **31** (2003), 19–28.

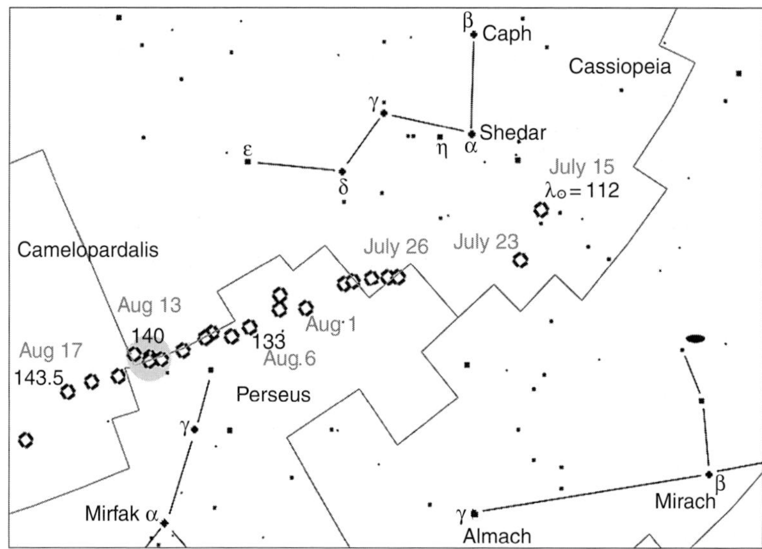

Fig. 4.2 The daily drift of the Perseid radiant from a synopsis of video observations by Rainer Arlt, IMO.

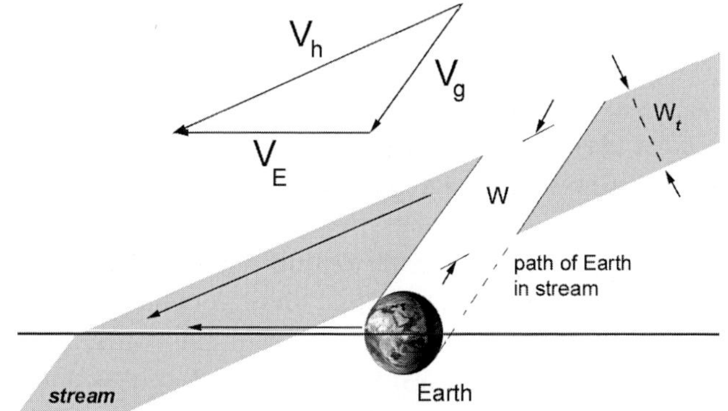

Fig. 4.3 The path of Earth through the meteoroid stream.

4.2 The structure of our atmosphere

After Brandes and Benzenberg, meteors were no longer a meteorological phenomenon. The old atmospheric inflagration explanation of Aristotle still popped up as late as 1892, but by that time it was commonly followed by a swift dismissal (ironically a century later lightning was discovered to reach in the upper atmosphere to the meteor layer in a phenomenon called "sprites." In one instance lightning was observed to travel from the cloud tops to the upper atmosphere, in part, along the ionized trail of a meteor ...[7]).

[7] E. M. D. Symbalisty, R. A. Roussel-Dupré, D. O. Revelle *et al.*, Meteor trails and columniform sprites. *Icarus* **148** (2000), 65–79.

Nevertheless, meteors are observed because the meteoroids collide with the atmosphere. As such, meteors are sensitive probes of the structure of the atmosphere in a range of altitudes that is not easily studied by other means. The first researchers to use this fact were *F. A. Lindemann* and *G. M. B. Dobson* at Oxford University in 1922.[8] At that time, the structure of the atmosphere above 35 km altitude was unknown. They used the height and speed measurements obtained by Denning to show that the air temperature begins to rise again above 50 km and that the density of the upper atmosphere was much higher than previously thought. That temperature rise is on account of warming by ultraviolet (UV) (<200 nm) light from the Sun.

Small variations of meteor rates can occur if the density scale-height at these altitudes alters by a small amount, for example when the atmosphere expands in response to solar activity.[9] Perseid shower rates have been found to be up by 20% in years of low sunspot activity, but the effect has never been established beyond doubt, in my opinion, partially because stream rates may also vary with the orbital period of Jupiter due to planetary perturbations on the same time scale (Chapter 11).[10]

The brightest Leonid fireballs are first seen as high up as 200 km (Fig. 4.4), in a region of our atmosphere called the *thermosphere*.[11] During daytime, this region absorbs the Sun's hard-ultraviolet light, which warms the air (at ground level 78.1% N_2, 20.9% O_2). UV light also breaks molecules and dislodges electrons to create atoms and charged particles, consisting of negatively charged *electrons* and their counterpart, the positive *ions*. This charged, or ionized, part of the atmosphere is called the *ionosphere* (Fig. 4.5).[12]

For a solid particle to slow down significantly, it has to meet more than its own mass in air. Hence, meteoroids penetrate deeper into the atmosphere than the Sun's hard-UV light, down below 120 km and into the cold *mesosphere*. The mesosphere is warmed by the longer wavelength UV sunlight that is absorbed by ozone molecules (protecting us from sunburn). Most ozone molecules are in a layer at about 37 km altitude. Because of this, air temperatures are highest here. Only the largest fireballs penetrate that deep.

The top of the mesosphere, where most meteoroids are stopped, is called the *mesopause*. This is the coldest place on Earth, a frosty ~180 K (−90 °C), with gale winds of 10s of m/s. Fast protons and electrons from the Sun will also sometimes penetrate into the mesopause and cause *aurora* in a circular region around the magnetic poles. For that reason, the mesopause is often considered to be the boundary between the Earth's atmosphere and the Sun's atmosphere. Here on the edge of space, the air density is about 5×10^{-9} g/cm^3 and the pressure about 0.003 mbar. Between 70 and 120 km, the density increases in an exponential manner by a factor of 10 every

[8] F. A. Lindemann and G. M. B. Dobson, A theory of meteors, and the density and temperature of the outer atmopshere to which it leads. *Proc. R. Soc., London* **102** (1922), 411–437.

[9] C. D. Ellyett and J. A. Kennewell, Radar meteor rates and atmospheric density changes. *Nature* **287** (1980), 521–522; B. A. Lindblad, Meteor radar rates, geomagnetic activity and solar wind sector structure. *Nature* **273** (1978), 732–734.

[10] P. Jenniskens, Meteor stream activity I. The annual streams. *Astron. Astrophys.* **287** (1994), 990–1013.

[11] P. Spurný, H. Betlem, K. Jobse, P. Koten and J. van't Leven, New type of radiation of bright Leonid meteors above 130 km. *Meteoritics Planet. Sci.* **35** (2000), 1109–1115.

[12] M. C. Kelley, *The Earth's Ionosphere. International Geophysics Series.* (New York: Academic Press, 1989).

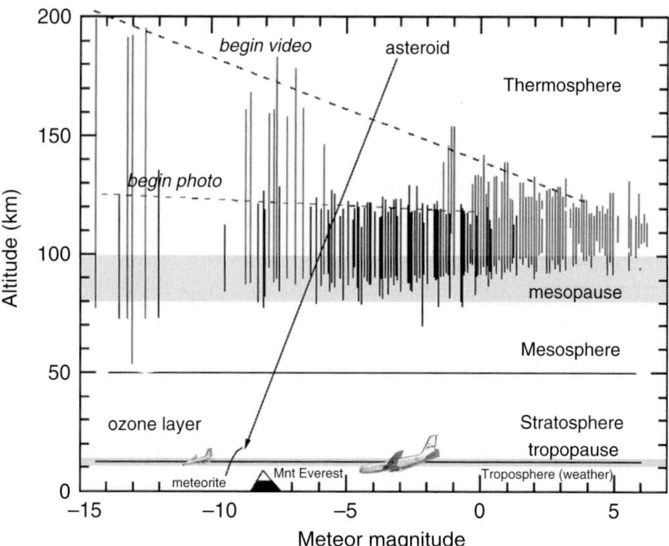

Fig. 4.4 Typical beginning and end heights of Leonid meteors in Earth's atmosphere from recent video and photographic observations by the Dutch Meteor Society.

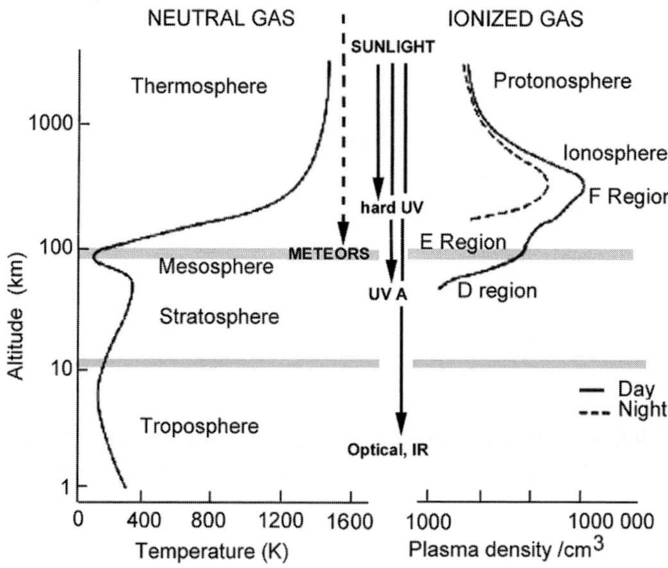

Fig. 4.5 Structure of the Earth's atmosphere for neutral and ionized molecules and atoms (dashed line: night-time). The ionized layers reflect how deep hard-UV light penetrates into the atmosphere.

13 km. Due to this very steep increase in air density, small meteoroids penetrate and deposit most of their mass into a narrow zone at around 89 km altitude. And material deposited higher up will settle down to these denser layers.

The meteors add a sprinkle of metal atoms to the atmosphere, creating a metal atom debris layer. The metal atoms help bring together ozone (O_3) and oxygen (O) atoms created during daytime, and in doing so give off a chemiluminescent glow called *airglow*, which includes the orange glow of sodium atoms. At lower altitudes, the atoms react with atmospheric carbon dioxide (CO_2), and in doing so can attract water molecules and form the core of the tiny ice crystals that make up the *noctilucent clouds* at ~84 km. Ice particles in noctilucent clouds scatter sunlight over the horizon, especially its blue colors, and thus stand out in a bright bluish glow against the dark background of twilight. The ice particles also help sustain ionization and cause sporadic *E layers*, responsible for unusual disturbances in radio and TV reception.

4.3 The mass of a meteor

The high altitudes measured by Brandes and Benzenberg led Professor Olmsted in 1834 to believe that the meteoroids had to be light weight and combustible to explain the high speed and brilliant light.[13] Here we still recognize Aristotle's view on meteors.

However, chemical energy released during burning is not the only potential source of the energy of radiation we call "light." The man who discovered nature's law that the sum of all energy is conserved, *James Prescott Joule* (1818–1889) – his name rhyming with cool, reminded his students to imagine the effect of a cannonball shooting through the classroom.[14] The exploding gun powder causes a cannonball of a mass M to move fast, achieving energy of motion called *kinetic energy*: $E = 0.5\, M V^2$, possibly achieving a speed of $V = 360$ kilometers per hour (km/h). That energy fuels the explosion at the other end of its trajectory. Leonids of the same mass move at 720 times that speed, causing a much bigger explosion, no matter what the makeup.

> NASA learned this the hard way when, in January, 2003, a piece of foam hit the reinforced carbon–carbon wing edge of the Space Shuttle Columbia during takeoff. That lightweight piece of foam with a mass of $M = 0.75$ kg (1.7 lb) hit with a velocity of about $V = 850$ km/h (530 mph), or with a kinetic energy of $E = 0.5 \times 0.75$ $(850 \times 1000/3600)^2 = 20\,906$ J. In comparison, Leonid meteoroids move at an astounding $V_\infty = 71.6$ km/s (258 000 km/h). Because of that, a fast Leonid meteoroid as light as 0.008 g will pack the same punch as the slower moving large piece of foam! Such a tiny Leonid meteoroid could have caused the same disaster should it have hit the Space Shuttle's wing edge. Because of that impact hazard, no Shuttle flights were executed during the earlier Leonid storms.

[13] D. Olmsted, Observations of the meteors of November 13, 1833. *Am. J. Sci.* **25** (1834), 354–411.
[14] D. W. Hughes, James Joule and meteors. *Vistas Astron.* **33** (1990), 143–148.

Even if only 1% of that energy is converted into visible light (actual *luminous efficiency* $\tau \sim 0.1-1\%$), then a tiny 0.008 g meteoroid moving at 71.6 km/s would shine like two hundred 100 Watt (= Joule/s) lamps for a whole second. At a distance of 100 km, it would make for a naked eye $+3^m$ Leonid, which solves Olmsted's dilemma.

The mass of a given meteor is usually calculated from a simple scaling equation introduced by Luigi Jacchia and Fred Whipple,[15] valid for the photographic film used in the Harvard Super-Schmidt cameras. This equation follows simply from luminosity being proportional to kinetic energy and an assumption about how the luminous efficiency depends on speed and how "magnitude" is defined. Appendix C gives a corresponding equation for visually observed meteors from a distance of 100 km (m_v), strictly for $V_\infty > 25$ km/s and $M < 1$ kg:

$$\log M\ (\mathrm{g}) = 6.31 - 0.40 m_v - 3.92 \log V_\infty\ (\mathrm{km/s}) - 0.41 \log(\sin(h_r)) \qquad (4.1)$$

A zero magnitude Leonid of pre-atmospheric speed $V_\infty = \sqrt{(V_g^2 + 11.2^2)} = 71.6$ km/s with the radiant at an elevation $h_r = 45°$ would have a mass of $M = 0.13$ g and $\tau = 0.21\%$.

During the 1998 Leonid Multi-Instrument Aircraft Campaign mission, we made an effort to measure the mass of a *Leonid* meteor by probing the neutral iron atom debris left in the path of a meteor, using the University of Illinois at Urbana resonant Fe Boltzmann lidar, in a project led by Chester S. Gardner and Xinzhao Chu (Fig. 4.6), and simultaneously filming the meteors with a high-definition intensified TV camera operated by the Japanese Broadcasting Service (NHK), in a project led by Hajime Yano. The lidar sends pulses of near-UV laser light up to the meteor layer, which are absorbed and re-emitted by the iron atoms in the trail. The time it takes the light pulse to travel up to the meteor layer and back down is used to measure the distance to the trail and its vertical width. The intensity of the scattered light is proportional to the iron atom (Fe) density.

Twenty atom debris trails were detected. In only one case could we identify the meteor that caused the trail: at 17:05:58 UT on November 17, when a -2.9 ± 0.3^m Leonid passed by the lidar beam just ahead of the direction of flight (Fig. 4.7). A 10 s signal (during which the aircraft moved 1640 m, suggesting a trail FWHM ~1263 m) was detected at 101.14 km altitude peaking at 17:06:59 UT, right when the aircraft was below the train, for which Chu calculated a peak Fe atom density 3.27×10^4 cm^{-3}.[16] One minute after deposition, the trail still had a

[15] F. Verniani, Meteor masses and luminosity. *Smithsonian Contrib. Astrophys.* **10** (1967), 181–195; L. G. Jacchia, F. Verniani and R. E. Briggs, An analysis of the atmospheric trajectories of 413 precisely reduced photographic meteors. *Smithsonian Contrib. Astrophys.* **10** (1967), 1–139.

[16] X. Chu, W. Pan, G. Papen, *et al.* Characteristics of Fe ablation trails observed during the 1998 Leonid meteor shower. Geophys. Res. Lett. **27** (2000), 1807–1810.

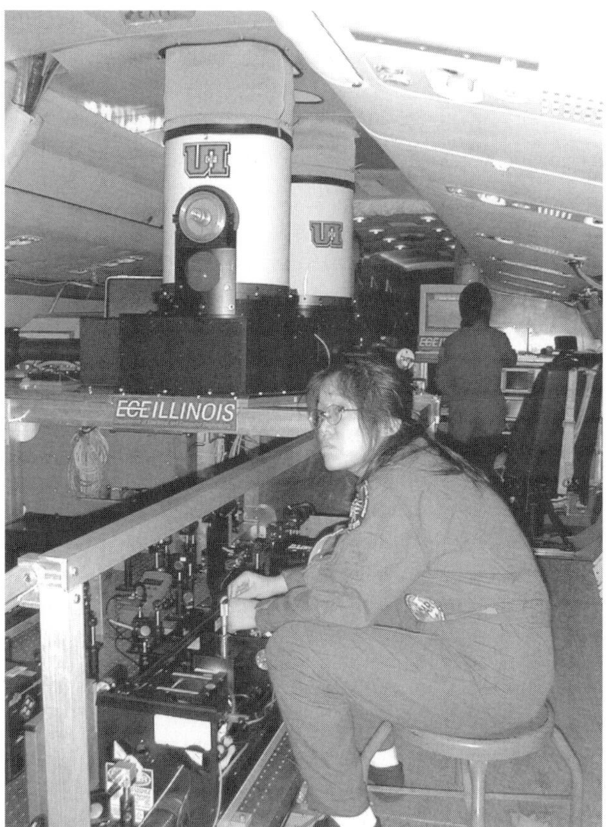

Fig. 4.6 Xinzhao Chu (foreground) and Weilin Pan operate the University of Illinois at Urbana two-beam Fe Boltzmann lidar installed in the NSF/Electra aircraft during the 1998 Leonid MAC mission. Photo courtesy: Chet Gardner, UIU.

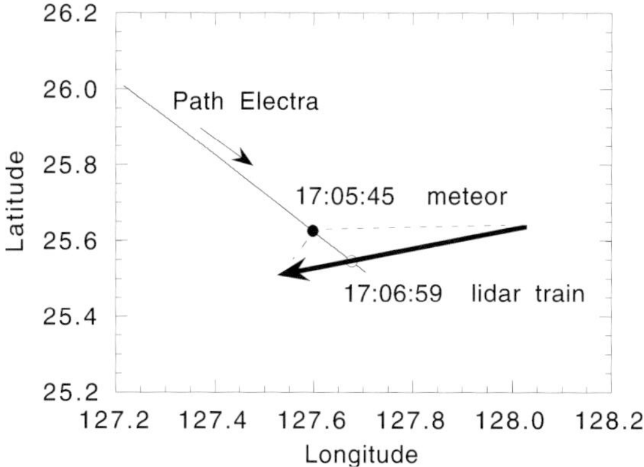

Fig. 4.7 Detection of the neutral atom debris train (open circle) compared to the observed meteor track (line) and moment that the meteor was observed (dot).

vertical rms width of only 63 ± 5 m, but was already ~1640 m dispersed in horizontal direction. From 17:07:15 to 17:07:45 (peaking at 17:07:30 UT), a second detection was made at 100.20 km altitude with only a factor of two lower $Fe = 1.66 \times 10^4 \, cm^{-3}$ density, compensated for by the factor of three wider horizontal width. The vertical rms width $= 78 \pm 19$ m. This was probably the same trail, distorted in the upper atmosphere winds.

From the measured Fe density and an expected Fe (55.8 g/mole) abundance fraction by mass of 6.2%, I calculate that the Leonid meteor of -2.9^m at 100 km distance moving at 71.6 km/s deposited 0.25 ± 0.04 g/s matter in atomic form at 101.14 km and 0.48 ± 0.07 g/s at 100.20 km. That mass represents a kinetic energy deposition of 0.64 and 1.2×10^6 J/s, respectively. The visible light output of the meteor is 13 800 W, so that $\tau = 2.2 \pm 0.4\%$ and $1.2 \pm 0.2\%$, respectively.

According to Eq. (4.1), 0.21% of kinetic energy is transferred into light. If this estimate is correct, then much mass is not counted. The lidar does not detect the iron atoms that are in ionized or solid form. At ambient temperatures, all iron atoms should be in neutral form, albeit that the recombination process takes some time and starts from a high $Fe^+/Fe \sim 3600$ in the meteor plasma itself. If most of the ions had recombined by the time of the lidar measurement, as expected, then I conclude that as much as 90% of a Leonid meteoroid ended up as solid debris instead of atoms! This may explain why other researchers have found that neutral atom debris trains can vary strongly in the relative composition of the expected meteoric metal atoms.[17]

The meteor light is a combination of emission lines from metal atoms from the meteoroid itself and broad emission bands from the collisionally excited atmospheric molecule N_2 (Fig. 4.8). By putting a so-called *transmission grating* in front of the lens of a camera, astronomers disperse the light into all colors of the rainbow and can thus distinguish the contribution of each atom or molecule. Each color of light creates a separate image of the point-like meteor. The result is called a spectrum (Fig. 4.8). Under normal conditions, that spectrum is surprisingly independent of meteor mass and speed,[18] although the relative contributions of air plasma and metal atoms vary a lot. Note that equation (4.1) does not take into account such changes in the ratio of air plasma and metal atom emissions that are responsible for most observed color changes in meteors.

[17] U. Von Zahn, M. Gerding, J. Höffner, W. J. McNeil and E. Murad, Iron, calcium, and potassium atom densities in the trails of Leonids and other meteors: strong evidence for differential ablation. *Meteoritics Planet Sci.* **34** (1999), 1017–1027.

[18] P. Jenniskens, C. O. Laux, M. A. Wilson and E. L. Schaller, The mass and speed dependence of meteor air plasma temperatures. *Astrobiology* **4** (2004), 81–94.

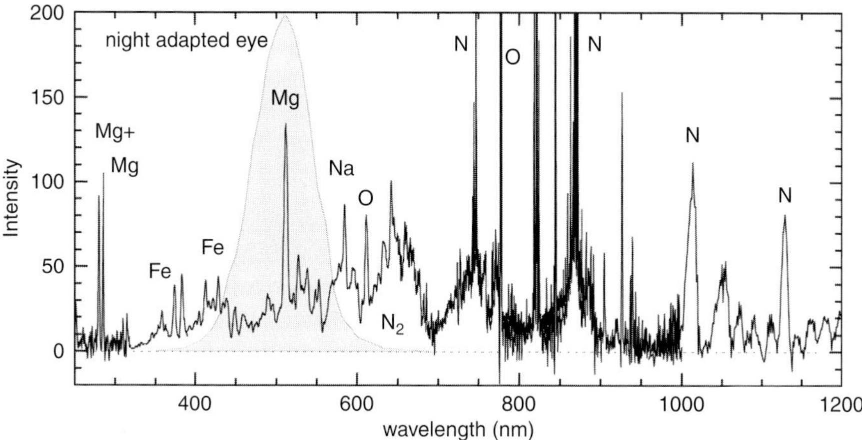

Fig. 4.8 Spectrum of a typical -2^{m} Leonid meteor from ultraviolet (left), over violet, blue, green, yellow, and red, into the near-infrared. Compilation of observations from the *Leonid MAC* campaign in different wavelength regimes (and at different spectral resolutions).

4.4 How meteors emit light

The particulars of how a meteor emits light and what compounds and debris are deposited in Earth's atmosphere are not important to understand the rest of this book and the uninterested reader may want to skip to Chapter 5. These details do matter, however, when meteor showers are used to study the delivery of organic compounds at the time of the origin of life, the supply of metal atoms to the upper atmosphere, or how efficiently the meteoroids of different streams are detected by high-aperture radar, just to mention a few good reasons to read on.

What matters is not just how much energy is available, but how that energy is used. If the energy is only used to break the atomic bonds of the meteoroid, then there is enough kinetic energy to break every single bond in a Leonid meteoroid some 50 times. Indeed, with all those metal atom lines shining bright, it seemed obvious that the whole of the meteoroid was atomized.

In reality, much can remain in solid or molecular form. Most energy goes into heating the air and the process of evaporation can carry away much of the heat imposed on the meteoroid. Because of that, for example, some of the rocky matter of asteroids survives the impact if the impact speed is less than about 22 km/s. The recovered pieces are called *meteorites*. Meteorites have only a thin crust of molten rock and stay at their original hand-warm temperature inside. That crust is a black opaque glass, made dark by sub-micrometer-sized inclusions of magnetite (Fe_3O_4). In this case, the heat is lost by rapid evaporation.

In the same manner, small grains can lose heat efficiently by radiation, if the meteoroids are larger than the wavelength of infrared light. At sizes smaller than the wavelength of infrared light, that radiation process is not efficient. Because of this, tiny 5–50 μm dust grains arriving at a slow ~11 km/s can survive the impact nearly intact, creating *micro-meteorites*, but even smaller grains do not survive.

The classical 1958 book by Ernst Julius Öpik: *Physics of Meteor Flight in the Atmosphere*,[19] and the more recent 1983 book by Vitalij Aleksandrovich Bronshten, *Physics of Meteoric Phenomena*,[20] describe the basic processes of meteor entry in terms of how hot a meteoroid can become and how much matter is lost. I will not repeat this. The reader is referred to these excellent books and that of McKinley.[21]

The recent Leonid storm observing campaigns provided more insight into the actual physical conditions and the fate of the meteoric matter. We now know that at high altitudes (up to 250 km) *sputtering and the subsequent collision cascade* of metal atoms and air molecules with the ambient environment is the dominant luminous mechanism. Colliding air molecules cause meteoric metal atoms to be ejected at speeds in excess of the entry velocity of the meteoroid. The subsequent cascading collisions with the atmosphere result in a broad V-shaped glow, which becomes narrower the deeper the meteoroid penetrates into denser air layers.[22] Sputtering does not significantly depend on the surface temperature of the grain.

Below about 136 km, rapid evaporation adds to the sputtering, the latter accounting for no more than 10% of the total ablation for a 0.01 g grain. The meteoroid surface starts to warm up to the point where minerals near the surface of the grains melt and evaporate in the form of atoms and molecules. An *ablation vapor cloud* is formed that travels along with the meteoroid and surrounds the meteoroid out to a size comparable to the distance an ambient air molecule can travel before hitting another molecule (Fig. 4.9). This *mean free path* is of the order of 1 m at 111 km, 10 cm at 96 km, 1 cm at 83 km, and 1 mm at 68 km altitude. Each impact with an air molecule will evaporate up to 80 atoms and molecules from the surface of the meteoroid. The most important effect of the vapor cloud is to greatly expand the surface area now exposed to collisions, increasing the rate of collisions and hence the brightness of the meteor. The lightcurve of the meteor will show a rapid increase at the onset of this rapid evaporation.

It is only when the meteoroid is larger than this mean-free path, that the molecules start colliding with each other in front of the meteoroid and a shock wave is formed. It is said that the airflow changes from *rarified* to *continuum flow*. Very rarely, this shock wave penetrates deep enough in the atmosphere that it can be heard as a distant rumble long after the fireball is seen. In such exceptional cases, the sound traveling at about 270–350 m/s takes several minutes more to reach us than the nearly instantaneous visible light. Meteors do not normally cause audible sounds. Hence, they will pass by unnoticed if not seen. But hissing sounds ("crackling," "rushing,"

[19] E. Öpik, *Physics of Meteor Flight in the Atmosphere* (New York: Interscience, 1958); B. Yu. Levin, *Physikalische Theorie der Meteore und die meteoritische Substanz im Sonnensystem*, vol. II. *Scientia Astronomica* 4. (Berlin: Akademie-Verlag, 1961), 330 pp.

[20] V. A. Bronshten, *Physics of Meteoric Phenomena*. (Dordrecht: Reidel, 1983), 356 pp.

[21] D. W. R. McKinley, *Meteor Science and Engineering* (New York: McGraw-Hill, 1961), Chapter 7.

[22] Presentations by: D. Vinkovic, O. Popova, R. L. Hawkes *et al. Meteoroids 2004 Conf.*, London, Ontario, Canada.

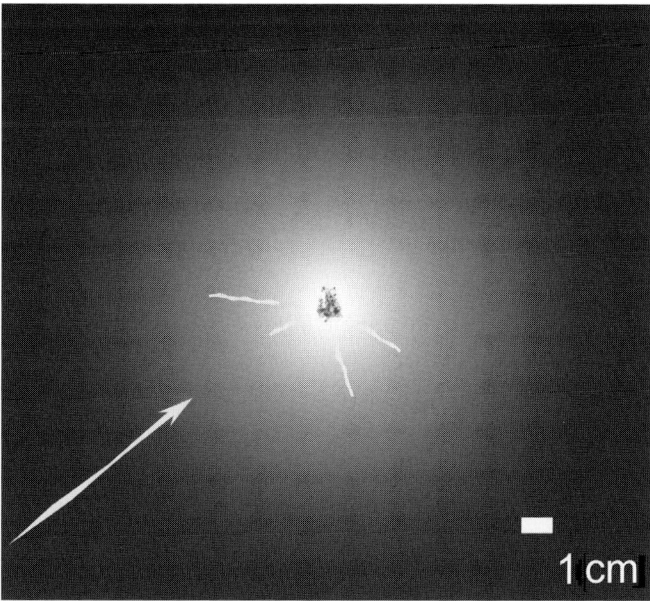

Fig. 4.9 The creation of an ablation vapor cloud upon bombardment of a meteoroid by air molecules.

"popping," "vits," and "sharp clicks") have been reported for very bright meteors. Although still a contentious issue, these sounds are thought to be due to very low frequency (VLF) radio waves interacting with metal, paper, or other electrically conducting materials in the environment.[23]

When air molecules hit the meteoroid or vapor cloud, they speed up relative to the ambient air, but not as fast as the meteoroid, and immediately lag behind. At the same time, the vapor cloud atom is slowed down relative to the meteoroid, lagging behind as well. Both will continue to bounce off many ambient air molecules before slowing down (Fig. 4.10). This cascade of collisions with ambient molecules is called the *cascade phase*, and is responsible for impact excitation of a "hot" component in meteor spectra, dominated in the visual region by light from ions of the elements magnesium, calcium, and silicon. This hot component is most clearly seen in high-velocity or bright meteors.

When the molecules and atoms finally slow down, a column of warm *air plasma* is created with an initial radius of the order of a few meters, mainly determined by the mean free path. This column quickly expands to tens of meters diameter (*expansion phase*) to establish pressure equilibrium with the surroundings (Fig. 4.11). At higher elevations, the original mean-free path and the subsequent dilution is large and the

[23] C. S. L. Keay, Anomalous sounds from the entry of meteor fireballs. *Science* **210** (1980), 11–15; M. Beech and L. Foschini, Leonid electrophonic bursters. *Astron. Astrophys.* **367** (2001), 1056–1060.

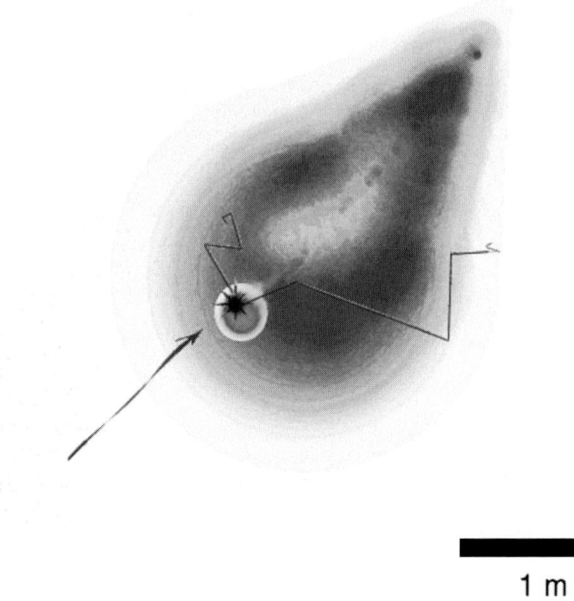

Fig. 4.10 The cascade of collisions in the region behind the meteoroid in the cascade phase.

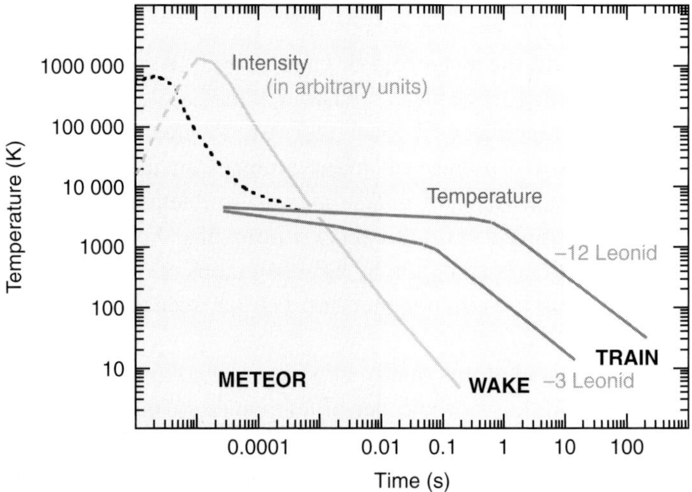

Fig. 4.11 Temperature and intensity in the path of the meteoroid since the meteoroid passed by. Measured temperatures are relative to the ambient atmospheric temperature. The intensity of the meteor is in arbitrary units. A summary of results from the Leonid MAC observations. The dashed lines are results from theoretical models.

electrons generated in the impacts diffuse more rapidly.[24] There is a height above which no radar meteor echoes are detected, which is known as the *echo height ceiling* in back-scatter radar.

The light of meteors originates mostly from this warm *air plasma* due to processes similar to those in discharge lamps such as the familiar low pressure yellow–orange sodium lamps. The plasma is warm, at about 4400 K, and energetic collisions are common. In collisions, electrons bound to those metal atoms such as sodium are knocked into distinct excited states with orbits at larger mean distances from the nucleus of the atoms. When the electrons fall back to their rest positions, light is emitted at very specific wavelengths (orange in the case of sodium, blue–green for magnesium). Light is also emitted from the air molecules that take part in those collisions. Rather than specific emission lines, molecules create bands of light spread over a range of wavelengths that are a reflection of all the vibrational and the rotational states induced by the collisions (Fig. 4.8). The total intensity of light (and the electron density: this is also the region from which the moving radar signals bounce that are called *meteor head echoes*) is proportional to the amount of kinetic energy (meteor mass) deposited in the air at any given moment.

During the 2001 Leonid Multi-Instrument Aircraft Campaign, Hans Stenbaek-Nielsen of the University of Alaska at Fairbanks discovered that Leonids brighter than -3^{m} show a halo around the meteoroid, with a bite-out that is shaped like a shock wave (Fig. 4.12). This halo is now interpreted as the result of hard ultraviolet light generated in collisions in the vapor cloud. This light is absorbed by trace molecules in the air and re-radiated as visible glow.[25]

The size of that glow reflects the decreasing intensity of the UV light source away from the meteoroid. This is an astounding several hundreds of meters around a 1 cm meteoroid (the air pressure is only one millionth of that at the surface). In those same meteors, a shock-like feature has been observed, the source of which is possibly a shadow from the vapor cloud.

The classical lightcurve of a meteor (dashed lines Fig. 4.13) is one in which the meteor brightens exponentially due to the rapidly increasing air density deeper down in the atmosphere, then peaks and fades when the meteoroid becomes smaller.[26]

In practice, there is usually fragmentation of the meteoroid early in its trajectory, after which the meteoroid consists of a number of individual fragments. These

[24] From the decay of radar echoes at wavelength λ with time constant t_d, the electron diffusion coefficient, defined as $t_d = \lambda^2/(16\pi^2 D)$, was measured to increase exponentially from $D = 1\ m^2/s$ at 84 km to $D = 400\ m^2/s$ at 120 km altitude. The classical rate of diffusion of the trail radius is $r = r_0(1 + 4Dt^2)$, with initial radius $r_0\ \rho_\alpha^{-0.25} V^{0.6}$ is 63 cm at 20 km/s increasing to 123 cm at 60 km/s at 100 km altitude, 22 cm at 20 km/s increasing to 42 cm at 60 m/s at 75 km altitude. W. G. Elford, Radar observations of meteors. In *Meteoroids and Their Parent Bodies, Proc. IAS Symp. Meteoroids and Their Parent Bodies*, ed. J. Stohl and I. P. Williams (Bratislava: Inst. Slovak Acad. Sci., 1992), pp. 235–244.

[25] H.-C. Nielsen and P. Jenniskens, A "shocking" Leonid meteor at 1000 fps. *Adv. Space Res.* **33** (2004), 1459–1465; P. Jenniskens and H.-C. Stenbaek-Nielsen, Meteor wake in high frame-rate images – implications for the chemistry of ablated organic compounds. *Astrobiology* **4** (2004), 95–108.

[26] The classical light curve can be expressed in terms of air density (ρ_a) (from McKinley 1961):

$$I/I_{max} = \frac{9}{4}\rho_a/\rho_a^{max} \times (1 - \rho_a/3\rho_a^{max})^2 \qquad (4.2)$$

Fig. 4.12 A -3^{m} Leonid meteor in 1 ms snapshots by Hans Stenbaek-Nielsen of the University of Alaska at Fairbanks. This meteor developed a shock-like feature in a halo of light.

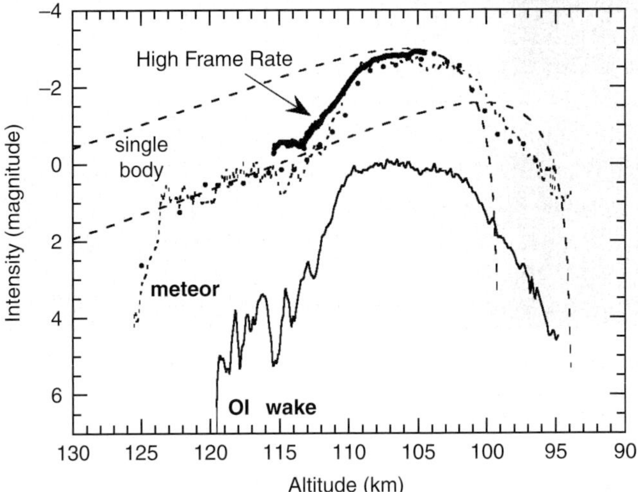

Fig. 4.13 The lightcurve of the Leonid meteor of Fig. 4.12 and that of its OI wake. The dashed lines show the expected lightcurve for a single body.

fragments come down together because the low air density is not efficient at stopping them. This is called the *dust-ball* model of meteor light curves. Deeper into the atmosphere, the smallest fragments are slowed down most and form a wake of debris particles, with the larger fragments penetrating most deeply. If a single fragmentation event creates a spray of tiny particles, each weighing no more than 1 millionth of a gram, a brief *flare* may be observed.[27] Meteor flares show a very abrupt onset. The

[27] H. J. Smith, The physical theory of meteors. V. The masses of meteor-flare fragments. *Astron. J.* **119** (1954), 438–442.

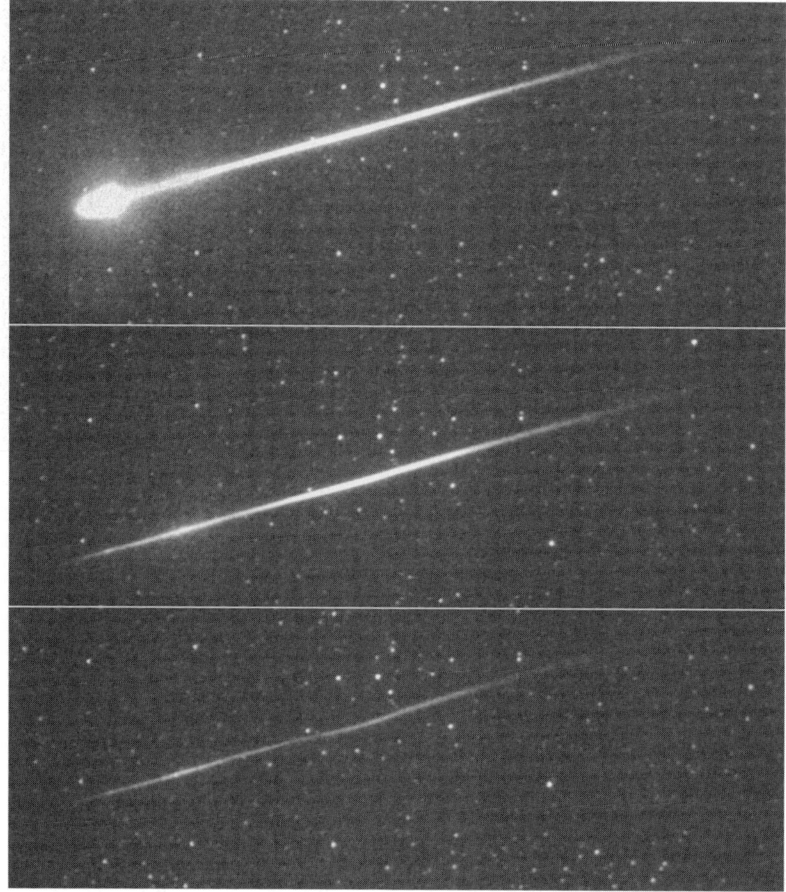

Fig. 4.14 Forbidden green line luminescence in the wake of the 02:15:39 UT (Nov. 18, 1999) Leonid, filmed by an NHK Hivision HDTV-II camera onboard the 1999 Leonid MAC mission. Photo courtesy NHK and Hajime Yano, ISAS.

increased ablation rate will cause an increase of metal atom line emission, often causing the meteor to turn green. Moreover, the large ablation vapor cloud will cause more cascade-phase radiation from the "hot component."

The warm air plasma stretches 5–50 m behind the meteoroid, then fades when it cools and collisions become less frequent. The cooling is gradual enough (and inhibited by continued secondary ablation from debris) to sometimes cause a brief *afterglow* in the meteor images, usually lasting less than a second or two.

Sometimes there is also a slightly longer lasting *wake* due to emission generated by free electrons that find their way back to ions, called *recombination line emission*.[28]

[28] J. Borovicka and P. Koten, Three phases in the evolution of Leonid meteor trains. *ISAS SP* **15** (2003), 165–173.

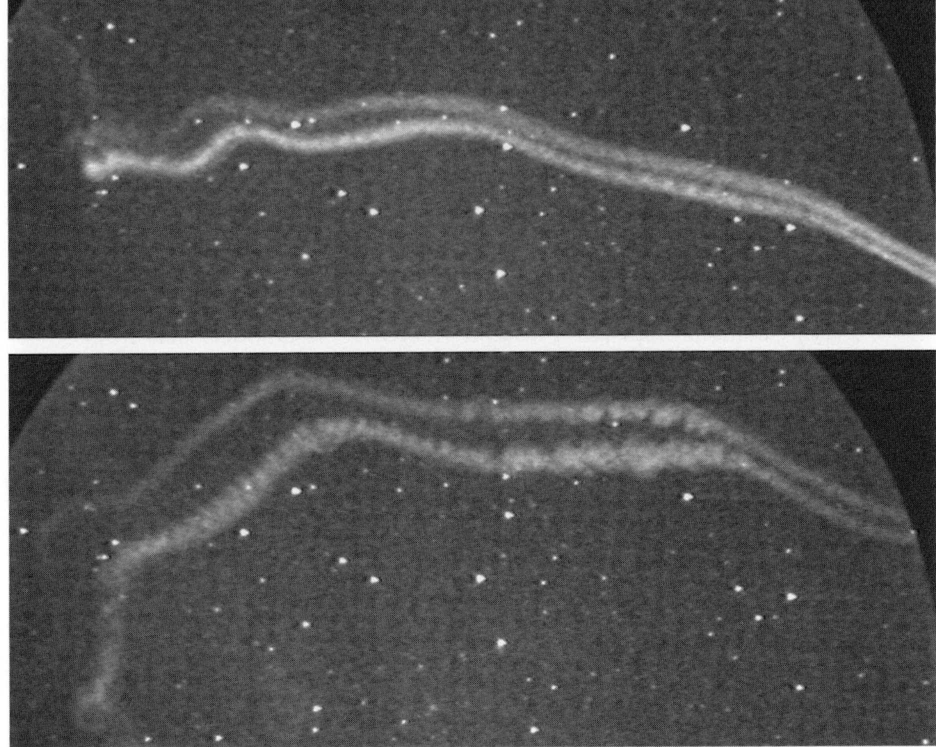

Fig. 4.15 Persistent train 40 and 72 s after a Leonid fireball passed by from right to left. This train lasted 9 min. Photo by Kouji Maeda (Miyazaki Astronomical Group).

This emission can last up to several tens of seconds. The ionization efficiency per gram varies with entry speed $\sim V^5$, making this a more dominant phenomenon in fast Leonid meteors.[29] This emission is very different from that of the meteor itself or its afterglow, containing high energy transitions that do not occur as a result of thermal collisions.

Visual observers can see that fast meteors have a 1–10 s *wake* caused by the "forbidden" 557 nm green line luminescence of oxygen atoms (Fig. 4.14). Forbidden, because the final relaxation step in the energy diagram of oxygen atoms is not allowed by quantum mechanical rules.

Bright fragile meteoroids continue to leave a *persistent train*. Only during the recent Leonid MAC missions was it discovered what is responsible for the eerie glow. Its light derives from chemiluminescence of iron oxide (FeO) molecules and sodium atoms

[29] F. L. Whipple, The physical theory of meteors. VII. On meteor luminosity and ionization. *Astrophys. J.* **121** (1955), 241–249.

(Na), because meteoric iron and sodium atoms participate as catalysts in the recombination reactions of oxygen atoms and ambient ozone molecules.[30]

The glow can persist for many minutes at 95–75 km altitude and thus trace the prevailing winds at those altitudes. Variations in air pressure and density called *gravity waves* are caused by hot air bubbles that buoyantly rise from the troposphere and expand with increasing altitude. These bubbles have wind directions that change rapidly with altitude, causing the familiar corkscrew patterns in windblown trains.

Many persistent trains show two parallel lanes of billows, whereby the amount of billowing can vary (Fig. 4.15). John Zinn (LANL) at the 2003 Leonid MAC workshop explained that the cylinder of low density hot-air in the path of the meteoroid displaces the overlaying column of air more so at the center of the cylinder. Hence, the center will rise more rapidly by buoyancy than the edges, which results in a breakup of the hot air column into two oppositely rotating cylindrical line vortices.[31]

[30] P. Jenniskens, M. Lacey, B. J. Allan, D. E. Self and J. M. C. Plane, FeO "Orange Arc" emission detected in optical spectrum of Leonid persistent train. *Earth, Moon Planets* **82–83** (2000), 429–438.

[31] J. Zinn and J. D. Drummond, Observations of persistent Leonid meteor trails: 4. Buoyant rise/vortex formation as mechanism for creation of parallel meteor train pairs. *J. Geophys. Res. Lett.* **110** (2005), A04306.

5

Comet and meteoroid orbits

The forecasting of meteor storms is all about knowing the orbit of the meteoroids and their parent body in the past, present, and future. When I first learned about meteor orbits in space, I found it very hard to imagine the orbit in three dimensions. Modern computer games and 3D software tools have made it easier to visualize a comet orbit in space, but there remains a need to express the shape and orientation of an orbit with numbers, the so-called *orbital elements*. The astronomical language of orbital elements is the $qae\omega\Omega i$-system[1].

5.1 Orbital elements

Theoretical astronomers still like to give position and motion by three positional coordinates X, Y, Z, which give the location of a comet or meteoroid at a given time, and three velocity coordinates V_x, V_y, V_z, which describe the direction of its motion. They tell us, for example, that the comet on January 1, 2005, is at −1.223 15, +0.352 52, +0.025 33 AU (1 AU = the Earth–Sun distance) and moves at a speed of +22.5251 km/s towards the X-direction, at +12.1523 km/s towards Y, and at +2.1523 km/s towards Z. However, from that it is hard to imagine what type of orbit the comet is in, or how it will move in the future.

Everyone else uses the fact that comets and their offspring move in elliptical orbits around the Sun. The orbits are more elongated than those of the planets and tilted out of the plane of Earth's orbit called the "*ecliptic plane*". The orbital elements describe the shape and orientation of the ellipse (Fig. 5.1) by a system of six numbers (and a variety of derivatives): the size and shape of the orbit (two numbers: q and a), the orientation of the orbital plane (three numbers: ω, Ω, i), and the position in the orbit (one number: T_p).

The shape and size of the ellipse – is described by two numbers: the distance from the Sun (strictly the focal point of the ellipse) at closest approach, called *perihelion distance*, usually assigned the letter "q," and the size of the ellipse, expressed in terms of the *semimajor axis* (letter "a"), with $2 \times a$ being the length of the longest axis of

[1] C. D. Murray, S. F. Dermott, *Solar System Dynamics*. Cambridge University Press, Cambridge (1999), 592 pp.

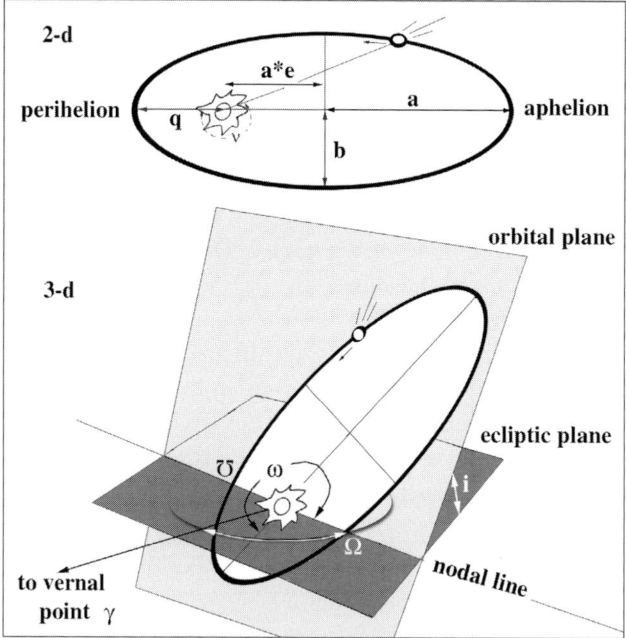

Fig. 5.1 A definition of the orbital elements of a comet or meteoroid orbit in the solar system in two and three dimensions. The line from perihelion to aphelion is called the line of apsides.

the ellipse. Derivative numbers include the point furthest from the Sun, which is called the *aphelion distance*, assigned the letter "Q". The distance between q and Q is twice the semimajor axis. Another derivative is the time it takes to complete one revolution, which is called the *orbital period* (symbol $P = \sqrt{a^3}$). The shape of the ellipse can also be defined by the *eccentricity* (letter e), defined as ae is the distance from the Sun to the midpoint of the ellipse. The eccentricity is usually derived from the equations:

$$q = a(1 - e) \quad \text{or} \quad Q = a(1 + e) \tag{5.1}$$

The eccentricity is $e < 1.0$ for an elliptic orbit, $e = 1.0$ (and $1/a = 0$) for a parabolic orbit, and $e > 1.0$ ($1/a < 0$) for a hyperbolic orbit.

The orientation of the ellipse in space – relative to the ecliptic plane – is defined by three angles. The angle between comet orbital plane and the ecliptic is called the *inclination* of the orbit (i), measured between 0° and 180° (360° being a full circle). The *nodal line* is the intersection of these two planes and its orientation is called the *longitude of the ascending node* (Ω), or *node* for short. The angle is measured between 0° and 360°, and measured from the direction of the spring point (= *vernal equinox*) in the constellation of Pisces. A derivative is the descending node (☋ = Ω + 180°). Finally, the orientation of the ellipse (anchored at the focal point where the Sun is), or direction of perihelion in the plane of the comet orbit, is expressed with the *argument of perihelion* (ω), measured from the direction of the ascending node. The derivative parameter "*longitude of perihelion*" $\Pi = \omega + \Omega$ gives the direction of perihelion measured from the vernal point. The pronunciation of Greek letters is given at the end of this book.

The positions of the comet and the meteoroid in their orbits can be expressed as the moment in time when the comet passes by perihelion, the *perihelion time* (T_p), or alternatively as an angle (*anomaly*) measured from perihelion in the direction of motion along the orbit.[2] A derivative number used for meteor showers is the moment when passing by Earth (the node of the orbit). The *eccentric anomaly* is the angle measured around the center of the ellipse, the *true anomaly* (v) is the angle measured with the Sun at the center.

The motion along the ellipse is not constant. In an eternal swapping of kinetic and potential energy, a comet speeds up while falling in the gravity well of the Sun, then loses that speed again when moving back out. The effect is similar to the roller coaster ride that comes to a near stop at the top of the hill before rushing down again. As a result, comets and meteoroids spend much of their time close to aphelion (the top of the hill).

The *Vernal*, or *Spring equinox* is the direction where the equatorial plane intersects the ecliptic plane. Because the Earth's spin axis changes direction in space continuously like the axis of a tilted spinning top, the orientation of the equator changes and the vernal equinox moves with about +0.013 96°/yr towards Capricorn. Because of this spin axis *precession*, all orbital element angles need to be expressed in a coordinate system that is valid for a given position of the *equinox*, such as on January 1 11:58:55.816 UTC UT in the year AD 2000 (J2000). And because the orbital elements change with time even over one orbit due to the gravitational pull of the planets, they are really valid only for a given *epoch*, or moment in time. Instantaneous orbital elements, describing the unperturbed elliptical orbit based on speed and position at any given moment, are called *oscular orbital elements*, from the Latin verb *osculare*, meaning "to kiss". Throughout this book, I will use J2000 and the epoch being the time of perihelion passage of the comet or the moment the meteoroid is observed, unless otherwise specified.[3]

[2] *The position of the comet or meteoroid* – in the ellipse can be expressed in terms of the time since perihelion passage (T), when the *mean anomaly* $\mathcal{M}$ is defined as: $\mathcal{M} = 360° (t - T)/P$. The comet will speed up near perihelion and slow down at aphelion. The *true anomaly*, the angle v = perihelion – Sun – comet, gives the actual angular position of the planet in its orbit. Calculating v, is an iterative process. First, calculate the *eccentric anomaly* E = the angle perihelion – midpoint ellipse – comet: $E = \mathcal{M} + e \sin(\mathcal{M})\ (1 + e \cos(\mathcal{M}))$. Then iterate using $E' = E$, $E = E' - (E' - e \sin(E') - \mathcal{M})/(1 - e \cos(E'))$, until the magnitude of $E - E'$ is sufficiently close to zero. Finally, the true anomaly v relates to E as: $v = 2 \tan^{-1} \{\sqrt{[(1 + e)/(1 - e)]} \tan(E/2)\}$.

It is useful to realize that when a meteoroid is observed at Earth, it is in either the ascending or the descending node. If it is in the ascending node, then $v = 360 - \omega$ (Fig. 5.1). If it is in the descending node, then $v = 180 - \omega$.

[3] Here, the "J" stands for "Julian," meaning that the year is defined as having precisely 365.25 d. In the old form of "B1950.0", for example, the "B" meant "Besselian," with the year being the tropical year of 365.2421988 d. This affects the definition of the date of the epoch and therefore the exact position of the meteoroid or comet at a given time, but only by a very small amount. The coordinate system adopted by the *International Astronomical Union* is that of the *International Celestial Reference System*, with its origin in the Solar System barycenter, and in which use of the "mean" equator and equinox of J2000 means that nutation of the Earth's spin axis is averaged out or omitted altogether.

5.2 Miss distance

For a meteoroid to hit Earth, the heliocentric distance of the nodes has to be close to the heliocentric distance of Earth. The heliocentric distance at any point along the orbit, for a given position with *true anomaly* (v) is given by:[4]

$$r = q(1 + e)/[1 + e\cos(v)] \tag{5.2}$$

Substituting $v = \omega$, for the ascending node, or $\omega + 180°$ for the descending node, gives the heliocentric distance (r) of that node, respectively.[5]

Compare this to the heliocentric distance of Earth (r_E) at the point of intercept which follows from Eq. 5.2 with $q = 0.983\,289\,90$ AU, $e = 0.016\,710\,22$, and $v = \Omega - 91.686\,55°$ for the ascending node or $v = \Omega + 88.313\,4537°$ for the descending node. The *nodal miss-distance* is defined as: $\Delta r = r - r_E$. This is not always the shortest distance between the Earth's orbit and the comet orbit (δ in Fig. 5.2), especially for low-inclination orbits. This difference plays a role in calculating the maximum time of a shower and to some extent the theoretical radiant, if the orbit does not intersect that of the Earth. For low inclination showers with a node far from Earth's orbit, the difference can be quite large ($\Delta\Omega$).

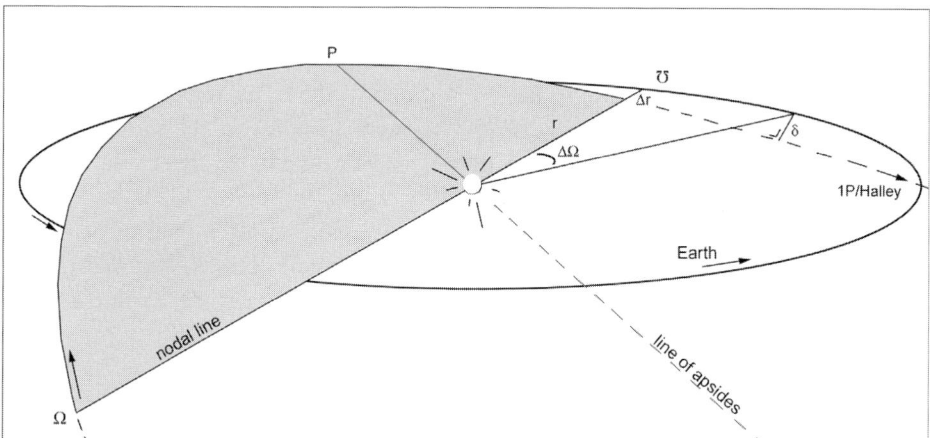

Fig. 5.2 Illustration of the difference between the nodal miss-distance (Δr) and the shortest distance (δ) between comet Halley's orbit and Earth's orbit.

[4] *The position of the comet or meteoroid* – in the ellipse can be expressed in terms of the time since perihelion passage (T), when the *mean anomaly* $\mathcal{M}$ is defined as: $\mathcal{M} = 360°\,(t - T)/P$. The comet will speed up near perihelion and slow down at aphelion. The *true anomaly*, the angle v = perihelion – Sun – comet, gives the actual angular position of the planet in its orbit. Calculating v, is an iterative process. First, calculate the *eccentric anomaly* E = the angle perihelion – midpoint ellipse – comet: $E = \mathcal{M} + e\,\sin(\mathcal{M})\,(1 + e\,\cos(\mathcal{M}))$. Then iterate using $E' = E$, $E = E' - (E' - e\,\sin(E') - \mathcal{M})/(1 - e\,\cos(E'))$, until the magnitude of $E - E'$ is sufficiently close to zero. Finally, the true anomaly v relates to E as: $v = 2\tan^{-1}\{\sqrt{[(1+e)/(1-e)]}\tan(E/2)\}$.

It is useful to realize that when a meteoroid is observed at Earth, it is in either the ascending or the descending node. If it is in the ascending node, then $v = 360 - \omega$ (Fig. 5.1). If it is in the descending node, then $v = 180 - \omega$.

[5] G. S. Hawkins, R. B. Southworth and F. Steinon, Recovery of the Andromedids. *Astron. J.* **64** (1959), 183–188.

There are different ways to adjust the comet orbit so that it intersects with Earth. Long-time director of the B.A.A., *John Guy Porter*,[6] first described a method, used later by Jack Drummond and Duncan Olson-Steel, where the direction of the comet orbit at the node is simply rectilinearly transposed to the orbit of Earth. This is not how a meteoroid orbit differs from a comet orbit. *Ishiro Hasegawa*[7] proposed a technique whereby the line of apsides of the comet orbit is rotated until the orbit intersects Earth's. This assumes a particular mechanism of evolution (precession, see later), which does not apply to young showers, and assumes that the shape of the orbit does not change, which is not the case for old showers. Another way is to change q and ω in unison, but keep Π constant, or to change the semimajor axis a. For low-inclination streams, the results can differ significantly. The only good method would be to calculate a theoretical radiant from the history of the meteoroid orbit since ejection until it can hit Earth. That is a lot of work, but there are relationships between orbital elements that can sometimes be used to get a good guess of how the orbit changes over time.[8] These methods were summarized in a software program by *Lubos Neslusan* of the Astronomical Institute of the Slovak Academy of Sciences,[9] which was used here to calculate theoretical radiants.

5.3 Reservoir of Jupiter-family comets: the Kuiper Belt

With this system of orbital elements, it is possible to study the origin and evolution of comet orbits. Comets originate in large reservoirs in the outer regions of our solar system. Many short-period comets ($P < 20$ yr) tend to move in the same direction (prograde) as the planets. Because they do not survive long in the inner solar system, there has to be a continuous supply. *Gerard Peter Kuiper*[10] (1905–1973), and before him *Kenneth Essex Edgeworth*[11] (1880–1972), proposed that a remnant of planetesimals just outside of Neptune's orbit could be the reservoir. This *Kuiper Belt* (sometimes called the *Edgeworth–Kuiper Belt*) are planetesimals that never grew into a larger planet.

The first comet still in the Kuiper Belt, was found in 1992 by *David Jewitt* and *Jane Luu* (1992 QB_1, nicknamed "Quebewan"), now called a *Kuiper Belt object* (KBO). These are also called *Trans-Neptunian Objects*, which are all objects that permanently reside outside of Neptune's orbit. Most found so far are between 30 and 50 AU from

6 J. G. Porter, *Comets and Meteor Streams* (London: Chapman and Hall, 1952).

7 I. Hasegawa, Y. Ueyama and K. Ohtsuka, Predictions of the meteor radiant point associated with an Earth-approaching minor planet. *Publ. Astron. Soc. Japan* **44** (1992), 45–54.

8 P. B. Babadzhanov, Formation of twin meteor showers. In *Asteroids, Comets, Meteors III*, ed. C.-I. Lagerkvist, *et al.* (Uppsala: University of Uppsala, 1990), pp. 497–503.

9 L. Neslusan, J. Svoren and V. Porubčan, A computer program for calculation of a theoretical meteor-stream radiant. *Astron. Astrophys.* **331** (1998), 411–413.

10 G. P. Kuiper, On the origin of the solar system. In *Proc. Topical Symposium, Commemorating the 50th Anniversary of the Yerkes Observatory and Half a Century of Progress in Astrophysics*, ed. J. A. Hynek. (New York: McGraw-Hill, 1951), p. 357.

11 K. E. Edgeworth, The evolution of our planetary system. *J. Br. Astron. Assoc.* **53** (1943), 181–188; K. E. Edgeworth, The origin and evolution of the solar system. *Mon. Not. R. Astron. Soc.* **109** (1949), 600–609.

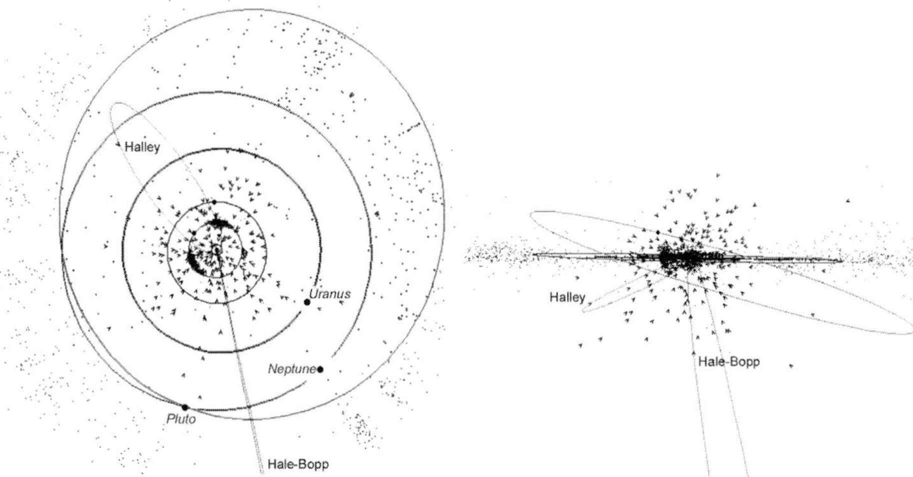

Fig. 5.3 Kuiper Belt: position of all known comets in the inner solar system (V) and in the Kuiper Belt (•) with a semimajor axis larger than 5 AU on July 1, 2003. In both diagrams, the vernal equinox is to the right along the horizontal axis. Images by Paul Chodas (NASA/JPL).

the Sun, not much further than Neptune (Fig. 5.3). Many have sizes of 100 to 500 km, but the biggest so far is 2004 DW at 1400–1600 km. Indeed, Pluto, at 2200 km, was soon recognized to be simply one of the largest of these objects. There are now thought to be about 70 000 KBOs larger than 100 km diameter in the observable region of the Kuiper Belt.[12]

At the other end of the size range, there may be as many as 2.8 billion $D > 1$ km sized comets in the Kuiper belt with a steep differential size distribution $\alpha \sim 4.0 \pm 0.5$,[13] enough to replenish the population of short-period comets in the inner solar system. Most mass is in the smaller 0.2–20 km sized objects.

The Kuiper Belt has a group of objects trapped in the 2:3 mean-motion resonance with Neptune (Fig. 5.4). As a result, they never approach the planet close enough to be ejected. These are called *Plutinos*, after its main object Pluto. The resonance is also an efficient dust particle trap.[14]

> Gerard Kuiper himself came to the idea of a Kuiper belt by believing that Pluto may be a giant comet. Although this is true, Kuiper based his hypothesis on the incorrect data that the density of Pluto was a low 0.1 g/cm^3, the same low density as he expected comets would have. We now believe that both are closer to 1 g/cm^3.

[12] D. Jewitt, From Kuiper Belt Object to cometary nucleus. *ESA SP* **500** (2002), 11–19.

[13] W. F. Bottke, A. Morbidelli, R. Jedicke *et al.*, Debiased orbital and absolute magnitude distribution of the near-Earth objects. *Icarus* **156** (2001), 339–433.

[14] E. K. Holmes, S. F. Dermott and B. Å. S. Gustafson, Dynamical evolution of dust particles in the Kuiper disk. *ESA SP* **500** (2002), 43–46.

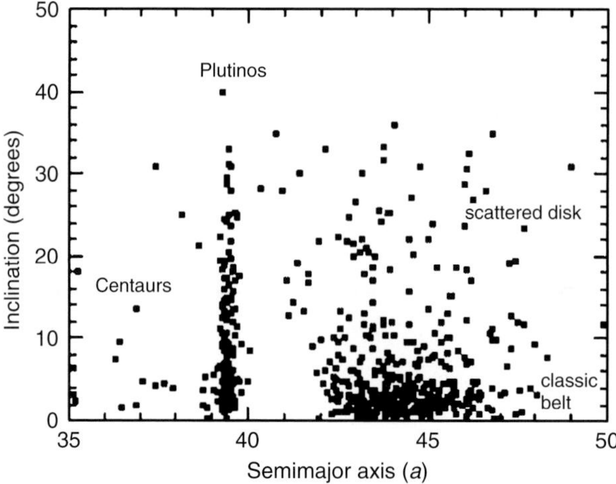

Fig. 5.4 KBOs discovered prior to July, 2004 in a diagram of *i* versus *a*. Different populations are marked.

> Pluto is so large that it is spherical with a top layer of volatile ices (nitrogen and methane), overlaying the less volatile water ice. Like Pluto, many Kuiper-belt (and Oort cloud) objects are binaries. Pluto's moon Charon is smaller and does not have volatile ices on top.[15]

Many comets have been found outside the reach of Neptune, between 48 and 52 AU, in a structure that resembles Kuiper's original remnant of a planetary disc. This is called the *Classical Kuiper Belt* containing some 47 000 comets. Because the main planets have not much influence on these objects, they are in relatively stable orbits.

The source of the short-period comets is a structure further out: the *scattered disk*, a thick torus with an inner edge near 35–40 AU ($q \sim 35$ AU), containing more than 1.2 billion comets >1 km in size, and some 30 000 larger than 100 km. The "scattered disk objects" (SDO) have elongated orbits ($a > 50$ AU, high e) that can be unstable. The disk evolved as a consequence of perturbations by Neptune, either in a progressive process of order 1 billion years, or during planet formation. Passing stars or an early proto-planet could also scatter comets and might better explain the outer edge of the classic Kuiper Belt at $a = 48$ AU. This is being investigated. It is not known how far the Kuiper Belt extends beyond this distance. Comets beyond 48 AU appear to be on perturbed noncircular orbits.

The Kuiper Belt was initially 100 times more massive than today, with planetary perturbations gradually eating away the inside parts, causing a steady stream of

[15] Z. Sekanina, Detection of a satellite orbiting the nucleus of comet Hale–Bopp (C/1995 O1). *Earth, Moon Planets* **77** (1997), 155–163.

comets to come our way. Collisions, too, can perturb the comets inward, and can lead to much dust being created.

Comets are called *Centaurs* once they are found with a perihelion just outside of Jupiter's orbit and an aphelion just inside of Neptune's orbit. There may be some 10 million Centaurs with diameters larger than 100 km, most of which are scattered out of the solar system by the giant planets within a million years. One of those every thousand years makes it into the inner solar system to become a Jupiter-family comet.

5.4 Reservoir of long-period comets: the Oort cloud

It was long known that long period comets arrived at Earth on very elongated orbits from all directions, including orbits that moved against the direction of motion of the planets (retrograde). After correcting the orbital elements measured near Earth for the earlier perturbations by the planets, *Elis Strömgren*[16] published the first list of "original" orbits, which lacked the hyperbolic orbits that were predicted by the then popular theory of *Pierre Simon Laplace* (1749–1825) who had thought that comets were captured from interstellar space. Instead, the comets moved on elongated elliptical and (nearly) parabolic orbits. *Ernst Öpik*[17] found that elongated elliptical orbits are stable against the perturbations from nearby stars, unless their aphelion was beyond $Q = 100\,000$ AU. It was Dutch astronomer *Jan Hendrik Oort* (1900–1992) who in 1950 first formulated the idea of a Sun-bound reservoir continuously supplying new comets (Fig. 5.5).[18] Oort found that all well observed "new" comets used to have their farthest point at a distance of around $Q = 50\,000$ AU, ranging from 6000 to 90 000 AU, or out to one third the distance of the nearest star.[19] He also found no preferential direction of arrival and concluded there had to be a spherical cloud of comets that supplied a steady stream of new objects to the inner solar system (Fig. 5.6). This is now called the "Oort cloud" of comets.

There may be 1 trillion comets in that cloud with $D > 1$ km with a total mass of a few times that of Earth, originally evenly distributed between the inner and outer Oort cloud. The first comet still fully in the Oort cloud was discovered on March 15, 2004, it is called "Sedna" after the Inuit goddess who rules over the seas. Sedna was discovered close to perihelion when it was at 90 AU from the Sun. It is a slow rotator, between 20–50 d. This comet is big, 1200–1600 km in diameter. Long-period comets can be very big!

"New" comets represent only the easily perturbed outer regions of a much larger reservoir. The tidal force of our Galactic plane is the main cause of those perturbations and is effective in perturbing comets only for those in elongated orbits with semimajor

[16] E. Strömgren, *Publ. Obs. Copenhagen* **19** (1914), 187.

[17] E. J. Öpik, Note on stellar perturbations of nearly parabolic orbits. *Proc. Am. Academy Arts Sci.* **67** (1932), 169–183.

[18] H. Rickman, Dynamics of meteoroid parent bodies: a conceptual history. In *Meteoroids and Their Parent Bodies*, ed. J. Stohl and I. P. Williams, 1992, pp. 83–92.

[19] J. H. Oort, The structure of the cloud of comets surrounding the Solar System and a hypothesis concerning its origin. *Bull. Astron. Inst. Netherlands* **11** (1950), 91–110.

Fig. 5.5 Jan Hendrik Oort at his 90th birthday, as I remember him. Photo: Sterrewacht Leiden.

axis $a > 10\,000$ AU, leaving the inner Oort cloud of objects with $a < 10\,000$ AU untouched. That inner Oort cloud supplies the outer Oort cloud with new comets over the age of the solar system.[20]

The origin of the Oort cloud dates back to the time of the origin of the solar system in the region where the giant planets were formed, where planetesimals accumulated into comets and planets. During close encounters, about 3%–10% of comets in the Uranus–Neptune region were thrown in elongated orbits, while a smaller fraction of comets survived collisions with early versions of Jupiter and Saturn. Because there were more comets closer to the Sun, most Oort cloud comets may have come from near Jupiter. Once in an elongated orbit, small perturbations by nearby stars and

[20] J. A. Fernández, Dynamics of comets: recent developments and new challenges. In *Asteroids, Comets, Meteors 1993*, ed. A. Milani *et al.* (Dordrecht: Kluwer, 1994), pp. 223–240.

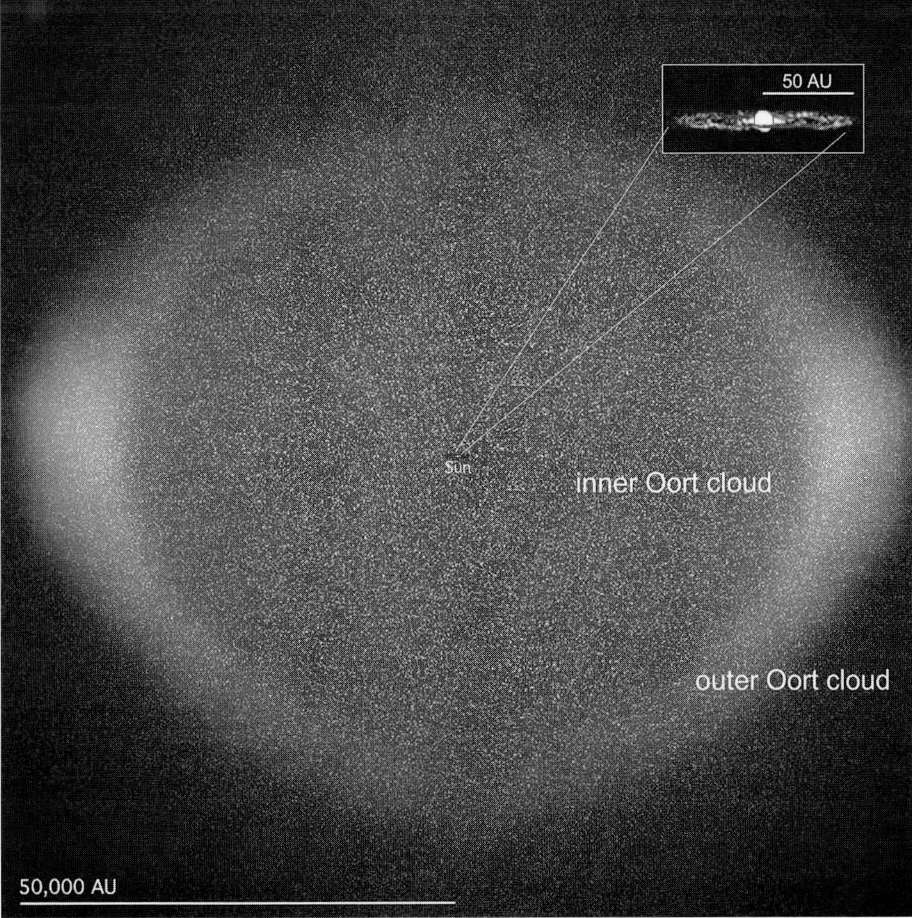

Fig. 5.6 Artist impression of the Oort cloud. This diagram shows a region that is 100 000 AU across. At this scale, the Kuiper Belt is very small.

molecular clouds could increase their perihelion to safe distances and put the comet in cold storage for the next 4.5 billion yr until, one day, the orbit was perturbed enough to bring it back to those now fully grown giant planets and ultimately into the inner solar system. There, ices would evaporate and a new long-period comet would excite observers on Earth.[21]

[21] J. A. Fernandez, The formation of the Oort cloud and the primitive galactic environment. *Icarus* **129** (1997), 106–119.

Part II

Parent bodies

6
Long-period comets

Long-period comets arrived recently from cold storage in the Oort cloud. They are of particular interest because they can still have a high content of volatile ices, impact craters, and other mementos from the origin of the solar system. Their crust is not yet fallen-back debris, but not freshly exposed ice either. The crust has been irradiated by energetic cosmic rays in the past 4.55 billion years and is a potential source of unusual meteoroids.

The comet is called "long period" if $P > 200$ years or, more correctly, when its orbital evolution is not tightly controlled by the major planets. As I said there are some 1 trillion long-period comets (D > 1 km) in the Oort cloud.[1] They have a magnitude distribution index:[2] $\chi_c = N(H_{10} + 1)/N(H_{10}) = 1.80$, where H_{10} is the comet magnitude when located 1 AU from the Sun and 1 AU from Earth (the *absolute magnitude*). This corresponds to a mass distribution index of $s = 1.64 \pm 0.02$. There is a conspicuous lack of comets smaller than about 3 km in size.[3]

According to *Bill Bottke* and colleagues, about 415 long period comets with $H_{10} < +16.8^m$ ($D > 0.2$ km) cross Earth's orbit each year (strictly having $q < 1.05$ AU), whereby ten are brighter than $H_{10} < 11^m$ ($D > 1$ km). The impact rate on Earth is about once per 150 million years for long period comets larger than 1 km in size.

Jan Oort calculated that most "new" comets return to their perihelion only once every 4 million years. The average of all observed comets is about once every 100 000 years, on account of the *intermediate long-period comets* ($P = 200$–10 000 yr), abbreviated as "ILPCs," which are more efficiently detected on account of their more frequent visits.

There are far fewer comets in intermediary orbits than expected (Fig. 6.1). This mystery is called the *fading problem*. The planets change the orbits in predictable ways

[1] W. F. Bottke, A. Morbidelli, R. Jedicke *et al.*, Debiased orbital and absolute magnitude distribution of the near-Earth objects. *Icarus* **156** (2002), 399–433.

[2] D. W. Hughes, The magnitude distribution, perihelion distribution, and flux of long-period comets. *Mon. Not. R. Astron. Soc.* **326** (2001), 515–523; K. J. Meech, O. R. Hainaut, B. G. Marsden, Comet nucleus size distribution from HST and Keck telescopes. *Icarus* **170** (2004), 463–491.

[3] P. Lamy, I. Toth, Y. R. Fernandez and H. A. Weaver, The sizes, shapes, albedos and colors of cometary nuclei. In *Comets II*. (Tucson, Az: University of Arizona Press, 2004), 223–264.

continuous scans of the sky looking for heat radiation. On April 25, he detected an object (Fig. 6.3) that was a little too bright and, after checking for technical problems, asked some ground-based telescopes to have a look. All he had was a stream of numbers on a piece of paper, no images to interpret easily. *What happened next is a bit of a saga of confusion, missed messages and chance conversations*, wrote Davies.[11] The news got out on April 27, but early confusion relating to the discovery prevented any formal announcements of the comet from being mailed until May 4. By then, it was independently discovered on approximately May 3 by two amateur observers: the Japanese observer Genichi Araki, and the British observer George Alcock. Alcock, who started meteor observing with Prentice at age 18 following a July, 1931 BAA Meteor Section meeting,[12] used no more than a pair of 15 × 80 binoculars to spot this comet, while observing indoors through a closed window . . .

The record close encounter made unusual observations possible. The comet was the first to be studied by radar with the Arecibo radio telescope. From the echo signal, a projected diameter of 5–16 km was derived, depending on surface reflection

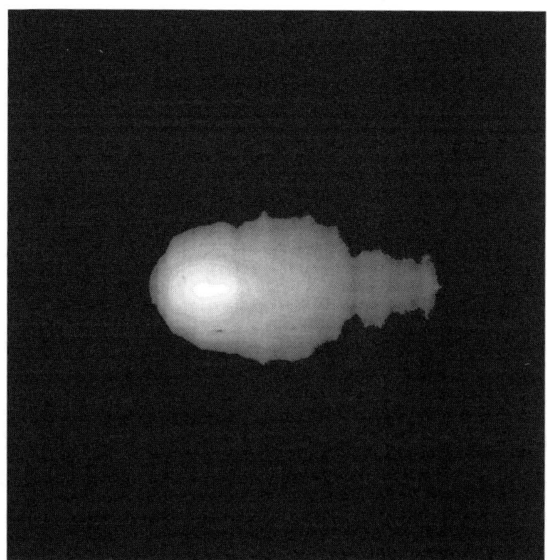

Fig. 6.3 Comet C/1983 H_1 IRAS–Araki–Alcock as seen in mid-infrared light by IRAS. Image: IPAC/Caltech.

[11] J. K. Davies, S. Green, S. B. Meadows and H. H. Aumann, IRAS and the search for fast moving objects. *Nature* **309** (1984), 315–319; *Sky Telescope* July (1983), pp. 26–29.

[12] M. Mobberley, George Alcock remembered. The comet's tale. *Newslett. Comet Section BAA* **8** (2001), 1–4.

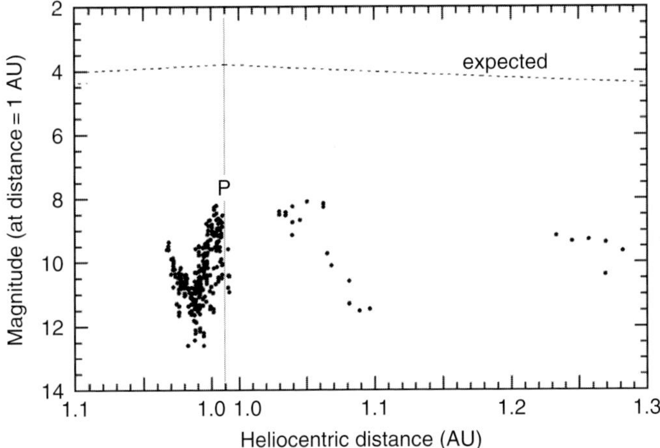

Fig. 6.4 The lightcurve of comet IRAS–Araki–Alcock from IQC data. The dashed line is the expected brightness according to Eq. (6.1).

properties, which *Zdenek Sekanina* refined to 9.3 ± 0.5 km. Combined with optical observations, Sekanina derived for the nucleus a size of 16.0 by 7.1 by 7.0 km, a volume of 416 km^3, a surface area within 15% from 298 km^2 and a mass of 8.3×10^{13} kg, spinning with a sidereal period of 51.3 ± 0.3 h in a prograde manner.[13]

That size was much larger than expected from the observed brightness of the comet ($H_{10} = +8.6 \pm 0.2^{m}$, Fig. 6.4). This comet was similar in size and rotated at a similar rate to the comet 1P/Halley, but the comet was five magnitudes fainter than Halley, emitting only half the amount of dust and a hundred times less water vapor.

The reason for the low activity is now thought to be because the spin axis of the comet nucleus was tilted by $62 \pm 2°$ (=*obliquity*) to almost in the plane of the comet orbit. As a result, only one side of the nucleus was pointed to the Sun during approach. The *northern pole*, from which the nucleus is seen to spin in clockwise direction, was directed to R.A. $= 255 \pm 3°$, Decl. $= -15 \pm 3°$. About 0.2–1% of the surface of the comet was active, with three jets emanating from within a 40° region near the sunlit southern pole. Jets A and B were about 150° apart near latitude 50°, and a third vent was located halfway between those two at a latitude of 60°.

While the comet produced less water vapor than expected (only 1×10^{28} molecules/s in May 6–13), the close approach permitted the detection of a host of other molecules. This was the first time that molecular sulfur (S_2) was detected in a celestial object, by the *International Ultraviolet Explorer* satellite. Carbon monosulfide (CS) was also visible in the spectrum. In addition, a host of other small molecules were detected by

[13] J. K. Harmon, D. B. Campbell, A. A. Hine, I. I. Shapiro and B. G. Marsden, Radar observations of Comet IRAS–Araki–Alcock 1983d. *Astrophys. J.* **338** (1989), 1071–1093; Z. Sekanina, The nucleus of comet IRAS–Araki–Alcock (1983 VII). *Astron. J.* **95** (1988), 1876–1894.

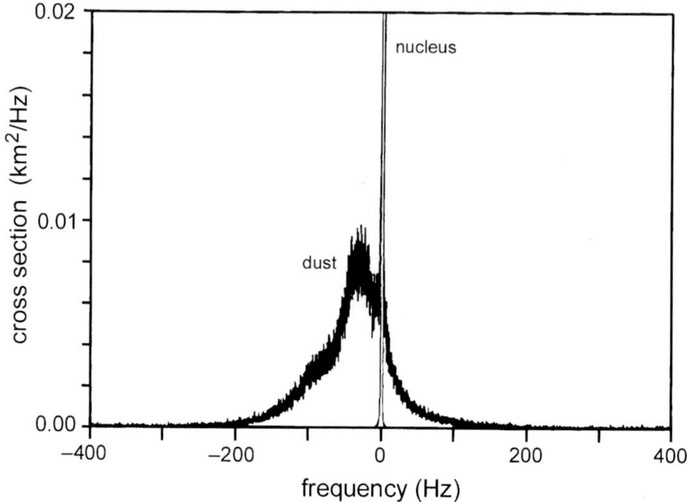

Fig. 6.5 Large dust grains surrounding the nucleus of comet IRAS–Araki–Alcock bounced back Arecibo radar signals, changing the frequency of the radio waves in doing so when approaching or moving away from the radar observatory (called Doppler frequency shift). Image courtesy of John K. Harmon (Arecibo Observatory).

filling the beam of radio telescopes (including OH, CO, CS, HCN, HCO^+, CN, CH_3C_2H, NH_3, H_2O, HC_3N, and CH_3CH_2CN).[14]

At the time, the surface reflected only $3 \pm 1\%$ of visible light and was heated to 310 K, much higher than the ~250 K needed to sublimate water ice. Hence, most of the surface was already covered in dust. The dust mantle appears to be a poor thermal conductor, because sub-mm radiation at 1.3 and 2.0 cm wavelength originating deeper inside the mantle came from colder ~180 and <100 K dust grains, respectively.[15]

Comet IAA lost dust at a rate of 10^6 g/s for 1 cm dust grains. The radar echo contained a diffuse nondepolarized signature caused by large dust grains with sizes less than about 3 cm in the comet coma near the nucleus, with a total surface area of $0.80 \pm 0.16\,km^2$ (compared to $32 \pm 10\,km^2$ for comet 1 P/Halley) (Fig. 6.5). A dust density of $0.7 \pm 0.2\,g/cm^3$ was derived.[16] The amount of dust[17] decreased with distance from the nucleus as $r_N{}^{-1}$. IRAS detected a 200 000 km long tail of warm large

[14] W. M. Irvine, Z. Abraham, M. A'Hearn *et al.*, Radioastronomical observations of comets IRAS–Araki–Alcock (1983d) and Sugano–Saigusa–Fujikawa (1983e). *Icarus* **60** (1984), 215–220.

[15] I. de Pater, C. M. Wade, H. L. F. Houpis and P. Palmer, The nondetection of continuum radiation from Comet IRAS–Araki–Alcock (1983d) at 2- to 6 cm wavelengths and its implication on the icy-grain halo theory. *Icarus* **62** (1985), 349–359.

[16] R. M. Goldstein, R. F. Jurgens and Z. Sekanina, A radar study of comet IRAS–Araki–Alcock 1983d. *Astron. J.* **89** (1984), 1745–1754.

[17] M. C. Festou, T. Encrenaz, C. Boisson, H. Pedersen and M. Tarenghi, Comet IRAS–Araki–Alcock (1983 VIII) – Distribution of the dust and of gaseous species in the vicinity of the nucleus. *Astron. Astrophys.* **174** (1987), 299–305.

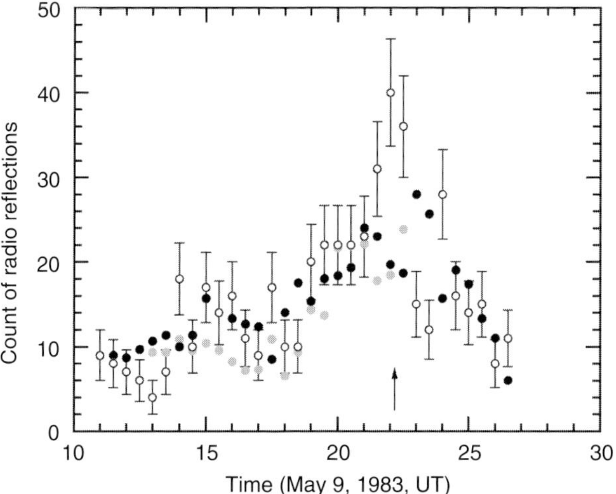

Fig. 6.6 Increase in radio forward meteor scatter counts close to the node of IRAS–Araki–Alcock by Chikara Shimoda (Asahi Village): • = average rates May 5–12 in 1982 (gray) and 1983; ○ = counts for May 9, 1983.

dust grains stretching behind the comet (Fig. 6.3).[18] The comet was observed only when it was close to Earth and we do not know how it brightened approaching the Sun.

Jack Drummond (Stewart Observatory) was first to point out that meteor activity was expected on May 10.1 ($\lambda_\odot = 49.1°$) from a radiant at R.A. = 289°, Decl. = +44°, near η-Lyra.[19] Japanese observers proceeded to observe the region at the time of the closest encounter, but no believable elevated rates were noticed.[20] 2.4 h after sunrise (07.2 JST, 22.2 UT, $\lambda_\odot = 48.96$), close to passing the orbit node, a narrow rate hike of meteor counts was detected by *Chikara Shimoda* of NMS (Fig. 6.6), but this has not been confirmed.[21] In later years, it was found that the comet has an associated annual shower. This discovery was the topic of my first research paper, submitted to a professional astronomical journal when I was still an undergraduate student, and cordially rejected by the journal as irrelevant (and badly written). All beginnings are difficult. Later, veteran amateur meteor astronomer *Katsuhito Ohtsuka* of the Tokyo Meteor Network understood the importance of adding to the very short list of known long-period comet parent bodies and pointed out the photographed meteors in a paper published in the *Tokyo Meteor Network Report*.[22]

[18] R. G. Walker, H. H. Aumann, J. Davies *et al.*, Observations of comet IRAS–Araki–Alcock 1983d. *Astrophys. J.* **278** (1984), L11–L14.

[19] B. Marsden, Comet IRAS–Araki–Alcock (1983d). *IAU Circular No. 3805* (1983 March 9).

[20] P. Jenniskens, Meteoren van 1983-d? *Radiant, J. DMS* **7** (1985), 31–33.

[21] C. Shimoda and K. Ono, Iras–Araki–Alcock outburst with FRO. *Proc. Meteor Conf.* (Nippon Meteor Society, 1984), p. 62.

[22] K. Ohtsuka, The association of the eta-Lyrids with comet IRAS–Araki–Alcock. *Tokyo Meteor Network Rep.* **8** (1989), 44–49; K. Ohtsuka, Eta-Lyrid meteor stream associated with comet IRAS–Araki 1983 VII. *Astrophys. Space Sci. Library* **173** (1991), 315.

On the nomenclature of meteor showers

There is much confusion about the naming of meteor showers. The general rule is that a shower should be named after the then current constellation that contains the radiant, specifically using the possessive Latin form. The possessive Latin name for the constellations end in one of seven declensions: "-ae (e.g. Lyrae), -is (e.g. Leonis), -i (e.g. Ophiuchi), -us (e.g. Doradus), -ei (e.g. Equulei), -ium (e.g. Piscium), or -orum (e.g. Geminorum). Custom is to replace the final suffix for "-id," or plural "-ids." Hence "Quadrantid" after the now defunct constellation the wall quadrant *Quadrans Muralis* (possessive: Quadrantis Muralis), introduced in the 1801 atlas by Johann Elert Bode (1747–1826) *Uranographia* (Gdansk, 1690). Meteors from Aquarius (Aquarii) are Aquariids, not Aquarids. Meteors from Canes Venatici (Canum Venaticorum) would be Canum Venaticids or Canids.

If a higher precision is needed, then the shower is usually named after the nearest (if in doubt: brightest) star with a Greek letter assigned, as first introduced in the *Uranometria* atlas by Johann Bayer (1603), or one with a later introduced Roman letter. If in doubt, the radiant position at the time of the peak of the shower (in the year of discovery) should be taken. Hence, the meteors of comet IRAS–Araki–Alcock would be named "η-Lyrids."

One may add the name of the month to distinguish among showers from the same constellation. In this case, one would call the shower from comet IRAS–Araki–Alcock the "May Lyrids," in order to differentiate from the more familiar "April Lyrids." For daytime showers, it is custom to add "Daytime," hence the name for the "Daytime Arietids" in June as opposed to the Arietids in October (which are really an extension of the Southern Taurids).

South and North refer to *branches* of a shower south and north of the ecliptic plane, resulting from meteoroids of the same parent body that are perturbed by secular perturbations (Chapter 9). Because they have nearly the same longitude of perihelion at a given solar longitude, the South and North branches have ω and Ω reversed and are active over about the same time period.

If the meteoroid stream is encountered at the other node, it is customary to speak of "twin showers." The Orionids and η-Aquariids are twin showers, even though each represents dust deposited at different times and are now in quite different orbits. As a matter of custom, twin showers and the north and south branches of showers from the same stream carry different names.

Sometimes, the meteor showers are named after their parent body. Examples are the *Giacobinids* of comet 21P/Giacobini–Zinner and the *Bielids* of comet 3D/Biela. My objection to that is, that the names of comets tend not to be Latin, raising the question what part of the name is to be replaced by "-id." As a result, these name constructs are arbitrary (e.g., Biela-ids or Bielids?).

The size of comet nuclei from comet activity

The comet brightness depends on the gas production rate and that can vary a lot. The comet apparent brightness (m_c, in magnitudes) varies with the square of the distance to the Earth (Δ, in AU) and more so with the distance to the Sun (r, in AU). To isolate the fading due to the decreasing activity of the comet away from the Sun, the inherent brightness of a comet is the brightness when placed at 1 AU from the Sun and 1 AU from the Earth. This *absolute brightness* (H, in magnitudes) is traditionally expressed as:[23]

$$H_y = m_c - 5 \log \Delta - y \log r$$

where $y = 2.5n$. Here "n" reflects how fast the comet loses brightness away from the Sun ($\sim r^{-n}$). If n is unknown, then a typical value of $n = 4$ is assumed and the absolute magnitude is expressed as H_{10}, while otherwise the comet absolute brightnesses is called H_0, with n provided.

Before many diameters of comet nuclei were measured, David Hughes[24] derived a tentative relationship between the diameter (D) of a long-period comet nucleus and its absolute brightness:

$$\log_{10} D\,(\text{km}) = 1.543 - 0.20\, H_{10}$$

which assumed that the rate of gas released by a comet (Q, derived from: $H_{10} \sim 2.5 \log_{10} Q$) is proportional to its surface area, or the amount of sunlight received. In that case: $\log D \sim 1/2 \times 1/2.5\, \text{H}_{10}$ (shown as a dashed line in Fig. 6.7). In order to fit all data, Hughes adopted different proportionality factors for different comet groups.

Dismissing comet IRAS–Araki–Alcock, I find from recent visits and radar measurements that the gas production rate is proportional to the volume (or mass) of a comet instead (solid line in Fig. 6.7, which has a slope $1/3 \times 1/2.5$):

$$\log_{10} D\,(\text{km}) = 1.48 - 0.133\, H_{10} \qquad (6.1)$$

This relationship appears to hold for both short-period (at their most active) and long-period comets and goes against most existing comet models.

The possible reason for this unexpected result could be that the efficiency of transforming heat into water sublimation proportionally decreases for smaller comet nucleus sizes. Small nuclei lose their water more easily, but in doing so decrease the gas production rate for given nucleus size in later returns.

[23] S. K. Vsekhsvyatskii, *Physical Characteristics of Comets*. 1964, Israel Program for Scientific Translations, Jerusalem.
[24] D. W. Hughes, The magnitude distribution, perihelion distribution, and flux of long-period comets. *Mon. Not. R. Astron. Soc.* **326** (2001), 515–523.

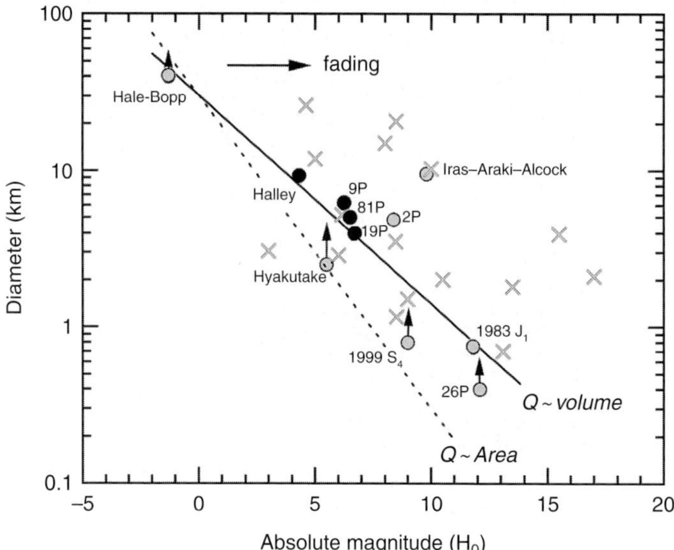

Fig. 6.7 Diameter of observed comet nuclei versus the absolute magnitude of the comet: • = visited comets; gray circles radar measurements; × = photometric. The dashed line has the slope expected if the gas production rate is proportional to the surface area (amount of energy received from the Sun), the solid line if it is proportional to the volume of the comet nucleus.

6.2 Other identified parents

Only one month after IRAS–Araki–Alcock, on June 12, 1983, there was a second comet that made a very close approach to Earth (at 0.0628 AU), currently number 16 on the list of all-time close approaches. This comet *C/1983 J_1 (Sugano–Saigusa–Fujikawa)* was also observed by radar on 11 June, 1983. It was a lot smaller, only $D = 0.74$ km, and is one of the smallest comets on record. *In the telescope, the comet appeared simply as a ghostly mist drifting across the sky*, according to then amateur astronomer Maurice Clark. Japanese visual observers reported seeing faint meteors from the predicted shower radiant close to the encounter and radar rates were said to be enhanced in 1984.[25] Both results are now in question. C/1983 J_1 passed Earth on a very elongated orbit on a path to never come back.

Comet *C/1861 G_1 (Thatcher)*, parent of the April Lyrid shower is second on my list (Fig. 6.8). The orbital period of this comet (~415 yr) is very close to the limit between short-period and long-period comets. It is an intrinsically bright comet, $H_{10} = +5.5$. Not much else is known, as it was seen last in 1861. According to my Eq. (6.1), the diameter of comet Thatcher is 5.6 km. That is, unless one side of the comet nucleus is

[25] M. Simek and P. Pecina, On the meteor activity associated with the comet Sugano–Saigusa–Fujikawa 1983 V. *Bull. Astron. Inst. Czechoslov.* **37** (1986), 103–107.

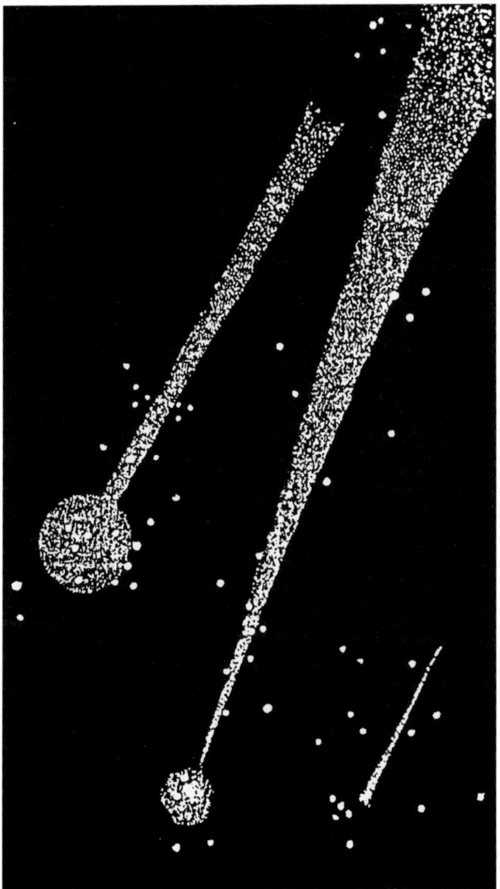

Fig. 6.8 Comet C/1861 G_1 (Thatcher), parent of the Lyrid shower, seen on May 5, 1861 (left), together with comet C/1863 Y_1 Respighi (January 5, 1864), and comet C/1873 Q_1 Borrelly (September 3, 1873). This drawing was published in 1898.[26]

always facing the Sun, as with comet IRAS–Araki–Alcock, or unless the comet has an episode of eruption and leaving boulders in its wake as did comet Hyakutake (Fig. 6.7).

The comet was discoved by amateur astronomer A. E. Thatcher of New York on April 5, 1861, who described it as a "tailless nebulosity, 2 arcmin in diameter, with a central condensation." It was a $+7.5^m$, but brightened slowly during April and was discovered independently by Bäker of Nauen, Germany, with the unaided eye. It brightened to $+2.5^m$ with a tail of 1° on May 9 and 10, when approaching Earth to only 30 million km. It passed perihelion on June 3, now moving away from Earth. Until 1983, Thatcher's comet was the only long-period comet that could be linked with

[26] A. Freiher von Schweiger-Lerchenfeld, *Atlas der Himmelskunde*. Vienna, (Wien: Hartleben; 1898), p. 195.

certainty to a meteor shower. However, the relatively short orbital period left open the possibility that the dust evolved in a manner more typical for that of Halley-type comets.

A comet of unambiguous long-period character, and the third comet on my list, is *C/1911 N_1 (Kiess)*, parent of the *Aurigids* (also known from the historic name of α-Aurigids), and named after astronomer *Carl Clarence Kiess* (1887–1967),[27] who at that time was a fellow at the Lick Observatory working towards his Ph.D. He discovered the comet while searching for Encke's comet with the Crocker Photographic Telescope on July 6.[28] This was only five days after the July 1, 1911 earthquake, which had moved the 90 cm refractor on its supporting pier until it came in contact with the observing floor in the dome. Chimneys of buildings had broken off and one building was so badly damaged that it had to be rebuilt. Kiess would later spend much time documenting the effects of that earthquake, which came in the wake of the more famous earthquake of April 18, 1906 that leveled nearby San Francisco.

The comet was observed for only 70 d. It was a large diffuse nebula of $+7^m$ with a vague central condensation and a 0.2°– 0.5° tail (Fig. 6.9, image by Wilhelm Högner and Nikolaus B. Richter[29]) with $H_{10} = +7.9^m$ ($D \sim 2.7$ km).[30] The association with Aurigids was recognized by Cuno Hoffmeister and colleague Arthur Teichgraeber of the Sternwarte Sonneberg,[31] following the 1935 Aurigid outburst. However, Hoffmeister[32] later pointed out that the orbit of the comet was nearly parabolic, making the meteor outburst 24 yr after the perihelion passage of the comet difficult to explain. He believed incorrectly that an isolated cluster of dust in the comet orbit was responsible. Since 1994, we know that the agreement is very good and there is no doubt anymore that this association is genuine.

The parent of the October *Leonis Minorids* is probably comet *C/1739 K_1(Zanotti)*, number four on my list. It was discovered with the naked eye on May 27.9, 1739 by *Eustachio Zanotti* of Bologna, Italy, as a $+3^m$ star surrounded by nebulosity.[33] The orbital period is unknown and the orbit passes Earth at a relatively large $\Delta r = 0.049$ AU, but this comet is intrinsically bright, with $H_{10} = +3.3^m$ ($D \sim$ 10.9 km), and meteors radiate from R.A. = 162.4°, Decl. = +36.3° on October 24.2, 2000, which is close to the radiant measured by the photographic and the video observations of the Dutch Meteor Society: R.A. = 161.4 ± 0.2°, Decl. = +37.3 ± 0.5° on October 24.2, 2000.[34] The October Leonis Minorids was one of seven showers

[27] C. C. Kiess, S. Einarsson and W. F. Meyer, Discovery and observations of comet b, 1911 (Kiess). *Lick Observatory Bull.*, vol. 198 (Berkeley, CA: University of California Press, 1911), pp. 138–139.

[28] Anonymous, *Biennial Report of the President of the University on Behalf of the Regents to His Excellency the Governor of the State [1910–12]* (Berkeley, CA: University of California Press, 1912).

[29] W. Högner and N. Richter, *Isophotometrischer Atlas der Kometen*, Teil I, mit 90 Bildtafeln erläuterungen und Tabellen. 2. Auflage. (Leipzig: Johann Ambrosius Barth, 1979).

[30] E. Hahr, On the orbit of comet 1911 b (Kiess). *Stockholms Observatoriums Ann.* **10** (1912), 1–27.

[31] A. Teichgraeber, Unerwarteter Meteorstrom. *Sterne* **15** (1835), 277; V. Guth, Über den Meteorstrom des Kometen 1911 II (Kiess). *Astron. Nachr.* **256** (1936), 27–28; C. Hoffmeister, Unerwarteter Meteorstrom. *Astron. Nachr.* **258** (1936), 25.

[32] C. Hoffmeister, *Meteorströme*. (Weimar: Werden and Wirken, 1948), 286 pp.

[33] G. W. Kronk, *Cometography, a Catalogue of Comets. Vol. I: Ancient–1799* (Cambridge: Cambridge University Press, 1999), p. 402.

[34] M. C. de Lignie and H. Betlem, A double-station video look on the October meteor showers. *WGN* **27** (1999), 195–201.

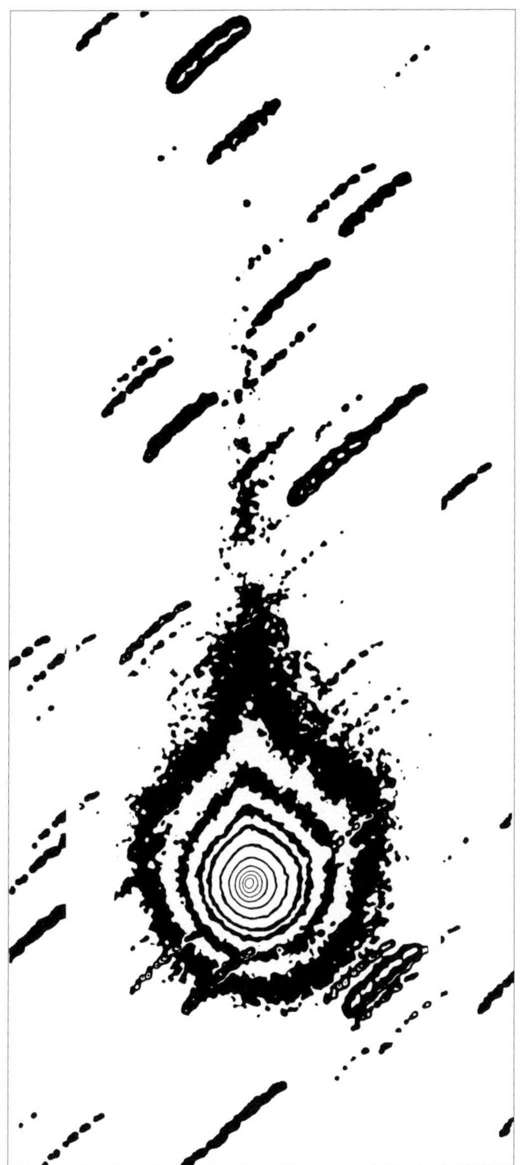

Fig. 6.9 C/1911 N₁ on August 5, 1911 at $\Delta = 0.469$ AU, $r = 0.998$ AU (blue-sensitive plate 360–490 nm).[29]

discovered by *Dick McCrosky* and *Annette Posen* from the Harvard Meteor Program.[35] They also recognized the association. The shower may be variable in intensity. *Peter V. Bias* of California reported unusually high rates of Leonis Minorids on several nights in 1967 and 1968, but particularly on October 23, 1967, when meteors were rising up from

[35] R. E. McCrosky and A. Posen, New photographic meteor showers. *Astron. J.* **64** (1959), 25–27.

the north-eastern horizon at a rate of 12 per hour.[36] In a recent e-mail he added: *I remember thinking that I had discovered a new annual shower after seeing it again in 1968 – only to have it more or less disappear after that. I can tell you that, strength-wise, watching the Leonis Minorids in 1967 was quite comparable to watching the Orionids or the 1970s-Taurids at maximum.*

Number five on the list is comet *C/1852 K_1 (Chacornac)*. On May 22, B. A. Gould wrote a letter to the still young *Astronomical Journal*, published in volume 2, saying in full that: *It is stated in the Boston Evening Traveler of May 19 and 20, that a new comet has been detected at the Cambridge Observatory, within 15° of the pole; – in right-ascension about 22.5 hours, declination 75° north. I have, however, received no information to this effect from the astronomers of the Observatory.*[37] Even in those days, professional astronomers would sometimes get their information from the newspaper. Gould later learned that the comet had been observed by Mr. G. P. Bond at the observatory on May 18, but that was three days after it was discovered in Marseille by the French astronomer Jean Chacornac (1823–1873). Chacornac described it as: *une petite comète, très-faible, assez diffuse, sans queue ni noyau.*[38] The comet was faint, without a central condensation. This retrograde comet moved rapidly along the sky following the path of a parabolic orbit, but was lost around June 14. The observations permit a shorter Halley-type orbit. This unassuming comet ($H_{10} = +9.8$ or $D \sim 1.5$ km) was recently associated with meteor activity from η-Eridani in mid August during the Perseid season by *Katsuhito Ohtsuka* and coworkers.[39]

C/1874 G_1 (Winnecke) has a theoretical radiant at R.A. = 18.5°, Decl. = −12.0° ($\lambda_\odot = 92.2°$), four degrees southeast of the third magnitude star η-Ceti, which is close to the observed radiant at R.A. = 20°, Decl. = −14° of the June π-Cetids (solar longitude 95.0 ± 0.7). However, there is a significant discrepancy in the peak time and the separation distance is a large 0.064 AU. Based on the width of the shower, this could be a Halley-type comet.

C/1947 F_2 (Becvár) is a good match to the February δ-Serpentids, except that the shower itself is not that well defined. I found two meteoroid orbits in the Harvard Super-Schmidt program that have a different perihelion distance, bracketing the comet. Again, the comet is intrinsically faint and passes wide.

C/1943 W_1 (Van Gent–Peltier–Daimaca) is a good match to the November Hydrids, but again the shower itself is not well defined. As before, the comet is intrinsically faint and passes wide.

C/1862 N_1 (Schmidt–Tempel) is a match to the August ζ-Arietids, suspected of fireball activity in Medieval times. The comet orbit passes close enough to

[36] P. V. Bias, *Meteors and Meteor Showers, an Amateur's Guide to Meteors.* Miracle Publ. Co., Cincinnati, OH (2005), 238 pp.
[37] B. A. Gould, Comet. *Astron. J.* **2** (1852), 123.
[38] A. C. Petersen, Comet and asteroid observations. *Astron. J.* **2** (1852), 142–143.
[39] K. Ohtsuka, T. Tanigawa, H. Murayama and I. Hasegawa, The new meteor shower η Eridanids. *ESA SP* **495** (2001), 109–111.

Earth's orbit, but the shower itself is not well known. This is also the case for **C/1976 D_1 (Bradfield)**, which can have dust trails that wander in Earth's orbit (Chapter 13). The February/March β-Tucanids are at high southern latitude and there is no solid evidence of the existence of the shower. The author saw only one possible β-Tucanid on March 1, 2003 between 09:20 and 10:22 UT, in intermittent clouds, haze, and interuptions to relocate to an area with better observing conditions, during a dedicated observing campaign in South Africa.

The status of the other comets is more uncertain. Comet **C/1723 T_1 (Kegler-Crossat-Saunderson)** was independently discovered at such remote sites as Peking in China (*Ignatius Kegler*), Cayenne in French Guiana (*Crossat*), and Bombay in India (*William Saunderson*) in the week of October 10–17, 1723. It became a naked eye object, described as a small "nebulous star" of third magnitude with a tail of ~1°, moving at over 5°/d. It was intrinsically bright ($H_{10} = +5.5^m$)[40] and passed very close to Earth's orbit. Evidence for its associated shower, the October Monocerotids, is weak. As a further word of caution, the orbital period of the comet remains unknown, with some solutions finding a hyperbolic orbit.[41]

Drummond pointed out that comet *C/1964 N_1 (Ikeya)* missed Earth by 0.12 AU during the October ε-*Geminid* shower and other long-period comets might pass at similar large distances from Earth's orbit. However, recent searches for meteor showers from long-period comets that passed Earth's orbit at >0.012 AU have always come up empty, even for those where Earth passed the node only days after the comet. In the case of C/1964 N_1, comet and meteor shower are not a good match either. There should be a stronger shower in July, which is not observed. When *C/1987 B_1 (Nishikawa–Takamizawa–Tago)* was discovered, Duncan Steel[42] argued that this orbit matched the shower better. He predicted a possible outburst in 1987, which was not observed. The association remains in doubt. Comet NTT has a nearest encounter on October 7 with a theoretical radiant at R.A. = 93°, Decl. = +28°, when it passes Earth's orbit at $\Delta r = 0.050$ AU, while the ε-Geminid shower peaks on October 19, with a radiant at R.A. = 101.6°, Decl. = +26.7°. In comparison, comet Ikeya has a radiant at R.A. = 107°, Decl. = +27° on October 23.

6.3 What makes long-period comets unique

Long-period comets were created in the giant planet region, in more violent accumulation processes than those that led to Kuiper-belt comets, but they can be brought to

[40] S. K. Vsekhsvyatskii, *Physical Characteristics of Comets*. (Jerusalem: Israel Program for Scientific Translations, 1964), p. 51.

[41] G. W. Kronk, Cometography, a catalog of comets. Vol. 1: ancient–1799 (Cambridge: Cambridge University Press, 1999), pp. 392–394.

[42] D. Olsson-Steel, Comet Nishikawa–Takamizawa–Tago (1987c) and the Epsilon Geminid meteor shower. *Mon. Not. R. Astron. Soc.* **228** (1987), 23P–28P; D. Olsson-Steel, Prospects for an enhanced epsilon-Geminid shower in 1987. *WGN* **15** (1987), 109–111.

us in a more pristine form, unchanged by lack of earlier returns to the solar system. Some do sport a pristine comet crust from 4.55 billion years of exposure by cosmic rays. Many fall apart into transient meteoroid streams that can contain large boulders not normally expected from Whipple-type ejection. Karen Meech of the University of Hawai'i has pointed out that Oort cloud comets systematically possess dust comae out to large heliocentric distances (>3 AU), whereas short-period comets do not (unless there is an outburst). Long period comets eject dust from a larger fraction of their surface, rather than from a few concentrated jets.

C/1999 S_4 (LINEAR) ejected a brief puff of small particles on July 14, 2000,[43] and unraveled over a period of a few days starting on July 19, shortly before its closest approach to the Sun on July 26. It started brightening in an irregular manner on June 26, and showed the first sign of a splitting nucleus,[44] shedding large fragments in the process. During amateur astronomy night at Fremont Peak Observatory (CA) in late July, I happened to see the comet when it had just broken apart and was spread out into an elongated blur. I had not expected that and was happily surprised. Soon, the comet disappeared from view and the *Hubble Space Telescope* found 26 tiny comets in its path (Fig. 6.10). The largest fragment measured 100 m across, and all the fragments together weighed about 2 million tons. The original nucleus was only about $D = 0.8 \pm 0.1$ km across.

Because this comet fell apart rather gently, these fragments may still be much the same as the *cometesimals* that accumulated by their own gravity to build a "Rubble Pile" nucleus[45] as the end result of a process called *hierarchical accretion*, whereby attogram grains build dust particles, dust particles build pebbles, pebbles build cometesimals and cometesimals build comets, each step dictated by different forms of attraction and sticking. The accumulation of cometesimals is driven by gravity.

Long-period comets originated from the giant-planet formation region of the solar system, most from the Uranus–Neptune region. Billions of cometesimals moved at high relative speed in close proximity to each other. The final stages of accretion were rather violent and there could be a wide variety in nucleus properties among long-period comets, from highly unstable and loosely bound rubble piles to relatively consolidated conglomerates.

It was calculated that comet C/1999 S_4 was so loosely bound at the time of breakup that it could be pulled apart by the strength of bare hands. In this breakup some 200 times more mass was shed in the form of mm-sized dust and cm-sized pebbles than large fragments.[46]

[43] Six papers on Comet LINEAR (C/1999 S4) were published in the 18 May 2001 issue of *Science* magazine. Also: T. Bonev, K. Jockers, E. Petrova, *et al.*, The dust in comet C/1999 S4 (LINEAR) during its disintegration: narrow-band images, color maps, and dynamical models. *Icarus* **160** (2002), 419–436.

[44] R. Schulz and J. A. Stüwe, The dust coma of comet C/1999 S4 (linear). *Earth, Moon, Planets* **90** (2002), 195–203.

[45] P. Weissman, Are cometary nuclei primordial rubble piles? *Nature* **320** (1986), 242–244.

[46] W. J. Altenhoff, F. Bertoldi, K. M. Sievers *et al.*, Radio continuum observations of comet C/1999 S4 (LINEAR) before, during, and after break-up of its nucleus. *Astron. Astrophys.* **391** (2002), 353–360.

Fig. 6.10 Comet C/1999 S_4 on August 5, 2000 (Hubble WFPC2). Photo: Harold Weaver, NASA HST-VLT-LINEAR Team.

Interestingly, the breakup did not result in a great brightening of the comet, as would be expected if gas-laden regions of the comet nucleus were suddenly freshly exposed to sunlight. This has since been interpreted as meaning that the comet was unusually poor in ices at the time of breakup, only 0.4–10% of its mass. Unlike other Oort cloud comets, the ice of this comet was poor in volatile ices such as methane and carbon-monoxide, with no difference before and after breakup. Perhaps, because the comet was formed further in towards the Sun than other Oort cloud comets,[47] with more of the volatiles having evaporated early on. C/1999 S_4 could have been formed in the region between Jupiter and Saturn, or on the outer parts of the asteroid belt where about 8% of all objects 3.3–5.0 AU from the Sun are ejected into the Oort cloud.[48] Unfortunately, the heavy water content of comet LINEAR, a potential tracer of the region of formation, could not be measured.

It could also be that the size of the comet nucleus was the deciding factor. It has been suggested that the comet breakup occurred because it ran out of ices, which perhaps holds the cometesimal pieces together like the pins in a wood construction.

[47] H. Levison and M. Duncan, From the Kuiper belt to Jupiter-family comets: the spatial distribution of ecliptic comets. *Icarus* **127** (1997), 13–32.

[48] M. C. de la Fuente and M. R. de la Fuente, On the origin of comet C/1999 S4 LINEAR. *Astron. Astrophys.* **395** (2002), 697–704.

7
Halley-type comets

Over time, about 1% of the long-period Oort cloud comets evolve into shorter orbits and a class of comets called "Halley-type," the remainder being ejected from the solar system. *Halley-type comets* are defined as comets with orbital period $P < 250$ yr and encounter velocities with Jupiter larger than Jupiter's orbital velocity.[1] This amounts to about one new Halley-type comet every four years, resulting in a steady state of about 300 such comets being present in the solar system at any given time.[2] Halley-type comets are responsible for some of the most exciting meteor showers because they are in relatively stable orbits that can accumulate dust from many returns.

The Oort-cloud origin of Halley-type comets is still evident in the proportion of *retrograde orbits*, those that go around the Sun in a direction opposite to that of the planets. Near head-on collisions result in fast meteors. Examples are the Orionid and η-Aquariids of comet 1P/Halley itself (Fig. 7.1). Normal *prograde* orbits (moving in the same direction as the planets) are also common among Halley-type comets (Ursids of 8P/Tuttle, for example). Slightly more are prograde than retrograde due to the progressive capturing processes by the planets.

Their Oort cloud origin also carries the possibility that individual Halley-type comets may originate from quite different regions of the solar system: close to Jupiter or as far out as Neptune. Comet observers are looking for such differences in the composition of ices.[3]

7.1 Orbital resonances

The meteoroid streams of Halley-type comets differ dynamically from those of long-period comets mainly because the comet and dust can avoid close encounters with the

[1] H. F. Levison and M. J. Duncan, From the Kuiper belt to Jupiter-family comets: the spatial distribution of ecliptic comets. *Icarus* **127** (1997), 13–32.

[2] J. A. Fernandez and T. Gallardo, The transfer of comets from parabolic orbits to short-period orbits: numerical studies. *Astron. Astrophys.* **281** (1994), 911–922; J.-L. Zhou, Y.-S. M. Sun, J.-Q. Zheng and M. J. Valtonen, The transfer of comets from near-parabolic to short-period orbits: map approach. *Astron. Astrophys.* **364** (2000), 887–893.

[3] E. L. Gibb, M. J. Mumma, N. dello Russo, M. A. DiSanti and K. Magee-Sauer, Methane in Oort cloud comets. *Icarus* **165** (2003), 391–406; M. J. Mumma, I. S. McLean, M. A. DiSanti *et al.*, A survey of organic volatile species in comet C/1999 H1 (Lee) using NIRSPEC at the Keck Observatory. *Astrophys. J.* **546** (2001), 1183–1193.

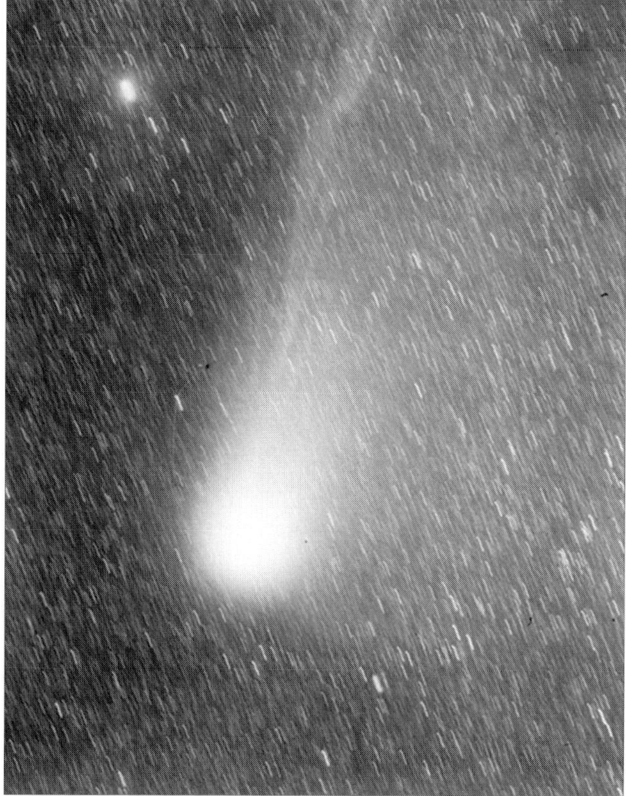

Fig. 7.1 Comet 1P/Halley on April 3, 1986. A sheet of thin dust particles stretches to the right, with Earth about to cross the orbital plane in early May. Photo by Richard Wainscoat with the Uppsala Schmidt telescope at Siding Spring Observatory on Kodak IIIa film. Photo: Research School of Astronomy and Astrophysics, Mount Stromlo Observatory.

planets by getting trapped in (and librate about) a mean-motion resonance with Jupiter.[4] The analogy in music would be called a "harmonic." Trapping in a *mean motion resonance* means that the bodies orbit the Sun in a simple integer ratio to the orbital period of a perturbing planet. For example, the 1:11 mean-motion resonance is when a comet completes one orbit around the Sun in the same time it takes Jupiter to complete 11 orbits.

Those trappings are never permanent, because a third party can change the orbital period of the comet or meteoroid. A meteoroid in resonance with Jupiter can suffer a perturbation by Saturn, changing the orbital period to bring it closer to the planet Jupiter in the next return. In the case of Halley-type comets, however, such close

[4] J. E. Chambers, Why Halley-types resonate but long-period comets don't: a dynamical distinction between short- and long-period comets. *Icarus* **125** (1997), 32–38.

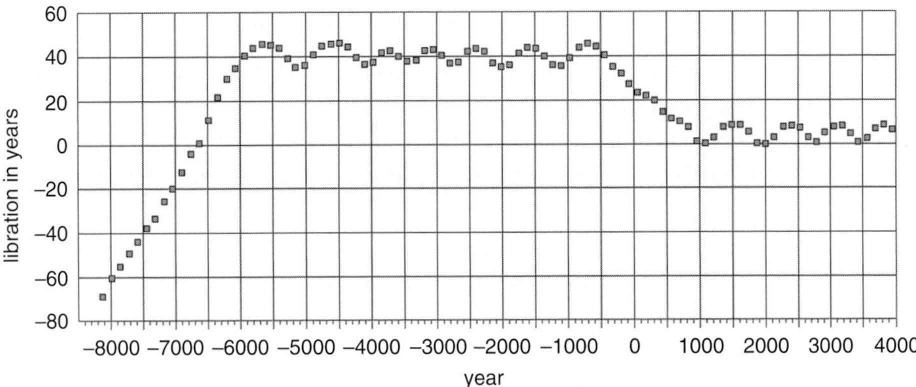

Fig. 7.2 The deviations of comet 109P/Swift–Tuttle from the 1:11 resonant orbit $P = 11 * 11.862$ yr. The graph represents the difference of perihelion timing as compared with the computed period back from the 1992 return.

encounters with Jupiter tend to be brief and inefficient, and subsequent perturbations by Saturn can perturb the comet into another resonance, where it then stays safe again for a long time.

When trapped, the orbital periods tend to oscillate around the resonant value (Fig. 7.2), in a process called *libration*.[5] The reason for this is that the relative position of the comet and the planet during an encounter oscillates. For example, when the comet is on its way to the nearest point when the planet is there and is accelerated in the encounter, then the orbital period increases, causing a delay in the next return, conveniently keeping the comet away from the planet. After a while, the comet will be delayed by over half an orbit of Jupiter so that it sees the planet after it reaches the nearest point. As a result, the comet is slowed down, returns earlier, and sees its orbital period further decreased, until half a cycle later the process reverses. And so on, and so on. The effect is that the orbital period increases and decreases periodically. Long-period comets do not have stable enough orbital periods to be trapped in this manner. In addition, orbital resonances tend to work against stability if the orbits are too short, by quickly making the orbits more elongated (see the next chapter).

For comets with $q < 2$ AU, the average physical lifetime as a Halley-type comet is about 300–500 revolutions. Sooner or later Halley-type comets are either ejected from the solar system, or will impact the Sun. Comet 1P/Halley itself is expected to evolve into a sungrazing orbit in the next ~200 000 years.[6]

[5] A. Carusi, L. Kresák and G. B. Valsecchi, *Electronic Atlas of Dynamical Evolutions of Short-Period Comets* (IAS Computing Centre, 1987) (website); A. Carusi and G. B. Valsecchi, Dynamical evolution of short-period comets. In *Proc. 10th European Regional Astronomy Meeting of the IAU*, Prague, Czechoslovakia, August 24–29, 1987 (Prague: Czechoslovak Academy of Sciences, 1987), pp. 21–28.

[6] M. E. Bailey and V. V. Emel'yanenko, Dynamical evolution of Halley-type comets. *Mon. Not. R. Astron. Soc.* **278** (1996), 1087–1110.

7.2 Comet Halley and the breakup of meteoroids

Dust accumulates over time and oscillations of the comet orbit can lead to detectable showers over large volumes of space. Because of that, it is necessary to consider much more liberal criteria in deciding which Halley-type comet can be the parent of a shower. That is clearly demonstrated by 1P/Halley itself, number one on another very short list. It is the most famous comet, because it is bright and because it is periodic on a human timescale, just frequent enough to appear in each generation.

> Author *Mark Twain* was born in 1835, only fourteen days after comet Halley had started a new orbit, and diagnosed with heart disease he said in 1909, according to his biographer Albert Bigelow Paine: "I came in with Halley's Comet in 1835. It is coming again next year, and I expect to go out with it. It will be the greatest disappointment of my life if I don't . . .".[7] Twain died on April 21, 1910, a day after comet Halley completed its 76-year orbit.

Comet Halley was seen at each return since the first record in 239 BC (1P/-239 K_1). The orbital period varied between 75 and 80 yr, while its rocket-like outgassing delayed the return to perihelion consistently by 4 d in each return. Halley's outgassing rate and the position of the spin axis must have remained nearly constant.[8] From the Giotto encounter in 1986, we know that it has a relatively large 14.4 by 7.4 by 7.4 km sized nucleus. Vega 1 and 2 measured the size at 15.3 by 7.2 by 7.2 km, with a rotation axis ratio of 1:3.5, rotating in a complicated manner (in about 175 h about the long axis and in 88 h in a precessing motion around the moment of inertia). The comet weighs about 2×10^{14} kg (200 trillion tons), with a volume of 420 km^3, and has a surface area of 294 km^2, with a mean bulk density of about 0.55 g/cm^3 (Chapter 2).

Halley is the parent of two meteor showers, the May *η-Aquariids* and the October *Orionids*. In both cases, the orbit of the comet is now far from Earth's orbit (up to 0.155 AU in the case of the Orionids), demonstrating that dust is distributed over a large volume. Halley revolves in an orbit in between the 1:6 and 1:7 resonances with Jupiter. Even though Halley does currently not librate about a mean-motion resonance, there will be no close encounters within 0.50 AU from outer planets, or within 0.05 AU from inner planets in the next centuries, because the nodes of Halley's orbit have moved away from Jupiter's orbit (Chapter 27).

[7] A. B. Paine, Personal memoranda. *Mark Twain: a biography* (New York: Harpers, 1912), Ch. 282.
[8] D. K. Yeomans, Ancient Chinese observations and modern cometary models. *Bull. Am. Astron. Soc.* **27** (1995), 1286–1286.

Relationships of particle size, mass, and magnitude

The Giotto and VeGa probes observed the formation of these meteoroid streams close up. The satellites counted dust particles in small bins of mass (M).[9] Because mass relates to magnitude as $\log_{10}(M) \sim -0.4\,\mathrm{m}$, each bin of $\log_{10}(M)$ corresponds directly to a range in visual magnitude m of a corresponding meteor. If the density is constant, this would also be equivalent to counting the meteoroids in bins of size (d). The distribution of meteoroid masses can often be summarized by a single parameter, the *differential mass distribution index* (s):

$$\Delta N(M) \sim M^{-s}\, \Delta \log(M) \tag{7.1}$$

The magnitude distribution of Orionids and η-Aquariids can be compared to the measured dust distribution at the comet itself (Fig. 7.3), because the differential mass distribution index relates to the *magnitude distribution index* $\chi = N(m+1)/N(m)$ according to:

$$s = 1 + 2.5 \log(\chi) \tag{7.2}$$

The corresponding *differential size distribution index* is:

$$\alpha = 3s - 2 \tag{7.3}$$

Plotting the number $N(M)$ of meteoroids in a bin of logarithmic mass $\log M$ (or versus a bin of magnitude) gives a slope $-s$. When $s = 2.0$ (or $\chi = 2.51$), then the same amount of mass is contained in each bin of magnitude. In the same way, if $s = 1.67$ (or $\chi = 1.85$), an equal combined cross sectional area for the meteoroids is contained in each magnitude bin. Fig. 7.4 shows an example of what these particle mass distributions look like.

From the counts by their "Dust Impact Detector System (DIDSY)," *Tony McDonnell* and coworkers derived $s = 1.61 \pm 0.17$ ($\chi = 1.75 \pm 0.28$) for the measured incremental mass distribution index (s) of large grains in excess of 1 billionth kg, the slope in Fig. 7.3, for the period -40 to $+40$ s from encounter, while $s = 0.30 \pm 0.09$ for the period -300 to -40 s.[10] The corresponding *differential size distribution index* $\alpha = 2.8 \pm 0.5$ and -1.1 ± 0.3, respectively.

More recently, Marco Fulle *et al.* applied a dust outflow model and derived from the same data $\alpha = 3.5 \pm 0.2$ ($s = 1.83 \pm 0.07$) for grains larger than 20 μm. They also found a most probable outflow velocity of 50 ± 5 m/s for $d = 1$ mm sized grains scaling

[9] N. Divine, H. Fechtig, T. I. Gombosi, *et al.*, The comet Halley dust and gas environment. *Space Sci. Rev.* **43** (1986), 1–104.

[10] J. A. M. McDonnell, P. L. Lamy and G. S. Pankiewicz, Physical properties of cometary dust. In *Comets in the Post-Halley Era*, ed. R. L. Newburn Jr., M. Neugebauer and J. Rahe. (Dordrecht: Kluwer, 1991), pp. 1043–1073; J. A. M. McDonnell, G. C. Evans, S. T. Evans *et al.*, The dust distribution within the inner coma of comet P/Halley 1982i – Encounter by Giotto's impact detectors. *Astron. Astrophys.* **187** (1987), 719–741.

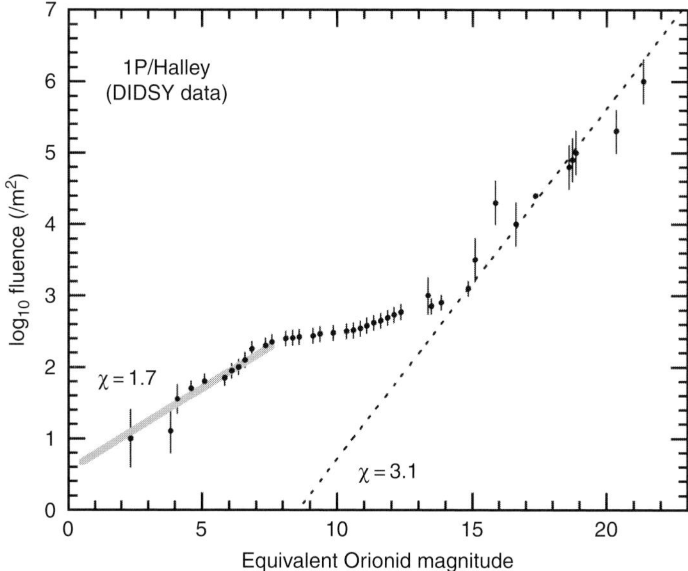

Fig. 7.3 Dust number distribution per mass interval measured by the DIDSY sensors during the period −450 to −169 s before the closest approach of 1P/Halley. I have compared the counts to the magnitude distributions observed during Orionid outbursts ($\chi = 1.7$) and annual shower activity ($\chi = 3.1$).

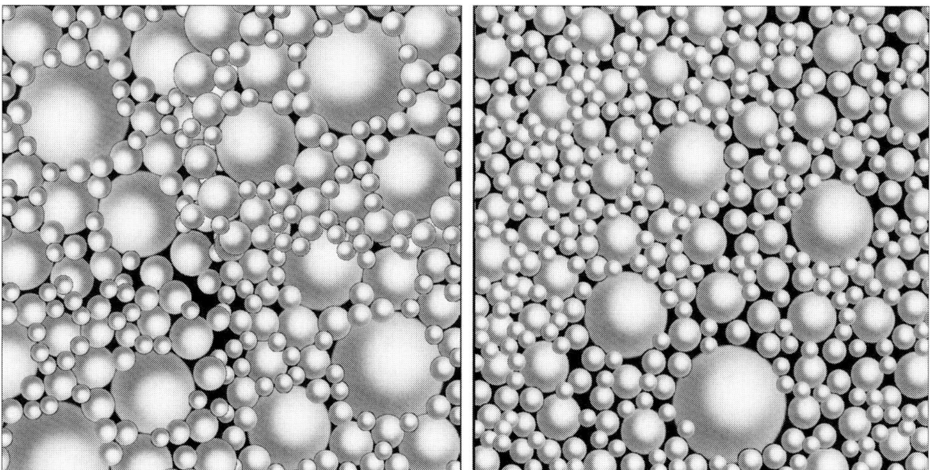

Fig. 7.4 Dust mass distribution covering seven magnitudes for $\chi = 1.85$ (left, equal cross section per magnitude interval) and $\chi = 2.51$ (right, equal mass per magnitude interval).

with $1/\sqrt{d}$, dispersed by 35 ± 5 m/s.[11] A typical 0.01 g (2.2 mm sized) Orionid meteoroid of magnitude +2.8 would have a relatively high ejection speed of 33 ± 24 m/s.

In comparison, the annual Orionid meteor stream has $s = 2.23$, while $s \sim 1.6$ during the 1993 Orionid shower outbursts (see Chapter 18). Hence, the large meteoroids in the coma of comet Halley had a size distribution equal to that observed in the 1993 Orionid outburst, while the small grains have a distribution similar to that of the annual Orionid shower.

7.3 Types of fragmentation

Catastrophic fragmentation, for example by the impact of a fast meteoroid, results in a fragment mass distribution $s = 2.1 \pm 0.1$ ($\chi \sim 2.5$).[12] This has been explained in many ways, but the exact reason is not clear. Such collisions tend to lead to only slightly ellipsoidally shaped meteoroids with a ratio of orthogonal diameters: B/A and C/A of on average 0.7 and 0.5, where A is the long axis. Collisions with other interplanetary dust particles can lead to catastrophic fragmentation, which is perhaps why many annual showers have $\chi \sim 2.5$.

In the gas outflow of a comet, mutual collisions between meteoroids are typically not catastrophic. The number of particles increases steeply with mass and thus a meteoroid is most likely to be broken by another meteoroid that has just enough mass (and hence energy) to do so. The collisional fragments will have ejection velocities that are almost identical to those of the original particle. This peculiar property is the reason why there are asteroid families (Chapter 10).[13] This may also account in part for the peculiar dust distributions in comet dust trails (see Chapter 15).

Hence, if meteoroids fall apart after the ejection of the comet, they can create a *collisional cascade*, where bigger particles break up into smaller pieces, and then those smaller pieces break and become the parents of even smaller pieces, etc, ... The size distribution that results from such a cascade is $s = 1.83$ ($\chi = 2.15$), if all meteoroids have the same strength against impacts.[14] Asteroid families have been measured to have $s = 1.78 \pm 0.02$,[15] suggesting a small effect of changing strength with increasing mass. Such things can occur in asteroids because of gravitational overburden or strain-rate effects. If the collisional cascade has fully developed and the meteoroid strength against impacts does not change with size, then we would expect the meteors in meteor storms to have a shallow $\chi = 2.15$.

[11] M. Fulle, L. Colangeli, V. Mennella, A. Rotundi and E. Bussoletti, The sensitivity of the size distribution to the grain dynamics: simulation of the dust flux measured by GIOTTO at P/Halley. *Astron. Astrophys.* **304** (1995), 622–630.

[12] A. Fujiwara, P. Cerroni, D. Davis *et al.*, Experiments and scaling laws for catastrophic collisions. In *Asteroids II*, ed. R. P. Binzel, T. Gehrels and M. S. Matthews. (Tucson, AZ: University of Arizona Press, 1989), pp. 240–269.

[13] S. F. Dermott, K. Grogan, D. D. Durda *et al.*, Orbital evolution of interplanetary dust. In *Interplanetary Dust*, ed. E. Grün, B. Å. S. Gustafson, S. F. Dermott and H. Fechtig. (Berlin: Springer, 2001), pp. 569–639.

[14] J. S. Dohnanyi, Collisional model of asteroids and their debris. *J. Geophys. Res.* **74** (1969), 2531–2554.

[15] D. D. Durda and S. F. Dermott, The collisional evolution of the asteroid belt and its contribution to the zodiacal cloud. *Icarus* **130** (1997), 140–164.

A third type of size distribution is called *collisionally relaxed.* If the meteoroids (or minor planets) frequently collide among each other, then the size distribution will evolve towards having equal cross section per magnitude interval, which means $s = 1.67$ and $\chi = 1.85$.[16]

Once the largest grains have disappeared from the observed sample of grain sizes, the distribution steepens. Interestingly enough, the sporadic background has a steeper size distribution index: $\chi = 4.3$ ($s = 2.58$), on the high-mass slope of the mass influx curve (Chapter 32). Indeed, the peak of the sporadic meteoroid mass influx is at smaller masses, around 10^{-5} g, or meteoroid sizes of ~200 μm.[17]

7.4 Comet Halley and the dispersion of meteoroids

It is also possible to compare the dust number density measured along the path of Giotto with the activity profile of meteor storms. Giotto encountered comet Halley on March 13, 1986, and came within 596 km of the nucleus. It had been helped to locate the nucleus by the Russian VeGa 1 and VeGa 2 spacecraft that had passed on March 6 and 9 at 8890 and 8030 km distance. The dust number density[18] for grains from about 10^{-16} up to 0.01 g (mostly the smaller grains) shows a $1/r_N{}^n$ dependence, with $n = 2.7 \pm 0.2$, away from the comet nucleus. Very similar distributions are measured from the flux profile of meteor storms when Earth travels through the extended atmosphere of a comet.[19]

At the time, the comet was visible from Earth as a +1st magnitude object with a 30° tail. The comet was observed well and its brightness compared to the stars thousands of times. With the brightness of a comet depending on the square of distance from Earth (Δ, AU) and some power n of heliocentric distance (r, AU) for small phase angle ϕ (°):

$$m = H_0 + 5 \log (\Delta) + 2.5\, n\, log(r) - 0.04\, \phi \tag{7.4}$$

1P/Halley's brightness was well described by $H_0 = 4.69$ (± 0.35) and $n = 4.0$ during the preperihelion part, and $H_0 = 3.71$ (± 0.12) and $n = 3.0$ during the postperihelion part (Fig. 7.5).[20] This implies that the mass loss of the comet increased as r^{-4} towards perihelion and fell off as r^{-3} after perihelion. That means that half of all matter was ejected close to perihelion between January 21 and March 04, 1986, when the comet was north of the ecliptic plane and far inside Earth's orbit. The water outflow speed

[16] I. S. Dohnanyi, Collisional model of asteroids and their debris. *J. Geophys. Res.* **74** (1969), 2431–2554.

[17] E. Grün, H. A. Zook, H. Fechtig and R. H. Giese, Collisional balance of the meteoritic complex. *Icarus* **62** (1985), 244–272.

[18] S. Nappo, J. A. M. McDonnell, A. C. Levasseur-Regourd *et al.*, Intercomparison of Giotto DIDSY/PIA and HOPE data. In *Asteroids Comets Meteors III.* (Uppsala: University of Uppsala, 1989), pp. 397–400.

[19] J. A. M. McDonnell, S. F. Green, E. Grün *et al.*, *In situ* exploration of the dusty coma of comet P/Halley at Giotto's encounter – flux rates and time profiles from 10 to the –19th Kg to 10 to the –5th Kg. *Adv. Space Res.* **9** (1989), 277–280.

[20] G. W. Kronk, *Comets, a Descriptive Catalog.* (Hillside, NJ: Enslow, 1984), 331pp (and website).

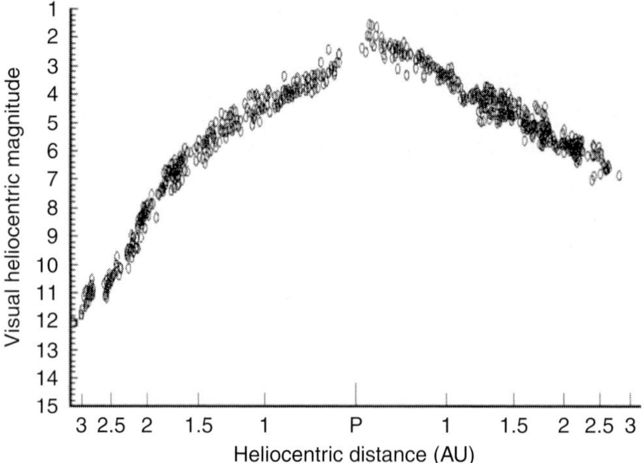

Fig. 7.5 The Halley visual light curve as measured by J. E. Bortle, R. Bouma, D. W. E. Green, A. Hale, R. Keen, C. S. Morris, and D. A. J. Seargent. Adapted from: Morris and Hanner (1993).[21]

was measured at $V\,(\mathrm{km/s}) = 2.33 \pm 0.07 - 0.98 \pm 0.05\, r$ preperihelion ($0.6 < r < 2.2$ AU) and $V\,(\mathrm{km/s}) = 2.63 \pm 0.18 - 1.19 \pm 0.14\, r$ postperihelion ($0.6 < r < 1.6$ AU).[22]

The total dust and gas production rate was estimated to be equivalent to the comet nucleus losing about 1/2000th of its mass per apparition (or 100 billion kg/return). Halley's meteor shower suggests that since capture by Jupiter, the comet has lost about 10% of its mass (Chapter 27), consistent with the survival of craters on the surface of the nucleus. Giotto recorded three major jets (and perhaps four fainter ones) from active areas that measure about 10 km^2 each. From these active areas, which cover 10% of the comet surface, about 6 m depth of dust and ice is lost in each apparition. The dust is ejected not directly towards the Sun, but 30° south of the comet–Sun line. This direction changed slightly during the encounter, possibly because of the rotation of the nucleus. Over the 3 h period of the encounter, the dust production rate did not change by more than 1%. Dust was ejected at a rate that was about twice as much as vapor from ices, with a dust to gas ratio of 2, excluding boulders.

The nongravitational force parameters, $A_1 = +0.1130 \pm 0.0013$ and $A_2 = +0.015\,5501 \pm 0.000\,000\,24 \times 10^{-8}$ AU/d^2, amount to an increase of the orbital period by about 4 d with each revolution.

[21] C. S. Morris and M. S. Hanner, The infrared light curve of periodic comet Halley 1986 III and its relationship to the visual light curve, C_2 and water production rates. *Astron. J.* **105** (1993), 1537–1546.

[22] D. Bockelée-Morvan, J. Crovisier and E. Gérard, Retrieving the coma gas expansion velocity from the 18-cm OH line shapes. In *Asteroids Comets Meteors III*, (Uppsala: University of Uppsala, 1998), pp. 267–270.

Fig. 7.6 Comet 109P/Swift–Tuttle on December 1, 1992, composite of four images taken by Jim Scotti, Spacewatch, LPL, University of Arizona.

7.5 Perseid comet 109P/Swift–Tuttle

The best known meteor shower is that of the Perseids in summer. Parent *109P/Swift–Tuttle* (Fig. 7.6), number two on the list, was identified as such shortly after the return of 1862. The comet did not return at the predicted time in the early 1980s. Then it was found that comet 1737 II (Kegler) could be this object and the new predicted return date in December of 1992 was very close to the actual value ($P = 135$ yr). It was then found that the comet was seen before in ancient China in AD 188 and 69 BC (109P/−68 Q_1). During that time, the absolute magnitude of the comet remained nearly constant and its orbital motion was not affected by the rocket-like forces of outgassing, presumably because of its weight.

Because nongravitational forces appear to be negligible, the future orbital evolution can be investigated until an exceptionally close encounter with Earth in AD 4479, when according to *John Chambers* the chance of collision will be 1 in 1 million.[23] The descending node is likely to stay within 0.1 AU of Earth's orbit for the next 20 000 years. The comet will continue to librate about the 1:11 mean-motion resonance with Jupiter (Fig. 7.2), mainly due to indirect rather than direct planetary perturbations.

[23] J. E. Chambers, The long term dynamical evolution of comet Swift–Tuttle. *Icarus* **114** (1995), 372–386.

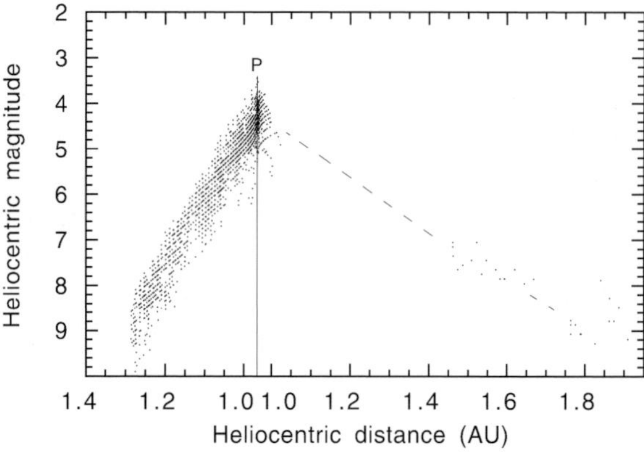

Fig. 7.7 Lightcurve of the 1992 return of comet 109P/Swift–Tuttle from data collected by the International Comet Quarterly.

This is a relatively large comet, with a nucleus of 26 ± 3 km[24] diameter, much bigger than comet 1P/Halley, reflecting $3 \pm 1\%$ of red light. It was found that the nucleus rotates in a more simple manner than comet Halley every 67.3 ± 2.6 h[25] with a rotation axis that is tilted by $45 \pm 10°$ (obliquity).

The gas production rate of Swift–Tuttle is comparable to that of Halley at the same heliocentric distance. The preperihelion brightness of the comet was well represented by $H_0 = 4.56$, $n = 7.58$, while postperihelion by $H_0 = 4.6 \pm 0.3$ and a less steep $n = 6.7 \pm 0.2$ (Fig. 7.7).[26] The comet ice is rich in hydrogen cyanide (HCN) and methanol. Small dust grains (submicron in size) were observed to be ejected with a speed of >660 m/s, comparable to the gas ejection speed.[27] Larger 0.6–10 μm sized grains observed at infrared wavelengths imply a size–distribution index $\alpha = 2.5 \pm 0.5$, not quite as steep as $\alpha = 3.7$ measured for comet 1P/Halley.[28] The dust to gas ratio of Swift–Tuttle is comparable to, but slightly larger than, that of comet Halley.

Over the past 2000 years, the location of the active areas on the nucleus can not have changed much, although the direction of the spin axis may be precessing. During the 1992 return, the comet had one strong jet close to its north pole, with an opening angle of about 20° and small-grain dust ejection speeds of 300 m/s.[29] In an earlier interpretation of observations of the 1862 return, Sekanina[30] found that the siderial

[24] Higher limit given in JPL database. Lower limit by A. Fitzsimmons from nuclear R-magnitude at 5.3 AU ($m_R = +19.3$). D. P. O'Ceallaigh, A. Fitzsimmons, I. P. Williams, CCD photometry of comet 109P/Swift–Tuttle. *Astron. Astrophys.* **297** (1995). L17–L20; H. Boehnhardt, K. Birkle, M. Osterloh, Nucleus and tail studies of comet P/Swift–Tuttle. *Earth, Moon, and Planets* **73** (1996), 51–70.

[25] L. Jorda, F. Colas and J. Lecacheux, The dust jets of P/Swift–Tuttle 1992t. *Planet Space Sci.* **42** (1994), 699–704.

[26] M. P. Mobberley, Swift–Tuttle: the 1992 apparition. *J. Brit. Astron. Assoc.* **104** (1994), 11–26.

[27] Y. Goldberg and N. Brosch, Imaging polarimetry of the comet P/Swift–Tuttle. *Mon. Not. R. Astron. Soc.* **273** (1995), 431–442.

[28] J. Sarmecanic, M. Fomenkova and B. Jones, Modeling of mid-infrared dust emission from P/Swift–Tuttle. *Planetary Space Sci.* **46** (1998), 859–863.

[29] L. Jorda, F. Colas and J. Lecacheux, The dust jets of P/Swift–Tuttle 1992t. *Planet. Space Sci.* **42** (1997), 699–704.

[30] Z. Sekanina, Distribution and activity of discrete emission areas on the nucleus of periodic comet Swift–Tuttle. *Astron. J.* **86** (1981), 1741–1773.

nuclear rotation period was 66.5 h, the pole was tilted by 80°, and oriented 60° from the solar direction. The axis of the north pole of the comet was pointed at ecliptic longitude 217°, latitude −76°. Longitude was measured from subsolar meridian at perihelion. He found that as many as eight discrete active regions (covering less than 1% of the surface) were responsible for the observed jets, envelopes, and tail bands. Waning dust jets developed into envelopes, while old envelopes in turn became the observed tail bands. A jet located at about latitude 60° N and longitude 90° W was the only one persisting over the 7 week observing period and the one with the highest gas production rate. As pointed out by Sekanina, this could have been the same active region seen in the 1992 return.

7.6 Leonid comet 55P/Tempel–Tuttle

Number three on my list of Halley-type parent comets is *55P/Tempel–Tuttle*, parent of the Leonids. Orbiting the Sun every ~33.23 yr, the comet was first seen by Chinese astrologers in 1366 during a rare encounter with Earth, passing at 0.023 AU on October 26.4, when it became sufficiently bright to be seen by the naked eye ($m_v < +3.5^m$). This is the second closest approach of a comet on record (after comet D/1770 L_1 Lexell). The identity of C/1366 U_1 as 55P/Tempel–Tuttle was first suggested by *John Russel Hind*[31] in 1866 after computing a parabolic orbit from the Chinese observations. During the next Leonid return, *Shigeru Kanda*[32] of Tokyo Observatory calculated an elliptical orbit with an assumed period of 33.35 yr and confirmed the identity of the comet. As head of the Astronomical Society of Japan, Kanda was always motivating amateur astronomers and as such is sometimes considered to be the Japanese counterpart of Olivier in the West.

In 1965, *Joachim Schubart*[33] in a formal study of planetary perturbations confirmed the link with C/1366 U_1 and also identified a single sighting by Gottfried Kirch at Güben, Germany, on the morning of October 26 in 1699 (C/1699 U_1), when the comet passed only 0.064 AU from Earth. The comet was visible with the naked eye in the morning sky but moved southwards at 24° per day, so fast it was not seen again.

Schubart recovered the object on plates taken by M. J. Bester at Bloemfontein in South Africa on June 30, 1965 and found that his prediction was off by only 5 d. That year, the comet stayed far from Earth and was observed at no brighter than +16 magnitude.

During its approach in 1997, the comet was discovered as a faint star-like object long before any dust was ejected. From the brightness of the nucleus at that time, a size of 3.5 ± 0.4 km was derived and a nuclear axis ratio of larger than 1.5. From the

[31] J. R. Hind, *Observatory* **9** (1886), 282–284.
[32] S. Kanda, On the orbits of the comet of 1366 A.D. and of 868 A.D. *Japan J. Astron. Geophys.* **10**, (1932) 30.
[33] J. Schubart, E. Geyer, Periodic Comet Tempel–Tuttle (1965); J. Schubart, M. J. Bester, E. Geyer, Periodic comet Tempel–Tuttle (1965). *IAUC* 1926 (1965), O. Gingerich (ed.); J. Schubart, Comet Tempel–Tuttle: recovery of the long-lost comet of the November meteors. *Science* **152** (1966), 1236–1237.

Fig. 7.8 Comet 55P/Tempel–Tuttle, while at 1.067 AU from the Sun and 0.680 AU from Earth in a 13 by 21 arcmin wide image by Paul Gitto, Arcturus Observatory, Whiting, NJ. The photograph shows the comet as it appeared through a telescope.

motion of a jet near the northern pole (Fig. 2.8), a spin period of 15.33 ± 0.02 h was derived.[34]

The comet passed near Earth on January 17, 1998, when it reached $+7.5^{m}$ (Fig. 7.8). A thin tail was seen in the near-infrared R-band images, which was not seen in visual V-band images, implying some amount of fine dust grains being ejected with the ice. Only $5 \pm 1\%$ of visible sunlight is reflected. The comet passed perihelion on February 28, after which it was lost in the glare of the Sun after May 10.

During this return, the heliocentric dependence of activity of the comet peaked steeply near perihelion. Postperihelion observations were scarce, but suggest that the profile may have been symmetric. For $H_0 = +8.5$, the comet brightness increased with a steep exponent $n = 10.7 \pm 0.6$, after an onset of activity around 1.40 AU from the Sun (Fig. 7.9). The comet is not active at 3.5 AU.[35] This trend is much steeper than the average value of $n = 4$ for all other comets.[36] In general, Halley-type comets have higher values of around $n = 6.3$, while for Jupiter-family comets the mean is as high as 7.4.[37] Hence, the gas (and dust) outflow of comet 55P/Tempel–Tuttle is concentrated near perihelion. The nongravitational parameters are presently $A_1 = +0.0996 \pm 0.0023$ and $A_2 = +0.009\,196 \pm 0.000\,017 \times 10^{-8}$ AU/d^2 (1865–1998).

The observed water production rate is ~10 billion kg/orbit derived from SOHO/SWAN data (Fig. 7.10) by Teemu Mäkinen of F.M.I. The comet does not emit many

[34] L. Jorda, J. Lecacheux, F. Colas *et al.* (Cambridge, MA: IAU), Comet 55P/Tempel–Tuttle. *IAU Circ.* **6816**, ed. D. W. E. Green 1998.

[35] O. R. Hainaut, K. J. Meech, H. Boehnhardt and R. M. West, Early recovery of comet 55P/Tempel–Tuttle. *Astron. Astrophys.* **333** (1988), 746–752.

[36] J.-I. Watanabe, H. Fukushima and T. Nakamura, The activity profile of comet 55P/Tempel–Tuttle in 1998 return: meteoroid release concentration on the perihelion. *ESA SP* **495** (2001), 175–178.

[37] L. Kámel, On the photometric indices of periodic comets. *Icarus* **122** (1996), 428–431.

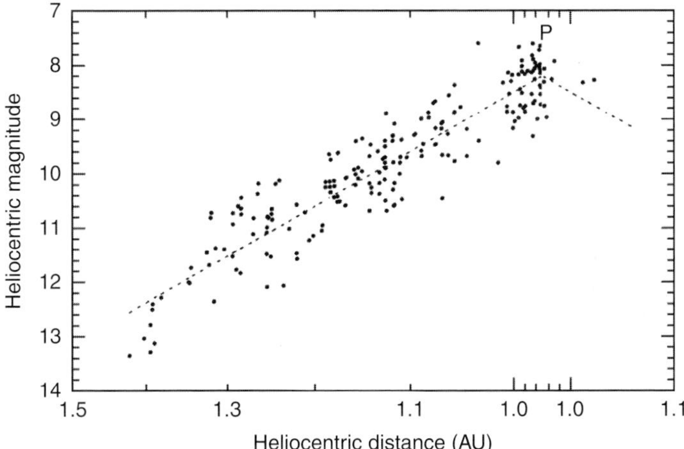

Fig. 7.9 The comet 55P/Tempel–Tuttle light curve. Uncorrected magnitude estimates collected by the International Comet Quarterly.

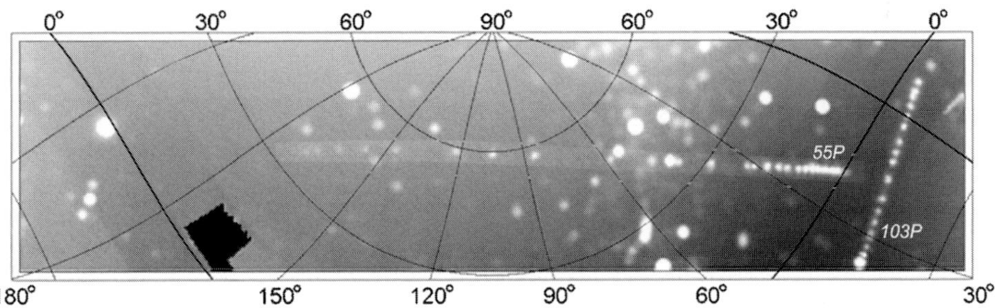

Fig. 7.10 SOHO/SWAN observations of the hydrogen cloud of comet 55P/Tempel–Tuttle in Lyman alpha emission. The sequence to the right shows comet 103P/Harley 2. Image courtesy Teemu Mäkinen, Finnish Meteorological Institute – Geophysical Research Division.

small dust grains. Based on the scattered light from ~1 μm grains, the dust to ice ratio is low. However, based on the total mass of dust in the dust trail of the comet from meteor observations, about 13 ± 4 billion kg/orbit (Chapter 15),[38] the dust to ice ratio is a typical 2.3 ± 0.9.

The comet is close to, but does not librate about the 4:15 mean-motion resonance with Jupiter. The nodal line gradually rotates with respect to the line of apsides. A close encounter with Saturn occurred on March 05, 1630 (0.337 AU), while the comet is near Uranus on July 31, 2317 (0.292 AU) and near Earth on November 29, 2397 (0.024 AU). Due to the high relative speed in these encounters, the orbital elements will not be affected much.

[38] P. Jenniskens, Forecast for the remainder of the Leonid storm season. *ESA SP* **495** (2001), 83–90.

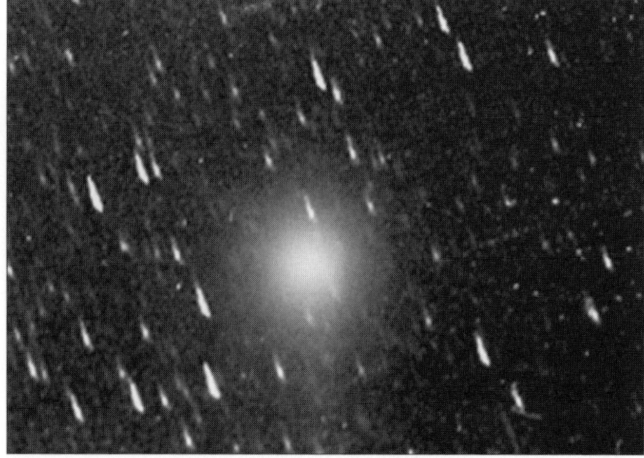

Fig. 7.11 Comet 8P/Tuttle on January 11, 1981. Five minute exposure on Kodak Tri-X (5 arcsec f/5 refractor prime focus). Photo: Maurice Clark at Yericoin, Western Australia.

7.7 Ursid comet 8P/Tuttle

Comet *8P/Tuttle* is the parent of the Ursid shower and number four on this list (Fig. 7.11). When it was first seen by Pierre A. Méchain at Paris in 1790 (1790 A_2), H_{10} was about $+7.7^m$. When Horace P. Tuttle observed a faint comet on January 4, 1858, (1858 A_1) and his brother Charles calculated the parabolic orbit, the striking similarity with the orbit of the 1790 comet was soon noticed by Gould and others.

Because of its high inclination ($i = 55°$) and argument of perihelion ($\omega = 207°$), the comet has only weakly perturbing encounters with Jupiter. The most recent occurred on December 29, 1900 and December 29, 1995. At first sight, the comet does not appear to librate[39] about a mean-motion resonance. There are oscillations of the orbital period, but those are due to the similar periods of Jupiter and the comet. The semi-regular oscillation in aphelion distance reflects a nearly constant excess over the 1:1 resonance. These oscillations, however, occurred around the 7:6 resonance in the period AD 1585–1805 and around the 15:13 resonance since. During the 1994 return, q dipped just below 1 AU, but most of the time its perihelion is just outside of Earth's orbit.

Modern comet catalogs imply that 8P/Tuttle has decreased in brightness over the years, but a good fit to the 1980 curve gives $H_0 = +8.0^m$, $n = 6.0$ (Fig. 7.12), not unlike the $H_0 = +7.7^m$ (for $n = 4$) when the comet was first discovered. From the brightness of the nucleus when it still looked stellar, $H_N = +13.3^m$, the diameter is $D = 14.9$ km (for an albedo of 0.04).[40] This is the size of comet 1P/Halley! Even the nongravitational parameters from observations in the period 1967–1992 are almost

[39] G. B. Valsecchi, *Electronic Atlas of Dynamical Evolutions of Short-period Comets*. LTEP website, Istituto di Astrofisica Spaciale e Fisica Cosmica.

[40] J. Licandro, G. Tancredi, M. Lindgren, H. Rickman and R. G. Hutton, CCD photometry of cometary nuclei, I: Observations from 1990–1995. *Icarus* **147** (2000), 161–179.

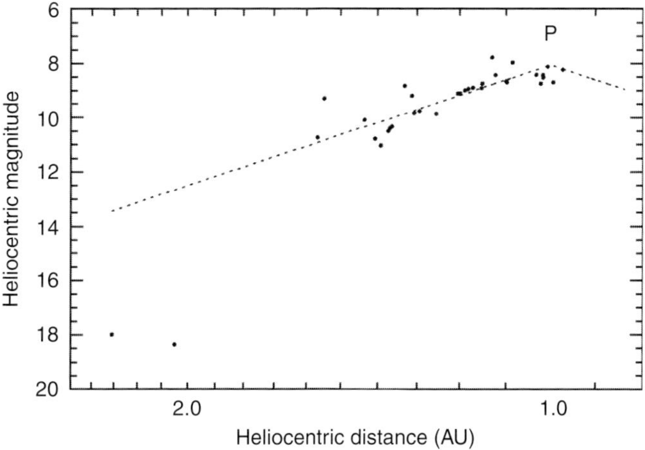

Fig. 7.12 Light curve of comet 8P/Tuttle during the 1980 return. From: IAU Circulars.

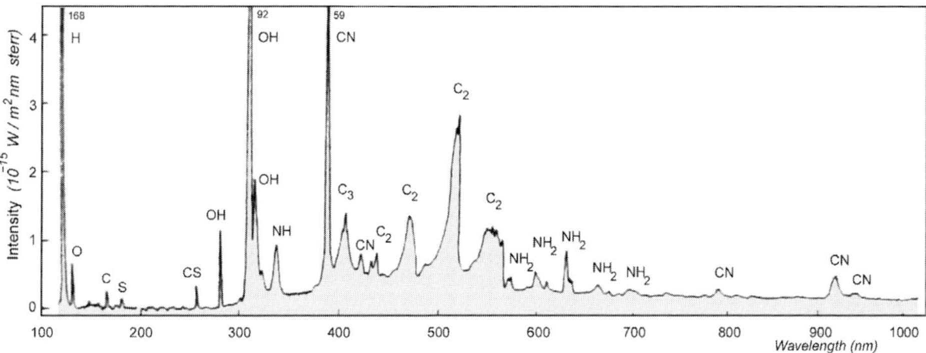

Fig. 7.13 Spectrum of comet 8P/Tuttle in November, 1980 in a compilation of IUE satellite and Arizona Catalina Station telescope data by S. Larson and J. Johnson, courtesy of Hal Weaver, The Johns Hopkins University. Emission bands from molecules are identified.[41]

the same as those for comet Halley: $A_1 = +0.090 \pm 0.051 \times 10^{-8}\,\mathrm{AU}/d^2$ and $A_2 = +0.011\,57 \pm 0.000\,96 \times 10^{-8}\,\mathrm{AU}/d^2$, stable at least since 1858.[42]

Why then is comet 8P/Tuttle not as intrinsically bright? The water evaporation rate was measured using ultraviolet emissions of OH, observed with the IUE satellite on December 7, 1980, when the comet was at 1.02 AU from the Sun and close to perihelion (Fig. 7.13).[43] In the optical spectra, comet Tuttle had strong vapor emissions from molecules of NH_2, C_2, H_2O^+, and CS^+ that originate from the

[41] W.-H. Ip, U. Fink and J. R. Johnson, CCD observations of comet Tuttle 1980 XIII – The $H_2O(+)$ ionosphere. *Astrophys. J.* **293** (1985), 609–615; J. R. Johnson, U. Fink and S. M. Larson, Charge coupled device (CCD) spectroscopy of comets – Tuttle, Stephan-Otema, Brooks 2, and Bowell. *Icarus* **60** (1984), 351–372.

[42] V. V. Emelianenko, Study of the motion of comet Tuttle from 1858 through 1967. *Kazansk. Gorodsk. Astronom. Observ.* **42–43** (1978), 203–210 (in Russian).

[43] H. H. Weaver, P. D. Feldman, M. C. Festou, M. F. A'Hearn and H. U. Keller, IUE observations of faint comets. *Icarus* **47** (1981), 449–463.

evaporating ices. In the ultraviolet, there were strong OH, CS_2, CO_2^+, and hydrogen atom emissions, along with C, S, CS_2, CO_2^+ emissions, both showing evidence of much gas and few fine dust grains that scatter light efficiently. The amount of water evaporating from 8P/Tuttle was only a factor of 4 less than that of comet Halley, at about 5.7×10^{28} molecules of water per second, or about 1700 kg/s.

Unlike Halley, 8P/Tuttle appears to eject dust in the form of large grains. The production rate of small meteoroids was only about 3.12 kg/s at 1.92 AU (mean diameter 0.1 μm), 5.51 kg/s at 1.75 AU (0.174 μm), 23.85 kg/s at 1.35 AU (1.2 μm), and 36.3 kg/s at 1.19 AU (1.32 μm). The ratio of fine dust to gas appears to have been independent of heliocentric distance between 1.92 and 1.19 AU.[44]

The mass-loss rate of large meteoroids can be estimated from the density of Ursid shower dust trails, should more be known about the 3D distribution of dust in the trails. It is possible that 8P/Tuttle lost most dust in the form of larger grains, but that is not likely because the Ursid showers are not particularly rich in bright meteors.

7.8 C/1917 F_1(Mellish)

Number five is *C/1917 F_1 (Mellish)*, parent of the December *Monocerotid* shower.[45] It has a perihelion inside the orbit of Mercury ($q = 0.190$ AU), the second smallest perihelion distance after comet 96P/Machholz. The comet passed Earth's orbit at 0.06 AU at the descending node. It has an orbital period of 145.8 ± 0.8 yr,[46] but the orbital period and its status as a Halley-type comet is more uncertain than that (more in Chapter 18). It is due back at perihelion in 2061. With $H_0 = +7.4$ ($n = 3.9$), it may have a diameter $D \sim 3.1$ km.

The comet was discovered on March 20, 1917 by John E. Mellish at the Harrold Observatory in Leetonia, Ohio, but tracked only until June 25. It was 5th magnitude on March 27, then rapidly grew to 2nd magnitude on April 3. In South Africa, the comet reached +1st magnitude, the brightest object since 1P/Halley's return in 1910, and was reported in the Cape Town newspapers. James Francis Skjellerup (discoverer of comet 26P/Grigg–Skjellerup) observed the comet on April 19, noting: *5.30 am Tail of comet visible above trees on horizon. The nucleus cleared the horizon about 5.45 am tail about 10° long nearly straight up towards the zenith but with a slight curve towards the N. Nucleus very bright about 3rd mag. Coma small and diffused, tail narrow but widened out a little towards the end. Visible to naked eye vision until 6.25 am (Sunrise 7.13)*. By April 25, it was still clearly visible

[44] R. L. Newburn and H. Spinrad, Spectrophotometry of seventeen comets. II – The continuum. *Astron. J.* **90** (1985), 2591–2608.

[45] F. L. Whipple, Photographic meteor orbits and their distribution in space. *Astron. J.* **59** (1954), 201–217; B. A. Lindblad and D. Olsson-Steel, The Monocerotid meteor stream and comet Mellish. *Bull. Astron. Inst. Czech.* **41** (1990), 193–200.

[46] S. Asklöf, Orbit of comet 1917 I (Mellish). The definitive orbit of comet 1917 I from visual and photographic observations separately. *Arkiv För Matematik, Stronomi och Fysik* 23A (Stockholm: Almqvist & Wiksells, 1932), 16pp.

but considerably fainter.[47] On April 28th, the tail measured 9° long. By June 16 and 19, the comet was a very faint, nebulous mass, without condensation.[48]

7.9 What makes Halley-type comets unique?

Halley-type comets are special because they are from the Oort cloud, not from the Kuiper Belt. They originate from regions near Jupiter or as far out as Neptune. Materials from the inner solar system may have mixed in with other dust, more so if they originate closer to Jupiter. On a larger scale, the rubble pile of cometesimals may have a wider range of compactness, because collisions in this region were more violent than in the Kuiper Belt. This could mean a larger range in tensile strength between the grains and perhaps also mineralogical differences. Unlike long-period comets, Halley-type comets spend significant time in the inner part of the solar system and may have long ago lost their pristine crust.

The Halley dust experiments onboard Giotto and VeGa 1 and 2 captured 5000 individual particles with the three instruments, all very small 0.2–2 μm in diameter, with a total mass of no more than a few ng. From this data, it was found that the comet dust was rich in *CHON*: particles consisting of carbon, hydrogen, oxygen, and nitrogen. The element carbon was present at 20 weight %.[49] Some small grains consisted only of CHON, others only of rock forming elements (Si, Mg, Fe, ...; mostly magnesium silicates with ~10% iron sulfides). Only about half were intimately mixed. Rock-only particles were found mostly in the outer reaches of the coma, while CHON particles were found more in the inner coma,[50] implying that some organics were lost while leaving the nucleus.

High in the Earth's atmosphere, dust particles are found that originate outside of Earth, called *Interplanetary Dust Particles* (IDP). The IDPs suspected of being cometary in origin are porous aggregates built from thousands or millions of smaller ~0.3 μm sized units (Fig. 7.14). There is a wide range of porosity and a large amount of organics, which assists in binding them together. Surfaces are irregular and give direct access to the interior of the dust grains. Particle densities of 10 μm sized IDPs were measured in the range 0.3–6.2 g/cm^3, with an average of 2.0 g/cm^3 and 75% are in the range between 1 and 3 g/cm^3. These values compare well to the densities estimated indirectly for the Halley dust.[51]

[47] W. Orchiston, J. F. Skjellerup: a forgotten name in South African cometary astronomy. *Mon. Not. Astron. Soc. SA* **62** (2003), 56–75.
[48] J. Voûte, Observations of Mellish's comet (1971a) made at the Royal Observatory, Cape of Good Hope. *J. Observ.* **2** (1918), 3–4.
[49] E. K. Jessberger and J. Kissel, Chemical properties of cometary dust and a note on carbon isotopes. In *Comets in the Post-Halley Era 2*, ed. R. L. Newburn Jr., M. Neugebauer and J. Rahe. (Dordrecht: Kluwer, 1991), pp. 1075–1092.
[50] M. N. Fomenkova, S. Chang and L. M. Mukhin, Carbonaceous components in the comet Halley dust. *Geochim. Cosmochim. Acta* **58** (1994), 4503–4512.
[51] S. G. Love, D. J. Joswiak and D. E. Brownlee, Densities of stratsopheric micrometeorites. *Icarus* **111** (1994), 227–236; D. Maas, F. R. Krueger and J. Kissel, Mass and density of silicate and CHON type dust particles released by comet P/Halley. In *ACM III*, ed. C.-I. Lagerkvist *et al.* (Uppsala: Reprocentralen HSC, 1989), pp. 389–392.

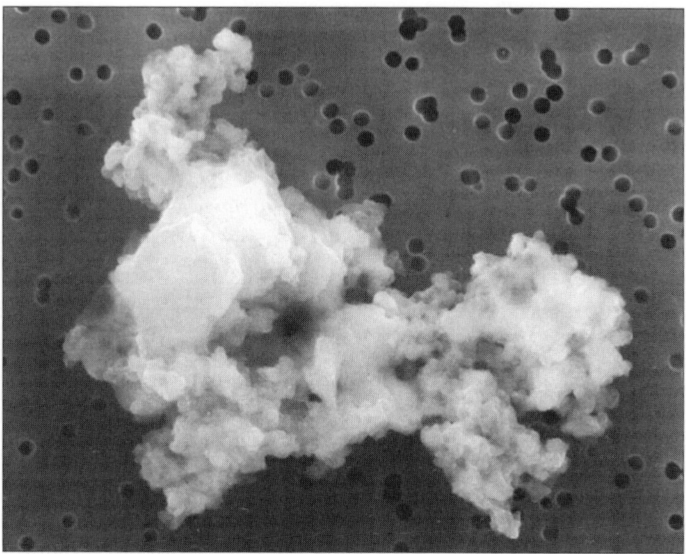

Fig. 7.14 A 10 μm interplanetary dust particle with a large (foreign?) clay inclusion. This is U2012C-II, collected by a NASA U2 aircraft, with a mean density of 1.8 g/cm^3. The open structure was once filled in with ice. It entered Earth's atmosphere at 19 km/s. Photo: NASA Stardust, courtesy Donald Brownlee, University of Washington.

IDPs tend to have elemental compositions very similar to those of (primitive) C I type chondrites, but for a factor 2–10 less of the elements calcium, sulfur, and, nickel. The eleven elements from copper to zinc are enriched by factors of ~1.5–3.[52] There is also less iron by about 30%. Such differences are large enough to be observed in meteor spectra if they also apply to the much larger-cm sized grains.

Just like the interstellar grains from which they are formed, amorphous silicates dominate IDPs. The rocky material is found mostly in the form of beads of GEMS: *glass with embedded metals and metal sulfides* (Fig. 7.15), which are usually a few 100 nm in diameter and are composed of silicate glass with large numbers of 10 nm size rounded grains of iron–nickel (FeNi) metal and metal sulfide such as troilite (FeS). Individual GEMS show evidence of a long bombardment with energetic particles, consistent with exposure in the interstellar medium.

IDPs contain on average about 10% carbon in weight, mostly as CHON. The CHON is thought to be the product of two processes. Tiny components have been identified as the gas phase condensations from the carbon-rich atmospheres of evolved stars. The bulk, however, originates from ice mantles frozen on the grains when they were still dust particles in dense interstellar clouds. The UV light inside the clouds generated by cosmic rays would break the oxygen-rich ice molecules and molecular fragments could then react into more complex (and thus less volatile) organic

[52] E. K. Jessberger, T. Stephan, D. Rost *et al.*, Properties of interplanetary dust: information from collected samples. In *Interplanetary Dust*, ed. E. Grün, B. Å. S. Gustafson, S. F. Dermott and H. Fechtig. (Berlin: Springer, 2001), pp. 253–294.

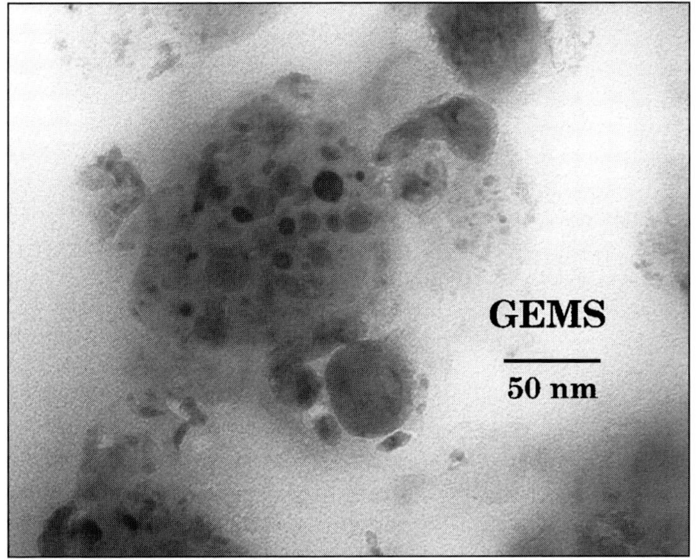

Fig. 7.15 GEMS in an interplanetary dust particle. Photo: NASA Stardust.

molecules.[53] This mechanism was studied in progressively more detail in the astrophysics laboratory of Mayo Greenberg at Leiden University, each Ph.D. student adding more information. My contribution focussed on how to make that "yellow stuff" dark brown or black, a requirement of Mayo's interstellar dust model. With the help of Gianni Strazzulla and Giuseppe Baratta at Catania Observatory, we discovered that cosmic rays can quickly turn yellow stuff black, but it is UV light outside dense interstellar clouds that does so in space. When leaving the protection of the cloud, that organic residue would be exposed. That organic residue would be exposed to more UV light for tens of millions of years, removing much of the hydrogen, oxygen, and nitrogen in a process called *carbonization*.[54]

Larger crystalline silicate grains, clays, and large chunks of CHON are often mixed in (Fig. 7.15), which do not come from interstellar space. These "hard bits" may have been created during the formation of the solar system, possibly further in towards the Sun, in active regions of planet formation, or at the extremes of the planetary disk above and below the region of comet formation during grain–grain collisions. If so, their relative abundance and size could vary among comets. Evidence for mixing also comes from mid-infrared spectra of Halley-type and long-period comets from the Oort cloud, which sometimes show the sharp features of crystalline silicates. In meteors, such hard bits could manifest as fragments penetrating deeper into Earth's atmosphere.

[53] J. M. Greenberg, A. J. Yencha, J. W. Corbett and H. L. Frisch, *Mem. Soc. R. Des Sci. de Liège* 6e ser. **3** (1972), 425; J. M. Greenberg, Radical formation, chemical processing, and explosion of interstellar grains. *Astrophys. Space Sci.* **39** (1976), 9–18.

[54] P. Jenniskens, G. A. Baratta, A. Kouchi *et al.*, Carbon dust formation on interstellar grains. *Astron. Astrophys.* **273** (1993), 583–600.

8
Jupiter-family comets

Jupiter-family comets (JFC) have their origin in the Kuiper Belt, a remnant of the solar nebula from which the planets formed, on the outskirts of our planetary system beyond Neptune. There, they were colliding with typical speeds of 1.5 km/s, sometimes losing speed and falling into orbits closer to the Sun. Encounters with the giant planets caused many to be tossed out, but some ended up crossing Jupiter's orbit. In a close encounter with Jupiter, those comets were finally slowed down enough to fall into an orbit towards Earth, circling the Sun in the same direction as the planets and close to the ecliptic plane in less than 20 yr (typically around 6 yr), with an aphelion at Jupiter's orbit (Fig. 8.1).

There are about 20–100 JFCs of $D \geq 1$ km (with $\chi_c \sim 2.2$, with a deficit of comets below 1.6 km diameter) that have $q < 1.3$ AU at any given time,[1] but only some of these will come close to Earth's orbit. Of all known Jupiter-family comets, only 13 have crossed Earth's orbit in the recent past and seven of those are known parents of meteor showers.

8.1 The Tisserand parameter

Most active Jupiter-family comets were only recently captured. In most cases, this is not a one-step process but rather a cascade.[2] Close encounters with Jupiter can both speed up and slow down the comet. The retrograde and highly inclined orbits of the Oort cloud comets, and the stochastic nature of this process more often tends to lead to ejection rather than capture.[3]

Because their aphelion is near Jupiter's orbit, these comets spend much time in that neighborhood and pass by Jupiter at speeds of less than that of the planet itself. As a result, the perturbations from Jupiter's gravitational pull last long and the orbits can

[1] W. F. Bottke, A. Morbidelli, R. Jedicke *et al.*, Debiased orbital and absolute magnitude distribution of the near-Earth objects. *Icarus* **156** (2002), 399–433.
[2] E. I. Kazimirchak-Polonskaya, *The Motion, Evolution of Orbits, and Origin of Comets*, ed. G. A. Chebotarv, E. I. Kazimirchak-Polonskaya and B. G. Marsden. (Dordrecht: Reidel, 1972), p. 373.
[3] M. Duncan, T. Quinn and S. Tremaine, The origin of short period comets. *Astrophys. J.* **328** (1988), L69–L73.

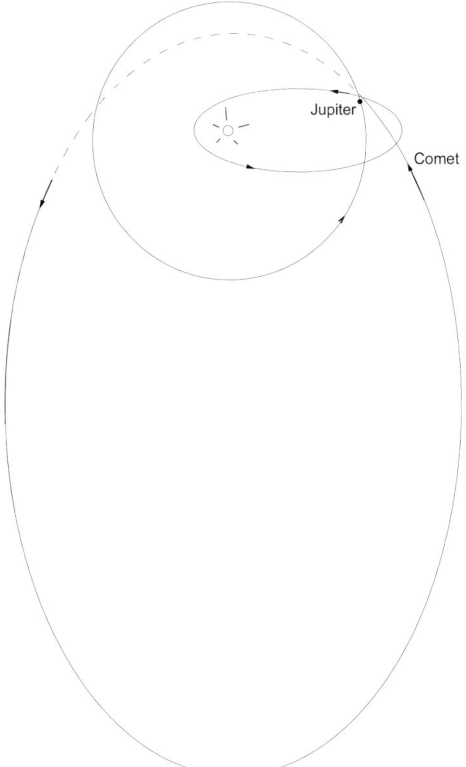

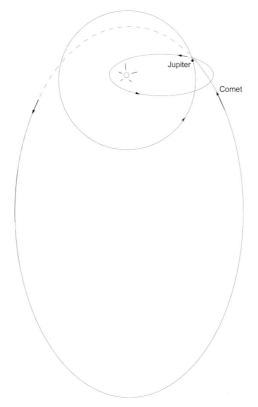

Fig. 8.1 The capture of a comet into a Jupiter-family comet orbit. North polar view.

vary dramatically from one return to the next. Another way of saying this is: the *Tisserand parameter* is larger than 2. The Tisserand parameter is defined as:[4]

$$T_J = a_J/a + 2 \cos i \sqrt{[a/a_J (1 - e^2)]} \quad (8.1)$$

Where a_J refers to the semi major axis of Jupiter ($a_J \sim 5.203\,36$ AU) and a, e, and i are the orbital elements of the comet (or meteoroid). The relative speed between comet and Jupiter is about $V_{rel} = V_J\sqrt{(3 - T_J)}$, with $V_J = 13.06$ km/s being the average value for a circular orbit of Jupiter. The Tisserand parameter is a convenient way of discriminating Jupiter-family comets from asteroids. Jupiter-family comets tend to have T_J between 2 and 3. Halley-type comets have $T_J < 2$, while asteroids (Chapter 10) have $T_J > 3$.

Not only does Jupiter perturb the orbits in frequent close encounters, but the planet also wields its power through weaker long-distance perturbations. Especially, when those are in any way nonrandom. The particular orientation, shape, and size of the

[4] F. Tisserand, *Traité de Méchanique Celeste (Paris)* **IV** (1896), 203.

orbit can lead to resonances but, unlike Halley-type comets, comets never stay long in a resonance. What initially seemed a stable situation, with Jupiter far from the node when the comet passed by, can turn quickly into an extra dangerous situation with Jupiter being near the node every time when the comet is there.

Due to the low inclination prograde nature of their orbits, they approach Earth at small speeds, arriving from the antihelion or the helion directions (Chapter 28). Their meteoroids tend to be slow (11–35 km/s, typically around 20 km/s) and the meteor showers are best seen just after midnight, while Halley-type and long-period type comet showers are best seen just before dawn. The frequent and effective perturbations cause these showers to disperse rapidly and vary greatly in activity from one return to the next.

Ultimately, $5.5 \pm 0.8\%$ will collide with the Sun, most of the rest being ejected. Comets can survive in Jupiter-family orbits for about 450 000 years,[5] but long before that time they will have lost most of the ices that drive the creation of meteoroid streams, once their perihelion distance dips below $q \sim 2.7$ AU. *David Hughes* estimated that 10, 20, and 30% of comets decay away after 1000, 2000, and 3000 yrs, respectively.[6] After that, they are mostly dormant.

8.2 Draconid comet 21P/Giacobini–Zinner

Confirmed and potential future parents of meteor showers are listed in Fig. 8.2. Here, I will discuss the seven comets with known showers. The first on my list is comet *21P/Giacobini–Zinner*, the cause of the spectacular 1933 and 1946 Draconid storms (Fig. 8.3). More recent outbursts occurred in 1985 and 1998 (Chapter 19). There is also an extended low-level shower activity in the years when the comet is near the Sun.

The orbit changes erratically, even though the orbital period remains in between the 2:1 and 3:2 resonances with Jupiter.[7] After a pair of close encounters with Jupiter in 1626 and 1628, the inclination decreased and the orbit aligned closely with that of the planet. The inclination stayed around a high 30° and other orbital elements vary now slowly, at a near constant rate, with q remaining close to Earth's orbit at least until AD 2400.

The comet nucleus is the fastest spinning on record. Lacking imaging of the nucleus itself, Zdenek Sekanina (J. P. L.)[8] modeled the nucleus as an oblate spheroid with an equatorial diameter of 2.5 km, a ratio of about 8.3:1 between the equatorial and polar diameters, and a rotation period of 1.66 h. More recently, the nuclear absolute

[5] H. F. Levison and M. J. Duncan, The long-term dynamical behavior of short-period comets. *Icarus* **108** (1994), 18–36.

[6] D. W. Hughes, The variation of short-period comet size and decay rate with perihelion distance. *Mon. Not. R. Astron. Soc.* **346** (2003), 584–592.

[7] A. Carusi, L. Kresák and G. B. Valsecchi, *Electronic Atlas of Dynamical Evolutions of Short-Period Comets* (IAS Computing Centre, 1987) Website: http://www.rm.iasf.cnr.it/ias-home/comet/catalog.html.

[8] Z. Sekanina, Precession model for the nucleus of periodic comet Giacobini–Zinner. *Astron. J.* **90** (1985), 827–845.

Comet	P (yr)	q (AU)	i (°)	H_{10}	H_N[9]	D_N (km)	P_N (h)	Comments:	Shower:
3D/Biela	6.62	0.87	12.3	+7.5	-.-	~3.0	-.-	Now dormant	**Andromedids**
6P/d'Arrest	6.16	1.04	9.3	+9.5	+16.7	3.0	6.67		1780–1786: Sept. Bootids
7P/Pons–Winnecke	5.89	0.97	18.3	~+9	+18.1	1.5	-.-		**June Bootids**
15P/Finlay	6.41	0.94	6.8	+13.5	+17.9	1.8	-.-		Future showers
21P/Giacobini–Zinner	6.61	1.00	30.7	+8.9	+17.7	2.0	1.66	Low C_2/CN	**Draconids**
26P/Grigg–Skjellerup	5.08	0.99	21.1	+12.1	+19.4	0.8	>12	Normal C_2/CN	**π-Puppids**
45P/Honda–Mrkos–Pajd.	5.43	0.60	12.0	+13.1	+19.3	0.7	-.-		Future showers
73P/SW-3	5.41	1.01	17.4	+10.5	+17.7	2.0	-.-	Fragmented	**τ-Herculids**
103P/Hartley 2	6.20	0.96	13.2	+8.5	>+14.7	1.1	-.-		Future showers
D/1770 L1 Lexell	5.60	0.67	1.6	+7.7	-.-	~2.8	-.-	Now lost	μ-Sagittariids?
D/Blanpain	5.12	0.92	9.6	+8.5	-.-	≤2.2	-.-		**Phoenicids**
D/1978 R1	5.97	1.10	5.9	+12.5	-.-	~0.7	-.-		**October Capricornids**
P/2004 CB (Linear)	5.09	0.97	21.2	+17.0	+17.4	2.1	-.-		Future showers

Fig. 8.2 A list of Jupiter-family comets that are certain (bold) or probable parents of meteor showers. The orbital period, the perihelion distance, and the inclination of the comet orbit is given, as well as the total absolute magnitude near perihelion of the comet, the absolute magnitude of the nucleus, and the nucleus diameter and spin period.

magnitude of $H_N = +17.7^m$ was measured and an approximate diameter of 2.0 km derived.[10]

Sekanina estimated from the density of Draconid meteoroids a bulk density of the comet of 0.7 g/cm^3. From the comet brightness at large heliocentric distances he estimated that 5–6% of red (600–700 nm) sunlight is reflected back into space (the geometric albedo) and about 4% of yellow light. In the 1959 return, the tilt of the spin flipped from retrograde to prograde motion. The spinning north pole following the 1959 apparition was tilted by 64° and pointed to ecliptic longitude 152° and latitude +6°. This all changed the nongravitational forces on the comet orbit over time. The comet light curve (Fig. 8.4) peaked at different times either before or after perihelion.

The rapid rotation can be the result of spin-up: a reaction force from the ejection of gas and dust. Soon the comet may spin too fast and break. Already, the surface stress at the equator (a measure of tensile strength) is 200 N/m^2. Draconids in the 1946 and 1952 outbursts had a sudden brightening at the beginning of their trajectories, when the meteoroids experienced a pressure of about 100–500 N/m^2 (measuring compression and

[9] From amongst others: J. A. Fernández, G. Tancredi, H. Rickman and J. Licandro, The population, magnitudes, and sizes of Jupiter family comets. *Astron. Astrophys.* **352** (1999), 327–340; G. Tancredi, J. A. Fernández, H. Rickman and J Licandro, A catalog of observed nuclear magnitudes of Jupiter family comets. *Astron. Astrophys. Suppl. Ser.* **146** (2000), 73–90.

[10] G. Tancredi, J. A. Fernández, H. Rickman and J. Licandro, A catalog of observed nuclear magnitudes of Jupiter family comets. *Astron. Astrophys.* **146** (2000), 73–90. In contrast, Meech *et al.* have D = 6.0 km. K.J. Meech, O.R. Mainaut, B.G. Marsden, comet nucleus size distributions from HST and Keck telescopes. *Icarus* **170**, 463–491.

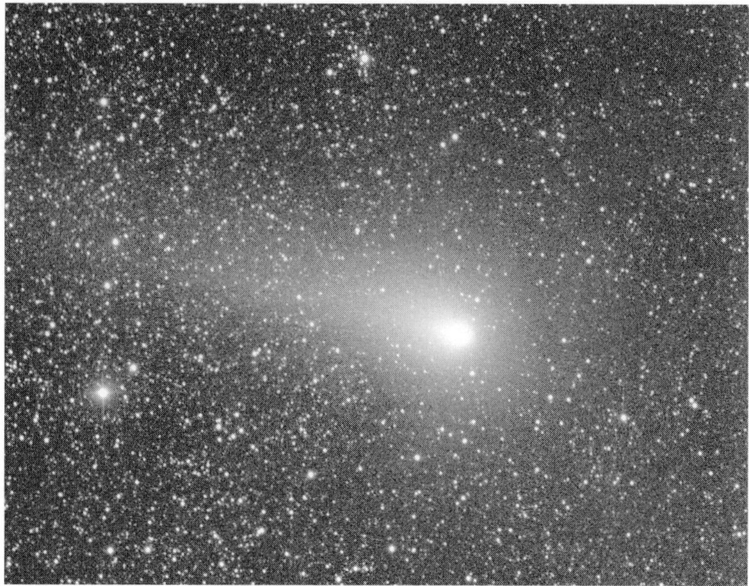

Fig. 8.3 Comet 21P/Giacobini–Zinner photographed on November 1, 1998 at the Kitt Peak 0.9 m telescope. Trailing of stars is removed. North is up, East is to the left. Photo: N. A. Sharp (NOAO/AURA/NSF).

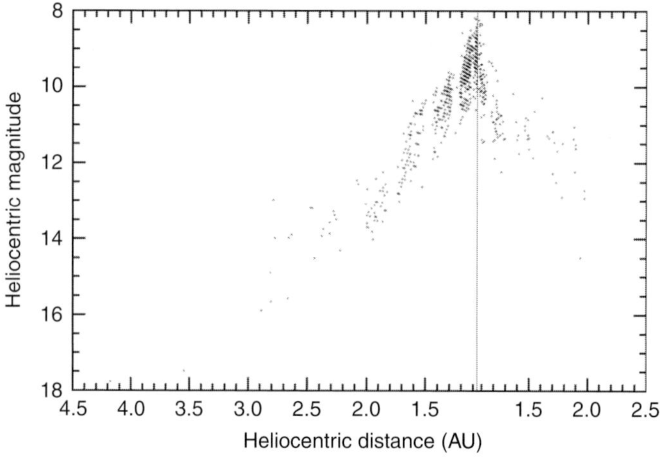

Fig. 8.4 Light curve of the 1998 return of comet 21P/Giacobini–Zinner, from data gathered by the International Comet Quarterly (D. W. Green, ed.).

the shear strength of the material), compared to the 2000–20 000 N/m^2 for other cometary meteoroids and about 6000–260 000 N/m^2 for meteorites.

The comet was visited by NASA's *International Cometary Explorer* in September, 1985 and was the first comet to be studied by a probe. It passed through the plasma tail at a distance of about 7800 km downstream from the nucleus at a relative speed

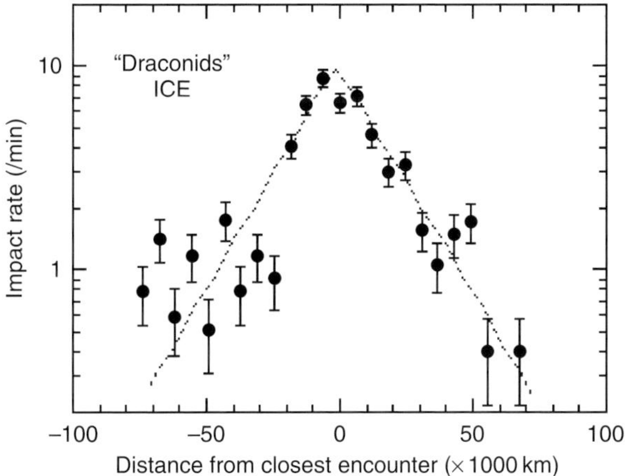

Fig. 8.5 The count rate for 1–3 μm sized dust impacts by ICE, for particle sizes corresponding to an instrumental detection threshold of 1 mV (Gurnett *et al.* 1986).

of 21 km/s. Among the many ions measured in the plasma tail were those of sodium,[11] presumably originating from small dust grains. The spacecraft was repeatedly hit by meteoroids only 1.0–2.8 μm in size (much smaller than typical Draconids), most within 30 000 km of the nucleus, falling off with an $1/r_N{}^2$-dependence (Fig. 8.5).[12] Ground-based observations revealed that these small dust grains were ejected with speeds of ~1.19 km/s, while the water vapor outflow speed $V_{H_2O} = 0.85\ r^{-0.5}$ had an inverse-root dependence with heliocentric distance.[13]

Dust was ejected at a rate of 1000 kg/s at perihelion, a factor of 100 less than comet Halley at $r = 1.03$ AU.[14] A dust tail of large grains was detected in infrared images on August 4.4, 1985 by *Ed Tedesco* and coworkers, when the comet was at 1.12 AU from the Sun. They concluded that these were typical Draconid meteoroids: larger than 100 μm if 1 g/cm^3 in density (syndyne for $\beta < 5 \times 10^{-4}$) and ejected with speeds less or equal to 3 m/s.[15] The tail was 18 000 km wide at a distance of 26 000 km from the nucleus.

During the 1998 return, the comet reached a brightness of $H_0 = +8.9$, with $n = 5.4$ (Fig. 8.4). Again, most of the gas is ejected near and approximately symmetrically around perihelion. This comet was unusual in the composition of its ices. Compared to comets Hale–Bopp and Hyakutake, Giacobini–Zinner was a factor of 5–15 poorer in

[11] K. W. Ogilvie, M. A. Coplan and L. A. McFadden, Sodium near the tail of comet Giacobini–Zinner. *Icarus* **134** (1998), 249–252.

[12] D. A. Gurnett, T. F. Averkamp, F. L. Scarf and E. Grün, Dust particles detected near Giacobini–Zinner by the ICE plasma wave instrument. *Geophys. Res. Lett.* **13** (1986), 291–294.

[13] L.-M. Lara, J. Licandro, A. Oscoz and V. Motta, Behaviour of comet 21P/Giacobini–Zinner during the 1998 perihelion. *Astron. Astrophys* **399** (2003), 763–772.

[14] M. S. Hanner, G. J. Veeder and A. T. Tokunaga, The dust coma of comet P/Giacobini–Zinner in the infrared. *Astron. J.* **104** (1992), 386–393.

[15] C. M. Telesco, R. Decher, C. Baugher *et al.*, Thermal-infrared and visual imaging of comet Giacobini–Zinner. *Astrophys. J. Lett.* **310** (1986), L61–L65.

C_2, C_3, and ethane. NH_2 is depleted as well.[16] The comet is one of about one-third of all Jupiter-family comets sampled by *Michael A'Hearn* and colleagues that are carbon-chain depleted.[17] Many authors have remarked on the lack of C_2 and C_3 (carbon chains) versus CN and ethane versus carbon monoxide.[18] The production of CN is normal with respect to OH. Surprisingly, other molecules such as methanol are in the same (albeit low end of the) range of other comets, with methanol/water production rates at about 0.9–1.4%.[19]

A lack of carbon chains in the gas phase does not mean that the cometary matter is necessarily poor in organics. It could be that the carbon chain depletion is due to the organics being locked in large grains. Excess emission at 3.43 μm wavelength was detected that could not be explained by methanol fluorescence,[20] possibly from complex organic molecules instead (Chapter 34). The comet does not show the 3.28 μm feature indicative of polycyclic aromatic hydrocarbons.

8.3 π-Puppid comet 26P/Grigg–Skjellerup

A very different Jupiter-family comet is *26P/Grigg–Skjellerup* (Fig. 8.6), number two on the list, parent of the *π-Puppid shower*. The comet is named after the singing teacher and amateur astronomer *John Grigg*, who discovered the comet on July 23, 1902, and after *J. Frank Skjellerup*, an Australian telegraphist working at the Cape of Good Hope in South Africa, who rediscovered the comet on May 17, 1922. A link was suspected and *Gerald Merton*'s 5 yr orbit[21] panned out when the comet was recovered in 1927. Sixty years later, *Lubor Krésak* identified an earlier sighting by Pons in 1808.

The comet first approached Earth's orbit at the end of the sixteenth century, when the orbital period dropped below the 2:1 resonance with Jupiter. The comet continued to have frequent close encounters. In one encounter on March 17, 1964, it passed only 50 million km from the giant planet and gained energy, changed orbital inclination and lengthened the orbit so that its perihelion moved outwards in the path of the Earth. Impressive π-Puppid showers were seen in 1972, 1977, and 1982, with low level activity in the years following.[22] In the near future, the perihelion distance will increase

[16] J. E. Beaver, R. M. Wagner, D. G. Schleicher and B. L. Lutz, Anomalous molecular abundances and the depletion of NH_2 in comet P/Giacobini–Zinner. *Astrophys. J.* **360** (1990), 696–701.

[17] M. F. A'Hearn, R. L. Millis, D. G. Schleicher, D. J. Osip and P. V. Birch, The ensemble properties of comets: results from narrowband photometry of 85 comets, 1976–1992. *Icarus* **118** (1995), 223–270.

[18] A. L. Cochran and E. S. Barker, Comet Giacobini–Zinner – a normal comet? *Astron. J.* **93** (1987), 239–243; M. J. Mumma, M. A. DiSanti, N. Dello Russo, K. Magee-Sauer and T. W. Rettig, Detection of CO and ethane in comet 21P/Giacobini–Zinner: evidence for variable chemistry in the outer solar nebula. *Astrophys. J.* **531** (2003), L155–L159.

[19] H. A. Weaver, G. Chin, D. Bockelée-Morvan *et al.*, An infrared investigation of volatiles in comet 21P/ Giacobini–Zinner. *Icarus* **142** (1999), 482–497.

[20] G. Tancredi, J. A. Fernández, H. Rickman and J. Licandro, A catalog of observed nuclear magnitudes of Jupiter family comets. *Astron. Astrophys.* **146** (2000), 73–90.

[21] G. Merton, Comet 1922 b (Skjellerup) *Astron. Nachr.* **216** (1922), 91.

[22] B. A. Lindblad, The meteor stream associated with comet P/Grigg-Skjellerup. *Astron. Astrophys.* **187** (1987), 931–932.

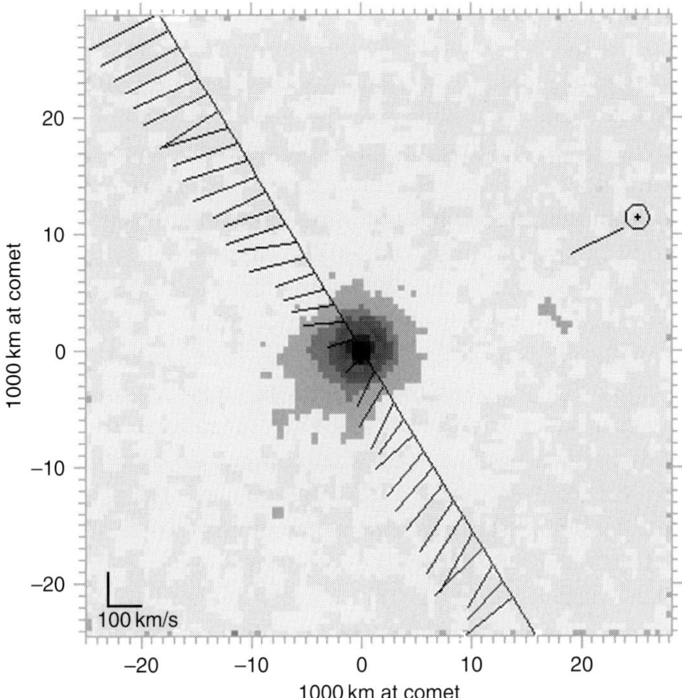

Fig. 8.6 26P/Grigg–Skjellerup 7 h before the encounter with Giotto. Overlaid are the path of Giotto and the measured solar wind vectors. Photo: Keith Mason, Mullard Space Science Lab., University College London and the Anglo-Australian Telescope.

and the orbit will evolve towards the 9:4 resonance, returning back to $q \sim 1$ AU only in the twentysecond century.[23]

The shimmering light reflected from the nucleus far from the Sun suggests a diameter of less than 2.4 ± 0.4 km,[24] and a body axis ratio of 0.9 or (probably) less.[25] Meech *et al.* even give D = 5.8 km. But that result may be contaminated by activity beyond 2.7 AU. Radar observations measured $D = 0.8 \pm 0.2$ km, one of the smallest known comets.[26] Gregorz Sitarski concluded that the nucleus of comet Grigg–Skjellerup had the shape of a prolate spheroid rotating around its longer axis, with a ratio of equatorial to polar diameters of 10:14.[27] The surface is unusually bluish. The nucleus rotation period is unknown, but appears to be longer than 12 h.

[23] A. Carusi, L. Kresák and G. B. Valsecchi, *Electronic Atlas of Dynamical Evolutions of Short-Period Comets* (IAS Computing Centre, 1987) (website http://www.rm.iasf.cnr.it/ias-home/comet/catalog.html).

[24] C. Lowry, A. Fitzsimmons, I. M. Cartwright and I. P. Williams, CCD photometry of distant comets. *Astron. Astrophys.* **349** (1999), 649–659; Licandro J. *et al.*, CCD photometry of cometary nuclei, I. *Icarus* **147** (2000), 161–179.

[25] H. Boehnhardt, N. Rainer, K. Birkle and G. Schwehm, The nuclei of comets 26P/Grigg–Skjellerup and 73P/Schwassmann–Wachmann 3. *Astron. Astrophys.* **341** (1999), 912–917.

[26] P. Kamoun, D. Campbell, G. Pettengill and I. Shapiro, Radar observations of three comets and detection of echoes from one: P/Grigg–Skjellerup. *Planet. Space Sci.* **47** (1998), 23–28.

[27] G. Sitarski, On the rotating nucleus of comet P/Grigg–Skjellerup. *Acta Astron.* **42** (1992), 59–65.

There is a rapid onset of activity around 1.4 AU (Fig. 8.7).[28] In 1982, outbound activity fell off less quickly. Despite the small size and asymmetric light curve, non-gravitational forces are small: $A_1 = +0.0067 \times 10^{-8}$ and $A_2 = -0.0011 \times 10^{-8}$ AU/d^2, which implies a spin axis orientation near the pole of the orbit.[29]

The comet released gas and dust at a rate ~300 kg/s. This comet is not poor in carbon chains. C_2 and CN were emitted at normal ratios when compared to the amount of scattered light from dust grains, but in a ratio three times higher than for comet 1P/Halley.[30] The dust ejection rate was about 200 kg/s on June 10–July 10, 1992. Over 65% of the surface area may be active. In the 1987 return, there was a broad 50°–100° wide jet, at intermediate latitudes on the northern hemisphere.[31] Marco Fulle and coworkers determined non-Whipple-like dust ejection speeds of only 15 m/s for 0.15 mm diameter grains in the comet coma 2000–10 000 km from the nucleus, surprisingly independent of particle size, with a size distribution index of $\alpha = 3$ to 4 over the range 0.004–1.2 mm.[32]

The comet was visited by the Giotto satellite when it was 1.01 AU from the Sun, after its successful close encounter with comet Halley. The spacecraft was put in hibernation on April 2, 1986, to be reactivated in April of 1990. On July 10, 1992, the now blind Giotto was aimed directly at the nucleus of comet 26P/Grigg–Skjellerup and came within a distance of 100–200 km at a relative speed of 14 km/s (Fig. 8.6). The camera had been damaged in the Halley flyby and there are no pictures of the nucleus.

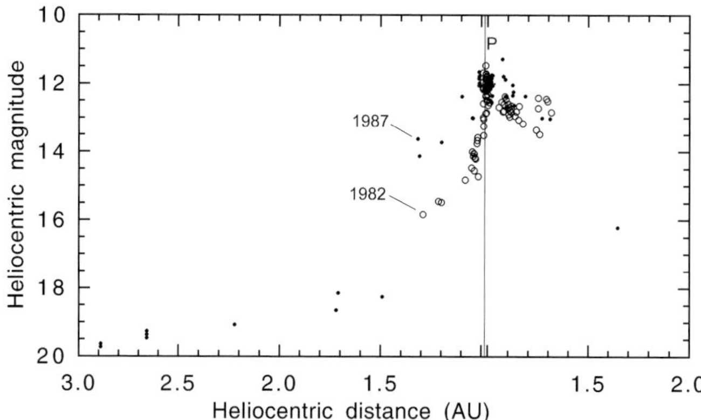

Fig. 8.7 Light curve of comet 26P/Grigg–Skjellerup during the 1982 and 1987 returns. Data: International Comet Quarterly.

[28] K. Birkle and H. Boehnhardt, Observations of the Giotto target comet P/Grigg–Skjellerup at the Calar Alto Observatory. *Earth, Moon, Planets* **57** (1992), 191–201.

[29] D.K. Yeomans and P.W. Chodas, An asymmetric outgassing model for cometary nongravitational accelerations. *Astron. J.* **98** (1989), 1083–1093.

[30] K. Jockers, N.N. Kiselev, H. Boehnhardt and N. Thomas, CN, C_2, and dust observed in comet P/Grigg–Skjellerup from the ground eight hours after the Giotto encounter. *Astron. Astrophys.* **268** (1993), L9–L12.

[31] Z. Ninkov, A near stellar occultation by P/Grigg–Skjellerup. *Astron. J.* **107** (1994), 1182–1188; K. Birkle and H. Boehnhardt, Observations of the Giotto target comet P/Grigg–Skjellerup at the Calar Alto Observatory. *Earth, Moon, Planets* **57** (1992), 191–201.

[32] M. Fulle, V. Mennella, A. Rotindi, L. Colangeli and E. Bussoletti, The radial brightness dependence in the dust coma of comet P/Grigg–Skjellerup. *Astron. Astrophys.* **289** (1994), 604–606.

Only three dust particles impacted on the front shield of Giotto.[33] Two of the particles were rather heavy, 1×10^{-6} g (0.12 mm diameter) and 2×10^{-7} g, the other one being 2×10^{-10} g. The lack of smaller grains means that the dust cloud near the nucleus was dominated by large meteoroids. Such masses give meteors of only $+18^{m}$ and $+20^{m}$, respectively. Another 1.4×10^{-6} g meteoroid was detected by the Giotto Radio-Science Experiment.[34]

Interestingly, *Neil McBride et al.* proposed the presence of actively dust-emitting boulders near the path of Giotto at the time of the encounter, to explain three increases in the intensity of scattered sunlight (from very small dust particles) measured away from the nucleus by the spacecraft during the flyby.[35] Each spike had a rapid onset and fell off as $1/r^2$, like the main peak.

8.4 7P/Pons–Winnecke

The orbit of comet *7P/Pons–Winnecke* (Figs. 8.8, 8.9), number three on my list and the parent of the June Bootid shower, is dominated by an irregular libration about the strong and broad 2:1 mean motion resonance with Jupiter, punctuated by sets of close encounters with Jupiter. The result is that q moves periodically in and out of Earth's orbit (Fig. 8.10), at least for the next three centuries. Meteoroids, too, can get trapped in this 2:1 resonance and survive as trailets for extended periods of time.

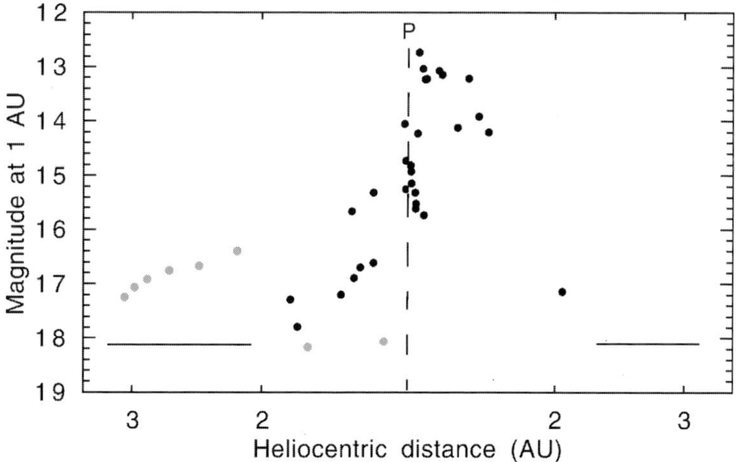

Fig. 8.8 7P/Pons–Winnecke light curve during the 2002 (and 1996, gray) return. Data: ICQ.

[33] J. A. M. McDonnell, N. McBride, R. Beard *et al.*, Dust particle impacts during the Giotto encounter with comet Grigg–Skjellerup. *Nature* **362** (1993), 732–734.

[34] M. Paetzold, M. K. Bird and P. Edenhofer, The change of Giotto's dynamical state during the P/Grigg–Skjellerup flyby caused by dust particle impacts. *J. Geophys. Res.* **98** (1993), 20 911–20 920.

[35] McBride N. *et al.*, The inner dust coma of comet 26P/Grigg–Skjellerup: multiple jets and nucleus fragments? *Mon. Not. R. Astron. Soc.* **289** (1997), 535–553; see also: T. Le Duin, J. F. Crifo, D. Le Queau and F. Crifo, A quantitative interpretation of the *in-situ* observations of the dust coma of comet P/Grigg–Skjellerup by the OPE photopolarimeter. *Astron. Astrophys.* **308** (1996), 261–272.

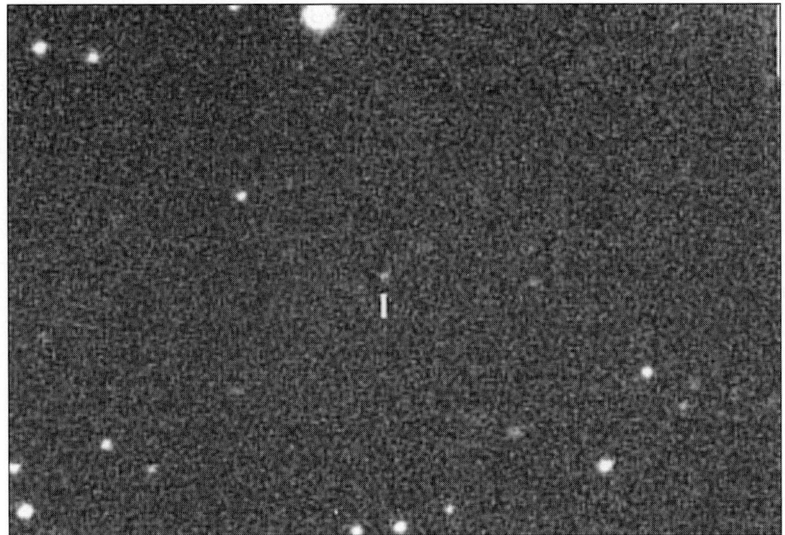

Fig. 8.9 Comet 7P/Pons–Winnecke on May 14.052, 2002 UT in a photo by Michael Jäger.

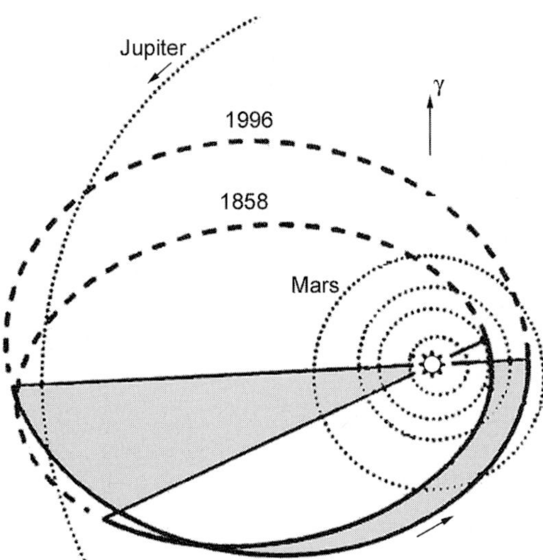

Fig. 8.10 The changing orbit of comet 7P/Pons–Winnecke. Graph from Takema Hashimoto.

Kresák[36] pointed out that the comet may have been responsible for meteor storms around June 24, 1752 and 1771 at the other node from a radiant in Corvus. They were not observed, perhaps because the comet was less active in the past. It has been claimed that it is fading over time, but I find that the comet has varied between

[36] L. Kresák, Meteor storms. In *Meteoroids and Their Parent Bodies*. ed. J. Stohl and I. P. Williams (Bratislava: Inst. Slovak Acad. Sci., 1992), pp. 147–156.

about $H_{10} = +7.2^m$ and $+12^m$, seemingly with the faintest values when the lightcurve peaked at perihelion. In most returns, the lightcurve is asymmetric (Fig. 8.8). *Kresák* and his wife *Margita Kresákova* noticed that the peak oscillated from around −20 d before perihelion for five to seven returns and then abruptly to +20 d after perihelion in the next five returns.

From the high value of H_{10}, I have $D = 3.3$ km, while the low value only gives $D = 0.77$ km (Eq. 6.1). *Lowry* and *Fitzsimmons* derived from the nuclear magnitude at $r = 5.58$ AU a diameter $D < 5.2 \pm 0.2$ km. Lower nuclear magnitudes have been measured by others. *Tancredi et al.* give $H_N \sim +16.8$, corresponding to $D = 3.0$ km, but that may still be too large (Fig. 8.8).[37] The spin period remains unknown.

The comet passed only 0.040 AU from Earth on June 26, 1927 when it was seen as a $+3.5^m$ object. At that time, a fan-shaped nebulosity extended northward[38] suggesting this comet, too, has a jet on the northern hemisphere. Kresák and Kresákova estimated a gas loss rate of 5×10^8 kg/orbit.[39] A 40 ± 14 thousand km wide dust trail was detected by IRAS, from which *Mark Sykes* and *Russell Walker* calculated $V_{ej} = 3.0$ m/s and a total dust mass of 1.0×10^9 kg, finding that the dust had accumulated in ~1 orbit.[40]

8.5 Andromedid comet 3D/Biela

Number four, comet *3D/Biela*, was first observed in 1772 and seen again in 1805. Suspected by Bessel, Gauss, and others to be the same comet, it was the recovery in 1826 by Austrian officer *Wilhelm, Freiherr von Biela* (1782–1856)[41] at Josephstadt in Bohemia that established the connection. A fear of impact created some concern when *M. Charles Theodorde Damoiseau* calculated that the comet would pass by Earth's orbit at a distance of only 35 000 km a month before the Earth would be at that point in 1832. This inspired the lyrics of a French poet: *We shall not escape this great impact. I feel our planet crumbling already . . .*[42] Earth survived the ordeal unharmed.

When the comet returned in 1845/46, it had broken apart in two pieces that were 250 000 km apart (Fig. 8.11). The companion grew in brightness, developed a tail, and both continued to shed fragments. At the next return in 1852, the smaller piece had moved ahead of the larger by about 2 million km, corresponding to a difference in

[37] G. Tancredi, J. A. Fernández, H. Rickman and J. Licandro, A catalog of observed nuclear magnitudes of Jupiter family comets. *Astron. Astrophys. Suppl. Ser.* **146** (2000), 73–90.

[38] H. E. Burton, Observations of comet Pons–Winnecke *Astron. J.* **37** (1927), 190.

[39] L. Kresák and M. Kresáková, The mass loss rates of periodic comets. In *Symposium on the Diversity and Similarity of Comets. ESA SP* **278** (1987), 739–744; L. Kresák and M. Kresáková, The absolute magnitude of periodic comets. *BAC* **41** (1990), 1–17.

[40] M. V. Sykes and R. G. Walker, Cometary dust trails. I. Survey. *Icarus* **95** (1992), 180–210.

[41] W. T. Lynn, Wilhelm von Biela. *Observatory* **21** (1898), 406–407; W. T. Lynn, Biela and his comet. *Observatory* **28** (1905), 423–425; K. Hujer, On the history of Wilhelm von Biela and his comet. *J. R. Astron. Soc. Canada* **77** (1983), 305–309.

[42] P. J. de Béranger La comète de 1832. *Musiques des chansons de Béranger*. 7th edition (Paris, Perrotin, 1858), p. 207.

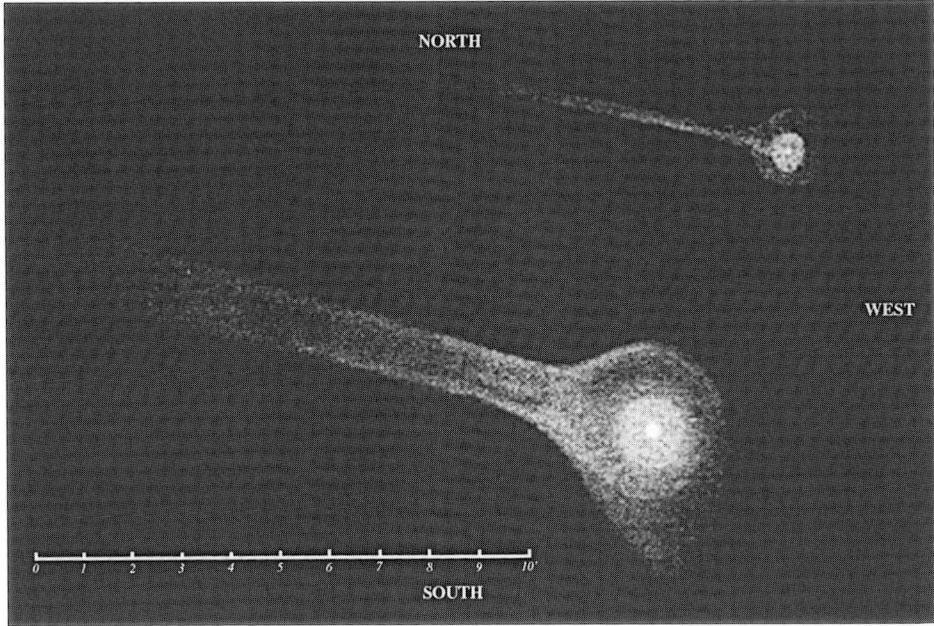

Fig. 8.11 Comet Biela in a drawing by Otto Wilhelm Struve of Pulkovo Observatory in Russia in the night of February 19, 1846. Struve was the last to see the comet on September 29, 1852.[43]

speed at perihelion of 60 cm/s.[44] Marsden and Sekanina traced the fragments back to a breakup in early 1842 up to mid 1843, when the comet was near aphelion at ~6 AU from the Sun and a distant ~1 AU from Jupiter." The comet was not seen again in later returns, but a spectacular storm of Andromedids occurred on November 27, 1872, and again in 1885. Not much is known about the nuclear properties. The brightest fragment would have been about $D = 3.0$ km.

Periodic interest has persisted in trying to find the comet. Marsden and Sekanina studied the nongravitational effects and calculated a close approach with Earth (0.05 AU) in November, 1971, but no comet was found. While searching, *Lubos Kohoutek* discovered his now infamous long-period comet 1973 XII at Hamburg Observatory in Bergedorf, Germany.[45] 3D/Biela should have been detected on several occasions in the late eighteenth and early nineteenth centuries.

It is likely that at least one fragment survived, now an extinct comet nucleus. The size of that fragment remains unknown, but a clue is provided by the nongravitational forces $A_1 = +2.22 \pm 0.33 \times 10^{-8}$ and $A_2 = -0.16228 \pm 0.00038 \times 10^{-8}$ AU/d^2, which Marsden and Sekanina found to be constant between 1832 and 1852. Dynamically, the comet remains deep within the zone between the 2:1 and 3:2 resonances with

43 O.W. Struve, Observations of Biela's Comet. *Mon. Not. Astron. Soc.* **16** (1856), 137–138.
44 B. G. Marsden and Z. Sekanina, Comets and nongravitational forces. IV. *Astron. J.* **76** (1971), 1135–1151.
45 L. Kohoutek, On the discovery of comet Kohoutek /1973 f/. *Icarus* **23** (1974), 491–492.

Jupiter.[46] Close encounters with Jupiter are few. A recent encounter (<0.5 AU) was on June 4, 1794 (0.38 AU), while the next is not due until June 7, 2163. More distant encounters (even though more frequent) do not efficiently change the orbit due to a high relative speed of comet and Jupiter near aphelion. The comet may have librated about the 7:4 mean-motion resonance with Jupiter in the period AD 1653–1866. The inclination has gradually decreased from 28° in 1585 to 8° in 1997 and is now increasing again. The node has changed from 271.271° in 1585 to 213.868° in 1997, and will continue to change to 127.812° in AD 2406.

8.6 73P/Schwassmann–Wachmann 3

When comet *73P/Schwassmann–Wachmann 3* (SW-3, Fig. 8.12), number five on my list, missed Earth by only 0.062 AU in 1930, Japanese observers were on alert for a possible meteor outburst. One saw a shower of faint meteors from a radiant near the star τ-Herculis.[47] In later years, a weak annual shower was recognized in photographic orbit surveys.

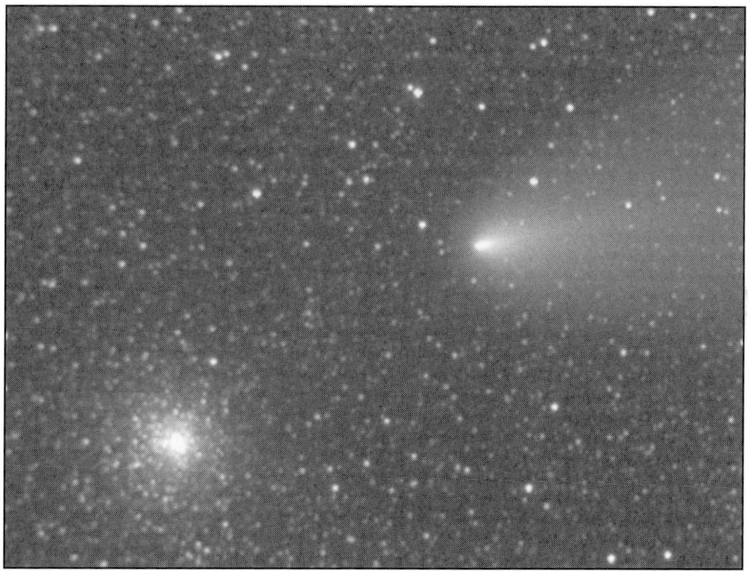

Fig. 8.12 Comet 73P/SW-3. The image was taken by Steven Lee from Australia on October 21, 1995 prior to the breakup, while the comet was near the globular cluster M62.

[46] A. Carusi, L. Kresák and G. B. Valsecchi, *Electronic Atlas of Dynamical Evolutions of Short-Period Comets* (IAS Computing Centre, 1987) (website http://www.rm.iasf.cnr.it/ias-home/comet/catalog.html).

[47] K. Ohtsuka, The Meteor Shower Associated with P/Schwassmann–Wachmann 3, *Tokyo Meteor Network Rep.* **1** (1986), 5–7; H. B. Ridley, Meteors associated with Comet P/Schwassmann–Wachmann 3. *J. Brit. Astron. Assoc.* **100** (1990), 92–93.

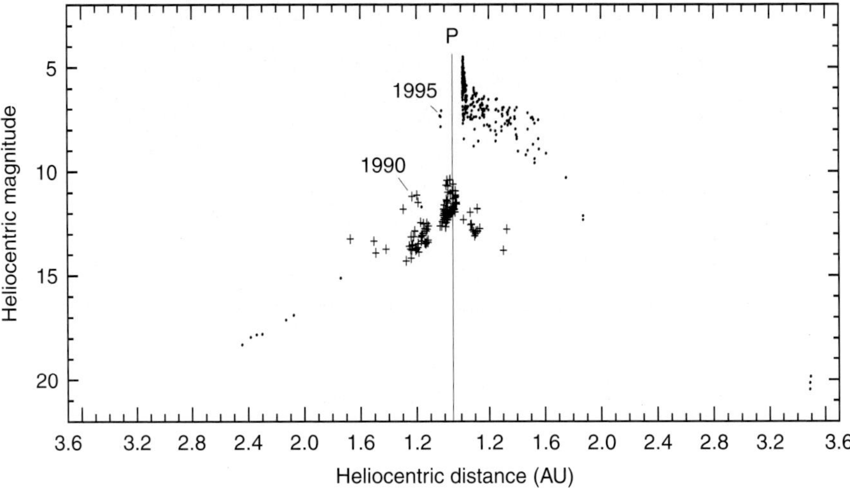

Fig. 8.13 The light curve of comet 73P/Schwassmann–Wachmann 3 in the returns of 1990 and 1995. Data: International Comet Quarterly.

Just weeks prior, the comet had been discovered by Arnold Schwassmann and *Arthur Arno Wachmann* of the Hamburg Observatory in Germany, on May 2, 1930. It first came near Earth's orbit after a pair of close encounters (~0.22 AU) with Jupiter in 1882 and 1894. The orbital period dropped to below the 2:1 resonance and q decreased. Since that time, the perihelion of the orbit has remained close to Earth's orbit and will continue to do so until an encounter with Jupiter on November 25, 2167 (0.066 AU), when q will increase abruptly.[48]

It was weakly active at 3.0 AU on the inbound leg in December of 1994 and subsequently developed in brightness much as during the previous return in 1990. Then, on September 8, 1995, comet astronomer Jacques Crovisier noticed during radio observations of the OH radical emission (a product of water vapor photolysis), that SW-3 was emitting enormous amounts of water. This breakup caused it to brighten from $+13^{\mathrm{m}}$ to $+8^{\mathrm{m}}$ (Fig. 8.13). Another outburst followed in early October and on October 22nd. By mid October, the comet had reached $+6^{\mathrm{m}}$ and was faintly visible with the naked eye.

In December, there was a train of five fragments (Fig. 8.14). The main fragment "C" was recovered after aphelion, and continued to be active at 4 AU on the inbound leg.[49] Sekanina[50] concluded that Fragment B broke from C in the timeframe November 7–15, with a relative velocity of 1.7 m/s. Fragment "B" was observed by Hubble upon

48 A. Carusi, L. Kresák and G. B. Valsecchi, *Electronic Atlas of Dynamical Evolutions of Short-Period Comets* (IAS Computing Centre, 1987) (website http://www.rm.iasf.cnr.it/ias-home/comet/catalog.html).

49 H. Boehnhardt, S. Holdstock, O. Hainaut *et al.*, 73p/Schwassmann–Wachmann 3 – one orbit after break-up: search for fragments. *Earth, Moon, Planets* **90** (2002), 131–139.

50 Z. Sekanina, Comet 73P/Schwassmann–Wachmann 3. *IAU Circ.* **7541** (2000), 1, ed. D. W. E. Green.

Fig. 8.14 Fragment train of 73P/S-W 3 in December, 1995. Photo: Michael Jäger.

its return in 2001 and was 1.36 ± 0.08 km in diameter (assuming 4% light reflected), with an axis ratio larger than 1.16. This size is sufficiently large for the comet to remain a conglomerate of primordial building blocks. The fragment was observed outbound until 2.75 AU. The other three fragments appear to have been less than 400 m in diameter. Fragment "E" broke off from "C" in mid December 1995, 85 ± 7 d after perihelion, with a relative speed of <1 m/s.[51]

Before the breakup, 73P was smaller than 2.2 km diameter (for an albedo of 0.04) and heavily crusted.[52] Kresák and Kresákova estimated that the dust production of this comet was a factor of 7 less than that of 21P/Giacobini–Zinner.[53] Lately, this comet has had a brightness outburst nearly every year. Unlike other Jupiter-family comets, it must still be laden with ices. The breakup in 1995 was not induced by tidal forces, but perhaps by the loss of ice.

8.7 Phoenicid comet D/1819 W_1 Blanpain

Number six is *D/1819 W_1 (Blanpain)*, parent of the *Phoenicid* shower, a shower seen periodically in 1887, 1938, 1956, and 1972.[54] The orbital period ($P = 5.10$ yr) is uncertain by several months. The comet was observed just a few years after a close

[51] I. Tóth, P. L. Lamy and H. A. Weaver, Hubble Space Telescope observations of the nucleus fragment 73P/Schwassmann–Wachmann 3-B. *Am. Astron. Soc., DPS meeting* #35, #38.05 (2003).

[52] H. Boehnhardt, N. Rainer, K. Birkle and G. Schwehm, The nuclei of comets 26P/Grigg–Skjellerup and 73P/Schwassmann–Wachmann 3. *Astron. Astrophys.* **341** (1999), 912–917; G. Tancredi, J. A. Fernández, H. Rickman and J. Licandro, Catalog of observed nuclear magnitudes of Jupiter family comets. *Astron. Astrophys. Suppl. Ser.* **146** (2000), 73–90.

[53] L. Kresák and M. Kresáková, The mass loss rates of periodic comets. In *Diversity and Similarity of Comets*, ed. E. J. Rolfe and B. Battrick, *ESA SP* **278** (1987), 739–744.

[54] Anonymous, The Phoenicids. *Mon. Not. Astron. Soc. Southern Africa* **16** (1956), 2.

encounter with Jupiter in 1817. Because of that, the orbital evolution in past and present is uncertain. The comet was perturbed into a near-Earth's orbit only during a series of close encounters in the seventeenth century and will stay with us for some time into the future. The comet was identified by Kresák as a potential source of meteor storms in the period 1770–1817, with a radiant in Gruis at Decl. $= -41°$. More recently, the comet is responsible for outbursts of Phoenicids at its other node in early December. In the twenty-second and twenty-third centuries, it is expected to librate around the 9:4 resonance with Jupiter. The nodal line rotates rapidly and in an irregular manner. There is no information about the comet nucleus. Orbit and $H_{10} = +8.5$ suggest a fairly large $D \sim 2.2$ km and a modest mass loss of about 1×10^9 kg/orbit, unless the comet was in eruption. This object was expected to be found as a dormant nucleus. I checked the list of NEO discoveries and found 2003 WY_{25}. More about that in a later chapter.

8.8 October Capricornid comet D/1978 R_1 (Haneda–Campos)

Number seven on the list is the parent of the *October Capricornid* shower: comet *D/1978 R_1(Haneda–Campos)*. A series of four close encounters with Jupiter (in 1945, 1957, 1969, and 1981) put this comet in a libration about the 2:1 resonance with Jupiter, where it will be trapped for the foreseeable future. Dust may be trapped there as well.

The perihelion distance lowered dramatically to $q \sim 1.10$ AU and the inclination decreased to $i = 6°$, making meteor showers possible. The comet was very diffuse in 1978, fading on approach to perihelion. The absolute magnitude $H_{10} \sim +12.5$ (Fig. 8.15), suggests that $D = 0.65$ km, but that is a lower limit. The nucleus was clearly intermittently active in 1978 and the comet was lost after that. The comet may have lost most ices long ago.

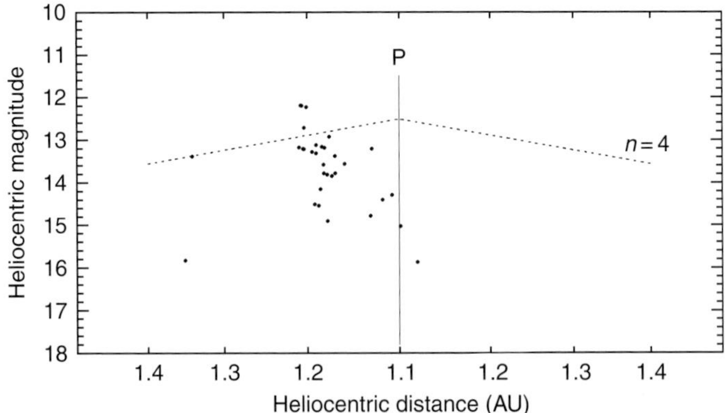

Fig. 8.15 Light curve of D/1978 R_1. Data: International Comet Quarterly.

8.9 A brief visit by D/1770 L_1 Lexell

A really bad case of a rapidly changing comet orbit and a brief encounter with Earth was that of comet *D/1770 L_1 (Lexell)*, discovered by Charles Messier. When the Swede *Anders Johan (Andrei Ivanovich) Lexell*,[55] then professor of astronomy at the St. Petersburg Academy of Sciences, determined an orbital period of only 5.6 yr, he had found the first short-period comet. He explained why the comet had not been seen before by pointing out that there had been a close encounter with Jupiter in 1767 and that another one was forthcoming in 1779. Prior to AD 1767, the orbit was ~9.23 yr and the comet did not approach the Sun to less than 2.90 AU. The comet was captured in a shorter Jupiter-family orbit on March 27, 1767, when it passed Jupiter at a distance of 0.020 AU. This changed the orbit into one intersecting Earth's orbit in 1770 and 1776 (Fig. 8.16).

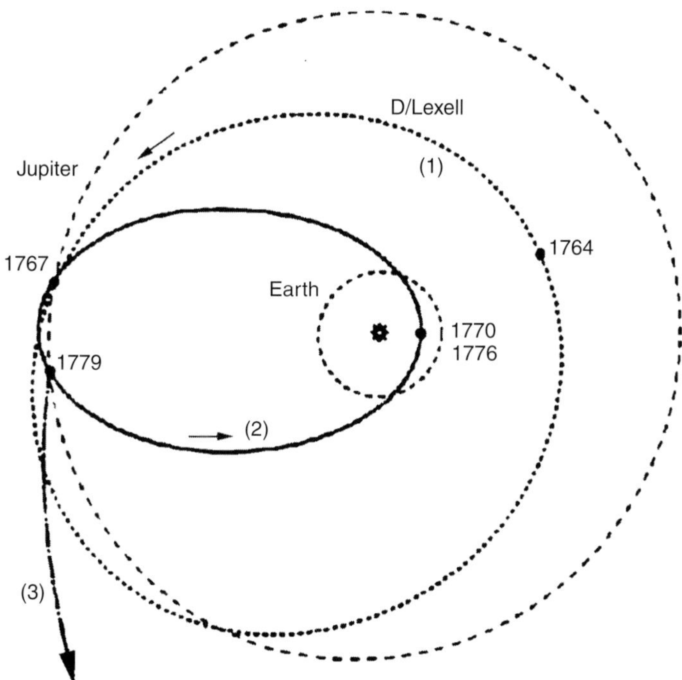

Fig. 8.16 The abrupt changes in the orbit of D/1770 L_1. The initial orbit (1), after a perturbation in 1767 (2), and after a second perturbation in 1779 (3).

[55] A. J. Lexell, *Réflexions sur la Temps Périodique des Comètes en Général et Principalement sur Celui de la Comète Observée en 1770* (St. Petersbourg: Imprimerie de l'Academie Imperiale des Sciences, 1772).

The July 1, 1770 encounter with Earth has the distinction of being the closest on record, only 0.015 AU (3426 000 km). The comet reached magnitude −1. Now having a period of 5.60 yr, the orbit was of very low inclination (1.55°), and meteor showers from dust ejected in 1770 may have occurred at both nodes in 1776.

The proposed association of Lexell with the μ-Saggitariids[56] is very doubtful, but some (historic) shower activity is not out of the question. Because of the low inclination, the Earth is nearest to the orbit of the comet (as it was in 1770) as much as 62 days before passing the ascending node.[57] The association is highly uncertain, because only the 1770 and 1776 returns could have contributed dust. The shower appears to contain more mass than the $\sim 8 \times 10^9$ kg of dust released at that time.

Lexell discovered that the comet would approach Jupiter again on July 27, 1779, to less than three and a half radii from the center of the planet (0.002 AU). The new orbit (Fig. 8.16) takes comet Lexell out to beyond the orbit of Uranus while $q \sim 5.2$ AU is now at Jupiter's orbit. When *William Hershel* discovered Uranus on March 13, 1781, it was Lexell who calculated the orbit and determined it was a planet, rather than a comet as Hershel had at first thought.

8.10 What else makes Jupiter-family comets unique

While long-period comets tend to emit dust of small 0.1–10 μm grain sizes that scatter light efficiently, with high albedos of ~0.1–0.4 and at a high rate of ~10 000 kg/s, Jupiter-family comets tend to emit >10 μm dust grains with albedos of only 0.05–0.07, low optical polarization per unit mass, and at a much lower rate of ~100–1000 kg/s.[58] Halley-type comets fall in both classes, with 1P/Halley emitting many fine dust grains, while 55P/Tempel–Tuttle emits mainly large grains. Comets such as Tempel–Tuttle have relatively faint dust tails and emit infrared radiation more typical of blackbody spheres.

While long-period comets may still have their original crust exposed, many Jupiter-family comets are a mere kernel of the original comet. Fragments have broken off. Whatever terrain survived years of abuse is now covered by large meteoroids that have fallen back. It appears that the dust fills bowls in the topography (Fig. 2.2). It is likely that what remains are the parts with the highest tensile strength and many of the physical properties of Jupiter-family comets may be related more to their dynamical history in the inner solar system, rather than to the processes that occurred in the Kuiper Belt at the time of their formation.

Some Jupiter-family comets have only recently wandered into the inner solar system within $q < 2.7$ AU where water ice sublimates. These are *dynamically young*, such as

56 J. Wood, The October-Capricornids: a new stream. *NAPO-MS Bull.* 200 (*Radiant* **10** (1988), 71).
57 J. Porter, *Comets and Meteor Streams* (New York: Wiley, 1952), 123 pp.
58 A. C. Levasseur-Regourd, E. Hadamcik and J. B. Renard, Evidence for two classes of comets from their polarimetric properties at large phase angles. *Astron. Astrophys.* **313** (1996), 327–333.

81P/Wild 2. In contrast, comet 2P/Encke is old, while 26P/Grigg–Skjellerup, 21P/Giacobini–Zinner, and 7P/Pons–Winnecke are in their middle age. Sadly, no dynamically young comets are a known source of meteors, although the lost comet D/Blanpain and 73P/Schwassmann–Wachmann 3 are candidates.

Jupiter-family comets of past and future showers

Some other comets worth mentioning include **6P/d'Arrest**. Its orbit is chaotic, moving between three different values for the perihelion distance: $q = 1.17$, 1.28, and 1.38 AU.[59] The comet was in the $q \sim 1.17$ AU situation in the 1678 return, during discovery in 1851 and 1857, and recently in 1970 and 1976. Of particular interest are two orbits in 1780 and 1786, when q decreased to 1.037 and 1.035 AU, respectively. Meteors at that time radiated from R.A. = 233.0, Decl. = +16.3 ($V_g = 11.03$ km/s) around September 1 ($\Delta r = 0.033$ AU).

Comet **15P/Finlay** only came into Earth crossing orbits after perturbations in 1600. A series of close encounters with Jupiter, including that of February 24, 1767 (0.168 AU), decreased the inclination to <4°. A close encounter on May 8, 2004 raised i to 6.8°. Meteors are now expected to radiate from R.A. = 251–257°, Decl. = −47 to −49°, October. Time of expected activity is uncertain.

41P/Tuttle–Giacobini–Kresák has been associated with the Virginid shower, but this association is unlikely. Its perihelion hovers just outside of Earth's orbit at a miss-distance of ~0.3 AU. This is a small $D \sim 1.4$ km comet.

Since an encounter with Jupiter on March 27, 1983, **45P/Honda–Mrkos–Pajdušáková** has a low inclination (4.2°). Meteors would radiate from R.A. = 329°, Decl. = −16° ($V_g = 24.7$ km/s), around February 16, when the comet passes Earth's orbit at 0.062 AU, and from a radiant R.A. = 325°, Decl. = −11° ($V_g = 24.7$ km/s) around August 15, passing at 0.060 AU. This comet is tentatively associated with the August δ-Capricornids (Table 7).

Following an encounter with Jupiter on April 10, 1972 (0.28 AU), and again on February 26, 1984 (0.47 AU), the perihelion distance of tiny ($D \sim 1.2$ km, $P \sim 7.6$ h) **46P/Wirtanen** decreased from $q = 1.6$ to 1.08 AU. The comet continues to pass just outside of Earth's orbit until a close encounter on November 25, 2042 (0.56 AU), when q will increase to 1.22 AU. Until then, meteors should radiate from R.A. = 9°, Decl. = −47° ($V_g = 11.0$ km/s) on December 14, with a miss-distance of ~0.075 AU. The comet itself was close to Earth on December 16.5, 2018 (0.078 AU).

$D \sim 1.6$ km sized comet **62P/Tsuchinshan 1** had a perihelion close to 1 AU and very low inclination (1.4°) before a close encounter with Jupiter in August 21, 1759 (0.05 AU). The encounter increased the semi major axis, moving the perihelion outwards. Before that time, meteors may have radiated from R.A. ~ 32°,

[59] K. Kinoshita, Comet orbit home page (website http://www9.ocn.ne.jp/~comet/

Decl. ~ +7° (V_g = 8.8 km/s), on January 30, with a miss-distance of 0.003 AU. The comet eventually settled in the libration about the 2:1 resonance and is not expected to cause future meteor showers.

Comet **67P/Churyumov–Gerasimenko** is the target of the Rosetta comet landing mission. It measures $D \sim 4.0$ km, rotating once every 12.3 h. Following a close encounter with Jupiter on January 6, 2280 (0.057 AU), this comet will come close to Earth's orbit for a period of time until the next close encounter on October 30, 2338.

72P/Denning–Fujikawa is another proposed (but unlikely) parent of Capricornid activity. The inclination increased from 4.7° in 1802 to 11.2° in 2059 due to numerous shallow encounters with Jupiter. The orbital period is relatively high at 9.0 yr. In the next 50 years, meteors would be expected from R.A. ~ 271°, Decl. ~ −38° (V_g = 19.0 km/s), on December 5 (passing at 0.075 AU, epoch 1979-01-07), with the August radiant missing Earth at ~0.15 AU.

79P/du Toit–Hartley is a potential source of historic showers. With a perihelion just outside of Earth's orbit, its motion is relatively stable. The comet diameter is estimated to be 2.8 ± 0.6 km. Between perturbations in 1702 and 1761, the comet was relatively close to Earth's orbit, passing at ~0.09 AU, and meteors might have appeared on April 6.6 around R.A. = 92.5°, Decl. = +69.7° (V_g = 10.9 km/s).

103P/Hartley 2 has dramatic orbit changes, sweeping past Earth's orbit on occasion. It measures about 1.4 km. The comet was first put in the current orbit after a close encounter with Jupiter (0.085 AU) on April 28, 1971, which decreased q from 1.62 to 0.90 AU. Over the next two centuries, the orbit remains relatively unchanged with q oscillating between 0.90 and 1.07 AU. The comet will approach Earth to 0.121 AU on October 20.8, 2010. After a perturbation on March 18, 2054, the orbit will move close to Earth and meteors may radiate from R.A. = 304°, Decl. = +23° (V_g = 11.8 km/s), on October 12.8 (miss-distance 0.0033 AU, $\lambda_\odot$ = 199.75).

141P/Machholz 2 is a potential source of historic showers. The current miss-distance is around 0.1 AU.

P/1999 RO$_{28}$ (LONEOS) is a possible source of historic showers with a radiant in 1966 at R.A. = 288°, Decl. = +17°, V_g = 12.9 km/s (miss-distance 0.073 AU) on August 30.5. A close encounter with Jupiter on October 2, 1966 (0.13 AU) moved q from 1.049 to 1.217 AU. In June of 2028, q will decrease again to 1.090 AU.

P/2000 G$_1$ (LINEAR) is a possible source of modern daytime *Lepusids* on March 31, with meteors radiating from R.A. = 77°, Decl. = −16° (V_g = 11.1 km/s), with a miss-distance of only 0.02 AU, since Jupiter perturbed the orbit from i = 4° to 10° in 1987. Before that time, the radiant was at about R.A. = 73°, Decl. = +8°, V_g = 11.4 km/s on April 17, missing Earth by only 0.007 AU.

In the same way, **P/2001 Q$_2$ (Petriew)** is a possible source of daytime *β-Cygnids*, following a close encounter with Jupiter in July 4, 1982 (0.15 AU), when q decreased from 1.37 to 1.00 AU. Meteors radiate from R.A. = 295°,

Decl. = +30° (V_g = 12 km/s), on October 27, with a miss-distance around 0.06 AU.

P/2004 CB (LINEAR) is a possible source of future *June Lyncids*. As a result of infrequent shallow close encounters with Jupiter, the inclination of the orbit of this comet is evolving from 15° in 1898 to a projected 21° in 2044. An encounter on February 18, 2012 (0.59 AU) brings the perihelion distance up to 0.969 AU, at which time the miss-distance with Earth's orbit becomes very small. The orbit evolves outside Earth's orbit after the encounter of September 3, 2047. In the mean time, meteors may radiate from R.A. = 123°, Decl. = +79° (V_g = 16 km/s), on May 24. The miss-distance is exceedingly small: ~0.002 AU during AD 2014–2037. Earth itself has a close encounter with the comet (0.051 AU) on May 29, 2014.

P/2004 R_1 (McNaught) is a possible source of future showers. A close encounter with Jupiter on November 5, 1990, changed q = 2.3 to 1.07 AU and the inclination from 23.8° to 5.0°. q continues to hover around 1 AU until 2061. Meteors might be expected around June 26–July 5 from R.A. = 211°, Decl. = −32° (V_g = 10 km/s), and from August 2–9 from R.A. = 184°, Decl. = −22° (V_g = 10 km/s).

The preliminary orbit for **P/2005 JQ_5 (Catalina)** suggests that it has had no strong perturbations from Jupiter in recent years, nor will it have in the near future, but the nodal line rotates rapidly. Meteors may radiate from R.A. = 243, Decl. = −9° (V_g = 16.1 km/s) around June 12, with a decreasing miss-distance (−0.04 to −0.01 AU from 1900 to 2100).

D/1766 G_1 (Helfenzrieder) was only observed in 1766. It has a very low inclination, decreasing from 7.4° to 1.7° during a close encounter following the 1844 return, reversing nodes after the 1928 return. It will do so again following the 2034 return. Theoretical radiants are very uncertain, perhaps around R.A. = 165°, Decl. = +6° (V_g = 30 km/s) on February 23 (0.02 AU) and R.A. = 187°, Decl. = −2° (V_g = 30 km/s) on October 11 (0.012 AU).

D/1892 T_1 (Barnard 3) was observed only from October 17 to December 08, 1892, suggesting a 30° inclined orbit. If the orbit is correct, then a close encounter in 2041 would bring the perihelion inwards to 0.98 AU, at which time meteors may radiate from R.A. = 291°, Decl. = +57° (V_g = 20 km/s), passing at 0.014 AU on October 12.

9

Fading comets of the inner Solar System

Perhaps 30 active comets exist that are in orbits decoupled from Jupiter with a perihelion distance $q < 1.3$ AU. They evolved into high inclination or short period orbits by the gradual push of outgassing or as a result of close encounters with the terrestrial planets Mercury, Venus, Earth, or Mars. After an episode of activity, these so-called *Encke-type comets* tend to become dormant in about 12 000 years.[1]

9.1 Dancing with the planets

It is still Jupiter, through long-range interactions, that determines much of the change in the instantaneous orbit (oscular orbital elements) over time. For that reason, Encke-type comets are usually classified among the Jupiter-family comets. Unlike other Jupiter-family comets, they do not have close encounters with Jupiter and their orbital evolution is determined mostly by the long-range interactions.

Changes from long-range interactions are often grouped in two types:[2] *short-period oscillations* depending on the position of the comet or Jupiter in its orbit, which cancel out over longer periods of time, and the long-period oscillations are called *secular variations*. Secular variations, from the Latin verb *saeculum* meaning "century" or "long period", are those changes that persist over periods of many orbits. Only when there are *resonances* (matching periods in the system) do persistent changes occur over shorter time intervals and we no longer speak of secular changes.

The long-term change in the orbital elements of comet 2P/Encke was first calculated by Fred Whipple in 1940 (Fig. 9.1).[3] In this case, the effect of Jupiter is to slowly change the orientation and shape of the orbit (eccentricity, semi-major axis, and longitude of perihelion), and to rapidly change the nodal line relative to the line of apsides. This creates a change in unison of the node, the argument of perihelion, and the inclination.

[1] H. F. Levison and M. J. Duncan, From the Kuiper Belt to Jupiter-family comets: the spatial distribution of ecliptic comets. *Icarus* **127** (1997), 13–32; M. J. Duncan and H. F. Levison, A scattered comet disk and the origin of Jupiter-family comets. *Science* **276** (1997), 1670–1672; H. F. Levison, M. J. Duncan, K. Zahnle, M. Holman, and L. Dones, NOTE: planetary impact rates from ecliptic comets. *Icarus* **143** (2000), 415–420.

[2] C. D. Murray, S. F. Dermott, *Solar System Dynamics*. Cambridge University Press, Cambridge (1999), 592.

[3] F. L. Whippe, Photographic meteor studies. III. The Taurid shower. *Proc. Am. Phil. Soc.*, **83** (1940), 711–745.

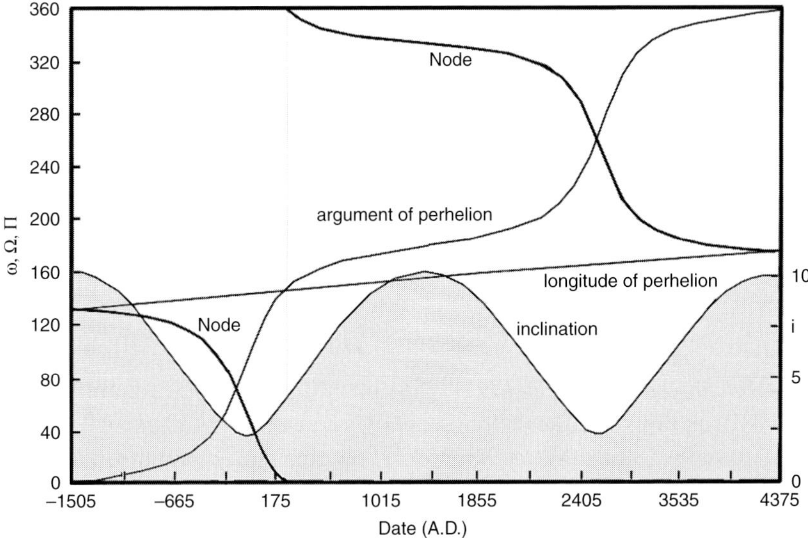

Fig. 9.1 Secular changes of the orbital elements of comet Encke over time during a full rotation of the nodal line relative to the line of apsides, as calculated by Fred Whipple.[4]

The range of orbits created in this manner will intersect Earth's orbit four times in the rotation cycle, as will be shown in Chapter 25, Fig. 25.3. The resulting meteor shower will have *northern and southern branches* (e.g., the Northern and Southern Taurids), as well as *twin showers* in daytime (the daytime ζ-Perseids and the β-Taurids in June). In some cases, the changes in eccentricity and longitude of perihelion are important too, in which case as many as eight showers can originate from a single parent comet.

The secular variations are not always such that the nodal line rotates over 360°. In some cases, such as for comet 1P/Halley, the node and argument of perihelion vary over a smaller range of values (Chapter 27). They do still tend to vary more rapidly than the longitude of perihelion and eccentricity.

All these secular changes are called *precession*, but just as with the motions of the Earth's spin axis, the name "precession" is usually preserved for the slow changes, while more rapid secular changes are sometimes called *nutation*. Note that precession and nutation are simply different frequency components of the same physical effect. In this book, I will restrict the use of "precession" only to the slow changes of longitude of perihelion (changes in the direction of the line of apsides). I will call the more rapid rotation of the nodal line relative to the line of apsides a *nutation cycle*. I find that the most relevant evolutions of meteoroid streams recognized in meteor showers are those working on the short time scale of nutation (Chapters 20–25), not on that of precession (Chapter 27).

[4] C. D. Murray, S. F. Dermott, *Solar System Dynamics*. Cambridge University Press, Cambridge (1999), 592.

Over longer periods of time, the orbits tend to evolve to states that either avoid close encounters with the terrestrial planets, or where the encounters occur in a random manner.[5] For example, the orbit of comet 2P/Encke has evolved to where the dangerous $\omega = 0°$ and $180°$ occur during the highest inclination (Fig. 9.1), which is the most stable phase of the nutation cycle. In contrast, when the inclination of the orbit is high and the orbital period is still similar to that of Jupiter, then the best solution is $\omega = 90°$ or $270°$, which puts aphelion and perihelion as far out of the ecliptic plane (away from Jupiter) as possible.

9.2 Dying comet 2P/Encke

Professor Johann Franz Encke (1791–1865) himself persistently referred to the first comet of 1819 as Pons' Comet, after its discoverer, but it was Encke who recognized that Pons' comet was the same as those found by Méchain in January, 1786, Caroline Herschel in November, 1795, and Thulis in October, 1805. Comet *2P/Encke* (Fig. 9.2, 9.3) has the shortest known orbital period among regularly active comets, $P = 3.5$ yr,

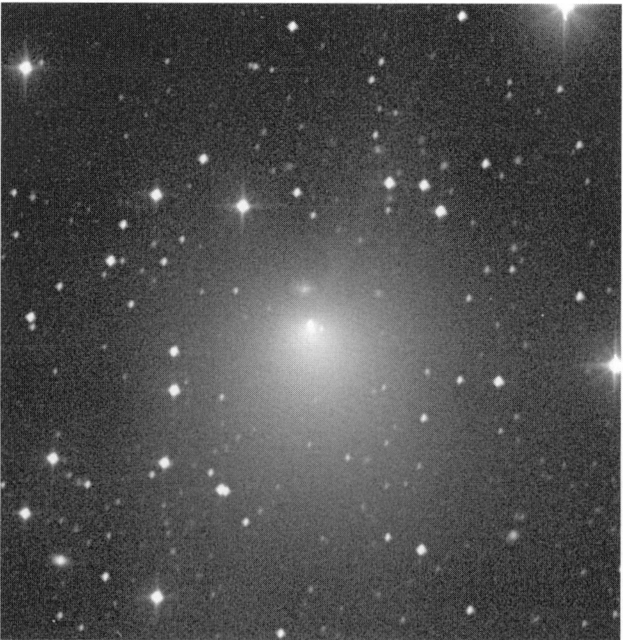

Fig. 9.2 Comet 2P/Encke on January 5, 1994. The fan of dust is towards the Sun. Photo: Jim Scotti, Spacewatch, LPL, University of Arizona.

[5] Milani A., Carpino M. *et al.*, Dynamics of planet-crossing asteroids: classes of orbital behavior. *Icarus* **78** (1989), 212–269.

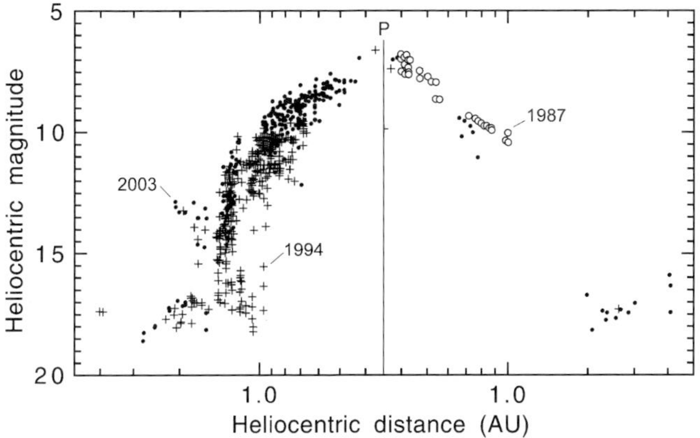

Fig. 9.3 The 2P/Encke lightcurve from three returns in 1987, 1994, and 2003. Data: International Comet Quarterly, courtesy of Dan Green.

with a Tisserand invariant $T_J = 3.027$ in the asteroid domain! The orbit is completely within Jupiter's orbit. In the far future, Encke is expected to collide with the Sun, unless it hits one of the terrestrial planets.[6]

Encke passes far from Earth's orbit. The association with the Taurid shower was not established until Fred Whipple pointed out that at some time in the past the rapidly changing node, the argument of the perihelion, and the inclination of the comet orbit (Fig. 9.1) could have been a much better match with the Taurid stream, if the comet and the meteoroids evolved at a different rate. For example, because the meteoroids were in a wider orbit (Chapter 25).[7]

The dust trail seen near the nucleus in mid-infrared[8] and optical images[9] (Fig. 9.4) consisted of large ~5 cm in size ($\beta < 10^{-5}$) meteoroids, according to Bill Reach (IPAC) and coworkers. Such large meteoroids would create bright -2^{m} Taurids! The trail stretched at least 80° behind Encke and some distance in front. It was 20 000 km wide when Encke was outbound 1.15 AU from the Sun. The trail is about 10 orbits old and weighs ~400 billion kg.

The comet is inactive at aphelion and looks cometary only for about 200 d around the perihelion passage. About 100 d prior to the perihelion passage, the comet suddenly starts to brighten $\sim r^{-10}$. Once activity has started, it develops predictably, following an $r^{-2.8}$ dependence in recent returns (1990–2003). Most dust is ejected 10 d inbound until 30 d outbound (Fig. 9.3). After that, the activity falls back steeply

[6] G. B. Valsecchi, From Jupiter-family comets to objects in Encke-like orbit. Evolution and source regions of asteroids and comets. In *Proc. IAU Coll.* 173, (Tatranska Lomnica: Astronomical Institute of the Slovak Academy of Sciences, 1999) p. 353.

[7] F. L. Whipple, Photographic meteor studies. III. The Taurid shower. *Proc. Am. Phil. Soc.*, **83** (1940), 711–745.

[8] W. T. Reach, M. V. Sykes and J. K. Davies, The formation of Encke meteoroids and dust trail. *Icarus* **148** (2000), 80–94.

[9] S. C. Lowry, P. R. Weissman, M. V. Sykes and W. T. Reach, Observations of periodic comet 2P/Encke: physical properties of the nucleus and first visual-wavelength detection of its dust trail. *LPSC* **34** (2003), 2056 (abstract).

Fig. 9.4 The dust trail of comet 2P/Encke (Encke is the bright dot in the center) detected in red light (coaddition of 195×300 s R-band exposures) at the Kitt Peak National Observatory (2.3 m Bok telescope of Steward Observatory) in September 2002, after subtraction of stars. Photo: Steve Lowry, NASA JPL.

according to $r^{-6.3}$. At perihelion, the comet produces >1000 kg/s of dust and some 600 kg/s of water vapor. From optical imaging, *Marco Fulle* of Trieste Observatory found that large ~2.4 mm sized grains are ejected at speeds of 5 m/s far from perihelion, increasing to 40 m/s at perihelion ($q = 0.34$ AU). The initial time averaged size distribution index $\alpha = -3.2 \pm 0.2$ ($\chi = 1.96$) steepened to -4.0 ($\chi = 2.51$) at perihelion. Dust is lost at a rate of $>5 \pm 1$ billion kg/orbit.[10]

The dust tail of comet Encke contained few small, <200 μm, grains ($\beta > 0.0015$). The dust coma, a peanut-shaped region of freshly emitted dust with a diameter of about 100 000 km consisted of mm-sized particles ($\beta < 0.001$). Two lobes of dust could have been created by a single jet positioned at latitude $-15°$, assuming a pole at R.A. = 102.8°, Decl. = +32.6°, so that Encke's pole is in the orbital plane. All of this mass adds up to 40 ± 20 billion kg/orbit from the 1997 return alone.

The presence of large grains makes the interpretation of optical observations tricky, creating some uncertainty in the pole position. According to Zdenek Sekanina,[11] the nucleus is 4.6 ± 0.6 km in diameter with a flatness of less than 4%. The geometric albedo is 0.03 at a *phase angle* (Sun–comet–Earth) of 55°. The total mass is about

[10] M. Fulle, Meteoroids from short period comets. *Astron. Astrophys.* **230** (1990), 220–226; E. Epifani, L. Colangeli, M. Fulle *et al.*, ISOCAM imaging of comets 103P/Hartley 2 and 2P/Encke. *Icarus* **149** (2001), 339–350.

[11] Z. Sekanina, Outgassing asymmetry of periodic comet Encke. II – Apparitions 1868–1918 and a study of the nucleus evolution. *Astron. J.* **96** (1988), 1455–1475.

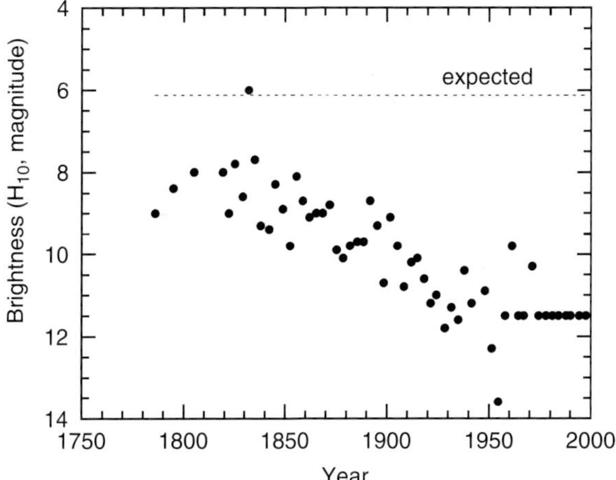

Fig. 9.5 Brightness decay of 2P/Encke, from absolute magnitudes listed in the MPC Comet Orbit Catalogue.

10^{13} kg (10 trillion tons), of which 0.09% is lost by sublimation in each orbit, amounting to about 9 billion kg/orbit.

The northern hemisphere is facing the Sun as the comet approaches perihelion. Only 5% of the surface is active, with two jets at latitudes of $-75°$ and $+55°$ with a pole that would have precessed to R.A. = 206° and Decl. = +3° in 1997, again at a high obliquity of 70°. The point of maximum outgassing lags 45° behind the subsolar point. In this case, the NW spike in the ISO image would have been created solely by the $-75°$ jet, the southern lobe solely by the $+55°$ jet, according to Reach *et al.*[12] Sekanina assumed that the nucleus rotated in a retrograde fashion every 6.55 h. More recently, Lisse *et al.*[13] found a similar nucleus diameter of 4.8 ± 0.6 km (confirmed by radar observations) and geometric albedo 0.05 ± 0.02, but put the spin period instead at 15.2 ± 0.3 h, with one axial ratio ≥ 2.6, and a migrating pole making a circle in ≤ 81 yr. The dust-to-gas ratio was 2.3 and the dust albedo was 0.06 ± 0.02. Others favor rotation periods of $P \sim 22$ h, assuming that other periodicities are overtones caused by shape and surface irregularities.

The small amount of ice evaporated is consistent with the fact that the comet only once lived up to its potential $H_{10} = +6.1^m$ for a 4.6 km sized nucleus (Fig. 9.5). Given the lack of prior sightings at favorable returns, it seems that Encke's activity started rather abruptly around 1786, and has steadily declined since. In recent years, the activity appears to have leveled off at a faint $H_{10} = +11.5^m$.

[12] W. T. Reach, M. V. Sykes and J. K. Davies, The formation of Encke meteoroids and dust trail. *Icarus* **148** (2000), 80–94.

[13] C. M. Lisse, Y. R. Fernández, M. F. A'Hearn *et al.*, A tale of two very different comets: ISO and MSX measurements of dust emission from 126P/IRAS (1996) and 2P/Encke (1997). *Icarus* **171** (2004), 444–462; Y. R. Fernández, C. M. Lisse *et al.*, Physical properties of the nucleus of comet 2P/Encke. *Icarus* **147** (2000), 145–160.

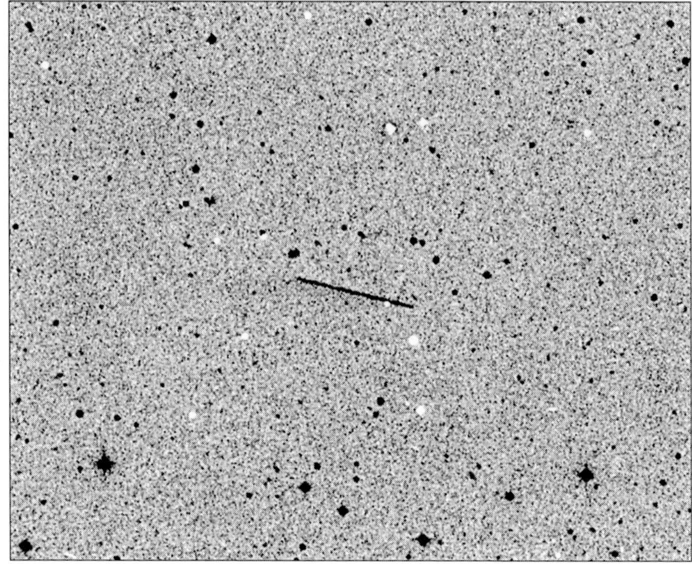

Fig. 9.6 Comet 107P/Wilson–Harrington on the Palomar Sky Survey print of November 19.1, 1949 in a 45 min exposure on red photographic emulsion. A faint tail is pointed to the left. Photo: Palomar Observatory Sky Survey.

9.3 Comet 107P/Wilson–Harrington

A more extreme case of fading is that of comet *107P/Wilson–Harrington*, which was a weakly active comet in 1949,[14] when it was discovered on the Palomar Observatory Sky Survey plates. It had a tail and was logged as comet 1949 III (Fig. 9.6). At the time, the magnitude of the comet was only $H_{10} = +16$, almost a nuclear magnitude. Analysis of the gas tail showed that CO and H_2O vapor were contributing about the same amount.[15] When the comet was finally recovered in 1979, it was dormant. It received the designation 1979 VA and an asteroid number: 4015.

The comet has a perihelion just outside Earth's orbit ($q = 1.0039$ AU) and a low $i = 2.8°$. Comet 107P/Wilson–Harrington was linked to the Corvid shower, but I find that another minor body, *2004 HW*, is a more likely candidate. Instead, this comet may be responsible for the September Sagittariids (Table 7). The comet was scheduled to be visited by NASA's Deep Space 1 in January, 2001, but the star tracker failed on November 11, 1999 and only Borrelly was visited. The comet nucleus is about

[14] S. C. Lowry and P. R. Weissman, CCD observations of distant comets from Palomar and Steward observatories. *Icarus* **164** (2003), 492–503.

[15] Y. R. Fernández, L. A. McFadden, C. M. Lisse *et al.*, Analysis of POSS images of comet–asteroid transition object 107P/1949 W_1(Wilson–Harrington). *Icarus* **128** (1997), 114–126.

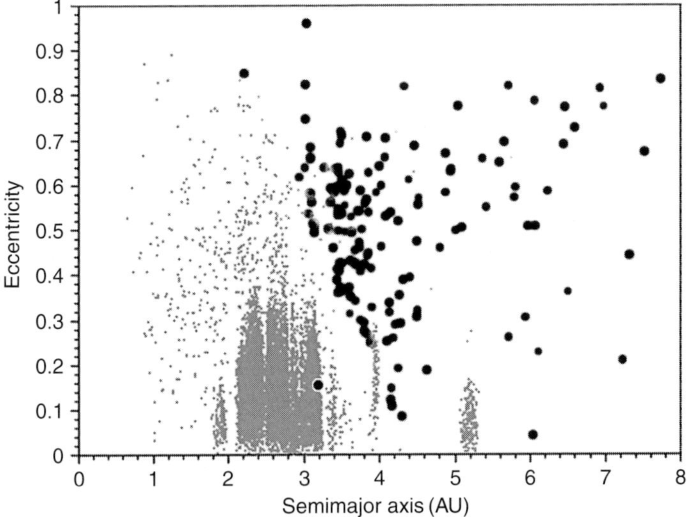

Fig. 9.7 The distribution of Jupiter-family comets and suspected extinct comet nuclei (•) in relation to asteroids (gray points) in a Kresák diagram of semimajor axis and eccentricity space. Earth is at 1 AU, Mars is at 1.52 AU and Jupiter at 5.2 AU. Diagram from Andrea Milani, University of Pisa.

3.9 ± 0.5 km in diameter, rotating every 6.1 ± 0.05 h (obliquity $\sim 90°$),[16] and reflecting $5 \pm 1\%$ of near-infrared light.[17]

If a fraction A (albedo) of light is reflected, then nuclear magnitude and size relate approximately as:[18]

$$\log D\,(\mathrm{km}) = 3.2 - 0.205 H_{\mathrm{N}} - 0.5 \log A \qquad (10.1)$$

which would give a diameter of $D = 4.2$ km for a typical albedo $A = 0.04$. Clearly, even in 1949 this comet was much less active than expected for its size.

Of course, the nature of the body does not change whether it is active or not. This remains a comet. Fortunately, it is possible to recognize dormant comets (*extinct comet nuclei*) because they tend to be in wider orbits with higher eccentricity (Fig. 9.7). Quantitatively, they have different *Tisserand parameters*, with most asteroids having $T_{\mathrm{J}} > 3$ and most Jupiter-family comets (•) have T_{J} between 2 and 3. Only very few Jupiter-family comets have wandered among the asteroids in (a, e) space. Among

[16] D. J. Osip, H. Campins and D. G. Schleicher, The rotation state of 4015 Wilson–Harrington: revisiting origins for the near-Earth asteroids. *Icarus* **114** (1995), 423–426.

[17] H. Campins, D. J. Osip, G. H. Rieke and M. J. Rieke, Estimates of the radius and albedo of comet-asteroid transition object 4015 Wilson–Harrington based on infrared observations. *Planet. Space Sci.* **43** (1995), 733–736.

[18] L. Kresák, Short-period comets at large heliocentric distances. *Bull. Astron. Inst. Czech.* **24** (1973), 264–283; S. C. Lowry, A. Fitzsimmons and S. Collander-Brown, CCD photometry of distant comets. III. Ensemble properties of Jupiter-family comets. *Astron. Astrophys.* **397** (2003), 329–343.

those are these Encke-type comets with 2P/Encke at $T_J = 3.027$ and 107P/Wilson–Harrington at $T_J = 3.084$.

It is possible to give a probability for a particular minor body to have evolved from an initial Jupiter-family comet orbit. From this, *Bill Bottke* of the Southwest Research Institute calculated upon request the probability for all known near-Earth objects. All those that may have evolved from a Jupiter-family comet are listed in Table 2. Interestingly, comet Wilson–Harrington has only a 7% probability of being a Jupiter-family comet on such dynamic grounds.

Once they are extinct, comets may still be distinguished from asteroids because they have slightly different surface reflection properties, and lower tensile strength.

9.4 3200 Phaethon

It was Ernst Öpik who first proposed that some of the "asteroids" that approach Earth's orbit are in fact degassed cometary nuclei.[19] In reality, only a small fraction is now suspected to be so. There may be 60 such extinct comets of sizes larger than 1 km with $q < 1.2$ AU. It is estimated that extinct comet nuclei stay dormant for about 40% of the time, but that could perhaps be less.[20] Even as an extinct nucleus, planetary

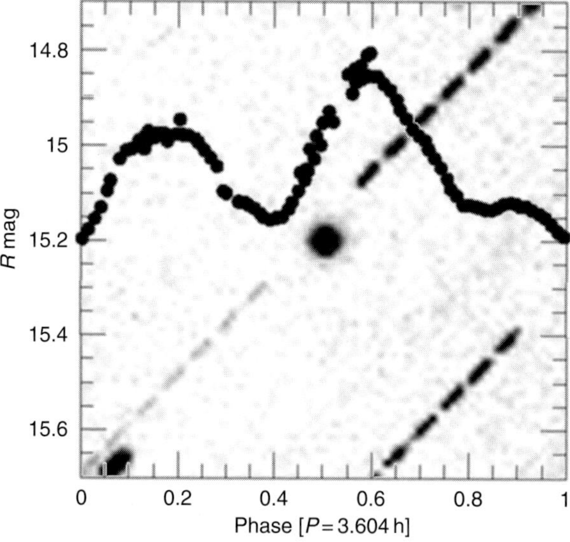

Fig. 9.8 3200 Phaethon in a CCD image with the University of Hawai'i 2.2 m telescope, on January 5, 1995 at $r = 1.09$ AU and a graph showing its brightness variation. Photo: Karen Meech and Oliver R. Hainaut, University of Hawai'i.

[19] E. J. Öpik, Survival of cometary nuclei and the asteroids. *Adv. Astron. Astrophys.* **2** (1963), 219–262.

[20] J. A. Fernandez, T. Gallardo and A. Brunini, Are there many inactive Jupiter-family comets among the near-Earth asteroid population? *Icarus* **159** (2002), 358–368.

perturbations will throw these comets out of the solar system within 100 000 years from when they first entered the inner solar system.[21]

A good example of an extinct or dormant comet nucleus is *3200 Phaethon* (Fig. 9.8), but it is also an unusual one because of its short orbital period of $P = 1.59$ yr. Because of that, it has an asteroidal $T_J = +4.51$, but a meteor shower to prove it a comet nucleus. Discovered in the IRAS fast moving object catalog,[22] it was Fred L. Whipple[23] who realized that this asteroid possessed an orbit almost identical to that of the Geminid meteoroid stream, which identified this object as a comet. Dynamical calculations confirmed the link.[24]

The nucleus of 3200 Phaethon has a diameter of ~5.1 km and rotates in 3.604 h, based on the periodic brightness variation of Fig. 9.8. Other estimates range from 3.57 to 4.08 ± 0.08 h.[25] The comet approaches the Sun close enough for silicates at its surface to melt ($q = 0.14$ AU). Perhaps as a result, the object is not as dark as a typical comet nucleus. 3200 Phaethon has a high geometric albedo of 11%, and a color slightly bluer than sunlight, with the spectral shape of B-type asteroids, which is similar to that of slightly metamorphosed carbonaceous-chondrites (Chapter 10).[26] It loses less than 0.01 kg/s of dust and gas.[27] It is remarkable that this comet has managed to survive the harsh conditions of its orbit.[28]

[21] W. F. Bottke, A. Morbidelli, R. Jedicke *et al.*, Debiased orbital and absolute magnitude distribution of the near-Earth objects. *Icarus* **156** (2002) 399–433.

[22] S. Green, *International Astronomical Union Circular* 3878, October (1983); J. K. Davies, S. F. Green, B. C. Stewart, A. J. Meadows and H. H. Aumann, The IRAS fast-moving object search. *Nature* **309** (1984), 315–319.

[23] F. L. Whipple, *International Astronomical Union Circular* 3881, October 25 (1983).

[24] B. Å. S. Gustafson, Geminid meteoroids traced to cometary activity on Phaethon. *Astron. Astrophys.* **225** (1989), 533–540.

[25] W. Z. Wisniewski, T. M. Michalowski, A. W. Harris and R. S. McMillan, Photometric observations of 125 asteroids. *Icarus* **126** (1997), 395–449.

[26] W. K. Hartmann, D. J. Tholen and D. P. Cruikshank, The relationship of active comets, 'extinct' comets, and dark asteroids. *Icarus* **69** (1987), 33–50.

[27] H. H. Hsieh and D. Jewitt, Search for activity in 3200 Phaethon. *Astrophys. J.* **624** (2005), 1093–1096.

[28] C. M. Birkett, S. F. Green, J. C. Zarnecki and K. S. Russell, Infrared and optical observations of low-activity comets, P/Arend–Rigaux (1984k) and P/Neujmin 1 (1984c). *Mon. Not. R. Astron. Soc.* **225** (1987), 285–296.

10

Asteroids as parent bodies of meteoroid streams

All the minor bodies of the solar system originating from beyond Jupiter's orbit are *comets*, even though their identity may long be hidden. In order to receive a comet designation, it is required that a cloud of dust or gas is detected. That means detecting a fuzziness in the image or other proof that a stream of cometary meteoroids is caused by the minor body.

Asteroids are the source of meteorites. Like comets, they are a remnant of the formation of the solar system, but they formed (and most spend all their life) closer to the Sun where ices did not remain frozen on dust grains and matter was compacted into rock. The name *asteroid* means "star-like" (Fig. 10.1). They are rocks larger than 1 m in diameter, somewhat arbitrarily discriminated from meteoroids because they are observed by telescope. The size limit has shrunk in recent years from ~50 to ~5 m and is expected to go further down with the increase of sophistication in asteroid detection.

Asteroids have also long been championed as a source of meteoroid streams.[1] Not the asteroids that make up the main belt between the orbit of Mars and Jupiter (Fig. 10.2), but the small fraction of asteroids that cross the orbit of Mars. Those that do not cross Earth's orbit are said to be of *Amor-type*, after namesake 1221 Amor (Fig. 10.3). Those that do cross Earth's orbit, but have an aphelion outside of Mars' orbit are called *Apollo-type*, after 1862 Apollo. Finally, those that cross Earth's orbit and have their aphelion inside the orbit of Mars are called *Aten-type*, after 2062 Aten. *Near-Earth asteroids* (NEA, $q < 1.2$ AU) return frequently to Earth and are about 80% of impacting bodies on Earth with diameter $D > 1$ km.

Most of these NEAs originate from the asteroid belt as a result of collisions between asteroids. Collisions can change the orbital period of a fragment into one that resonates with Jupiter. The result is a rapid increase of eccentricity and a decrease of perihelion distance until a close encounter with one of the terrestrial planets changes the orbital period again and puts it temporarily out of Jupiter's grip. Of all the asteroids that approach Earth, about 40% come from the ν_6 resonance, 25% from

[1] C. P. Olivier, *Meteors*, (Baltimore, MD: Williams & Wilkins, 1925), p. 272.

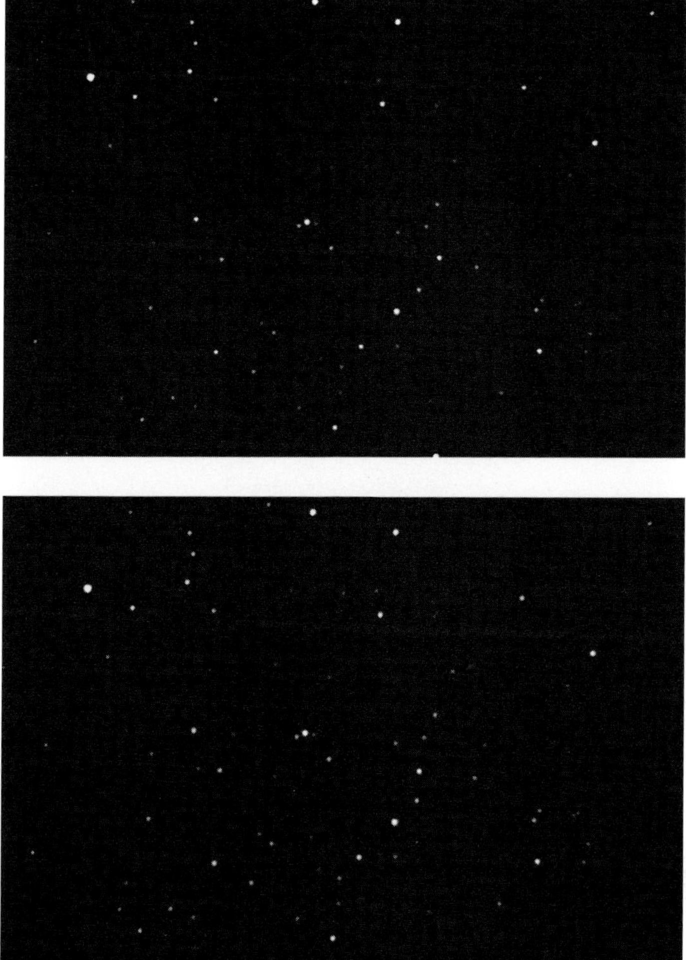

Fig. 10.1 Asteroid 433 Eros in Taurus on March 22, 2003, at times 21:26:43 (top) and 21:53:46 UT. Can you find it? Photo: Paul Smeets, Belgium.

the 3:1 resonance and the remaining 35% from numerous diffuse resonances across the main belt.[2]

10.1 Taxonomic classes of asteroids

Whether or not a particular minor planet is a suitable candidate parent body for a meteor shower depends on its taxonomic class. Asteroids are classified in groups from

[2] G. H. Stokes, D. K. Yeomans, W. F. Bottke *et al.* (2003) *Study to determine the feasibility of extending the search for Near-Earth Objects to smaller limiting diameters*. Report of the Near-Earth Object Science Definition Team. NASA Office of Space Science Report.

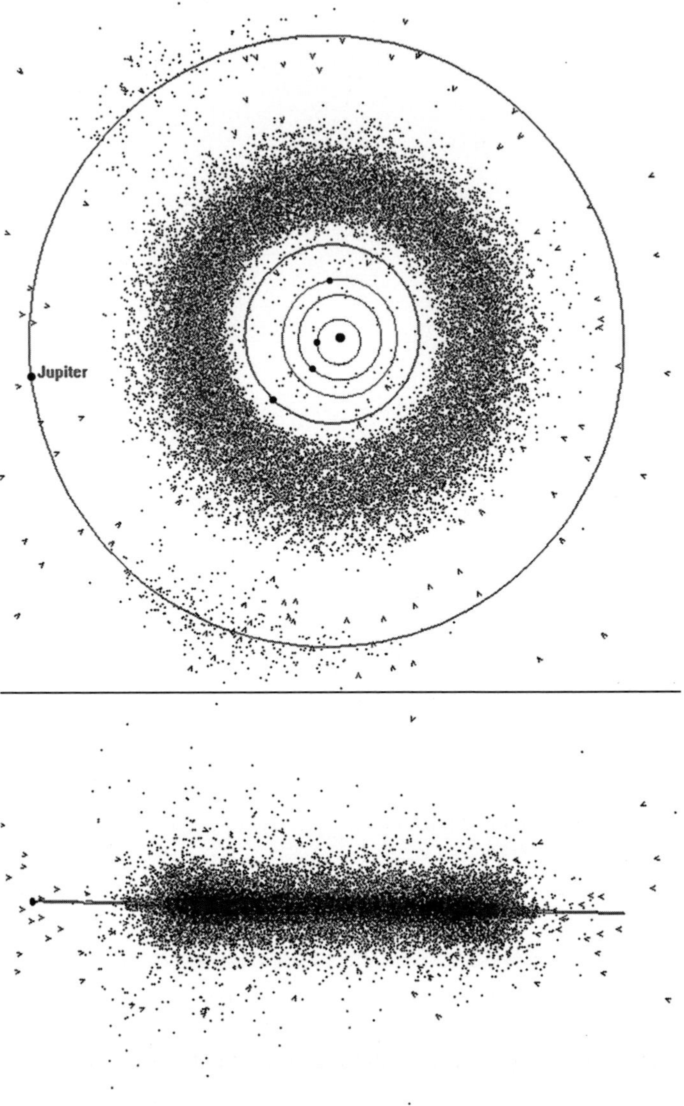

Fig. 10.2 The asteroid belt on January 1, 2005, showing the position of all known asteroids with well-determined orbits, with the vernal equinox to the right. Comets are shown by "V". Figure by Paul Chodas (NASA-JPL).

the color (spectral shape) of their reflected light and from the albedo, or percentage of reflected light. Some have a broad absorption band at 1 μm from mafic sillicates. These also tend to be more reflective and fall in the "S-complex" (for "stony"). Those that do not are darker, with a relatively flat spectrum, sometimes with an absorption

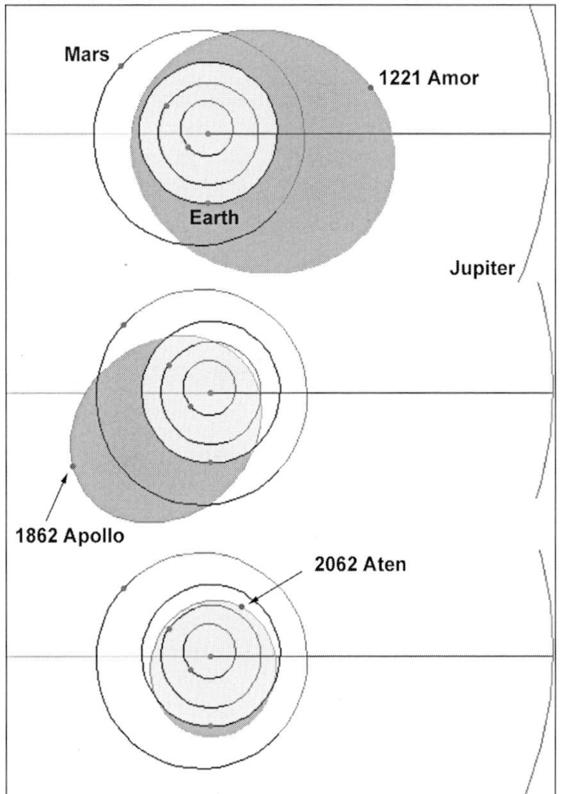

Fig. 10.3 Near-Earth asteroid orbit archetypes.

band at 0.7 μm, and fall in the "C-complex" (for carbonaceous).[3] The 0.7 μm band is due to the presence of oxidized iron in phyllosilicates resulting from aqueous alteration. Since these first taxonomic classes were introduced by Bus and Tholen, many outliers have been found and the complexes themselves have been cut into different taxonomic types (notably "C" into "C" and "X"). Each has been assigned a letter in the alphabet.[4] These groupings are not unique and there are at least six taxonomic classification systems used by different researchers.[5] Fig. 10.4 attempts a synthesis from the different taxonomy schemes.

[3] C. R. Chapman, D. Morrison and B. Zellner, Surface properties of asteroids: a synthesis of polarimtry, radiometry, and spectrophotometry. *Icarus* **25** (1975), 104–130.

[4] S. J. Bus and R. P. Binzel, Phase II of the small main-belt asteroid spectroscopic survey. A feature-based taxonomy. *Icarus* **158** (2002), 146–177.

[5] D. J. Tholen, Asteroid taxonomy from cluster analysis of photometry. Doctoral thesis, University of Arizona (1984); D. J. Tholen and M. A. Barucci, Asteroid taxonomy. In *Asteroids II*, ed. R. Binzel, T. Gehrels and M. Shapley Matthews. (Tucson, AZ: University of Arizona Press, 1989); E. S. Howell, E. Merényi and L. A. Lebofsky, Using neural networks to classify asteroid spectra. *J. Geophys. Res.* **99** (1994), 10 847–10 865; M. A. Barucci, M. T. Capria, A. Coradini and M. Fulchignoni, Classification of asteroids using G-mode analysis. *Icarus* **72** (1987), 304–324; E. F. Tedesco, J. G. Williams, D. L. Matson *et al.*, Three-parameter asteroid taxonomy. *Astron. J.* **97** (1989), 580–606; S. Xu, R. P. Binzel, T. H. Burbine and S. J. Bus, Small main-belt asteroid spectroscopic survey: initial results. *Icarus* **115** (1995), 1–35.

Asteroid taxonomic class:		Albedo:	NEO%	Example:	Suspected meteorite class:
Primitive asteroids:					
B	linear, blue/flat slope, dry	(0.09)	-.-	3200 Phaethon	-.-
(+F)	blue to flat slope	(0.05)	-.-	4015 Wilson–Harrington	-.-
T	red, flattening >850 nm	(0.06)	-.-	2001 SK_{162}	-.-
D (+P)	slighty red <550 nm, + featureless very red >550 nm	0.042	17	2001 SG_{286}	Tagish lake carbonaceous chondrite (carbon, organic-rich silicates)
C	flat/slightly reddish, wet Main belt:	0.10 0.06 ± 0.04	10	14827 Hypnos	CI/CM carbonaceous chondrites (hydrated silicates + organics)
X	featureless, positive slope	0.07	34	3362 Khufu	CI/CM carbonaceous chondrites
K (+L)	steep UV slope <750 nm	(0.14)	-.-	1991 XB	CV/CO carbonaceous chondrites
(+G)	UV band, slightly reddish	(0.09)	-.-	(1 Ceres)	altered carbonaceous
Asteroids with metamorphic surfaces:					
Q	deep rounded 1000 nm	0.25	14	1866 Apollo	H, L, and LL chondrites
(+U)		0.30	0.4	1993 UC	-.-
O	UV slope, deep 1000 nm	0.52	0.5	(4034) 1986 PA	L6 and LL6 ordinary chondrites
Differentiated (as in subjected to highest level of heating):					
S	moderate to strong absorption >700 nm	0.52	22	433 Eros	pallasites with accessory pyroxene stony-iron ureilites, and primitive achondrites
A	very steep UV slope	0.20	0.2	246 Asporina	olivine achondrite or pallasite
R	deeper 1 μm band	0.34	0.1	1990 MU	pyroxene–olivine achondrites
V (+J)	strong band >750 nm	0.42	1	1998 WZ_6	basaltic (HED) achondrites
(+E)	flat to slightly reddish	(0.55)	-.-	5751 Zao	enstatite achondrite
M	featureless, flat	(0.18)	-.-	6178 (1986 DA)	iron meteorites (metal, trace silicates)

Fig. 10.4 Asteroid taxonomic classes and suggested associated meteorite types. Albedo and percentage of NEOs were derived by Stuart and Binzel after correcting for observing bias.[6]

[6] J. S. Stuart and R. P. Binzel, Bias-corrected population, size distribution, and impact hazard for the near-Earth objects. *Icarus* **170** (2004), 295–311.

It is an ongoing effort to find out what the relationships are between these taxonomic classes and the meteorites that are recovered on Earth. Surface weathering affects the reflection spectra of asteroids differently than the bulk that survives as meteorites. This effort has been frustrated even more by finds that a single asteroid may contain different types of materials. Therefore, asteroidal meteoroid streams may also consist of different materials.

The letters D, (P), C, (X), and K are reserved for the dark primitive asteroids, least affected by heating. These asteroids are found dispersed throughout the asteroid belt for $a > 2.3$ AU. D-types include the Hildas. About 75% of known main belt asteroids are type C, as are most near-Earth objects,[7] thought to be the source of carbonaceous CI- and CM-type chondrites. Many C-type asteroids show a 3 μm OH-stretch vibration feature and other evidence of mineralogical changes due to the melting of ice and the incorporation of liquid water into the mineral structure of stony compounds to form new minerals called clays. This process is called *aqueous alteration*. This group includes the asteroids of the Veritas family. Types K contain olivine and pyroxene in their reflection spectra. It has been claimed that either CV or CO meteorites, but not both, are derived from the EOS family.[8]

The types B, G, and F appear to be altered carbonaceous chondrites and include the large asteroids 1 Ceres (G) and 2 Pallas (B), as well as suspected extinct comet nuclei. At 487 by 455 km, Ceres is the largest known asteroid. F-types are rare, but common in the Nysa family, Nysa being an M-type interloper.

Asteroids with metamorphic surfaces are named types Q and O. Q types (such as near-Earth asteroid 1866 Apollo) are the source of ordinary H, L, and LL chondrites. This is a rare class among main belt asteroids, but appears to represent the most abundant class of meteorites found on Earth![9] The O-class was initially created after finding that the spectrum of 3628 Boznemcovna resembled that of L6 and LL6 ordinary chondrites.[10]

A third class of asteroid surfaces appear to be igneous, as in the materials most subjected to heating. These are types S, A, (E), R, V, J, and M. In the main belt, they tend to be clustered in families, groups of which are dispersed throughout the belt. They are associated with differentiated meteorites: achondrites, pallasites, and metals. In the main belt, about 17% are type S. These "stony" asteroids dominate the inner side of the asteroid belt (<2.3 AU) and the Flora family. Most of the remainder are type M ("metallic"), such as the asteroid Psyche. The reality of "E" types as a class is debated, a name reserved for the source of the enstatite achondrites. V are

[7] *Ibid.*

[8] C. R. Chapman, P. Paolicchi, V. Zappala *et al.*, *Asteroids II*, ed. R. P. Binzel, T. Gehrels and M. S. Matthews. (Tucson, AZ: University of Arizona Press, 2001), pp. 386–415.

[9] M. J. Gaffey, J. F. Bell and D. P. Cruikshank, Reflectance spectroscopy and asteroid surface mineralogy. In *Asteroids II*, ed. R. P. Binzel, T. Gehrels and M. S. Matthews. (Tucson, AZ: University of Arizona Press, 1989), pp. 98–127.

[10] R. P. Binzel, S. Xu, S. J. Bus *et al.*, Discovery of a main-belt asteroid resembling ordinary chondrite meteorites. *Science* **262** (1993), 1541–1543.

"4 Vesta-type" basaltic (HED) achondrites.[11] The Apollo and Aten objects in the V class appear to have been produced in the ν_6 secular resonance on the inner edge of the asteroid belt, which delivers asteroids rapidly to an orbit intersecting Earth's and is thought to produce 70% of all near-Earth objects with a <2 AU. This is thought to be why they keep their fresh (highly reflective) surfaces. The distribution of S-type NEOs is similar to that produced by the intermediate-source Mars-crosser population of asteroids, which is a much slower pathway.[12]

10.2 Meteoroid streams from asteroids

Rocky asteroids can cause meteoroid streams as a result of frequent small impacts with asteroid boulders or during a rare catastrophic collision in which both asteroids break into fragments. The meteoroids and fragments will disperse much like a dust trail, but differ from cometary streams in being rare, transient in nature, and composed of rocky material.

In a small impact with sufficient kinetic energy, all the boulders and dust from previous collisions can be shaken loose, but only when the object is a solid rock and not a rubble pile. The asteroid Eros (Fig. 10.5), an S-type asteroid with a mean density of $2.4\,g/cm^3$, was studied by the NEAR-Shoemaker mission. The asteroid surface was covered by about 30 000 boulders, 6760 of which were bigger than 15 m.[13] The distribution of diameters (D) of these fragments should be close to that expected from a catastrophic collision $N(D) \sim D^{-\alpha}$, with $\alpha = 4.3$ ($s = 2.1$).[14] Most of the mass is in the smallest objects. Based on the estimated volume of regolith,[15] the corresponding meteoroids need to have a smaler $\alpha = 3.5$ ($\chi \sim 1.85$), or much of it was lost when the boulders were created. If all debris is shaken loose, I estimate the total amount in sizes of 1 mm to 10 cm to be $\sim 2 \times 10^{13}$ kg, 20–2000 times the dust loss from a single return of a short period comet. The escape speed from a spherical asteroid is about $V_{esc} = 59$ (m/s) ($D/100$) km, with D the diameter of the asteroid. Once released from the asteroid, most fragments will have relative velocities less than about twice this value. In cratering models, it is often assumed that the fragments leave with speeds that have a power-law distribution

[11] T. B. McCord, J. B. Adams and T. V. Johnson, Asteroid Vesta: spectral reflectivity and compositional implications. *Science* **178** (1970), 745–747.

[12] T. H. Burbine and R. P. Binzel, Asteroid spectroscopy and mineralogy. In *Asteroids, Comets, Meteors 1993*, ed. A. Milani *et al.* (Dordrecht: Kluwer, 1994), pp. 255–270.

[13] P. C. Thomas, J. Veverka, M. S. Robinson and S. Murchie, Shoemaker crater as the source of most ejecta blocks on the asteroid 433 Eros. *Nature* **413** (2001), 394–396.

[14] A. Fujiwara, P. Cerroni, D. R. Davis *et al.*, Experiments and scaling laws on catastrophic collisions. In *Asteroids II*, ed. R. P. Binzel, T. Gehrels and M. S. Matthews. (Tucson, AZ: University of Arizona Press, 2001), pp. 240–265.

[15] C. R. Chapman, W. J. Merline, P. C. Thomas *et al.*, Impact history of Eros: craters and boulders. *Icarus* **155** (2002), 104–118.

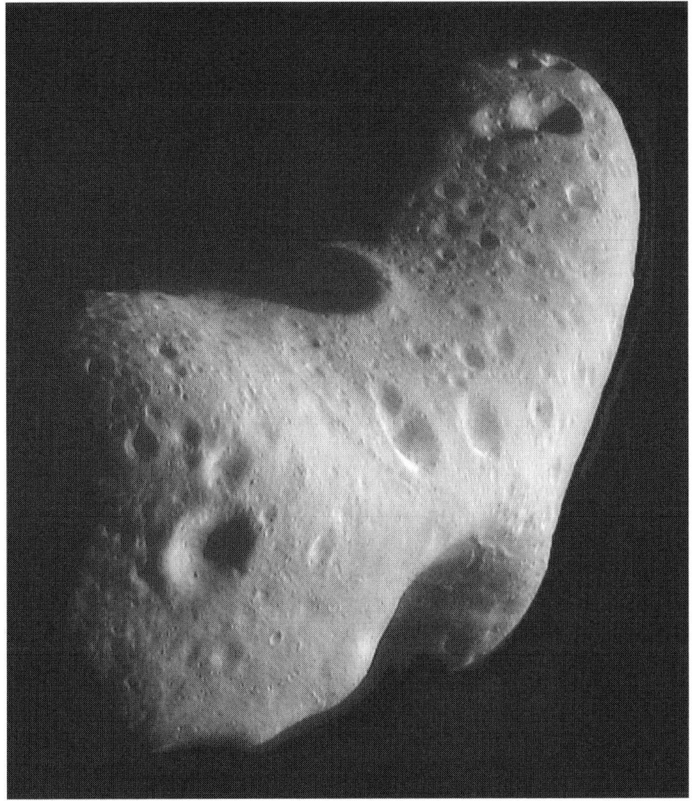

Fig. 10.5 Asteroid Eros as imaged by the NASA NEAT mission. Photo: NASA/JPL.

$\Delta N/\Delta V_{ej} \sim V_{ej}^{-3.25}$, with a lower limit cutoff,[16] which has a peak at $V_{ej} = 0.55\ V_{esc}$.[17]

Smaller collisions are more frequent, but generate less debris. Hypervelocity impacts on asteroids by a 1 kg boulder will generate about 1000 kg ejecta, while a similar impact on a comet might generate some 3200 kg. Half the kinetic energy of the impacting body is shared between the fragments leaving the impact site. The fragments have a mass distribution close to $s = 2.1 \pm 0.1$, about equal mass per log mass interval.[18] Larger fragments tend to have smaller ejection speeds, about proportional to $\Sigma M\ (> V_{ej}) \sim V_{ej}^{-2}$.

[16] R. Greenberg and C. R. Chapman, Asteroids and meteorites: parent bodies and delivered samples. *Icarus* **55** (1983), 455–481; J. M. Petit and P. Farinella, Modeling the outcomes of high-velocity impacts between small solar system bodies. *Celest. Mech.* **57** (1993), 1–28.

[17] P. Farinella, C. Froeschlé and R. Gonczi, Meteorite delivery and transport. In *Asteroids, Comets, Meteors 1993*, ed. A. Milani *et al.* 1994, pp. 205–222.

[18] D. W. Hughes, The largest asteroids ever. *Q. J. R. Astron. Soc.* **32** (1991), 133–145; I. P. Williams, The dynamics of meteoroid streams. In *Meteoroids and Their Parent Bodies*, ed. J. Stohl and I. P. Williams. (Bratislava: Astronomical Institute, Slovak Acad. Sci., 1993), pp. 31–40.

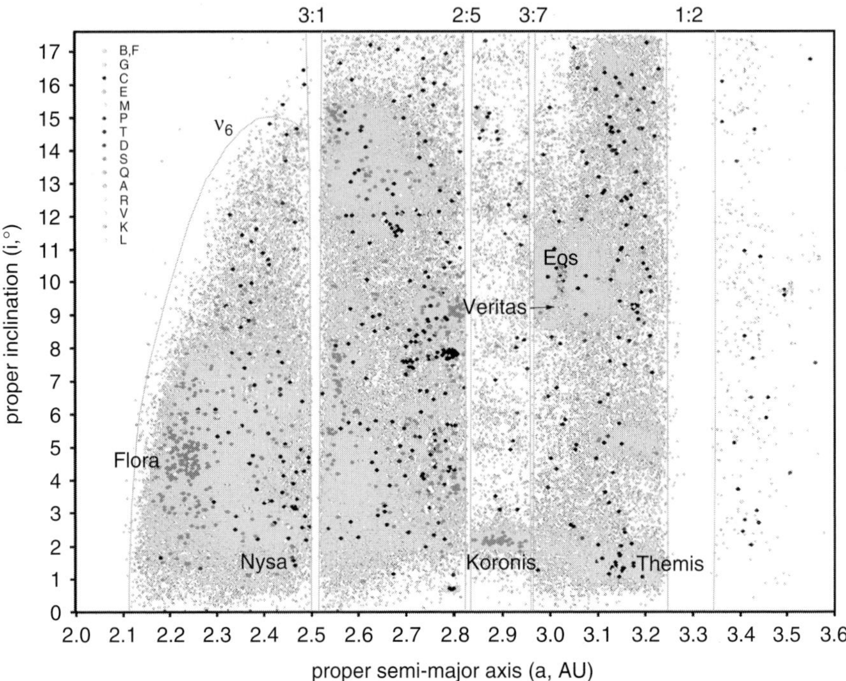

Fig. 10.6 5335 main belt asteroids in proper elements *i* versus *a*. The most prominent asteroid families are marked. Gaps are due to mean-motion resonances with Jupiter, and due to the secular resonance (ν_6) where the line of apsides of asteroids and that of Saturn rotate at the same rate.

Small grains have speeds of about $V_{\mathrm{ej}} \sim D^{-0.75 \pm 0.25}$.[19] Thus, most of the large fragments of an impact event will have low relative velocities and may fall back on to the asteroid, while small dust grains >1 mm in size escape with high speeds >50 m/s.

Catastrophic collisions of near-Earth asteroids with main-belt asteroids are rare. On the other hand, there are many examples of mutual collisions in the main asteroid belt. These relatively gentle collisions result in a (Kiyotsugu) *Hirayama asteroid family*[20] (Fig. 10.6).[21]

Asteroids in the main belt move around the Sun at speeds of $V_{\mathrm{h}} \sim 20$ km/s, and relative speeds of a few km/s. In a typical collision between two asteroids of similar size, the biggest fragment has a relative speed of 10–80 m/s, and other larger fragments leave at 80–130 m/s.[22] The ratio of the largest piece to the original mass can

[19] T. Mukai, J. Blum, A. M. Nakamura, R. E. Johson and O. Havnes, Physical processes on interplanetary dust. In *Interplanetary Dust*, ed. E. Gün, B. Å. S. Gustafson, S. F. Dermott and H. Fechtig, (Berlin: Springes 2001), pp. 445–507.

[20] K. Hirayama, Groups of asteroids probably of common origin. *Proc. Phys.-Math. Soc. Japan*, Series 2 (1918), 354–361.

[21] A. Milani and Z. Knezevic, Asteroid proper elements and the dynamical structure of the asteroid belt. *Icarus* **107** (1994), 219–254.

[22] P. Michel, P. Tanga, W. Benz and D. C. Richardson, Formation of asteroid families by catastrophic disruption: simulations with fragmentation and gravitational reaccumulation. *Icarus* **160** (2002), 10–23.

range from 0.04 to 0.69. The breakup mirrors the internal structure. Many asteroids are a loosely consolidated rubble pile, judging from the fact that few asteroids >150 m rotate faster than once every two hours.

The principal Hirayama families are *Hungarias, Floras, Phocaea, Koronis, Eos, Themis, Cybeles,* and *Hildas* (named after the main asteroid in the group). Members have relative velocities of a few hundred of m/s and a size distribution typical of a catastrophic breakup, with the Koronis group $\alpha = 3.55 \pm 0.34$ and the Flora group $\alpha = 3.85 \pm 0.17$. The dispersion is larger than measured in the laboratory, due to planetary perturbations. Most known asteroid families are very old and have therefore undergone significant dynamical (and collisional) evolution since their formation. The large Koronis and Themis families, for example, are about 2 billion years old.[23] About 90% of all asteroids appear to be part of such families.[24]

[23] F. Marzari, D. Davis and V. Vanzani, Collisional evolution of asteroid families. *Icarus* **113** (1995), 168–187.
[24] Z. Ivezic, R. H. Lupton, M. Juric *et al.*, Color confirmation of asteroid families. *Astron. J.* **124** (2002), 2943–2948.

Part III

Young streams from water vapor drag

11

Forecasting meteor storms from what planets do to dust trails

Meteor storm forecasting became feasible in 1867, when it was discovered that comets are responsible for meteor storms. By tracking the changing comet orbit in the gravity field of the Sun and planets, it was possible to predict when and where the comet itself would be near Earth's orbit. It took four more Leonid seasons, before meteor storm forecasting became a reality with the capability of tracking the path of the meteoroids instead.

11.1 The 1872 and 1885 Andromedids

The first successful meteor storm prediction based on the behavior of the parent comet was that for the Andromedid storm of 1872. Following Schiaparelli's work, both Edmond Weiss in Vienna and *Heinrich Ludwig d'Arrest* at Leipzig Observatory announced in 1867 that Biela's comet was in the same orbit as the Andromedids observed by Brandes on the evening of December 6, 1798. D'Arrest noticed that there had been six complete revolutions by 3D/Biela from the initial report of Brandes to the shower of December 6, 1838, when rates in Europe and the USA were four times as high as in a normal night. These showers occurred in sync with the short 6.6 yr orbit of the comet. However, his prediction of another shower on December 6, 1878, did not pan out.

Weiss found the reason why.[1] By comparing past orbits, he discovered that Jupiter had rotated the comet orbit so that now Earth passed its node a week earlier. Extrapolating, he predicted that a shower in 1872 should occur on about November 28, and not on December 6 when Brandes saw the meteors a century earlier. Weiss' result was quoted, but not believed. Herschel wrote: *Should a clear night permit a watch to be kept for these meteors during the last week in November, and the first week in December next, and especially on the nights of the 4th and of the 7th of December, both in this and in next year* (1872 and 1873 – PJ), *a favourable opportunity exists for verifying practically these ingenious suggestions of Dr. Weiss.*[2] Close to the prediction by Weiss, a spectacular storm of slow meteors was observed on November 27, 1872.

[1] E. Weiss, Beiträge zur Kenntniss der Sternschnuppen. *Astr. Nachr.* **72** (1868), 81–102.
[2] A. S. Herschel, Meteor-showers supposed to be connected with Biela's Comet. *Mon. Not. R. Astron. Soc.* **32** (1872), 355–359.

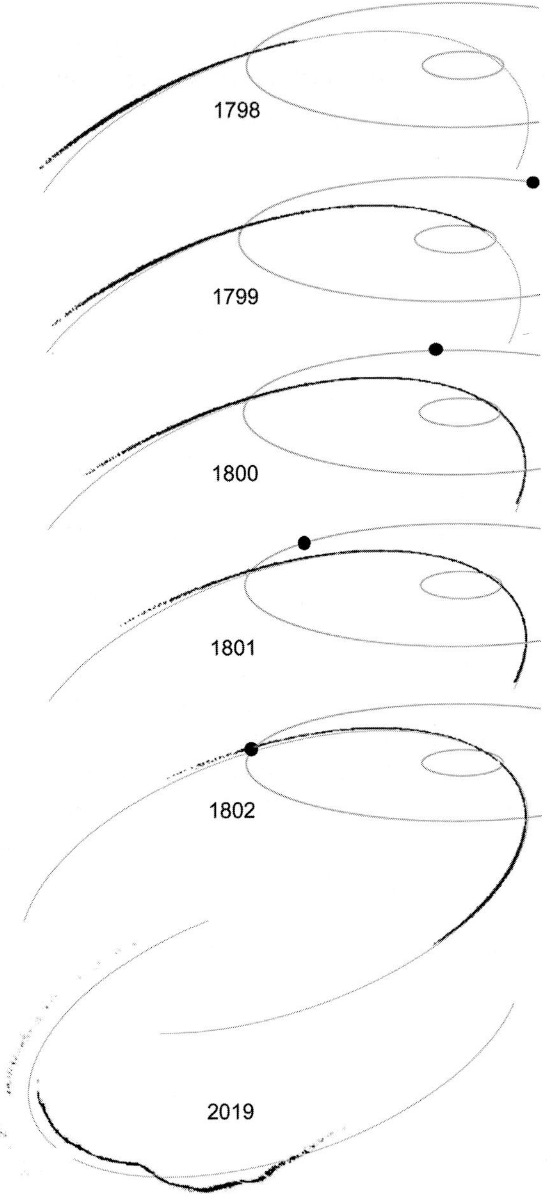

Fig. 11.1 The path of Leonid meteoroids in the 1767-dust trail of 55P/Tempel–Tuttle and the passage of Jupiter by the trail over a period of 5 yr during the 1799 return. The date in each frame is November 17. The bottom frame shows the dust trail in 2019 at aphelion. Calculations by Jérémie Vaubaillon, IMCCE.

A second Andromedid storm in 1885 under somewhat similar encounter conditions was highly anticipated and resulted in the first photographed meteor by *Ladislaus Weinek* at Prague.[3] No spectacular picture, just a single 7 mm long track on a glass plate. A first attempt at multi-station photography from a second site at Jena was spoiled by clouds.

11.2 The 1899 Leonids

These early successes created great expectations for the next anticipated Leonid return in 1899. Perturbations by the planets were clearly important, but had to be calculated by hand, a very tedious effort. Back in 1867, LeVerrier[4] realized that the gravity of Jupiter, orbiting the Sun, would affect meteoroids in a different way than the comet because the meteoroids were in different parts of the comet orbit.

Modern computers can visualize this effect by plotting the position of the dust trail (of 55P/Tempel–Tuttle) in different years (Fig. 11.1). Note how Jupiter is at different positions in its orbit when the particles pass by Jupiter's orbit. Only when the planet is nearby do the biggest perturbations occur. Those perturbations are clearly seen when the trail arrives at aphelion some centuries later. The result is a periodic distortion.

All planets perturb the orbit of the dust grains in this manner, but Jupiter is the heaviest planet in the solar system at 318 times Earth's mass and a mean distance of 5.2028 AU from the Sun.[5] Jupiter completes one orbit every 11.867 yr. Fig. 11.2 shows the relative importance of the planets in terms of diameter (pictures) and in terms of mass (filled circles, middle row). Mass is what determines the force of gravity and Jupiter is clearly king of its domain. The next most significant influence is Saturn at 95 Earth masses, traveling around the Sun in 29.46 yr. Together, they cause a periodic perturbation repeating roughly once every sixty years (five orbits of Jupiter and two of Saturn).

Uranus and Neptune have a relatively small gravitational field, but they are in the outer solar system where the Sun's gravity is weak. Because of that, they exert influence over a wider region. The radius of the zone of influence is expressed by Laplace's formula:

$$R_P = a_P(M_P/(M_{sun} + M_P))^{0.40} \qquad (11.1)$$

with R_P the radius of the zone of influence (Fig. 11.2), M_P and a_P the mass and semimajor axis of the planet, respectively. This has Jupiter's zone of influence at 0.32 AU distance from the planet, while Earth only reaches out to 0.0062 AU.

[3] L. Weinek, *Astronomische Beobachtungen an der K. K. Sternwarte zu Prag, in then Jahren 1885, 1886 und 1887* (Prague: Sternwarte, 1888) Before Weinek, a young German photographer Hermann Krone (1827–1916) is rumored to have taken a daguerrotype of a shooting star in 1848, while at the observatory of Breslau. I can not confirm that.

[4] U. Le Verrier, Sur les étoiles filantes de 13 November et du 10 Août *Comptes Rendes* **64** (1867), 94.

[5] H. N. Russell, On the origin of periodic comets. *Astron. J.* **33** (1920), 49–61.

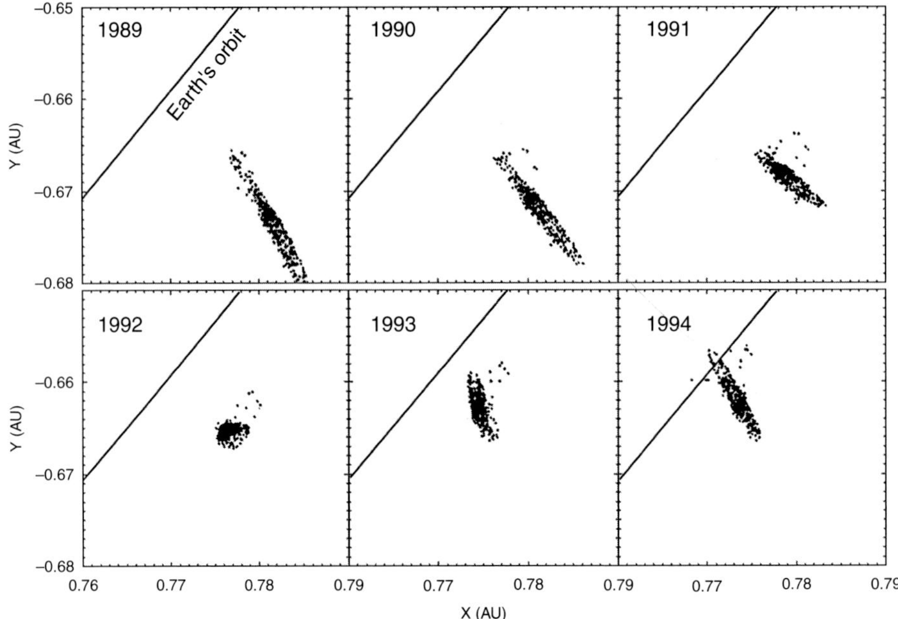

Fig. 11.3 The one-revolution 1862-dust trail of comet 109P/Swift–Tuttle near Earth's orbit in August of the years 1989–1994, in an early calculation by Zidian Wu and Iwan Williams.[11]

orbit. Because the planets perturb the meteoroids in different parts of the trail in a different manner, that point moves around from year to year in a cyclic manner with the 12 yr period of Jupiter. This is clearly seen in the early Perseid stream model by *Zidian Wu* and *Iwan Williams* of Queen Mary College at the University of London, shown in Fig. 11.3. Each dot marks where the orbit of a Perseid meteoroid, ejected in 1862 from 109P/Swift–Tuttle at a speed of ~50 m/s, crossed the ecliptic plane during the time when Earth arrived at the node. The reason the cross section is elongated is due to a lack of particles in the model, Wu and Williams had to show a relatively long section of the trail, the shape of the assembly reflecting how the trail moves about in the days and weeks before and after the time of Earth passing by the node.

Predicting the time of a meteor storm boils down to calculating the position of Earth at the time when we are nearest to the trail center. The direction of the Sun at that moment, the *solar longitude* ($\lambda_{\odot}$), is traditionally used to mark that position irrespective of the day count, which is off by one-quarter of a day each orbit. More precisely, the solar longitude is the direction of the Sun's center as seen from the geodetic center of the Earth.

In Chapter 13 I will discuss the fact that the Sun moves around the barycenter of the Solar System. The longitude of that barycenter (λ_b) would be more suitable for

[11] Z. Wu and I. P. Williams, The Perseid meteor shower at the current time. *Mon. Not. R. Astron. Soc.* **264** (1993), 980–990.

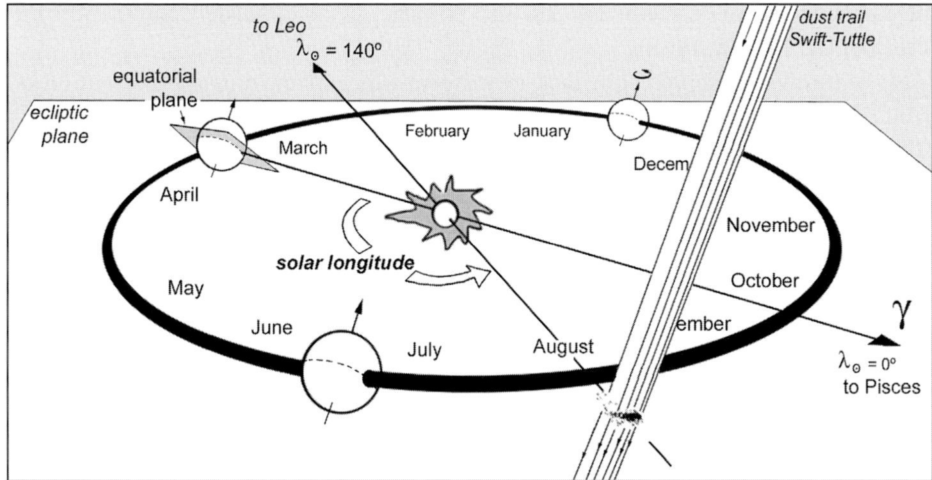

Fig. 11.4 The definition of solar longitude ($\lambda_{\odot}$). View from the north, looking down on the ecliptic plane. The Earth moves in an anticlockwise direction.

marking the position of Earth, which can differ from the solar longitude by up to 0.6°. Nevertheless, predictions and models can be compared in terms of both longitudes and time can be translated into either.

The longitude is measured relative to the direction where Earth's equator and ecliptic plane intersect, the Sun having a northerly motion, called the *vernal equinox* (Fig. 11.4). That direction moves around due to precession of the Earth's spin axis, so that solar longitude has to be given for a certain choice of the equinox. As said before, the angles of the orbital elements in this book are given in the FK5 system at the *equinox* of 2000 January 1.5: J2000.0.

Expectations for the 1899 Leonids were immense and the predictions of the peak time were eagerly greeted. In the mean time, Stoney and Downing were having second thoughts about the earlier results and tackled again the close approaches to Saturn in 1870 and to Jupiter in 1898. The Jupiter encounter had the effect of moving the stream further inside the Earth's orbit.[12] Stoney's latest calculation shifted Adam's orbit 800 000 km nearer to the Sun (−0.014 AU) than where the Earth would be . . .,[13] a result that now agreed with Berberich. Foreshadowing the habit of forecasters a century later, the early optimistic predictions were revised downward in the months before the expected storm.

The general public knew nothing of those misgivings, recalled *Charles P. Olivier*, founder of the American Meteor Society and author of the 1925 book *Meteors*. There was a sense that man had mastered nature and all that one needed to know about

[12] G. Johnstone Stoney and A. M. W. Downing, Next week's Leonid shower. *Nature* **61** (1899), 28–29.
[13] C. P. Olivier, *Meteors*. (Baltimore, MD: Williams & Wilkins, 1925), 276 pp.

meteor showers was known. Indeed, the popular media predicted confidently the return of the Leonid storms.

No storm was observed. The lack of a storm was explained as a perturbation of orbits by Jupiter and a storm was predicted for 1900. No storm was seen. Again it was explained and a storm predicted for 1901. None was seen. Olivier called the lack of a meteor storm that season "the worst blow ever suffered by astronomy in the eyes of the public". A humiliation that took a century, and much better computational tools, to overcome.

12
Meteor storm chasing

Theory and observations go hand in hand in our efforts to understand nature, but observers are not always remembered in the same way when discussing the gain made over the years. Their bold ventures into the cold and dangerous world painstakingly paved the path to wisdom, and those before us, many now forgotten, found that path much less traveled. This section recalls some of past travels in the pursuit of meteor showers.

The anticipated 1899 Leonid return prompted some astronomers to rise above the clouds. In France, *Jules Janssen (1824–1907)*, first director of the Observatory of Meudon, and his colleague *M. W. de Fonvielle*, organized a balloon flight to bring visual observers above the ground fog, with support of the French Society of Aerial Navigation. Before that, meteor showers had often been seen from hot air balloons in the wind-still early morning hours before sunrise (Fig. 12.1) and de Fonvielle had earlier viewed the Leonid shower of November 13/14, 1867, above clouds over Paris in what was probably the second airborne astronomical expedition (the first being a total solar eclipse observation that year). Now, everybody wanted to ascend in a balloon to view the Leonids. A German balloon launched from Strasbourg fell at Fanxault, causing one serious injury, while a British balloon was nearly lost at sea.[1] Janssen's 1898 balloon mission was flown by Russian astronomer *Gavriil Adrianovich Tikhov*, then stationed at Meudon. Rates had been high, but there was no storm that year. In 1899, there were five balloon flights. On the night of November 15/16, it was a woman, 38 yr old San Francisco born astronomer *Dorothea Klumpke* (1861–1942), then working at the Observatoire National de Paris, who was chosen to be the observer. "I do not know what good fairy overheard my wish to take a trip in the blue sky," Klumpke wrote of her voyage in the balloon called *Le Centaure*.[2] As Klumpke waited to go aloft, she knew of the disappointing reports from the previous night. Undaunted that the Leonids failed to appear when the peak had been expected, she joined the pilot of the balloon M. de Fonvielle and secretary M. Mallet into their tiny basket and

[1] M. Beech, The makings of meteor astronomy: part XVIII: the free-flying balloon. *WGN* **27** (1999), 45–51.

[2] D. Klumpke, A night in a balloon. *Century Mag.* **60** (1900), 276; D. Klumpke, Les Léonids. *Bull. Soc. Astron. France* **14** (1900), 34; R. Aitken, Dorothea Klumpke Roberts. *Proc. Astron. Soc. Pac.* **54** (1942), 217–222; J. H. Reynolds, Obituary notices: Fellows: Roberts, Dorothea Klumpke. *Mon. Not. R. Astron. Soc.* **104** (1944), 92–93.

Fig. 12.1 Shooting stars seen from a balloon, 1870 – Drawing by Albert Tissandier. From: Glaisher (1871).[3]

around 1 a.m. they lifted off. The sight of the city of Paris from the air was glorious. They tossed off ballast sacks, rose to an altitude of 500 m (1600 feet), and drifted westward over Normandy towards the English Channel. Before landing at the crack of dawn, they recorded 30 meteors during five hours of observing, of which only 12 were Leonids.

With the advent of technology, so increased the ambition to relive the experiences of past Leonid storm observers. In 1933, Czech astronomer *Vladimir Guth* (1905–1980) flew above clouds in a three engine Fokker FVII aircraft,[4] equipped with 720 HP Castor engines, in a mission supported by the Ministry of Transport. The threat of icing from fog and low clouds at night demanded a three-engine aircraft for a possible emergency landing. Past midnight, in the morning of November 17, Guth joined chief pilot K. Brabenee, radio telegrapher J. Kupka, astronomers Dr. E. Buchar and Dr. Hubert Slouka, and amateur astronomer M. Bláha, who was a clerk of the airlines. An earlier test flight had shown that the cloud deck was at 800 m. In a 2 h flight from 00:15 to 01:30 UT, with legs of 20 min, Guth saw 12 meteors (10 Leonids), Bláha saw five (five Leonids), and Slouka saw four (three Leonids). Guth would go on to become director of the Slovakian *Skalnaté Pleso Observatory* (1951–1956), during which time numerous visual observations of meteor showers were made and analyzed. The resulting 1966 paper

[3] J. Glaisher, C. Flammarion, W. de Fonvielle and G. Tissandier, *Travels in the Air* (London: Richard Bentley, 1871), 398 pp.

[4] V. Guth, The Leonids in the last years. In *The Realm of the Stars* 15 (1, Feb. 1934), (in Czech, translation by Milos Weber).

by Margita Kresáková on meteor stream activity[5] was still the most comprehensive study of its kind in the early 1980s, when I first became interested in meteor observing.

A spectacular storm of slow-moving *Draconids* in October, 1933 did much to keep the fascination with meteor showers alive.[6] The Draconids had been under scrutiny ever since *M. Davidson* calculated[7] that this comet might produce meteors on October 10, 1915. Indeed, W. F. Denning reported a handful of Draconids in 1915 and 1922, but when parent comet 21P/Giacobini–Zinner (with a 6.61 yr orbital period) evolved rapidly into orbits intersecting that of Earth in 1926, expectations were high. In Stowmarket, England, *J. P. M. Prentice* had succeeded M. Davidson as director of the British Astronomical Association's Meteor Section (1923–1954) and observed a definite outburst at *zenith hourly rate*, ZHR ~14/h. The zenith hourly rate is the hourly rate of meteors seen in an unobstructed sky by a visual observer if only the radiant had been in the zenith and the sky was clear and dark with stars visible up to $+6.5^{\mathrm{m}}$. Even so, the 1933 10 000/h level storm of slow Draconids came as a big surprise to many. A rising 70% illuminated Moon had discouraged many observers. Nonetheless, a photographic radiant was determined by *Peter MacKenzie Millman* in Canada.[8]

In 1937, astronomer *Cuno Hoffmeister* (1892–1968) of Sonneberg Observatory in Germany and author of *Die Meteore* (1937) and *Meteorströme* (1948),[9] took the long journey by ship to Southwest Africa to observe Southern Hemisphere meteor showers. The journey paid off in unexpected ways. During his routine visual observations, Hoffmeister saw 109 slow meteors in the period June 25–30, 1937, from a radiant at celestial coordinates R.A. = 192° and Decl. = −19°. This *Corvid* shower has not been seen again.

Earlier, Hoffmeister had published a catalog with *G. von Niessl*[10] that contained 611 fireball orbits determined from visual triangulations. 79% of those orbits were hyperbolic, implying they arrived from outside the solar system. This inspired Estonian astronomer *Ernst Julius Öpik* (1893–1985), author of *Physics of Meteor Flight in the Atmosphere*.[11] Öpik is best known for his theoretical work, including an early discovery in 1932 that the Sun could keep together a cloud of comets against perturbations from passing stars over the lifetime of the solar system.[12] He founded the meteor groups at Tashkent, Harvard, and Armagh Observatory and was the first to describe the physical processes of meteor ablation. His early "Oort-cloud" concept

[5] M. Kresáková, The magnitude distribution of meteors in meteor streams. *Contr. Astron. Obs. Skalnaté Pleso* **3** (1966), 75–112.

[6] W. F. Denning, A new cometary meteor shower (1926 October 9). *Mon. Not. R. Astron. Soc.* **87** (1927), 104–106; J. P. M. Prentice, The great meteor shower of comet Giacobini–Zinner. *J. Brit. Astron. Assoc.* **44** (1934), 108–111.

[7] M. Davidson, Meteor radiants from debris of comets. *J. Brit. Astron. Assoc.* **25** (1915), 292–293.

[8] P. M. Millman, Meteor news – note on the Perseid meteors in 1936; a meteor radiant photographically determined. *J. R. Soc. Canada* **30** (1936), 293–295.

[9] C. Hoffmeister, *Die Meteore* (Leipzig: Akademische Verlag, 1937), C. Hoffmeister, *Meteorströme* (Leipzig: Barth Verlag, 1948).

[10] G. von Niessl and C. Hoffmeister, Catalog der Bestimmungsgrößen für 611 Bahnen Großer Meteore. *Denksch. Akad. Wiss., Wien, Math. Naturw.* (1925), p. 100.

[11] E. J. Öpik, *Physics of Meteor Flight in the Atmosphere* (New York: Interscience, 1958).

[12] E. J. Öpik, Note on stellar perturbations of nearly parabolic orbits. *Proc. Am. Acad. Arts Sci.* **67** (1932), 169.

Fig. 12.2 Fred L. Whipple.

was mostly forgotten because Öpik believed that comets leaked from the clouds of other stars and arrived from interstellar space instead. As a contribution to practical astronomy, Öpik invented a rocking mirror technique to measure meteor angular speeds from visual observations during the 1931–1933 *Arizona Expedition for the Study of Meteors*. Plagued by large observational errors, he found incorrectly that 60% of meteoroids were arriving from interstellar space, an idea that agreed well with his picture of comets.[13] That result led to much controversy, Sir Arthur Stanley Eddington said that he would "*as lief belief in ghosts as in hyperbolic meteors*."

Öpik's visit to Harvard in 1931–32 inspired Ph.D. student *Fred Lawrence Whipple* (1906–2004).[14] Whipple (Fig. 12.2) started his career by calculating the radiant of meteors arriving from the star Sirius. When electricity companies started to precisely control the 60 Hz AC frequency of the commercial net in 1936, Whipple got his hands on newly available small synchronous motors for rotating shutters. He teamed with

[13] H. E. Shapley, E. J. Öpik and S. L. Boothroyd, The Arizona expedition for the study of meteors. *Proc. Nat. Acad. Sci. Wash.* **18** (1932), 16.

[14] F. L. Whipple, The incentive of a bold hypothesis – hyperbolic meteors and comets. *Int. Conf. on Education in the History of Modern Astronony*, New York, August 31, *Ann. NY Acad. Sci.*, **198** (1971), 219.

Fletcher Watson in a project of meteor multi-station photography, the Harvard meteor patrol (1936–1951),[15] modeled on early work by Yale Observatory astronomer *William Lewis Elkin* in the period 1893–1909.[16] An astrometic method to reduce such data had been published by H. H. Turner in 1907.[17] By offering a reward of $1 to visual observers for reporting the time of any photographed meteors, Whipple and colleagues obtained the first precise orbits of Perseids, Geminids, and Taurids and identified comet 2P/Encke as the parent of the Taurid shower.[18] By calculating the drag on the meteoroids, the work also provided information on the air density of the upper atmosphere.[19]

These observations grew into the photographic multi-station *Harvard Meteor Program* (1951–1957) at the Smithsonian Astrophysical Observatory. The early 35 mm cameras were replaced with two Baker–Nunn Super-Schmidt cameras in the often clear New Mexico skies of Soledad and Doña Ana (separated by only 18 miles),[20] boasting *f*/0.85, 12.25 inch aperture, 20 cm focal length, and a 55° field of view. These cameras (Fig. 12.3) caught slow +4 and fast +2 magnitude meteors at a rate of 3–5 per hour, ten times more than small cameras.[21] Film was exposed for periods of 12 minutes. Two other cameras were placed elsewhere. Elements of the Super-Schmidt cameras were tested with normal cameras during the October 9, 1946 Draconid storm. Apart from photographic and visual observations at the ground stations of Cambridge and Oak Ridge, Whipple organized for a group of six visual observers to go aloft in a Coast Guard PBY during the evenings of October 8 and 9. As it happened, the ground stations were completely overcast, while the airborne observers reported disappointing rates of only 20/h at around 03:50 UT October 9. Rates peaked at ZHR = 10 000/h elsewhere.

Based on meteors recorded by these Super-Schmidt cameras in the early 1950s, *Luigi Jacchia et al.* found that nearly all meteoroids were made up of fragile material from comets, and that none came with certainty from beyond the solar system. These large meteoroids were bound to our solar system, just like comets. As a by-product in trying to understand the physics of meteoroid ablation and atmospheric drag, Whipple studied the jet action of ablation on a rotating meteoroid and ended up applying this physics to the concept of a cometary nucleus. As said, this is how he found an explanation for nongravitational forces and arrived at his dirty-snowball model of a comet nucleus.

The Super-Schmidt cameras were the basis for a worldwide satellite-tracking network that went into full operation in June of 1958. Other types of meteor cameras were

[15] F. L. Whipple, Of comets and meteors. *Science* **289** (2000), 728; F. L. Whipple, *Proc. Am. Phil. Soc.* **79** (1938), 499.

[16] W. L. Elkin, Photographic observations of the Leonids at the Yale Observatory. *Astrophys. J.* **9** (1899), 20–22; W. L. Elkin, Results of the photographic observations of the Leonids, November 14–15, 1898, at the Yale Observatory. *Astrophys. J.* **10** (1899), 25–28; W. L. Elkin, The Velocity of meteors as deduced from photographs at the Yale Observatory. *Astrophys. J.* **12** (1900), 4–7.

[17] H. H. Turner, Measurement of a meteor trail on a photographic plate. *Mon. Not. R. Astron. Soc.* **67** (1907), 562–565.

[18] F. L. Whipple and S. El-Din Hamid, On the origin of the Taurid meteors. *Astron. J.* **55** (1950), 185–186.

[19] F. L. Whipple, Upper atmosphere densities and temperatures from meteor observations. *Popular Astron.* **47** (1939), 419–425.

[20] F. L. Whipple, The Harvard photographic meteor program. *Sky Telescope* **8** (1949), No. 4 (February).

[21] L. G. Jacchia and F. L. Whipple, The Harvard photographic meteor programme. *Vistas Astron.* **2** (1956), 982–994.

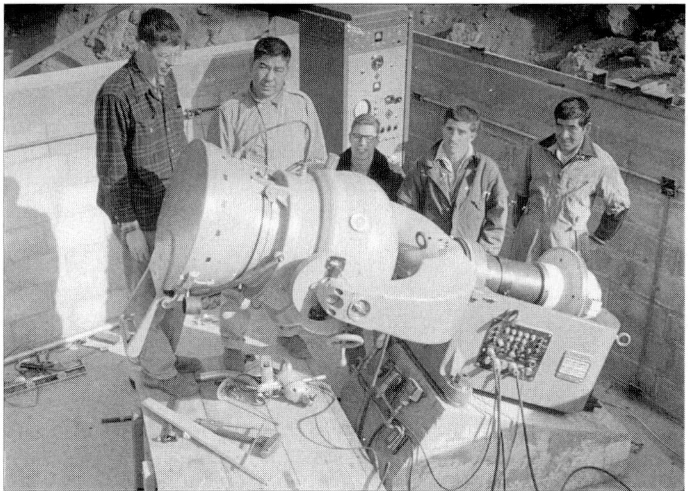

Fig. 12.3 Richard (Dick) McCrosky of the Harvard-Smithsonian Observatory (left) with one of the Baker–Nunn Super-Schmidt cameras, shortly after the camera was moved to the new observatory at Haleakala for satellite tracking in early 1958. Next to him is Mr. Ed Ige, a local contractor who hauled the telescope up the mountain. Next to him is Mr. Walter Webb, McCrosky's assistant and observing technician, and next to him are two employees of Mr. Ige. Photo courtesy of Walter Steiger, University of Hawai'i.[22]

distributed in the southwestern states of the United States to form the *Prairie Network*, photographing bright fireballs in search of meteorite falls, resulting in the recovery of the *Lost City* meteorite following its fall on January 3, 1970, by McCrosky and others.

When Earth approached comet 21P/Giacobini–Zinner on 1946 October 9, Canadian astronomer and Whipple's fellow graduate student at Harvard in 1932 (then completing a thesis on meteor spectra) *Peter MacKenzie Millman* (1906–1990) measured a photographic radiant. To do so, he had to use his connections from serving in the Royal Canadian Air Force during the war years to secure a military aircraft with permission to land anywhere in Canada. Only thus he escaped an ominous cloud bank over the Ottawa area, in a frantic last-minute dash to North Bay, Ontario. After landing there, he made single-station photographic and spectroscopic observations from the ground at the airport, first learning the fragile nature of Draconid meteoroids.[23]

In 1969, *K. Stuart Clifton* of Marshall Space Flight Center participated in the NASA *Airborne Auroral Expedition* for the purpose of making low-light level TV observations of the aurora and meteor flux. He deployed an image orthicon and a

[22] W. Steiger, *Origins of Astronomy in Hawai'i* (2001) (University of Hawai'i website).
[23] D. W. R. McKinley, *Meteor Science and Engineering* (New York: McGraw-Hill, 1961), p. 157.

secondary electron conduction vidicon intensifier.[24] His intensified video technique would later prove popular in the Leonid Multi-Instrument Aircraft Campaign. Later, Millman tried in vain to observe the Draconids from an airplane in 1972, and used the NASA Ames Lear Jet aircraft for pioneering intensified video spectroscopy of the Quadrantids in 1976.[25]

The 1946 Draconid storm was also the first to be observed by radar, a new technology that saw great use in the Second World War. Results of the early radar work led to the books *Meteor Astronomy* by *Sir Bernard Lovell*[26] and *Meteor Science and Engineering* by *Don McKinley*.[27] McKinley gives a nice history of the early radio astronomy, illustrating the great influence of the 1946 Draconid storm (also called *Giacobinids*) on the development of this field. Among the new discoveries with radar was that of daytime annual showers, the most active being the *June Arietids*.

Lovell (1913–), a cosmic ray physicist at the University of Manchester, was set to work on the use of radar during World War II. When the war ended, he was hoping to get back to studying cosmic rays by means of a mobile army radar unit, which he obtained with the help of colleague *James Stanley Hey*. Electric trams in the city of Manchester caused interference, and they moved to an open field outside the city called *Jodrell Bank*, now a famous radio observatory, where the first daytime meteor showers were discovered.

An Australian theoritician, *Thomas Reeve Kaiser* (1924–1998) (editor of *Meteors*)[28] joined the group in 1950 and went on to write the defining paper on meteor trail ionization and radar reflections.[29] Kaiser, a committed communist (until Hungary), ran into a conflict with Lovell over a protest telegram that Kaiser had written to President Truman (and over Lovell's subsequent fear of losing US funding). After the fall-out, Kaiser found new employment at the University of Sheffield in 1956, where he introduced astronomy and started a meteor radar program on a nearby hill at the Bradfield site. There, his work spawned a group of researchers (David Hughes, Iwan Williams, Jim Jones, Alan Webster, and Jack Baggaley) that would dominate meteor (shower) research in the next decades worldwide.

[24] K. S. Clifton, Airborne meteor observations at high altitudes. NASA TN D-6303, NASA, Washington DC (1971), 49pp.

[25] I. Halliday, Obituary – Peter MacKenzie Millman, 1906–1990. J. *R. Astron. Soc. Can.* **85** (1991), 67–78; P. Millman, Quadrantid meteors from 41,000 feet. *Sky Telescope* **51** (1976), 225–228; P. Millman, Airborne observations of the 1972 Giacobinids. *J. R. Astron. Soc. Canada* **67** (1973), 35–38; P. Millman, L. G. Jacchia and Z. Kopal, A photographic study of the Draconid meteor shower of 1946. *Astrophys. J.* **111** (1950), 104–133.

[26] A. C. B. Lovell, *Meteor Astronomy* (Oxford: Clarendon Press, 1954), 463pp.

[27] D. C. W. McKinley, *Meteor Science and Engineering* (New York: McGraw-Hill, 1961), 309pp.

[28] T. R. Kaiser (ed.), *Meteors* (*J. Atm. Terr. Phys.- Suppl.*) (London: Pergamon Press, 1955), 204pp.

[29] T. R. Kaiser and R. L. Closs, Theory of radio reflections from meteor trails. *Phil. Mag.* **43** (1952), 1–32; T. R. Kaiser, Radio echo studies of meteor ionization. *Phil. Mag. Suppl.* **2** (1953), 495–544.

In the 1960s, radar studies were also pursued at Springhill Observatory in Canada (*Bruce McIntosh*), at Ondrejov Observatory in the Czech Republic (*Milos Simek*), by *Graham Elford* at Adelaide in southern Australia, by *Clifford Ellyet* and *Colin Keay* in New Zealand, and at the Ukranian Kharkov Polytechnical Institute in the former Soviet Union by *B. L. Kashcheyev (1920–2004)* and *V. N. Lebedinets*. The latter deployed a 36.7 MHz radar in Somalia (Mogadishu) in 1968–1970 to measure the equatorial meteoroid streams, resulting in over 6000 meteor orbits with masses $>5 \times 10^{-6}$ g ($<+12^{m}$). The Smithsonian Astrophysical Observatory also included a radar meteor orbit survey, a project driven by *Richard B. Southworth* and *Gerald S. Hawkins*,[30] from which hundreds of meteor showers were identified by *Zdenek Sekanina*.[31] These are listed in the annual stream list of Table 7. The radar surveys were plagued by big observational biases, resulting in an underestimation of the influx of fast meteoroids and an overestimation of the activity from high northern latitudes, the true extent of which was recognized only in recent years.

When Fred Whipple semi retired in 1977, the very active Smithsonian meteor group was all but disbanded. For years, meteor astronomy was a field almost exclusive to radar observers and astronomers pursuing the recovery of meteorites. In western Canada, *Ian Halliday* continued the *Meteorite Orbit and Reovery Project* (MORP), with sixty fireball cameras operated from 1971–1985, while in Europe *Zdenek Ceplecha* pursued optical observations at Ondrejov Observatory from 1951, behind the iron curtain in what was then called Czechoslovakia. In 1963, Ceplecha established the European Network of allsky fireball cameras, which included stations at the other side of the iron curtain in Austria and Germany. That network is still in operation.

12.1 My pursuit of meteor outbursts

Once an amateur himself, Ceplecha reached out to amateur meteor astronomers, including an ambitious young physics student at Leiden University called *Hans Betlem*. In 1979, Hans joined photographer *Casper ter Kuile*, visual observer *Rudolf Veltman*, and other Dutch observers to found the *Dutch Meteor Society* (DMS). Hans visited Ceplecha in 1981, and hand-carried a printout of a meteor astrometry reduction program developed at Ondrejov. The print was so bad that it was impossible to discriminate "O" from "0". In those days, computer codes such as Ceplecha's "FIRBAL" were entered by means of a huge stack of punch cards. Hans punched in those cards, succeeded in debugging the code, and for the next 24 years pursued a program of multi-station photography in the Netherlands (Fig. 12.4),[32] while editing

[30] G. S. Hawkins, The Harvard radio meteor project. *Smithsonian Contrib. Astrophys.* **7** (1963), 53–62.

[31] Z. Sekanina, Statistical model of meteor streams. III. Stream search among 19303 radio meteors. *Icarus* **18** (1973), 253–284; Z. Sekanina, Statistical model of meteor streams. IV. A study of radio streams from the synoptic year. *Icarus* **27** (1976), 265–321.

[32] H. Betlem, C. R. Ter Kuile, M. de Lignie *et al.*, Precision meteor orbits obtained by the Dutch Meteor Society – Photographic Meteor Survey (1981–1993). *Astron. Astrophys.* **128** (1998), 179–185.

Fig. 12.4 Amateur astronomers Hans Betlem (foreground) and Marc de Lignie of the Dutch Meteor Society in the process of reducing photographic negatives at Leiden Observatory.

Radiant, Journal of the Dutch Meteor Society. All of that along side a demanding full-time high-school physics teaching career.

I met Hans at the first meeting of the newly founded Dutch Meteor Society at the Kamerlingh Onnes Laboratory in Leiden in 1980, and he encouraged me to become a photographer and visual observer for the DMS small-camera program. When I thus took up meteor observing as a hobby, I noticed that professional meteor astronomy appeared to be a dead field, with few active workers from the 1950s and 1960s remaining. Still, there was much to learn. It was widely believed that meteor showers were unpredictable. A wide range of processes had been proposed to influence the dynamical evolution of meteoroid streams,[33] but no reliable forecasts of meteor shower activity were available.

I joined other members of the DMS and we set out to observe meteor showers systematically and in the following years watched this field evolve rapidly. At the same time, the amateur community became better organized. From a shifting alliance of amateur meteor organizations in the early 1980s, initially called the *Federation of European Meteor Astronomers* (with the same acronym as the U.S. Government's Federal Emergency Management Agency – pun intended), the *International Meteor Organization* (IMO) was founded at a meeting in the Belgium town of Hingene in 1986, which I attended. The IMO was an initiative by Paul Roggemans, leader of the

[33] V. A. Bronshten, *Physics of Meteoric Phenomena* (Dordrecht: Reidel, 1983), 356pp.

Fig. 12.5 Modern day meteor storm chasers. Just back from the pursuit of the 2002 Leonid storms, the researchers and the crew of the Leonid Multi-Instrument Aircraft Campaign pose at Offutt AFB (Nebraska). Photo: Eric James, NASA Ames.

Belgian *Werkgoep Meteoren* of the *Vereniging Voor Sterrenkunde* (VVS), and Juergen Rendtel, the leader of the very active East-German *Arbeitskreis Meteore*. The Dutch language journal *Werkgroepniews* became *WGN,* the Journal of the IMO. This West meets East group ended up uniting scattered observers throughout the world, later joined by the already well organized active groups. In recent years, the IMO has taken over the lead in organizing amateur meteor observers worldwide.

In the late eighties, personal computers became available and initially drew interest away from the tedious task of meteor observing. That gradually changed in the early nineties, when the internet emerged as a powerful tool of organization and information exchange, and greatly helped the education of new observers. The computer became a tool to share experiences and find support from fellow meteor observers.

I first became interested in meteor outbursts in 1985–1986, when a spate of such events were reported. In 1992, after ten years of meteor observing, I finally saw my first meteor outburst: seven bright Perseids during a "crash campaign" to northern Switzerland.

Later that year, I completed my astronomy studies at Leiden University on a thesis called *Organic matter in Interstellar Extinction*. I moved to NASA Ames Research Center as a National Research Council Associate and two years later became a research scientist with the SETI Institute in California. There, I continued the pursuit of meteor outbursts with the help of San Francisco Bay Area amateur astronomers led by San Jose Astronomical Society president *Mike Koop*, and my friends of the Dutch Meteor Society back in Europe. This resulted in successful observations of a 1994 Perseid outburst, the return of the Leonids in 1994, and the spectacular and very rewarding 1995 α-Monocerotids. Step-by-step, we became more adept at anticipating the next meteor outburst, and started traveling to have the best observing conditions. And just in time.

Typing "Leonid" on the young internet in early 1996 would give references to famous Russians called "Leonid". That all changed when satellites in orbit were perceived to be in danger of "being sandblasted" in the 1998 Leonid storm. A NASA Shuttle launch was postponed. My small ground-based observing program of meteor outbursts evolved into the four 1998–2002 missions of the NASA- and USAF-sponsored *Leonid Multi-Instrument Aircraft Campaign* (Fig. 12.5).[34] These missions encouraged and facilitated the deployment of modern instruments for the observation of meteors, many techniques for the first time.

[34] P. Jenniskens, S. J. Butow and M. Fonda, The 1999 Leonid multi-instrument aircraft campaign – an early review. *Earth, Moon, Planets* **82–83** (2000), 1–26; P. Jenniskens, The 2002 Leonid MAC airborne mission: first results. *WGN* **30** (2002), 218–224.

13
Meteor outbursts from long-period comets

I started chasing meteor storms in earnest when I realized that there is a simple method for predicting when a dust trail is in Earth's path: just wait until the planets Jupiter and Saturn are back at the same location in their orbit. In 1994, I was in hot pursuit of a mysterious class of showers that occurred with no comet in sight. Persistent anecdotal accounts reminded me of ancient days when "stars fell like rain" (Table 1). What follows is a personal account of an exciting storm chase that resulted in the first proof that these showers are due to long-period comets and, for that matter, that some long-period comets have dust trails.

13.1 Far-comet type outbursts

When I first looked into this, there was only one known cause of such outbursts: comet C/1861 G_1 (Thatcher), parent of the April Lyrid shower (Fig. 13.1). It was not a very good example, because the orbital period of ~415 yr is only a factor of two higher than the Halley-type limit of ~250 yr. As recently as 2001, Vladislav Emel'Ianenko, pointed out that the Lyrid meteoroids may get trapped in the 1:10, 1:11, and 1:12 mean-motion resonances with Jupiter, with orbital periods of 113–150 yr.[1] In his scenario, the Lyrid outbursts could be caused by a cloud of dust trapped in a shorter 60 yr orbit.

The annual Lyrid shower, peaking in the days around April 21/22, has always been my favorite. After the low rates in the cold months of February and March, this shower is the proverbial swallow of spring for observers in the northern hemisphere. The meteors radiate with a medium speed from a direction at R.A. = 272° and Decl. = +33°, in the constellation of Lyra. They appear to be relatively old, the meteoroids are not very fragile and the magnitude distribution is relatively steep at $\chi = 2.36 \pm 0.11$.[2] From the Harvard Meteor Survey data, Franco Verniani[3] measured the particle density as 0.41 ± 0.12 g/cm^3, with a median value of 0.30, the same as that

[1] V. V. Emel'Ianenko, Resonance structure of meteoroid streams. *ESA SP* **495** (2001), 43–45.

[2] A. Dubietis and R. Arlt, The Lyrids in 2003. *WGN* **31** (2003), 97–98; P. Jenniskens, Meteor stream activity. I. The annual streams. *Astron. Astrophys.* **287** (1994), 990–1013.

[3] F. Verniani, Meteor masses and luminosity. *Smithsonian Contrib. Astrophys.* **10** (1967), 181–195.

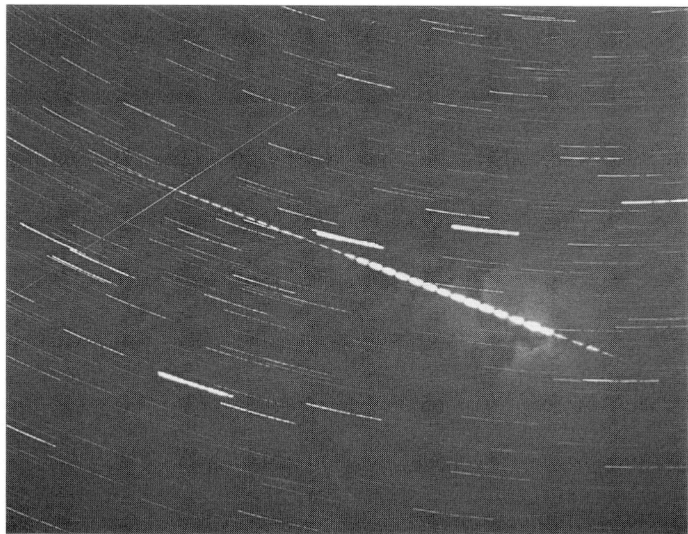

Fig. 13.1 The – 5 Lyrid fireball of 01:51:31 UT, April 22, 1996, as photographed from Biddinghuizen, the Netherlands. A satellite track is also seen in this photograph by Casper ter Kuile, DMS.

of sporadic meteoroids. In most years, the Lyrids are good for a peak ZHR = 12.8 ±0.7/h and that is enough to enjoy a great visual observing session.

In the spring of 1982, BAA observer Jonathon Shanklin found himself on a troop transport to the Falkland Islands when, in the dark southern Atlantic (28° S, 26° W), he witnessed one of his best showers ever between 06:00 and 07:10 UT, his Lyrid count in 10 min intervals was 0, 4, 2, 10, 12, 7, 0, respectively. His report was one of only a handful of this spectacular meteor outburst.

Another account of the 1982 Lyrid outburst is that of Paul Jones, then a Florida member of the *American Meteor Society* (Fig. 13.2), who wrote: "Just before midnight, I arrived at my observing site at St. Augustine Waterway. Although the sky was clear, I saw not a single Lyrid in a period of more than 40 min. My concentration lapsed and I started to fall asleep. Finally, I did see a few Lyrids. Between 05:25 and 06:25 UT, I saw 12. At the end of that first hour, I did notice an increasing rate, increasing to the point that it was obvious that something unusual was going on. Soon, short faint Lyrids flashed in all directions at an alarming rate. I estimate to have seen at least 50 between 6:20 and 6:40 UT. At a certain moment, I saw three at the same time and as many as six in one active minute! I was exalted, could hardly believe what I saw!" While the shower was still going on, Paul ran for help and recalls: "Desperate, I ran down the street to wake up Dave and Brenda Branchett, so that at least somebody would confirm this spectacle. When I returned, the rate had decreased a little, but not much. Between 6:25 and 7:25

Fig. 13.2 Current leading members of the *American Meteor Society*, the *North American Meteor Network*, and the important "meteorobs" mailing list, at the AMS Staff Spring Workshop 2000 in Talahassee, Florida. From left to right: (top) Kim Youmans, Pete Gural, Bob Lunsford, Jim Richardson (with guide dog, Ivy), Norman & Joan McLeod; (bottom) Lew Gramer, Wayne Hally, Mike Morrow, and Cathy Hall (Photo). Norman W. McLeod III of Miami, Florida, was a witness of the 1982 Lyrid outburst.

UT, I saw the astounding number of 74 Lyrids! Dave and Brenda arrived at 7:00 UT. The next hour gave a gradual decay in activity. Brenda still saw 50 while I saw 43. At 8:01 UT, we were surprised by a beautiful −4 Lyrid that illuminated the nebulosity on the eastern horizon. In the hour between 8:25 and 9:25 UT, the Lyrids returned to normal rates with a count of 23. What had happened? Why such an unexpected outburst with no apparent cause?"

Earlier Lyrid outbursts had occurred in 1803 and 1922, and the comet had returned in 1861, an obvious 60 yr pattern. Many astronomers assumed that this was because a clump of dust was orbiting in a 60 yr orbit, much less than the 415 yr it took the comet to complete one orbit.[4] Some less well documented outbursts, including a rate of ~100/h in 1945 seen in Japan,[5] led others to believe that this cloud of dust might orbit in only 12 yr. There was great difficulty in understanding how such a clump could exist. Vladimir Guth[6] thought that the meteoroids clumped in "some kinds of resonant

[4] V. Porubčan, J. Stohl and J. Svoren, On the origin of the 1982 Lyrids burst. *Contrib. Astron. Obs. Skalnate Pleso* **22** (1992), 25–31.
[5] K. Komaki, *Circ. Kii Astron. Soc.* No. 25 (1945), April 24.
[6] V. Guth, On the periodicity of the Lyrid showers. *Bull. Astron. Inst. Czechoslovakia* **1** (1947), 1–4.

nodes" along the orbit. Emel'Ianenko[7] believed that grains were trapped in the 1:5 mean-motion resonance with Jupiter, without explaining how the grains could end up there efficiently.[8] At the beginning of 1995, theoretical astronomers Terrance Arter and Iwan Williams summed up the state of affairs by stating that the most likely scenario was one of a recent breakup of Thatcher, leaving a fragment in a 60 yr orbit, which had now disintegrated.[9]

In the same manner, the Aurigid shower had two identical outbursts. In 1935, Cuno Hoffmeister and Artur Teichgraeber at the *Sonneberger Sternwarte* in Germany[10] and visual observers from the Stefanik Observatory in Prague reported an outburst of meteors in the predawn hours of the morning of September 1 during regular observations. A number of bright meteors radiated from a point in the constellation of Auriga. The German and Czech observers nicely confirmed each others' reports and, together, documented the event well. Hoffmeister's colleague Teichgraeber immediately realized that the radiant of the Aurigid stream (86.3°, +40.5°) was not far from that of comet C/1911 N_1 (Kiess) (91.3°, +39.2°), seen only in 1911 while passing close to Earth's orbit.

This scenario was hard to believe. Comet Kiess was last near the Sun about 2500 yr ago, definitely longer than the 24 yr between the return of the comet in 1911 and the meteor outburst of 1935. To confuse matters, Hoffmeister also suggested that "maybe there was a second radiant at R.A. = 86.7°, Decl. = +59.0°," some 20° from the theoretical radiant. Although no unusual activity from the area was noticed in the hours before, nor the next night, in later years other observers noticed fast meteors radiate from this general area in early September and suggested these "δ-Aurigids"[11] were somehow related to the outburst of 1935 (Fig. 13.3). Lovell, for example, did not include the association in his discussion on cosmological relationships of meteors in *Meteor Astronomy*.[12]

The confusion was considerably reduced when another outburst in 1986 confirmed activity from the radiant. The Hungarian amateur astronomer *Istvan Teplickzky* of the *Magyar Meteor- és Tüzgömbészlelö Hálózat* (MMETH) saw a flurry of bright meteors radiating from the constellation Auriga that was in many respects similar to the event in 1935.[13] These were predominantly bright meteors. There was no independent observation that year, but the fact that the Aurigids had returned was quickly accepted on account of Teplickzky's experience as a visual observer. Again, the peak of the outburst was close to the comet node. And this time, the radiant from his plots

[7] V. V. Emel'Ianenko, A resonance model of a meteoroid shower. *Kinematika I Fizika Nebesnykh Tel.* **6** (1990), 58–65; V. V. Emel'Ianenko, Dynamics of the Lyrids meteoroid shower. *Astronomicheskii Vestnik* **24** (1990), 308–313 (in Russian).

[8] V. V. Eml'Ianenko, Dynamics of the Lyrids meteoroid shower. *Astronomicheskii Vestnik* **24** (1990), 308–313 (in Russian).

[9] T. R. Arter and I. P. Williams, The April Lyrids. *Mon. Not. R. Astron. Soc.* **277** (1995), 1087–1096.

[10] A. Teichgraeber, Unerwarteter meteorstrom. *Sterne* **15** (1835), 277; V. Guth, Über den meteorstrom des kometen 1911 II (Kiess). *Astron. Nachr.* **256** (1936), 27–28.

[11] A. Dubietis and R. Arlt, Annual activity of the alpha Aurigid meteor shower as observed in 1988–2000. *WGN* **30** (2002), 22–31; A. Dubietis and R. Arlt, The current delta Aurigid meteor shower. *WGN* **30** (2002), 168–174.

[12] A. C. B. Lovell, *Meteor Astronomy* (Oxford: Clarendon Press, 1954), Ch. 21, pp. 413–434.

[13] I. Tepliczky, The maximum of the Aurigids in 1986. *WGN* **15** (1987), 28–29.

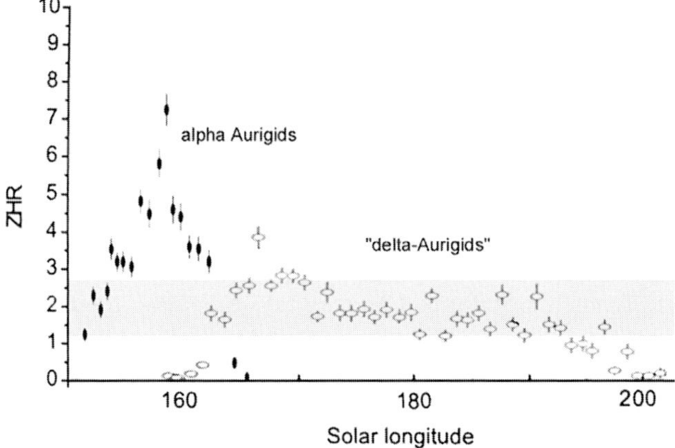

Fig. 13.3 Annual α-Aurigid activity stands out by $ZHR_{max} \sim 4$ ($B \sim 0.22$) over background activity of fast meteors from the apex. Data calculated by Audrius Dubietsis and Rainer Arlt[14] from observations collected in the IMO Visual Meteor Database. The population index of all Aurigids was measured as $\chi = 2.6 \pm 0.1$.

(R.A. = 90.5°, Decl. = +34.6°) was again close to the Kiess position (Fig. 13.4). There was no doubt now that comet Kiess was the parent.

Following the August, 1994 Perseid outburst (my second, see Chapter 17), I decided that I would monitor all nights that in recent times had produced meteor outburst activity. My first opportunity was approaching rapidly: the 1994 Aurigid return. In California, the radiant would only just be starting to rise at the time of the nodal crossing. A low radiant normally means low meteor rates, even if the ZHR is high. Hence, I settled in for a visual watch starting at around 8:00 UT, an hour after radiant rise, with the radiant somewhat higher in the sky.

Since graduating and moving to California, my visual meteor observing days were mostly in the past. The most active observers in the world now included Dutch Meteor Society's Koen Miskotte and the head of the American *Lunar and Planetary Observers – Meteor Section*, *Bob Lunsford*, and his observing partner and co-founder of N.A.M.N., the *North American Meteor Network*: *George Zay*. Bob and George live near San Diego, a thousand kilometers further south from my location. They started early, and right at around 06:55 UT, when the Aurigid radiant rose above their local horizon, there was a sudden appearance of long bright meteors from the Eastern horizon. The meteors were described as white tinkered with blue or green and leaving a smoky trail of at least 45° long. Most were of first magnitude (Fig. 13.5). With the radiant low on the horizon, these were *grazing meteors*, entering Earth's atmosphere above the observing site at a very shallow angle. They traveled long distances before penetrating deep in the atmosphere and seemed to move slowly and last a long time,

[14] A. Dubietis and R. Arlt, Annual activity of the alpha Aurigid meteor shower as observed in 1988–2000. *WGN* **30** (2002), 22–31; A. Dubietis and R. Arlt, The current delta Aurigid meteor shower. *WGN* **30** (2002), 168–174.

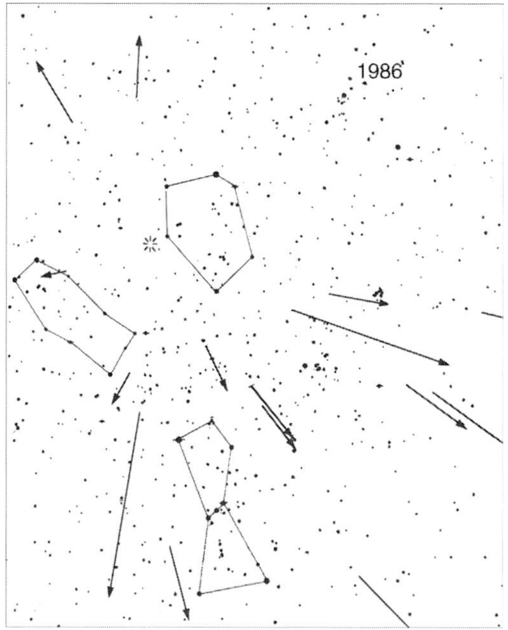

Fig. 13.4 Plotted meteor tracks during 1986 Aurigid outburst point away from the radiant of comet C/1911 N_1 (Kiess). Diagram by Istvan Teplickzky, MMETH.

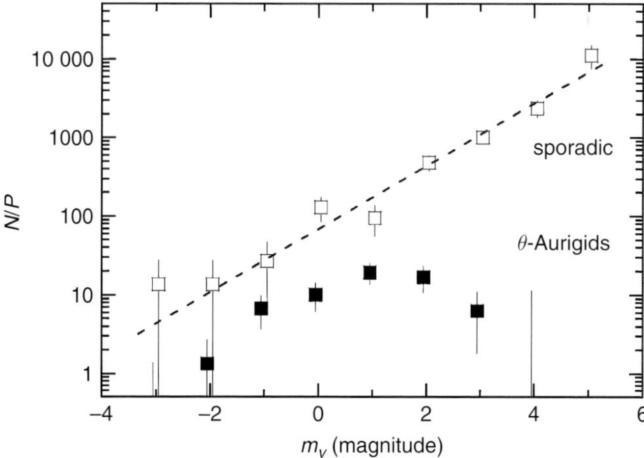

Fig. 13.5 The magnitude distribution of the 1994 Aurigids outburst after correction for detection efficiency per unit area (*P*) by Bob Lunsford and George Zay.

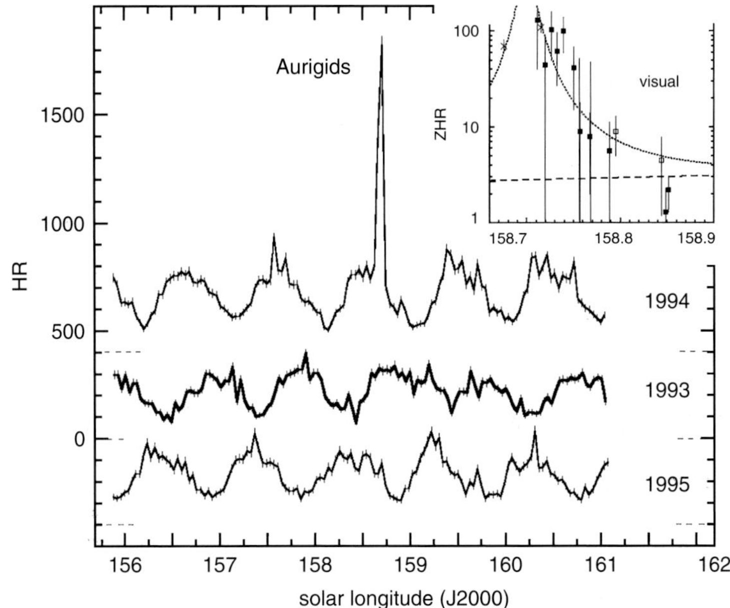

Fig. 13.6 The 1994 Aurigid shower outburst detected in radio-forward-MS by Ilkka Yrjölä.

some up to 2 seconds. With a climbing radiant, the rate went down, and when I finally set myself to start the meteor observing further north near San Francisco, the event was over.[15] I missed the outburst that year by 20 minutes.

That was in the beginning of September. The Leonids came in November that year, and shortly thereafter I met *Ilkka Yrjölä*, a Finnish radio amateur from Kuusankoski. He used his narrow-band radio receiver to listen for distant commercial TV and radio stations, the signal of which would bounce off ionized meteor tracks. By counting those reflections, it is possible to monitor the meteor activity on the sky. When Ilkka's counts of meteor reflections arrived, I was so excited. Until that point there had been no independent confirmation of Bob and George's observations. Downloading the file took long minutes. But there it was: a whopping peak in the counts on the day of September 1 (Fig. 13.6).

The first thing to check was whether the time corresponded to Bob and George's observing interval. The time matched. The counts were high in two consecutive hours, at 06:30 and 07:30 UT. Only the second hour had been observed by the two visual observers.

[15] P. Jenniskens, Meteor stream activity. IV. Meteor outbursts and the reflex motion of the Sun. *Astron. Astrophys.* **317** (1997), 953–961.

13.2 The Sun's reflex motion

In July of 1994, big chunks of comet Shoemaker–Levy 9 impacted Jupiter and comet impacts were no longer an academic exercise. It hit me that meteor showers can serve as an early warning for the impact of long-period comets, or at least betray the presence of the comet from its dust trail. It turned out that NASA was willing to fund research into why "stars fell like rain" and in the beginning of 1995, I set out to prove that such outbursts of meteors with no comet in sight were due to long period comets.

Where to start? None had been observed before in an organized manner and none had been predicted. The 1994 return of the Aurigids provided an important clue: there was no simple periodic pattern in the years of return: 1935, 1986, and 1994. I focused on the idea that dust trails may on occasion wander into Earth's orbit by planetary perturbations and examined the position of Jupiter and Saturn at the time of the showers. I discovered that each time the Sun was displaced from the barycenter of the solar system by about the same amount (Fig. 13.7).

The solar system *barycenter* is the center of mass, the point in space where the Greek god "Atlas" would have kept his finger to balance the weight of both Sun and planets. This point is located inside of the Sun, but displaced from its center. In a manner of speaking, the Sun always balances the planets. The resulting motion of the star has been used successfully to discover planets around other stars.

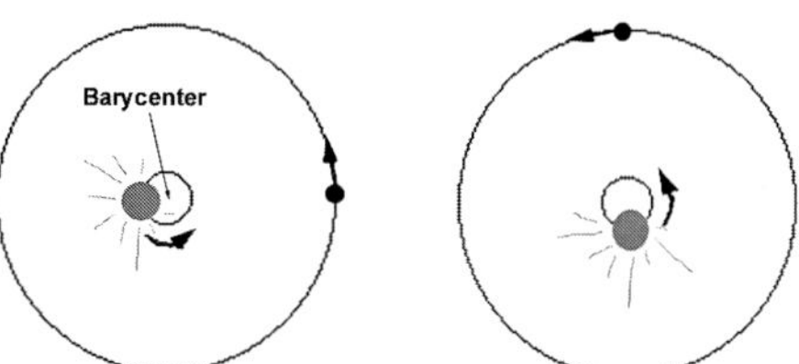

Fig. 13.7 The Sun's reflex motion. This shows the motion of a star in response to that of its planet.

The **Sun's reflex motion**: the displacement of the Sun from the barycenter of the solar system is given by:

$$\Delta r = -\Sigma_i \, (m_i/(M_\odot + m_i))r_i \qquad (13.1)$$

where the summation is over the nine planets, m_i is the mass of the planet, $M_\odot$ is the mass of the Sun, and r_i is the heliocentric distance of planet i at the time of nodal passage. The amplitude of this motion is: 0.0101 AU. As much as ± 0.00496 AU of that displacement is due to Jupiter alone, another

±0.002 72 AU from Saturn, Uranus contributes ±0.000 84 AU and Neptune ±0.001 55 AU. Earth has a negligible 3×10^{-6} AU. An easy way to calculate the Sun's reflex motion, is to use a planetarium program to find the year (y_i) when the giant planets (Jupiter: $i = 5$, Saturn $i = 6, \ldots$), are at the same solar longitude as Earth during the shower, and then calculate the perpendicular displacement (Δr) and parallel displacement ($\Delta\Omega$) for that point in time:

$$\begin{aligned} \Delta r\ (\mathrm{AU}) &= -0.004\,96\ \cos(360^\circ(y - y_5)/11.863) \\ &\quad -0.002\,72\ \cos(360^\circ(y - y_6)/29.447) \\ &\quad -0.000\,84\ \cos(360^\circ(y - y_7)/84.017) \\ &\quad -0.001\,55\ \cos(360^\circ(y - y_8)/164.791) \end{aligned} \tag{13.2}$$

$\Delta\Omega$ (AU) follows by replacing *cosine* by *sine*. This gives the displacement of the trail perpendicular and parallel to the Earth's orbit relative to an unknown central position, which can be determined approximately from past outbursts, when the trail was known to be in Earth's path. Keep in mind that the solar longitude is the direction to the Sun's center, not to the barycenter. The difference can amount to $\pm 0.6^\circ$.

13.3 No luck with the 1995 Lyrids and κ-Pavonids

My new hypothesis was that the Sun's reflex motion is a good mimick of the motion of the trail near Earth's orbit.[16] The maximum amplitude, when all planets are lined up on one side of the Sun, is close to the miss-distance of comet and Earth's orbits in the case of comets Thatcher and Kiess. I discovered that the 1-revolution dust-trail for comet 109P/Swift–Tuttle (Fig. 11.3), calculated by Wu and Williams in a failed attempt to explain the 1992–1993 Perseid outbursts, moved in and out of Earth's orbit in a pattern with the same amplitude and in a manner following the Sun's reflex motion (Chapter 17).

I set out to prove this hypothesis by attempting to observe far-comet type outbursts when the Sun's reflex motion was similar to that of past outbursts and the dust trails should again be in Earth's path. The project did not start very hopefully. Earlier, Kresák[17] had predicted that the Lyrids could produce a meteor storm over the North Atlantic in 1994 (around 4.8 UT, $\lambda_\odot = 31.899^\circ$), based on the earlier observation by Guth that high rates seemed to have a 12 yr periodicity: on April 22, 1945 between 18:00 and 19:07 UT ($\lambda_\odot = 32.053^\circ$), *Kojiro Komaki* of Japan had seen 103 Lyrids and

[16] *Ibid.*
[17] L. Kresák, Meteor storms. In *Meteoroids and Their Parent Bodies*, ed. J. Stohl and I. P. Williams. (Bratislava: Astronomical Institute, Slovak Acad. Sci., 1992), pp. 147–156.

Fig. 13.8 The Lyrid campaign on Big Island, Hawai'i, April, 1995. Members of the AMS affiliated *Hawaiian Meteor Society* led by Mike Morrow, included Steven O'Meara (right, with camera platform) and Paul Sears (in background).

nine sporadics in 67 min of very clear skies, with magnitudes (-4^m up): 2, 4, 7, 6, 13, 9, 19, 12, 10, 5 ($\chi = 1.4$).[18] Kresák's prediction did not come true. The Sun's reflex position in 1995 was still similar to that during past Lyrid outbursts, and I traveled to the big island Hawai'i with two sets of mobile camera platforms (Fig. 13.8) to position myself in the path of the dust with a radiant high in the sky at night time. Cameras at the remote site were operated by members of the *Hawaiian Meteor Society* led by Mike Morrow. Sadly, there was no Lyrid outburst that year.

There were very few chances left during the brief duration of my NASA project. The remaining showers were the α-Monocerotids, with an expected outburst in November, and two little-known southern hemisphere showers that had been observed only once before. The 1985 β-Hydrusids were to reappear during the full Moon. The 1986 κ-Pavonids were expected to reappear in the evening of July 17, 1995 in a moon-less sky. I set out in the footsteps of Cuno Hoffmeister and traveled to South Africa.

Back in 1986, Jeff Wood had written: *While carrying out a routine meteor watch on the evening of July 17, 1986, N.A.P.O. – M. S. observers Neil Ingwood and Paul Stacey started noticing a number of bright yellow–orange meteors radiating out from a point near the star kappa-Pavonis. The meteor shower was of a very short duration starting at 11:50 UT and finishing by 13:00. During this time Neil and Paul saw 26 and 30 kappa-Pavonids, respectively.*[19]

Members of the *Astronomical Society of South Africa* (ASSA) – Meteor Section, led by *Tim Cooper*, stepped in to operate the camera platforms at two sites. South Africa had just passed a difficult political transition, and smoke from grass fires from the traditional clearing of old growth reminded us, instead, of an uneasy truce. After days of haze and clouds in the Transvaal, the sky cleared and in the late afternoon of July 16,

[18] K. Kojiro, *Nippon Meteor Soc. Astron. Bull.* No. 25 (April 24, 1945).
[19] J. Wood, The kappa-Pavonid meteor stream in 1986. *WGN* **14** (1987), 129–130.

Fig. 13.9 Members of ASSA in Pretoria inspecting one of the camera platforms of the mobile station. Left is Tim Cooper.

Tim and I met the other observers at Pretoria observatory for a short introduction (Fig. 13.9). We split into two groups, with one team headed for the tracking station at Hartebeesthoek, west of Pretoria. I was guest at a farm east of Pretoria, where I thought myself back in the colonial era. The farmhouse had white painted walls, a reed roof, and small cozy rooms, surrounded by a lush garden. When we entered, a large table was covered with food . . .

After a practice night that saw many unforeseen problems, the observers were ready to face the hoped-for meteor shower. The sky was clear and the center of our galaxy was above us, as were the bright stars of the southern cross. Our host brought soup and cakes. Meteors were routinely logged. Any minute now, nature was expected to open the curtain on the κ-Pavonid shower . . .

No such meteors were seen. That night in the Transvaal, the κ-Pavonid activity was ZHR$<$0.5/h. An observing campaign in Brazil the next year gave a handful of tentative κ-Pavonids. I now suspect that this shower was not caused by a long-period comet.

13.4 The night it rained stars

The key to solving the puzzle was another such shower, the α-Monocerotids of November 22, which seemed to peak every ten years: in 1925, 1935, and 1985, each observed by few observers. The accounts from different years were very similar: all reports mentioned an activity profile lasting less than half an hour! Rates peaked above ZHR = 1000/h (a storm!) and all showers lacked bright meteors (Fig. 13.10).

The α-Monocerotids had not returned in 1994, and the Sun's reflex motion in 1995 was similar to that during the prior sighting of the shower in 1985 (Fig. 13.11). The Moon would be new and the complete period of opportunity between 0:00 and 6:00 UT could be observed from a single site in Western Europe. Kresák[20] had high hopes for this stream as well, but based on an assumed 10 yr periodicity. He put the peak time around 0:00 UT at the node of prior returns. Time to call home and involve my friends.

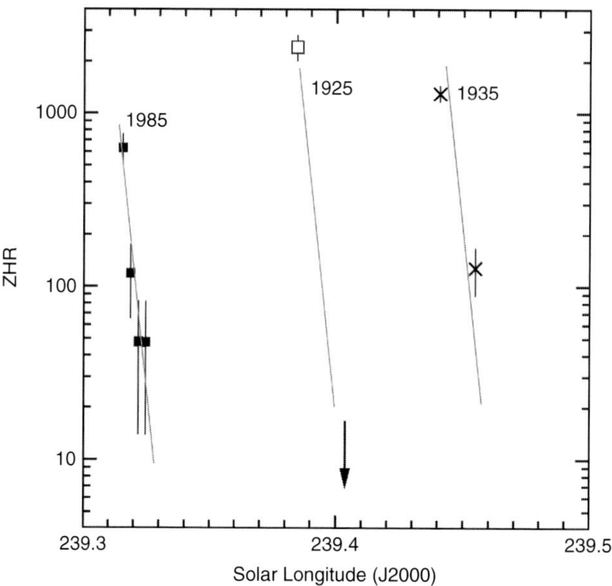

Fig. 13.10 The α-Monocerotid activity curves during their returns in 1925, 1935, and 1985.

[20] L. Kresák, Meteor storms. In *Meteoroids and Their Parent Bodies*, ed. J. Stohl and I. P. Williams (Bratislava: Astronomical Institute, Slovak Acad. Sci., 1992), pp. 147–156.

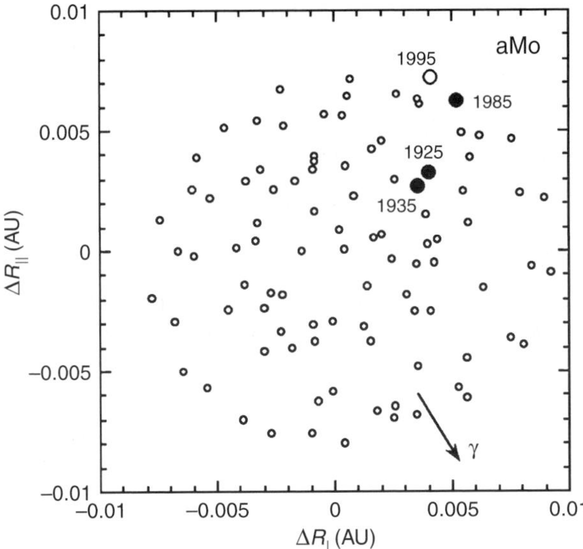

Fig. 13.11 The position of α-Monocerotid shower dust trail in a diagram of the Sun's reflex motion at that time.

Hans Betlem was way ahead of me. As soon as he learned I had seen the return of the Leonids in 1994 (Chapter 14), he made plans to observe the 1995 Leonids from abroad. Clouds would surely spoil the view in the Netherlands. His choice fell on Andalucia in southern Spain. The observing sites were chosen at Almedinilla, north-west of Granada, at Zaffaraya to the south, and at Alcudia-de-Guadix to the east, in an arid area north of the Sierra Nevada. Further support came from Spanish observers of S.O.M.Y.C.E., the *Sociedad de Observadores de Meteores y Cometas Españoles*, under the guidance of Luis Bellot Rubio and Josep Trigo-Rodríguez, who would gather at Chirivel, north-east of Alcudia-de-Guadix, at a suitable distance for multi-station meteor photography.

The α-Monocerotids were due on November 22, 4 d after the Leonids. It took the advocacy of Marc de Lignie and Marco Langbroek to convince Hans to extend the observing interval to include the night of November 22. It would force the whole team to stay in Spain for four more days. Observers included students from Hans' high school (Fig. 13.12). I arranged to join the DMS observers in the days just after the return of the Leonid shower, which Mike Koop and I observed from California for better coverage.

Arriving in Alcudia-de-Gaudix in the company of my girlfriend Charlie Hasselbach in the evening of November 20, I found the DMS observers dispirited by days of bad weather. Only at the peak of the Leonids had the clouds let up for a few hours. As expected (Chapter 14), but against better hopes, Leonid rates had been elevated but not to the level of the 1994 outburst.

Fig. 13.12 The Dutch Meteor Society team that observed the 1995 α-Monocerotid outburst. From left to right: Casper ter Kuile, Hans Betlem, Marco Langbroek, Jos Nijland, Vera Pijl, Petrina van Tongeren, Engelien Geerdink, Guus Docters van Leeuwen, Olga van Mil, Hans Klück, Iris Ooms, Annemarie Zoete, Wendy van Mil, Ruud de Voogt, Fieke Mol, Marc de Lignie, Michiel van Vliet, Jaap van 't Leven, Klaas Jobse, and Koen Miskotte. Photo: Hans Betlem/DMS.

During the evening of our arrival, the skies cleared for the first time and in twilight under a clear sky we talked about the pending observations the next night. The rented house at Alcudia-de-Gaudix was carved into local rock and only the facade was of masonry. The rooms were small as in a monastery but pleasantly warm in the cold fall nights. The DMS photographers had set up their equipment in the yard, while the visual observers had found themselves a nice spot on top of the hill above the house. Hopes were high, but few were confident that the predicted outburst was worth the wait. Luis visited us in the late afternoon of November 21, to pick up a camera battery. He was plainly disbelieving. *There is no chance at all,* he said, *that we will see an α-Monocerotid outburst. But, then again, I would hate to miss it.*

To make sure that others were aware of the imminent return of the α-Monocerotids, I had published a small paper in *WGN, the Journal of IMO* in June of 1995:[21] *This year is one of the best opportunities to try and observe an outburst of alpha-Monocerotids,*

[21] P. Jenniskens, Good prospects for alpha-Monocerotid outburst in 1995. *WGN* **23** (1995), 84–86.

because of a near New Moon, a favorable radiant position for Europe, and a similar relative position of the major planets. In August, I met various professional meteor astronomers at a conference in Florida and was able to convince astronomer Milos Simek at Ondrejov Observatory to continue his radar observations a few days beyond the Leonid maximum. Colleagues Jiri Borovicka and Pavel Spurný were also aware.

I had set my mind on establishing a fifth observing site at Calar Alto Observatory, because past accounts of outbursts had very few meteors brighter than zero magnitude. I suspected that the shower had a large magnitude distribution index: few bright meteors and many faint ones. More photographic sites would increase our chances of measuring a precise orbit. Intensified TV camera systems, first pioneered in DMS circles by Klaas Jobse in 1987,[22] would obtain orbits for fainter $+2$ to $+6^{m}$ meteors, albeit not as precisely. The visual observers would be able to pick up meteors over a wide magnitude range and would provide rate measurements and the times of the bright meteors. From the past accounts of α-Monocerotids, we also knew that the event would be very brief. The peak rate was expected to fall back to 14% of that level in only 6 min. This demanded a large team of observers to gather precise enough counts. All observers were urged to keep a count in one-minute intervals. The period of opportunity would start at around 11 p.m., with the rising of the α-Monocerotid radiant.

Early in the evening, the Alcudia team dined in a restaurant located somewhat on our way to Calar Alto. Dinner was good, conversation was light. Expectations were high, in spite of all attempts to laugh those away. I was surely going to be tarred and feathered if the shower did not show. After dinner, we said good-bye and went on our way. Marco and Charlie joined me at Calar Alto to give visual support. When we finally arrived at the 2160 m top, it was brilliantly clear with only a few clouds on the horizon. The stars were many. We found an empty parking lot near the entrance to the observatory and chose that as our observing site. The camera platforms were installed and I sat down to accept whatever would happen. The temperature had dropped to just below freezing and it was uncomfortably windy. We wrapped ourselves in blankets and tried to keep warm while scanning the sky for meteors.

All went routinely. We were not too certain about the radiant position of the α-Monocerotids. Periodically, sporadic meteors would get us all excited. I got my hopes up a few times and quickly started a new exposure in order not to need to do that during the outburst. Each time in vain. Then, at 00:50 UT, clouds came in. A lenticular cloud, it seemed, as it sat at rest over our head in the strong mountain wind. Thick enough to lower the limiting magnitude by at least one magnitude. Any moment, we expected the clouds to thicken and spoil the view completely. Faint meteors could now no longer be observed.

Suddenly, around 01:10 UT, three meteors radiated from a point on the border of the constellations of Monoceros and Canis Minor, 15° away from the position given in past accounts. And this time it did not stop after just a few. Meteors started pouring

[22] K. Jobse and M. C. de Lignie, Een TV-systeem voor meteoren. *Radiant, J. DMS* **9** (1987), 38–41.

Fig. 13.13 The α-Monocerotid outburst on November 22, 1995, 01:25–01:50 UT. Photo by Hans Betlem at station Almedinilla, Dutch Meteor Society.

out of Canis Minor, falling left and right, up and down. Bright meteors too, many of magnitude zero, laughing through the whimsical deck of clouds. The magnitude distribution index was not high at all.

While Marco and Charlie were counting, I decided to change my observing strategy. Nobody was marking the exact location of all those bright meteors, some of which were bright enough to be photographed. I took my star charts and pencil and did the good old plotting work, thinking of the Chinese astronomers who had reported for so long: *stars fell like rain at midnight*. Stars do fall at night. You just have to know when and where.

> And meteors fright the fixed stars of heaven;
> The pale faced moon looks bloody on the earth . . .
> These signs forerun the death or fall of Kings.
>
> – William Shakespeare, Richard II, Act 2, Scene IV

By 01:29 UT, meteors were falling at a rate of five or more per minute (Fig. 13.13). By 01:40 the cirrus started to dissipate, but the meteor rate also decreased. Twenty minutes later it was all over. We had seen the "signs" of lore.

Emotions were strong when the stream peaked that night on Calar Alto, and we broke down in tears. Out of happiness, or perhaps out of frustration because of the clouds. As at other stations, we allowed the weather to clear before ending the last exposure so that the end points of the star trails were well defined. We were cold. It was time to go home. Wow!

When we arrived back in Alcudia an hour later, it was still deep in the night. At the gate, Casper put up his thumbs. An enormous burden fell off my shoulders. They had seen it too! That meant the multi-station effort had a chance of success. Photographer Robert Haas was all smiles! He had captured one of the α-Monocerotids on color film. After sharing our experiences, we continued observing until morning dawn, charting the annual α-Monocerotid activity. When twilight came and the final stars of the night of November 22 faded, it was time to open the bottle of champagne.

While the Sun put the sleepy town of Guadix in a warm morning light, a group of excited Dutchmen were gathered around a telephone booth. Robert and Marco were delegated to be our spokespersons with the task to brief the world of what had happened, and they brought the news live on breakfast radio.

When we all met at Malaga airport the next day on our way home, there was not just one, but twenty radiant faces. I asked all the observers to plot on a star chart the radiant position of the α-Monocerotids as they recalled it from visual observations. Faces beaming with pride, the high-school students and observers from all walks of life signed off on an activity that went against common beliefs about what one ought to be doing in the middle of the week in the middle of November. The person that marked the radiant nearest to the photographic position was Marc de Lignie (Fig. 13.14). The radiant position (and time of the peak) is the most important piece of information for predicting the future return of the shower. The map shows that even experienced visual observers have some difficulty securing this data.

The meteoroids moved in almost the same orbit, as in a trail of dust. The recurrence of the shower at the predicted epoch and the duration of the event left no doubt: this was a dust trail that was steered into Earth's path by the planets.

After reducing the orbits of photographed and filmed meteors, we measured that the α-Monocerotid orbits were of long orbital period ($P > 149$ yr). A ten-year period could be completely ruled out! A great success.

The outburst peaked at $ZHR = 500/h$, perhaps the strongest meteor shower since 1966, lasting only 40 min.[23] From the counts, we found that the meteor shower profile was symmetric, with exponential slopes and a rounded top, well matched by a Lorentz profile with $FWHM = 0.090 \pm 0.005^\circ$ with an annual shower background of about ZHR $\sim$5/h (Fig. 13.15). To our surprise, there was not much difference in the dispersion of faint and bright meteors.

Most surprising was that the magnitude distribution was dominated by large meteoroids, but a cut-off at about -1^m created the impression that the shower was poor in bright meteors (Fig. 13.16). That is why past observers had spoken of relatively few bright meteors. Interestingly, the lenticular cloud over our site could not have tempered the rates by much. Other observers reported similar rates to those that we saw from Calar Alto Observatory.

[23] P. Jenniskens, H. Betlem, M. C. de Lignie and M. Langbroek, The detection of a dust trail in the orbit of an Earth-threatening long-period comet. *Astrophys. J.* **479** (1997), 441–447.

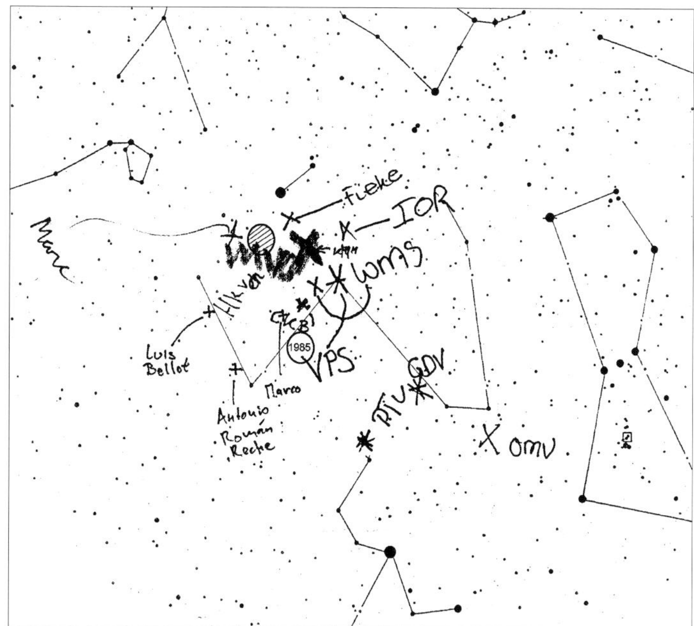

Fig. 13.14 Signing off on a very successful campaign: experienced visual observers of the DMS plotted the radiant position from memory one day after the outburst. The open circle is the position reported in the past. The shaded circle is the measured position from multistation photography. Orion is shown for scale.

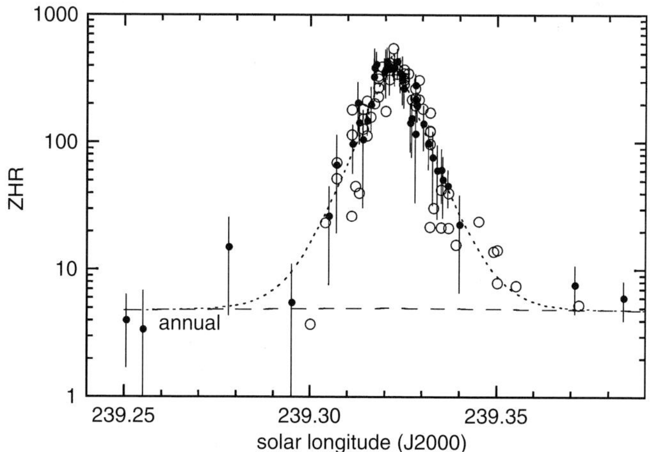

Fig. 13.15 The activity curve of the 1995 α-Monocerotid dust trail crossing.

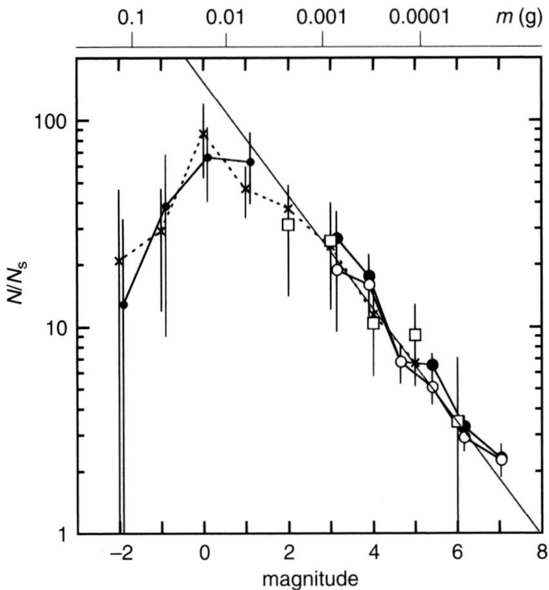

Fig. 13.16 The magnitude distribution.

Initially, I thought that the large grains had been ejected in hyperbolic orbits because of radiation pressure, but the example of Fig. 13.17 shows that most meteoroids are ejected in bound orbits even if the semimajor axis is 50 AU ($P = 353$ yr). This trail seen 1.5 orbits after ejection at aphelion also shows the perturbations induced by Jupiter (broad) and Earth (narrow spikes).

We now understand that the peculiar magnitude distribution comes on account of the larger grains not being able to make it all the way out to where Earth encounters the meteoroids. This could help determine where the comet is in its orbit.

From the multi-station orbits, we found also that the α-Monocerotid meteoroids themselves are peculiar. They penetrated 5 km deeper into the atmosphere than Perseids and Orionids of similar brightness and speed, suggesting that these grains had a higher density. They lacked the flares common for Perseids and Quadrantids.

At the same time at the Ondrejov Observatory, astronomers *Jiri Borovicka* and *Pavel Spurný* had observed the outburst comfortably from behind their office window and succeeded in capturing five low-resolution optical spectra. All lacked the element sodium. Sodium is normally associated with the most volatile minerals in cometary matter and could have been lost in a variety of ways, including by heating or by radiation. The lack of sodium emission in the Ondrejov spectra suggested to me that these grains were perhaps representative of the pristine cometary crust of the long-period parent comet. The α-Monocerotids do not come close enough to the Sun to be

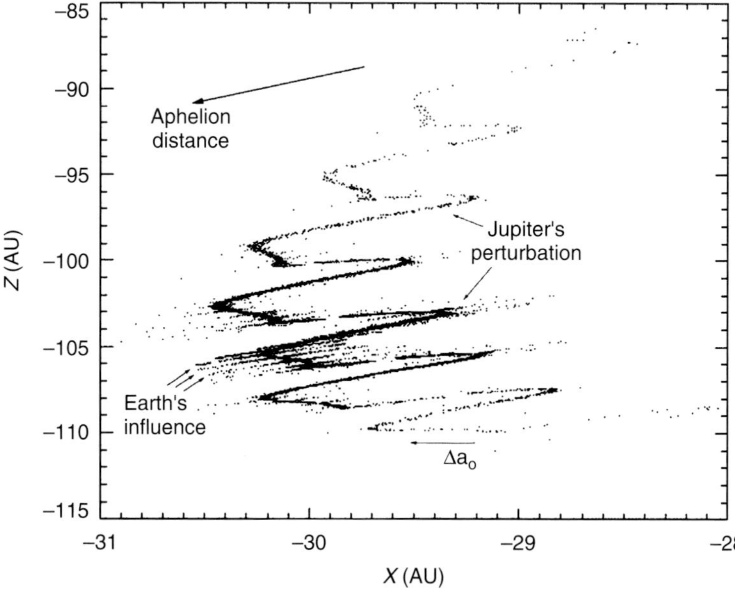

Fig. 13.17 A one-revolution dust trail of a long-period comet back at aphelion. The case of an $a = 50$ AU comet in the orbit of C/1854 L_1 (Klinkerfues). Calculations by Vaubaillon.

significantly heated and they have completed only one orbit since there were ejected (see below).

Several observers reported the event to the *Minor Planet Center* of the Smithsonian Institution of Astrophysics, which is a clearing house for meteor outburst announcements. Brian Marsden prepared an IAU Circular informing the community about what had happened. He sent his congratulations with the following message: *I have been meaning to congratulate you (for the past 24 hours, which have been very busy here) on the great success of your prediction for the alpha Monocerotid meteors. I presume you have seen IAUC 6265 by now. I know how you must be feeling, for it was, after all, rather a long shot, and I know that you had difficulty convincing everybody (including yourself) that it might really occur.*

In hindsight, it is not surprising that long-period comets may show coherent dust trails and those trails wave around in a predictable manner reflecting the motion of the main planets. It does not matter how long the orbital period of the comet is for predicting the return of a long-period comet dust trail. The perturbations by the planets of different parts of a trail occur only on the inward leg after the dust has spread due to different orbital periods (Fig. 3.5).

I teamed with Finnish amateur astronomer *Esko Lyytinen*, who had developed an early dust trail model to explain the Leonid storms, to verify this. Fig. 13.18 shows the calculated miss-distance from Earth's orbit for particles ejected in the previous return

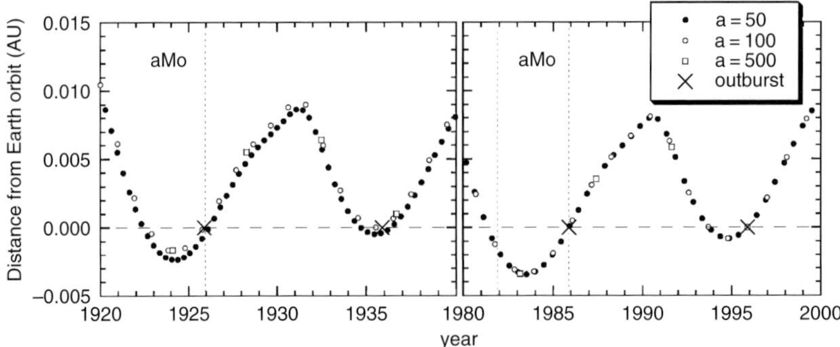

Fig. 13.18 The position of the α-Monocerotid shower dust trail near Earth's orbit (dashed line) for different semimajor axes (in AU). Observed showers are marked by a cross. The distance from Earth's orbit is the difference of the heliocentric distance of Earth's orbit and the node: $r_E - r_C$.

of the comet. Crosses mark the time of observed meteor showers. Each symbol represents an ejected particle returning at slightly different times. It makes no difference for the position of the trail whether the semimajor axis of the initial comet orbit was 50, 100, or 500 AU.

Esko was also able to confirm my hypothesis that dust trails follow the Sun's reflex motion. Fig. 13.19 shows the displacement of the trail from Earth's orbit (solid line) and the Sun's reflex motion (dashed line). There is a small phase lag of about 1.5 yr, corresponding approximately to the time taken since the particle was near Jupiter's orbit.

We were also able to show that these outbursts are caused only by one-revolution old dust trails. After a second revolution, the trail is immediately widely dispersed because of the induced changes in the orbital period. As a result, parts of the trail catch up on each other and a broad distribution of dust results that has a nodal dispersion of about ±0.0101 AU (±0.59 d) wide. This has traditionally been called the "Filament" of the shower (Chapters 14 and 17). Long-period comets differ from Halley-types, because a "Filament" already forms after one revolution (Fig. 13.20).

All this demonstrates that the return of a shower can be predicted even if the comet orbital period is not known. Esko Lyytinen and I have since examined the dust trails in the orbit of all known long-period comets that pass close to Earth's orbit and all historic accounts of meteor showers that may have been due to unknown long-period comets (Table 3).[24] The predictions critically depend on the accuracy of the radiant determination. A slightly different position can lead to a closer encounter with Jupiter.

The most promising return will be the 2007 Aurigids, the encounter conditions of which were calculated from a full planetary perturbation calculation and are

[24] E. Lyytinen and P. Jenniskens, Meteor outbursts from long-period comet dust trails. *Icarus* **32** (2003), 51–53.

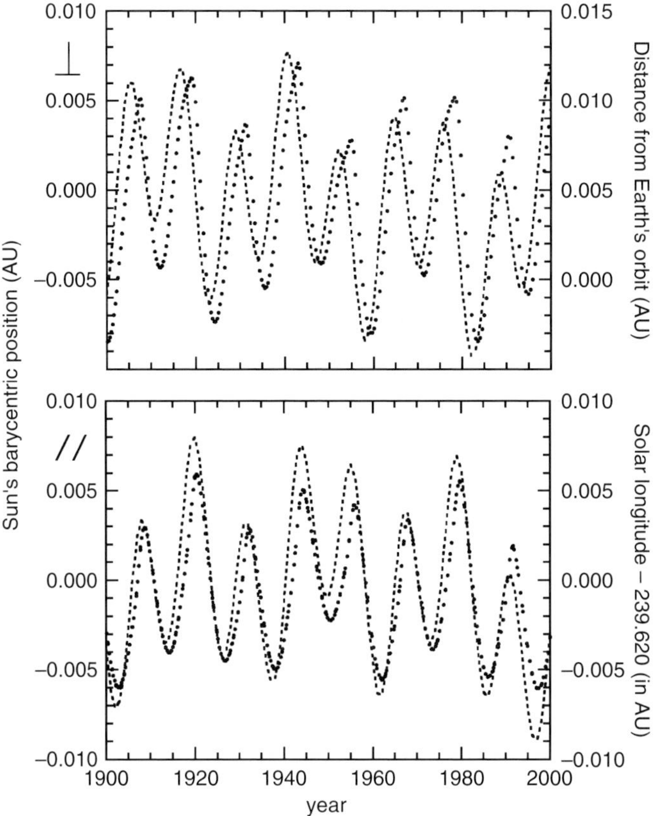

Fig. 13.19 A comparison of the Sun's reflex motion (dashed line) and the motion of the α-Monocerotid dust trail (dots).[25]

summarized in Figure 13.21. Note how in 2007 the dust trail is expected to be at a similar distance from Earth's orbit as during prior Aurigid outbursts. This shower can again be rich in bright meteors, with high peak rates. With a peak at 11:37 UT, California is well positioned this time!

Sample of other potential long-period comet dust trails

β-Leonis Minorids

December 5, 1921, about 19:15–20:10 UT ($\lambda_{\odot} = 254.26$). Astronomer *Shiro Inouye* of Tokyo Observatory observed 56 meteors (Fig. 13.22), of which 44 had a common radiant at R.A. = +160°, Decl. = +37° and the following distribution

[25] *Ibid.*

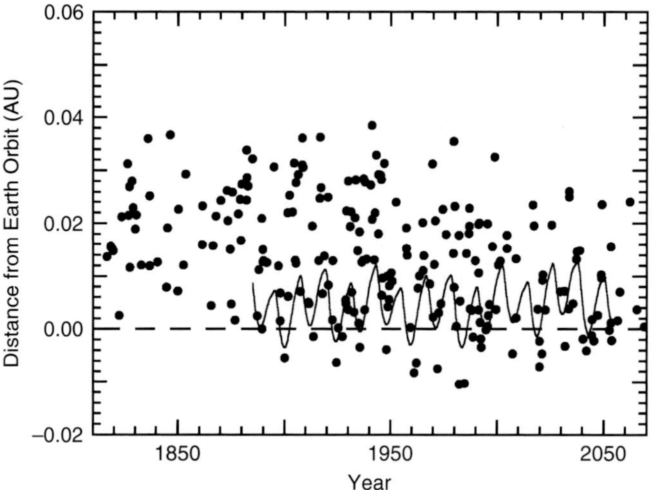

Fig. 13.20 A dust trail of a long-period comet after one (line) and two revolutions (dots).

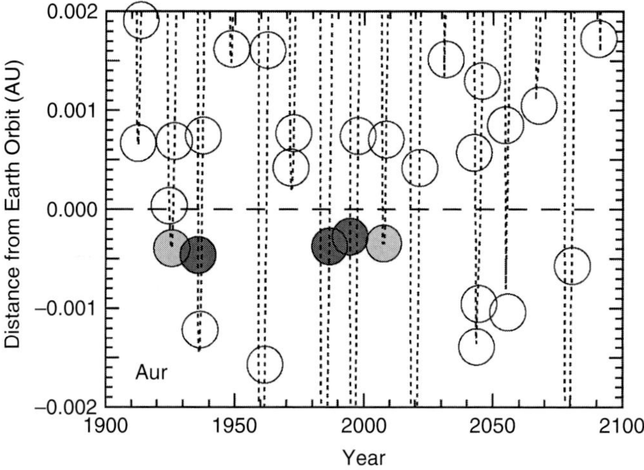

Fig. 13.21 The position of the α-Aurigid stream near Earth's orbit in the past and the future.

of magnitudes from $+2.0^{\mathrm{m}}$ up in steps of 0.5^{m}: 2 ($+2.0^{\mathrm{m}}$), 2, 6, 10, 16, and 8 ($+4.5^{\mathrm{m}}$).[26]

γ-Delphinids

June 11, 1930 ($\lambda_{\odot} = 80.42$). A total of 51 γ-Delphinids were observed between 02:15 and 02:45 UT, and none in the period 02:45 to 04:00 UT, by *Paul S. Watson*,

[26] I. Shiro, *Mon. Rep. Tokyo Obs.* **15** (1922), 12; S. Kanda, A new meteoric shower. *Kyoto Bull.* **1** (1922), 2, 3.

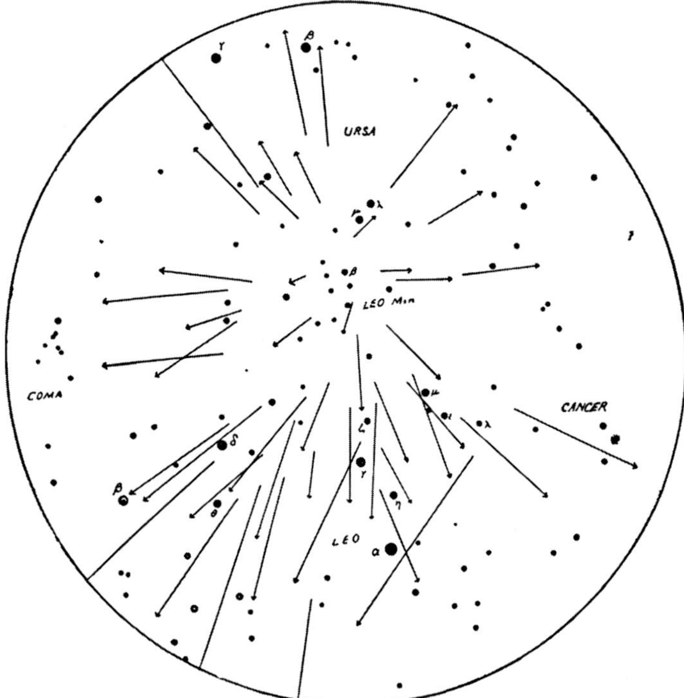

Fig. 13.22 Plot of the 1921 β-Leonis Minorids by Shiro Inouye.

Frank Oertle, and *Joseph Field* near Baltimore, Maryland.[27] The radiant was near R.A. = 312°, Decl. = +17°. The meteors had short trails and brief duration and where yellowish white, 39 were of magnitude +1, 10 of magnitude +0. The radiant elevation was only 12°, with a full Moon in the sky, but *the sky was one of the most transparent we have had at Flower Observatory*, wrote C. P. Olivier. He himself observed no γ-Delphinids between 02:45 and 03:15 UT.

β-Perseids

August 7, 1935 ($\lambda_\odot = 135.358$). *S. Holm* from Silkeborg, Denmark a student at the Polytechnikum in Copenhagen, reported seeing a large number of very faint 5–6th magnitude meteors with only 6–8 meteors within each 5 min count brighter than $+4^{\mathrm{m}}$. The radiant was determined some time after the observations at R.A. = 52°, Decl. = +40° (B1935?). He counted 548 such meteors in a 5 min interval centered at 22:15 UT, others between $\lambda_\odot = 135.31$ and 135.43. An

[27] K. Simmons, *Meteor News* **51** (1980), 2–3.

observer 35 km from Silkeborg, at Brabrand near Aaarhus, noted *a large intensity of weak meteors at 22 h UT, but not at all as many as reported by Holm.*[28]

γ-Cetids

October 19, 1935, ~19 h UT ($\lambda_{\odot} \sim 206.36$) rates in excess of 100/h from R.A. = 40°, Decl. = −05° observed by *M. Wickom* (?). Short duration shower. Medium fast meteors, yellow in color, many leaving trains.[29]

ω-Orionids

November 25, 1964 ($\lambda_{\odot} = 244.12$). In a period of 10–15 min around 20:30 UT, 25 meteors of magnitude +0 to +1 were observed to radiate from the constellation of Orion by B. Warner at Radcliffe Observatory, Pretoria. The radiant was approximately at R.A. = 85°, Decl. = +04° (B1950). Some meteors left trails for more than a second. *Later during the night, the shower was still in action.*[30]

June Lyrids

June 15, 1966, evening ($\lambda_{\odot} \sim 85.12°$). American Meteor Society observer *Stan Dvorak* in California plotted 16 meteors during a 1.5 h watch from the San Bernardino mountains, of which 13 originated from Lyra at R.A. = 278°, Decl. = +30° (B1950). A few hours later, *F. W. Talbot* in Cheshire, U.K., independently discovered the radiant and placed it at R.A. = 275.5°, Decl. = +30°, with an hourly rate of 9/h.[31] More recently, the shower may have been observed again at a similar rate on June 15, 1996 by *Marco Langbroek* of the Dutch Meteor Society during routine visual observations (Fig. 13.23). The highest rates occurred at 23:35 UT ($\lambda_{\odot} = 85.167°$), with a peak ZHR ~17/h and a FWHM = 0.017°. From 13 observed June Lyrids with $\chi = 2.7$ (amidst 25 sporadics with $\chi = 3.8$), a radiant was determined near the head of Draco at R.A. = 280°, Decl. = +55° (B1950), which is significantly different. This is suspected to be the same shower because of the good agreement in the node.[32]

[28] A. V. Nielsen, *Himmelswelt* **46** (1936), 34; A. K. Hindley, *Observatory* **59** (1936), 26; A. King, The display of Aug. 7. *Observatory* **59** (1936), 62–67.
[29] As reported in: V. Guth and F. Link, *Czech Astronomical Year-book for 1951* (1951); Also: *Nippon Meteor Society Astron. Circular*, No. 614, Sept. 1993, p. 2.
[30] B. Warner, A possible re-observation of the Monocerotid meteor shower. *MNASSA* **24** (1965), 126–127.
[31] S. Dvorak, Unexpected meteor shower from *Lyra & Sky Telescope* **32** (1966), 237; K. B. Hindley, The June Lyrid meteor stream. *J. Brit. Astron. Assoc.* **79** (1969), 480.
[32] M. Langbroek, Een kleine meteoren uitbarsting op 15/16 Juni 1996. *Radiant, J. DMS* **18** (1996), 64–67; P. Jenniskens, Juni Lyriden? *Radiant, J. DMS* **18** (1996), 67–68.

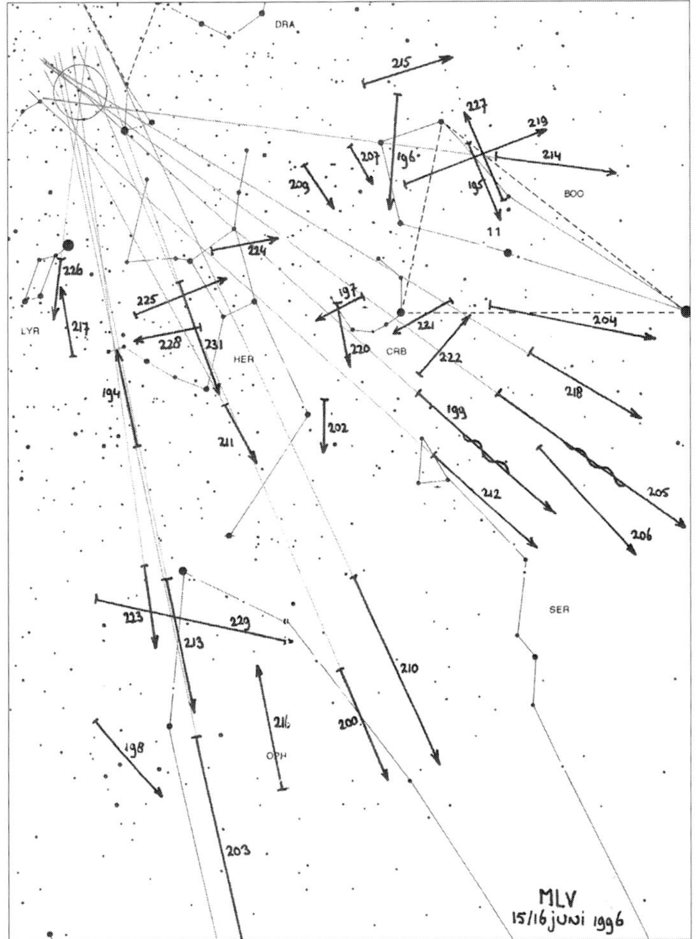

Fig. 13.23 Plot of the 1996 June Lyrids by Marco Langbroek.

β-Aurigids

September 21, 1968 ($\lambda_{\odot} = 179.258$). Observations from Japan ($\lambda = 140^{\circ}\,55'$, $\phi = +38^{\circ}\,13'$) by a three-person group of observers: *Mr. Saasaki*, *Mr. Abe*, and *Mr. Imaizumi*. 19 meteors (+1 to +4) were observed around 03:22 JST = 18:22 UT, radiating from R.A. $= 86 \pm 2^{\circ}$, Decl. $= +43 \pm 2^{\circ}$ +1(2), +2(6), +3(4), +4(4). They were medium fast and mostly yellow/white in colour (Fig. 13.24).[33]

[33] M. Sendai, *Nippon Meteor Soc. Astron. Circ.* No. 316, pp. 4–5 (November 1968, in Japanese).

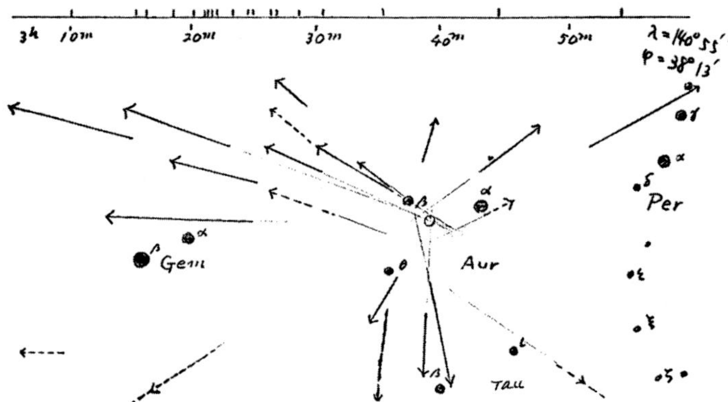

Fig. 13.24 Plot of the 1968 β-Aurigids, with a time bar indicating the moment of arrival of each meteor.

α-Lyncids

December 20, 1971 ($\lambda_{\odot}$ <= 268.78). *Malcolm Currie* of the *B.A.A. Meteor Section* in the U.K. first noticed activity from a radiant at about R.A. = 138°, Decl. = +44° (B1950) on December 14 under very transparent sky conditions while, setting up equipment. After 5–10 min, he started a formal watch and saw eight (perhaps nine) Lyncids between 22:24 and 23:08 UT (Lm = +6.8), with a mean magnitude of +3.0. On December 19/20, the limiting magnitude was +6.6 and he saw 14 Lyncids ($\langle m \rangle$ = +3.5) from 18:15–23:40 and from 00:01–02:55 UT, as well as 42 sporadics and 44 Ursids. The watch on December 20/21 (Lm = +6.6) started with a +1 Lyncid with a train of 1.5 s, followed by one of +0^{m}, but the remainder of the alleged Lyncids were +2^{m} to +5^{m}. Currie noticed seven Lyncids in the first 5 min interval at 21:55 UT (4 min effective observing time). In the next 5 min intervals he saw 4, 3.5, 1, 2, 2, 1, 0, and 1, in what seemed to be an outburst with decreasing rates. From 22:48 to 22:56 UT there were three, and from 23:56 to 00:23 four more, all for a mean magnitude of +3.47. In the same period he saw 14 sporadics and 19.5 Ursids.

α-Circinids

June 4, 1977, 08:30–09:20 UT ($\lambda_{\odot}$ = 73.92). *Belinda Bridge* of Queensland, Australia, reported "large numbers of meteors around 20:30 local time, with rates falling off around 21:00, and activity over by 21:20″. 15 Circinids came from a radiant near R.A. = 218.6°, Decl. = −70.2°.[34]

[34] K. Hindley, *Meteor News*, No. 40 (1977); K. B. Hindley and B. Bridge, Meteors from near alpha Circini. *IAUC* 3146 (1977), ed. B. G. Marsden, 5.

π-Cetids

June 28, 1977 ($\lambda_\odot = 96.63$). During morning sweeps for comets at the Satara and Olifants camps in the Kruger National Park, J. C. Bennett noticed meteors radiate from π-Ceti. 20 were counted between 02:45 and 03:30 UT (4:45–5:30 S.A.S.T.). *One was of first magnitude, leaving a short-lived train.* Some nonshower meteors were also seen. The sky was transparent and dark. Three bright meteors were seen from 02:45–03:30 UT the next year (1978), from a radiant at R.A. = 30°, Decl. = −15° (B1950).[35]

March α-Pyxidids

March 6, 1979, 19:00–21:15 UT ($\lambda_\odot \leq 345.91$, Lm = +5.5), 13 Pyxidids and three sporadics were seen by *Tim Cooper* of the *A.S.S.A.-Meteor Section* near Johannesburg. In the first 10 min interval, four were seen of magnitudes +3, +3, +1, and +3. The next appeared at 19:36 (+1), 19:41 (+3), 19:46 (+2), 20:00 (+4), 20:06 (+4), 20:23 (+4), 20:39 (+1), 20:59 (+1), and 21:11 (+3), in what seemed to be a decreasing rate. In comparison, on March 5, 1980, four meteors were identified as possible Pyxidids from 17:50 to 19:50 (Lm = +5.2), and the next night seven Pyxidids were seen from 17:30 to 20:00 UT (Lm = +5.2). The radiant was determined at R.A. = 135.3°, Decl. = −35.2°.[36]

α-Bootids

April 28, 1984 00:00–02:30 UT ($\lambda_\odot = 38.168$). A young but active visual observer of the Dutch Meteor Society, *Frank Witte* of Hengelo, the Netherlands, reported an outburst of telescopic meteors seen through a 6 cm f12–700 mm refractor (23 mm Huygens ocular) with a field of view of about 1° pointed just next to the star Arcturus. A total of 433 meteors and point meteors from a radiant of diameter less than 1° at R.A. = 214.4°, Decl. = +19.3° were classified as α-Bootids, 12 others as sporadics. The activity curve is that of a typical dust-trail crossing. No α-Bootids were seen by visual observer Klaas Jobse in that same period.[37]

February Canis Majorids

February 19, 1985 ($\lambda_\odot = 330.60$), *Jesús Otero* and *Pablo Silveira* of Caracas, Venezuela noticed fast and bright meteors from a radiant near R.A. = 104.3°, Decl. = −25.6° in Canis Major. In a dedicated watch, at one point seven

[35] J. C. Bennett, A possible new meteor stream. *Mon. Not. Astron. Soc. Southern Africa* **36** (1977), 110.
[36] J. C. Bennett, *Meteors* No. 9 (1989); T. Cooper, private correspondence.
[37] P. Jenniskens, Meteor stream activity. II. Meteor outbursts. *Astron. Astrophys.* **295** (1995), 206–235.

meteors were seen in 12 min in relatively poor sky conditions. Between 01:47 and 02:47 UT, 20 meteors were observed, 15 from the suspected radiant, for a peak ZHR $\gg$ 70/h. A total of 14 meteors were plotted. The next night, some meteors continued to radiate from the region, for a peak ZHR $\leq$ 19/h (J. Otero).[38]

December Canis Minorids

The night of December 3, 1988, just after midnight ($\lambda_{\odot}$ ~252.4). At a meeting of SOVAFA at one of the members private homes, *Carolina Munzi* of Caracas, Venezuela, noticed a swift and faint meteor radiate from near Procyon and then another and then another. Five (?) others joined in the observation, and with a transparent sky, they observed for two hours and plotted some 93 Canis Minorids, with a peak rate of ZHR = 54/h. The radiant was determined at R.A. = 111.2°, Decl. +7.7°.[39]

There are numerous other accounts from the last century that need confirmation before we can be certain that they pertained to a meteor outburst.

[38] J. Otero and P. Silveira, Fuerte actividad de las Canis Majoridas en 1985. *Universo* **7** (1987), 154 (in Portuguese).
[39] C. Munzi, A. Castillo, C. Soto *et al.*, LIADA, *Proc. 1ra Convention* (Caracas: Liga Iberoamericano de Astronomía, 1990), 18–19.

14

Trapped: the Leonid Filament

At the start of the Leonid season in November, 1994 the weather had worsened moving into late fall and, because of a full Moon that year, the interest in the shower was low. A full Moon would wipe out the faint annual Leonids (ZHR $\sim 13/\text{h}$), but perhaps not an outburst of bright meteors like that of 1961. An observing campaign was organized in the San Francisco Bay Area with the help of local amateur astronomers led by *Mike Koop* of the SJAA and *Mike Wilson* of NASA Ames. In 1961, rates peaked in the night prior to the nodal passage and, expecting the same, all of us went out in the night of November 16/17, 1994. Unfortunately, it rained water instead of meteors that night.

There were no volunteers for another campaign the next night. Work beckoned and it was deep in the night when I returned home around 12:30 UT. Before entering my apartment, I noticed the sky was clear. Still blinded by the car lights of opposing traffic, I quickly adapted to night vision by staring at my feet and blinking my eyes repeatedly while standing in the dark. I glanced up at the light polluted sky. Above my head was the constellation Lion in a clearing between trees. Well, there was one nice bright Leonid and not far from the radiant either! There was another one! I rushed inside to get my lawn chair, observing forms, and gnomonic star charts, and by 12:34 UT I was back out observing (Fig. 14.1).

From the start, I noticed many bright Leonids, falling left and right to the horizon at a rate I would have expected only for a good return of the August Perseids. One bright Leonid crashed in the zenith with a flare in the middle. Twilight set in at 14:00 UT. I phoned my friends in the Netherlands and then drove back to NASA/Ames to warn others by electronic mail. If this was the same broad stream encountered in 1961, then the shower would be visible at least for another day or so.

Confirmation came quickly from visual observers Josep Trigo-Rodríguez and *Francisco Reyes-Andres*, members of the Spanish *S.O.M.Y.C.E.*, who had been out that night as well, and from an unexpected source: the forward reflections of radio signals from faraway radio stations that were counted by *Ilkka Yrjölä* at Kuusankoski, Finland. His results (Fig. 14.2) led us to found *Global-MS-Net*, a collaboration of amateur radio-MS observers to monitor meteor activity. That same year, *Christian Steyeart* in Belgium independently initiated the *Radio Meteor Observers Bulletin*, which has served to provide feedback to participants.

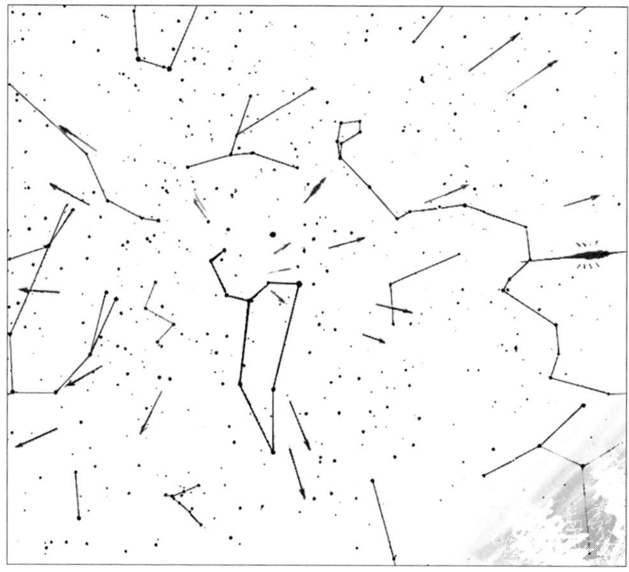

Fig. 14.1 The first-in-the-season Leonid outburst of 1994 as seen by the author from Mountain View, California.

The Moon and the city smog made the analysis of the visual data difficult. Late the next night, I set out to compare the observing conditions near my apartment with those out of town. The sky was clear again and I noticed that the faintest stars were about half a magnitude fainter than at downtown, possibly on account of the brown layer of smog visible during daytime, which contains particles that scatter the moon and city light. I also counted sporadic meteors to establish my personal perception. Allowing for that half a magnitude, and taking into account the Spanish data, I concluded that Leonid rates had peaked at ZHR $= 75 \pm 15$/h at about 10:00 UT on November 18. Except for being late, the shower was as wide, with as much activity, and as rich in bright meteors as the historic shower of 1961. The Leonids were back![1]

The comet was still beyond Jupiter's orbit at the time, hence the dust responsible for the outburst must have stretched at least a billion km along its orbit. In contrast, it took Earth only 0.8 d (FWHM) to cross the Filament, a distance of about 2 million km, while passing about 1 million km outside the orbit of the comet. These measures of length, width and depth ($1000 \times 2 \times 1$) are those of a narrow sheet or ribbon of dust, which I call the Leonid *Filament*, adopting an established name for a similar structure in the orbit of 109P/Swift–Tuttle.

Now it was official that the Leonids were back, ambitious plans were made for the next Leonid campaigns. It all started modestly. In 1962, rates had been down from

[1] P. Jenniskens, Meteor stream activity. III. Measurement of the first in a new series of Leonid outburst. *Meteoritics Planet. Sci.* **31** (1996), 177–184.

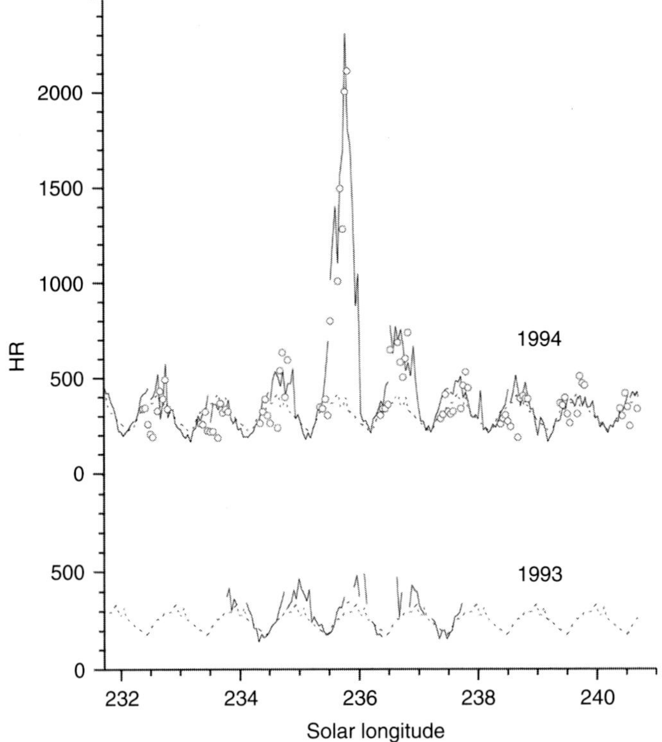

Fig. 14.2 Radio MS counts during the 1994 Leonid outburst by Ilkka Yrjölä of Finland.

those in 1961, and 1995 proved no different with a peak ZHR = 35/h.[2] We obtained the first outburst Leonid orbits that year during a concerted effort of the Dutch Meteor Society and the California Meteor Society in the south of Spain and in California.[3] The radiant dispersion was small, but resolved.

In 1995 a spectacular fireball was seen over California at 11:45:22 UT, which left a *persistent train* that was visible to the naked eye for six minutes. I captured the decaying luminosity of the train in a series of 1 min exposures (Fig. 14.3). For some time in 1996, this Leonid portrait was the first on the internet without Russian relatives.

This image made the point that it was possible to aim a telescope at the path of a meteor. In November, 1996, I flew to Chile to study the cause of this luminous glow at the European Southern Observatory (ESO). I saw the outburst peak at

[2] J. Rendtel, Leonids. Short summary: no outburst. *IMO Meteor Shower circular* (November 1995); P. Jenniskens, A second Leonid outburst in 1995. *WGN* **23** (1995), 198–200.

[3] H. Betlem, C. ter Kuile, J. van 't Leven *et al.*, Precisely reduced meteoroid trajectories and orbits from the 1995 Leonid meteor outburst. *Planet. Space Sci.* **45** (1997), 853–856.

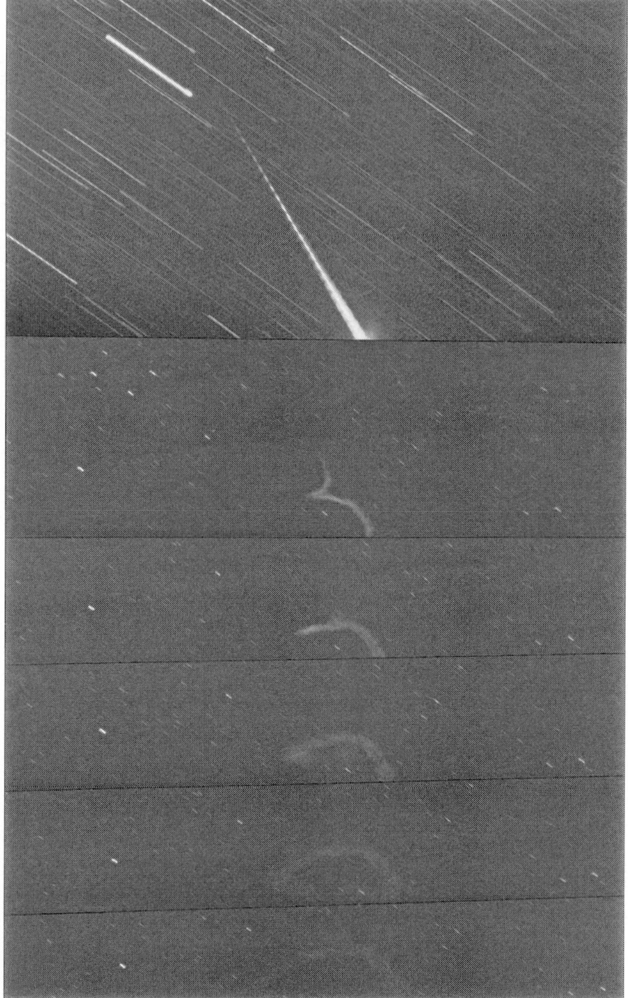

Fig. 14.3 The persistent train of this Leonid was photographed over California by Peter Jenniskens at the San Luis Reservoir on November 18, 1995.

ZHR = 80/h that year and was rewarded by a spectacular fireball in the zenith that left the most impressive persistent train in my memory: two straight bands, separating and spreading gradually. The telescope operator aimed the CAT telescope on my direction, but sadly did not recognize the train in the viewfinder and never started the integration. In the next year, seven telescopes participated in a repeated effort (Fig. 14.4), but no trains appeared.

When it became clear that an airborne lidar could not be made pointable, Leonid MAC participant Chet Gardner agreed to install one of the University of Illinois sodium lidars in a steerable telescope at the *Starfire Optical Range* at

Fig. 14.4 Pointing telescopes at persistent trains. The Leonid observing team at ESO, LaSilla, Chile, in November, 1997. Collaborator Giacomo Mulas from the *Osservatorio Astronomico di Cagliari* in Italy is kneeling in center, while Jiri Borovicka of *Ondrejov Observatory* is standing behind him.

Kirtland AFB in New Mexico in an effort led by *Jack Drummond* and *Mike Kelley*.[4] This resulted in some of the most spectacular images of trains yet, and the first probe of a persistent train by lidar. It took until the Leonid MAC mission of 1999, however, before, atmospheric chemist, *John Plane* and I discovered what was the likely source of the yellowish light: "orange arc" emission from iron oxide molecules (FeO) that are created in an excited state from the chemical reactions of ambient ozone and meteor included oxygen atoms, whereby meteoric iron atoms act as a catalyst: $Fe + O_3 \rightarrow FeO^* + O_2$; $FeO + O \rightarrow Fe + O_2$.[5]

That year, Japanese amateurs used intensified cameras to pick up luminosity from a −6 Leonid above 130 km altitude.[6] This was later recognized as the V-shaped collisional cascade.

The same wide shower rich in bright meteors returned in 1997 with peak rates of order ZHR = 100/h. These returning showers permitted more ambitious observing efforts. My collaborator, *Ed Tedesco*, and I obtained a spectacular image of the shower from space (Fig. 14.5) and the first UV spectrum of a meteor with the help of the

[4] M. C. Kelley, C. Gardner, J. Drummond *et al.*, First observations of long-lived meteor trains with resonance lidar and other optical instruments. *Geophys. Res. Lett.* **27** (2000), 1811–1814.

[5] P. Jenniskens, M. Lacey, B. J. Allan, D. E. Self and J. M. C. Plane, FeO "orange arc" emission detected in optical spectrum of Leonid persistent train. *Earth, Moon Planets* **82/83** (2000), 429–438.

[6] V. Fujiwara, M. Ueda, Y. Shiba *et al.*, Meteor luminosity at 160 km altitude from TV observations for bright Leonids meteors. *Geophys. Res. Lett.* **28** (1998), 285–288.

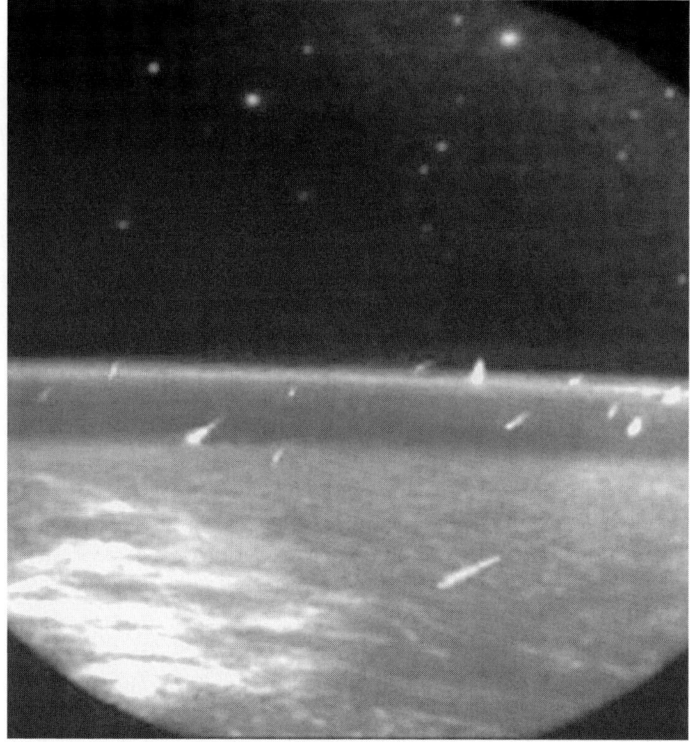

Fig. 14.5 The 1997 Leonid shower from space (15:20–15:59 UT, November 17). Different directions of motion are due to the changing perspective of the shower as seen from the orbiting satellite.[7]

Midcourse Space eXperiment (MSX) satellite.[8] Although its infrared sensors had long lost their coolant, the visible sensors were still operational. The telescope was aimed at the limb of the Earth and repeatedly scanned for meteors. During that time, 29 meteors were seen from an ever changing perspective, while the spacecraft circled Earth.

That year, observers of the California and Dutch Meteor Societies deployed a three-station photographic network in southern California, near Edwards Air Force Base, where Canadian meteor astronomers *Peter Brown* and *Robert L. Hawkes* had deployed a radar and intensified cameras, respectively.[9] The shower returned much as expected: broad and bright. The measured Leonid radiants were dispersed and displaced relative to 1995 (Fig. 14.6).

[7] P. Jenniskens, E. Tedesco, J. Murthy, C. O. Laux and S. Price, Spaceborne ultraviolet 251–384 nm spectroscopy of a meteor during the 1997 Leonid shower. *Meteoritics Planet. Sci.* **37** (2002), 1071–1078; P. Jenniskens, D. Nugent, E. Tedesco and J. Murthy, 1997 Leonid shower from space. *Earth, Moon Planets* **82/83** (2000), 305–312.

[8] *Ibid.*

[9] P. Brown, M. D. Campbell, K. J. Ellis *et al.*, Global ground-based electro-optical and radar observations of the 1999 Leonid shower: first results. *Earth, Moon Planets* **82/83** (2000), 167–190.

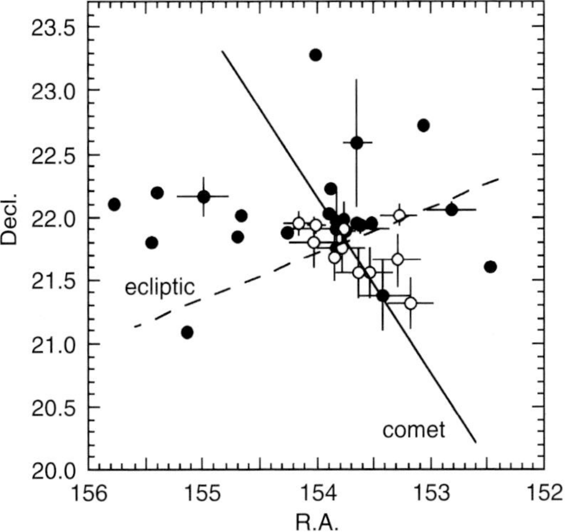

Fig. 14.6 1997 measured radiant positions (•) compared to those of 1995.

Early that summer, contacts had been made with the 418th Test Wing of the U.S. Air Force at Edwards Air Force Base (then the 452nd Flight Test Squadron), when collaborator *Captain Steven J. Butow* and I received a chance to deploy an aircraft for meteor shower observations. The Perseids were expected to have another outburst that year, also due to a Filament (Chapter 17). While I joined other CMS observers on the ground, Steve ran photographic, film, and video cameras and observed the meteor outburst from the air. We learned that intensified video cameras record meteors well at altitude.[10]

Results were presented at the April, 1998 *Meteoroid Storm and Satellite Threat Conference* in Manhattan Beach, organized by later Leonid MAC participant *David Lynch* of the Aerospace Corporation. The first of a series of Leonid MAC workshops (this one more of a 1 d business meeting) was held in conjunction with that meeting.

14.1 The 1998 shower of bright fireballs

We had discovered a massive Filament of dust in the orbit of the comet. That Filament had dominated shower activity in the years prior to and during the 1965 return of the comet and was doing so again.[11] In 1965, observers in Dushanbe, Tadjikistan, recorded six Leonids from a compact but resolved radiant at R.A. = 153.6 ± 0.1°, Decl. = +21.3 ± 0.4°. The same component must have been detected again in 1966, when three photographed Leonids radiated from the same R.A. = 153.5 ± 0.1°, Decl. = +21.3 ± 0.2°. The storm could not be seen from Tadjikistan that year. In other years, the annual Leonids radiated from a different R.A. = 154.0 ± 1.2° and

[10] P. Jenniskens and S. Butow, Successful Leonid airborne mission validation flight during August 1997 Perseids. *WGN* **25** (1997), 215–217.

[11] P. Jenniskens and H. Betlem, Massive remnant of evolved cometary dust trail detected in the orbit of Halley-type comet 55P/Tempel–Tuttle. *Astrophys. J.* **531** (2002), 1161–1167.

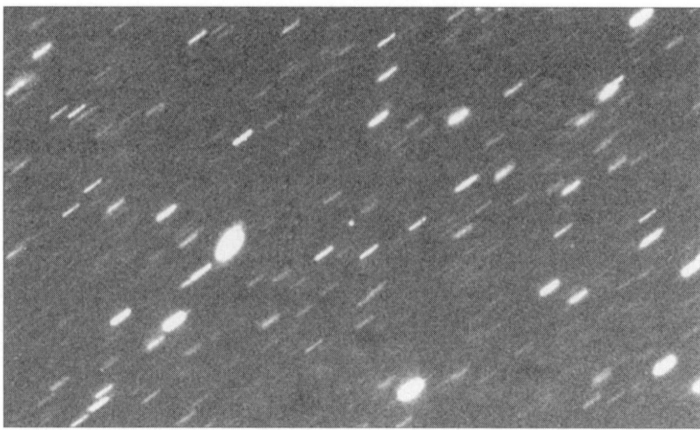

Fig. 14.7 Recovery image of comet 55P/1997 E_1 (Tempel–Tuttle) taken at the Keck II 10 m telescope on Manua Kea. The LRIS CCD camera was used.[12]

Decl. = +21.7 ± 0.6°. From these observations, *Pulat Babadzhanov* calculated that these fresh Leonid meteoroids, with some annual meteoroids mixed in, had only 60% of the density of the annual Perseids of comet Swift–Tuttle and 48% of the density of Geminids.

The comet was due in February, 1998. Guided by an improved orbit,[13] the parent comet was spotted approaching the Earth from afar with the Keck II 10 m telescope on Manua Kea (Fig. 14.7). The recovery image was taken by a team of astronomers including *Karen J. Meech*, *Oliver R. Hainaut* and *James M. Bauer* on March 4.6 UT, 1997.[14] By November, the comet was still a simple $+18^m$ point of light in even the biggest telescopes and was lagging brightness predictions. A coma was detected only on December 24, the first sign of activity. After that, the brightness of the comet rapidly increased and peaked at around $+7.7^m$ at the time of closest approach with Earth. When the comet passed Earth on January 17th, I could recognize the apparent motion among the stars easily in a telescope by glancing away for a minute and looking again. A faint dust tail developed in mid-February. While approaching the Sun and moving away from Earth, the comet coma became less diffuse. The comet reached the point nearest to the Sun (perihelion) on February 28th and rapidly faded in March before being lost in the glare of the Sun.

On November 17, 1998, the Filament showers peaked in a glorious rain of fireballs over Europe.[15] The zenith hourly rate increased to ZHR = 300/h (Fig. 14.8). At the observatory of Xing Long, near Beijing, Hans Betlem recalled the night of November

[12] O. R. Hainaut, K. J. Meech, H. Boehnhardt and R. M. West, Early recovery of Comet 55P/Tempel–Tuttle. *Astron. Astrophys.* **333** (1998), 746–752; O. R. Hainaut, 55P/1997 E1, *IAUC* 6579, ed. B. G. Marsden (Cambridge, MA: IAU Minor Planet Center, 1997).

[13] D. K. Yeomans, K. K. Yau and P. R. Weissman, The impending appearance of comet Tempel–Tuttle and the Leonid meteors. *Icarus* **124** (1996), 407–413.

[14] O. R. Hainaut, K. J. Meech, H. Boehnhardt and R. M. West, Early recovery of Comet 55P/Tempel–Tuttle. *Astron. Astrophys.* **333** (1998), 746–752; O. R. Hainaut, 55P/1997 E1, *IAUC* 6579, ed. B. G. Marsden (Cambridge, MA: IAU Minor Planet Center, 1997).

[15] R. Arlt, Bulletin 13 of the International Leonid Watch: the 1998 Leonid Meteor Shower. *WGN* **26** (1998), 239–248.

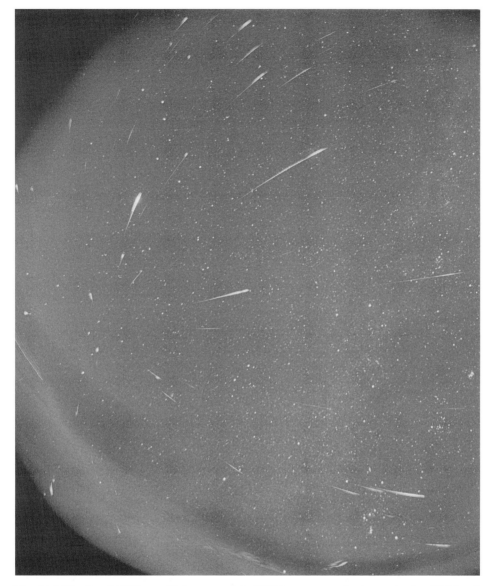

Fig. 14.8 The Leonid "Filament" shower peak in 1998 in a spectacular image by Juraj Tóth, of Modra Observatory, Slovakia. The photograph was exposed from November 16 23:33:00 UT until November 17 03:37:10 UT and shows 156 trails of Leonids brighter than magnitude −4.[16]

[16] J. Tóth, L. Kornovš and V. Porubčan, Photographic Leonids 1998 observed at Modra observatory. *Earth, Moon Planets* **82/83** (2000), 285–294.

16/17 as follows: *And then it happens. A long, very slow meteor of magnitude 0 or −1 glides from Gemini to Orion, through Eridanus and Cetus to extinguish low on the horizon after 8 seconds. "Earth grazing fireball!", yells Pavel Spurný, who is present too and observes the last part. A few minutes later, yet another. Then it hits me: the Leonids of lore have returned. Later that night we would write down seeing yet another −8 fireball without much second thought.*[17]

The Filament was broad and visible at all locations in the world at about the same total rate, albeit usually spread over two nights. In Slovakia, a brief flurry of fireballs peaked at about 01:40 UT ($\lambda_\odot = 234.42°$). The DMS teams, located in China, measured from double station meteors, a hollow circular radiant distribution centered at R.A. = 153.56 ± 0.02° and Decl. = +22.07 ± 0.02° (after correcting all radiant positions to the epoch corresponding to solar longitude of $\lambda_\odot = 235.0$, with ΔR.A./$\Delta\lambda_\odot = +0.659°$, ΔDecl./$\Delta\lambda_\odot = -0.325°$, as shown in Fig. 15.11 in the next chapter),[18] with meteoroid orbits spread by a large (intrinsic) ±0.09 (± 0.02)°. This compares well to the 1995 result, when the mean radiant was at R.A. = 153.41 ± 0.11° and Decl. = +22.05 ± 0.03° at that epoch.

14.2 The cause of the Filament

The exact cause of these Filament showers remains unknown. We do know that it is not the 1333-dust trail crossing proposed by *David Asher* and *Mark Bailey* of Armagh Observatory and *Vacheslav Emel'Ianenko*, of South Ural University in Chelyabinsk, in the Russian Federation. They suggested that one particular dust trail from comet 55P/Tempel–Tuttle alone was responsible for the 1998 outburst.[19] Unlike other dust trails, this dust was ejected when the comet orbit was close to a mean-motion resonance and much of the dust ejected at ~2.4 m/s in the forward direction of motion at perihelion would be trapped in that resonance.[20] In the calculations by Asher *et al.*, the dust stayed together as a well defined *trailet* (Asher's "resonant arc") and moved into Earth's path in 1998. The peak of that dust trail encounter occurred at $\lambda_\odot = 234.47 \pm 0.03°$, close to the fireballs seen at Modra.

The problem with this idea is that the assumption that the resonance keeps the dust density high enough, also implies that the dispersion of nodes stays small and much the same as that of other dust trail encounters (see the next chapter). A narrow shower would be expected, a miniature meteor storm, rather than a 1 d wide outburst, as was observed (Fig. 14.9). The model predicted a narrow dispersion of orbital elements,

[17] H. Betlem, Leoniden 1998 te Xing Long, People's Republic of China. *Radiant, J. DMS* **21** (1999), 1–14.

[18] H. Betlem, P. Jenniskens, J. van 't Leven *et al.*, Very precise orbits of 1998 Leonid meteors. *Meteoritics Planet Sci.* **3** (1999), 979–986.

[19] D. J. Asher, M. E. Bailey and V. V. Emel'Ianenko, The resonant Leonid trail from 1333. *Irish Astron. J.* **26** (1999), 91–93.

[20] The semi-major axis that corresponds to an orbital resonance for a particle of given beta (strictly speaking only in a restricted three-body system): $a_r = a_J\,(1-\beta)^{1/3}\,(k/l)^{2/3}$, where J denotes Jupiter and (k, l) are the numbers that characterize the ratio of orbital periods. From: J. Liou and H. A. Zook, Evolution of interplanetary dust particles in mean motion resonances with planets. *Icarus* **128** (1997), 354–367.

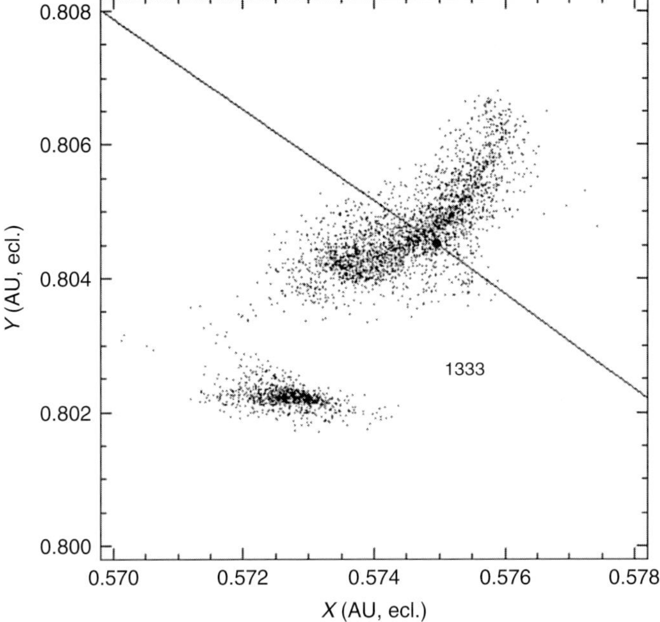

Fig. 14.9 The 1333 ejecta of comet 55P/Tempel–Tuttle in 1998. The dot in Earth's orbit marks Earth's position on November 17.0 UT. Calculations by Jérémie Vaubaillon.

while the photographic observations clearly show the orbital elements to be significantly dispersed, much more than in the calculated 1333-dust trail. Asher's calculated orbits had a very particular spread in inclination and argument of perihelion not observed in the photographed meteors.[21] Moreover, the total amount of mass in the Filament component would also demand that the comet was much more active in 1333 than in other years, coincidentally at the same time when dust apparently got preferentially trapped in a mean-motion resonance. There is no evidence for that. Hence, the 1998 fireballs were not from the 1333 trail alone.

Peter Brown[22] thought the 1998 Filament component was the product of numerous recent dust trails, with the most significant contributions from the trails ejected in AD 855–1600 (Fig. 14.10). He used reasonable ejection velocities in his model 10–45 m/s at perihelion (for $\beta = 0.1$) and 6–24 m/s for $\beta = 10^{-4}$. This way, he accomplished a surprisingly good fit to the activity curve, except for narrow peaks at $\lambda_\odot = 234.51$ and 234.74, and a more rapid fall-off after $\lambda_\odot = 234.8$ than observed.

Jérémie Vaubaillon calculated the dispersion of the recent dust trails near Earth's orbit for the Crifo ejection model (Fig. 14.11). Indeed, many trails cluster at encounter

[21] H. Betlem, P. Jenniskens, J. van't Leven *et al.*, Very precise orbits of 1998 Leonid meteors. *Meteoritics Planet Sci.* **34** (1999), 979–986.

[22] P. Brown and R. Arlt, Detailed visual observations and modelling of the 1998 Leonid shower. *Mon. Not. R. Astron. Soc.* **319** (2000), 419–428; R. Arlt and P. Brown, Bulletin 14 of the International Leonid Watch: visual results and modeling of the 1998 Leonids. *WGN* **27** (1999), 267–285.

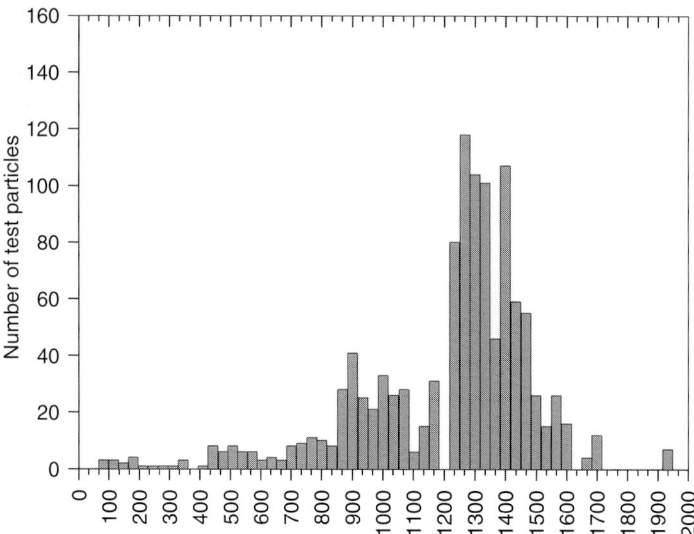

Fig. 14.10 Contribution of ejecta to the 1998 Leonid shower according to the calculations by Brown and Arlt (2000).[23]

times between November 16.5 and 18.5 UT, when the fireballs were observed. However, the result is not yet the diffuse 1 d wide stream recognized from the observations. As in Brown's model, there is still much clumpiness from the contribution of individual trails, each of which are active for about an hour or so.

In contrast, I have proposed that the Filament is older. It is a group of many dust trails that have slowly and inexorably blended together over the years due to planetary perturbations on different parts of the trails that catch up with each other, and the superposition of different dust trails.[24] If the orbital periods are perturbed within the range of periods allowed to stay trapped, then the particles of different parts of the trail will catch up with each other. Over long time scales, after avoiding close encounters with Jupiter, Saturn, or Earth this should result in a diffuse dust stream that has about the same width as the dispersion of trails near Earth's orbit, about 0.022 AU in diameter. The observed FWHM = 0.014 AU. That dust could have been ejected long before AD 855 when the comet first passed inside Earth's orbit.

A nice illustration of the final effect of this dispersion is shown by Jérémie's calculation of the evolution of the AD 608 ejecta integrated forward until the year 2984 (Fig. 14.12). The dispersion has the correct width, narrower than that of the

[23] *Ibid.*

[24] P. Jenniskens and H. Betlem, Massive remnant of evolved cometary dust trail detected in the orbit of Halley-type comet 55P/Tempel–Tuttle. *Astrophys. J.* **531** (2002), 1161–1167.

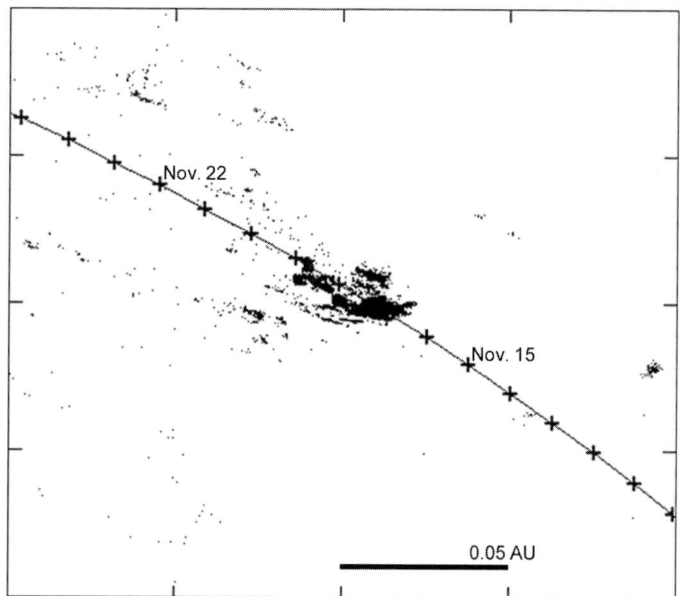

Fig. 14.11 Nodes of the 604 – 1965 dust trails (1–42 revolutions) of 55P/Tempel–Tuttle as seen in 1998. Calculations by Vaubaillon.

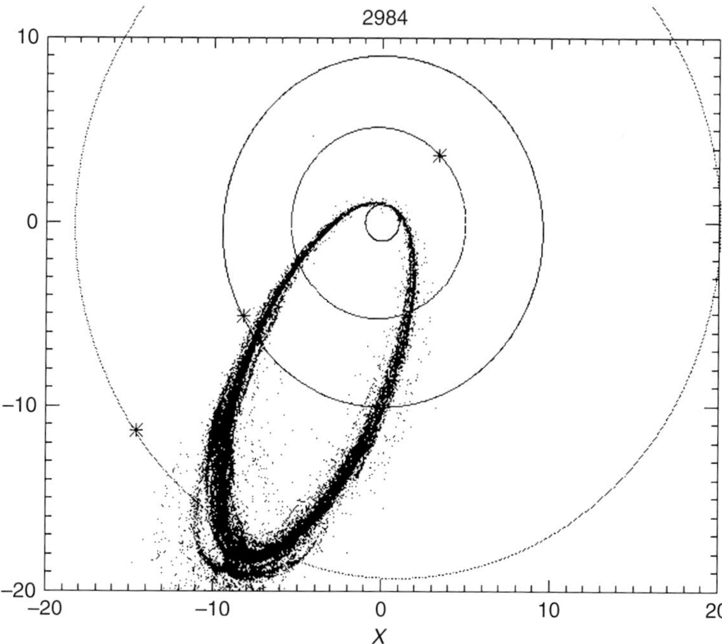

Fig. 14.12 Distribution of Leonid meteoroids ejected in 604, as seen in AD 2984. The outermost circle is the orbit of Uranus, the innermost circle is that of Earth. Calculations by Vaubaillon.

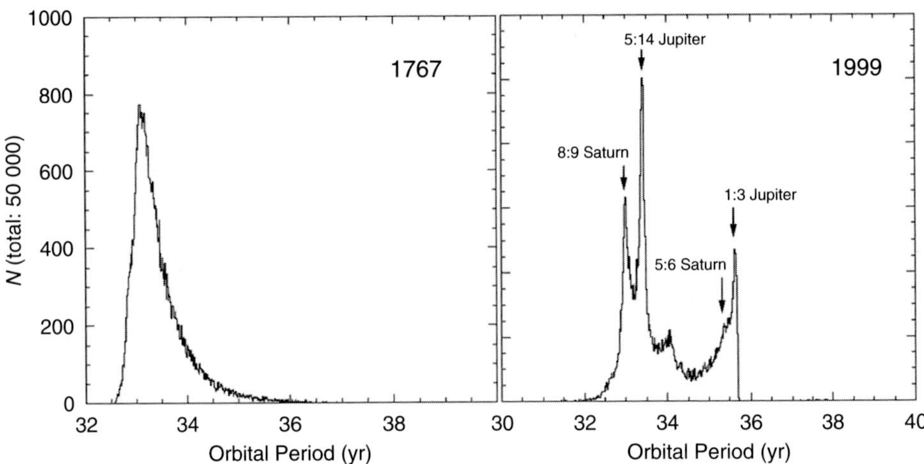

Fig. 14.13 The 1767-dust trail shortly after ejection and seven revolutions later, when meteoroids have accumulated in mean-motion resonances.

annual shower, but broader than a recent dust trail. Near aphelion, there is still some evidence of the individual trailets, but most residual structure has been lost.

The graph also illustrates the problem with this scenario: the meteoroids are spread along the comet obit and are no longer concentrated near the position of the comet, as the Leonid Filament seems to be. For dust to stay concentrated near the position of the comet for such a long time, it has to be moving in mean-motion resonances, or have been ejected with a very low speed.

Vaubaillon was able to demonstrate that trapping in mean-motion resonances is a rapid process. Fig. 14.13 shows a graph from his Ph.D. thesis, showing the orbital periods of particles ejected in 1767 shortly after ejection. After seven revolutions, planetary perturbations have changed the orbital period of the particles sufficiently to both smaller and larger values. At that time, the orbital periods have accumulated around three values that correspond to the 4:11, 5:14, and 1:3 resonances with Jupiter. Hence, the long survival of these Filaments does appear to be due to meteoroids being trapped in mean-motion resonances, preventing close encounters with Jupiter.

Jupiter's past perturbations may have been responsible for the sudden onset of this component in 1994, and the variations in peak activity observed in later years. I expected that the dust component would remain visible post perihelion for at least slightly less than one orbit of Jupiter (<12 yr), thus until 2004 or 2005.

I saw this validated in 2002, when the Filament component was detected for the first time after the perihelion passage of the comet, underlying two very narrow Leonid storm profiles (Fig. 14.14). The observed shift in the peak time and constant width over the years from 1994 to 2002 (Table 4) confirms that this component moves about Earth's path much like individual dust trails in reflection to the ever changing

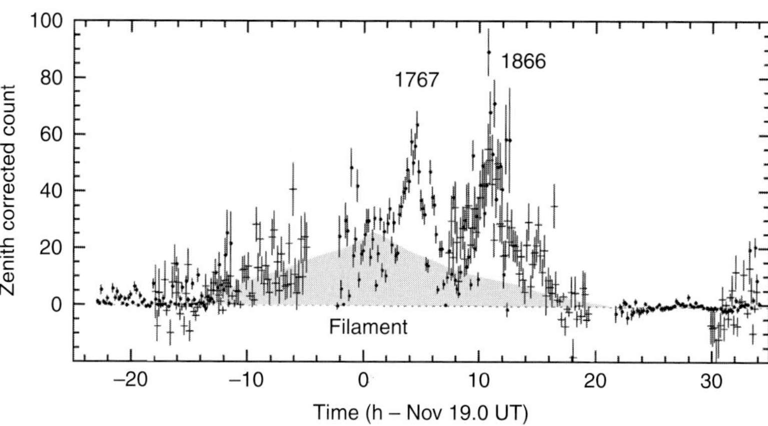

Fig. 14.14 The Filament component in 2002 as detected by radio MS counts by the Global-MS-Net stations of Jeff Brower (USA, •) and Ilkka Yrjola (Finland, +).

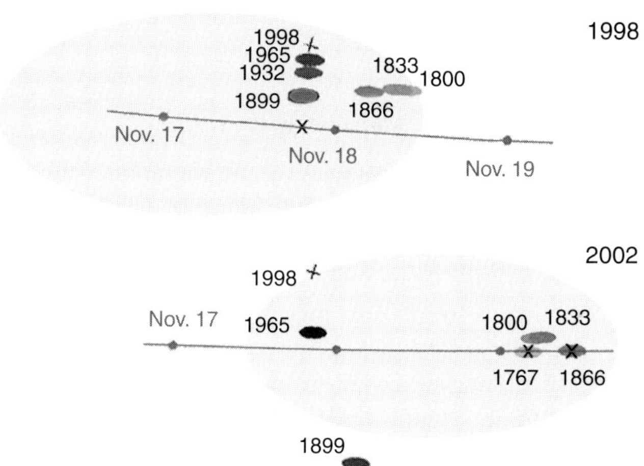

Fig. 14.15 The approximate position of the Filament component (gray ellipse) in relation to that of individual young dust trails.

gravitational field of the planets (shaded area in Fig. 14.15). Again, more or less following the Sun's reflex motion. If the meteoroids can evolve to lag the comet by more years than a period of Jupiter (as shown in Fig. 14.12), then the Filament component is expected to return on occasion in the Leonid off-season (Table 4b)!

15

The Leonid storms

In the years leading up to the 1998 Leonid maximum, anticipation was growing among amateur astronomers hoping to see a repeat of the 1966 storm, and among satellite operators hoping to avoid one. A meteor storm of that magnitude had not occurred since the beginning of the space age. This generated widespread concern culminating in testimony to the Subcommitte on Space and Aeronautics of the 105th US Congress that satellites were going to be *sand blasted*.[1]

15.1 Anticipating the storm

Peter Brown and *Jim Jones* of the University of Western Ontario and *Martin Beech* of Regina University in Canada, took a leading role in raising awareness about the impending danger.[2] They organized a campaign to provide meteor rate data to the satellite operator community from radar and intensified video observations, but achieved best flux estimates from visual meteor observations at sites spread around the globe. Rates were collected by a Canadian company called *CresTech* and normalized, before being distributed as e-mail reports. In later campaigns, this task was delegated to Marshall Space Flight Center in an effort led by Bill Cooke.

Brown and Jones[3] also provided the most elaborate theoretical models of expected Leonid activity (Fig. 15.1). They predicted an onset of high activity in 1997 with the highest rates on November 17, 1998, at 18 ± 05 h UT, and November 18, 1999, at 00 ± 05 h UT. Based on these predictions, the global observing effort was started in 1997, with the deployment of a meteor radar at Edwards AFB. In an early setback, no storm was observed that year.

[1] W. Ailor, The upcoming Leonid meteoroid storm and its effect on satellites. *Testimony to the US House of Representatives. Committee on Science. Subcommittee on Space and Aeronautics.* Hearing on "Asteroids: Perils and Opportunities" (May 21, 1998).

[2] M. Beech, P. Brown and J. Jones, The potential danger to space platforms from meteor storm activity. *Q.J.R. Astron. Soc.* **36** (1995), 127–152; M. Beech, P. Brown, J. Jones and A. R. Webster, The danger to satellites from meteor storms. *Adv. Space Res.* **20** (1997), 1509–1512.

[3] P. Brown and J. Jones, Evolution of the Leonid meteor stream. In *Meteoroids and Their Parent Bodies*, ed. J. Stohl and I. P. Williams (Bratislava: Astronomical Institute, Slovak Acad. Sci, 1992), pp. 57–60; P. Brown, The Leonid meteor shower: historical visual observations. *Icarus* **138** (1999), 287–308.

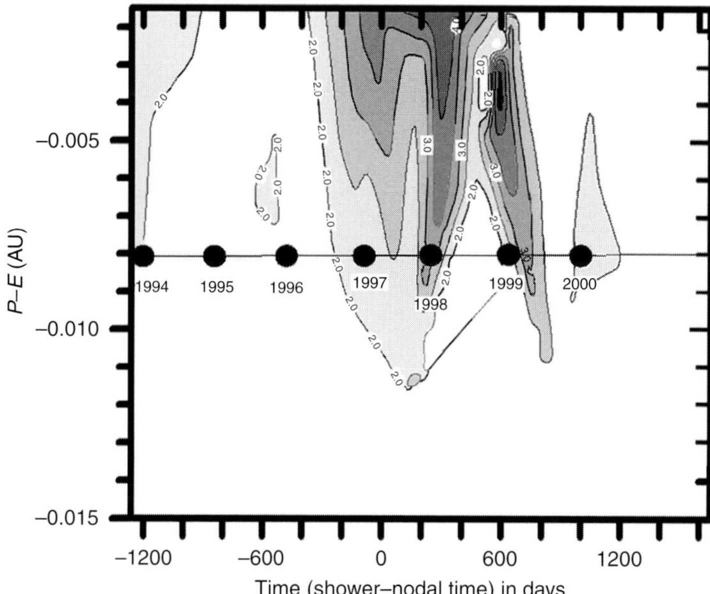

Fig. 15.1 Dust density near Earth's orbit at the node of the comet calculated by Brown and Jones.[4]

In England, *Zidian Wu* and *Iwan P. Williams* at Queen Mary and Westfield College of the University of London[5] predicted that the best rates would be in both 1998 and 1999, with either year being the strongest, depending on over how large an area near Earth's orbit the meteoroids in the model were averaged. Assumed ejection speeds were of order 600 m/s. They were pessimistic that the showers would be very intense, calculating that they would not rise much above the level of the weak showers in 1899 and 1932. Fortunately, that was not so.

Earlier, *Igor Stanislavovich Astapovich* and *Alexandra Konstantinovna Terentjeva* of Kiev State University had calculated the perturbations on meteoroids in an orbit similar to that of 1866, and expected high rates on November 17, 1997 20:19 UT, November 18, 1998 03:19, November 18, 1999 10:04, and November 19, 2000 17:02 UT.[6] But this was not so either.

Other astronomers made predictions based on the orbital evolution of the comet orbit and historic sightings of the Leonid storms. *Lubor Kresák*[7] used the delay in time between the return of comet and meteoroids to infer that the most likely storms would occur on November 17, 1998 (8.4 UT, ZHR = 10 000) and November 17, 1999 (14.4

[4] *Ibid.*

[5] Z. Wu and I. P. Williams, Leonid meteor storms. *Mon. Not. R. Astron. Soc.* **280** (1996), 1210–1218.

[6] I. S. Astapovich and A. K. Terentjeva, Conditions of an encounter of the Leonid meteor stream with the Earth in apparitions 1898–2000. *Problemy kosm. Fiziki* **7** (1972), 100–107.

[7] L. Kresák, Meteor storms. In *Meteoroids and Their Parent Bodies*, ed. J. Stohl and I. P. Williams (Bratislava: Astronomical Institute, Slovak Acad. Sci, 1992), pp. 147–156.

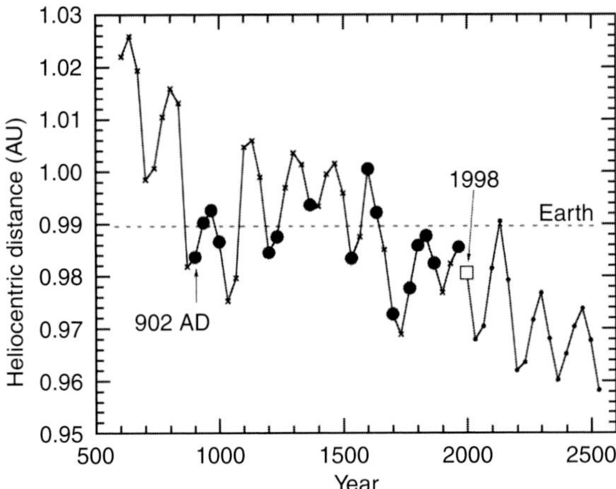

Fig. 15.2 The node, that is the point where 55P/Tempel–Tuttle crosses the ecliptic plane, moves inward to the Sun. Large dots indicate when a meteor storm was reported in the past (Table 4). After: Yeomans *et al.* (1996).[8]

UT, ZHR = 100 000). He added disappointedly: ... *but none of them should be visible from Europe*. That, too, turned out to be wrong.

The only solid lead seemed to be the evolution of the parent comet orbit, which was now much better known since 55P/Tempel–Tuttle was recovered in March of 1997. Donald Yeomans and colleagues at the NASA Jet Propulsion Laboratory extended the calculations of the comet orbit by Joachim Schubart (Chapter 7) into the past to AD 1 and into the future to AD 2530.[9] They confirmed that the descending node had steadily moved inwards, crossing Earth's orbit first in AD 865, and oscillated back and forth after that (Fig. 15.2). They confirmed that in the 1998 return the minimum distance was going to be larger than in 1932, when no meteor storm was reported. They did not dismiss the possibility that high rates might be observed. Five returns from now, Leonid storms would be a thing of the past.

Leonid storms were first recorded in AD 902, by observers in Egypt, and in Italy, on the flanks of Mount Etna in the Italian town of Taormina, now a nice tourist town.[10] Other recorded storms in Chinese, Arabic, and European annals are given in Table 4. Modern accounts date from 1799, 1832, 1833, 1866, 1867, and 1966.[11]

Another way of mapping the dust near the comet orbit is to plot Earth's position relative to the comet orbit and the position of the comet at the time of these past

[8] D. K. Yeomans, K. K. Yau and P. R. Weissman, The impending appearance of comet Tempel–Tuttle and the Leonid meteors. *Icarus* **124** (1996), 407–413.

[9] O. R. Hainaut, K. J. Meech, H. Boehnhardt and R. M. West, Early recovery of comet 55P/Tempel–Tuttle. *Astron. Astrophys.* **333** (1998), 746–752.

[10] I. Hasegawa, Further comments on the identification of meteor showers recorded by the Arabs. *Q. J. R. Astron. Soc.* **37** (1996), 75–78.

[11] P. Jenniskens, Meteor stream activity. II. Meteor outbursts. *Astron. Astrophys.* **295** (1995), 206–235; P. Brown, The Leonid meteor shower: historical visual observations. *Icarus* **138** (1999), 287–308.

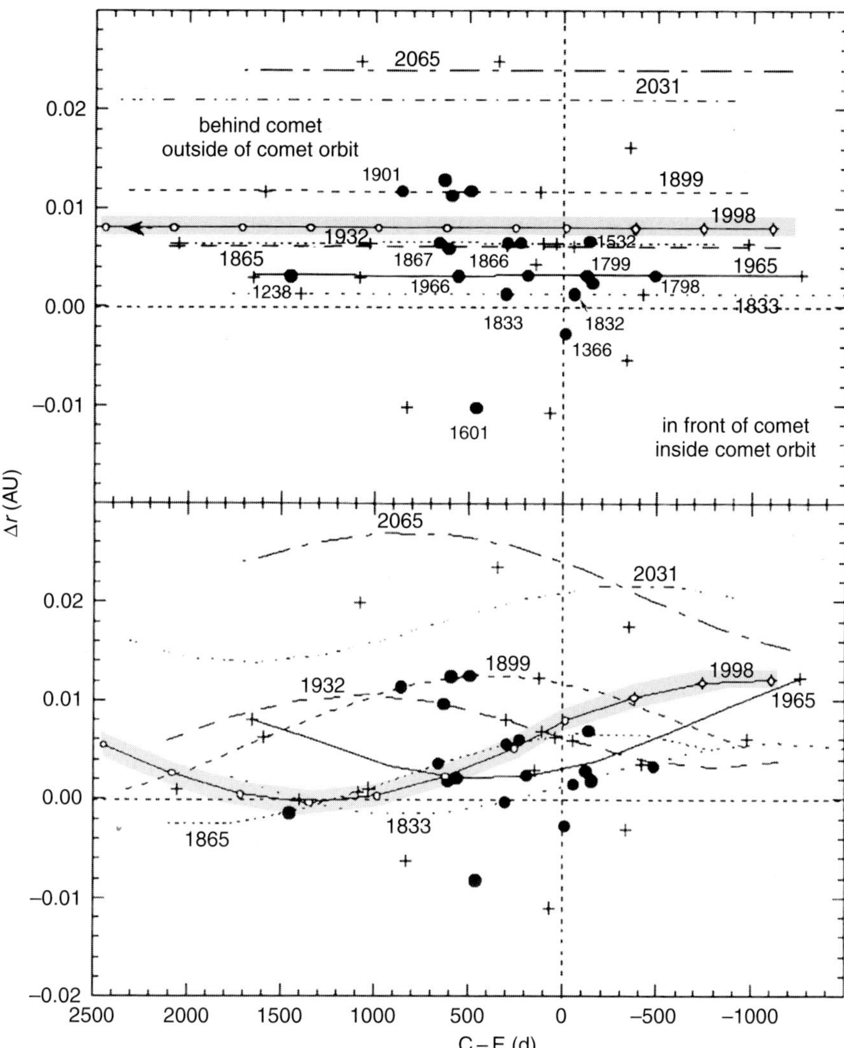

Fig. 15.3 Encounter conditions during past Leonid storms (•) and outbursts (+), expressed as miss-distance Δr (in AU) and as a time delay since the comet passed by the node = position along the dust trail (Comet – Earth, in d). Top diagram, from Yeomans (1981); bottom diagram by the author after correcting the miss-distance for Sun's reflex motion.[12]

storms (top of Fig. 15.3). From this, Zdenek Sekanina[13] and later Don Yeomans found that most rich showers (•) happened when Earth was just outside of the orbit of

[12] P. Jenniskens and S. J. Butow, An airborne mission to characterize the Leonid storms: results of validation efforts. *Proc. Leonid Threat and Impact Hazard Workshop*, ed. Manhattan Beach, CA, D. Lynch (El Segundo, CA: The Aerospace Corporation, 1998).

[13] Z. Sekanina, Meteoric storms and formation of meteoroid streams. In *Asteroids, Comets, Meteoric Matter*, ed. C. Cristesu, W. J. Klepczynski and B. Millet. (Bucharest: Editura Academici Repullicii Socialiste Romania, 1975), pp. 239–267.

the comet and behind the comet, even though a storm was not always reported when that was the case. In the upcoming 1998 return (gray line), the miss-distance would be relatively large, and Yeomans concluded that Earth might well encounter another storm in 1998 or 1999, but not at the level of 1966.

After including the Sun's reflex motion to estimate the miss-distance from the trail, rather than the miss-distance from the 1998 comet orbit (bottom of Fig. 15.3), I thought that 1999–2002, in fact, looked promising. Trails would then be moving into Earth's path.[14]

Publicity in advance of the 1966 storm was nothing like that leading up to the 1998 return, perhaps because few people expected anything really sensational after the failures of 1899 and 1932.[15] Peak activity was expected (with some reservation) over Europe.[16] Observing efforts were hampered by bad weather. The storm was detected by radar, but photographic and visual observations were poor. Those that succeeded produced photographs with numerous tracks (Fig. 15.4). A group at Kitt Peak in Arizona reported counts that translated to ZHR = 150 000 per hour.[17]

In the early 1990s, I found that there was a problem with the high-end ZHR estimates from 1966.[18] To go from a raw count to the zenith hourly rate, after correcting for the known radiant altitude, involves some scaling factor that is multiplicative. It includes the condition of the sky and observer perception, as well as all other factors that define the detection probability. The strategy is to plot the calculated ZHRs on a logarithmic scale (Fig. 15.5). If an error is made in the detection probability, then the whole graph will shift up or down. When plotting the 1966 Leonid rates on a logarithmic scale, the Springhill meteor patrol radar results reported by *Zdeňka Plavcová*,[19] *Bruce McIntosh*, and *Peter Millman*[20] showed a linear rise, indicative of an exponential increase or rates. The counts from Kitt Peak followed the exponential increase of the radar data, except right at the peak, when they started counting meteors by opening their eyes for one second only and then recalling each meteor from memory (Fig. 15.5). In doing so, they changed their detection probability by a factor of 7. Ignoring this peak, and scaling to early counts by experienced AMS observer *Karl Simmons* in Florida, I arrived at a ten times lower peak rate of only ZHR = 15 000 ± 3000/h.[21] Still, this is more intense than any storm since 1833 (Fig. 15.4).

[14] P. Jenniskens and S. J. Butow, An airborne mission to characterize the Leonid storms: results of validation efforts. *Proc. Leonid Threat and Impact Hazard Workshop*, ed. Manhattan Beach, CA, D. Lynch (El Segundo, CA: The Aerospace Corporation, 1998).

[15] G. Spalding and A. McBeath, Remembering the 1966 Leonids. *WGN* **26** (1998), 9–10.

[16] Anonymous, A good Leonid year? *Sky Telescope* **32** (1966), 251; J. Shubart, Comet Tempel–Tuttle (1866 I) *IAUC 1907*, ed. O. Gingerich (Cambridge, MA: IAU Minor Planet Center, 1965).

[17] D. Millon, Observing the 1966 Leonids. *J. Brit. Astron. Assoc.* **77** (1967), 89–93.

[18] P. Jenniskens, Meteor stream activity. II. Meteor outbursts. *Astron. Astrophys.* **290** (1995), 206–235.

[19] Z. Plavcová, Radar observations of the Leonids in 1965–66. In *Physics and Dynamics of Meteors*, ed. L. Kresák and P. M. Millman *IAU Symp*. 33. (Dordrecht: Reidel, 1968), pp. 432–439.

[20] B. A. McIntosh and P. M. Millman, The Leonids by Radar – 1957 to 1968. *Meteoritics* **5** (1970), 1–18.

[21] P. Jenniskens, Meteor stream activity. II. Meteor outbursts. *Astron. Astrophys.* **290** (1995), 206–235.

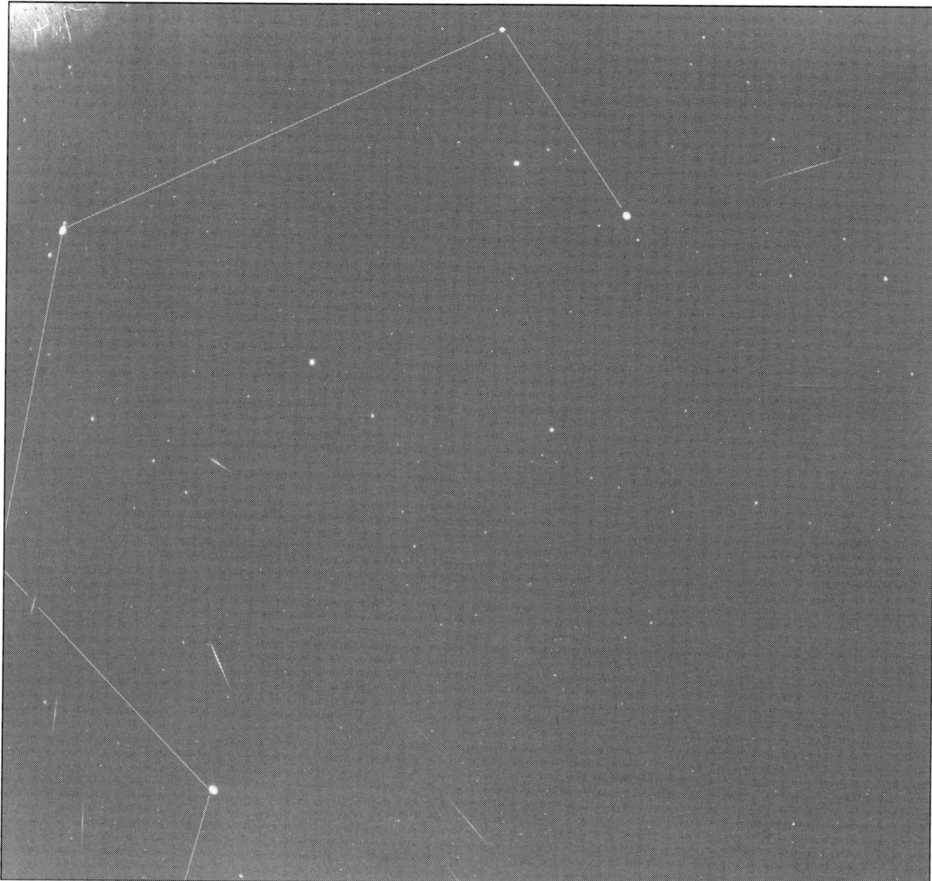

Fig. 15.4 The November 17, 1966 Leonid storm. Plate 76 A of the large format plates exposed by the astroscience program students at Pan American College in Edinburgh, Texas. Photo courtesy of Paul D. Maley of the NASA Johnson Space Center Astronomical Society.

15.2 The 1998 Leonid shower

If it was possible to improve the quality and variety of observations, a new storm could address difficult questions such as whether satellites were in danger of being hit. Did meteors contribute organic compounds to the prebiotic chemistry of life on the early Earth? And what could we learn about the comet itself? To answer such questions demanded the deployment of modern observing techniques. I set out to raise awareness among scientists that were using such instruments for other purposes.

The idea of fielding these instruments from aircraft came up during lunch at the NASA Ames Research Center cafeteria in 1995, in discussions with pilot *Captain Steven J. Butow* of the 19th RSQ at Moffett Air Field (Fig. 15.6), then also a research assistant at NASA Ames and fellow associate of the SETI Institute. What started as

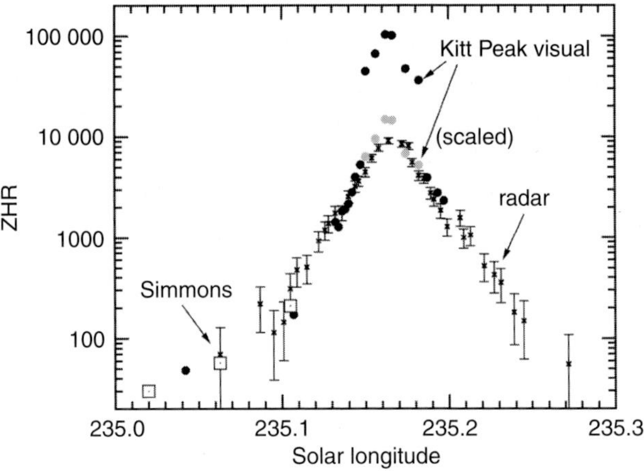

Fig. 15.5 A ZHR profile of 1966 storm. Radar data are scaled to visual results.[22]

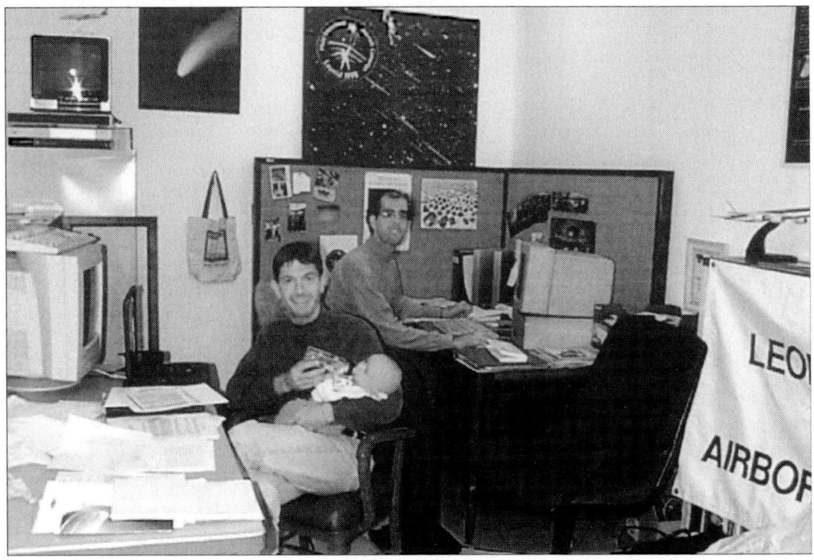

Fig. 15.6 A day at the office: Leonid MAC mission HQ at the NASA Ames Research Center. Front: Captain Steve Butow with his son Mason.

an opportunity to aim cameras out of the zenith port of the C-130 aircraft that Steve routinely piloted, quickly grew into a Leonid *multi-instrument aircraft campaign* ("MAC") when we discovered that NASA Ames itself had an active airborne observing program in atmospheric science.

[22] P. Jenniskens, Meteor stream activity. II. Meteor outbursts. *Astron. Astrophys.* **290** (1995), 206–235.

One of the biggest hurdles was that such MAC missions were a common tool in atmospheric science, but a novelty for peer reviewing astronomers. Fortunately, NASA was responsive to our proposals and funding was cobbled together from research programs in Planetary Astronomy, Astrobiology, and Suborbital MITM. Astrobiology had just been established as a new field of NASA research and, thanks to the efforts by the NASA Astrobiology Integration Office leads Lynn Harper and Greg Schmidt, ours would be NASA's first Astrobiology mission.

Ideal for the deployment of the instruments would be the NKC-135 "Flying Infrared Signatures Technology Aircraft" (FISTA), which was operated by the 418th Test Wing (at that time the 452nd Flight Test Squadron). The aircraft had 20 upward looking windows of 30 cm diameter. Experiments on that aircraft were mostly run by a program at Hanscom AFB, executed by Stewart Radiance Laboratory (SRL). A team led by Joe Kristl participated in the mission. The Aerospace Corporation, in an effort led by Ray Russell, would deploy newly developed mid-infrared sensors.

Our second aircraft was long unknown. Shortly before the campaign, the NASA Ames airborne program was transferred to NASA Dryden and, because of that, NASA's DC-8 Airborne Laboratory was no longer available. Also, the Kuiper Airborne Observatory had been grounded to facilitate the development of the next generation airborne astronomical observatory "SOFIA," and could not be used. When the DC-8 aircraft was no longer available, we identified a new opportunity during the '98 Leonid Meteoroid Storm and Satellite Threat Conference.' Chet Gardner, of the University of Illinois, would speed up his development of a 2-beam Fe Boltzmann Lidar if the National Science Foundation (with NASA funds) supported the deployment of its "Electra" aircraft stationed in Broomfield. Thus, three US Government agencies were involved.

The Leonid MAC website became a rallying tool among observers in the years leading up to the first airborne campaign. On one of the pages, I had summarized the available predictions. Most modelers put the 1998 storm around crossing the comet orbital plane, which was at solar longitude $235.258\,43 \pm 0.000\,11^\circ$ (epoch, March 8, 1998). This corresponded to 19:43 UT, November 17, which demanded an observing location in eastern Asia (Fig. 15.7). The storm could be anywhere in the time interval of ± 3 h centered around this time.

The Dutch Meteor Society eyed China as the ideal ground-based observing site, but initial contacts with Chinese astronomers led nowhere. Attending the May, 1998 IAU Colloquium 198 *Comets in Space and Time*, I had a chance to sit down at a restaurant with staff astronomers of Nanjing and Beijing Observatory. After an hour of small talk in broken English, I asked: "Who is in charge?" For the next fifteen minutes, only Chinese was spoken. It was decided then and there that Nanjing Observatory would fly one team out to Delingha in the Qinghai desert of upper Tibet, led by Casper ter Kuile and Marc de Lignie, while the Beijing Astronomical Observatory would drive a second team led by Hans Betlem to Xing Long

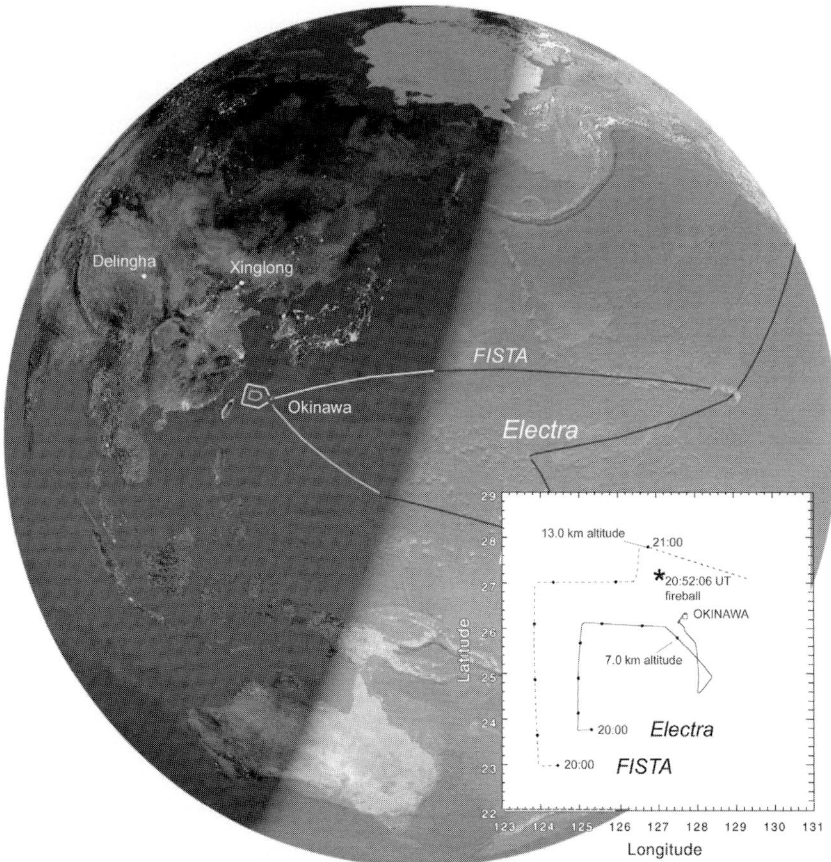

Fig. 15.7 The 1998 Leonid MAC mission route and a perspective of Earth from the viewpoint of the shower at the anticipated peak time.

Observatory. Each site had two stations for multi-station photography and observers from Ondrejov Observatory would assist the Xing Long station.

Okinawa was chosen as our best staging area for the airborne mission. A team of Japanese participants, led by Hajime Yano, included the Japanese Broadcasting Company NHK, who would deploy the new line of high-definition TV cameras, a technology that had emerged on the Japanese market and was on the brink of being introduced in the United States. With participants from Japan, Canada, the Czech Republic, the UK, the Netherlands, China, and the United States, this was truly an international campaign.

Observers at both sites in China could not believe their luck. Fireballs rained from the sky on the night of November 16/17, to be repeated at less high rates on November 17/18. In Okinawa, we had prepared the instruments for deployment into the wee hours and I was woken by a BBC reporter at noon on November 17 with the question: "How was it?" My stomach turned thinking I might have mistaken the time zone . . .

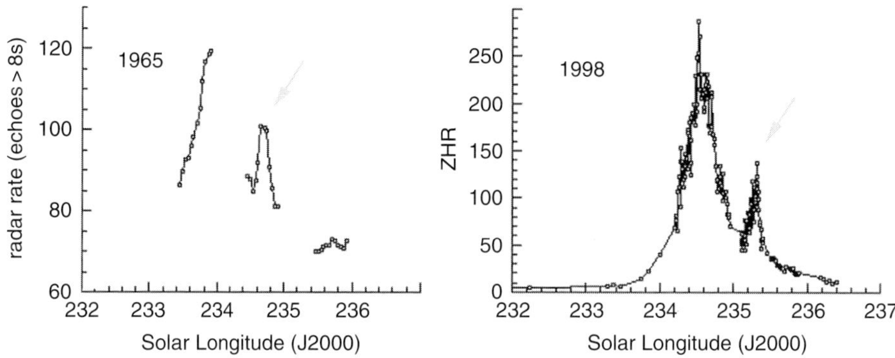

Fig. 15.8 Springhill meteor patrol radar rates in 1965,[23] compared to visual rates in 1998. A secondary peak of faint meteors was observed following the Filament in both years.

Sadly, we missed a good shower that night, but not because of an error in the date. Limited funding dictated the availability of crew, which prohibited us from flying two sorties after each other. I had chosen to deploy in the next night, when we hoped to see a storm. Both aircraft deployed that evening as planned, with all instruments operational and the scientists in great spirits. But as soon as we climbed through the cloud deck and saw the stars, it was clear: the shower was not as brilliant as reported from Europe earlier that day. Instead of a storm, we observed the rising branch of a small peak until twilight at 21:00 UT (arrow, Fig. 15.8). Fortunately, we discovered that observing near the horizon more than quadrupled the rate of meteors observed, permitting accurate measurements of flux. And one fireball just before dawn (20:52 UT, Fig. 15.7) created a spectacular persistent train.

The lack of a storm was a great disappointment. One European observer described the evening of November 16: *I opened the window before going to bed and immediately noticed a bright Leonid. I thought: 'Wow, this is going to be nice tomorrow', and went to bed*. Hopes had been pinned on a strong storm, the tail of which might still be observed in Europe at ~200/h. I had calculated the activity curve of Fig. 15.9 based on the observation that historic storm profiles appeared to be broader and have stronger wings ("background") when the Earth passed further away from the comet orbit. We now know that this is because Earth passed further from the dust trail center, instead.

In hindsight, it is surprising that we saw anything at all. Even though the secondary peak occurred at the predicted time, its cause remains a mystery. No dust trail was in Earth's path (Fig. 15.10). The peak occurred when Earth passed by the 1932-dust and 1899-dust trails, but the closer 1899 trail was depleted in grains from a recent encounter by Earth and the 1932 trail was far from Earth's orbit. From the model by Vaubaillon, it is not clear that the Earth's perturbation of the 1899-dust trail could

[23] B. A. McIntosh and P. M. Millman, The Leonids by Radar – 1957 to 1968. *Meteoritics* **5** (1970), 1–18; Z. Plavcová, Radar observations of the Leonids in 1965–66. *IAU Symp.* **33** (1968), 432–439.

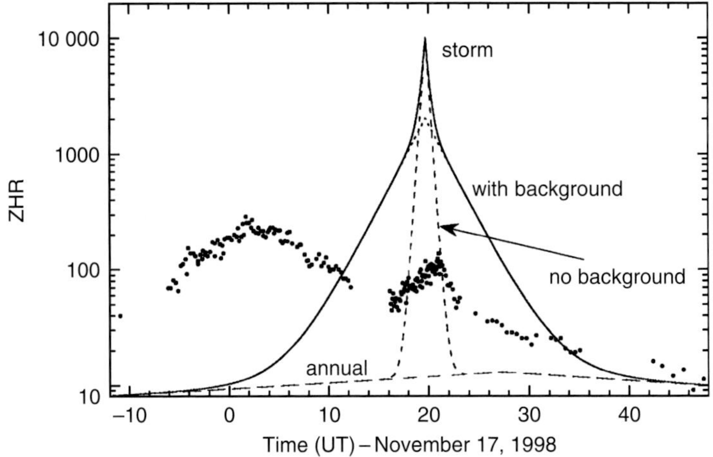

Fig. 15.9 Predicted and observed Leonid rates in November of 1998.

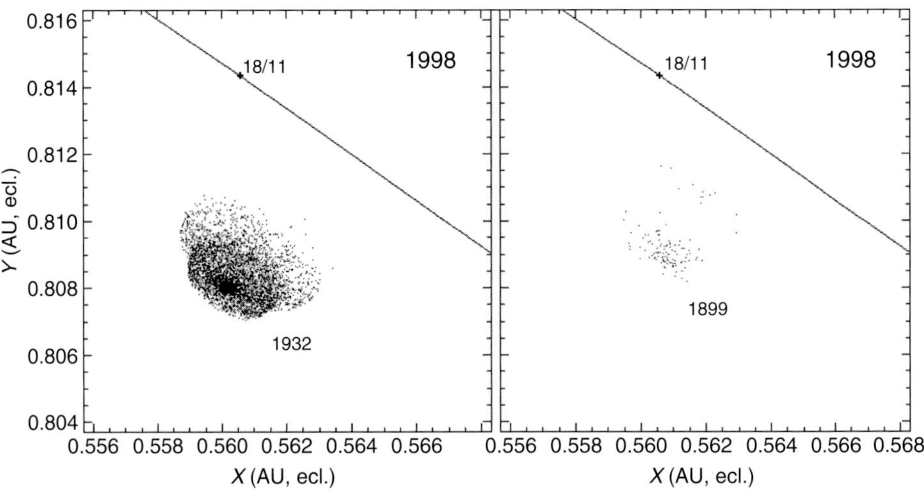

Fig. 15.10 The 1932 and 1899 dust trails inside Earth's orbit at the time of the "second peak." Calculations by Vaubaillon.

have produced sufficient particles to be in Earth's path. That leaves open the possibility that these were meteoroids with high ejection speeds in the right direction (~100 m/s) or unusually fluffy particles with low $\beta \sim 10^{-2}$ from the 1965- or 1932-dust trails, as proposed by Brown and Arlt.[24]

[24] P. Brown and R. Arlt, Detailed visual observations and modelling of the 1998 Leonid shower. *Mon. Not. R. Astron. Soc.* **319** (2000), 419–428; R. Arlt and P. Brown, Visual results and modeling of the 1998 Leonids. *WGN* **27** (1999) 267–285.

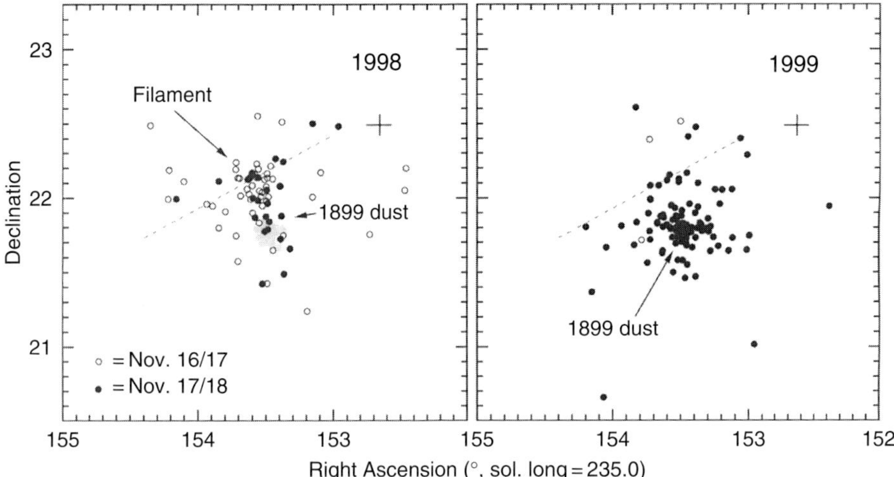

Fig. 15.11 Photographic radiants measured on November 16/17 and 17/18, 1998, compared to similar results from 1999, after correction of all radiant positions to the epoch corresponding to solar longitude 235.0° (J2000), with ΔR.A./$\Delta\lambda_\odot = +0.659°$ and ΔDecl./$\Delta\lambda_\odot = -0.325°$. Calculations: H. Betlem and M. C. de Lignie, Dutch Meteor Society.

Indeed, our measured radiant dispersion was large, spread out over ±0.5° in R. A., ±0.7° in Declination.[25] Most of that dispersion is due to meteors from the Filament component, which was richer in bright meteors (Chapter 14). The Filament had a hollow circular distribution of radiants, with some dispersion in a direction along the ecliptic plane (Fig. 15.11, dashed line). However, a few meteors were detected at the expected position of the 1899-dust trail (shaded area). This confirms that the second narrow peak was due to recently ejected meteoroids, sufficiently perturbed to impact Earth.

15.3 Forecasting meteor storms: the position of the dust trail

The approach to predicting meteor storms was not new, but first developed at the end of the nineteenth century, following the disappointing 1898 Leonid return.[26] Astronomers *G. Johnstone Stoney* and *Arthur M. W. Downing* in England and German astronomer Adolf Berberich showed, in principle, how to predict meteor storms.[27] The key was to

[25] H. Betlem, P. Jenniskens, J. van 't Leven *et al.*, Very precise orbits of 1998 Leonid meteors. *Meteoritics Planet. Sci.* **34** (1999), 979–986; M. C. de Lignie, M. Langbroek and H. Betlem, Temporal variation in the orbital element distribution of the 1998 Leonid outburst. *Earth, Moon Planets* **82/83** (2000), 295–304.

[26] D. J. Asher, Leonid dust trail theories. *Proc. Int. Meteor Conf.*, Frasso Sabino 1999, ed. R. Arlt. (Potsdam: International Meteor Organization, 2000), pp. 5–21.

[27] G. J. Stoney and A. M. W. Downing, Perturbations of the Leonids. *Astrophys. J.* **9** (1899), 203–210; G. J. Stoney and A. M. W. Downing, Leonids, ephemerides of two situations in the stream. *Mon. Not. R. Astron. Soc.* **59** (1899), 539–541; G. J. Stoney and A. M. W. Downing, Next week's Leonid shower. *Nature* **61** (1899), 28–29; G. J. Stoney and A. M. W. Downing, The Leonids – a forecast. *Nature* **63** (1900), 6.

calculate the planetary perturbation on the orbit of individual meteoroids and not only on that of the comet. After calculating the perturbations on dust ejected in 1866, Stoney and Downing predicted a peak time for November 16, 1899, 06:00 UT. Sadly, no meteor storm was observed that year, the miss-distance was just too big, which set back the development of models by a century.

A first partial success was achieved by *J. G. Davies* and *Wladyslaw Turski*[28] of the University of Manchester in 1962, who based their initial orbit on the parent comet rather than the stream. They showed that the Draconid storms of 1933 and 1946 could have been caused by dust ejected in 1894 (we now think that the 1900-dust and 1907-dust were more important, Table 6).

The effort was taken up again after the 1966 storm. A few particle orbits, sampled from the stream, were studied by *E. I. Kazimircak-Polonskaja* and *N. A. Beljaev* in Leningrad, in a collaboration with Astapovic and Terentjeva at Kiev State University.[29] They merely demonstrated how slightly different meteoroid orbits evolve dramatically differently.

By cleverly choosing the orbital period of the particle to match the delay between the return of the comet (in 1998, say) and the meteor shower (in 2001, say), the task was made manageable. In 1972, Kazan State University astronomers *Yu. V. Evdokimov* and *E. D. Kondrat'eva* in Cheliabinskii, Russia (in the radar group led by O. I. Bel'Kovich), found that dust ejected in 1800 could approach Earth in November 1833, the comet itself having crossed the ecliptic in January, 1833, and also found that the storm of 1966 was probably due to dust ejected in 1899.[30] This was a key result that led to the later break through paper in 1985 that tied meteor storms to dust ejected at specific returns of the comet.

Starting from a good orbit of the parent comet and using the tools developed by Kazimircak-Polonskaja and Beljaev, *E. D. Kondrat'eva* and *E. A. Reznikov* examined where a dust particle released at perihelion would cross the ecliptic plane near Earth's orbit a few revolutions later. The key is to find the one initial orbit that makes the particle hit the ecliptic plane when Earth is nearby. A wider orbit creates more delay so if, for example, the storm is expected four years after the comet return, from dust ejected three revolutions earlier, then the initial orbit is that of the comet three revolutions ago, and the initial difference in orbital period (ΔP) of the particle needs to be 4/3 years. If only radiation pressure causes the orbit to be wider, then according to Kondrat'eva and Reznikov the delay would be given by:[31]

[28] J. G. Daves and W. Turski, The formation of the Giacobinid meteor stream. *Mon. Not. R. Astron. Soc.* **123** (1962), 459–471.

[29] E. Kazimircak-Polonskaja, N. A. Beljaev, I. S. Astapovic and A. K. Terenteva, Investigation of perturbed motion of the Leonid meteor stream. In *Physics and Dynamics of Meteors*, ed. L. Kresák and P. Millman, *IAU Symp.* (Dordrecht: Reidel, 1968), pp. 449–475.

[30] Yu. V. Evdokimova and E. D. Kondrat'eva, Some observations about the origin of the Leonid meteor showers (1833 and 1966). *Trudy Kazan. Gor. Astron. Obs.* **38** (1972), 68–73 (in Russian).

[31] E. D. Kondrat'eva and E. A. Reznikov, Comet Tempel–Tuttle and the Leonid meteor swarm. *Solar Syst. Res.* **19** (1985), 96–101 (reprint from: *Astronomicheskii Vestnik* **19** (1985), 144–151).

$$\Delta P = \frac{3}{2}\,P\beta\,(1+e)/(1-e) + \frac{1}{2}\,P\beta \tag{15.1}$$

where P is the orbital period ($\sim$33.3 yr), e is the eccentricity, and β the ratio of radiation over gravitational forces. The first term in this equation describes the increase in the orbital period due to an adopted increase in the semimajor axis of the orbit (the biggest effect), while the second part describes the effect of seemingly reducing the Sun's mass by radiation pressure for the adopted semimajor axis. Because the second term can usually be neglected, the corresponding change in semimajor axis $a = P^{2/3}$ for given β is:

$$\Delta a = a\,\beta(1+e)/(1-e) = a\,\beta' \tag{15.2}$$

If a dust particle is released at perihelion from a comet with orbital elements (q_c, e_c) with no ejection speed, then the initial dust particle orbit (q_d, e_d, with $a_d = q_d/(1-e_d)$) is:

$$\begin{aligned} &a_d = a_c + \Delta a,\ q_d = q_c,\ \text{and} \\ &e_d = (e_c + \beta(1+e_c)/(1-e_c))/(1-\beta(1+e_c)/(1-e_c)) = (e_c+\beta')/(1-\beta') \end{aligned} \tag{15.3}$$

Choose β so that the orbit has the right ΔP, then start from this initial orbit and calculate the planetary perturbations. Vary β until the particle returns at the precise time.

Kondrateva and Reznikov then argued that it did not matter if the particles had ejection speed instead (for isotropic ejection): all particles seen in a storm arrived at Earth at the same time. If the particle had a different β (say that relevant for the size of a meteoroid causing a visible meteor) it could still return at the correct time if its ejection speed in forward direction of motion at perihelion was:

$$V_{ej} = (V_c/2)\beta + V_c/6\ \beta(1-e)/(1+e) \tag{15.4}$$

where V_c is the velocity of the comet at perihelion. In principle, similar equations can be written for a change in the angular elements as soon as ejection out of the plane is considered (Appendix B). Unless there is nonisotropic emission, it does not really matter what is responsible for delaying the particle, as long as the particle returns at the right time.

Usually, at least a series of test particles are calculated. In order to find the exact position of the trail (Δr, $\lambda_\odot$), the miss-distance and node are interpolated from those particles that returned closest to the appropriate date.

Nowadays, orbit integration programs (such as the *JPL/Horizons* software on the internet) can be used to quickly calculate the planetary perturbations of particles ejected with the right difference in orbital period to return at the time of the predicted encounter with Earth.

Kondrat'eva and Reznikov discovered that this very simple, perhaps naïve, approach resulted in dust trail positions that matched past observed storms. In particular, they found that dust ejected in 1899 was close to Earth's orbit in 1966 ($\Delta r = 0.000\,39$ AU minimum distance) and the approach time was November 17.50, exactly as observed for the 1966 storm. In contrast, the 1866-dust trail had a minimum distance of $\Delta r = 0.0036$ AU on November 18.16, and the 1932-dust trail did not come closer than $\Delta r = 0.0016$ AU on November 17.42.

They went on to predict that a meteor storm was due in November 18.09, 1999 from a trail ejected in 1899, for an ejection speed of (no radiation pressure effect, thus upper limits) $V_{\rm ej} < 13.7$ m/s, while the trail ejected in 1866 would be close to Earth's orbit in 2000, 2001, and 2002. Especially the miss-distances in 2001 and 2002 were very small (0.0002 and 0.0001 AU), for reasonable ejection speeds of 14.1 and 17.0 m/s, respectively. They followed up with a paper in 1997.[32] In 1977, Edward K. L. Upton[33] at Griffith Observatory independently found the trails responsible for the 1833 and 1966 storms and predicted the occurrence and cause of the 1999 storm, putting the time correctly close to 02:00 UT on November 18.

The problem all along had been one of taking a too complicated approach with too small computing power. Earlier models, such as that of Wu and that of Brown, assumed a wide range of relatively high ejection velocities, and allowed particles to be ejected at different directions and times along the comet orbit. Because few particles ended up in Earth's path from each set of calculations, particles passing far from Earth's orbit were included in making the predictions.

The successful timing of the 1998 Draconid outburst by Reznikov, just before the 1998 Leonids (Chapter 19),[34] spurred renewed efforts at predicting meteor storm occurrences in the more simple manner. By the time of the pending 1999 Leonids, two other teams had taken up this approach and prepared lists of dust trail encounters, finding more historic agreements. *Robert H. McNaught* (Siding Spring Observatory) and *David Asher* (Armagh Observatory) in Australia and the UK, and Finnish amateur astronomer *Esko Lyytinen.* David presented his predictions at the April, 1999 Leonid MAC workshop at NASA Ames. Results were published in *MNRAS* and in a nice paper in *WGN, Journal of the IMO*.[35] Lyytinen published his results in the more obscure journal *Metaresearch*,[36] edited by astronomer Tom Van Flandern, but results were presented at the Leonid MAC website.

Throughout this book, results are shown from calculations by Lyytinen, who worked out a procedure that considered only radiation pressure. He arrived at this simplification based on van Flandern's idea that a comet dust coma could consist of

[32] E. D. Kondrat'eva, I. N. Murav'eva and E. A. Reznikov, On the forthcoming return of the Leonid meteoric swarm. *Solar Syst. Res.* **31** (1997), 489–492.

[33] E. K. L. Upton, The Leonids were dead, they said. *Griffith Observer* **41** (1977), 3–9.

[34] E. D. Reznikov, The Giacobini–Zinner comet and Giacobinid meteor stream. *Trudy Kazan. Gor. Astron. Obs.* **53** (1993), 80–101 (in Russian).

[35] D. J. Asher, The Leonid meteor storms of 1833 and 1966. *Mon. Not. R. Astron. Soc.* **307** (1999), 919–924; R. H. McNaught and D. J. Asher, Leonid dust trails and meteor storms. *WGN* **27** (1999), 85–102.

[36] E. Lyytinen, Leonid predictions for the years 1999–2007 with the satellite model of comets. *Meta Res. Bull.* **8** (1999), 33–40.

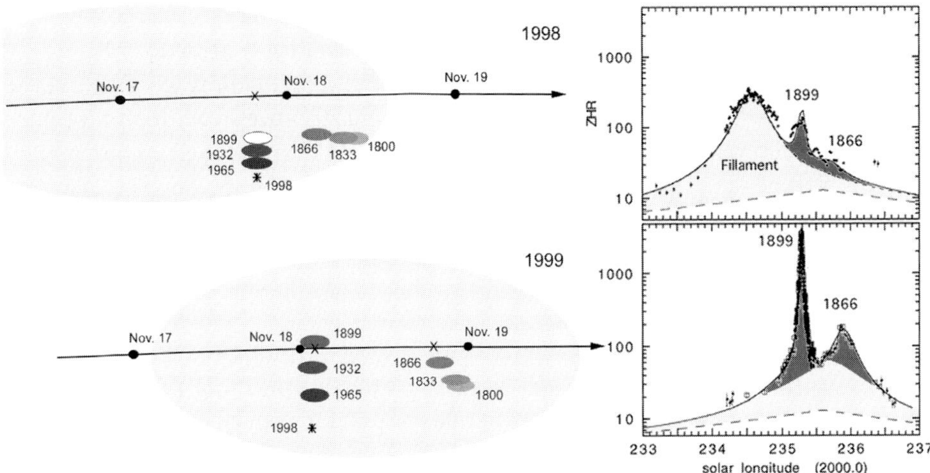

Fig. 15.12 The position of dust trails during the 1998 and 1999 Leonid showers from diagrams by David Asher, Armagh Observatory. To the right are the observed meteor rates from the Leonid MAC mission and visual counts by members of the IMO.

meteoroids orbiting the nucleus, which would "leak" dust away from the libration points. This idea goes back to the nineteenth century sandbox model of a comet. The idea is controversial. Marco Fulle calculated that only ~1% of large dust grains can be ejected in orbits around the nucleus.[37] However, when released at perihelion, the resulting orbit that matches the correct return date is identical and thus the method and the results are the same as those of Asher and Reznikov.

The dust trail positions calculated by McNaught and Asher[38] are shown in Fig. 15.12. In November of 1999, Earth was about to cross the 1899-dust trail on the inside.

Leading up to the 1999 storm, Rob McNaught drew attention to the fact that the trail positions were calculated relative to the center of the Earth. The peak time will be different if an observer is located north or south of the center of the Earth.[39] The expected storm would peak 12 min later for observers at the Earth's North Pole (where the observer is above the plane) and 12 min earlier if at the South Pole (where the observer is below).

15.4 Forecasting activity: the distribution of dust in the trail

Once the position of the dust trail is calculated, the next step is to estimate the activity of the shower. That depends on the distribution of the dust in the dust trail,

[37] M. Fulle, Injection of large grains into orbits around comet nuclei. *Astron. Astrophys.* **325** (1997), 1237–1248.
[38] R. H. McNaught and D. J. Asher, Leonid dust trails and meteor storms. *WGN* **27** (1999), 85–102.
[39] R. H. McNaught, Visibility of Leonid showers in 1999–2006 and 2034. *WGN* **27** (1999) 164–171; R. H. McNaught and D. J. Asher, Variation of Leonid maximum times with location of observer. *Meteoritics Planet. Sci.* **34** (1999), 975–978.

determined by the conditions of ejection and radiation pressure. The symbols in Fig. 15.12 could only be an approximation of the actual dust distribution.

Calculations of dust density distribution in the trail were out of reach. Instead, the expected peak rate (ZHR^{max}) for passing at different distances from the trail center (Δr), and at different distances behind the comet (Δa_0), were derived empirically by studying the encounters of historic Leonid storms, using simple functions to describe the dust distribution in the trail. The expected activity is:[40]

$$\text{ZHR} = \text{ZHR}^{\text{max}}\, f(\lambda_\odot) f(\Delta r)\, f(\Delta a_0) f_{\text{m}} \tag{15.5}$$

where "max" refers to the highest dust density in a one revolution trail at the peak of the dust distribution behind the comet and $f(\lambda_\odot)$ is the flux profile in Earth's path Δa_0 marks the initial change in the semi-major axis of oscular elements after ejection.

In each new orbit, the dust trail continues to stretch because neighboring particles continue to have slightly different orbital periods (Chapter 3). This stretching is included in what is called the "*mean anomaly factor*," f_{m}. For N revolutions since ejection:

$$f_{\text{m}} \sim 1/N \tag{15.6}$$

The mean anomaly factor also includes the stretching and compacting of the dust trail due to perturbations of the orbital period of the dust. This effect can be determined from the time delay between two test particles close to the moment of encounter.

In order to get the dust densities right, it is necessary to determine f_{m} in a very narrow time interval, which means having test particles close together. Close encounters with Earth create gaps in the trail (Fig. 15.13) only about 10 d wide, across which rates vary by over a factor 10 (Fig. 15.14). Jupiter causes similar depletions, but those tend to be more shallow (~factor 2) and have a longer duration (see Fig. 15.30 later).

The gaps were initially believed to be because the grains were scattered out of the dust trail. However, the majority of dust grains near Earth are only perturbed modestly, slightly accelerating the grains in the days when Earth approaches the stream and decelerating when moving away. That makes small changes to the orbital period. After one orbit, those particles that were accelerated come back later, piling up after the gap, while those slowed down come back early, piling up before the gap. Opening a curtain, so to speak, on a nearly dust free region. This is nicely demonstrated in the model by Vaubaillon (Fig. 15.14).[41] There is no change in the width of the dust trail. That is understood, because most particles are perturbed when Earth is relatively far from the node and all particles in the trail cross-section are perturbed to nearly the same amount.

[40] R. H. McNaught and D. J. Asher, Leonid dust trails and meteor storms. *WGN* **27** (1999), 85–102.
[41] J. Vaubaillon, Dynamics of meteoroids in the solar system. Application to the prediction of meteoric showers in general, and Leonids in particular. Ph.D. thesis. IMCCE, Observatoire de Paris, Paris, France (2002).

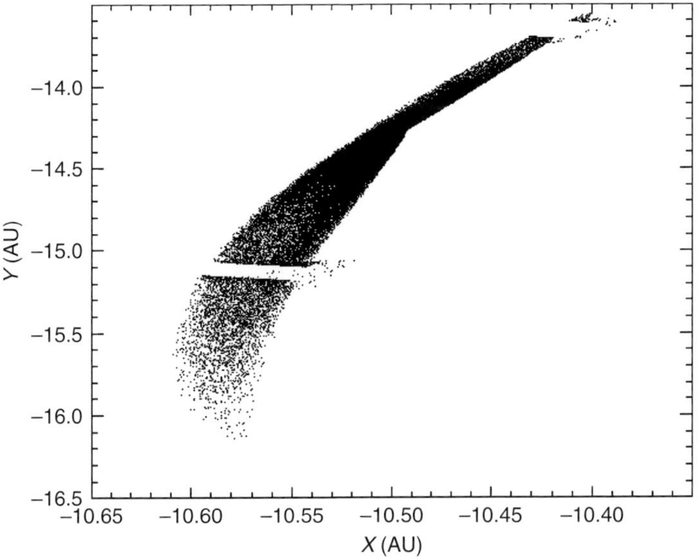

Fig. 15.13 The gap in the 1767 dust trail as seen in 1822 near aphelion. The gaps are from November 1799 and 1800, respectively.

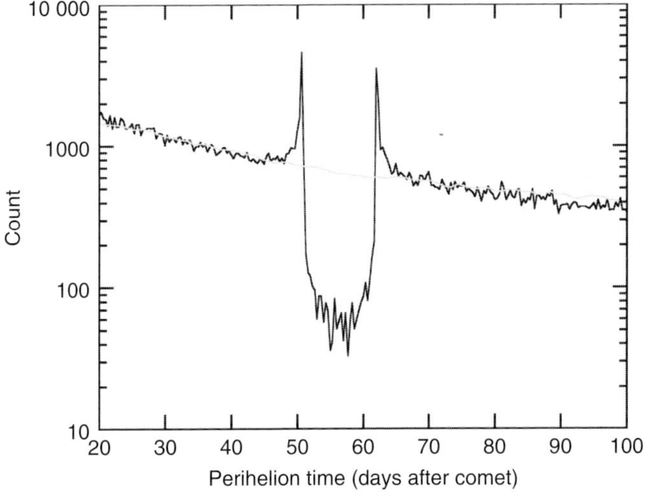

Fig. 15.14 Detail of the dust distribution resulting from a close encounter with Earth. The thin gray line is the unperturbed dust density.

15.5 The Leonid storm of 1999

Greatly encouraged, Steve Butow and I organized a second Leonid MAC mission to Israel, with the support of Tel Aviv University's *Dr. Noah Brosch*. Mark Fonda at NASA Ames acted as our program manager. Funding came together a bit more than just a month prior to the mission and we even had a chance to scout the facilities in

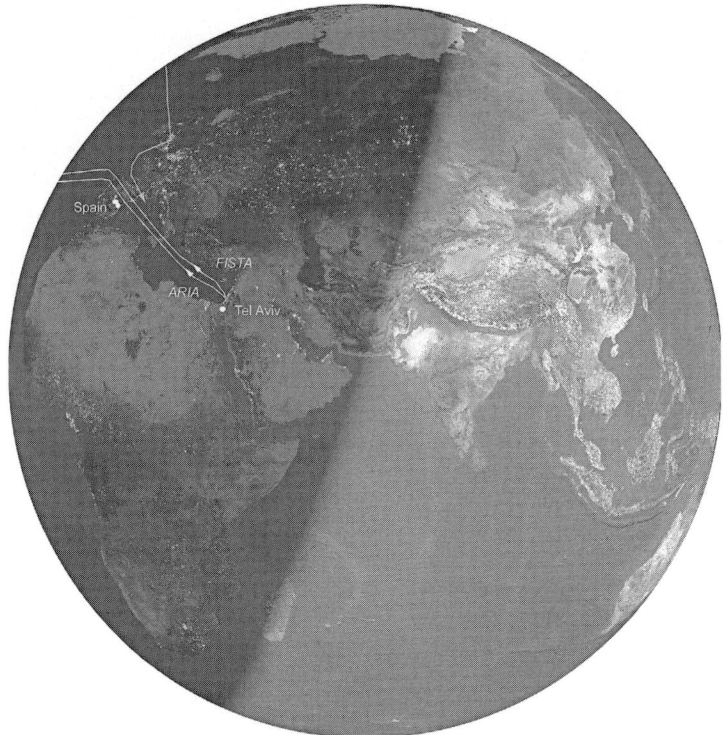

Fig. 15.15 The 1999 Leonid MAC trajectory and Earth at time of the Leonid storm.

Israel in advance. Noah arranged for ground-based support in Israel from visual, video, radar, and VLF/ELF technologies and DMS teams were supported by Josep Trigo-Rodríguez and S.O.M.Y.C.E. in Spain.

That year, 35 researchers and a wide range of sensors received a chance to observe the Leonid storm under ideal observing conditions onboard two USAF KC-135 aircraft, "FISTA" and "ARIA" (Fig. 15.15).[42] Those included mid-infrared spectrographs for the detection of organic matter in meteors and their trains. A C-130 plane flown by Steve helped transport the necessary liquid helium.

This time, we took part in the global effort to provide near-real time flux measurements to satellite operators by having a team of eight visual observers, with video headset displays, scan the output of video cameras mounted in front of the optical windows. A live S-band video uplink was established through NASA's TDRSS communication network to share our experiences with the world. US Air Force Colonel Simon (Pete) Worden, sponsor of the near-real time flux measurement effort,

[42] P. Jenniskens and S. J. Butow, The 1999 Leonid multi-instrument aircraft campaign: the storm from altitude. *WGN* **27** (1999), 305–307; P. Jenniskens, S. J. Butow and M. Fonda, The 1999 Leonid multi-instrument aircraft campaign – an early review. *Earth, Moon Planets* **82/83** (2000), 1–26.

Fig. 15.16 View of the 1999 Leonid storm from the ARIA aircraft. Compilation of high-definition TV images, courtesy of Hajime Yano, ISAS and NHK.

expressed the wish to join our mission after having helped to establish one of Brown's ground sites in the Negev.

My nervousness was on many levels. My program managers at NASA HQ and the USAF 452nd FTS at Edwards AFB had worked hard to secure support and accommodate our requirements. A huge preparation had led up to mission night. The previous night had shown no hint of a rate increase and, walking on the beach in Tel Aviv, I thought of Dorothea Klumpke's disappointing ascent in a balloon a century earlier.

Lack of knowledge of the radial dust distribution in the dust trail made the expected peak rate estimates uncertain. Earth would meet the 1899-dust near its inner edge in 1999. Asher adopted a Gaussian profile for $f(\Delta r)$ and $f(\Delta a_0)$. Lyytinen assumed $f(\Delta r) = (1 + \Delta r^2)^{-0.5p}$ for the radial distribution and an even more complex equation for $f(\Delta a)$ to account for the 1969 outburst. McNaught and Asher calculated a ZHR = 1200/h from a Gaussian distribution. Brown and Jones were expecting a heavy storm and I was quoted as saying: "I am optimistic ... we may get rates as high as 7000 per hour or so."[43]

Closer to the peak, Asher voiced similar doubts as his fellow countryman Stoney a century earlier: "It's marginal as to whether the meteor activity will reach storm level in 1999 ...." Wu and Williams insisted "only a few Leonids will be seen."

When the plane emerged from the cloud deck over Tel Aviv, I was tremendously relieved to watch the sky filled with meteors (Fig. 15.16). It was still early and rates

[43] J. Rao, Anticipation: the 1999 Leonid meteors. *WGN* **27** (1999), 177–194.

were climbing. Video games specialist *Chris Crawford* had designed a counting unit that calculated a mean of the instantaneous storm rates clicked off by the eight person strong flux measurement team, each equipped with a video headset display each hooked up to one of the intensified cameras behind the aircraft windows. Based on the smooth increase of rates, we could tell the satellite community to brace itself for half an hour prior to the peak, if the peak time held. The peak of the storm at 60 meteors/min was observed just west of Greece. Colonel Worden needed no more poignant images to stress how the storm increased the risk of impacts for satellites in orbit: *This is impressive if you are in the space business!* he said, while describing the storm to our audience at home.

Watching the video screens, I was only too glad that everything went as planned, and simply forgot to look out of the windows.

At the peak of the storm, lightning flashes were seen near the horizon and someone yelled: “Sprite”! We wondered if the meteors were causing the upward going lightning, which was observed for the first time over Europe. Sprites and elves emanated from a small storm complex with two centers over Croatia and Albania. Another highlight in the night was the capture of the 3.4 μm fingerprint of complex organics in the wake of a meteor train. This, and a lack of CN emission in meteors, was evidence that organic molecules are not efficiently broken down into small diatomic compounds, but instead more interesting complex compounds may have been sprinkled on the early Earth (Chapter 34). For a moment we saw the sky as it looked back in those days.

The predicted peak time was less than ten minutes off. One century after the biggest disappointment in astronomy, the precise timing of the 1999 Leonid storm was heralded as a great breakthrough. Instead, most predictions had been close that year, ranging from when Earth passed the comet orbit node at 01:48 UT, up to an hour later. Only Joe Rao had put the peak at 4:17 UT,[44] while Ignacio Ferrin of the University of the Andes in Venezuela had followed Kresák in putting the peak time at 08:48 UT, falsely raising hopes for observers in the Americas. The real breakthrough was that the dust responsible for the storm was now identified as dating from the 1899 return, providing a basis for interpreting the observations in a new way.

The storm was sharply peaked, but lasted most of the night, displaying strong tails to the activity profile (Fig. 15.17). The accurate measurements of flux from the aircraft closely matched that of a Lorentz profile defined by one width parameter (W):[45]

$$f(\lambda_\odot) = (W/2)^2 \;/\; ((\lambda_\odot - \lambda_\odot{}^{\max})^2 + (W/2)^2) \qquad (15.7)$$

W is the full-width-at-half maximum. Once the peak of the shower was matched, the wings were matched as well. Until that point, dust trail cross sections had been

[44] J. Rao, Anticipation: the 1999 Leonid meteors. *WGN* **27** (1999), 177–194; J. Rao, Awaiting the storm. *Sky Telescope* **97** (1999), 48–50.

[45] P. Jenniskens, C. Crawford, S. J. Butow *et al.*, Lorentz shaped comet dust trail cross section from new hybrid visual and video meteor counting technique – implications for future Leonid storm encounters. *Earth, Moon Planets* **82/83** (2000), 191–208.

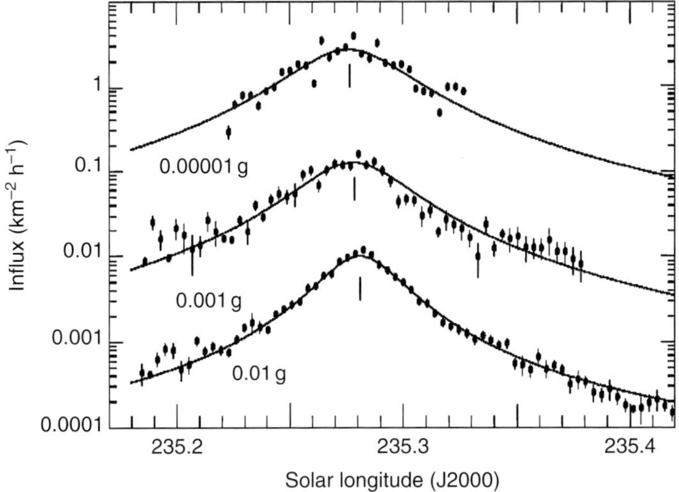

Fig. 15.17 Width of the storm profile for different particle sizes. Data from intensified cameras onboard the ARIA and FISTA aircraft aimed low and high in the sky.

fitted by a Gaussian profile or a set of two exponentials with exponent B. Now W $\sim$ 2 log (2)/B. This Lorentz function was later adopted by Lyytinen, improving the prediction model. If the profile were symmetric around a given heliocentric displacement Δr^{max}, then the full-width-at-half maximum (W, in astronomical units) would depend on the distance from the trail center:

$$(W/2)^2 = (W^{max}/2)^2 + (\Delta r - \Delta r^{max})^2 \quad (15.8)$$

Instead the observed distribution falls off slower than the curve matched by $W^{max} = 0.000\,42 \pm 0.000\,05$ AU, being the width of the trail at its center, and $\Delta r^{max} = +0.000\,77$ AU (see later Fig. 15.22).

A Lorentz function also described $f(\Delta a)$ behind the comet:[46]

$$f(\Delta a_0) = (W_a/2)^2 \ / \ ((\Delta a - \Delta a^{max})^2 + (W_a/2)^2) \qquad \Delta a_0 \gg 0 \quad (15.9)$$

where $W_a = 0.16 \pm 0.02$ AU describes the dispersion along a 1 revolution dust trail, and $\Delta a_0{}^{max} = 0.12 \pm 0.01$ AU is the mean increase in a from radiation pressure. The delay translates to $\beta = 5.8 \times 10^{-4}$ (Eq. 15.2) or a particle density of 0.97 ± 0.13 g/cm^3 for 0.003 g (+3.5^m) Leonids (Eq. 3.1).

With fewer meteors in each step of decreasing magnitude, the high rate permitted the measurement of the particle size distribution over a wide range in magnitude (Fig 15.18). During the storms, the rate of fireballs increased much like that of fainter

[46] *Ibid.*

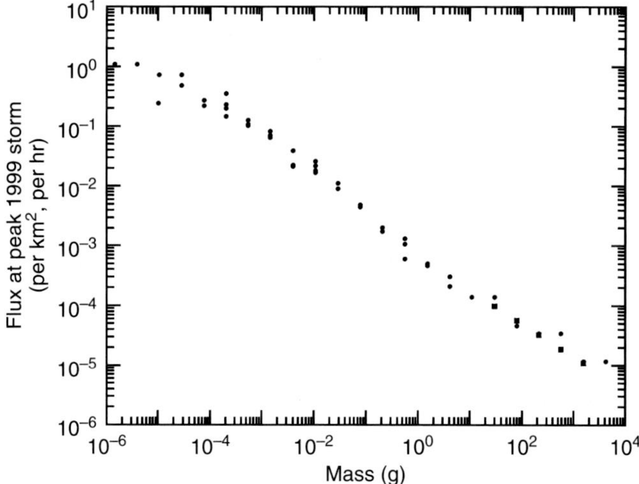

Fig. 15.18 The mass influx per unit of log mass at the peak of the 1999 leonid storm. Data are a compilation of results from 1999 and 2001.

meteors. However, even with the high rates, our counts ran out around magnitude -12^m, below which the observed impact flashes on the Moon became a way to count. The biggest Leonid fireballs associated with the 1998 Filament had a magnitude of $\sim -14.5^m$ and a mass of $\sim$5 kg according to Pavel Spurný (Eq. 4.1 gives 80 kg).[47]

The low cameras onboard the aircraft tended to observe brighter meteors than the high cameras and by plotting both sets of data, a small shift in the peak time is evident (Fig. 15.17). The peak of the shower was found to shift only slightly with particle mass and the trail broadens slightly, not inconsistent with the Whipple formula for dust ejection. Brown had earlier found that the radar observations of the width of the 1966 Leonid storm profile as a function of meteoroid mass followed the Whipple formalism.[48]

The video tapes showed an immense number of meteors. *Peter Gural* studied the statistical distribution in the stream and did not find an excess of pairs, but meteors did appear to occur at certain spots in the sky for periods of time, like mini-dust trails, each of which could result from the decay of boulders in the comet coma (Fig. 15.19).

The magnitude distribution was surprisingly flat and dominated by the brighter meteors, with $\chi = N(m+1)\,/\,N(m) = 1.8$.[49] Exhaustive observations of Leonid MAC video tapes by amateur astronomer *David Holman*[50] detected a deficiency in +5 and

[47] P. Spurný, H. Betlem, J. van 't Leven and P. Jenniskens, Atmospheric behavior and extreme beginning heights of the 13 brightest photographic Leonids from the ground-based expedition to China. *Meteoritics Planet. Sci.* **35** (1998), 243–249.

[48] P. G. Brown, Evolution of two periodic meteoroid streams: the Perseids and Leonids, Ph.D. thesis, UWO, Canada (1999).

[49] P. Gural and P. Jenniskens, Leonid storm flux analysis from one Leonid MAC video AL50R. *Earth, Moon Planets* **82/83** (2000), 221–247.

[50] D. Holman and P. Jenniskens, Leonid storm flux from efficient visual scanning of 1999 Leonid storm video tapes. *WGN* **29** (2002), 77–84.

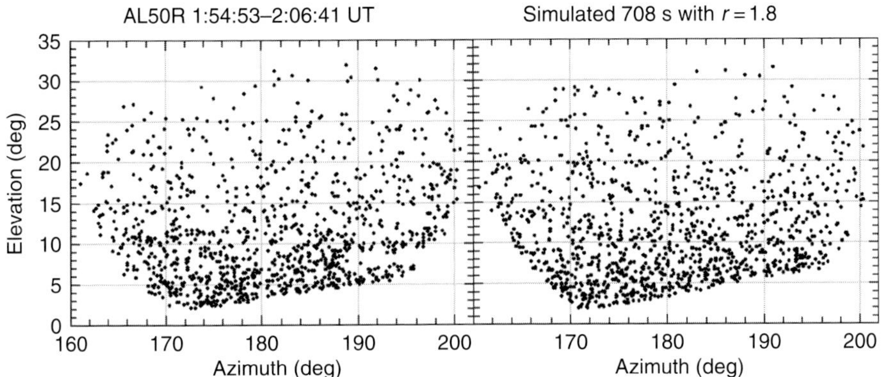

Fig. 15.19 The observed end-points of meteor trails in one of the video cameras, compared to a model of random meteor activity. Calculations by Peter Gural, SAIC. The gap in the lower right corner (left graph) is due to the time stamp on the video.

$+6^{m}$ Leonids over that expected for an exponential distribution. The trend was confirmed from telescopic CCD camera observations in Japan, which detected fewer Leonids than expected in the range +7 to $+11^{m}$. This result is in contrast to that reported by Brown,[51] based on very few meteors. Radar proved not to be an efficient technique for detecting fast Leonid meteors.

A disappointed Peter Brown was clouded out that year while trying to observe from the Canary Islands. Our teams in Israel and Spain were more fortunate and recorded a large number of multi-station orbits.[52]

As before, Brown modeled the activity profile of the 1999 storm and confirmed that the bulk of meteoroids were ejected with a low speed of 1–12 m/s and very close to perihelion, >90% between 1.5 and 0.98 AU. It seemed as if storm prediction models were converging and all there was to know about this topic was now established. There was a great sense of accomplishment. Behind the scenes, however, there was still much disagreement. For example, concerning the question of whether or not the 1932-dust trail contributed to the 1999 storm.[53] Moreover, the predictions for the 1866-dust and 1833-dust trails were far off. The Asian observations collected by the *International Meteor Organization*, and analyzed by Rainer Arlt, showed a peak at solar longitude $\lambda_{\odot} = 235.87 \pm 0.04^{\circ}$ and a broad FWHM $= 0.28^{\circ}$ profile, instead of the narrow predicted peak at $\lambda_{\odot} = 236.04^{\circ}$ (Fig. 15.20).

In order to understand this discrepancy, Lyytinen and Van Flandern introduced nonradial radiation pressure effects, called the "A_2-*effect*", analog to nonradial forces

[51] P. Brown, M. D. Campbell, K. J. Ellis *et al.*, Global ground-based electro-optical and radar observations of the 1999 Leonid showers: first results. *Earth, Moon Planets* **82/83** (2000), 167–190.

[52] H. Betlem, P. Jenniskens, P. Spurný *et al.*, Precise trajectories and orbits of meteoroids from the 1999 Leonid meteor storm. *Earth, Moon Planets* **82/83** (2000), 277–284.

[53] R. Artl *et al.*, First global analysis of the 1999 Leonid storm. *WGN* **27** (1999), 286–295.

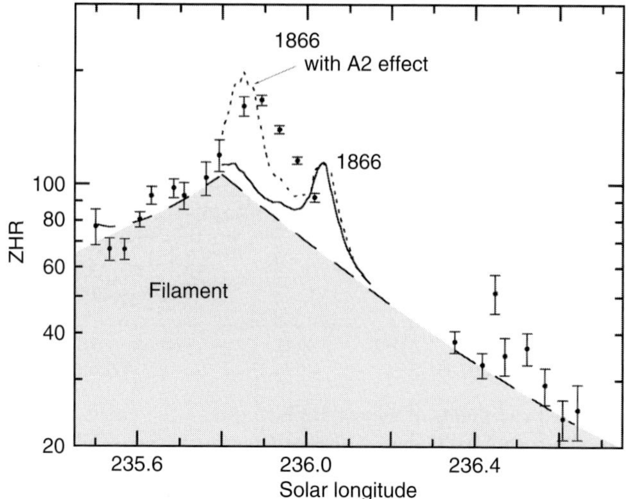

Fig. 15.20 The 1999 encounter with the 1866-dust trail as observed from eastern Asia (IMO observations). Calculations by Esko Lyytinen without (solid line) and with the "A_2" effect (dashed).

in comet dynamics.[54] Such forces were introduced in the model simply by artificially accelerating or decelerating particles when they passed perihelion. This results in small trail shifts, with dramatic effects when the dust trails are distorted. In 1999, one of the 1866-trail segments moved closer to Earth's orbit, while another moved farther away, changing the predicted peak time and rate dramatically (dashed line, Fig. 15.20). The problem with this proposed effect is that one value of A_2 does not always improve the predicted encounters of all dust trails in a model and the result of shifting the trail positions could have occurred for other reasons. In his models with A_2 included, Lyytinen assumed a distribution of A_2 values, centered on zero.

15.6 Multiple trail encounters in 2000

This set the stage for the next Leonid return of 2000, which would be badly hampered by a last quarter Moon close to the radiant. Three dust trail encounters were predicted: the 1932-dust trail on November 17, and the 1733 and 1866-dust trails on November 18 (Fig. 15.21).[55] The 1932 and 1866-dust trail encounters would occur at about the same time of the day and both would be visible from the eastern USA. All trail encounters would be at relatively large distances from the center of the trails and this made the 2000 returns very valuable for understanding how the dust is distributed away from the trail center.

[54] E. J. Lyytinen and T. Van Flandern, Predicting the strength of Leonid outbursts. *Earth, Moon Planets* **82/83** (2000), 149–166.

[55] D. J. Asher and R. H. McNaught, Expectations for the 2000 Leonids. *WGN* **28** (2000), 138–143.

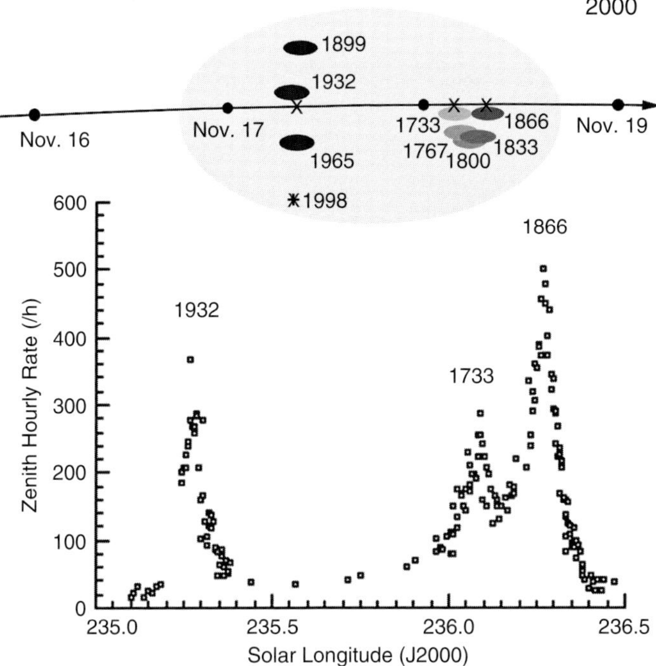

Fig. 15.21 Dust trail positions in 2000 and flux measurements.

After the 1999 Leonids, interest quickly waned and there was no funding for an intensive 2000 Leonid campaign. Ground-based observations were organized at I.S.T.E.F. on Cape Caneveral, Florida, and at Mount Lemmon Observatory in Arizona. In addition, a small airborne effort using a Cessna aircraft was made possible by astronomer *B. Å. S. Gustafson* of the University of Florida at Gainesville. Observers of the Dutch Meteor Society observed from Spain to cover the 1733-dust trail encounter.

The ground-based observations confirmed the return of three dust peaks at the predicted time of the dust trail positions, with the flux measurements from the Cessna aircraft establishing the flux profile of the otherwise poorly observed 1932-dust peak of faint meteors in the bright moonlight (Fig. 15.21). Results agreed best with the predictions by Lyytinen.[56] Still, the first maximum in 2000 was timed 12–17 min too early, the second peak was 16–20 min too late, and the final maximum ended up being 39–43 min late. The peak rates were also consistently off by up to a factor of two.

Our 2000 observations helped us to understand how the width of the shower varies with distance from the trail center, later complimented with data from 2001 and 2002

[56] P. Jenniskens and B. Å. S. Gustafson, The rare 1932-dust trail encounter of November 17, 2000, as observed from aircraft. *WGN* **28** (2000), 209–211; R. Arlt and M. Gyssens, Results of the 2000 Leonid meteor shower. *WGN* **28** (2000), 195–208.

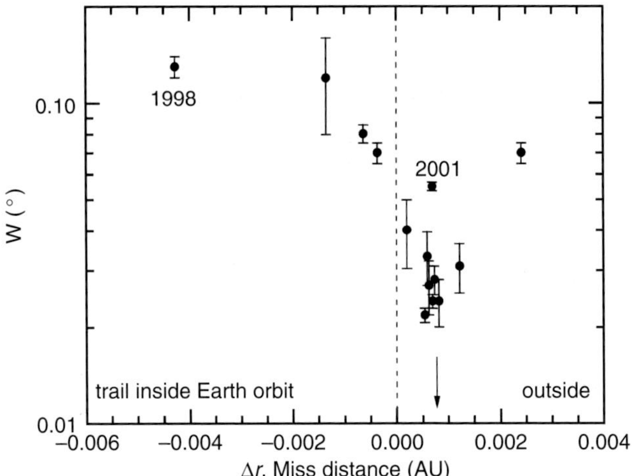

Fig. 15.22 The measured full-width at half maximum (W) of Leonid outbursts depending on how far we passed from the trail center (Δr) calculated by Vaubaillon. The 1998 and 2001 results are off, presumably due to an uncertain trail position.

(Fig. 15.22). The width increased exponentially away from the trail center, with apparent $W = 6.3 \pm 0.6 \times 10^4$ km at the center.

The distribution of dust in the heliocentric direction had to be derived in a similar manner, but the data showed significant scatter. I noticed that rates were seemingly too low when the trails were too wide to fit the overall pattern, which implied that the trail positions were not always correctly calculated. Permitting for some uncertainty in the calculated $\Delta r = r_C - r_E$ of the trail center, I found that the dust density, too, declined exponentially away from the trail center, described by:

$$f(\Delta r) = 10^{-B|\Delta r - \Delta r_{\max}|} \tag{15.10}$$

where $B = 1450$ AU^{-1}. This completes the formalism. The prominent Lorentz-wings in the path of Earth were not present in the perpendicular dust distribution towards the Sun.

The calculated positions of the dust trails tended to be slightly too far out from the center of the Sun (about $\Delta r_{\max} = 0.000\,25$ or $0.000\,77$ AU in Vaubaillon's model below). The trail center may circle around the expected center with an amplitude of about 0.0005 AU, but not as a function of the epoch of ejection as I thought early on.

15.7 The 2001 Leonid storms: where to go?

The year 2001 arrived with the prospect of a favorable encounter with the 1866-dust trail just inside Earth's orbit (Fig. 15.23), as foretold by Kondrat'eva and

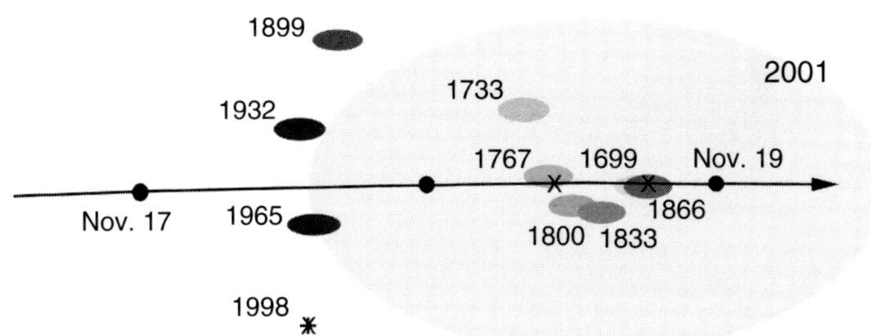

Fig. 15.23 Dust trail encounters in 2001.

Reznikov.[57] In addition, there was a trail ejected in 1767 just outside Earth's orbit. The 1866-dust trail was best seen from Asia and Australia, the 1767-dust trail from the Americas. The precise position of the trails and the heliocentric distribution of dust in the trails would decide which peak would be best. The issue was furher complicated because the 1767-dust trail was highly perturbed.

Asher and McNaught favored the east-Asia peak by a factor of ten (Fig. 15.24), including a contribution from the 1699-dust trail.[58] Esko Lyytinen[59] included the A_2-effect deduced from the 1932-dust trail observations in 2000 and the 1866-dust trail in 1999 to conclude that the 1767 trail would be later compared to earlier results given in Table 6a (10:28 UT rather than 10:03 UT) and the rate would be 2000/h, while the 1866-dust trail was not affected. That peak was expected at 18:27 UT, rather than 18:22 UT, with a peak ZHR of 5000. This would come on top of contributions from the 1699-dust trail (ZHR = 2600/h at 18:03 UT).

I suspected that both the 1767 and 1866-dust trails had been calculated incorrectly by 0.000 25 AU, putting the 1767-dust trail closer to Earth's orbit (higher rates) and the 1866-dust trail farther away (lower rates). Rates for both peaks could be the same, which implied that observing from the USA was a good option.[60]

Peter Brown and *Bill Cooke* further fueled the debate about whether the assumptions made in calculating the trail positions were correct. They concluded from their latest model results[61] that it was rather the 1799-dust trail that was going to be closest to Earth's orbit, while other trails would be far away (Fig. 15.24). They predicted a broad storm over Hawai'i.

The argument raised by Brown and Cooke was swiftly settled by observations. Loyal to his forecast, Cooke traveled to Hawai'i and missed the storm. Asher and

[57] E. D. Kondrat'eva, I. N. Murav'eva and E. D. Reznikov, On the forthcoming return of the Leonid meteoric swarm. *Astronomicheskii Vestnik* **31** (1997), 489 (in Russian).

[58] R. H. McNaught and D. J. Asher, The 2001 Leonids and dust trail radiants. *WGN* **29** (2001), 156–164.

[59] E. Lyytinen, M. Nissinen and T. Van Flandern, Improved 2001 Leonid storm predictions from a refined model. *WGN* **29** (2001), 110–118.

[60] P. Jenniskens, Forecast for the remainder of the Leonid storm season. In *Proc. Meteoroids 2001 Conf.*, 6–10 August 2001, Kiruna, Sweden, ed. B. Warmbein, *ESA SP* **495** (2001), 247–254.

[61] P. Brown and B. Cooke, Model predictions for the 2001 Leonids and implications for Earth-orbiting satellites. *Mon. Not. R. Astron. Soc.* **326** (2001), L19–L22.

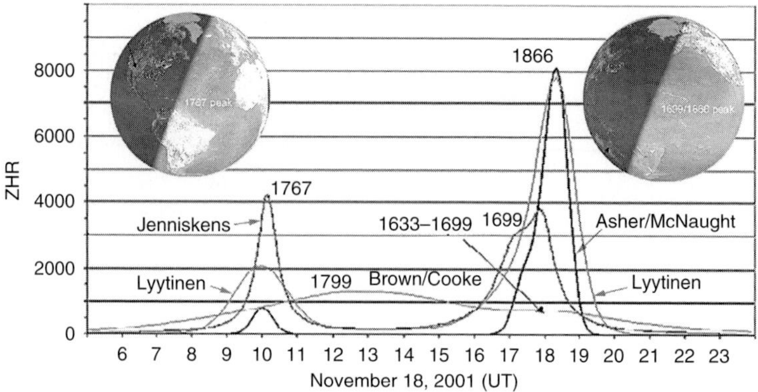

Fig. 15.24 Calculated trail positions (top) and predicted Leonid activity in 2001. After a graph by Bill Cooke, NASA Marshall Space Flight Center.

many other storm chasers traveled to Australia and witnessed a great storm at the predicted time. I stayed in the US and much to my pleasure saw rates increase to nearly half the Asian peak.

Bringing together the 2001 Leonid MAC mission was a long fight against all odds. Overall interest in the Leonids had decreased. Operators now assumed that the satellites would weather any future storm just fine. Simple mitigation efforts to decrease the risk were left in place. The scientific importance of future storms was the ability to apply lessons learned, deploy more modern instruments, and study how the dust ejecta of the comet were distributed in three dimensions. To keep momentum, a very successful *2000 Leonid MAC Workshop* was organized in Tel Aviv, Israel, where we were the guests of Noah Brosch and colleagues at Tel Aviv University. Results from the 1999 campaign were quickly published in a proceedings: a 606 page edited book *Leonid Storm Research* that was printed and available only seven months after the workshop in time for the 2000 Leonid season.[62]

The funding situation for further Leonid MAC missions was more uncertain than ever. Data reduction support did not come through, leaving me without a salary. I lost my office at Ames and moved to the SETI Institute (located just off base in Mountain View), but continued to be supported by the Ames Astrobiology Integration Office. The increased security measures from the notorious September 11 attack on the New York World Trade Center made an international campaign impossible. Instead, I worked towards a CONUS (Continental USA) mission to cover the 1st peak. My only spark of hope was my positive outlook for high rates over the USA.

When I hit rock bottom in mid October, funding for a mission did arrive from NASA's Astrobiology program, and we ended up flying a single plane (FISTA) from Alabama to California on a westward trajectory. DMS and CMS observers

[62] P. Jenniskens, F. Rietmeijer, N. Brosch and M. Fonda (eds), *Leonid Storm Research*, (Dordercht: Kluwer, 2000), 606pp (reprint from: *Earth, Moon Planets* **82/83**, 1–606).

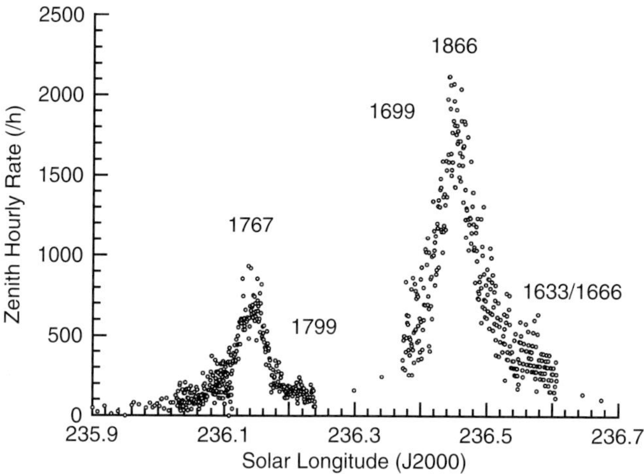

Fig. 15.25 Summary of Leonid MAC flux measurements of the 2001 Leonids.

established ground stations in Australia and China to cover the second peak. Even with funding in hand, the mission was canceled more than once due to the heavy use of FISTA in the post "9–11" world, because *FISTA* doubled as an in-air fueling tanker. In the end, the mission could be done only because it was a single-night deployment at the weekend, when no other programs were using the plane. We never got to fly a test flight that year.

Foreign nationals, including myself, were grounded, their paperwork for flight clearances having fallen victim to the late arrival of funding, the repeated cancellations, and the post "9–11" regulations. We organized alternative observing sites at Mount Lemmon Observatory in Arizona, centered on a very active group led by Jim Richardson and David Holman, while Japanese participants traveled to Mauna Kea, Hawai'i. *Morris Jones* and *Jane Houston-Jones* perfected the tools to tally meteor counts by mouse clicks in support of our contribution to the near-real time flux measurement program at Marshall Space Flight Center. They were sent to Alice Springs, Australia. Results of the near-real time counts from Mount Lemmon, Alice Springs, and the FISTA aircraft are summarized in Fig. 15.25.

US born Astrobiology Academy students Avi Mandell and Emily Schaller stepped in to operate my imagers and spectrographs onboard "FISTA," bringing back a rich harvest of data. I observed the 2001 Leonid storm from the ground at Red Rock Canyon in California, in the good company of *Ian Murray* of Canada.

Ian and I only left Edwards Air Force Base after all the other observers were settled on FISTA and the aircraft had departed. It was dark. We drove North, away from the city lights and into the hills, until 30 min before the Leonid radiant rose above the horizon. At that moment, we passed by the small entrance to the Red Rock Canyon wilderness area where a car was parked. To find our own private observing site, we drove into the park, turned a corner and found more cars parked along the road, with people camping in the bushes. Around us were hundreds of campers who specially came out into the wild to watch the meteor storm. Early in the night, we heard singing and later the oohs and aahs whenever a bright fireball appeared. The sky was clear at our site, completely clouded only 30 km further south. I was amazed by the bright red and orange colors of many Leonids, green on occasion. At one point, I stood facing the radiant and thought myself in the "10-forward" of spaceship Earth. Like Aristotle, pinching slippery seeds with my thumb and finger.

We were back at Edwards AFB at sunrise, just in time to see FISTA land and pose for a picture, and were delighted to hear the reports that had arrived from Mount Lemmon and Australia (Fig. 15.25).

After all that pain, it felt wonderful that the 2001 Leonid Meteor Storm was honored by *Astronomy magazine* as the "news story of the year 2001" (beating the spectacular landing of the "NEAR" spacecraft on the asteroid Eros and other astronomical discoveries).[63] The science mentioned was that conducted in the context of Leonid MAC. Among our most exciting results was the detection of a halo and "shock" in a -3^{m} Leonid by (Danish national) Hans Stenbeak-Nielsen of the University of Alaska at Fairbanks.

The storms had been spectacular from both the Americas and Asia. There are now numerous great Lenoid storm images on the internet. One example is shown in Fig. 15.26. This composite photograph from the 2001 Leonid return is from a location near Ayers Rock (Uluru) in Australia, by Fred Espenak. The composite is made from eight subsequent 9 min exposures on Fuji Superia 800 film, using a 16 mm full frame f/2.8 Nikon FE Nikkor fisheye lens. The individual frames were aligned in Photoshop™ and the meteor trails in each frame were then combined into one photo. To the lower right, the meteor shower radiates from the sickle shaped string of stars in Leo. Orion (center left) appears upside down in this southern hemisphere image with the "winter" Milky Way. Note the two bolides in Taurus (center left) and near the Large Magellanic Cloud (top right).

An interesting aspect of the Asia peak was the blend of both the 1866 and 1699 dust trails. This was recognized from the distribution of faint relative to bright meteors in the shower and from the radiant positions. Between 17:45 and 18:00 UT ($\lambda_{\odot} = 236.439° - 236.449°$), the magnitude distribution index was a low $\chi = 1.82 \pm 0.09$ from the 1699-dust trail, then it changed to a higher $\chi = 2.18 \pm 0.09$ from the 1866 dust[64] between 18:15 and 18:40 UT ($\lambda_{\odot} = 236.460° - 236.477°$). Kouji Ohnishi and coworkers,[65] also detected slightly brighter meteors on telescope-mounted CCD camera frames around 16:50 UT than around 18:00 UT. From those observations,

[63] S. Maran, Astronomy's top 10. The most significant cosmic news stories of the last year. In *Astronomy's Explore the Universe*, 9th edn, 2003 pp. 7–17.

[64] B. Suzuki, T. Sugaya, Y. Sato and T. Mizuno, Astro-HS management committee, The results of ASTRO-HS video observation network. *ESA SP* **500** (2002), 201–204.

[65] K. Ohnishi, T. Yanagisawa, K-I. Torii *et al.*, Leonid radiant project 2001. *ISAS SP* **15** (2003), 29–38.

Fig. 15.26 The 2001 return of the Leonids as seen from central Australia around 20 km north of Ayers Rock (Uluru) in a composite image. Photo © 2001 by Fred Espenak.

Yanasigawa *et al.*[66] derived two mean radiants separated by $0.157 \pm 0.059^\circ$, which agrees well with the predicted difference by McNaught and Asher.[67] The peak times were best predicted in the A_2-model by Lyytinen *et al.*

The stronger than expected (by some) showing of the 1767-dust trail, the unexpected abundance of bright meteors, and the longer than anticipated width all added to the enjoyment of the observers, but did not give much confidence in our ability to predict meteor storms accurately.

15.8 The 2002 Leonid storm: isolated storm profiles

Two further Leonid storms were predicted for 2002, with the Earth passing near the center of the trails and far from the comet position (Fig. 15.27). This time, there was no doubt that the storms would happen. An airborne mission in westward direction would enable us to observe both peaks 6.5 h apart at almost identical radiant elevations.

Leading up to the new campaign, a very succesful *2002 Leonid MAC workshop* was organized by participants *Hajime Yano*, *Shinsuke Abe*, and colleagues in Tokyo. Results were published in an ISAS Special Publication. Many Japanese high school students participated in this workshop, having performed their first research during the 2001 Leonid storm.

This time around, the organization of the campaign was supported early on and with the help of NASA's Dryden Flight Research Center's Airborne Astronomy program. The 2001 Leonid MAC success had generated considerable momentum. We flew NASA's DC-8 in combination with USAF's "FISTA" aircraft. 28 researchers from seven nationalities traveled to Spain, where we were the guests of the *Centro de Astrobiología* (CAB) in Torrejon de Ardoz near Madrid (hosted by Juan Pérez Mercader). The main mission night, November 19, took us from Spain to Omaha, Nebraska. DMS and SOMYCE ground based observers watched the stream from locations in southeastern Spain, but unfortunately were mostly clouded out by a low-pressure region that swept over Spain that night.

The meteor storms unfolded right on schedule. The Moon was low in the sky and the shower radiant hovered near 38° elevation throughout the mission. Moonlight and clouds made observations from the ground difficult, but our view was spectacular. At one point, I stood in the aisle of the plane surfing the air over the frozen moonlit tundra of Canada, and stared in awe at the monitors. An active aurora had developed just before the storm peak and now a barrage of faint meteors were seen against the backdrop of the fading aurora, truly a spectacular sight. This time, I did not forget to peek a glimpse through the windows.[68]

[66] T. Yanagisawa, K. Ohnishi, K.-I. Torii *et al.*, Separation of 1699 yr and 1866 yr radiant points by use of the line detection method. *ISAS SP* **15** (2003), 39–45.

[67] R. H. McNaught and D. J. Asher, The 2001 Leonids and dust trail radiants. *WGN* **29** (2001), 156–164.

[68] P. Jenniskens, The 2002 Leonid MAC airborne mission: first results. *WGN* **30** (2002), 218–224.

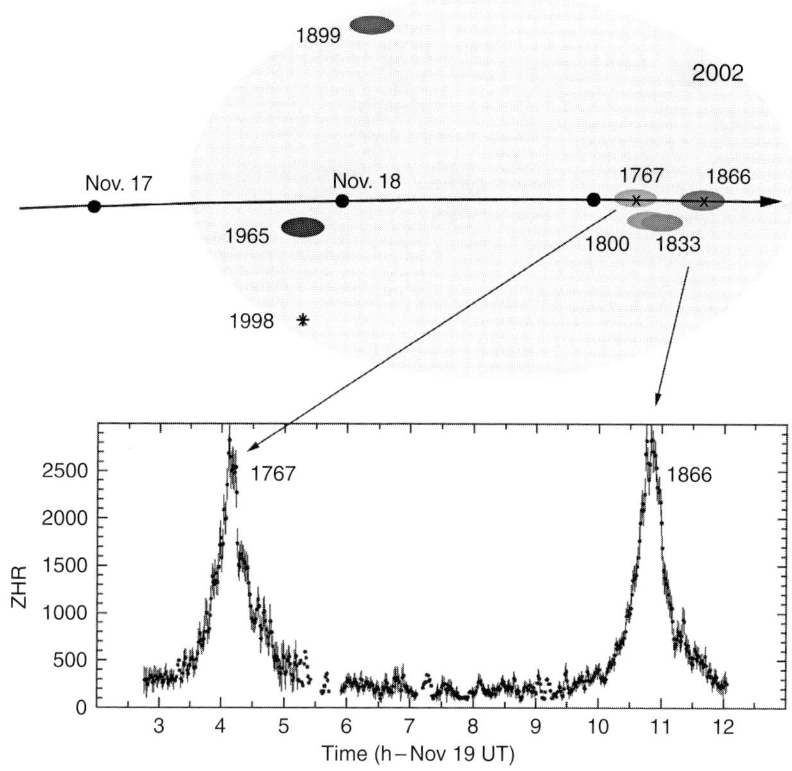

Fig. 15.27 Observed 2002 Leonid activity from near-real time counts. Note the high activity between the two storm peaks from the Filament component.

The most significant new result was confirmation that the dust trails do not broaden significantly with time and are most narrow when we cross through the trail center. No significant radiation pressure forces act on the grains other than the known radial force. An underlying background of activity that kept rates high between the two storm peaks was identified as a return of the Filament component. This was the first solid evidence that this component continues to be present after perihelion of the comet.

The observations also appeared to confirm the periodic error in the trail positions calculated. If indeed the 1767 trail was closer to Earth's orbit and the 1866 trail further away than calculated by Asher and McNaught, then the peak intensity ratio should be closer to 1:1 than the earlier estimate of 1:3. Indeed, the observed ratio was close to 1:1. Post-dicted peak times (04:05 and 10:46 UT) were also closer to the observed values (04:08 and 10:51 UT) than the no-shift results of 03:53 and 10:29 UT.[69]

[69] P. Jenniskens, Leonid MAC near-real time flux measurements and the precession of comet 55P/Tempel–Tuttle. *ISAS SP* **15** (2003), 73–80.

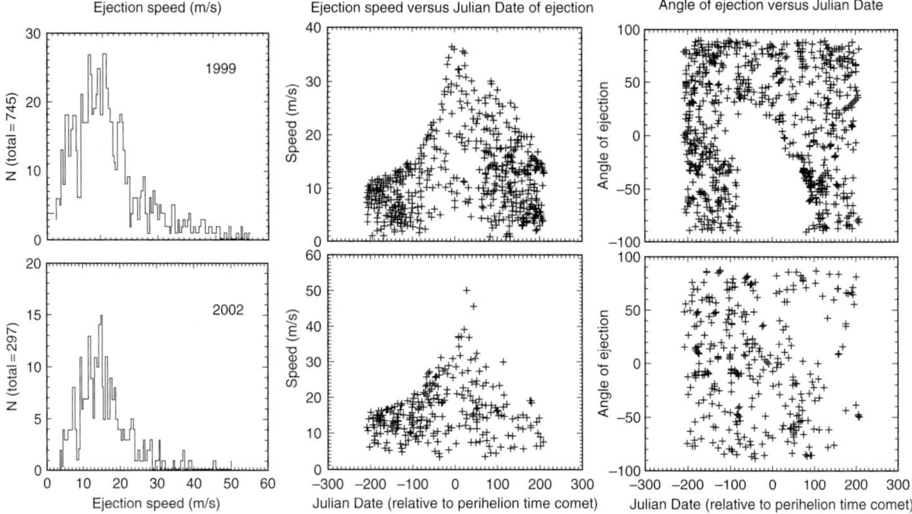

Fig. 15.28 Ejection conditions of meteoroids responsible for the 1999 and 2002 encounters with the 1866 dust trail. Left-hand diagrams are a histogram of ejection speed. Middle diagrams show the speed as a function of time of ejection. Right-hand diagrams show the angle of ejection in the orbital plane (towards Sun = 0), and as a function of the time of ejection. Model by Jérémie Vaubaillon.[70]

The best prediction of activity and peak time in 2002 was that of graduate student Jérémie Vaubaillon and thesis advisor François Colas of IMCCE,[71] who used, for the first time, massive parallel supercomputers to calculate the distribution of meteoroids in the stream from a full comet ejection model (that of Crifo[72]). Now it was possible to study the expected distribution of meteoroids in the dust trail and obtain statistical information about the distribution of ejection speeds and angles of ejection of the meteoroids seen during individual Leonid storms (Fig. 15.28).

Interestingly, in Vaubaillon's model most meteoroids are ejected at relatively high heliocentric distance because that is where the comet spends more time. Typical ejection speeds for grains arriving at Earth's orbit at the right time are of order 10 m/s in this model (for +3 to $+8^{m}$ Leonids) and nearly independent of position along the dust trail! The angle of ejection is more important than the speed itself in determining where in the dust trail the particle will end up.

[70] J. Vaubaillon, Dynamic of meteoroids in the solar system. Application to the prediction of meteoric showers in general, and Leonids in particular. Ph.D. thesis, I.M.C.C.E., l'Observatore de Paris, Paris, France (2004).

[71] J. Vaubaillon and F. Colas, Evolution of a meteor stream and Leonids 2002 Forecasting. *ESA SP* **500** (2002), 181–184.

[72] J. F. Crifo, A general physicochemical model of the inner coma of active comets. 1: Implications of spatially distributed gas and dust production. *Astrophys. J.* **445** (1995), 470–488; J. F. Crifo and A. V. Rodionov, The dependence of the circumnuclear coma structure on the properties of the nucleus. *Icarus* **127** (1997), 319–353.

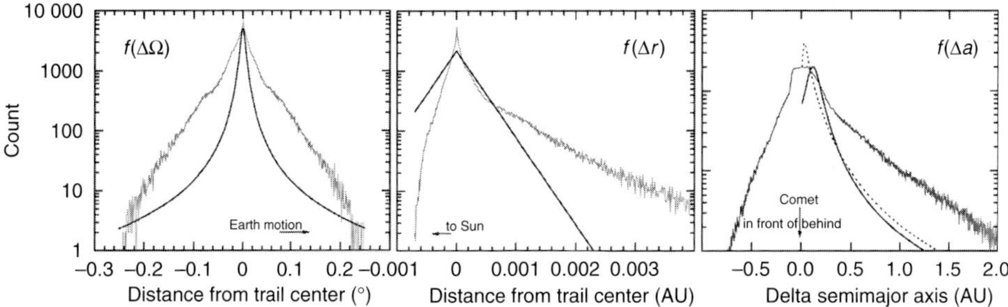

Fig. 15.29 Distribution of dust density along the dust trail in a one-revolution trail of 55P/Tempel–Tuttle, without planetary perturbations. Results for the adopted Crifo ejection model are shown in gray (for $0.1 - 0.5$ mm radius particles and $\Delta T = 200$–1500 d in the case of $f(\Delta\Omega)$ and $f(\Delta r)$). The black and dashed lines are the relationships derived from the observations.[73]

The one-revolution dust trail of comet Tempel–Tuttle

With support of NASA's Planetary Astronomy program, Jérémie was given the opportunity to visit the SETI Institute for a work visit in late 2004. We set out to make a close comparison between models and observations. The model handles planetary perturbations well (predicting the peak time), but Crifo's ejection model performs poorly. Calculated and observed profiles differ considerably (Fig. 15.29).

The problem in deriving the functional forms of $f(\Delta a_0)$ and $f(\Delta r)$ from observations is that both vary by orders of magnitude. In addition, perturbations of the orbital period of the meteoroids can change f_m from the expected value of $1/N$ by factors of up to 100 or more. The result is a nightmarish puzzle of dependencies.

We started from the assumption that the interpretation of the observations left much to wish for. The Crifo model produced a theoretical $f(\Delta a_0)$, $f(\Delta r)$, and f_m that can be used to disentangle these contributions from the observations if they have the correct functional form. That approach is more objective than starting from some assumed dependency on Δa_0 or Δr.

Our first step towards a better model of the one-revolution dust trail of 55P/Tempel–Tuttle was to figure out how to calculate the mean anomaly factor for a full-blown ejection model such as shown in Fig. 15.30. The separation between two particles is not enough to measure the change in particle density of the trail. In this case, any section of the dust trail is made up of particles with very different ejection conditions. We calculated the mean anomaly factor by comparing the dust density at the time of encounter with the original dust density of the model one-revolution trail just after ejection (with no planetary perturbations).

[73] P. Jenniskens, More on the dust trails of comet 55P/Tempel–Tuttle from 2001 Leonid shower flux measurements. *ESA SP* **500** (2002), 117–120.

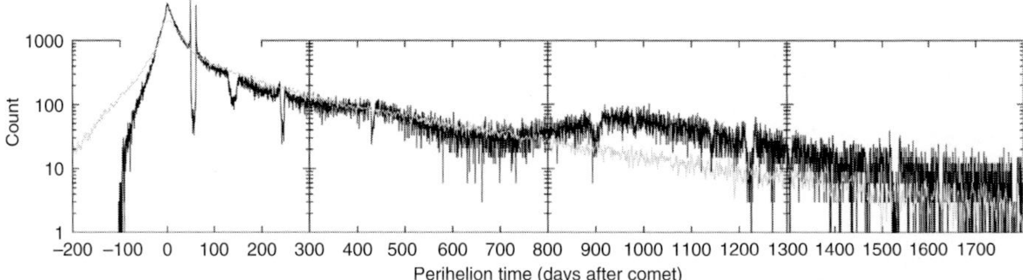

Fig. 15.30 Dust distribution in the comet orbit ejected in 1866 as encountered three revolutions later. The graph shows the distribution of the perihelion times of all meteoroids just after ejection (gray line), and after three revolutions (dark line) shown by compressing the X-scale by a factor of 3.

The problem then boils down to deciding what was the original section of the dust trail that was seen in a later year. For that, we used the initial (as in just after ejection) Δa_0 value of each particle. For each particle separately, we calculated one over the original density in the one-revolution trail for that Δa_0 and then summed up the contribution from all particles. In this way, the dust trail could be an isolated trail section, or a combination of several, and still give a good estimate of how the dust density has changed over time.

The miss-distance Δr (in AU) was determined not from where particles ejected at perihelion are found, but as a distance to the center of mass of the distribution of nodes. These miss-distances are systematically different from those calculated by Asher or Lyytinen by about +0.0005 AU.

In contrast to prior models, the trail cross section is made up of particles with different initial Δa_0. We simply calculated the mean, which agreed well with the ejection-at-perihelion values calculated by Asher or Lyytinen. This is because Δa_0 is mostly determined by the time of encounter. Results for Δr, Δa, and f_m are given in Table 4.

The distribution of dust grains in the trail

With the planetary perturbations taken care of, we can now compare the observed peak rate (ZHR), width of the profile in Earth's orbit (W), mass distribution index (s), and even the skew in the meteor light curves (F), for each dust trail encounter. First up, we examined the distribution of dust in the heliocentric direction (Δr) by removing the dependence on Δa_0. The value of Δa_0 is a given and can not be uncertain by any significant amount because it is determined by the time of encounter. Crifo's ejection model produces an exponential distribution along the dust trail, with higher density at the position of the comet. The slope of that distribution depends on meteoroid size.

Bin	Size (mm):	Mass (g):	m_v (magnitude)	Instrument
I	0.2–1.0	$4.2 \times 10^{-6} - 5.2 \times 10^{-4}$	+6.0 to +11.2	Radar
II	1.0–2.0	$5.2 \times 10^{-4} - 0.0042$	+3.7 to +6.0	Intensified video
III	2.0–10.0	0.0042 – 0.52 g	– 1.5 to +3.7	Visual Leonids
IV	10.0–20.0	0.52 – 4.2	– 1.5 to –3.8	PhotoGraphic Leonids
V	20.0–200	4.2 – 4200	<–3.8	Bright fireballs

Fig. 15.31 The Size and mass of meteoroids in each computational bin of the model by Vaubaillon.

The model by Vaubaillon assumed five classes of meteoroid sizes, represented with the same amount of 50 000 particles in each calculation (Fig. 15.31). The distribution of meteoroids in the various plots in this book are a sum of all the bins. In reality, meteors are counted by any given technique mostly in a small range of meteoroid size (and meteor magnitude).

As expected, the tail in the distribution of Δa_0 in Fig. 15.30 is mostly due to the smallest grains in bin I, which are most affected by radiation pressure. After correction for f_m and $f(\Delta a_0)$, Fig. 15.32 shows how the observations trace the dust distribution in the heliocentric direction. As expected, bin II gives the best result because our flux measurements are from video observations and visual observations normalized to a limiting magnitude of +6.5. Bin II has a slope of log $f(\Delta a_0) / \Delta a_0 = -4.2$/AU for the range 0.125–0.40 AU, bin III has a slope of –10.2/AU (for 0.049–0.21 AU), while bin I has a slope of –1.7/AU.

After correcting, the dust density peaks at $\Delta r = 0.000\,77 \pm 0.000\,05$ AU farther out than predicted by the model. The dust density falls off exponentially according to:

$$f(\Delta r) = 10^{(-1050\,|\Delta r\,(\mathrm{AU}) - 0.000\,77|)} \tag{15.11}$$

Interestingly, the model can be made to agree with observations if the whole curve is shifted by $\Delta r = +0.000\,77$ AU (Fig. 15.33). It is not clear at present why that is so. It is now possible to correct other data points as well, using the model $f(\Delta a)$ curve, instead of a simple exponential slope.

There are two or three discrepancies: the 1998 encounter with the 1899-dust trail (marked "1998") and the 2001 encounter with the 1767-dust trail. Presumably, distortions of the trail from close encounters with Earth are responsible. Model and observations agree if the calculated 1899-dust trail in 1998 is off by $\Delta r = -0.0017$ AU, and the 1767-dust trail in 2001 is off by $\Delta r = +0.0007$ AU. Again, the reasons for these shifts are not known.

I have included in the graph other peak ZHR estimates found in the literature, including those by Arlt and Brown(+). Note that they generally give a comparable or less good agreement. The high ZHR = 150 000/h estimate from the 1966 Leonid storm can not be right.

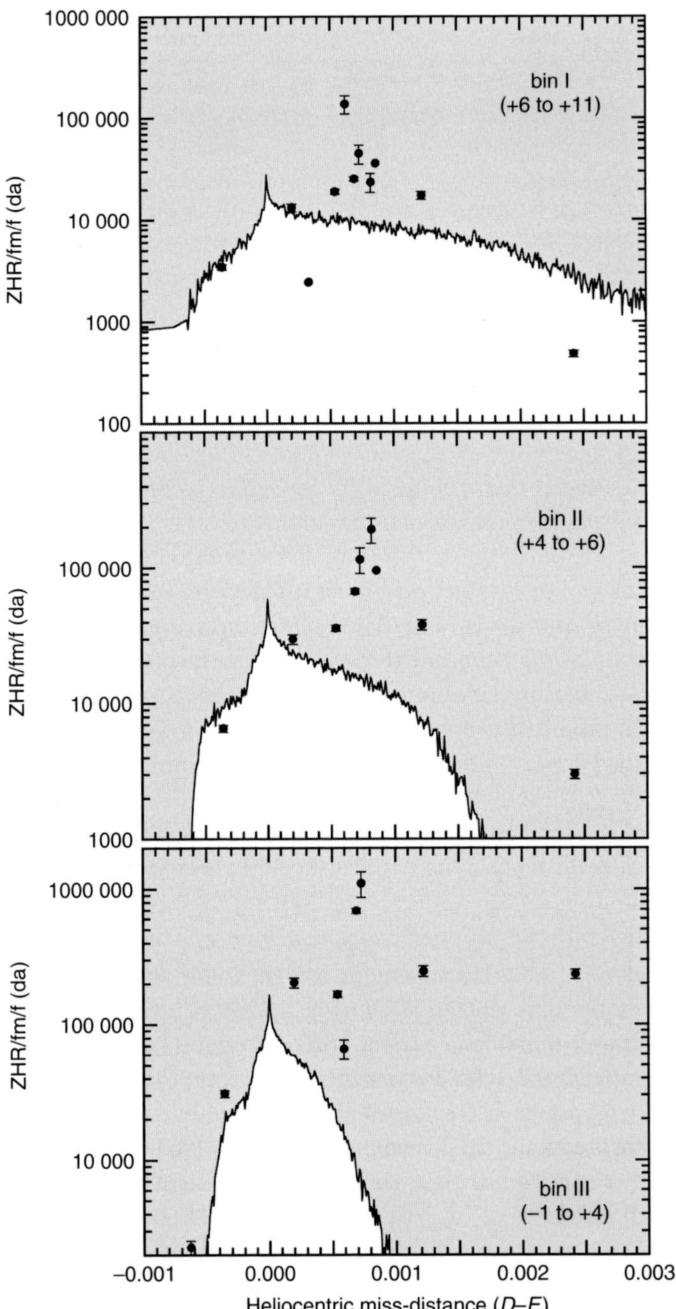

Fig. 15.32 Radial distribution of dust. Solid line: model, datapoints: observations if corrected for Δa-dependence typical of model particles.

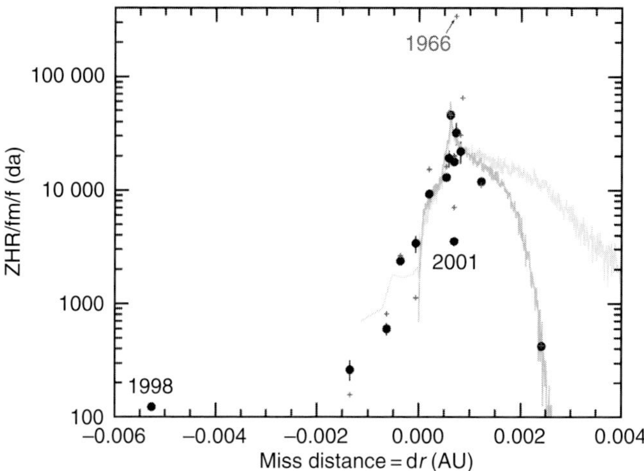

Fig. 15.33 As Fig. 15.32, with the model shifted and scaled to match observations. Crosses are flux estimates by Arlt and Brown. Gray lines are models for bins II and I.

The model predicts small meteoroids to have an outward tail (Fig. 15.33). Indeed, the 2000 encounter with the 1932 dust trail, responsible for the point $\Delta r = +0.002\,42$ AU, was well detected by radar.[74] This result is now understood. No such large dependence on meteoroid size is expected for the sunward size of the profile!

The best-fitting function for $f(\Delta r)$ has a shoulder on the anti-Sunward side. This shoulder is the result of a (sum of) Δr^{-2} – dependent (rather than exponential) drop off away from the trail center.

With this functional form for Δr, we can check how much variation remains in the distribution along the comet orbit (Fig. 15.34). The 1969 observations are consistent only with the distribution calculated for small +6 to $+11^{\mathrm{m}}$ Leonids, possibly consistent with the observation that most Leonids were faint. However, the rest of the data scatter around this model solution as well. The peak of the dust density is at $\Delta a_0 = 0.11 \pm 0.02$ AU.

The model predicted a peak around 0.061 AU. Hence, the meteoroids appear to be spread along the comet orbit more widely than in the model. In the same way, the stream cross section $f(\Delta\lambda_\odot)$ is not yet well described by the model (Fig. 15.35). The distribution of nodes in the model (= the ZHR curve) is a factor of three wider than observed, with $W \sim 0.09°$ instead of $W = 0.031°$.

More importantly, the dust density falls off in a different manner than observed, being evenly dispersed away from the trail center, with a rapid exponential decay beyond a certain range.

[74] P. Brown, M. Campbell, R. Suggs *et al.*, Video and radar observations of the 2000 Leonids: evidence for a strong flux peak associated with 1932 ejecta? *Mon. Not. R. Astron. Soc.* **335** (2002), 473–479.

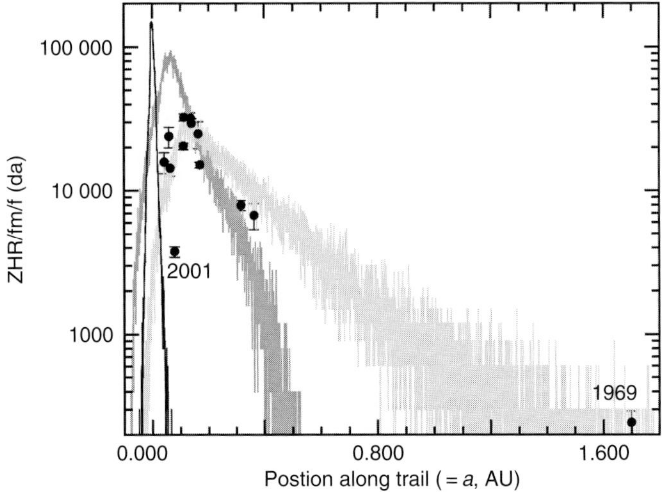

Fig. 15.34 Variation of dust density along the dust trail. Gray lines are model results for bins I (light), II, and V (dark).

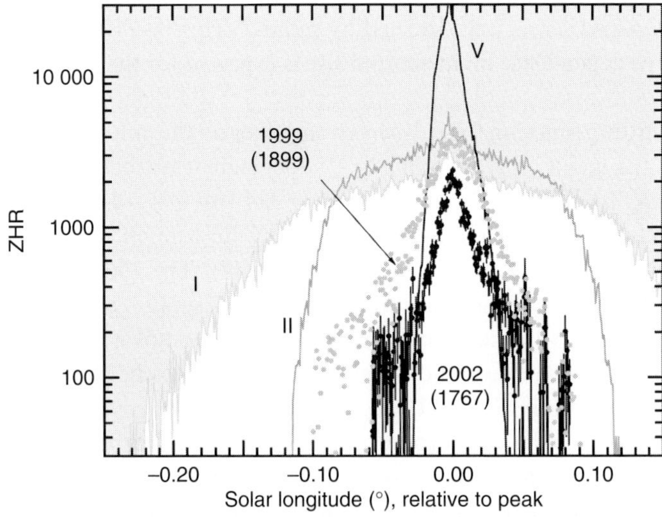

Fig. 15.35 Dust density in the Earth's path. ZHR curve of 1767-dust trail in 2002, 1899-dust trail in 1999, and three models for bins I, II, and V.

The size distribution of dust grains

According to the Crifo ejection model and that of Whipple, the small grains are ejected with higher speed and are pushed more by solar radiation forces. That should translate into smaller grains being found farther out from the trail center

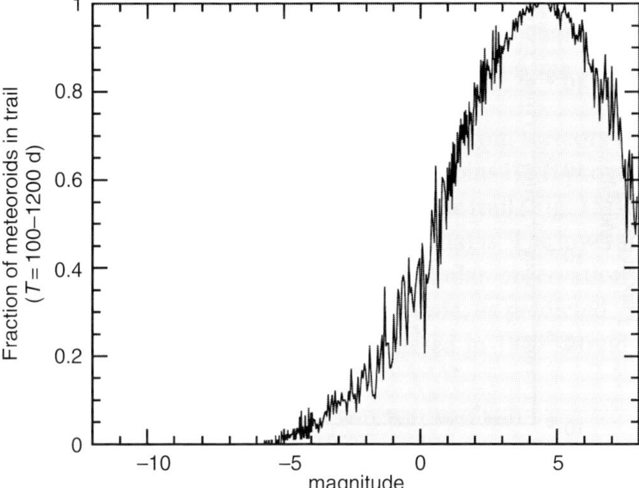

Fig. 15.36 The fraction of meteoroids ejected that are located in the trail segment observed during recent Leonid meteor storms ($\Delta T_{\mathrm{p}} = +100$ to $+1200$ d) for a density of 2.0 g/cm^3.

(compare bin V and bin I in Fig. 15.35). This is not observed. The magnitude distribution index χ is nearly constant across the ZHR profile.

The same applies to the distribution along the comet dust trail. Fig. 15.36 shows the fraction of all ejected grains from comet 55P/Tempel–Tuttle that end up in the dust trail segment that has perihelion time returns of $+100$ to $+1200$ d after the comet (after one orbit), relevant for many of the recent storms. No large grains are expected. The model mass distribution for masses smaller than 1 g is matched if the initial mass distribution index of the dust ejected by comet 55P/Tempel–Tuttle equals $s = 1.40 \pm 0.05$.

To explain the presence of large masses in the recent Leonid storms, it is possible to assume a low meteoroid density. That will help increase the ejection speeds, but it does not appear to be enough (Fig. 15.37). Pavel Spurný measured one of the -12^{m} fireballs over China to have a density of about 0.7 g/cm^3.[75] Even if the density of the meteoroids is 0.2 g/cm^3, there are more fireballs than expected.

It is possible to calculate the magnitude distribution index from the model by scaling the number of initially ejected meteoroids in each bin according to an assumed index of $s = 1.83$, as expected from a collisional cascade (Chapter 7.3). The observed ratio of meteors in neighboring bins can then be directly translated into the mass (and magnitude) distribution index at a given position in the resulting trail.

The model predicts that the magnitude distribution index changes as a function of position in the trail and as a function of meteor magnitude (Fig. 15.38). There

[75] P. Spurný, H. Betlem, J. van 't Leven and P. Jenniskens, Atmospheric behavior and extreme beginning heights of the 13 brightest photographic Leonids from the ground-based expedition to China. *Meteoritics Planet. Sci.* **35** (2000), 243–249.

than calculated from the extrapolated orbit. We were able to refine the comet orbit using the reported times in those old accounts of meteor storms. That pushed the known history of the comet orbit back by a factor two.

A confluence of dust trails led to a particularly intense storm in 1799. The young researcher and great narrator *Alexander von Humboldt* had set out with French botanist *Aimé Jacques Goujaud Bonpland* to explore the world, and had landed in the city of Cumaná (now Venezuela). Bonpland first noticed the shower at half past two in the night, when he had risen for some fresh air, and was quoted saying that: *from the first appearance of the phenomenon, there was not in the firmament a space equal in extent to three diameters of the moon, which was not filled every instant with bolides and falling stars.*[80] Humboldt interviewed eye witnesses during his exploration of the interior of the Spanish colony along the Orinoco river in the next months, and added: *A phenomenon analogous to that which appeared on the 12th of November at Cumana, was observed thirty years previously on the table-land of the Andes, in a country studded with volcanoes. In the city of Quito there was seen in one part of the sky, above the volcano of Cayamba, such great numbers of falling-stars, that the mountain was thought to be in flames. This singular sight lasted more than an hour. The people assembled in the plain of Exido, which commands a magnificent view of the highest summits of the Cordilleras. A procession was on the point of setting out from the convent of San Francisco, when it was perceived that the blaze on the horizon was caused by fiery meteors, which ran along the skies in all directions, at the altitude of twelve or thirteen degrees.*

Finally, Jérémie and I were pleased to identify the mythical storm that was seen "thirty years previously on the table-land of the Andes" reported by von Humboldt after seeing the 1799 storm, which helped establish the periodicity of Leonid storms. Without a precise date we can not be certain, but the encounter of November 9, 1771 with the 1600-dust trail (Table 4) was the only one that season capable of causing a meteor storm and it happened to be at 06:18 UT, well positioned for Southern America. In revealing the historic significance of the anecdote, we felt like the British Museum curator George Smith in 1872, when he first deciphered the cuneiform account on clay tablets of Noah's flood in the epic 2750 BC poem *Gilgamesh*. A long-hidden epic story of travel and fortune was revealed. Before our eyes, legendary accounts of "stars fell like rain" from the time of emperors and court astrologers became historic accounts of astronomical events.

[80] A. de Humboldt and A. Bonpland, Second abode at Cumana. Earthquakes. Extraordinary meteors. Personal narrative of travels to the equinoctial regions of America during the years 1799–1804 (1852, English translation of the 1815 Paris edition, ed. T. Ross (London: Bohn, 1852), Vol. I, Ch. 1.10, pp. 351–360 (Gutenberg Etext).

16
The Ursids

Shortly after the 1999 Leonid storm, Esko Lyytinen and I teamed up to tackle a long-standing mystery: why do Ursid showers (Fig. 16.1) tend to have outbursts when the parent comet is at aphelion?[1] Parent comet 8P/Tuttle, like 55P/Tempel–Tuttle, is a Halley-type comet and the same techniques could be expected to work.

When we started the investigation, even the identification of the Ursids with 8P/Tuttle was in doubt. The Ursid meteors radiated from R.A. = 219.4°, Decl. +75.3°, slightly different from the predicted radiant at R.A. = 214.5°, Decl. = +74.4°. More importantly, the orbit of comet 8P/Tuttle is far from Earth's orbit, with a minimum distance of 0.095 AU on December 22. In order to intersect Earth's orbit, the meteoroids have to evolve significantly to smaller perihelion distances.

We discovered that the Tuttle meteoroids can get trapped in a mean-motion resonance just outside of that which traps the comet. Because the orbital periods of the meteoroids and the comet are different, the dust gradually moves away from the position of the comet and after about 45 revolutions, the dust lags by half an orbit. And, guess what, it takes that long for dust to wander into Earth's path.

16.1 The Ursid Filament

The first clue to the cause of the Ursid aphelion outbursts came in 1994, when I discovered that Ursids have perihelion outbursts as well.

Ilkka Yrjölä's radio-MS data (Fig. 16.2) showed strong Ursid activity with varying intensity from year to year. Highest rates had occurred just before the comet reached perihelion in 1994. These enhancements stood out from the other aphelion Ursid outbursts by being significantly broader and about a day wide. I then realized that one such outburst had been seen in December, 1982 by Jos Nijland and Hans Breukers of the Dutch Meteor Society.[2] Word came that the 1994 Ursid outburst was spotted by visual observers in Japan.[3] The reason was that in

[1] P. Jenniskens, E. Lyytinen, M. C. de Lignie *et al.*, Dust trails of 8P/Tuttle and the unusual outbursts of the Ursid shower. *Icarus* **159** (2002), 197–209.

[2] R. Veltman, De herfst – and winterakties 1982. *Radiant, J. DMS* **5** (1983), 4–9 (in Dutch).

[3] K. Ohtsuka, H. Shioi and E. Hidaka, Enhanced activity of the 1994 Ursids from Japan. *WGN* **22** (1995), 69–72.

Fig. 16.1 A -1^{m} Ursid meteor photographed from Fremont Peak Observatory, CA, by the author on December 22, 1997.

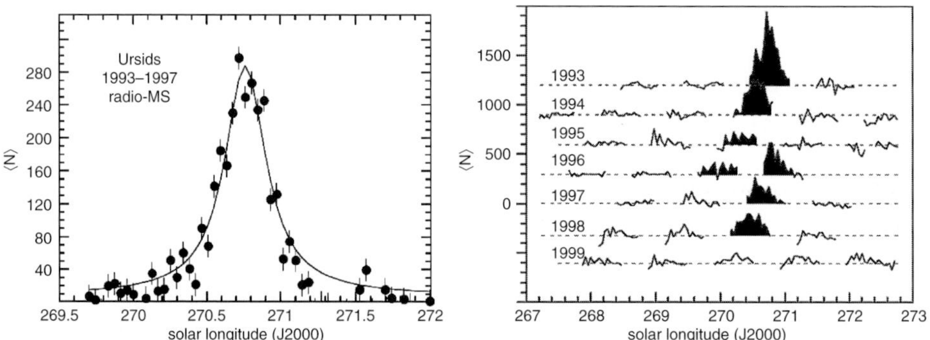

Fig. 16.2 The Ursid shower Filament. Radio-MS observations by Ilkka Yrjölä. The graph to the right shows raw counts, while the graph to the left is a summary of all the years corrected for observing effects (geometic dillution) and changing peak rates, to bring out the shape of the Filament flux profile.

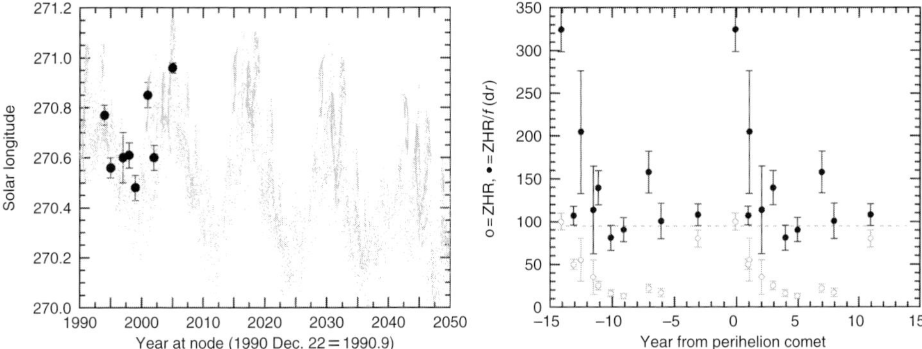

Fig. 16.3 The Ursid Filament shower peak ZHR (period 1993–2004) and the time of the peak compared to the solar longitude of several old dust trails, shifted by 0.1° to match recent peak times. The graph to the right shows the peak Ursid rates in each year.

1981 a similar outburst was seen, and Katsuhito Ohtsuka had suspected those enhanced rates might return in 1994 during the next visit of 8P/Tuttle.[4]

In 1997, the peak would finally be over California and volunteers of the California Meteor Society gathered to support a multi-station photographic and video campaign, the results of which were reduced by Hans Betlem and Marc de Lignie. In one night, the number of known Ursid orbits was increased from two to two dozen. The radiants were dispersed, consistent with a Filament.

The peak times of the Filament correlate with the position of nodes from old trails (about the seventh to the tenth century AD) when shifted by 0.1° (Fig. 16.3). They are likely to be dust that is even older. This pattern was used to predict the time of future Ursid Filament returns (Table 5b).

The peak rates correlate with the miss-distance from these older trails, with a radial dispersion three times that in the Earth's path. After correcting for this $f(\Delta r)$ dependence, the rates are constant along the orbit of the comet, except at the position of the comet (Fig. 16.3). There, we may find dust that is in the same resonance as the comet.

16.2 Dust trail encounters

The much narrower aphelion outbursts had to be something altogether different. Esko and I studied the evolution of the 8P/Tuttle trails and found that their perihelion takes at least 600 yr to evolve from the position of the comet to the position of Earth, mostly due to perturbations by Saturn, Uranus, and Neptune. If during that time the particles are getting trapped in the 7/6 orbital resonance with Jupiter, one further than that of the comet (15/13), then this will result in a gradual delay of the dust relative to the position of the comet, accumulating to $600 \times (13/15 * 7/6 - 1) = 6.67$ yr. That is half a

[4] K. Ohtsuka, *Tokyo Meteor Network Rep.* **14** (1994), 85–87.

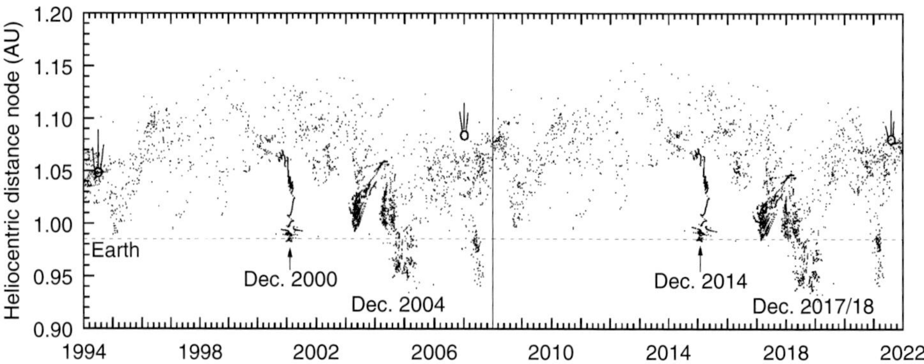

Fig. 16.4 The position of the node of 8P/Tuttle dust trails. Calculations by Esko Lyytinen.

$P = 13.6$ year orbit. We recognized that this is why Ursid shower outbursts are observed when the comet is at aphelion.

This explanation was quite different from the suggestion by Lubor Kresák, that a comet nucleus fragment had long lagged the main nucleus of 8P/Tuttle by six years and had now recently disintegrated.[5] Our explanation is explained in too simplified a manner, given that the comet is not librating about the 15/13 resonance but in a periodic oscillation about the 15/13 resonance due to the nearby 1:1 mean motion resonance with Jupiter (Chapter 7). Nevertheless, the effect is much the same and the idea is panned out by calculations (Fig. 16.4). The meteoroids remain so concentrated in a section of their orbit that the showers are only seen in one year.

In contrast to the Leonids, these ancient dust trails are far from well-behaved lines. The trails become highly distorted. Some parts diffuse into an ill-defined chaotic cloud of dust particles, but the lack of close encounters with Jupiter cause other parts of a trail to remain nearly undistorted. Those conserved *trailets* are expected to give the higher rates and are most easily recognized as a meteor outburst when in Earth's path.

It was found that only a handful of dust trails are near Earth's orbit after 600 yr of evolution, notably those ejected in 1378, 1392, and 1405, with a rapid worsening of the encounter conditions between the trail of 1405 and 1419 (Fig. 16.5).

The 1945 outburst, when the shower first received widespread attention, is now identified as due to a more or less well behaved dust trailet ejected in 1392. The shower was quite intense and narrow, with a peak ZHR of around 120, observed by Czech observers at Ondrejov[6] before clouds interfered. The calculated peak time of the shower was 18:29 UT, December 22, close to the time of the observations.

Later, observers from the *British Meteor Society* detected a radio enhancement on December 22, 1973 (11:00 UT) lasting for 1 h. This, too, may have been an 8P/Tuttle

[5] L. Kresák, Meteor storms. In *Meteoroids and Their Parent Bodies*, ed. J. Stohl and I. P. Williams (Bratislava: Astronomical Institute, Slovak Acad. Sci., 1992), pp. 147–156.

[6] Z. Ceplecha, Umids-Becvar's meteor stream. *Bull. Astron. Inst. Czech.* **2** (1951), 156–159.

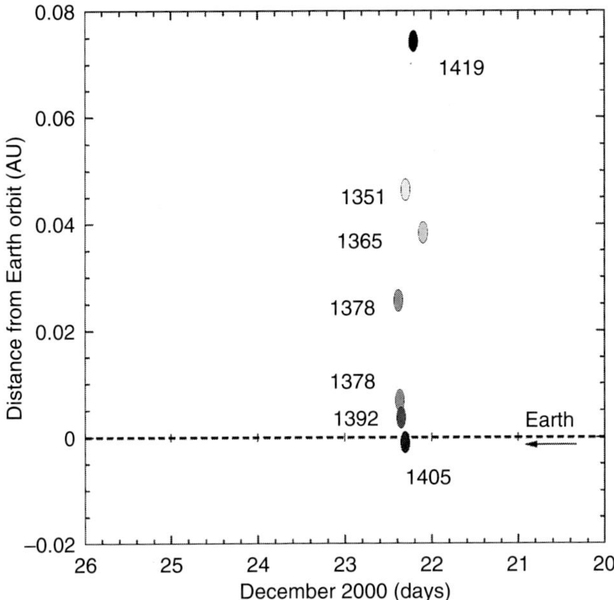

Fig. 16.5 The point of intersection of Ursid shower dust trails.

dust trail encounter, as the outburst followed the 1945 event by two orbits of the comet.[7]

Another orbit later, in 1986, an Ursid outburst was seen by *Kai Gaarder* and *Lars Trygve-Heen* of the Norwegian Astronomical Society – Meteor Section. This was one of a series of unusual accounts that first attracted my attention to meteor outbursts. We now calculated that this outburst was caused by the dust ejected in 1378, 44 revolutions earlier. The trail came close to Earth's orbit and the timing agreed to within 0.01°. At the time, there were two trailets near Earth's orbit, both of which may have contributed to the outburst.

The December 2000 encounter with the dust trails would bring Earth close to a well-defined trailet of dust ejected in 1405 (Fig. 16.6), which however did not quite extend to the crossing point in Earth's path. Some nongravitational forces, or slightly different initial conditions during ejection, were needed to bring the trailet into Earth's path.[8] If so, the trail would cause an outburst at 07:29 UT, observable from California. We found that the 1392 trail might contribute as well, with a predicted encounter time at 08:32 UT (Fig. 16.5). The calculations were later confirmed by *Hartwig Lüthen* following a similar technique but with an independent orbit integrator. He found a peak time at 07:27 UT for the 1405 trailet.

[7] R. A. MacKenzie, *Solar System Debris*, British Meteor Society, internal publication (New York: Dover, 1980), p. 30.

[8] P. Jenniskens, Ursid meteors 2000. *IAU Circ.* 7543, ed. D. W. E. Green; P. Jenniskens and E. Lyytinen, Possible Ursid outburst on December 22, 2000. *WGN* **28** (2000), 221–226.

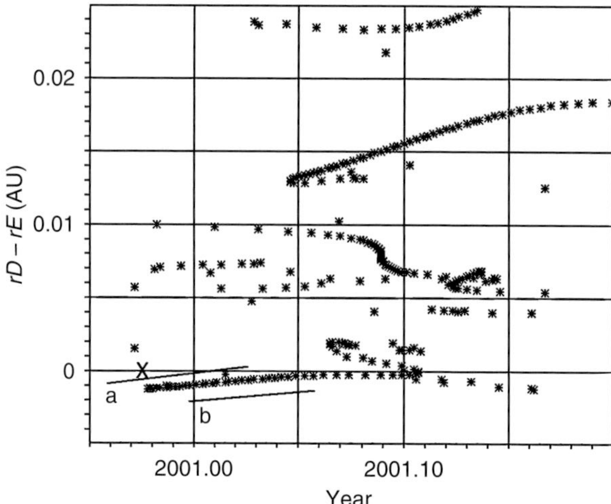

Fig. 16.6 The nodes of the 1405-dust trail in 2000 versus time at the node. "a" and "b" mark the position of the trail for different nongravitational effects by radiation pressure on the grains. Calculations by Esko Lyytinen.[9]

The notion to have a heavenly fireworks from dust ejected just before the time Columbus first set foot in the Americas was too much for the media to pass up on. Even *Phil Frank*, a cartoonist of the San Francisco Chronicle, picked up on the occasion.

Clouds were expected to cover the Bay Area, so we went to two sites further south than usual in California. The anticipated outburst was observed as predicted and clearly recognized in intensified video records (Fig. 16.7). The low radiant position and steep magnitude distribution index made this a difficult shower for visual observers, but DMS observers confirmed the increase leading up to the peak, while NMS observers recorded the descending branch.[10]

For the 1392 and 1405 dust trails to contribute about equal parts to the 2000 outburst, the dispersion perpendicular to Earth's path must be at least a factor of 3 larger than that in Earth's path, the same as we found for the Leonids. The apparent width of the shower, assuming a Lorenzian profile gave $W = 0.06 \pm 0.01^\circ$ for each trail, in good agreement with $W = 0.05 \pm 0.01^\circ$ derived from the 1986 1378-dust trail crossing. The trails may have been slightly more inward than calculated, by about −0.0013 AU. In comparison, the Leonid trails seem to be off by −0.000 25 AU.

There is now no longer doubt that the Ursids are from parent 8P/Tuttle. While the comet orbit itself has an inclination of $\sim$54.9° and $q \sim 1.002$ AU, different from that of

[9] L. Kresák, Meteor storms. In *Meteoroids and Their Parent Bodies*, ed. J. Stohl and I. P. Williams (Bratislava: Astronomical Institute, Slovak Acad. Sci., 1992), pp. 147–156.

[10] P. Jenniskens, Ursid meteors 2000. *IAU Circ.* 7547, ed. D. W. E. Green; P. Jenniskens and E. Lyytinen, 2000, Ursid outburst confirmed. *WGN* **29** (2001), 41–45; P. Jenniskens, E. Lyytinen, M. C. de Lignie *et al.*, Dust trails of 8P/Tuttle and the unusual outbursts of the Ursid shower. *Icarus* **159** (2002), 197–209.

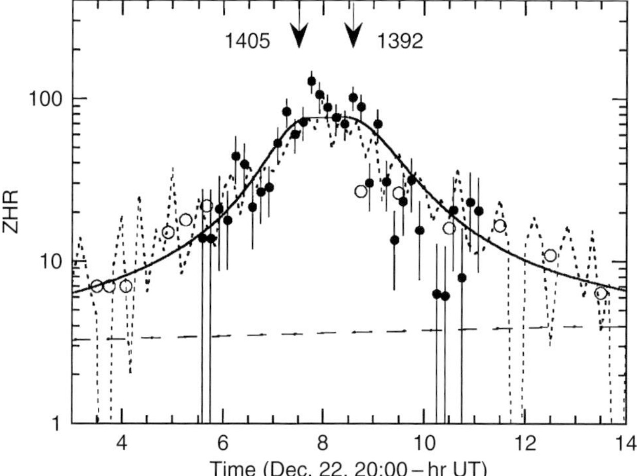

Fig. 16.7 The observed Ursid rates during the December 2000 outburst measured by video (●), visual (○), and radio forward meteor scatter observations (- - -).[11]

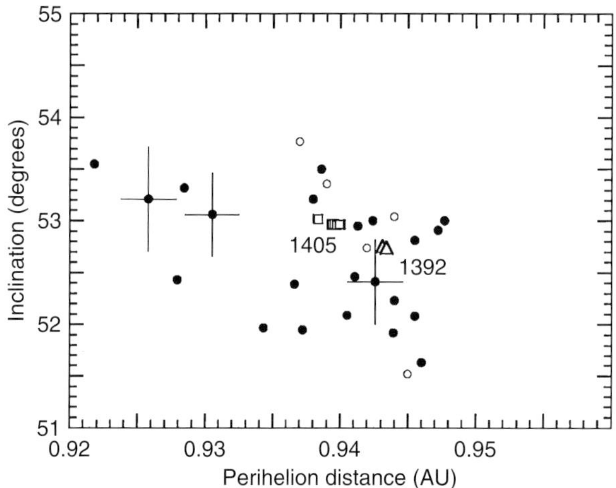

Fig. 16.8 Theoretical and observed Earth-intersecting 2000 (●) and 1997 (○) Ursid outburst meteoroid orbits. Crosses indicate typical observing errors, other symbols are theoretical values for the dust trails of 1392 and 1405.[12]

Ursids, the calculated theoretical orbits of Ursid meteoroids agree with the observed outburst Ursids which have $i \sim 52.8$ and $q \sim 0.941$ AU (Fig. 16.8).

From these video orbits, we did not find a more compact radiant for the narrow 2000 outburst as compared to the broader 1997 Filament due to observational errors.

[11] *Ibid.*
[12] *Ibid.*

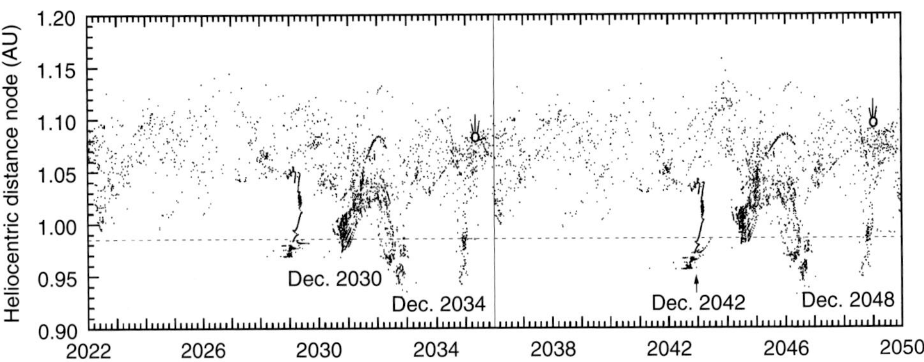

Fig. 16.9 Future (AD 600–1500) dust trail encounters. Calculations by Esko Lyytinen.

Several of the 2000 meteoroids are outliers with a similar slightly higher inclination as the 1997 Ursid Filament. Perhaps these are older trails of more than 1000 years old (Fig. 16.8).

The outburst Ursids were found to light up ∼5 km higher in the atmosphere than other meteors of similar speed (they end higher as well), and release sodium earlier than magnesium.

These observations are evidence of a relatively fragile morphology compared to annual shower grains and point at a relatively recent time of ejection, in agreement with the proposed scenario of their origin. A similar effect was only observed earlier for outburst Draconids and outburst Leonid meteors.[13] Older outbursts tentatively associated with the Ursids (Table 1) do in fact not fall in the current 6 yr delay pattern, nor do they occur at the time of the return of the comet. Hence, their identity is uncertain. Before the Ursid shower was recognized in 1945, only Denning's 1916 paper[14] in which he searched for meteors to be associated with 8P/Tuttle, made mention of the shower. He reported having seen *a meteoric shower from R.A. = 218°, Decl. = +76° on December 18–25 in various years, but the display has shown no special abundance*. This sounds like normal annual Ursid activity.

Future dust trail encounters (Fig. 16.9) are given in Table 5a. While the comet orbit moves farther away from Earth's orbit in the near future, the Filament remains just outside Earth's orbit. Table 5b gives anticipated peak time and activity for the Ursid Filament. Keep in mind that the given peak times of the Filament are based on Fig. 16.4 and they are very uncertain. These values are likely to improve once we better understand the cause of the Ursid Filament.

[13] J. Borovicka, R. Stork and J. Bocek, First results from video spectroscopy of 1998 Leonid meteors. *Meteoritics Planet Sci.* **34** (1999), 987–994.

[14] W. F. Denning, Mechain–Tuttle's comet of 1790–1858 and a meteoric shower. *Observatory* **39** (1916), 466–467.

17
The Perseids

The Perseids are the amateur astronomer's main entertainment on sultry summer nights. The Perseids have been around as long as there are records (Chapter 1). They are caused by the largest comet to frequent Earth for thousands of years past, 109P/Swift–Tuttle, which will continue to frequent Earth for thousands of years to come. When the comet returned in 1992, a series of meteor outbursts were observed that led to my first successful meteor storm chase. One year later, meteor outbursts were the astronomical theme for a Mediterranean cruise and all of America went out to look at the shooting stars.

These meteor outbursts sparked a debate on whether they were caused by young dust trails or the Filament of older trails. In this chapter, I will argue that they were mostly from older trails, making it appropriate that the name 'Filament' was first given to these outbursts.

17.1 The 1979–1981 Perseids

Perseid meteoroids approach Earth from a northern direction (Fig. 17.1) with a radiant in the constellation of Perseus at R.A. = 48°, Decl. = +58°, just below the "W" of Casseiopeia.

Parent comet 109P/Swift–Tuttle was last seen in 1862. In 1973, *Brian Marsden*, at the time director of the IAU's Bureau for Astronomical Telegrams, recalculated the orbit of the comet and predicted a return on September 16.9, 1981 (±1.0 yr), suspecting that a comet seen by astronomer Pehr Wargentin at Uppsala in 1750 was an earlier sighting.[1]

Although no comet was observed in those years, considerable excitement was generated by some accounts of high Perseid rates. In the summer of 1980, a group of experienced Belgian observers in the company of Hans Betlem drove to Switzerland to observe the Perseid shower for the first time in the high Alps.[2] Since 1976, individual

[1] B. G. Marsden, The next return of the comet of the Perseid meteors. *Astron. J.* **78** (1973), 654–662.
[2] P. Roggemans, Perseiden in Switzerland. *Radiant* **2** (1980), 107–108; H. Betlem. Perseiden 1980 (1). *Radiant* **2** (1980), 140–141.

Others had normal rates.[9] *A big difference with the previous nights was that the Perseids described long trajectories*, writes Joop D. Bruining at Appingedam.[10] *There were also many negative magnitudes recorded on our observing forms. However, that could be due to the fact that we were still rather inexperienced.* No Perseids were photographed from two stations. This was the beginning of the DMS multi-station photographic program. In October, the news came that Prentice had passed away.

17.2 The skeptic years of 1988–1990

Twelve years and one orbit of Jupiter later, there was a sense of *dejá vu* (Fig. 17.4). Puzzled by the fact that 109P/Swift–Tuttle had not been seen in the early eighties, Marsden took up the idea that comet *1737 II (Kegler)* might well be another apparition of P/Swift–Tuttle, a suggestion first made by William Lynn in 1902. Comet Kegler was discovered by the Jesuit missionary Ignatius Kegler, who was then the director of the Imperial Astronomical Bureau in Beijing, China. Although it was intrinsically bright, the comet kept its distance from Earth and was not well observed, except during the week from July 3 to July 10. Turning to the 1737 observations, Marsden had to allow some leeway with the 1862 observations, but found that it was possible to match the two sightings into one orbit. Marsden predicted a return in late November of 1992.

In 1988, a report[11] of a double peak in the Perseid profile by Paul Roggemans, founder of the *International Meteor Organization* (IMO) was received with skepticism. His results came from one of the first efforts to analyze visual observations on a global scale from groups that had little interaction and Roggemans had announced double peaks before in 1985 and 1986.

It was IMO's great achievement to unite visual observers in this manner. Earlier, the *Royal Astronomical Society*, the *British Astronomical Association*, the *British Meteor Society*, and the *American Meteor Society* had all provided a podium for reports from observers all over the world, but never was it attempted to systematically bring together visual observations in this manner. Because the Perseids can be seen from each location for only 5–6 h, a combination of counts from locations in Europe, the Americas, and East and West Asia were needed to cover the profile.

The problem was to normalize counts from different observers into the *zenith hourly rate*, proportional to the actual number of stream meteoroids hitting the Earth's atmosphere per unit area and time (Appendix G). The observed rates are lower if the star limiting magnitude (the faintest star visible by averted vision to a naked eye observer) is less than the standard $+6.5^{m}$, if the radiant altitude is lower than the zenith, and if the observer has a less than normal perception. High ZHRs were usually

[9] R. Veltman and H. K. Ploos van Amstel, Perseiden 1981: een normale terugkeer. *Radiant, J. DMS* **3** (1981), 116–123.
[10] J. D. Bruining, Perseiden 1981: een groot success!! *Radiant, J. DMS* **3** (1981), 95.
[11] P. Roggemans, The Perseid meteor stream in 1988: a double maximum! *WGN* **17** (1989), 127–137.

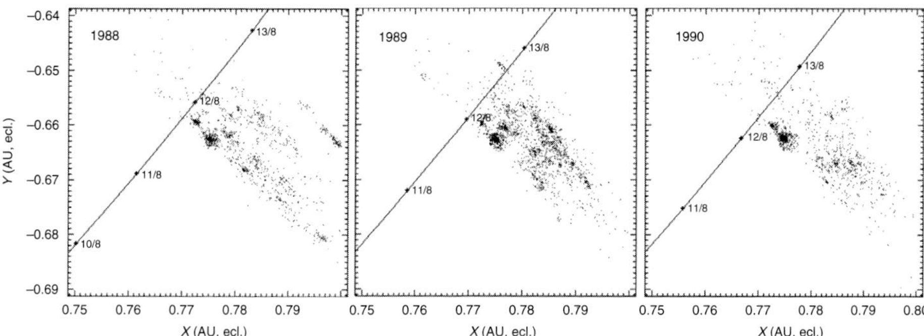

Fig. 17.4 Dust trails from comet Swift–Tuttle in 1988–1990.

the result of over correction, from underestimating the star limiting magnitude or the ease of spotting meteors in twilight or moonlight. Sometimes, observations from a whole continent would be dominated by a handful of observers with exceptionally good eyesight or a different approach to estimating the sky conditions.

In 1989, Roggemans claimed yet another double peak (Fig. 17.5). This time, Ralf Koschack of AKM assisted in the analysis of global visual observations.[12] That peak proved to be the real McCoy.

17.3 The prelude: the Perseid outburst of 1991

Even the most skeptic observers were convinced when Japanese observers reported a spectacular outburst of Perseids in August, 1991, with photographs to proof it (Figs. 17.6 and 7).[13] *Tatsuo Nakagawa* and *Maroshi Hayashi* of the Japanese Shinshu University Astro O.B. Club in Nagano captured 16 bright $<-2^{m}$ Perseids in a 1 h 8 min exposure.[14] The high rates were only for a short period of time, no longer than 1–2 h.

I first learned about these observations when I received a letter from Masahiro Koseki of the Nippon Meteor Society. In an astronomical telegram[15] on August 28, 1991, Brian Marsden had cited from the Yamamoto Circular no. 2170, where Mr. Y. Taguchi of Osaka told about the observations of a group of observers of the Kiso Observatory, which at the time were at a location at 1720 m altitude in the Japanese Alps. I remember thinking: "Ah, yes. Unusual observing site, excellent observing conditions, information out of third hand, probably another report of very high

[12] P. Roggemans and R. Koschack, The 1989 Perseid meteor stream. *WGN* **19** (1990), 87–98.

[13] P. Roggemans, M. Gyssens and J. Rendtel, One-hour outburst of the 1991 Perseids surprises Japanese observers! *WGN* **19** (1991), 181–184.

[14] J.-I. Watanabe, T. Nakamura, M. Tsutsumi and T. Tsuda, Radar observation of the strong activity of a Perseid meteor shower in 1991. *Publ. Astron. Soc. Japan* **44** (1992), 677–685.

[15] D. Levy, P. Jedicke, Y. Taguchi *et al.*, Perseid meteors and periodic comet Swift–Tuttle. *IAU Circ.* 5330 (August 28), ed. B. G. Marsden (1991); J.-I. Watanabe, Activity of Perseids 1991 and the parent comet P/Swift–Tuttle. *Proc. Int. Meteor Conf. Smolenice* (1992), pp. 82–85.

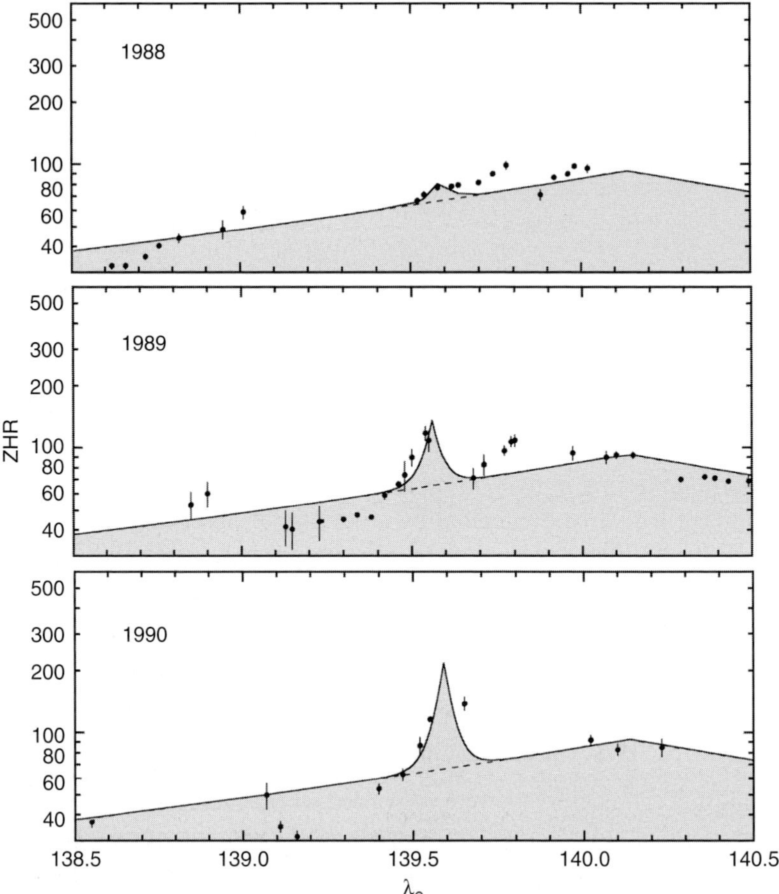

Fig. 17.5 Perseid shower rates in 1988–1990. The gray area is a fit of the normal annual shower activity in 1981–1991 plus a Filament component with constant width.[16]

meteor rates." And I was right, because the zenith hourly rate was supposed to have been ZHR = 352/h in the hour centered at 15.8 UT on August 12. Then again, rates before and after were quite normal, at about ZHR = 60/h. During this brief peak, one observer saw eight Perseids in 10 s.

At the time, the 1862-dust trail was in Earth's path (Fig. 17.8). Photographic observers participating in the Japanese Fireball Network, led by *Yasuo Shiba*, succeeded in recording nine multi-station fireballs, all of which emanated from a very compact radiant, shifted from the center of annual Perseid activity (Fig. 17.9).

[16] P. Jenniskens, H. Betlem, M. de Lignie *et al*., On the unusual activity of the Perseid meteor shower (1989–96) and the dust trail of comet 109P/Swift–Tuttle. *Mon. Not. R. Astron. Soc*. **301** (1998), 941–954; P. Brown and J. Rendtel, The Perseid meteoroid stream: characterization of recent activity from visual observations. *Icarus* **124** (1996), 414–428.

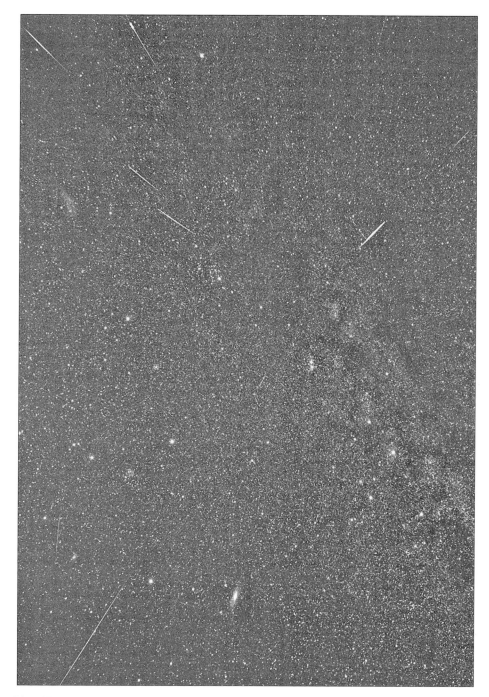

Fig. 17.6 Perseid 1991 outburst, photograph by Maroshi Hayashi of the Astronomical OB-Group of Shinshu University. The guided 30 min exposure on Sakura GX3200 film was taken from 15:45 to 16:16 UT, August 12, with an Asahi Pentax 67 camera and 55 mm f5.6 lens. The negative has 12 Perseids.

Fig. 17.7 Perseid outburst photograph by Tatsuo Nakagawa (Astronomical OB-Group of Shinshu University) at Takane Village on August 12, 1991. This is a 68 min exposure on RHP film, from 15:08 to 16:16 UT, with an Asahi Pentax 67 camera and 35 mm f4.5 lens. The negative has 26 Perseids. Photo: courtesy of Shinsuke Abe.

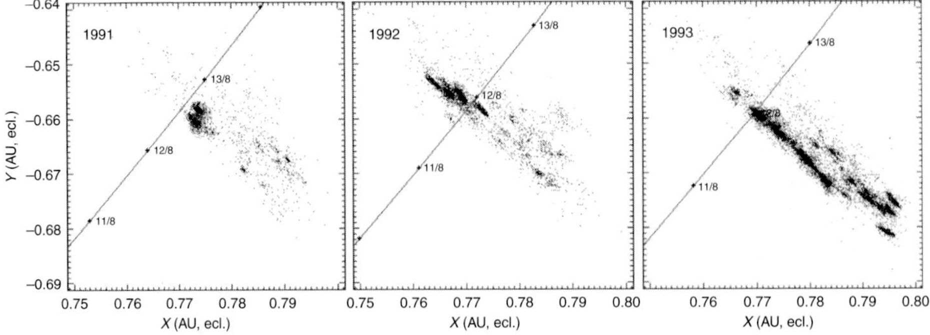

Fig. 17.8 Dust trails from comet Swift–Tuttle in 1991–1993: dust trail encounters.

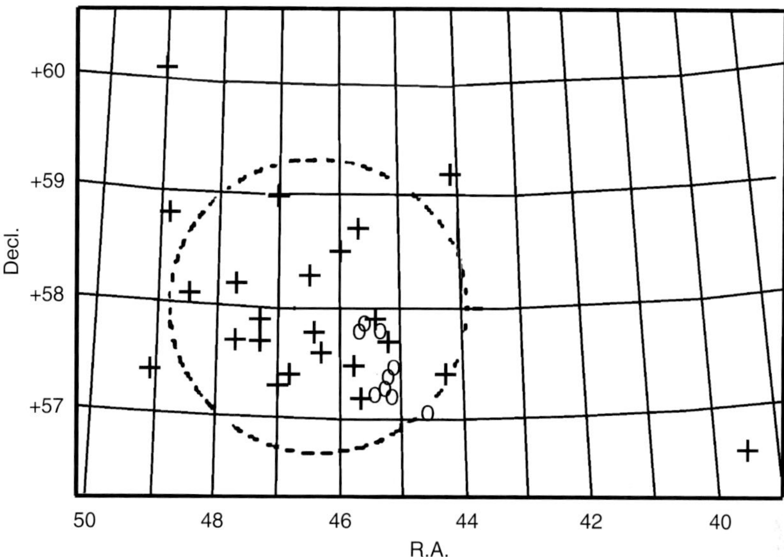

Fig. 17.9 Dispersion of meteor radiants during the Perseid outburst in 1991 (○), as compared to normal annual activity measured by the DMS (+). Image by Yasuo Shiba and coworkers, NMS.[17]

Marsden saw in those reports a confirmation of his prediction that comet 109P/Swift–Tuttle was on its way back to Earth. In his enthusiasm, he also cited meteor observations by comet observer David Levy, who reported high Perseid rates from a location in the United States. Bulgarian observers followed suit and observers from Belgium also reported high rates from an observing site in the Haute Provence. But one DMS observer in the Haute Provence, an archeology student and amateur astronomer called *Marco Langbroek*, did not notice anything out of the ordinary. Observers in the Netherlands, too, did not see higher than usual Perseid rates.[18] Marsden stepped in an age-old trap, it seemed. By quoting only the enthusiastic reports, the highest rate, the brightest meteor, or the flurry of meteors, the emerging picture was one of fantastic activity. For European and American observers, the annual Perseids were just as annual as other years. The Japanese record, on the other hand, could not be denied.

17.4 Chasing the anticipated outburst of 1992

Back in 1992 the comet orbit was thought to pass just outside Earth's orbit and I suspected that the dust was distributed in a sheet-like dust trail, extending ahead of the comet at locations inside the comet orbit (Fig. 17.10). If the comet had not been discovered by August of 1992, we might see another 1–2 h outburst.

[17] Y. Shiba, K. Ohtsuka and J.-I. Watanabe, Concentrated radiants of the Perseids outburst 1991. In *Meteoroids and Their Parent Bodies*, ed. J. Stohl and I. P. Williams. (Bratislava: Astron. Inst. Slovak Acad. Sci., 1993) pp. 189–193.

[18] P. Jenniskens and Y. Taguchi, Perseid meteors 1991. *IAU Circ.* 5340 (Sept. 10), ed. B. G. Marsden (1991).

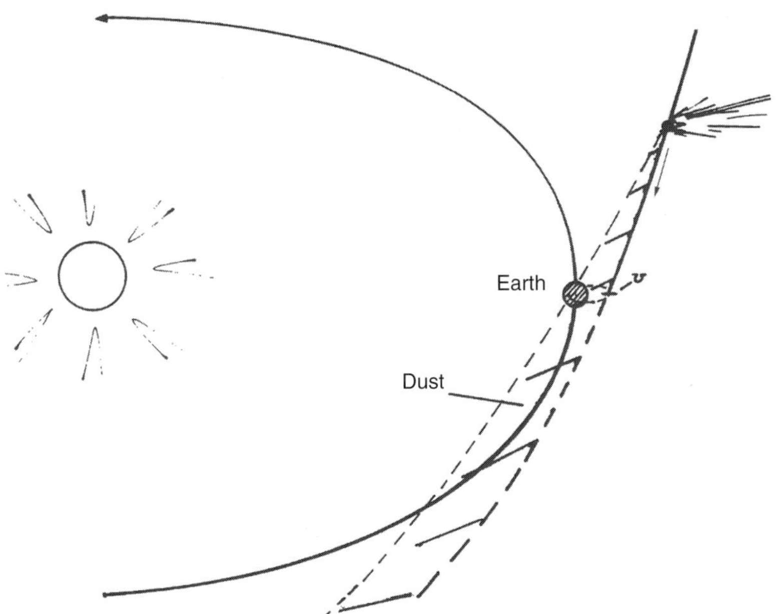

Fig. 17.10 A model of dust in a comet orbit. Drawing by Marco Langbroek.

The shower was expected to stay rich in bright meteors as in 1991,[19] because the meteoroids were expected to be brighter closer to the position of the comet. In 1992, Earth would arrive at the same location in solar longitude ($\lambda_\odot = 139.58°$) around 21.8 UT on August 11. Unfortunately, slight disturbances can move the trail by hours. From historic outbursts of the Leonids, I had found that the node of the sheet now known as the Filament tended to shift towards the node of the approaching comet. In 1991, the peak was some time after the passing of the node of comet 109P/Swift–Tuttle. Hence, I suspected that the time of peak activity could well be a little earlier than 21.8 UT. That was a problem, because evening twilight in the Netherlands ended only at about 21.5 UT.

By the summer of 1992, the comet had not yet been discovered. Convinced that there was enough ground to point out this opportunity, I wrote a paper with Marco Langbroek for the popular journal *Zenit* of the Dutch Association of Meteorology and Astronomy.[20] Because Marco was an associate editor, we had the space and control to measure out all the ifs and buts. Sadly, the article had little impact because observers were frustrated by bad weather and the near full Moon.

In the afternoon of August 10, Marco and I had settled at my parents' house in anticipation, but watched in frustration to see the deck of clouds over the Netherlands

[19] Y. Yabu, M. Ueda and K. Suzuki, Outburst of the Perseids 1991 observed in Japan. In *Meteoroids and Their Parent Bodies*, ed. J. Stohl and I. P. Williams. (Bratislava: Astron. Inst. Slovak Acad. Sci., 1992), pp. 201–204.

[20] M. Langbroek and P. Jenniskens, De Perseidenregen van 1991: herhaling op 11 augustus? *Zenit, J. NVWS* July–August, (1992), 328–330.

Fig. 17.11 1992 Perseids: Marco setting up at "Blauen."

thicken. I proposed a bold move: *let's drive south to clear weather, to Switzerland if necessary*. A crash campaign. A quick phone call enticed Casper ter Kuile of post "Pegasus" in de Bilt and Carl Johannink and Romke Schievink of post "Denekamp" to join us and provide photographic support for a multi-station effort. Five of us gathered near the border town of Venlo the next morning. After a 10 h drive, Carl and Casper manned a station in the French Les Vosges (pronounced "Vojs") mountains, while Marco, Romke, and I settled near the small town of Blauen in the Swiss Jura mountains.

At Blauen, farmer Siegfried Meury walked us to the perfect grass field for our cameras and offered gracious lodging in a garden house. Within minutes, we had set up the photographic cameras and an intensified video camera (Fig. 17.11). There were still some cirrus and cumulus clouds, but large parts of the sky were clear. It was 19:36 UT. The first stars were spotted when suddenly a very bright Perseid sailed

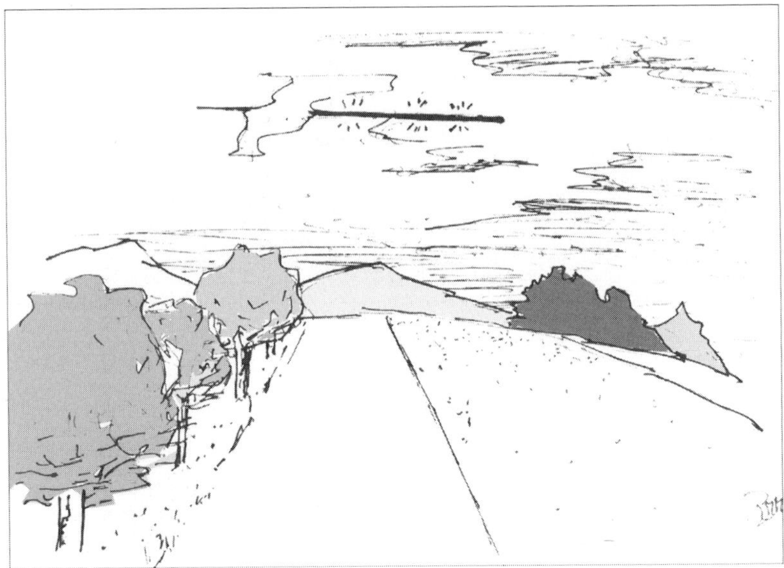

Fig. 17.12 Sketch of the grazing 19:36 UT August 12 Perseid over the Swiss Jura Mountains – the start of my first meteor outburst.

parallel to the mountains in the east, erupting in flashes, shooting behind the clouds and then emerging again (Fig. 17.12).

Adrenalin rushed and somewhere from my body sprang a loud yell. Here is how Marco described what happened next:

> We were still calming Peter when a −2 Perseid through the zenith caused another fierce scream. Romke and I reacted a little worried, thinking of angry farmers with forks and threshing tools who would come to check out why those crazy foreigners were making such a dreadful noise. We asked Peter kindly to be a little less enthusiastic. Because what is that, two meteors? Two? In the corner of my eye, I see a −3 Perseid, flaring, moving gracefully, cutting through the square of the big Dipper, whose stars are just barely visible. And one more: there is a −1 or −2 Perseid going in the direction of the star Arcturus . . . ! Now, we all were on fire. It must already be going on, it is too early! Or did it just start? Again a yell when another bright meteor appears. In the meantime, the cloud cover was getting worse and after seven semifireballs and some weak meteors in barely 35 minutes, the sky was completely overcast. That was truly unusual: so many bright meteors in such a short time, with a radiant altitude at merely 20°, lousy sky conditions, and increasing cloud cover. Absolute madness!

When it cleared again, 20 min later, rates were back to normal. Casper and Carl had seen the bright meteors too: 14 grazing Perseids in half an hour, at one time three in one minute. Unfortunately, the twilight did not permit photography. We were not disappointed. For the first time in our lives, we had seen a significant enhancement of rates above the normal annual Perseid activity (Fig. 17.13). That was a *meteor*

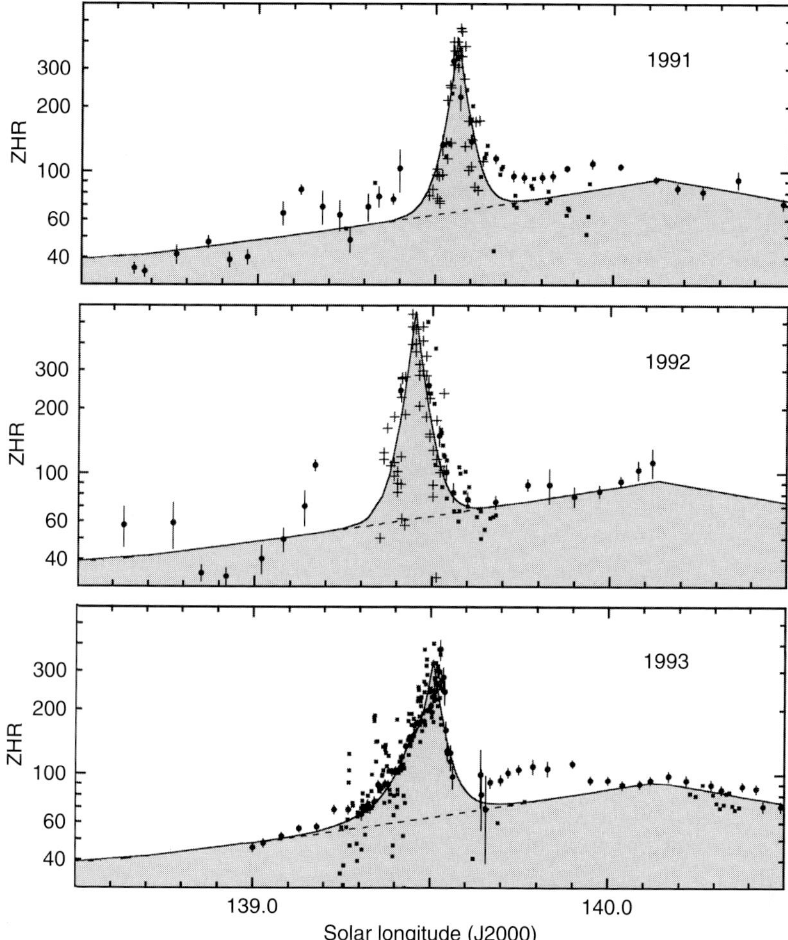

Fig. 17.13 Perseid activity 1991–1993. Visual (●) and radio MS (+) data.

outburst! The next day, we celebrated with a visit to the town of Ensisheim and saw its famous 500 yr old meteorite.

Brian Marsden was all too happy to learn about the new outburst,[21] raising further hopes for the return of the comet. Chinese observers confirmed the outburst[22] and 109P/Swift–Tuttle was recovered on September 26 that year as a bright magnitude +9 object by Japanese amateur astronomer *Tsuruhiko Kiuchi*[23] who, spurred by the meteor outburst, checked the predicted position using 15 cm binoculars. Perihelion passage occurred on December 12, 1992. Marsden had predicted November 25th, only 17 days earlier. This was the first time that a meteor outburst helped recover the parent

[21] A. Mizser, P. Jenniskens and J. Rao, Perseid meteors 1992. *IAU Circ.* 5586 (August 13), ed. B. G. Marsden (1992).

[22] X. Pin-xin, The 1992 Perseid outburst in China. *WGN* **20** (1992), 198.

[23] H. Kosai, T. Kiuchi, J. B. Tatum *et al.*, Periodic comet Swift–Tuttle (1737 II = 1862 III = 1992t). *IAU Circ.* 5620 (September 27), ed. B. G. Marsden (1992).

comet. Among the observers of the 1992 Perseid outburst was Lubor Kresák, who estimated from his own observations a peak ZHR = 800/h.[24]

17.5 Everybody catches on in 1993

Now Perseid outbursts were established, it was soon realized that the Perseids also had an outburst in 1862, during the previous return of the comet.[25] Chinese gazetteers reported about August 10, 1862,[26] at around $\lambda_\odot = 139.61 \pm 0.10$: Immediately after twilight, *At the hour hsu* (7–9 p.m.), *stars fell like rain during about two hours*, and further west: *At the hour kai* (9–11 p.m.) *numerous stars flew from the northeast to the southwest in confusion. They were too many to count.* Someone added: *They sounded as if glossed silk were torn.* The next year, in 1863 at $\lambda_\odot = 139.64$, an outburst of Perseids was observed in Europe and the eastern Americas, raising hope of a return in 1993 (Fig. 17.14). Other historic accounts of high Perseid activity (Table 1) do not coincide with the return of the comet to perihelion.

In the 1992 return, the comet orbit passed so close to Earth's orbit (0.001 AU) that some believed a meteor storm was on the cards. Marsden created further alarm by warning that a mere 15 d uncertainty in the time of perihelion of the comet (the 1992 return had been incorrect by 17 d) could cause the comet to collide with Earth during the next return in August 14, 2126.[27] That fear dissipated, when the orbit was traced back to sightings in AD 188 and 69 BC.[28] The more precise orbit has the approach to Earth of the comet at only 0.153 AU, in no danger of hitting Earth. A closer encounter will occur on September 2, 3044 (~0.1 AU), followed by an even closer one in September, 4479 (~0.04 AU).

There was huge anticipation in August of 1993, whipped up to a frantic hyperbole by *Joe Rao*'s popular articles in *Sky & Telescope* and *WGN*.[29] The possibility of Leonid-type storms made Martin Beech and Peter Brown decide to issue a warning that satellites in Earth's orbit might be in danger.[30] That led the astronomers operating the Hubble Space Telescope to decide to reorientate the HST for 6 h on either side of the potential storm period near 1:00 UT on August 12. Rao's enthusiastic *Sky & Telescope* article and the measures taken by the HST team subsequently alarmed journalists, who then asked NASA officials if the meteors might not endanger NASA's space shuttle program, because a shuttle mission was scheduled during the

[24] L. Kresák, Meteor storms. In *Meteoroids and Their Parent Bodies*, ed. J. Stohl and I. P. Williams. (Bratislava: Astron. Inst. Slovak Acad. Sci., 1992), pp. 147–156.

[25] D. W. Olson and R. L. Doescher, Meteor observations on August 10–11, 1863. *WGN* **21** (1993), 175–181.

[26] N. Nogami, Chinese local records of the 1862 Perseids and the 1885 Andromedids. *Earth, Moon Planets* **68** (1995), 435–441.

[27] B. G. Marsden, Periodic comet Swift–Tuttle (1992t). *IAU Circ.* 5636 (October 15), Minor Planet Center (1992).

[28] K. Yau, D. Yeomans and P. Weissman, The past and future motion of comet P/Swift–Tuttle. *Mon. Not. R. Astron. Soc.* **266** (1994), 305–316.

[29] J. Rao, Perseids 1993: the big one? *WGN* **21** (1993), 110–111; J. Rao, Storm watch for the Perseids. *Sky Telescope* **86** (1993), 43.

[30] M. Beech and P. Brown, Impact probabilities on artificial satellites for the 1993 Perseid meteoroid stream. *Mon. Not. R. Astron. Soc.* **262** (1993), L35–L36.

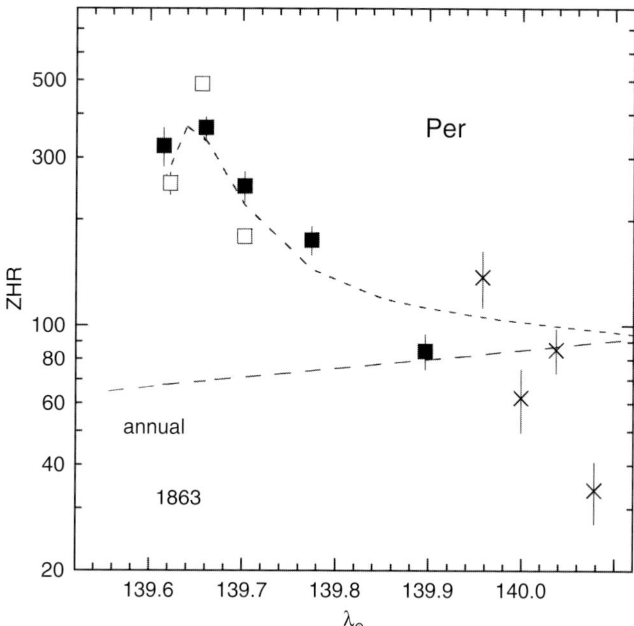

Fig. 17.14 An activity curve of the 1863 Perseid outburst, from J. F. Julius Schmidt at Athens Observatory, a group of observers in Germany as reported by Eduard Heis, and from observers Andres Poey and Ricardo Zenoz from Havana, Cuba, in that order.[31]

next Perseid shower. Again people did the math and found that the impact danger was negligibly small compared to the daily impact rate of meteoroids, even in a 1966 Leonid-like storm. Even so, the danger from large and fast meteoroids would dramatically rise. Better to keep your new car in the garage during rush hour! The Shuttle launch was postponed and all of America drove out to see the meteors. Against all odds, the ESA Olympus satellite lost one of its gyros at the time of the shower, due to a meteroid impact, and had to be parked in a graveyard orbit.[32]

An interesting perspective was offered by astronomers Zidian Wu and Iwan Williams. From a computer model, Wu and Williams[33] found that the planets moved the dust in and out of the Earth's path. They calculated that rates should continue to increase in 1993 and peak in 1994 (Fig. 11.3). Their paper came out just days before the 1993 Perseid peak and received wide-spread attention.

Journalists were having a field day. Not only Europe was on full alert, Rao had warned that a meteor storm might occur with rates increasing to over 100 000/h, perhaps even over the USA. He himself, however, joined the Perseid shower theme

[31] P. Jenniskens, Meteor stream activity. II Meteor outbursts. *Astron. Astrophys.* **295** (1995), 206–235.
[32] R. D. Caswell, N. McBride and A. D. Taylor, Olympus end of life anomaly – a Perseid meteoroid event? *Int. J. Impact Eng.* **17** (1995), 139–150.
[33] Z. Wu and I. P. Williams, The Perseid meteor shower at the current time. *Mon. Not. R. Astron. Soc.* **264** (1993), 980–990.

Fig. 17.15 1993 outburst Perseids in a compilation of photographs taken at post "Rognes" in the French Haute Provence between 01:37:00 and 02:37:00 UT, August 12. The brightest meteor (left) appeared at 02:23:37 UT. Star background is that of the first 01:37–01:47 UT exposure. Photo: Casper ter Kuile, DMS.

cruise to the Mediterranean Riviera organized by *Ted Pedas*. Dutch Meteor Society observers established a multi-station photography effort in the Haute Provence. I had resigned myself to staying in California, where I had moved earlier that year to start a job at NASA Ames Research Center. That August night, all the roads in the San Francisco Bay area were crowded. Thousands gazed up at the dark sky.

For US observers, the 1993 Perseids was as notorious as the ill-fated 1899 Leonid storm that set meteor astronomy back a century. East coast observers had slightly elevated rates at the tail of the outburst, but few noticed that the low rates were higher than they should be. West coast observers saw nothing unusual. *There were too few meteors to keep the kids quiet*, said David Blake, my new boss at NASA Ames. As expected, it was the European observers who watched a spectacular outburst of bright meteors. No storm, but rates did go up to ZHR = 350/h towards the early morning. The cries of enthusiastic observers were heard far and wide, which excited local canines. This night came to be known as the *night of the howling dogs* (Figs. 17.15 and 17.16).[34]

[34] M. Langbroek, Vuurwerk boven de Provence!!! De aktiviteiten van het 'dreamteam' Rognes. *Radiant, J. DMS* **15** (1993), 96–106.

Fig. 17.16 The same outburst of 1993 Perseids between 01:37:00 and 02:37:00 UT, August 12, showing the typical irregular light curves. The bright -3^{m} Perseids appeared at 01:38:48 (right) and 01:58:39 UT (left). The star background is that of the first exposure. Photo: Casper ter Kuile, DMS.

17.6 After the disappointment

The next year, I realised that my move to California had put me at the best location for watching the 1994 Perseid outburst. I tried to alert the local media, but nobody would listen. If the press is like a herd of buffaloes, this flock was grazing. Only astronomer Iwan Williams[35] created some headlines with his prediction that the series of outbursts might well peak in 1994.

At Fremont Peak Observatory, Henry Coe State Park, and at a site near Los Banos, a new team of amateur meteor observers gathered to assist a multi-station photography campaign. This was the start of the California Meteor Society.

Among those new observers was *Mike Koop* from San José, on whom I came to depend in future campaigns for technical support and as my man on the other hill (or the other plane) for stereoscopic observations. When late that afternoon we passed by Fremont Peak, on the way to our observing site further East, we saw, in alarm, that a billow of smoke rose over the state park, where park ranger Rick Morales (Fig. 17.17) had just spend much of the afternoon battling one of California's notorious wild fires. Only one other observer, lawyer Duncan M. McNiell, managed to weasel past the fire guard to join Rick in operating the station that night. The smoke dissipated just in time for the Perseid observations.

[35] I. P. Williams and Z. D. Wu, The current Perseid meteor shower. *Mon. Not. R. Astron. Soc.* **269** (1994), 524–528.

Fig. 17.17 Park ranger and amateur astronomer Rick Morales with one of the 1994 Perseids camera platforms.

At about 03:00, rates noticeably increased (Fig. 17.18). It was quite a spectacle: *+2, +3, +1. Wow! A −6 in the zenith! Time . . ., let's see, ... 09:32:14 UT!* A fireball crossed the zenith and exploded at the end of its path (Fig. 17.19). I remember looking up in amazement, enjoying myself immensely. Just imagine: the sky conditions were perfect, no Moon and a radiant high in the sky! While observing, I could see, almost feel, the rates go up. Bright meteors fell left and right. One particularly nice fireball low in the east left a persistent train that was seen for 8 s, while drifting rapidly towards the south, changing in shape as a thread of silk in the wind.

That night in 1994, a lonely observer in California's Sierra Nevada mountains relaxed from a busy day and, while his cameras were exposing the sky, he dozed off in the late hours of the night. Between 10:10 and 11:55 UT, *Kenton R. Parker* missed what could well have been the best meteor shower of his life, but fortune had it that he woke up to end the exposure just before twilight set in. Twelve Perseid trails had been recorded in what was arguably the most beautiful astronomy picture of the year (Fig. 17.20).

The peak was at about 04:10 a.m. local time (Fig. 17.18), with a ZHR ~190/h. After that, rates noticeably declined. When twilight came at 5:10 a.m., rates were back to normal. We were excited, we were happy, we were proud. A pattern had established,

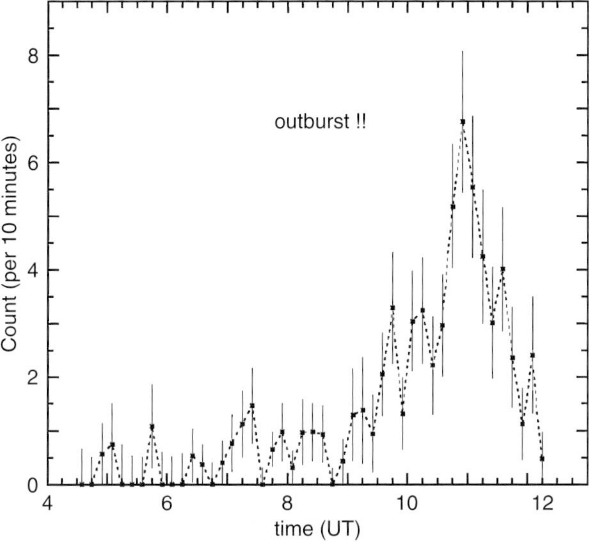

Fig. 17.18 The number of photographed Perseids during the night of August 12, 1994.

Fig. 17.19 Right: Perseid at 10:52:21 UT August 12, 1994. Photo: Paul Canfield at Springville, CA. Left: this -8^m Perseid at 09:32:14 UT (August 12) was photographed from Los Banos, CA, by Peter Jenniskens. This meteor was also photographed by Rick Morales at Fremont Peak State Park.

Fig. 17.20 Twelve photographed Perseids are visible in the original negative of this photograph by Kenton R. Parker, CMS.

but the showers did not behave as indicated by the model of Wu and Williams. The timing was off and the shower was not as strong as in 1992 and 1993 (Fig. 17.21).

Perseid outbursts would continue to be reported in the following years. The final well documented outburst in this series was that of the 1997 Perseids.[36] It was well observed by us in California. This was the year before the first Leonid multi-instrument aircraft campaign, and the outburst helped jump start the Leonid campaign.

Our photographic results were reduced by Hans Betlem and the video results by Marc de Lignie (Fig. 17.22). The 1997 Perseids had three photographic orbits and six video orbits clustered at R.A. $= 46.3 \pm 0.36°$ and Decl. $= 57.76 \pm 0.17°$, amidst significantly dispersed radiant positions from meteoroids that were clearly part of the annual shower.

17.7 The failure of dust trail models

First to study of the ejection of meteoroids from comet 109P/Swift–Tuttle was Salah E. Hamid,[37] using Fred Whipple's vision of an icy-snowball comet nucleus, emitting dust grains by water vapor drag. Computing methods of the time allowed only the calculation of mean perturbations, assuming the planet mass was distributed in a ring around

[36] R. Arlt, Global analysis of the 1997 Perseids. *WGN* **26** (1998), 61–77.
[37] S. E. Hamid, The formation and evolution of the Perseid meteor stream. *Astron. J.* **56** (1951), 126–127.

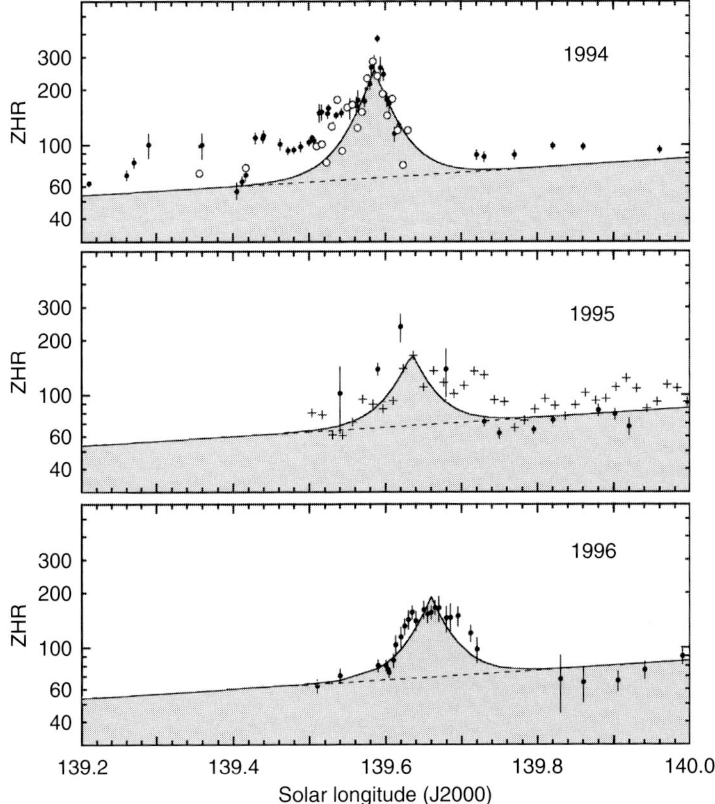

Fig. 17.21 Perseid shower activity in 1994–1996.[38]

the Sun. *Richard B. Southworth*[39] first calculated the orbit of individual particles in an effort to explain the large dispersion of meteoroid orbits in the annual shower and could not understand how planetary perturbations could scatter the meteoroids so much over periods of only a few thousand years (the annual stream is older than that). Sekanina[40] examined how the direction of ejection affects the particle position in the stream. From systematic variations of the time of historic Perseid returns, he concluded that the meteoroids were ejected near perihelion, over a period of only a few months.

Following the recovery of Swift–Tuttle in 1992, and using greatly improved computing facilities, Wu and Williams[41] calculated the orbit of 500 test meteoroids, using

[38] P. Jenniskens, H. Betlem, M. de Lignie *et al.*, On the unusual activity of the Perseid meteor shower (1989–96) and the dust trail of comet 109P/Swift–Tuttle. *Mon. Not. R. Astron. Soc.* **301** (1998), 941–954; P. Brown and J. Rendtel, The Perseid meteoroid stream: characterization of recent activity from visual observations. *Icarus* **124** (1996), 414–428.

[39] R. B. Southworth, Dynamical evolution of the Perseids and Orionids. *Smithsonian Contrib. Astrophys.* **7** (1963), 299–303.

[40] Z. Sekanina, Meteoric storms and the formation of meteor streams. In *Asteroids, Comets and Meteoric Matter*, ed. C. Cristescu, W. J. Klepczynski and B. Milet. (New York: Scholium, 1974), pp. 239–267.

[41] Z. Wu and I. P. Williams, The Perseid meteor shower at the current time. *Mon. Not. R. Astron. Soc.* **264** (1993), 980–990; I. P. Williams and Z. Wu, The current Perseid meteor shower. *Mon. Not. R. Astron. Soc.* **269** (1994), 524–528.

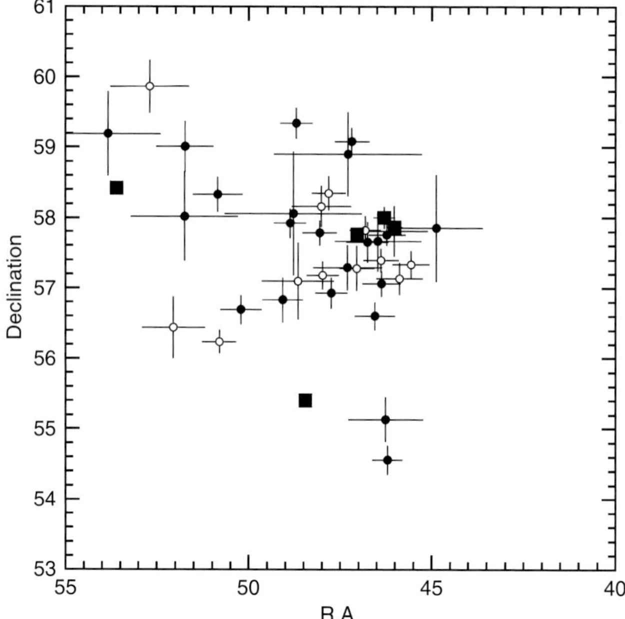

Fig. 17.22 Video (○) and photographic (■) Perseids on August 12, 1997, split between periods 8–9 h UT (● = outburst) and 9–11 h UT (○ = annual).

Whipple's ejection model (Fig. 17.23). They first noticed how the planets periodically moved the meteoroids into Earth's path. They held the 1862 dust responsible for the Perseid outbursts and predicted even higher rates in 1994, but we saw earlier that this did not pan out. Nor did the calculated variation of peak times (Fig. 17.24).

The methods pioneered by Wu and Williams were further developed by *Petr Pecina* and *Milos Simek* at Ondrejov Observatory[42] and in the Ph.D. thesis of *Peter Brown* at the University of Western Ontario in Canada. All adopted a range of particle densities and ejection along the comet orbit. Ejection speeds were typically high, resulting in a wide distribution of dust near Earth's orbit around the time of an expected outburst. A length of trail was averaged to derive the expected peak time and dust density.

Brown and Jones[43] ejected a total of 610 000 test particles and studied where those particles crossed the ecliptic plane near Earth's orbit. Following Wu and Williams, they concluded that the 1991–1994 Perseid outbursts could all have been caused by dust ejected in 1862 (with some contribution from 1610 dust), expecting the highest rates to be just behind the comet in 1993 or 1994. Instead, rates had peaked in 1992, ahead of the comet. They also expected the particles to become fainter along the trail.

[42] P. Pecina and M. Simek, The orbital elements of a meteoroid after its ejection from a comet. *Astron. Astrophys.* **317** (1997), 594–600.

[43] P. Brown and J. Jones, Simulation of the formation and evolution of the Perseid meteoroid stream. *Icarus* **133** (1998), 36–68.

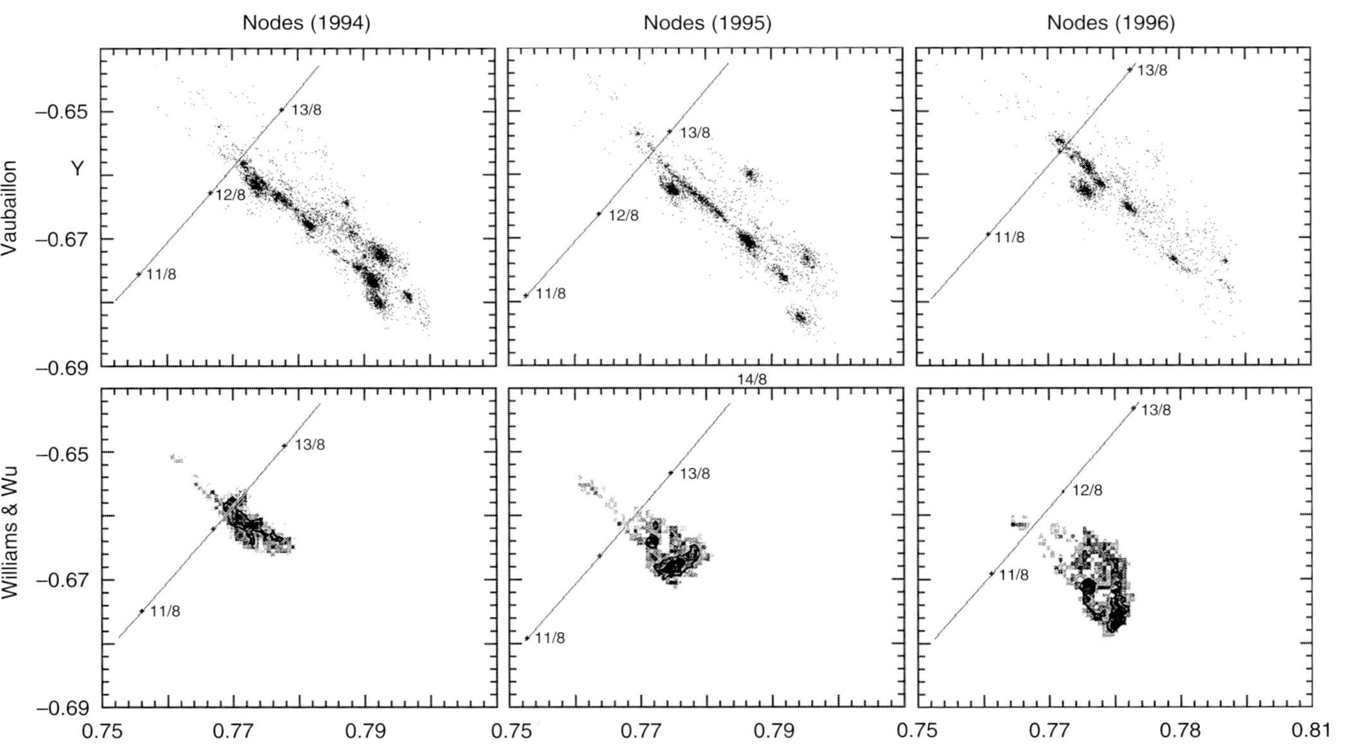

Fig. 17.23 Dust trails from comet Swift–Tuttle in 1994–1996: no trail encounters. Top: results by Jérémie Vaubaillon. Bottom: early results by Zidian Wu and Iwan Williams.

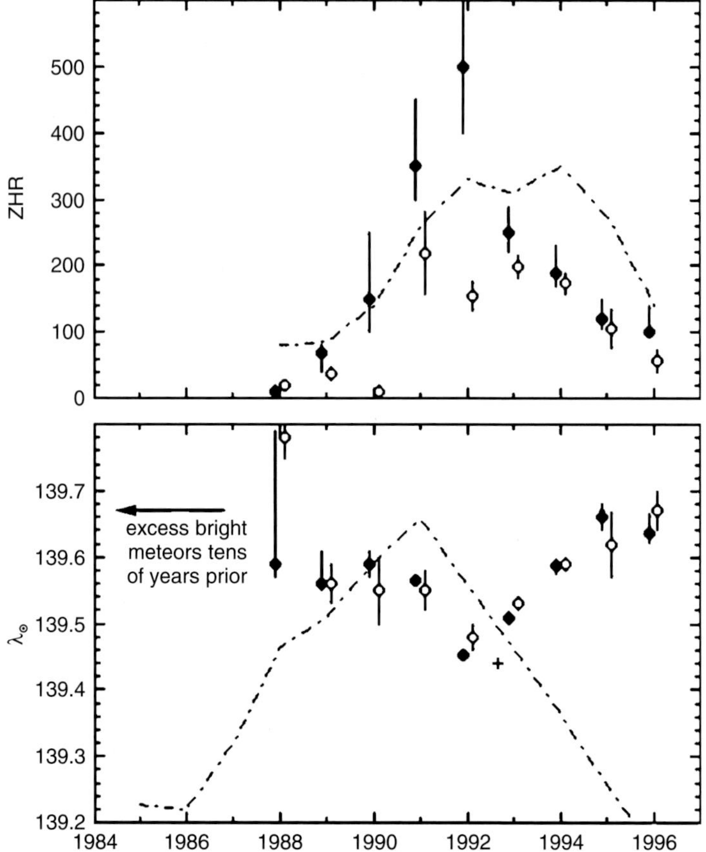

Fig. 17.24 The observed variation of the peak activity (top) and the time of maximum shower activity compared to the model calculations by Wu and Williams (dashed line).[44]

Instead, $\chi = N(m+1)/N(m)$ remained constant, changing from 1.86 ± 0.14 in 1991 to 2.12 ± 0.15 in 1992, to 2.02 ± 0.04 in 1993 and 1.9 ± 0.2 in 1994 (Table 5d).

Jupiter moved the pattern of trails periodically towards Earth's orbit, for a short period every 12 years when Jupiter approached the dust trail. The shifts induced in 1991 were thought to be responsible for the 1991–1994 outbursts. In this model, rates should also have been high in 1980, when no meteor storm was reported.

In order to explain the shift of the nodes following the 1992 outburst, Brown and Jones invoked that each outburst had contributions from more than a single dust trail (Fig. 17.23): the 1988 and 1990 returns consisted of 1610 and 1737-dust; the returns from 1991 through 1994 were thought to be 1862 and 1610-dust, while from 1995 through 1997 Earth met streams ejected in 1079, 1479, and 1862. The older dust trails tended to shift the node to a later time, but did not explain the fact that the peak time

[44] Z. Wu and I. P. Williams, The Perseid meteor shower at the current time. *Mon. Not. R. Astron. Soc.* **264** (1993), 980–990.

in 1993 was later than that of 1992. The result is sensitive to the adopted particle density. Very fluffy particles ($\rho = 0.10\,\text{g/cm}^3$) moved the 1862-, 826-, and 1079-dust trails into Earth's path in 1992 and 1993, but also had trails move into Earth's path at times when no outburst was observed. In 1997, the Perseids returned as postdicted $\lambda_{\odot} = 139.72 \pm 0.04$ and ZHR $= 137 \pm 7$ (Table 5), and were observed in California, but stronger predicted outbursts in 1998 and 1999 did not occur.

17.8 Young dust trail or old Filament?

What clues are there to decide if the Perseid outbursts were indeed due to recent dust trails, or instead from older dust accumulated in mean-motion resonances, now called a *Filament*?[45] Such a Filament would be narrower in Earth's path than the corresponding structure in the Leonid shower because of the higher inclination of the comet orbit.

The large dispersion of radiants measured in our photographic observations during the 1993, 1994, and 1997 Perseid campaigns (Fig. 17.25) points towards older dust. The radiants were more spread out than calculated in the model by Brown and Jones for recent dust trails, especially considering the large ejection speeds adopted in their model.

Interestingly, the radiant dispersion before (•) and after (o) solar longitude 139.46 is quite different. After $\lambda_{\odot} = 139.46°$, the radiants are significantly dispersed along the ecliptic plane, with a dispersion of $\pm 0.31°$ in inclination. Before that, seven out of 15 radiants cluster in a tight radiant at R.A. $= 45.88 \pm 0.08°$, Decl. $= 57.84 \pm 0.03°$ (defined for Earth being at solar longitude 139.5°), offset from that of the Filament and not far from the theoretical radiant calculated for the 1862-dust trail encounter by Brown and Jones.

The activity profile was asymmetric in 1993, with a more rapid decline than increase. This could be explained if there were two components to the activity profile, including what could have been an 1862-dust trail crossing centered close to the current comet node at $\lambda_{\odot} = 139.444°$ (Fig. 17.26). The magnitude distribution index was also lower at the onset: $\chi = 1.72 \pm 0.02$ instead of 2.20 ± 0.04 at the peak (after splitting into two components). The observed differential size distribution power index of dust ejected by 109P/Swift–Tuttle was $\alpha = 3.3 \pm 0.2$, which translates to $s = 1.77 \pm 0.07$, in good agreement with the value for this, what I called, the "nodal blanket."

I conclude that we did cross the 1862-dust trail in 1993 in the hours before the peak of the Filament outburst later that night. In light of the recent Leonid trail observations, the large width is now understood because we passed far from the trail center at that time. Because of that, the width of the fresh trail was as wide as the Filament. In more central crossings, the meteor outburst could be as narrow as FWHM ~ 0.52 h!

[45] P. Jenniskens, H. Betlem, M. C. de Lignie *et al.*, On the unusual activity of the Perseid meteor shower (1989–96) and the dust trail of comet 109P/Swift–Tuttle. *Mon. Not. R. Astron. Soc.* **301** (1998), 941–954.

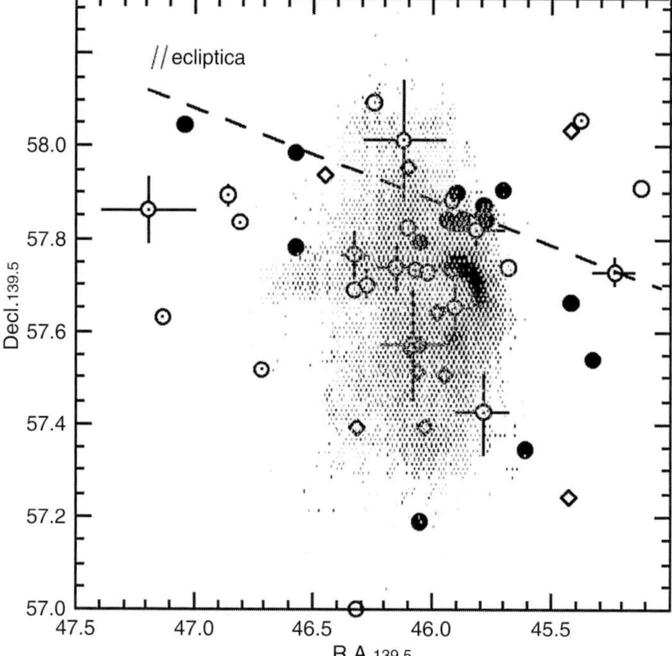

Fig. 17.25 The radiant of 1993 outburst Perseids before (●) and after (⊙) solar longitude 139.46°. Open circles are data for 1994. Also shown is a radiant derived during the 1991 outburst (◇).[46] Representative error bars are given. The gray scale is the distribution of radiants calculated by Brown and Jones[47] for dust trails ejected between 59 and 1862 AD. The dark spot is mostly dust that was ejected in 1862.

Esko Lyytinen confirmed that the one-revolution trail in 1993 passed at a relatively large $\Delta r = +0.0124$ AU at $\lambda_\odot = 139.433°$. Other such dust trail encounters are listed in Table 5c, predicted from the model by Vaubaillon, with my estimate for the expected peak activity. Earlier outbursts may also have had a contribution from yang trails. We crossed the 1610-dust trail in 1990, and 1991, while the 1862- and 826-dust trails may have been responsible for much of the 1992 outburst, in agreement with the results by Brown and Jones.

17.9 Future outbursts

Twelve years later in 2004, Jupiter steered the 1862-dust back into Earth's orbit (Fig. 17.27) and Lyytinen predicted an outburst of several hundred meteors per hour at 20:54 UT on August 11. The outburst was well observed in Eastern Europe (Fig. 17.28).

[46] Y. Shiba, K. Ohtsuka and J.-I. Watanabe, Concentrated radiants of the Perseids outburst 1991. In *Meteoroids and Their Parent Bodies*, ed. J. Stohl and I. P. Williams. (Bratislava: Astron. Inst. Slovak Acad. Sci., 1993), pp. 189–193.

[47] P. Brown and J. Jones, Simulation of the formation and evolution of the Perseid meteoroid stream. *Icarus* **133** (1998), 36–68.

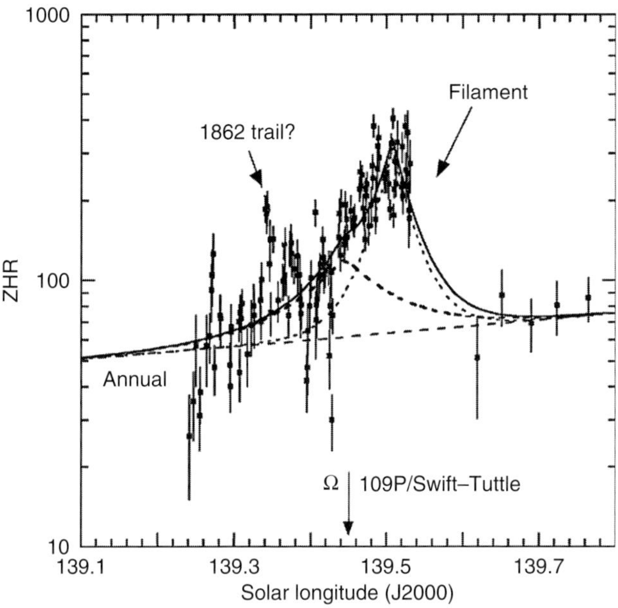

Fig. 17.26 The rate of 1993 Perseids with contributions from the annual shower, the Perseid Filament, and the 1862 dust trail identified.

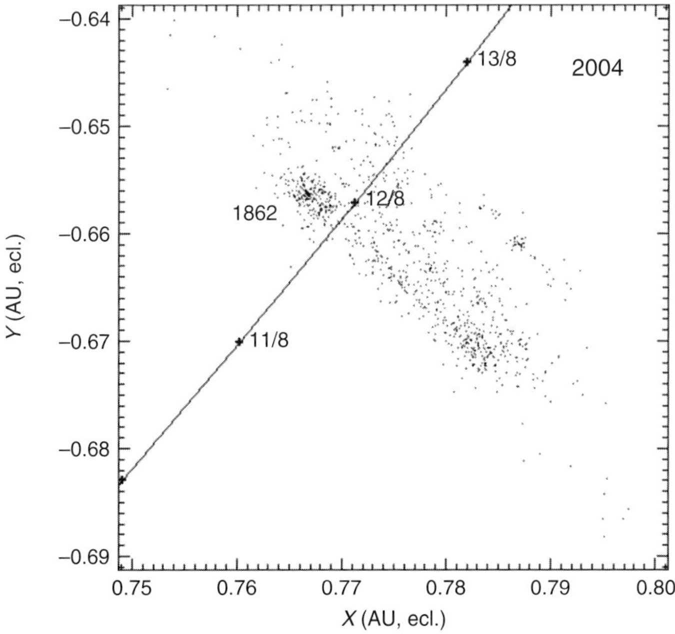

Fig. 17.27 The 1862-dust trail encounter in 2004. Calculations by Vaubaillon.

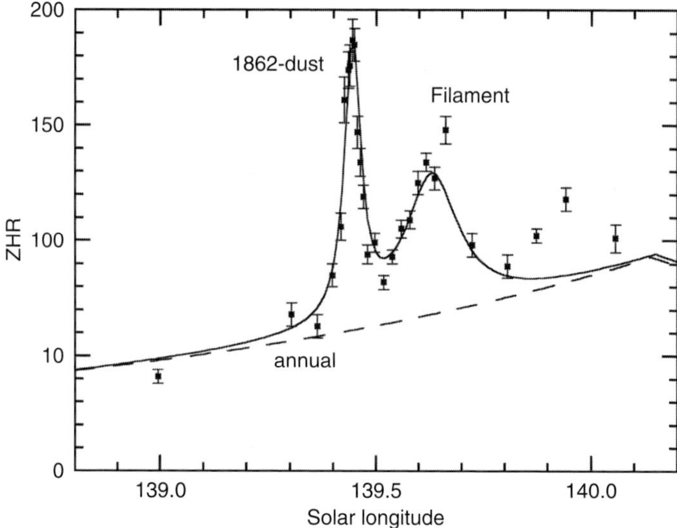

Fig. 17.28 Perseid activity in 2004 with the 1862 dust trail and Filament encounters identified. Visual observations gathered by IMO.

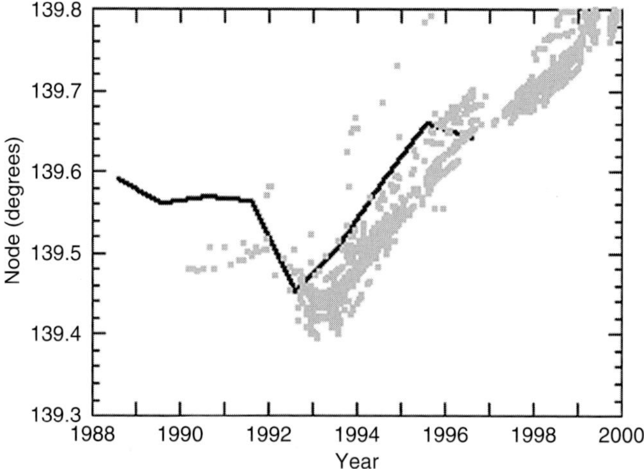

Fig. 17.29 The node of Filament (line) and old dust trails ejected in the eight revolutions following the AD 188 return of Swift–Tuttle in the model by Lyytinen.

A second broader peak was observed in western Europe, presumably the Filament albeit three times broader than past Filament encounters. I estimate the total amount of matter in the Filament component to be larger than 3×10^{11} kg, the mass of a few trails.

Lyytinen studied the orbital evolution of the more recent dust trails ejected since 69 BC (Fig. 17.29). The calculated node of the meteoroids ejected in the eight

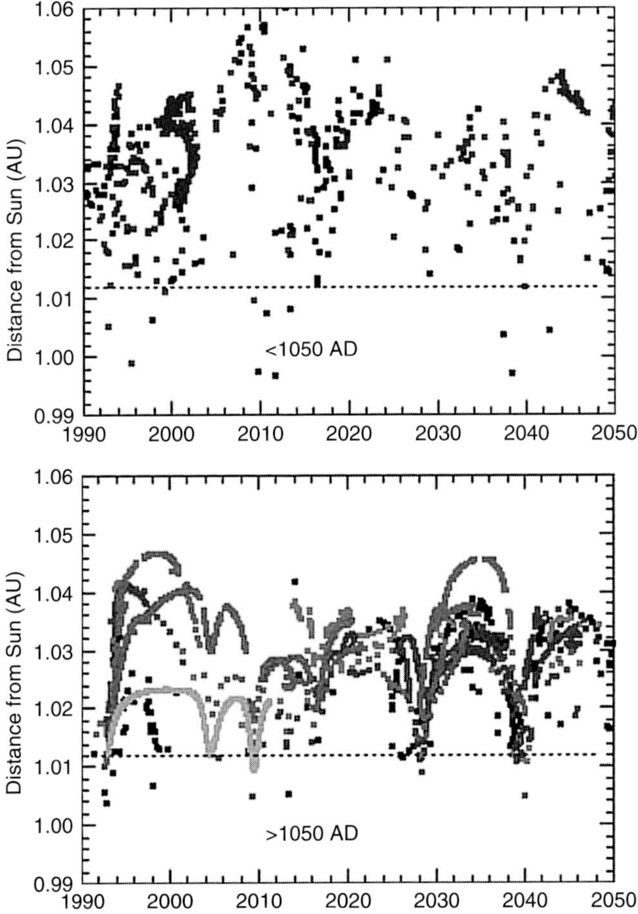

Fig. 17.30 Position of old (top) and new (bottom) trails of 109P/Swift–Tuttle near Earth's orbit. Calculations by Lyytinen.

revolutions following AD 188 (if no ejection velocity is assumed and beta values are low) closely follows the observed variation of peak times, but there is a discrepancy of −0.4°. This discrepancy may prove that ultimately older trails are responsible for the Perseid Filament. The meteoroids may date from long before the comet evolved in its current 1:11 mean-motion resonance window since about the 688 return.

The most likely years of Perseid shower outbursts are those when Jupiter steers the dust trails in Earth's path every 12 years. Already in 1923, Denning[48] concluded that historic records of Perseid displays dating back to AD 714 show a pattern with a period of 11.72 yr (counting back from the intense 250/h shower of AD 1921.16). In

[48] W. F. Denning, Radiant points of shooting stars observed at Bristol chiefly from 1912 to 1922 inclusive. *Mon. Not. R. Astron. Soc.* **84** (1923), 43–56.

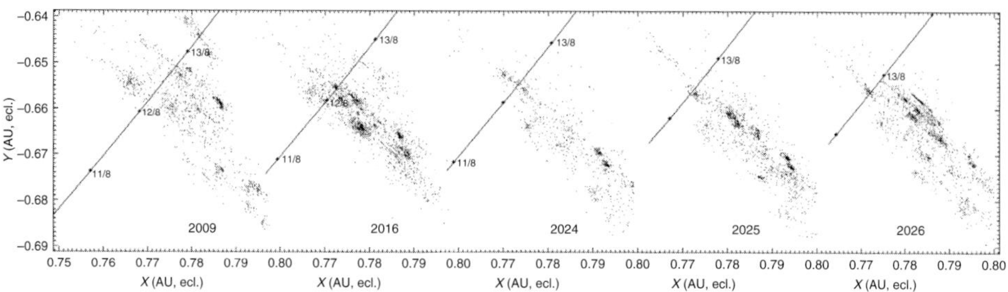

Fig. 17.31 Perseid dust trail encounters in 2008, 2016, and 2024. Calculations by Vaubaillon.

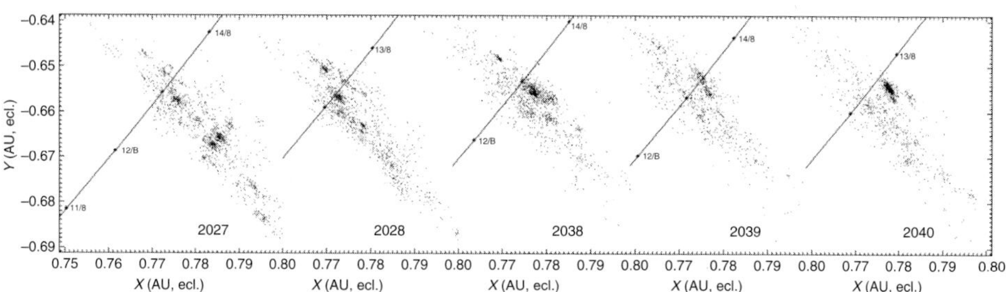

Fig. 17.32 Perseid dust trails in other favorable returns shown in Fig. 17.30.

that same sequence we can fit AD 1979.8, 1991.5, and 2003.2, as well as 2014.9, 2026.6, and 2038.4. These dates correspond roughly to the years when Jupiter moves the Perseid meteoroids into Earth's orbit (Fig. 17.30).

Both the Filament and the dust trails are affected by this, as well as the annual Perseid shower, albeit to a lesser extent.

A promising return is that of the 1479-dust trail in AD 2028 (Fig. 17.32), when rates could increase significantly. The one-revolution 1862-dust trail will pass just outside the Earth's orbit at that time. Further predictions are listed in Table 5c. It remains to be seen how well the model predicts the dust trail positions. The dust trail encounters in future years (Figs. 17.31 and 32) may help us to better understand the distribution of dust and predict whether or not future cousins of the ESA Olympus satellite face a same fate.

18
Other Halley-type comets

Apart from the Leonids, Ursids, and Perseids, there are a handful of other known Halley-type comet induced showers with past meteor outbursts, and some that are not so well known. In addition, I will report here on an ongoing investigation into a mechanism of comet ejection peculiar to Halley-type comets and how that may manifest as meteor showers on Earth.

18.1 The Halley streams

The Orionid shower was among those discovered in the years following the 1833 Leonid storm. It is a relatively strong annual shower with a peak of ZHR = 25 around October 22. The parent is comet 1P/Halley itself, now in an orbit passing a far +0.151 AU from Earth.

Hence it came as a surprise when, in 1993, the most active and experienced visual observer of the Dutch Meteor Society, amateur astronomer *Koen Miskotte*, reported an outburst of meteors from the Orionid shower (Fig. 18.1).[1] Observing under a good sky limiting magnitude $Lm = +6.6^{m}$ in the two nights of October 16/17 and 17/18 (peak at $\lambda_{\odot} \sim 204.5°$), rates were 2–3 times higher than normal. In that second night, Koen saw a -5^{m} Orionid fireball, another of -4^{m} and two of -3^{m} from R.A. = 90.3°, Decl. = 14.8°. This is very unusual since Orionids are quite faint (population index $\chi = 2.9$) and as Koen recalls *I had never seen Orionids brighter than −2 before in my career as a meteor observer. And I should know, because I watched this shower in seven previous years. Hans Betlem*, observing from Sinderen in the eastern part of the Netherlands, confirmed the high activity. Some of these bright meteors were photographed, but only one from two sites simultaneously. That Orionid with a -5^{m} end flare had a radiant at R.A. = 90.1 ± 0.2°, Decl. = +15.4 ± 0.2° and speed $V_g = 67.5 \pm 0.7$ km/s. Its orbit had a higher than usual perihelion distance at 0.613 ± 0.013 AU (Table 7).

The outburst happened several days prior to nearest approach with the comet orbit on October 22.2, $\lambda_{\odot} = 208.9°$, and prior to the annual shower peak at $\lambda_{\odot} = 208.6 \pm 0.4°$

[1] P. Jenniskens, Meteor stream activity II. Meteor outbursts. *Astron. Astrophys.* **295** (1995), 206–235.

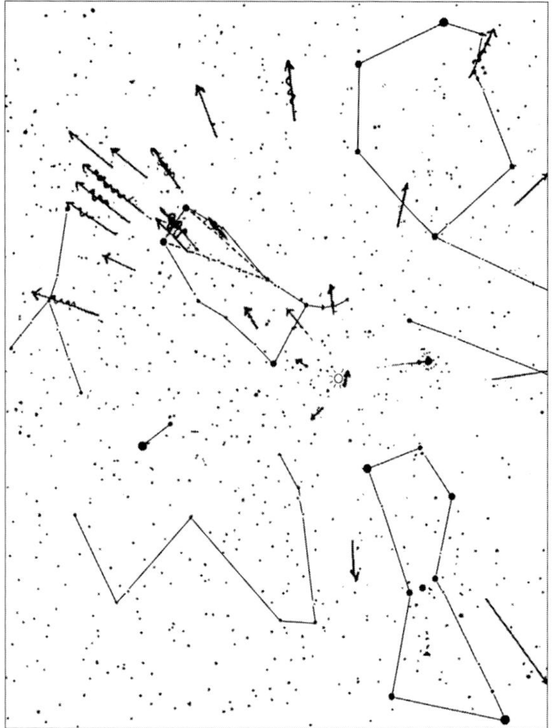

Fig. 18.1 Outburst Orionids from the night of October 17/18, 1993, plotted on a gnomonic map by Koen Miskotte, DMS.

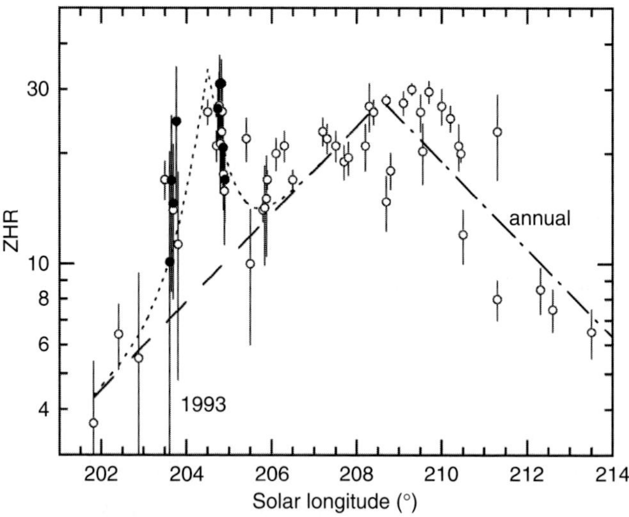

Fig. 18.2 The 1993 Orionid outburst from observations by Koen Miskotte (●), visual observations of IMO and AKM, and radio meteor scatter data by Mr. Ken-ichi Shibata of Sapporo, Japan (○), NMS.

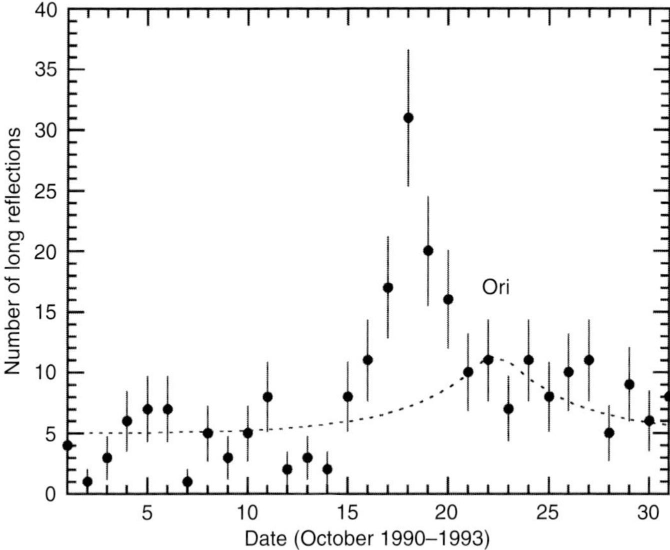

Fig. 18.3 The sum of long-duration radio forward meteor reflections during the Orionid season in 1990–1993 (October 18.0:$\lambda_\odot \sim 234.599$). Counts by Chikara Shimoda (Asahi, Japan), NMS.

(Fig. 18.2).[2] The 1993 Orionid outburst appears to have been limited in extent along the comet orbit, with no outburst reported in 1994. For that reason, it is almost certain that the meteoroids are trapped, possibly in the 13:2 mean-motion resonance with Jupiter.

The outburst was confirmed by the automatic radio forward meteor scatter stations of Ilkka Yrjölä (Kuusankoski, Finland) and Ken-ichi Shibata of Sapporo, Japan, both of which detected elevated rates by up to a factor of two in the period $\lambda_\odot = 202°–205°$ (Fig. 18.2).[3] Interestingly, elevated rates of long echo duration reflections were recorded by Shikara Chimoda, participant in Global-MS-Net, also in 1991 and 1992. The distribution of long-duration echoes peaks several days before the Orionid shower maximum (Fig. 18.3).[4]

The annual Orionids (Fig. 18.4) have been suspected of being variable before, with several "Ribbons" changing in position and peak activity from year to year. Instead, I find that the annual Orionid activity is very stable (Fig. 18.5).

η-Aquariids

If the Orionids have outbursts, then the *η-Aquariids* at the descending node in May should have them as well. At this time, Halley's dust approaches Earth's orbit

[2] J. Rendtel and H. Betlem, Orionid meteor activity on October 18, 1993. *WGN* **21** (1993), 264–268.
[3] P. Jenniskens, Meteor stream activity. II. Meteor outbursts. *Astron. Astrophys.* **295** (1995), 206–235.
[4] C. Shimoda, *Nippon Meteor Soc. Circ.* (Dec. 1993).

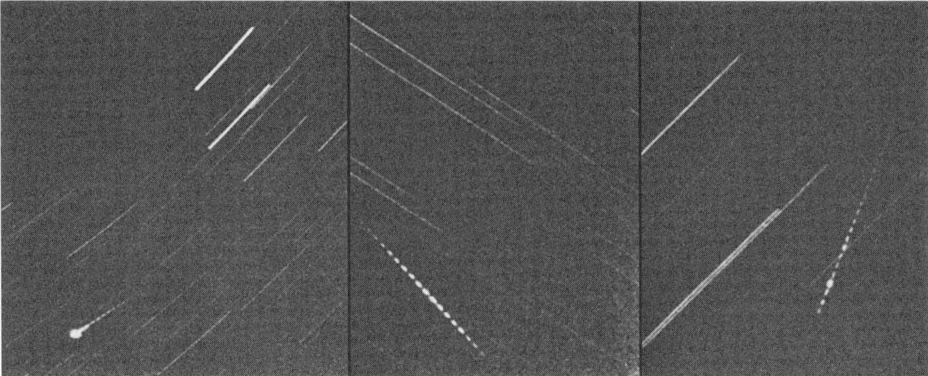

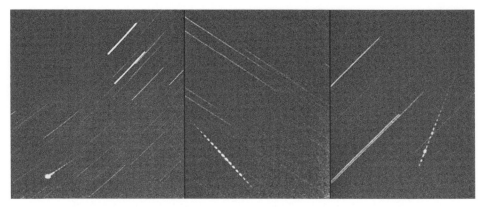

Fig. 18.4 Three Orionids photographed from the Netherlands on the night of October 18, 1993. The photo left is the −5 (flare) Orionid of 01:29:46 UT as captured at Oostkapelle by Klaas Jobse. The middle image is the Orionid in Pegasus at 01:07:47 UT, while the right image is the Orionid of 02:05 UT in the head of Hydra, both by Hans Betlem at Sinderen (rotating shutter with 25 breaks/s). Photo: DMS.

to −0.064 AU on May 6.4, ($\lambda_\odot = 46.1°$). From Ilkka Yrjölä's radio-MS counts, I find that the η-Aquariid rate is very constant from year to year, except perhaps between solar longitudes 43° and 45°, when counts varied by a factor of two.

This main southern hemisphere shower was first recognized only in 1870 by Lieutenant Colonel G. L. Tupman, a member of the Italian Meteoric Association, while cruising the Mediterranean in the course of his duties.[5] The year before, he had noted a radiant at R.A. = 331.2°, Decl. = −10.2°. Now, on April 30 and May 3 in the early morning, he observed a *fine shower* of swift meteors with trains,[6] 15 and 21 in total, and determined the radiant position at R.A. = 327.2°, Decl. = −2.2°. Later, William F. Denning found that 45 of his plotted meteors during April 29 to May 5 came from a radiant at R.A. = 337.2°, Decl. = −8.1°. He noticed the shower again on April 29 the next year. In 1876, astronomer Alexander S. Herschel (in a professional–amateur relationship with Denning) derived a theoretical radiant for comet Halley after working out a way to bridge the gap between the comet's and the Earth's orbit, and immediately noticed Tupman's shower to be close to the predicted radiant. No enhanced rates were observed during the 1910 and 1986 returns of Halley.

On closer inspection, I find that the profile of the η-Aquariids is asymmetric and consists of two components, the narrow one of which peaks at $\lambda_\odot = 44.4°$. I searched for possible variations of this first peak in the rate data collected by Tim Cooper of ASSA[7] and Jürgen Rendtel of IMO,[8] and found the peak present in 1993–1997, but

[5] W. F. Denning, Radiant points of shooting stars from Capt. Tupman's unreduced observations 1869–71. *Mon. Not. R. Astron. Soc.* **37** (1877), 349–352.

[6] G. L. Tupman, Catalogue of radiant points of meteors. *Mon. Not. R. Astron. Soc.* **33** (1873), 298–312.

[7] T. Cooper, A decade of visual eta-Aquariid meteor observations. *WGN* **24** (1996), 157–161.

[8] J. Rendtel, The eta-Aquariid meteor shower in 1997. *WGN* **25** (1997), 153–156.

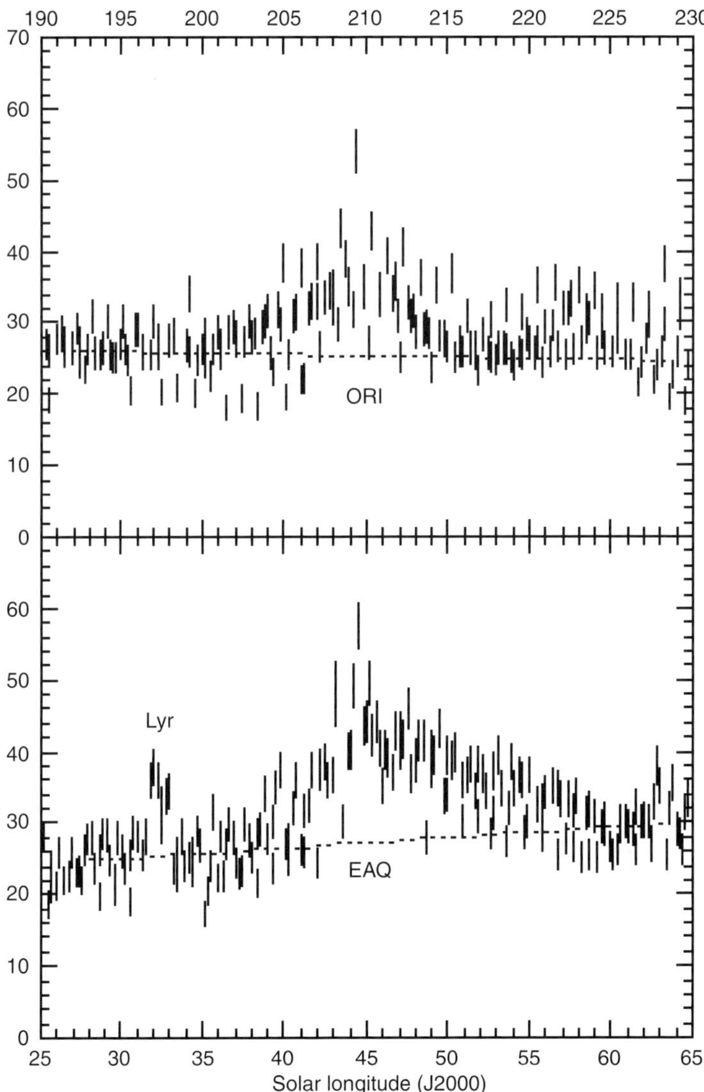

Fig. 18.5 Radio MS counts for Orionids and η-Aquariids in the period 1994–1998, as detected in forward meteor-scatter counts by Ilkka Yrjölä of Kuusankoski, Finland.

not in 1986–1992. During that time, Jupiter was on the same side of the Sun as Earth. Note, however, that the difference is not very large when averaging the years 1993–1997 on one hand and 1986–1992 on the other (Fig. 18.6). Some of the observed variation must come on account of under sampling.

According to Don Yeomans and Tao Kiang, who examined the orbit of Halley's comet back to 1404 BC, the descending node of the comet (η-Aquariids) was close to

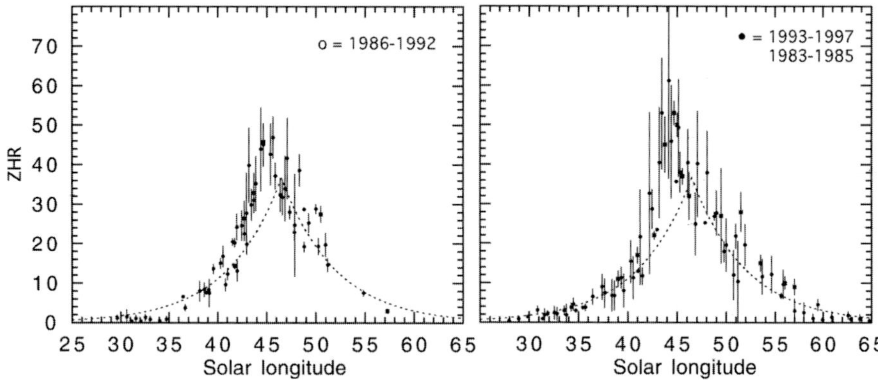

Fig. 18.6 Two components in the activity curve of the η-Aquariids. ZHR values calculated by Jeff Wood, NAPO-MS. Dashed curve by Jenniskens (1995).

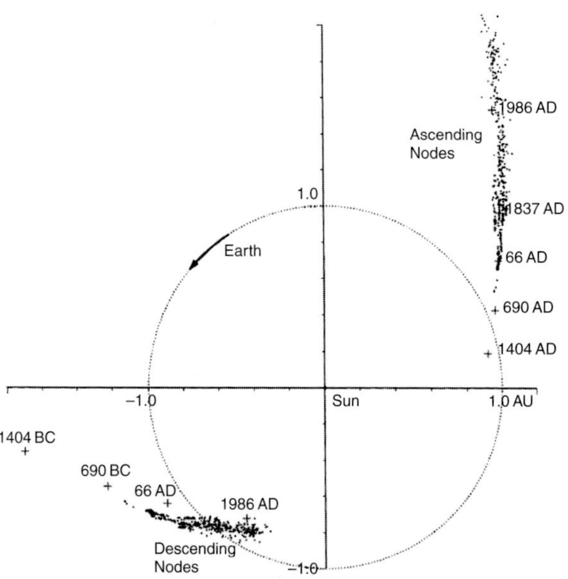

Fig. 18.7 The changing orbit of 1P/Halley and the distribution of 1404 BC-dust as seen in 1986. Calculations by Bruce McIntosh and Jim Jones, UWO.[9]

Earth's orbit in the AD 530 and 607 returns, while the ascending node (Orionids) was at Earth's orbit in the 836 and 763 BC returns. Bruce McIntosh and Jim Jones calculated how dust ejected in 1404 BC dispersed along the track of the comet orbit nodes (Fig. 18.7).

[9] B. A. McIntosh and J. Jones, The Halley comet meteor stream – Numerical modelling of its dynamic evolution. *Mon. Not. R. Astron. Soc.* **235** (1988), 673–693.

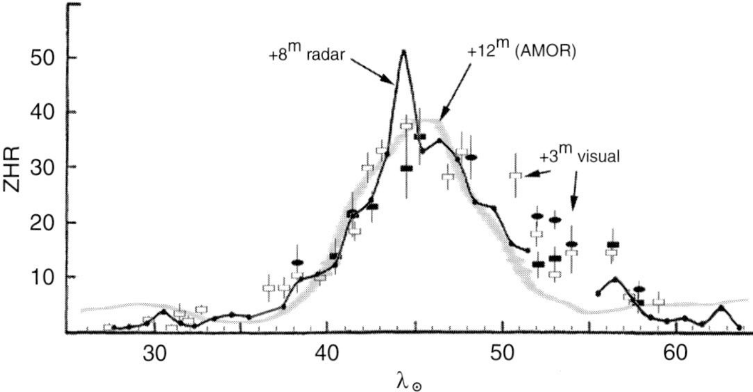

Fig. 18.8 Comparison of η-Aquariid activity profiles for meteors of widely different brightnesses. Whatever causes the nodal dispersion is not mass dependent.

Interestingly, the broad annual shower component peaks at about the expected moment of nearest approach to the orbit of Halley (46.1° and 208.9° for η-Aquariids and Orionids, respectively), while the second η-Aquariids peak lags by 1.7° and the Orionid outburst lagged by 4.7°, in proportion to the time since the comet was at the node. This deserves further study.

A remarkable feature of comet 1P/Halley is that it emits many small dust grains and in that respect behaves more like a long-period comet. It turns out that the η-Aquariid shower is one of the stronger showers detected by the AMOR radar, which is sensitive to very small meteoroids, most no bigger than 40–100 µm in size (+14 limiting magnitude). In contrast, the Leonid shower is not detected. The η-Aquariid activity profile (Fig. 18.8) has the same width and peak time as the visual activity curve. This implies that whatever is responsible for the nodal dispersion is not mass dependent.

The presence of small grains is also manifest in the breakup of the meteoroids in Earth's atmosphere. Although the Orionid meteoroid particle density $\rho = 0.25 \pm 0.05$ g/cm^3 from Harvard Meteor Survey Project data by Verniani[10] is not unlike that of the Perseids (0.32 g/cm^3), the meteoroids fragment differently. They have a high progressive fragmentation index, 0.46 ± 0.06 (see later), higher than that of the Perseids and Leonids, but comparable to the δ-Aquariids and Quadrantids. Jacchia *et al.* found that Orionids only need a relatively small amount of air to slow the meteoroids down, which end relatively high in the Earth's atmosphere (He = 98 km at $+5^{\mathrm{m}}$ and 85 km at -5^{m}). The meteors also have more wake and frequent end flares, suggesting the production of small fragments.

The small grains may be abundant, but they do not carry away the mass of Halley. Most of the mass is lost as the large meteoroids that cause visible meteors. Fig. 18.9 shows the mass-loss from comet 1P/Halley, in which I have plotted the Orionid meteoroid distributions during the outburst ($\chi = 1.7$) as a gray line, while a dashed

[10] F. Verniani, Meteor masses and luminosity. *Smithsonian Cont. Astrophys.* **10** (1967), 181–195.

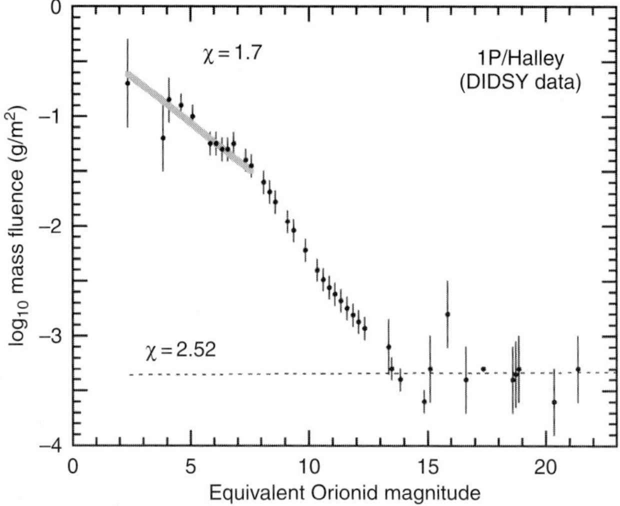

Fig. 18.9 The mass distribution of comet 1P/Halley dust during the 1986 flyby of Giotto as calculated by Tony McDonnell, University of Kent.

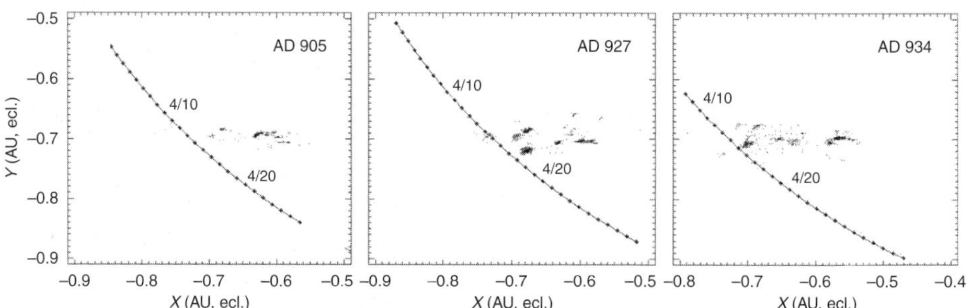

Fig. 18.10 Dust trails of comet 1P/Halley ejection between AD 566 and 934 near Earth's orbit. Calculations by Vaubaillon.

line shows the relationship for constant mass per log(mass) interval: $\chi = 2.52$. Most mass is lost in the form of boulders.

Did Halley cause meteor storms on Earth? The shower was hard to observe from China, only in the hours just before dawn with a radiant in the southeast. Several accounts from around AD 500 were linked to the η-Aquariids (Table 1), but none coincided with the return of Halley to perihelion. Jérémie and I studied all trails ejected between AD 66 and 934 and searched for close encounters in later years (Table 5). Fig. 18.10 shows how the trails would quickly separate due to the evolution of the comet. Again, I used a scaled Leonid dust trail model to estimate the expected meteor activity for each encounter. The result is that only on three occasions were meteor storms expected: on April 10, 531, April 10, 539, and April 12, 964 (Table 5e). None of those were favorable for eastern parts of Asia.

No such dust trail crossings would have been expected in recent history at the other node for the Orionid shower. On the other hand, the Orionids are well positioned for China and outbursts similar to that of 1993 may have been seen. Most historic accounts (Table 1) appear to relate to the normal annual activity. In AD 585 ($\lambda_\odot = 202.4 \pm 0.2°$), it was said that: *Hundreds of meteors scattered in all directions.*[11] In 903 ($\lambda_\odot = 204.2 \pm 0.2°$) *Many small meteors flew, crossing each other, and fell. A shower lasted all night.*

18.2 The December Monocerotids

The *December Monocerotids* (Fig. 18.11) of C/1917 F_1 (Mellish) were thought to be responsible for fireballs seen in AD 1038–1099.[12] I. S. Astapovich and A. K. Terent'eva found that 14 fireballs from the December 6–18 period in the eleventh century emerged from R.A. = 103°, Decl. = +26°, when the rapidly evolving Geminids were not yet in Earth's path.[13] This activity may have extended from AD 1006 to 1508.[14]

The shower now has a peak rate of only ZHR = 2/h,[15] but is clearly recognized in photographic orbit surveys.[16] The shower is detected between $\lambda_\odot = 258°$ and 264° and peaks on December 11 at $\lambda_\odot = 260.9 \pm 0.2°$, with a radiant at R.A. = 102.4°, Decl. = +8.1° ($V_g = 42.0$ km/s), not far from the active Geminid radiant.

Fred Whipple made the link with comet *C/1917 F_1 (Mellish)* in 1954, based on a theoretical radiant calculated by Porter. The closest approach is only on December 12.0, when the theoretical radiant R.A. = 102.0°, Decl. = +7.8° ($V_g = 41.9$ km/s) is identical to the observed radiant, if the radiant is calculated by rotating the orbit around the line of apsides.

Kresáková[17] pointed out that several meteors were photographed with a Monocerotid radiant at the peak of the shower, but with geocentric velocities less than those of the main stream (~41 km/s). She proposed that C/1917 F_1 (Mellish) was part of a Sungrazer-like family, with the Geminids (3200 Phaeton) being part of that family (see Chapter 23). There is also a nearby grouping of meteor radiants at Decl. ~+16° with lower inclination and lower $q \sim 0.13$ AU. It is quite possible, alternatively, that

11 S.-H. Ahn, Meteors and showers a millennium ago. *Mon. Not. R. Astron. Soc.* **343** (2003), 1095–1100.
12 K. Fox and I. P. Williams, possible origin for some ancient December fireballs. *Mon. Not. R. Astron. Soc.* **217** (1985), 407–411.
13 I. S. Astapovic and A. K. Terentjeva, Fireball radiants of the 1st–15th centuries. *IAU Symp.* no. 33 (1968), p. 308.
14 I. Hasegawa, Historical meteor showers – Geminids and December Monocerotids. In *Meteoroids 1998* (Bratislava: Slovak Acad. Sci., 1999), pp. 177–184.
15 P. Jenniskens, Meteor stream activity I. The annual streams. *Astron. Astrophys.* **287** (1994), 990–1013.
16 F. L. Whipple, Photographic meteor orbits and their distribution in space. *Astron. J.* **59** (1954), 201–217; B. A. Lindblad and D. Olsson-Steel, The Monocerotid meteor stream and Comet Mellish. *Bull. Astron. Inst. Czechosl.* **41** (1990), 193–200.
17 M. Kresáková, Meteors of periodic comet Mellish and the Geminids. *Bull. Astron. Inst. Czechoslov.* **25** (1974), 20–33.

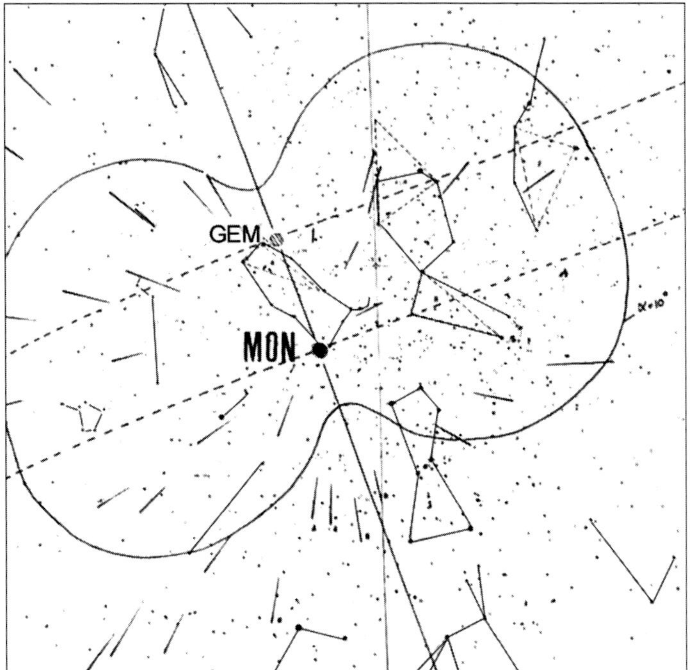

Fig. 18.11 December Monocerotids observed by the author during the 1990 Geminid campaign. Only in the butterfly-shaped area is discrimination with the more active Geminid shower feasible.

the slower meteors are due to measurement error and the lower q meteoroids are unrelated.

The bulk of the shower may be inside Earth's orbit at the moment. *Ken Fox* and *Iwan Williams*[18] found that comet Mellish has a stable orbit. During the period from AD 785 to 3185, the ascending node increased less than 5°. They also recognized that Mellish was further away from Earth's orbit in the Middle Ages and the spate of fireballs at that time was unexpected.

Esko Lyytinen studied (for this book) the generation of dust trails from this comet and found that integrating the comet back further in time than one orbit (about AD 1630) is difficult because of potential close encounters with Venus. The orbit is currently not in resonance with Jupiter, and at the limit for where it might become so. Over 2000 years, the orbit rotates around the line of apsides, with a node decreasing by about 1°/revolution and a corresponding increase of the argument of perihelion. As a result, the orbit will be closer to that of Earth in the future, with the nodal distance increasing from $r = 0.78$ today to 0.988 AU ($\lambda_\odot = 264.39°$) after six revolutions in about AD 2591, and to 1.014 AU ($\lambda_\odot = 264.47°$) around AD 3070. Dust trails

[18] K. Fox, The effects of planetary perturbations on observations of meteor showers. In *Asteroids, Comets, Meteors II* (Uppsala: Uppsala Observatory, 1986), pp. 521–525.

are not expected to wander into Earth's path until about that time, unless they evolve more rapidly than the orbit of the comet. It remains unknown whether or not fireball activity in the Middle Ages originated from this stream.

18.3 Comet C/1854 L_1 Klinkerfues and the ε-Eridanids

Comet *C/1854 L_1 Klinkerfues (1854 III)* could be one of those "long-period" comets not well enough observed to reveal that they are of Halley-type instead. Observations in 1854 only spanned a period of two months in June and July, when the comet was near perihelion. A parabolic orbit was calculated by *F. August T. Winnecke* and *C.F. Pape*,[19] who found no significant deviation from a parabolic orbit over the period that the comet was observed: *Die Bahn lässt nicht die geringste Abweichung von der Parabel mit Sicherheit erkennen, wodurch die Anfangs vermuthete Identität dieses Cometen mit dem von 960 umgestossen wird, ein Resultat, zu dem shon früher Oudemans durch Betrachtungen anderer Art ebenfalls gelangt ist.* Winnecke and Pape therefore dismissed the possible identity of comet C/962 B_1 as the same object, following the opinion of *Jean Abraham Chretien Oudemans* at Leiden Observatory (without providing a reference). Oudemans did observe comet 1854 III.

In 1981, observer *Murray Gayski* of WAMS reported high activity from Eridanus during the night of September 10/11 (Fig. 18.12). Jeff Wood writes about Gayski's observation: *Into his third hour which began at 14:00 UT, he started noticing fast bright trained yellow–orange meteors streaking out of Eridanus, which at this time was low on the eastern horizon.* There was a near-full Moon. The meteors had an average magnitude of +1.2 ($\chi = 2.0$), were mostly yellow or orange in color, and 44% left a brief train. Three 1 h counts were made starting at 14:00 UT (September 10): 3, 11, and 34 meteors ($\lambda_\odot > 168.12°$), with no meteors seen in the prior three hours, when the radiant was below the horizon.

Jeff Wood associated the event with C/1854 L_1 *Klinkerfues.* The comet is intrinsically bright, with $H_{10} = +6.0$ ($D \sim 4.8$ km), making this a good candidate for a meteor shower parent. In 1854, it just barely reached naked eye brightness. However, the miss-distance is a large +0.014 AU. Moreover, the outburst occurred a day before passing the comet node ($\lambda_\odot = 169.699°$), more than the usual discrepancy for the 1-revolution dust trails of long-period comets. A rotation around the line of apsides is needed to shift the node to the observed value. Also, the outburst was not as sharp as that expected for a 1-revoluton old long-period comet dust trail (Chapter 13). Finally, there is an annual shower active from a radiant near the star ε-Eridani, sometimes referred to as the "π-Eridanids" to discriminate from the ε-Eridanid outburst. Rates are about ZHR = 1.5/h at $\lambda_\odot = 167°$.[20] It is possible that some "π-Eridanids" are

[19] A. Winnecke and C.F. Page, Bahnbestimmung des Cometen 1854 III, von Herren Winnecke und Pape. *Astron. Nachrichten* **42** (1856), 113.

[20] P. Jenniskens, Meteor stream activity I. The annual streams. *Astron. Astrophys.* **287** (1994), 990–1013.

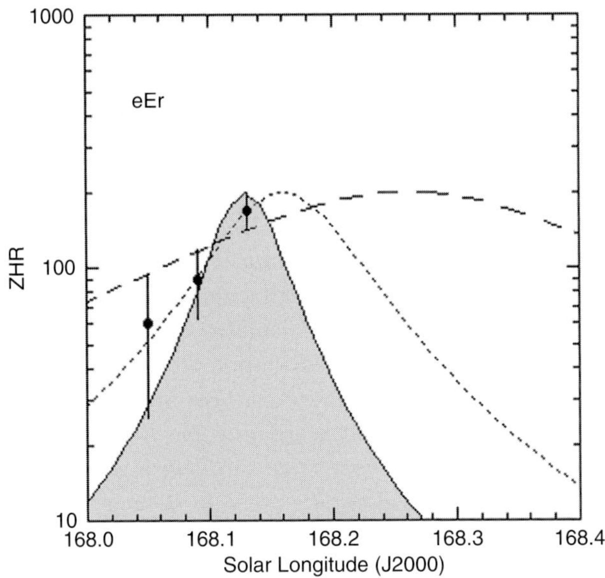

Fig. 18.12 The ε-Eridanid outburst as reported by Murray Gayski.[21] The stream profile may have been as broad as suggested by the dashed lines, or as narrow as indicated by the gray area.

sporadic apex shower meteors, but if not, then there is a dense stream just outside Earth's orbit. All these observations suggest that the parent comet is of Halley-type.

Jérémie Vaubaillon and I teamed up to investigate. First, we confirmed that the observations in 1854 permit an elliptic orbit for this comet. The comet was observed near perihelion and over too short an arc to make a significant difference between a parabolic and a long Halley-type orbit with period $P \sim 100$–250 yr. Next, we considered the possibility that C/962 B_1 could be the same object. From the two descriptions in the Chinese chronicles, comet C/962 B_1 was estimated to be similarly intrinsically bright, with an uncertain $H_{10} \sim +2.5$.[22] Matching a different number of orbits between 961 and 1854, solutions with $N = 4$–9 would make this a Halley-type comet with few recent returns. We discovered that $N = 7$ (or $P \sim 127$ yr) would have caused the comet to come back in 1981, when Gayski observed the outburst. The resulting orbit has a smaller miss-distance (0.0077 AU) and can explain the observed outburst by rotating around the line of apsides. We then integrated back the orbit of Klinkerfues until AD 961, by permitting a range of initial velocities at perihelion, to find the orbit with matching perihelion time. The resulting orbit is:

[21] J. Wood, The Epsilon Eridanid meteor stream. *WAMS Bull.* **167** (1981).

[22] S. K. Vsekhsvyatskij, *Physical Characteristics of Comets* (Jerusalem: Israel Program for Scientific Translations, 1964), p. 49.

Comet	T (UT)	a (AU)	q (AU)	i (°)	ω (°)	Node (°)
C/1854 L_1	961 Dec. 27.0	23.258 07	0.665 94	108.569	75.631	349.394
C/962 B_1	961 Dec. 28	Inf.	0.63	119	85	362

The inclination does not change much over time, and when plotting the calculated positions of the comet in AD 961, it appears to be higher in the sky than reported by the three accounts from China. No close encounters with planets occurred that could account for the difference. Oudemans may have been right, and the association of Klinkerfues and C/962 B_1 remains uncertain.

18.4 Comet 1913 I (Lowe) and the Comae Berenicids

In December and January, fast sporadic meteors radiate from Coma Berenices (Fig. 18.13). Formerly the tail of Leo, this constellation was introduced by Tycho Brahe (1546–1601) and named after the missing hair of Berenice, the wife of Ptolemy III of Egypt. On an important occasion her hair was ceremoniously clipped and laid out on the temple altar in honor of Aphrodite, when it went missing. It was the astronomer *Conon of Samos* who saved the lives of those responsible by proclaiming that Aphrodite had accepted the gift, now showing brightly in the sky as stars.

In January, Coma Berenices is close to the direction of Earth's apex. The meteoroids are in retrograde orbits. A center of activity in the period January 13–23 was recognized by Dick McCrosky and Annette Posen in 1959, the *January Comae Berenicids* (JCO). It has ZHR ~ 2/h. While others have extended this stream to start in December, the expected radiant drift excludes most, only six, and possibly eight, meteors are indisputable members.

McCrosky and Posen, following Whipple (1954),[23] tentatively linked this shower with the intrinsically bright comet *1913 I (Lowe)*, and I regard this identification as likely; if indeed the comet exists. Brian Masden considers this discovery to be doubtful.[24] This comet was so badly observed that the orbit is not even included in the official IAU list of cometary orbits. Comet Lowe was discovered by an Australian amateur astronomer B. Lowe on December 31, 1912, and observed on January 2, 4, 5, and 9, but he made an error in communicating the position to the Adelaide Observatory on January 7, so that no follow up observations were possible. The orbit calucated by M. Viljev[25] in 1913 is often adopted. This approximate orbit

[23] F. L. Whipple, Photographic meteor orbits and their distribution in space. *Astron. J.* **59** (1954), 201–217.

[24] B. Marsden, *Comae Berenicids, yes; comet connections, no.* Cambridge Conference Network 6/2001 (January 12, 2001). B. J. Peiser (Ed.). Electronic circular.

[25] M. Viljev, Zur Bahnbestimmung des Kometen 1912 d (Lowe). *Astron. Nachr.* **195** (1913), 107; M. Viljev, Komet 1912 d (Lowe), *Astron. Nachr.* **195** (1913), 415.

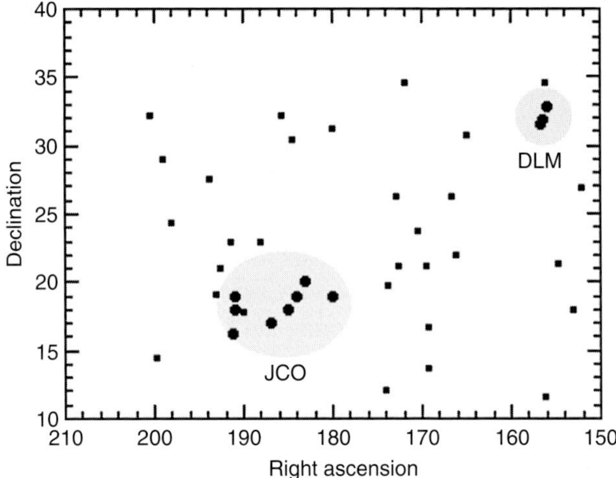

Fig. 18.13 The distribution of radiants in December and January in the region of activity of the "Comae Berenicid" showers, with R.A. = 150°–210° and Decl. = +10° to +40°, and $V_g > 50$ km/s. The December β-Leonis Minorids and January Comae Berenicids are marked (●).

suggests a radiant near R.A. = 187.9°, Decl. = +21.9° ($V_g = 59.4$ km/s) on January 25.1 ($\lambda_\odot = 304.4°$).

There is a small probability that the same comet was seen (briefly) in 1750 by the well-known Swedish astronomer Peter Wargentin on three nights, which would suggest an orbital period of 163 years (due back around AD 2076), or a fraction thereof. This was an intrinsically bright comet $H_{10} = +5.5$ and earlier mistaken as a possible sighting of 109P/Swift–Tuttle.[26] As with comet Lowe, comet C/1750 (Wargentin) was unconfirmed and therefore left out of most comet catalogs published in the nineteenth and twentieth centuries.[27] Marsden is more confident that this was a genuine comet on account of Wargentin's reputation, but considers the association with Lowe as unlikely.

Several historic outbursts (shower 32, Table 1) have been reported, that form a pattern with progressively increasing solar longitude matching the January Comae Berenicids. If all these reports of meteor activity relate to the same shower, then this may perhaps be a retrograde moving Halley-type shower, which would be active today at $\lambda_\odot \sim 304.4$ in late January. The node of the comets, too, fits the precession of the node of $\Delta\Omega/\Delta t = +0.0031°$/yr. As expected, the sightings in 609 and 764 are 155 years apart, not much different than the 163 years between 1750 and 1913, and they could fall in a regular pattern, allowing for outbursts a few years prior to and later than

[26] G. W. Kronk, *Cometography, a Catalog of Comets. Vol. 1: Ancient–1799* (Cambridge: Cambridge University Press, 1999), 563pp.

[27] R. Gorelli and A. McBeath, Call for observations, 2001 January 20–26: the January Comae Berenicids. *Meteorobs Mailing List Circular* (2001).

the return of the comet. If so, then the comet orbit is currently located outside Earth's orbit at the descending node (with the miss-distance progressively increasing) and the ascending node is inside Earth's orbit, with the miss-distance decreasing over time.

Two recent outbursts have been reported in the correct time frame in late January, but it is not clear if they are related. Two Slovenian observers (*Niko Stritof* and *Igor Grom*) reported a small visual outburst of ten bright meteors perhaps from near Corona Borealis on 1999 January 22 (01–04 UT, $\lambda_\odot = 301.63$). Finally, David Asher and Mark Bailey reported that a single UK witness observed a possible outburst on January 20/21, 2000. However, the radiant was said to be northwest of Ursa Major, not southeast of it.

18.5 Comet C/1742 C_1 = C/1907 G_1 Grigg–Mellish (?) and the δ-Pavonids

Another controversial identity is the parent of the *δ-Pavonid* shower, comet C/1907 G_1 *Grigg–Mellish*. It was poorly observed in 1907, permitting only a parabolic orbit fit, with a node at $\Omega = 189.719°$ and an inclination of $i = 110.06°$. It was later assumed by Weiss that this comet was identical to C/1742 C_1, with a node at $\Omega = 189.201°$ and an inclination of $i = 112.95°$. Based on that assumption, an orbital period of $P = 164$ yr was derived, making this a Halley-type comet. Then again, Brian Marsden also regards this link as "highly improbable."[28]

The shower was found by Michael Buhagiar of Perth, Australia, in response to the narrow miss-distance and theoretical radiant calculated from the comet orbit. During the period 1969–1980, he detected a weak shower, peaking at ZHR ~ 1.8 ± 0.2/h on April 5, 1980. The shower was seen between April 4 and 8 with a mean radiant at R.A. = 305°, Decl. = −65°. NAPO-MS observers later traced a broad FWHM ~ 9 d wide shower with a peak ZHR = 5.3 ± 0.7/h on April 1 at $\lambda_\odot = 11.1 \pm 1.5°$, from R.A. = 310°, Decl. = −63° (Fig. 18.14). This is close to the theoretical radiant of comet Grigg–Mellish.

The broad activity profile suggests that at least one of these two comets is of Halley type. Intrinsically faint $H_{10} = +9.7$ ($n = 8.4$) comet Grigg–Mellish passes very close to Earth's orbit. Its theoretical shower peaks at $\lambda_\odot = 10.4°$ from R.A. = 309.0°, Decl. = −60.4°, with $V_g = 58.72$ km/s. Intrinsically bright ($H_{10} = +3.7$) comet C/1742 C_1 has a miss-distance of 0.164 AU. Its theoretical radiant $\lambda_\odot = 9.2°$ is at about R.A. = 306°, Decl. = −59°, and $V_g = 60$ km/s. Both are candidate parent bodies. It is not impossible that both comets are independent objects, but related to each other from a breakup.

[28] B. G. Marsden, *Catalogue of Cometary Orbits*, 2nd edition, (Cambridge, MA: Smithsonian Center for Astrophysics, 1975), p. 83.

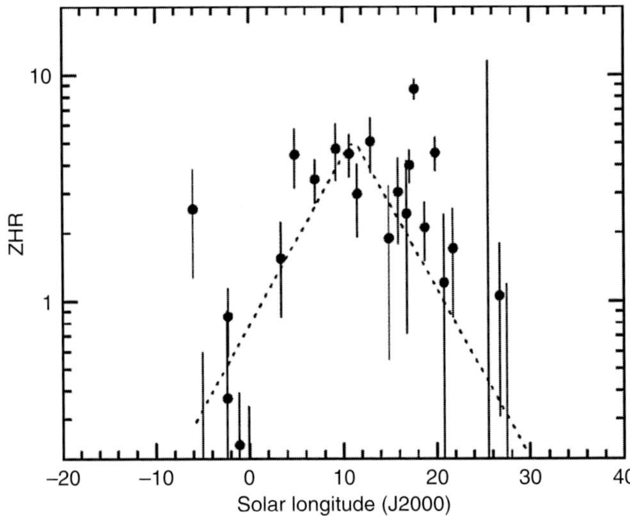

Fig. 18.14 The δ-Pavonid activity curve (zenith hourly rate) from NAPO-MS observations.[29]

18.6 Comet C/1991 L_3 (Levy) and the undiscovered Indids

Finally, I would like to point out that Halley-type comet C/1991 L_3 (Levy), with $P = 51.27$ yr, should be a shower parent. It had designations 1991 XI = 1991q, and is not to be confused with other comets called "Levy" that were discovered at this time. It is due back at perihelion on October 30, 2042. What makes this comet special is that it passed just outside of Earth's orbit at only +0.079 AU and is intrinsically bright (Fig. 18.15).

This is a relatively large comet, with a size of 11.6 km, based on its nuclear R-band magnitude, the same size as comet Halley, but with low activity. From a periodic variation of the nuclear brightness by about 1 magnitude, a spin period of 8.34 h was derived by *Alan Fitzsimmons* and *Iwan Williams*.[30] Most dust is ejected close to perihelion. It had a diffuse coma with a very weak central condensation, with $H_{10} = +7.8 \pm 0.1$ and $n = 10.9 \pm 1.0$ for the period outside the excess emission at perihelion.

Slow ($V_g = 18.3$ km/s) meteors should radiate on August 30.8 ($\lambda_\odot = 157.7°$) from R.A. = 320.2°, Decl. = −60.6°, near the star β-Indi. However, the theoretical radiant varies greatly with the method used. The shower could peak as early as August 22 ($\lambda_\odot = 149.4°$) from R.A. = 320°, Decl. = −58°, and $V_g = 19.8$ km/s, or as late as September 16 ($\lambda_\odot = 173.8°$) from R.A. = 314°, Decl. = −69°, and $V_g = 16.2$ km/s.

[29] P. Jenniskens, Meteor stream activity I. The annual streams. *Astron. Astrophys.* **287** (1994), 990–1013.
[30] A. Fitzsimmons and I. P. Williams, The nucleus of comet P/Levy 1991 XI. *Astron. Astrophys.* **289** (1994), 304.

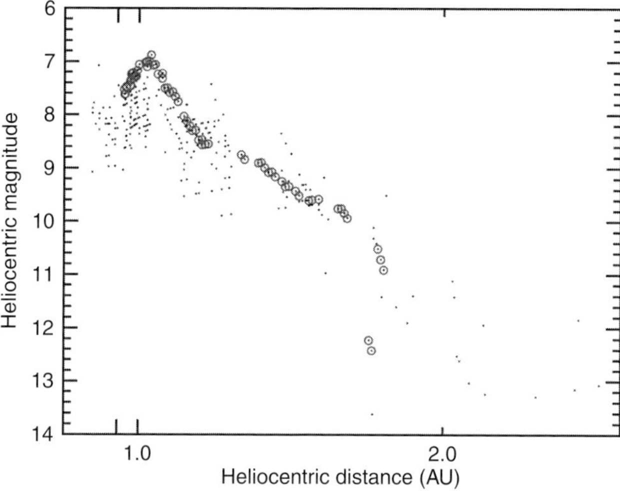

Fig. 18.15 Light curve of P/1991 L$_3$(Levy) during July–December, 1991. Data from International Comet Quarterly. Results by Valeriy L. Korneev, Zelenograd, Russia are highlighted.

Alexandra K. Terentjeva[31] lists a short-period shower ($a = 2.21$ AU) with a radiant at $\lambda_\odot = 164.8°$: R.A. = 342°, Decl. = −52° ($V_g = 14.1$ km/s). The perihelion distance ($q = 0.983$ AU for Levy), inclination ($i = 14.9°$ versus 19.2° for Levy), and argument of perihelion ($\omega = 43.1°$ versus 41.5° for Levy) are similar. According to *Giovanni Valsecchi* (LTEP), the comet moves close to the 4:7 resonance with Saturn and does not have close encounters between 1585 and 2406 that affect the orbit significantly.

It has been proposed that comet C/1499 Q$_1$ is the same ($i = 16°$, $q = 0.95$ AU).[32] This $H_{10} \sim +9^m$ comet became a naked eye object when it passed only 0.059 AU from Earth on August 17, 1499.

Meteor showers from comet activity at large heliocentric distances

Some Halley-type and long-period comets have eruptions at large heliocentric distances. Perhaps not a plentiful source of dust, this material does get into somewhat different orbits than dust ejected near perihelion, and can potentially result in unusual meteor outbursts.

Such comet activity was first seen while following comet Halley on its way out after the 1986 return. Halfway between Saturn and Uranus, at 14.3 AU from the Sun and about five years after passing perihelion, the comet suddenly became 300

[31] A. K. Terentjeva, Fireball streams. In *Asteroids Comets Meteors III.* (Uppsala: Astronomical Observatory, 1989), pp. 579–584.

[32] J. D. Shanklin, The comets of 1991. *J. Brit. Astron. Assoc.* **107** (1997), 186–199.

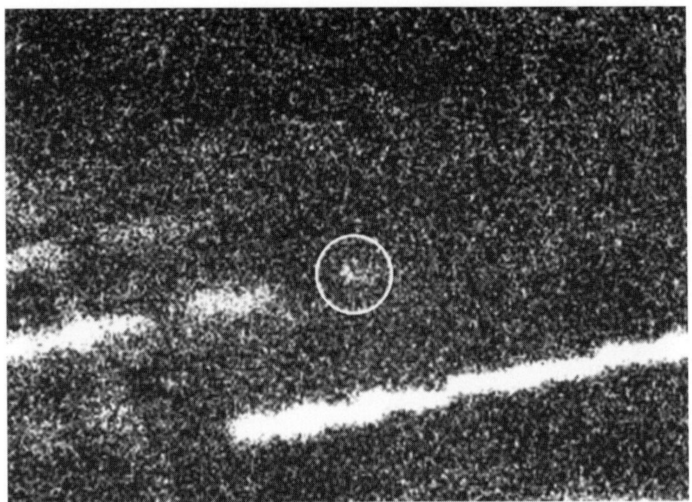

Fig. 18.16 Comet 1P/Halley flares up on its way out at 18.82 AU. This image was obtained with the ESO 3.58 m New Technology Telescope during the night of January 10–11, 1994. Halley had a brightness of $V = 26.5 \pm 0.2^{m}$. Photo: ESO.

times brighter than expected (Fig. 18.16).[33] A few months earlier, in December, 1990, dust had been released from a source on the sunlit hemisphere, which evolved into the shape of a fan, populating a segment of a conical surface. The ~1 μm sized grains that dominated the scattering of visible light were observed to expand at 45 m/s. A billion kg of dust was ejected. This is 100 times less than that ejected at perihelion in a single revolution of the comet.

Carbon monoxide (CO) gas is the suspected driver of these outbursts. Water remains solidly frozen on the ejected grains that far from the Sun. Most mass was lost in the form of large grains, the kind responsible for telescopic and naked eye meteors. Sekanina and colleagues concluded that the particle size distribution had to be relatively flat, with $\alpha = 3.7$ for particles larger than 14 μm. The dust to CO gas ratio was very high ($\gg 10$). Sixteen months later, when the comet reached 16.6 AU, the dust had dispersed and could no longer be detected.[34]

Marco Fulle of Trieste Observatory and the author (unpublished data) studied the possible occurrence of meteor showers from such outbursts near aphelion following the return of 1737 as a potential explanation for the 1861–1863 and 1989–1997 Perseid outbursts. The method used was to calculate the orbits of 82500 meteoroids ejected from the sun-lit hemisphere of the comet for a range of ejection

[33] Z. Sekanina, S. M. Larson, O. Hainaut, A. Smette and R. M. West, Major outburst of comet Halley at a heliocentric distance of 14 AU. *Astron. Astrophys.* **263** (1992), 367–386; R. M. West, O. Hainaut and A. Smette, Post-perihelion observations of comet P/Halley III: an outburst at $R = 14.3$ AU. *Astron. Astrophys.* **246** (1991), L77–L80.

[34] O. Hainaut, R. M. West, B. G. Marsden, A. Smette and K. Meech, Post-perihelion observations of comet P/Halley: IV. $r = 16.6$ and 18.8 AU. *Astron. Astrophys.* **293** (1995), 941–947.

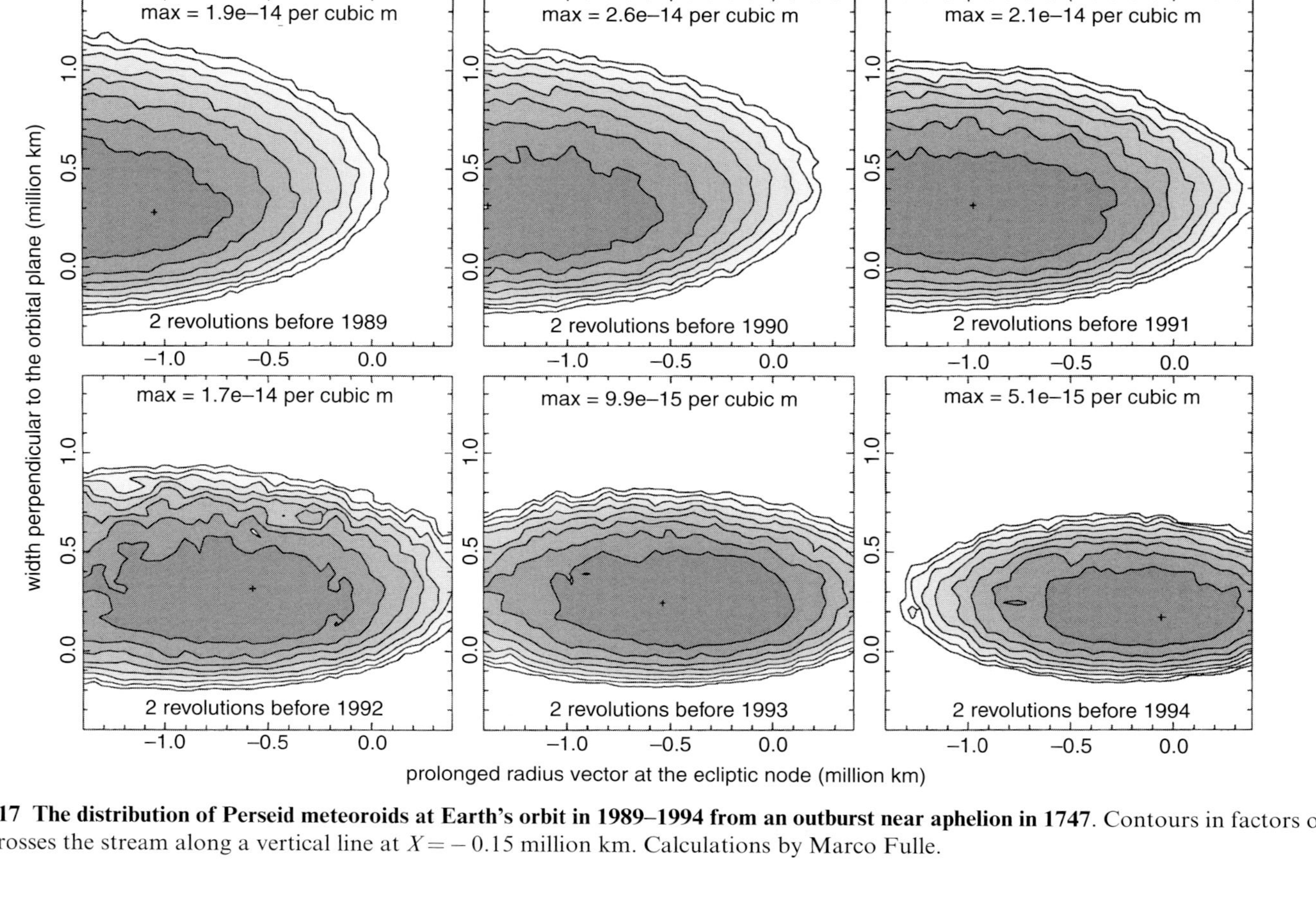

Fig. 18.17 The distribution of Perseid meteoroids at Earth's orbit in 1989–1994 from an outburst near aphelion in 1747. Contours in factors of 2. Earth crosses the stream along a vertical line at $X = -0.15$ million km. Calculations by Marco Fulle.

velocities and about 100 different values of β. We neglected the planetary perturbations that determine the position of the trail near Earth's orbit (Fig. 18.17).

In order to obtain a relatively thin sheet of dust with high dust density, the most favorable position of the cometary outburst would be close to the ascending node, exactly opposite to the Earth at the descending node: in this way, the effect of the dust orbit refocussing[35] builds up a thin meteor layer at the second orbital node.[36] Since meteoroids are sensitive to solar radiation pressure, the only way to obtain a stream with a high enough concentration close to the comet is to eject dust in a direction opposite to the motion of the comet. For sunward ejection, this implies ejection while the comet approaches aphelion. However, the ascending node of 109 P/Swift–Tuttle is placed after aphelion. We reached an acceptable compromise for a position on the outward part of the orbit at the anomaly of about $+162^\circ$, at a distance of 23–27 AU from the Sun, or an ejection around the year 1747.

The observed stream width constrains the cone of ejection to about 10° width, or somewhat less. A Gaussian ejection probability (sampled in 2800 points) resulted in a better reproduction of the stream structure than a Maxwellian profile. The time-dependence of the stream activity constrained the most probable velocity to be of order 25 m/s, consistent with the measured radiant dispersion. Given the low heliocentric velocity of the comet at 23 AU, that can easily account for the off-set of order 0.3° in the radiant position, in which case the active spot is close to the cometary plane. We obtain the best fit with a spot pointing 20° from the Sun direction, oriented according to the orbital motion of the comet and 8° south of the orbital plane of the comet. All of these parameters are similar to those required to fit Halley's outburst. However, the stream space density implied a dust mass released in the burst of 10^{10} kg (with an $\alpha = 3.5$ size distribution power index), a factor of 10 more than the Halley outburst. The model predicted a near-constant node of the dust, but the observed shift in the peak time can be explained by planetary perturbations.

In the end we decided against this mechanism as the cause of the Perseid Filament because the calculated stream profiles are not as sharply peaked as the observed profiles. Perhaps, this mechanism accounts for some Filament-like showers of other Halley-type comets.

[35] K. Richter and H. U. Keller, Density and brightness distribution of cometary dust tails. *Astron. Astrophys.* **171** (1987), 317–326.

[36] M. Fulle and G. Sedmak, Photometrical analysis of the neck-line structure of comet Bennet 1970II. *Icarus* **74** (1988), 383–398.

19

Dust trails of Jupiter-family comets

Jupiter's strong perturbations of the dust ejected by Jupiter-family comets creates odd-looking streams that, unlike the trails of long-period and Halley-type comets, can have widely varying cross sections at Earth's orbit. It is often the perturbations that cause the dust to spread along the orbit, more so than the initial ejection conditions or radiation pressure effects. Potential meteor storms are usually limited in time between a pair of close (<1 AU) encounters of the parent comet with Jupiter, before and after which the comet orbit can be quite different. Those also tend to break the trail in parts, and those *trailets* can catch up on each other.

19.1 The π-Puppid shower

Close encounters with Jupiter are not only more frequent, they are also more severe because of the low relative velocity between meteoroids and Jupiter. Their effects are best illustrated with the π-Puppid stream of comet 26P/Grigg–Skjellerup.[1] *Jérémie Vaubaillon*[2] has studied the dust trail evolution of this Jupiter-family comet for his Ph.D. thesis work, creating the dramatic Figs. 19.1 and 19.2.

The 1878-dust trail shown in Fig. 19.1 had a close encounter with Jupiter in 1881. Although it became distorted, it was not much spread out, continuing to evolve almost unharmed, being nearly continuous and stretching out for only about 2/3 of the orbit in 2003. This was possible because grains were moving close to the 5:12 resonance, escaping further encounters with Jupiter. Unlike Halley-type showers, the trappings in mean-motion resonances never last very long, and thus play a different role in the evolution of the streams and their manifestation as meteor showers.

Other trails are cut into *trailets*. Each trailet can survive for long periods of time. A nice example is the 1848-dust of comet 26P/Grigg–Skjellerup as seen in 1999, shown

[1] I. Yamamoto, A meteoric shower from Skjellerup's comet, 1927 K. *Mon. Not. R. Astron. Soc.* **88** (1928), 487; K. Nakamura, Observation of meteors from Skjellerup's comet, 1927 K. *Mon. Not. R. Astron. Soc.* **89** (1929), 141–143; A. King, Meteors from Skejellerup's comet. *Observatory* **51** (1928), 232–233; H. B. Ridley, A possible meteor shower associated with comet P/Grigg–Skjellerup. *J. British Astron. Assoc.* **82** (1972), 95–98; W. J. Baggaley, Observations of meteors associated with comet Grigg–Skjellerup. *Observatory* **93** (1973), 23–26.

[2] J. Vaubaillon, Dynamic of meteoroids in the solar system. Application to the prediction of meteoric showers in general, and Leonids in particular. Ph.D. thesis, IMCCE/l'Observatoire de Paris, Paris, France (2003).

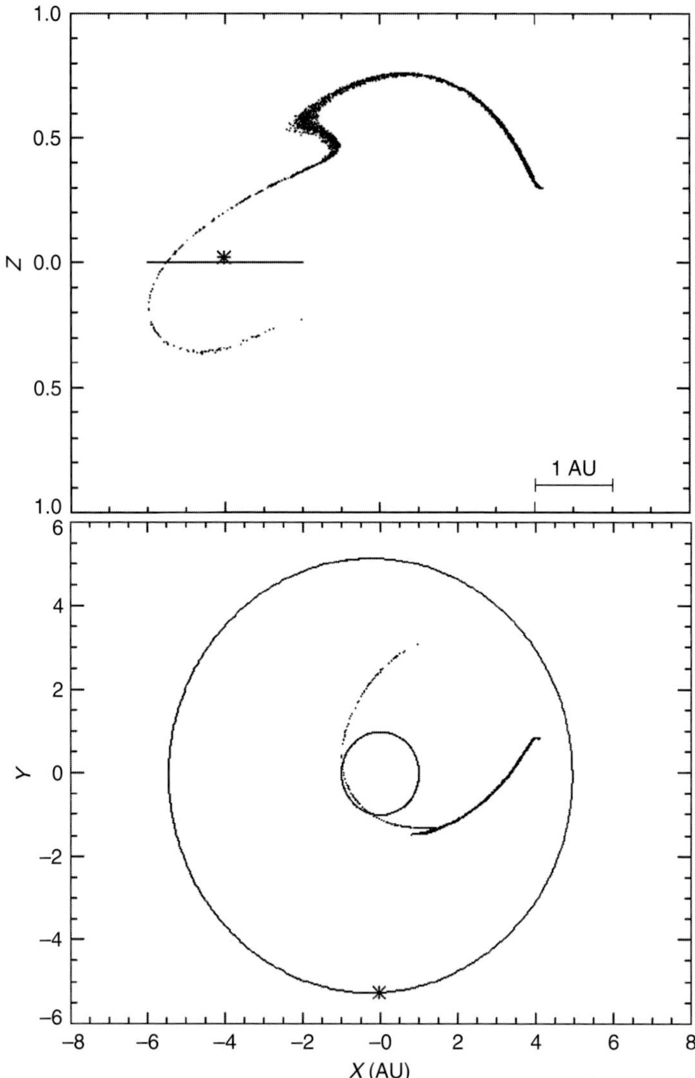

Fig. 19.1 Dust trail distortion from a close encouner with Jupiter in 1881. Shown are the 1848 dust ejecta of comet 26P/Grigg–Skjellerup as seen in 1889 from two perspectives: looking sideways (X, Z) and down upon the ecliptic plane (X, Y). The outer circle is Jupiter's orbit. The horizontal line in the upper graph is the orbit of Earth (diameter = 2 AU). Calculations by Jérémie Vaubaillon, IMCCE.[3]

[3] *Ibid.*

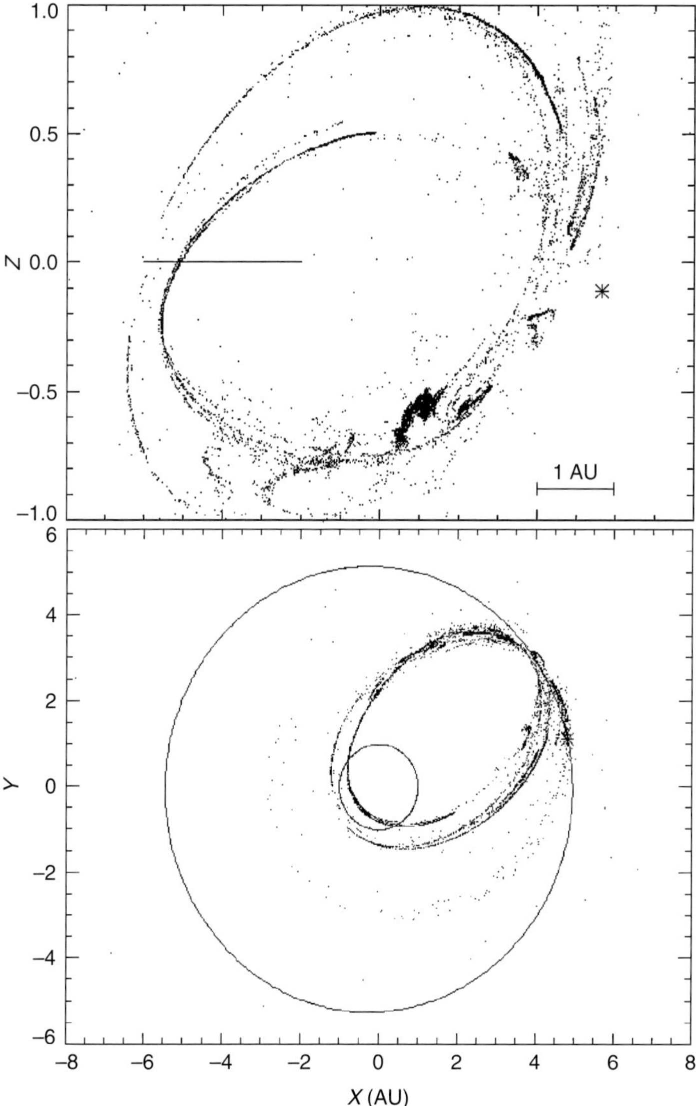

Fig. 19.2 The 1848-dust ejecta of Grigg–Skjellerup as seen in 1999. The graph shows two viewing angles.[4]

in Fig. 19.2. All the large separations and distortions of individual sections of the dust trail are caused by encounters with Jupiter near aphelion.

Over time, the effect of the planetary perturbations is to broaden the distribution of orbital elements, which can sometimes be used to estimate the age of the shower.

[4] *Ibid.*

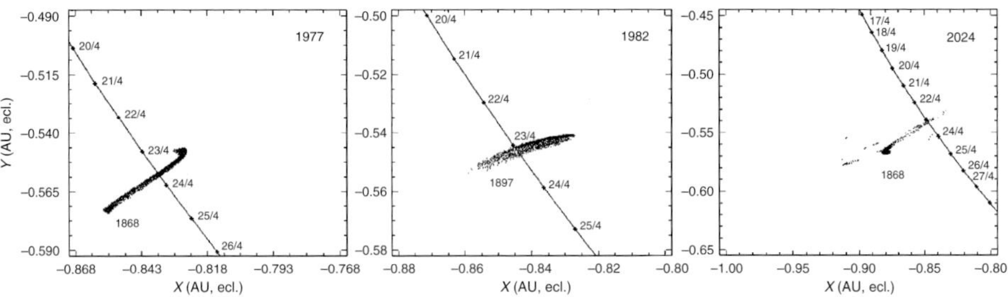

Fig. 19.3 Position of π-Puppid dust trails near Earth's orbit in 1977, 1982, and 2024.

At the same time, the distribution of nodes at Earth's orbit can take a wide range of shapes (Fig. 19.3).

These distortions tend to be similar for different dust trails during the same encounter, I believe because Jupiter's perturbations work on all trailets that are present at a small range of distances in a given cross section in the same amount. Over time, however, trailets catch up on each other after orbital periods are changed and their distortions become a reflection of their individual perturbation history. As a result, an assembly of trails near Earth's orbit can take a variety of shapes.

Predicting the peak time of a shower is exasperated by the low inclination of the orbit. Earth is closest to the center of the trail at a different time than when it passes the node of the meteoroid orbits in Earth's path. For example, the minimum distance to the center of the 1961-dust trail in 2003 was predicted to be on April 23, 13:05 UT, while the node was passed at 12:06 UT.

Finally, the radiant position of individual dust trailets can be significantly different. For example, the 2003 encounter with the 1961 dust would have given a radiant position at R.A. = 110.2°, Decl. = −45.0° (V_g = 15.0 km/s), while the 1957 dust radiated from R.A. = 112.0°, Decl. = −42.0° (V_g = 14.9 km/s). Other such values are given in Table 6.

Comet 26P/Grigg–Skjellerup was perturbed in an orbit close to Earth in 1964. *Harold Ridley*[5] first predicted a possible meteor shower when Earth would pass the comet by only 50.8 d on April 23.02, 1972, at only 0.004 AU distance. These were slow meteors, with an apparent speed of only 19.0 km/s, and visible in dark skies only in the evening hours. Following this prediction, US observers had a mean rate of 1.9/h with a very low radiant, while Australian observers observed on average 1.2/h. The stream was detected with a radar at Christchurch in New Zealand by *Jack Baggaley* for several days, before and after April 23.[6]

Subsequent meteor outbursts were observed in 1977 and 1982.[7] The outbursts were broad, about a day wide, with an even broader background of low-level activity (Fig. 19.4). The meteors tended to be bright (χ = 1.9) and yellow in color, with much fragmentation and persistent trains lasting 3–4 min for a -4^{m} π-Puppid.

[5] H. B. Ridley, A possible meteor shower associated with comet P/Grigg–Skjellerup. *JBAA*, **82** (1972), 95.
[6] W. J. Baggaley, Observations of meteors associated with comet Grigg–Skjellerup. *The Observatory*, **93** (1973), 23–26.
[7] D. Hughes, The Grigg–Skjellerupid meteoroid stream. *Mon. Not. R. Astron. Soc.* **257** (1992), 25–28.

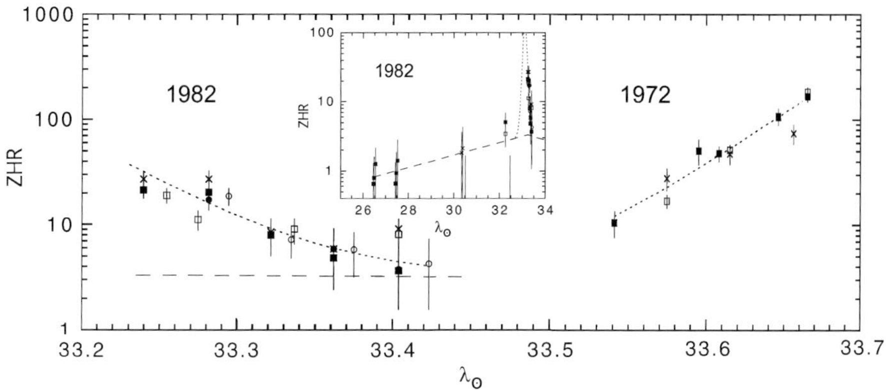

Fig. 19.4 Summary of Puppid observations in 1982 and 1977.[8]

Scott Messenger,[9] in searching for dust trail crossings that could offer an opportunity for the stratospheric collection of cometary dust from a known source, identified the 2003 π-Puppids as a possible next encounter, as did Kondrat'eva and others.[10] Dust from the 1952, 1957, and 1961 returns was expected to be in Earth's path. Amateur astronomer *Mikiya Sato* of the Fucho Astronomical Society Japan, found that relatively high ejection speeds of order 47 m/s were needed to bring dust particles far enough away from the comet position. Unfortunately, the 2003 return could not be confirmed, partially because of bad weather conditions and partially because of a lack of devoted meteor observers in parts of the southern hemisphere at that time. One observer, *Mike Begbie* of Zimbabwe, observed between 16:45 and 19:00 UT and saw nine π-Puppids between 17:05 and 17:15, six of which clustered at 17:10 UT. However, none were seen in the remainder of the observing period.

19.2 The Draconid shower

The 1946 Draconid storm (Fig. 19.5) is still remembered by some of us. Alasdair MacRaonuill,[11] for example, recalls: *I remember driving at night during this meteor storm – it was literally like a snow storm, with so many flakes coming at me; so many lights, many more than were possible to count. Sometimes I wonder why the Leonids get more attention in the news than the Draconids?*

Some did count the meteors and the activity curves constructed from that are shown in Fig. 19.6. Rates went up to ZHR = 10 000/h, both in 1933 and 1946. The observing

[8] P. Jenniskens, Meteor stream activity. II. Meteor outbursts. *Astron. Astrophys.* **295** (1995), 206–235.

[9] S. Messenger, Opportunities for the stratospheric collection of dust from short-period comets. *Meteoritics Planet. Sci.* **37** (2002), 1491–1506.

[10] E. D. Kondrat'eva, I. N. Murav'eva and E. D. Reznikov, On the return of comet Grigg–Skjellerup. *Solar Syst. Res.* **36** (2002), 348–352.

[11] A. MacRaonuill, Meteor shower. *Sideral Times*, J. *Austin Astron. Soc.*, January (2002), p. 4.

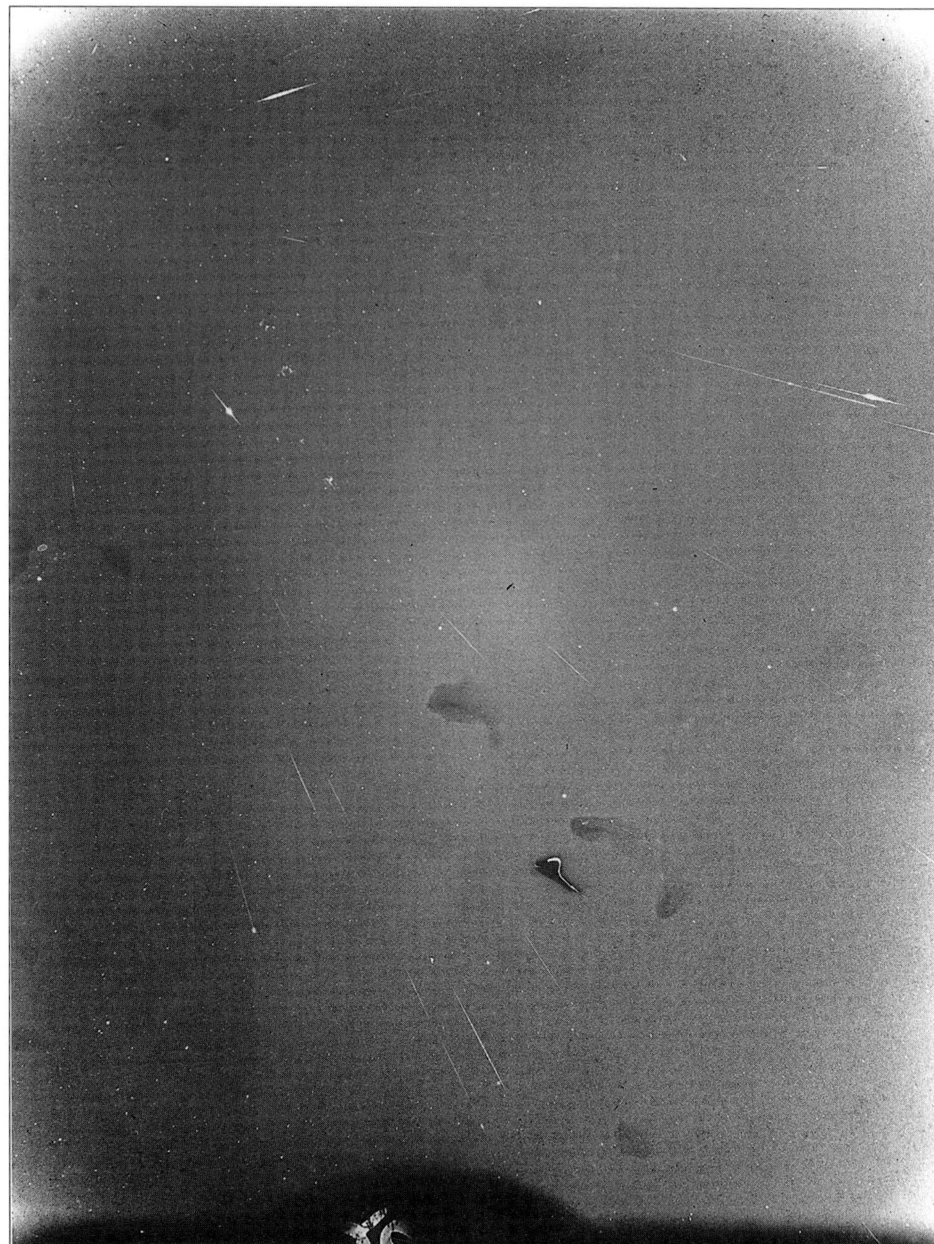

Fig. 19.5 The October 9, 1946 Draconid storm over Griffith Observatory, Los Angeles, California. A 13 min exposure ending at 4:00 UT by Paul Roques with the camera attached to the 30 cm telescope. Roland Michaelis and Karl Bouvier observed rates of 55 min between 3:45 and 4:02 UT. Photo from the original negative, courtesy of Tony Cook, Griffith Observatory.

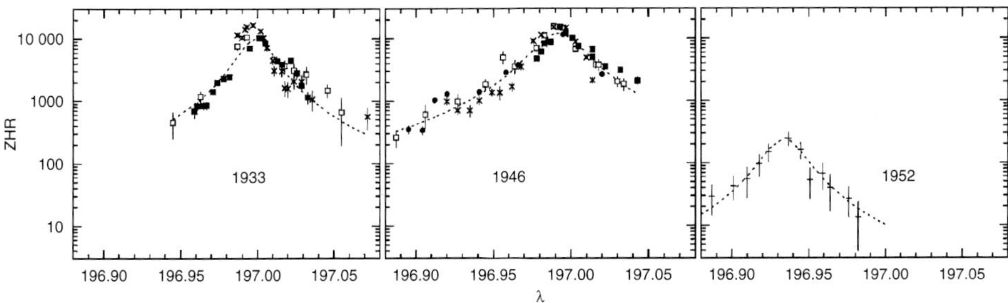

Fig. 19.6 Rate curves of Draconid outbursts of 1933, 1946, and 1952[12]. The dashed lines are Lorentz-shaped curves.

conditions were not great. In 1933, a 68% illuminated Moon was rising in the early evening. In 1946, there was a full Moon that interfered all night. Still, the high rate of very slow meteors ($V_g = 20.4$ km/s) caused many to be still visible when new ones appeared.

The shower is also called the *Giacobinids* after parent comet 21P/Giacobini–Zinner. Other names are the *October Draconids* and *γ-Draconids*. The anticipated 1946 return ushered in radar as a tool of meteor astronomy.[13] The Draconids flared up again in 1952, only detected by radar in daytime. The shower was weakly active in 1953 and 1972, and then nearly stormed in 1985 (Fig. 1.1), when many observers in Japan were on alert because the comet and Earth passed close to each other.[14] The 1985 activity curve (Fig. 19.7) had a Lorentzian profile with FWHM $W = 0.044° \pm 0.005°$. There was also a broader low-level activity, observed from the Netherlands (Fig. 19.7, right). The shower was active, but with rates below ZHR = 2, in 1986.[15]

Y. V. Yevdokimov[16] suggested that the 1946 Draconids were ejected only in 1940, with a tangential component of the velocity of about 13.7 m/s. In 1997, Wu and Williams[17] still believed that dust observed in the 1933 and 1946 Draconid storms was ejected in that same return earlier in the orbit on approach to the Sun with much higher ejection speeds.

These extremely young ages seemed to square with the excessively fragile nature of two stray Draconids photographed with the Super-Schmidt cameras one hour apart in 1953 by *Luigi G. Jacchia et al.*[18] They fragmented so readily that the shutter breaks near the

[12] P. Jenniskens, Meteor stream activity. II. Meteor outbursts. *Astron. Astrophys.* **295** (1995), 206–235.
[13] A. C. B. Lovell, C. J. Banwell and J. A. Clegg, Radio echo observations of the Giacobinids meteors, 1946. *Mon. Not. R. Astron. Soc.* **107** (1947), 64–175.
[14] K. Nagasawa and A. Kawagoe, Observations of the 1985 Giacobinid meteor shower in Japan. *Icarus* **70** (1987), 138–145; M. Koseki, Observations of the 1985 Giacobinids in Japan. *Icarus* **88** (1990), 122–128; B. A. Lindblad, The 1985 return of the Giacobinid meteor stream. *Astron. Astrophys.* **187** (1987), 928–930.
[15] M. Langbroek, The 1999 Draconids from the Netherlands and the Draconids of 1953. *WGN* **27** (1999), 335–338.
[16] Y. V. Yevdokimov, Masses of comet Giacobini–Zinner and the Draconid meteor stream. *Smithsonian Contrib. Astrophys.* **7** (1963), 297–298.
[17] Z. Wu and I. P. Williams, P/Giacobini–Zinner and the Draconid meteor shower. *Planet. Space Sci.* **43** (1997), 723–731.
[18] L. G. Jacchia, F. Verniani and R. E. Briggs, An analysis of the atmospheric trajectories of 413 precisely reduced photographic meteors. *Smithsoni Contrib. Astrophys.* **10** (1967), 1–139.

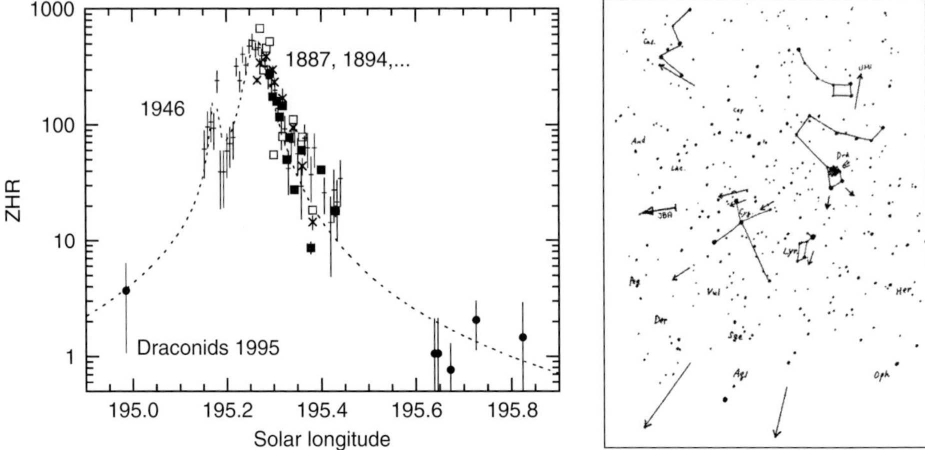

Fig. 19.7 The 1985 Draconid activity curve. The dashed line is a Lorentzian fit.[19] Right: drawings of the 1985 outburst, with one late Draconid recorded by Joop Bruining, DMS ("JBA").

end of the trajectory were no longer visible. A high fragmentation index of $\chi_f = 1.89 \pm 0.57$ was derived (Appendix D). No reliable meteoroid density could be determined.

Later, *Martin Beech*[20] argued that the meteoroid density was a normal 0.20–0.34 g/cm^3. He estimated that it needed only about 80 J/g to break up Draconids, compared to 100 J/g for Geminids, and 300 J/g to break up Perseid and Taurid meteoroids. The Draconids were detected first relatively deep in the atmosphere (Fig. 19.8), but the overall behavior of beginning and end heights is as expected for a "dustball:" all meteors start at about the same low height because they had already fragmented into their smallest sub units before reaching the point where the rocky grains started to evaporate. Only meteoroids larger than a critical mass (-2.5^{m}, or 107 g for Draconids) tend to delay fragmentation and penetrate deeper into the atmosphere. For other showers smaller masses resist fragmentation, down to critical masses of about 2.2 g (Taurids), 4.0 g (Geminids) and 2.0 g (Perseids). Interestingly, the more recent video observations of Draconids show normal end heights. Many meteors have light curves that peak early. This may help to explain why there is so much variation in beginning and end heights ($\pm$12 km) from photographed 1946 Draconids.[21]

The composition of the meteoroids is not to blame for their fragility. Peter Millman[22] analyzed 10 Draconid spectra recorded during his heroic campaign in 1946 and found that the meteoroid abundance was very similar to that of other meteoroids and not unlike carbonaceous or olivine–bronzite chondrite material

[19] P. Jenniskens, Meteor stream activity. II. Meteor outbursts. *Astron. Astrophys.* **295** (1995), 206–235.

[20] P. M. Millman, *Smithsonian Contrib. Astrophys.* **2**, 105; M. Beech, The Draconid meteoroids. *Astron. J.* **91** (1986), 159–162.

[21] L. G. Jacchia, A. Kopal and P. M. Millman, A photographic study of the Draconid meteor shower of 1946. *Astrophys. J.* **111** (1950), 104–133.

[22] P. M. Millman, Giacobinid meteor spectra. *J. R. Astron. Soc. Canada* **66** (1972), 201–211.

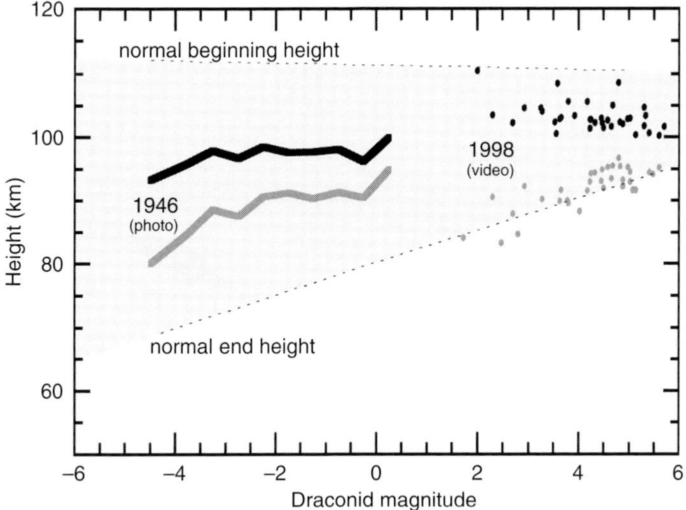

Fig. 19.8 The mean beginning and end heights of the 1946 Draconids[23] (solid line) and the 1998 Draconids[24] (dots) compared to other meteors with atmospheric speeds of 21–25 km/s in the IAU database.

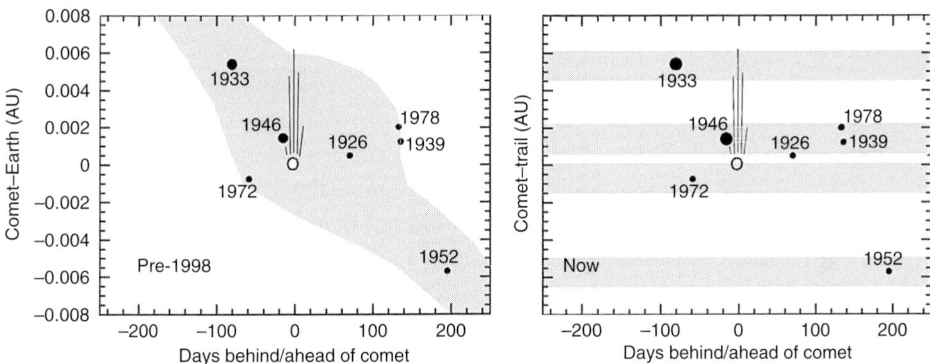

Fig. 19.9 View of the dust distribution in the neighbourhood of comet 21P/Giacobini–Zinner. Left: the picture prior to 1998 after Porter,[25] right: a schematic picture of the new dust trail model.

regarding the elements Mg, Fe, and Na. *Gale A. Harvey*[26] found similar results from two spectra recorded during the return of 1972.

Our view of how the dust is distributed around the comet has changed dramatically in recent years. Fig. 19.9 is that shown by *John Guy Porter* in "Comets and Meteor

[23] P. M. Millman, *Smithsonian Contrib. Astrophys.* **2**, 105; M. Beech, The Draconid meteoroids. *Astron. J.* **91** (1986), 159–162.

[24] Y. Fujiwara, M. Ueda, M. Sugimoto *et al.*, TV observations of the 1998 Giacobinid meteor shower in Japan. *ESA SP* **495** (2001), 123–127.

[25] J. G. Porter, *Comets and Meteor Streams* (London: Chapman & Hall, 1952), 123 pp; R. A. MacKenzie, *Solar System Debris* (Dover: British Meteor Society, 1981), p. 83.

[26] G. A. Harvey, Note on the 1972 Giacobinid Meteor Shower. *Publ. Astron. Soc. Pac.* **85** (1973), 474–475.

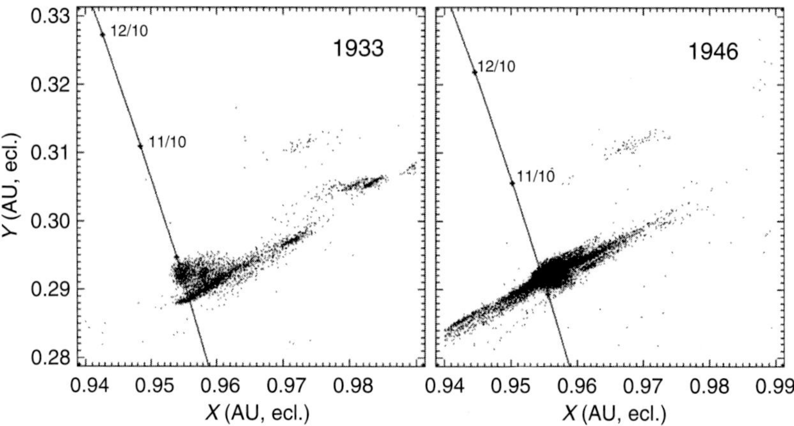

Fig. 19.10 A compilation of all recent dust trail positions in 1933 (left) and 1946, according to Vaubaillon's model. In both years, Earth passed through the center of a group of trails.

Streams" in 1952, outlining the distribution of dust as a function of where Earth sampled the stream relative to the position of the comet. The graph is similar to Yeomans' later Leonid graph Fig. 15.3. It does not take into account that meteoroids and comets are perturbed differently.

Davies and Turski calculated that the 1933 and 1946 Draconids were from dust ejected in 1894.[27] However, since Reznikov's paper,[28] the 1900- and 1907-dust trails are known to be responsible for the 1933 and 1946 storms (Fig. 19.10), each of which is a narrow expanse of dust now displaced from the current position of the comet.

The correctness of Reznikov's approach was not immediately obvious. His prediction for a return at 01:00 UT on October 8, 1988, did not pan out.[29] More importantly, he could not account for the strong 1985 outburst. Earlier, a fellow colleague astronomer at Engelhart Astronomical Observatory in Kazan, *Yu. V. Evdokimov*,[30] had forecast the time of the Draconid shower in 1985 at 05:00 UT, October 8, based on the orbital evolution of particles ejected from the cometary nucleus at perihelion between 1900 and 1959. This compared to Yeomans' forecast of 13:00 UT, when Earth would be at the node of the present comet. The observed time of October 8 at 09:36 UT was not in very good agreement with either of these predictions.

Despite the uncertainties, Reznikov boldly proceeded to predict the return of the 1998 Draconids at a time very different from the node of the current comet orbit and therefore quite different from other people's expectations.

[27] J. G. Davies and W. Turski, The formation of the Giacobinid meteor stream. *Mon. Not. R. Astron. Soc.* **123** (1962), 459–470.
[28] E. A. Reznikov, The Giacobini–Zinner comet and Giacobinid meteor stream. *Works Kazan' Town Astronomical Observatory* (1993), pp. 80–101.
[29] E. A. Reznikov, *Kometnyiy Circ.* No. 388 (1988), p. 4.
[30] Yu. V. Evdokimov, *Kometnyiy Circ.* No. 34 (1985), pp. 1–2.

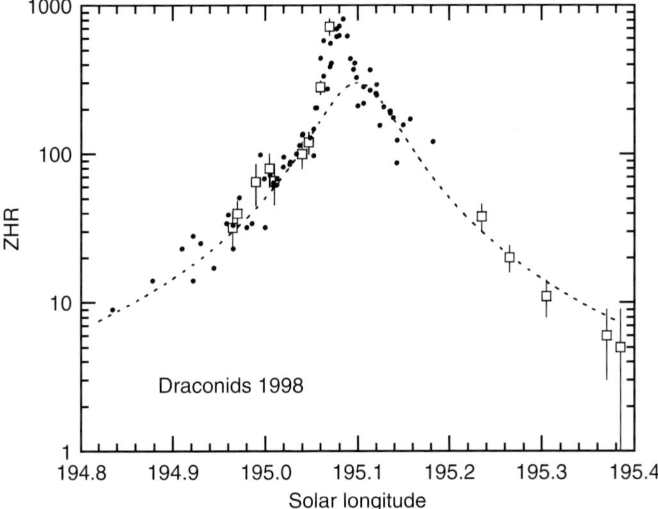

Fig. 19.11 1998 Draconid activity. A compilation of visual and radio-MS observations reported by Japanese and Dutch observers. The open squares are a summary by Rainer Arlt (IMO).

One popular announcement of the 1998 return read as follows: *Anyone hoping to see lots of Draconids must be alert for as long as possible, and also be rather lucky! Times that the highest rates have featured in the past have varied significantly. Thus the shower might be seen at any point from roughly 1998 October 6–10, most likely between October 8, 17 h to October 10, 12 h UT. We believe the most probable period for anything to happen in 1998 is on October 8, between 17 h–23 h UT.* Indeed, Kresák's guess, made in 1993, was 17:45 UT. There was no peak at this time; instead, the outburst peaked at 13:14 UT, close to Reznikov's prediction (Figs. 19.11 and 19.12).

Not aware of the earlier failed predictions, I took Reznikov's forecast seriously given it was based on the idea of a dust trail crossing, and organized a small observing campaign in California covering the hours leading up to the meteor outburst. A handful of Draconid meteors were captured on video, three of which were seen from both observing sites at Henry Coe State Park and Fremont Peak State Park. Two of these meteors at $\lambda_{\odot} = 194.6726°$ and $194.6802°$ are shown in Fig. 19.13 (•). Similar results from Japan are also shown (+).

The outburst was best seen further east over Japan, where rates increased to $\text{ZHR} = 720 \pm 90$.[31] *Yasunori Fujiwara* and coworkers[32] observed many Draconids simultaneously on intensified video cameras around $\lambda_{\odot} = 195.07°$ at the time of the

[31] R. Arlt, Summary of 1998 Draconid outburst observations. *WGN* **26** (1998), 256–259.

[32] Y. Fujiwara, M. Ueda, M. Sugimoto *et al.*, TV observations of the 1998 Giacobinid meteor shower in Japan. *ESA SP* **495** (2001), 123–127.

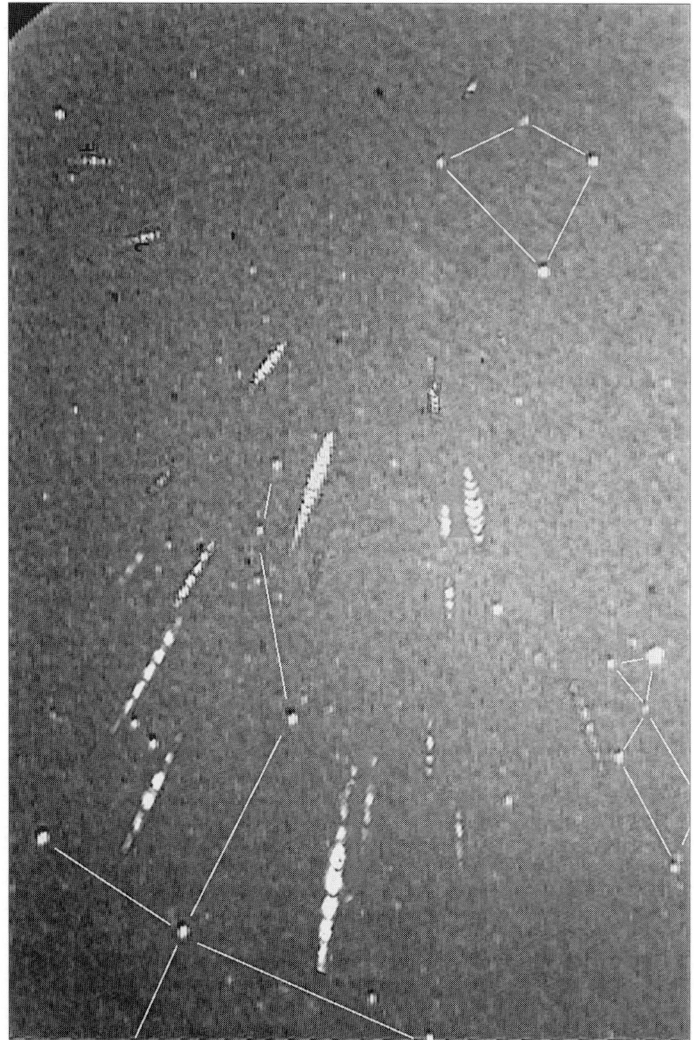

Fig. 19.12 The 1998 Draconid outburst (12:05–21:59 UT, October 8) recorded with an intensified video camera by Hideki Kondo at Narusawa Village, Japan.

peak (+). Error bars are representative for the whole data set. Open circles indicate the mean radiant position from photographic single-station meteors by Tomita *et al.*[33] and that derived from 20 station TV meteors by Suzuki *et al.* (1999).[34] The latter derived heights: H_b (km) = 107.3–0.99 m_v and H_e (km) = 84.4–1.45 m_v.

[33] M. Tomita, A. Murasawa, C. Shimoda *et al.*, On two double-station photographic 1998 Draconids. *WGN* **27** (1999), 118–119.

[34] S. Suzuki, T. Akebo, T. Yoshida and K. Suzuki, TV observations of the 1998 Giacobinid Outburst. *WGN* **27** (1999), 214–218.

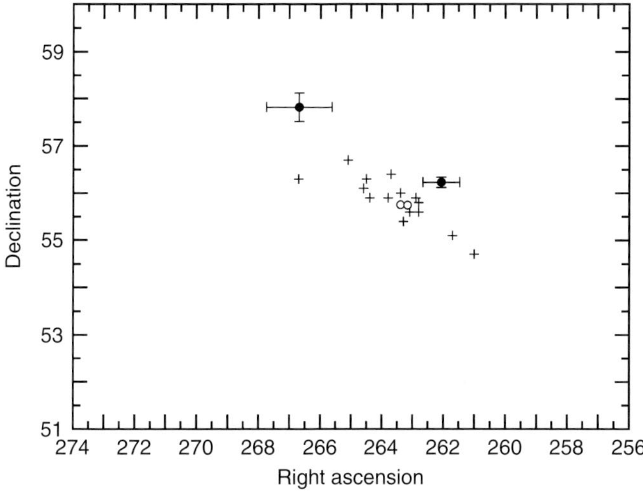

Fig. 19.13 Geocentric radiant positions from the 1998 Draconid outburst, from California (•) and from the most precise multistation video observations by Suzuki *et al.*, (+), NMS.

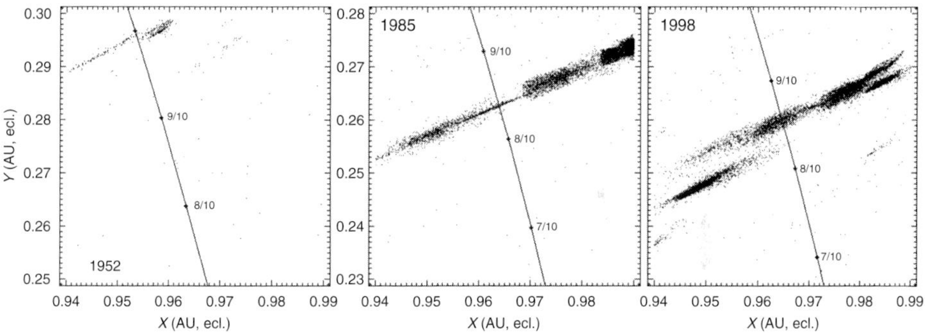

Fig. 19.14 Dust trail positions during the 1952, 1985, and 1998 Draconid outbursts. All trails from 1596 onwards.

It is interesting to find that the 1998 Draconid activity profile is composed of two components, well described by a Lorentz profile. One is a broad shower ($W = 0.090 \pm 0.007°$) centered on $\lambda_\odot = 195.10 \pm 0.02°$, with a peak rate of 300 ± 40/h, while a narrower ($W = 0.017 \pm 0.002°$) component is at $\lambda_\odot = 195.078 \pm 0.002°$ with a rate of 500 ± 100. I identify the narrow component with the thin trail of dust found in the simulation by Vaubaillon (Fig. 19.14). The broader outburst remains unexplained, but has probably suffered more planetary perturbations.

Sato continued to predict a possible encounter with the 1966-dust trail the next year in 1999, but he needed unusually high ejection velocities (of order 60–70 m/s) to bring

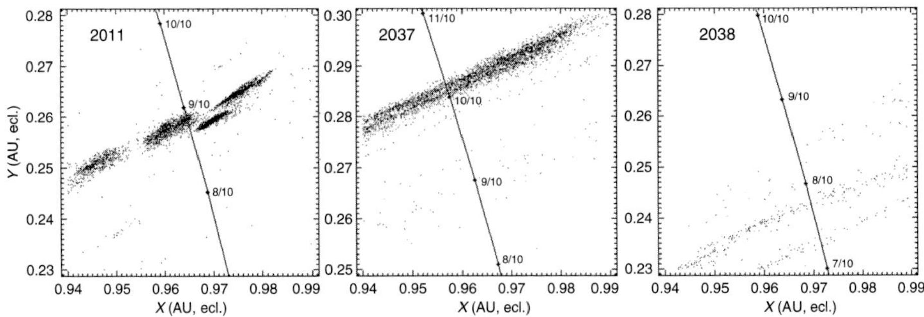

Fig. 19.15 The best dust trail encounters in future years (from left to right: 2011, 2037, and 2038). Rates in 2011 may increase to storm level.

the particles far enough out.[35] As predicted, a peak was observed around 11:00 UT (1959 trail) and 12:00 UT (1966 trail) on October 9, with an hourly rate of about 20–30 and 5–10/hr, respectively.

This established that dust trail models also work for short-period comets. More recent results by Vaubaillon are shown in Fig. 19.14. Vaubaillon, while at the SETI Institute, also worked out why the 1985 Draconid outburst could not be predicted by Reznikov: he simply did not include old enough dust trails. The 1887- and 1894-dust (and perhaps other older trails) were at the position of the bulk of activity. Note how trails of different perturbation history have a quite different width. The 1946-dust was a thin trail, the 1985 encounter of which is detected in the radar observations by *Bertil-Anders Lindblad* from Lund, Sweden.

Recent outbursts are insignificant compared to what can happen when Earth passes through the densest parts of a trail. The best return in future years will be the anticipated storm of October 8, 2011 (Fig. 19.15). Reznikov held the 2018 return promising as well because of the 1953-dust trail, but now it appears that past encounters of Earth left a gap in the trail. After 2028, conditions get worse. The perihelion distance of the comet shifts to 1.07 AU in 2037 and 1.10 AU in 2045. In the 2037/2038 encounter, there could still be nice outbursts. Dispersed low-level activity is expected in 2035 and 2044.

19.3 The June Bootids = ι-Draconids

The 1916 ι-Draconids

Comet 7P/Pons–Winnecke first surprised observers in 1916 (Fig. 19.16), when *William F. Denning* fondly recalled an outburst radiating from near ι-Draconids. He provided sufficient information to trace the activity curve over six 10 min time intervals, but I could not recognize much variation in rate.

[35] M. Sato, An investigation into the 1998 and 1999 Giacobinids by meteoroid trajectory modeling. *WGN* **31** (2003), 59–63.

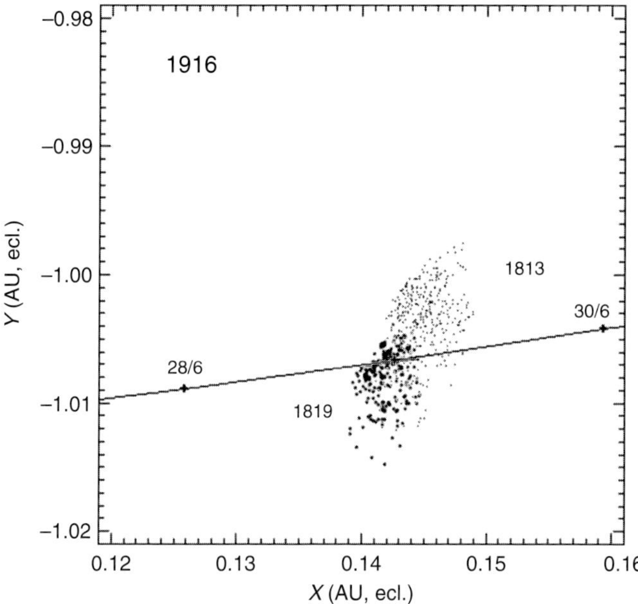

Fig. 19.16 The historic 1916 June Bootid encounter: the 1813 (top) and 1819 (bottom) dust in Earth's path during Denning's historic observation in 1916.

This case haunted me in my 1995 study of meteor outbursts, because I had a really hard time deciding on what curve would best fit the data (Fig. 19.17) and, therefore, if I should include this result in the analysis. Sadly, Denning's observations were truncated by clouds. In the USA, *Donald Brooks* of Washington, DC, was out that night and saw only a few meteors from this shower some hours later.

AMS president Charles P. Olivier[36] recalls how the unusually slow meteors from this shower were first noticed on May 27 that year, when he was first drawn to their appearance. From the observed radiant position (and assumed speed), he calculated the orbit of the meteoroids and found good agreement with that of the comet. Denning, too, recognized that 7P/Pons–Winnecke was the parent of the shower.

This event had historic significance, because it impressed upon the observers that new showers could occur when Jupiter threw a comet on an orbit nearer to us. The comet was first seen in 1809 by Pons and rediscovered in 1858 by Winnecke. In subsequent years, the orbit changed only gradually, bringing it to 0.03 AU from Earth's orbit in 1916.

Reznikov[37] first calculated that the 1819 ejecta were responsible for the 1916 ι-Draconid outburst, when Earth passed only −0.0005 AU from the trail center (Fig. 19.16). Particle ejection speeds of 8.2 m/s were needed. From results with the

[36] C. P. Olivier, *Meteors* (Baltimore, MD: Williams & Wilkins, 1925), p. 79–81.

[37] E. A. Reznikov, The Origin of the 1872 and 1855 Bielid meteor showers. In *Meteornoe Veshchestvo v Mezhplanetnom Prostranstve* (*Meteoric Material in Interplanetary Space*). (Kazan, 1982), pp. 151–152.

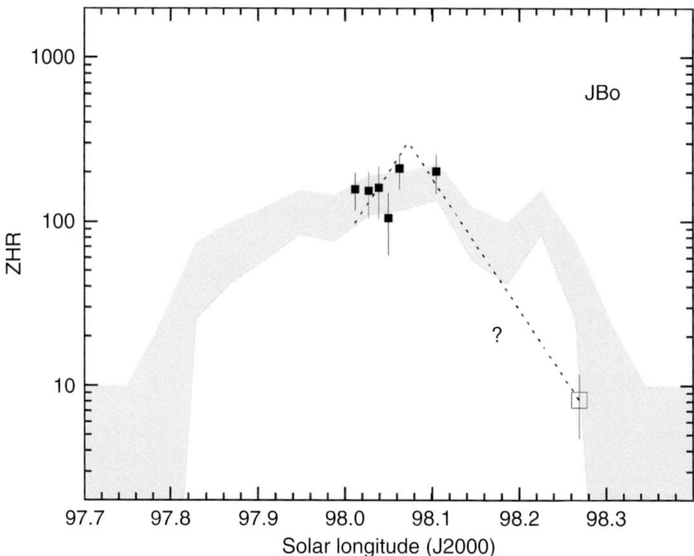

Fig. 19.17 The activity curve of the 1916 June Bootids, reproduction of the curve in Jenniskens (1995) with the model rate profile superposed.

new meteoroid stream modeling tools, we recognize why the flux profile did not permit a simple exponential fit: the particular trail cross section (both contributing dust trails) is unusually dispersed by the planetary perturbations and does not have the typical exponential increase and decrease of rates. Moreover, we find that two dust trails contribute to this profile, that of 1813 (just inside Earth's orbit) and the trail of 1819 identified by Reznikov (just outside Earth's orbit). In Figure 19.17, I reproduced my plot of the 1995 paper and show in gray the activity curve derived from the new model.

It was hoped that the next return in 1921 would see even higher rates, but that did not happen. A few meteors were observed in 1922 and 1927[38] and no more after that. As Olivier wrote: *the main group evidently had paid only a passing visit in 1916 and had then moved on. Again it is impossible to predict as to further returns, but for these meteors they seem less likely than for the Leonids.*

The 1998 June Bootids

The comet continued to evolve away from Earth's orbit, when in the weekend of June 27, 1998 a strong outburst of June Bootids was seen, radiating from: R.A. = 223°, Decl. = +48° (V_g = 14.1 km/s). The comet had been at perihelion in January 1996 passing 0.242 AU(!) from Earth's orbit. Olivier would have met the

[38] R. Arlt, The June Bootids – past and present records. *Proc. IMC Stará Lesná 1998*, ed. R. Arlt and A. Knöfel. (Potsdam: International Meteor Organization, 1998), pp. 29–42.

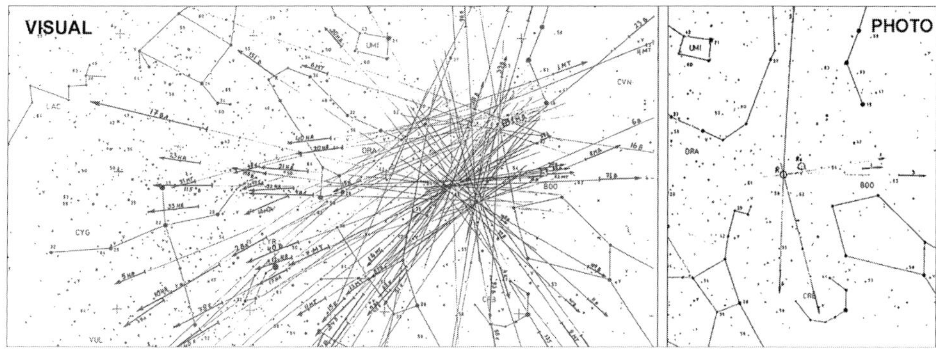

Fig. 19.18 Visually observed and photographed 1998 June Bootids in Bulgaria.[39]

announcement with disbelief. The outburst was first reported by Japanese observers[40] and in later hours also from Europe and the USA (Fig. 19.18). Many were bright ($\chi = 2.1$) and frequently there were two or three at once.

After the children were in their tents and most of the adults were sitting around the campfire, Dennis di Cicco – associate editor of *Sky & Telescope* – reported, *I was laying on a picnic table in Savoy State Forest, MA, looking at satellites and whatnot. In what was only a small hole in the canopy of trees I counted 15–20 meteors per hour*. Mark Taylor, a member of the California Meteor Society, reported seeing 20–25 meteors in the magnitude 1–2 range and another five at 3^m or dimmer in 2.5 h. Almost all of the bright meteors appeared greenish, were visible for 15°–20°, and at least half of them left a train of 1 s or more.

Two meteors photographed at Ondrejov observatory, confirmed the source as comet 7P/Pons–Winnecke.[41] One -4.5^m fireball with a -7.9^m flare (with an initial photometric mass of 0.14 kg) shown in Fig. 19.19 was found to be fragile, withstanding a maximum dynamic pressure of only 0.19 bar. With a speed of $V_g = 14.1 \pm 0.4$ km/s, it was first seen at 89.67 km and ended at 72.22 km from a direction R.A. $= 222.88 \pm 0.16°$ and Decl. $= +47.60 \pm 0.06°$.

The activity curve was again slightly more rounded than expected from a simple exponential increase and decrease, but not as severe as in 1916. The activity curve (Fig. 19.20) was broad, with $B = 2.5 \pm 0.3/°$ (the same as the Quadrantids), peaking at $\lambda_\odot = 95.69 \pm 0.01°$ and peak ZHR $= 250 \pm 50$/h. Some shadow-casting meteors were reported. Some -1 to -3 meteors left a wake that lingered for a few seconds.

[39] V. Velkov, The June Bootid outburst observed in Bulgaria. *Proc. IMC Stará Lesná, Slovak Republic*, 20–23 August 1998, ed. R. Arlt and A. Knöfel. (International Meteor Organization, 1998), pp. 84–101.

[40] K. Maegawa, HRO: a new forward-scatter observation method using a ham band beacon. *WGN* **27** (1998), 64–73.

[41] P. Spurný, EN270698 fireball: the first photographic data on a member of the June Bootid meteor stream. In *Meteoroids 1998*, ed. W. J. Baggaley and V. Porubčan. (Bratislava: Astron. Inst. Slovak. Acad. Sci., 1999), pp. 235–238; P. Spurný and J. Borovicka, *IAUC* 6973, ed. D. W. E. Green (July 21, 1998); P. Spurný and J. Borovicka (1998) Photographic observation of a June Bootid fireball. *WGN* **26** (1998), 177–179.

Fig. 19.19 June 27, 1998 21:23:04 UT June Bootid with flare. Photograph by Pavel Spurný, Ondrejov Observatory.

David Asher and *Vacheslav Emel'yanenko*[42] calculated that the 1998 shower was caused mainly by dust ejected in the 1825 return. The dust trails are relatively protected by being close to mean-motion resonances. Emel'yanenko[43] found that the June Bootid dust may well have librated about the strong 2:1 resonance with

[42] D. J. Asher and V. V. Emel'yanenko, The origin of the June Bootid outburst in 1998 and determination of cometary ejection velocities. *Mon. Not. R. Astron. Soc.* **331** (2002), 126–132.

[43] V. V. Emel'yanenko, Resonant structure of meteoroid streams. *Proc. Meteoroids 2001, Swedish Institute of Space Physics*, Kiruna, Sweden, 6–10 August 2001, *ESA SP* **495** (2001), 43–45.

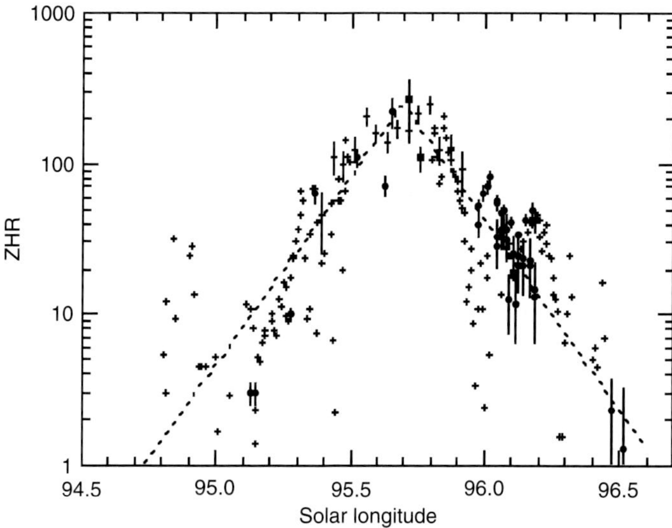

Fig. 19.20 The June 27, 1998, outburst of ι-Draconids, from visual observations collected by the author and by IMO (●), and the Global-MS-Net forward meteor scatter observations provided by Ilkka Yrjölä and Masayoshi Ueda (+).

Jupiter and nearly completed a libration cycle between 1825 and 1998 at the time of the June Bootid outburst. The 2:1 orbital resonance inhibits chaotic motion and keeps the stream compact enough over centuries to be observed as relatively strong meteor outbursts. Ejection speeds of 10–20 m/s are required. According to Asher and Emel'yanenko: *We pass the* resonant cloud *in 2004 again and may already hope for heightened rates in 2003 ....*

In research for this book, Jérémie and I noticed that several other years contributed to the outburst as well (Table 6b). The 1830 dust trail, for example, is shown in Fig. 19.21 and is superposed on the contribution from all other trails. Each trail has a profile much wider than an unperturbed one-revolution trail, significantly elongated in the radial direction towards the Sun and broadened along Earth's path. Hence, the width of the shower profile is not a good measure of the ejection conditions. The meteoroids may not be in resonance.

The 2004 June Bootids

The June Bootid story became even more interesting when *Sergey Shanov* from Russia and *Sergey Dubrovsky* from Byelorussia pointed out[44] that, on June 23, 2004, Earth was to meet a cluster of dust trails from comet 7P/Pons–Winnecke ejected in 1819, 1825, and 1830 with a predicted peak time at 09:30–13:00 UT. They used the same type

[44] S. Shanov and S. Dubrovsky, Personal website (2004) (in Russian).

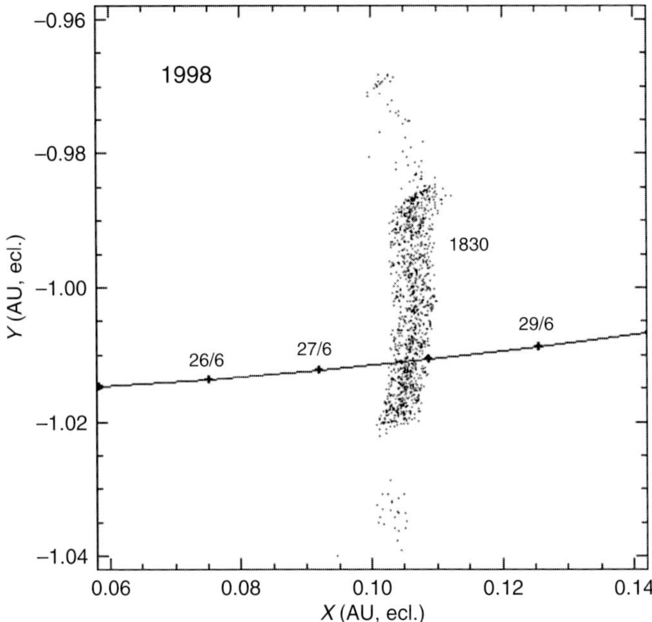

Fig. 19.21 The distribution of nodes of the 1830 ejecta of comet 7P/Pons–Winnecke, superposed on the contribution from other dust trails to the June 1998 outburst. Note the distortion of the dust trails.

of dust trail modeling as used by Reznikov, Asher, and Lyytinen for the Leonid predictions. Interestingly, these dust trails carried not the main brunt of dust near Earth at that time, but formed a separate cluster quite distinct from the main cluster of trails that had been responsible for the 1998 outburst.

The cause of that cluster was the result of a combination of perturbations by Jupiter and Earth. This was shown nicely by *Mikiya Sato* in Fig. 19.22. It shows the evolving 1819-dust trail after meeting Earth in 1910 ("A"). The Earth caused a part of the trail to be accelerated, another part decelerated, opening up a gap ("A1"–"A2"). A second encounter with the same trail during the 1916 ι-Draconid outburst, caused a further gap to form ("B1"–"B2"). The "B2" part is accelerated, and lags the earlier part of the trail, the other side "A1" now closing the gap to "B1". Jupiter then continues to warp the trail over a longer length scale. In 2004, Earth would cross the trailet A1–B2 close to the edge first perturbed in 1910.

Sato also pointed out how the position of the trails depended on the time of ejection. Fig. 19.23 shows the center of the dust trail for ejection times up to 100 d prior to perihelion (−) and after perihelion (+). Hence, the elongated distribution (seen in the model by Vaubaillon, Fig. 19.23) is not a direct consequence of the ejection process, but instead an effect of planetary perturbations being different for particles at different positions along the dust trail. Particles ejected at different times encounter Jupiter under slightly different circumstances and end up at different positions near Earth.

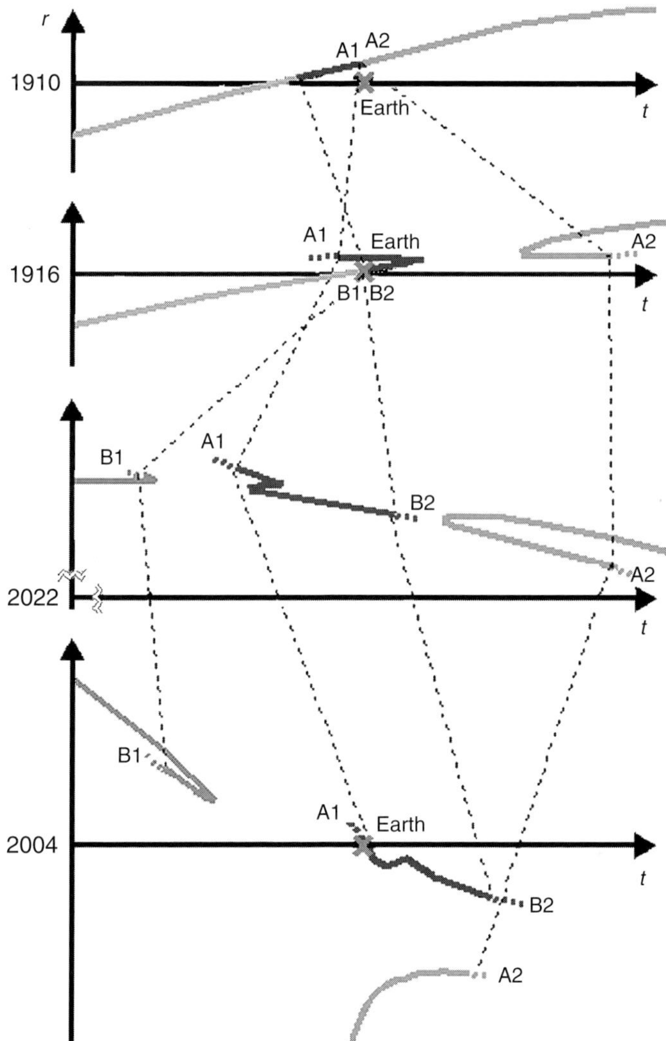

Fig. 19.22 The 1819 dust trail of comet 7P/Pons–Winnecke in the years 1910, 1916, 1922, and 2004. Calculations by Mikiya Sato, Japan.

Sato put the predicted peak between 12:30 (1830, 1825) and 19:30 UT (1813), depending on the dust trail approaching the Earth. The shower was expected to be less intense than that of 1998, but even a weak repeat provided a unique window on the properties of the dust from this comet. In a similar manner, Vaubaillon[45] predicted a broad distribution of dust with a peak time centered on 11:00 UT. Rates were expected to be less than the outburst of 1998.

[45] J. Vaubaillon, June Bootids forecastings (2004) (IMCCE website).

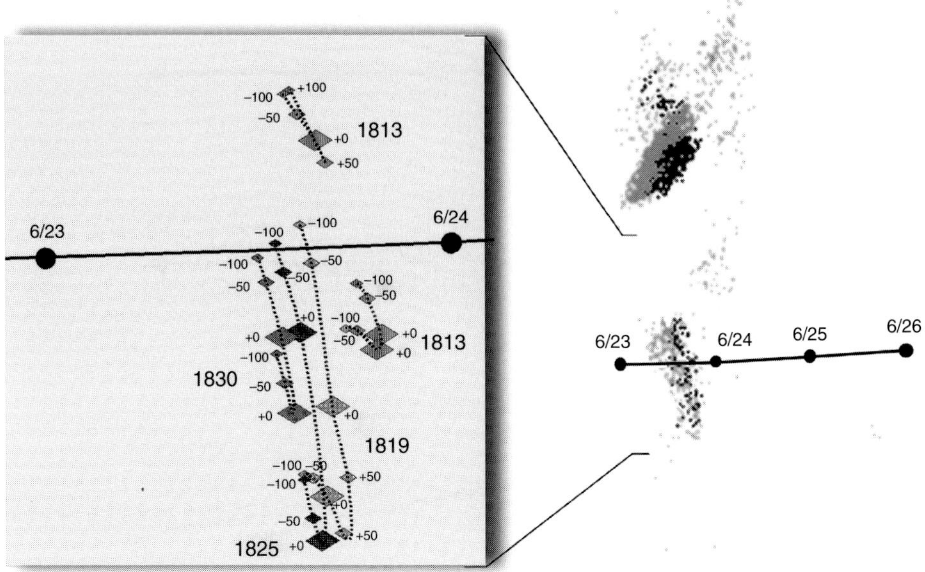

Fig. 19.23 Position of dust trails as a function of the point of ejection up to 100 d before (−) and after (+) perihelion. Right: dust at Earth in 2004, with dust from the 1819 dust trail shown as black dots. The previous figure pertains to the dust cloud at the bottom. Calculations by Sato.

In the night of June 23, visual observers[46] reported an outburst from about June 22, 20:00 UT, until June 23, 23:00 UT (ZHR > 3/h), peaking at a rate of ZHR = 18 ± 2/h at 10 ± 1 h UT, June 23, with a full-width-at-half maximum of 12 ± 2 h. The outburst was barely detected by the forward meteor scatter stations of Global-MS-Net, suggesting that the shower was not rich in faint meteors (Fig. 19.24). Indeed, the magnitude distribution index for the June Bootids was only $\chi = 2.3 \pm 0.2$, with $\chi_s = 3.2 \pm 0.3$ for sporadics.

I deployed two intensified cameras in Mountain View, California: an imager and a slit-less spectrograph, and detected six June Bootids of magnitudes between +0.4 and +2.3 (Fig. 19.25) and four sporadics of −0.9 to +3.0. There was a noticeable lack of June Bootids in the range +3 and $+4^m$, which should have been detected. This suggests that the distribution may not have been exponential towards fainter meteors.

The geocentric radiant is sharp at R.A. = 226.0°, Decl. = +47.9° (predicted 223.0°, Decl. +47.0°), mostly defined by the final June Bootid at 10:45:05 UT.

The predicted outburst produced one spectrum recorded by Toshihiro Kasuga and colleagues from Japan[47] and three by the author from California. My June Bootids appeared just outside the field of view at 07:50:07 (-0.3^m) and 10:23:03 UT (-1.7^m), while a partial spectrum was measured for a $+0.4^m$ meteor at 07:37:37 UT.

[46] R. Arlt (2004) June Bootids, Visual Data, *IMO Shower Circ.* June 25 (Potsdam: International Meteor Organization, 2004).
[47] T. Kasuga, J.-I. Watanabe, N. Ebizuka, T. Sugaya and Y. Sato, First result of June Boötid meteor spectrum. *Astron. Astrophys.* **424** (2004), L35–L38.

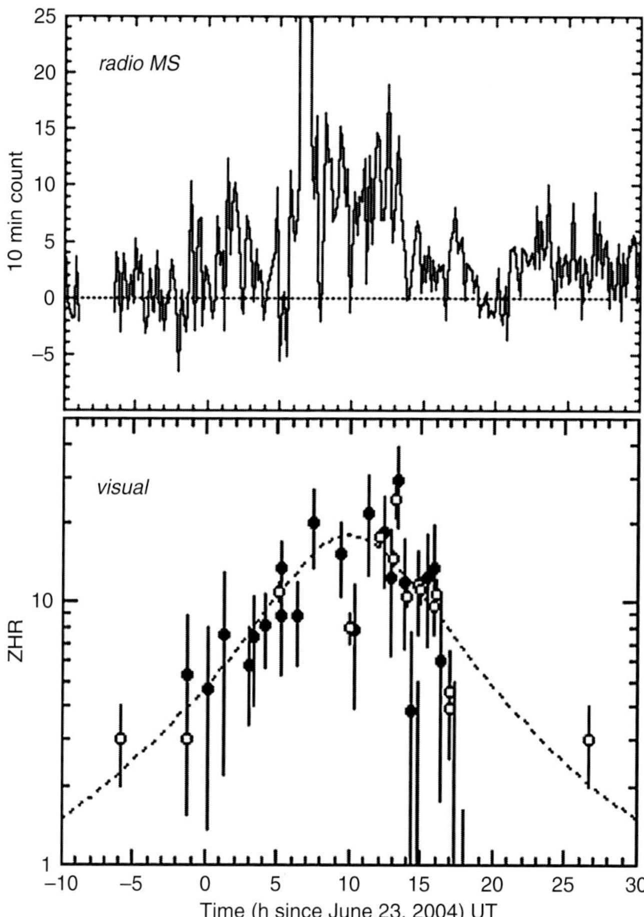

Fig. 19.24 2004 June Bootid rates. Top: raw 10 min counts of radio reflections by Esko Lyytinen, after subtraction of the (scaled) rates from the same day in 2003. Bottom: a compilation of visual rates by Bob Lunsord, Lew Gramer, Pierre Martin, Westley Stone, Roberto Haver, Mikiya Sato, Tomoko Sato, and Kazumi Terakubo (•) from IMO and NMS shower ciruclars (○).

The June Bootid spectra (Fig. 19.26) are dominated by strong sodium (Na) line emission. The sodium line has an earlier onset than the magnesium (Mg) line and, after an early peak, fades gradually when the meteor penetrates deeper into the atmosphere. This pattern is thought to be due to the more rapid loss of volatile minerals containing sodium and is only seen in meteoroids that fragment easily. Indeed, June Bootids do have the typical flat light curves of fragile meteoroids. There is also a strong continuum visible that has a relatively small contribution from the first positive system of N_2 (strong in Perseids).

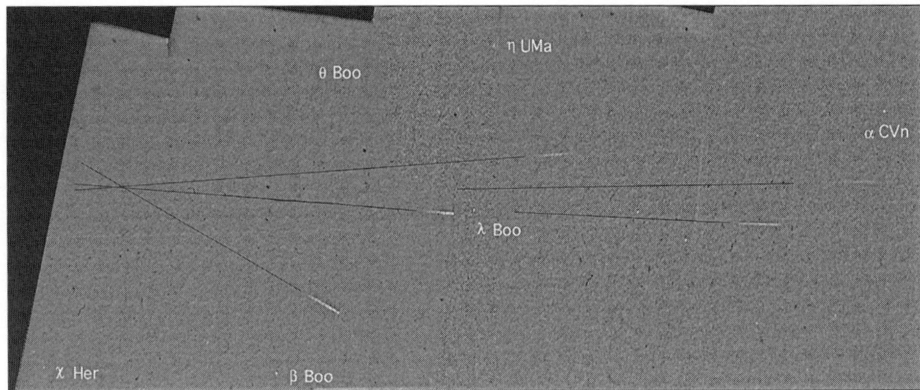

Fig. 19.25 A mosaic of video frames with June Bootid meteors obtained by the author in Mountain View, CA (07:30–10:45 UT on June 23, 2004). Each meteor image was compiled by pasting the track from four average frames, so that the meteors appear up to a factor of 4 less bright than the star background.

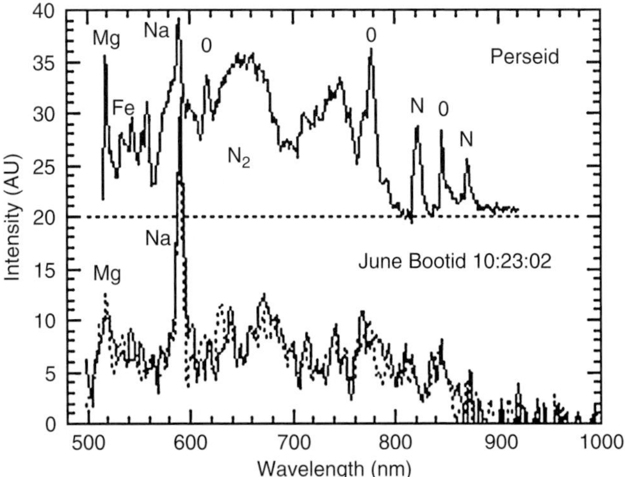

Fig. 19.26 Extracted optical spectrum of the June Bootid of 10:23:02 UT (June 23, 2004), without correction for instrument response curve. Dashed line = even fields, solid line = odd fields. A similar Perseid spectrum is shown for comparison, displaced upward by 20 units.

Sato studied the position of the dust trails in the next 50 years and showed that the trails will miss the Earth. Past returns may have occurred around July 1, 1917 and July 1, 1930, from a perturbed trail, and around June 21, 1939 (Fig. 19.27), in addition to the encounters listed in Table 6b found in a model by Vaubaillon.

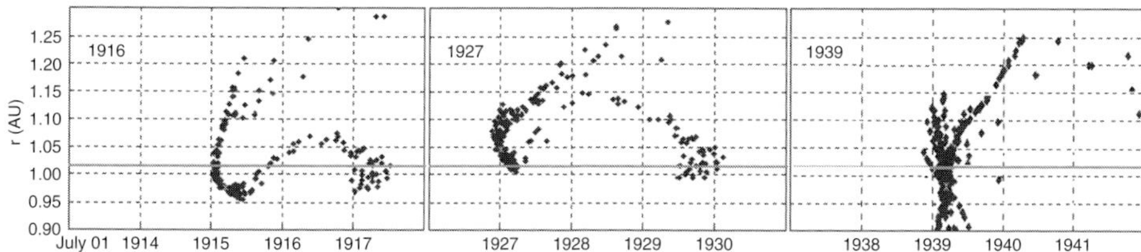

Fig. 19.27 The nodes of meteoroids at Earth's orbit as a function of time during three returns of comet 7P/Pons–Winnecke. Calculations by Mikiya Sato.

Fig. 19.28 Members of the Western Australian Meteor Society (After 1986: NAPO-Meteor Section) during an η-Aquariid campaign in 1986, under the guidance of Jeff Wood, the society was responsible for reports on many southern shower outbursts, including the κ-Pavonids and β-Hydrusids. Photo: Jeff Wood, NAPO-Meteor Section.

19.4 Lost comet D/1978 R1 /Haneda–Campos and the (October) ξ-Capricornids

Another Jupiter-family comet known for intermittent dispersed shower activity is comet *D/Haneda–Campos (1978 R1)*. In the grip of Jupiter, the comet had a gradually decreasing perihelion distance during the past century ($q = 1.57$ AU in 1900, 1.21 in 1954, 1.12 in 1962). In 1972, q had decreased to 1.101 AU, and two observers, *Dennis Rann* and *Derek Johns*, of the *Western Australia Meteor Society* (WAMS, later *NAPO-Meteor Section*, Fig. 19.28) noticed 10 bright and slow ξ-Capricornids in 35 min on October 2/3 that year ($\lambda_\odot = 189.90°$). This may have been a dust trail crossing.[48]

In the following years, it was noticed that slow meteors radiated from near ξ-Capricorni (R.A. = 302°, Decl. = −8.5°) for a few days around the time of the return of the comet (Fig. 19.29). The meteors are mostly yellow or orange, reflecting their low speed,

[48] P. Jenniskens, Meteor stream activity. II. Meteor outbursts. *Astron. Astrophys.* **295** (1995), 206–235.

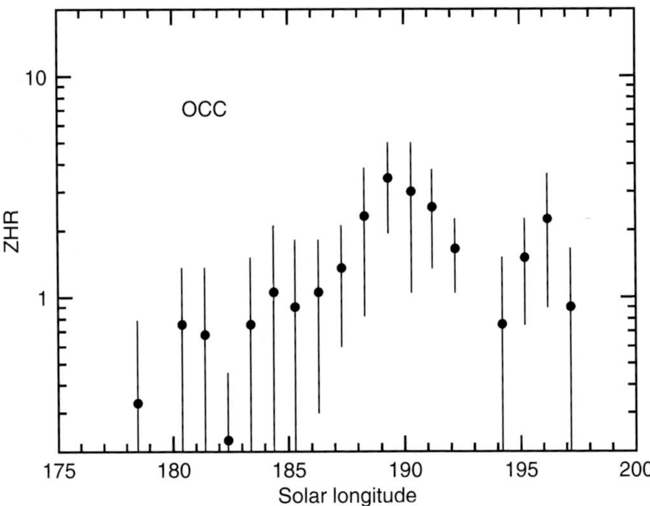

Fig. 19.29 Dispersed October Capricornid shower from observations by NAPO-MS in the period 1978–1987.

and only 4% leave a wake. From the magnitude distribution of all 138 observed meteors, Jeff Wood derived $\chi = 2.8$. The possible outburst of 1972 has not been identified yet.[49]

19.5 κ-Pavonids

Bright yellow–orange meteors radiated from a point near the star κ-Pavonis (R.A. = 275°, Decl. = −67°) on the night of July 17/18, 1986.[50] No 10 min counts were given, but the shower was said to have been of very short duration, starting at 11:50 UT and being finished at 13:00 UT (center at $\lambda_\odot = 114.828°$). With a near full Moon almost in the zenith, WAMS observers *Neil Inwood* and *Paul Stacey* from Karnet (116.1° E., −32.5° S) saw 26 and 30 slow κ-Pavonids and 4 and 6 sporadics, respectively, during this time interval. Their average magnitude was +0.71, ($\chi = 1.7$) and 14% left a train. My initial hypothesis that this could be a long-period comet dust trail (Chapter 13) did not pan out, and I now suspect that this shower was due to a short-period comet, based on the low inclination ($i = 24°$) prograde orbit of the meteoroids. No parent is known.

19.6 β-Hydrusids

The same argument suggests that the β-Hydrusid shower of August 16, 1985 was also caused by a Jupiter-family comet. No annual activity is known. The parent of

[49] J. Wood, *NAPO-MS Bull.* **200** (1988) (see also: *Radiant, J. DMS* **10** (1988), 71).
[50] J. Wood, Australia kappa-Pavonids a new shower. *WGN* **6** (1986), 186–187.

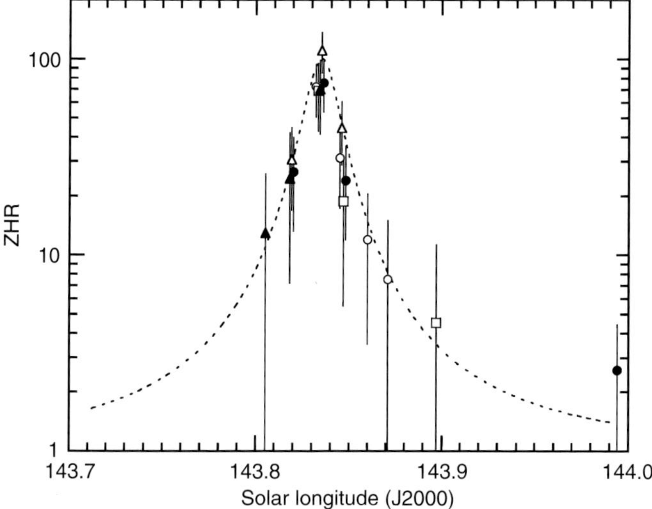

Fig. 19.30 The 1985 β-Hydrusids outburst rate curve. Data: NAPO-MS.

this shower remains a mystery. This dust trail crossing has been observed relatively well.[51] The rates shown in Fig. 19.30 are by *Brian Macauley* from Brickley, *Jason Tame* and *Simon Evans* at Kalumunda, *Paul Rawlings* at Belmont and *Joh-Ann Borrows* and *Megan Clay* from Byford, all in Western Australia.

The radiant of this shower was at R.A. = 23°, Decl. = −76° on the southern hemisphere, with a suspected entry velocity at or below $V_g = 24$ km/s. The meteors were relatively bright with $\chi = 2.1$ being a typical value. The ZHR curve is well matched by a Lorentz profile with a width of $W = 0.02°$ and a peak ZHR of 80 ± 20 at solar longitude, $\lambda_\odot = 143.833°$.

The approximate orbit of the parent comet is prograde.[52] No known extinct comet nucleus fits the bill. Again, it is unknown whether this shower was caused by a Jupiter-family, Halley-type, or long-period comet.

19.7 The α-Centaurids

In the summer months of February, southern hemisphere observers are treated on relatively fast meteors from the bright "pointer" stars β- and α-Centauri, in the general area of the (southern) Apex Source. The α-Centaurids, at R.A. = 210° and Decl. = −58° on $\lambda_\odot = 320°$, is the most active component, with a peak rate of ZHR ~7/h each year (Fig. 19.31). The shower was first seen by *Mr. V. Williams* at 4–5/h from Sydney, Australia, on February 10/11, 1889 ($\lambda_\odot = 323.6 \pm 0.1°$), from

[51] J. Wood, *NAPO-MS Bull.* **150** (1986) (see also: *Radiant, J. DMS* **8**, 75).
[52] P. Jenniskens, Meteor stream activity II. Meteor outbursts. *Astron. Astrophys.* **295** (1995), 206–235.

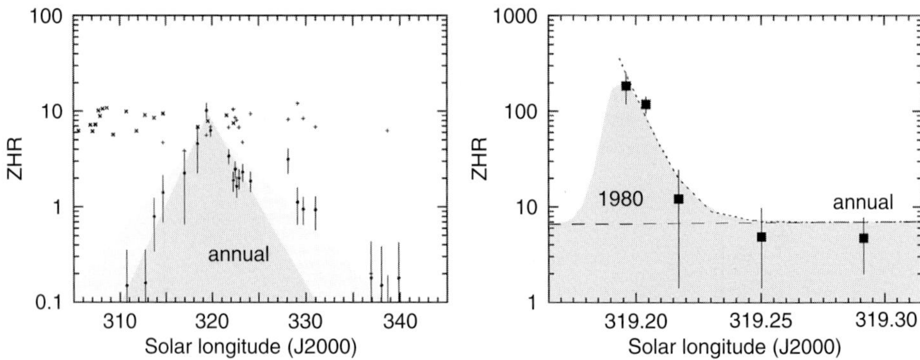

Fig. 19.31 Annual (left) and 1980 outburst activity of the α-Centaurids. Data: NAPO-MS.[53]

a radiant at R.A. = 216°, Decl. = −61.5°. Hoffmeister recorded a tentative radiant near α-Centauri, but the shower was first recognized as such only in the early seventies by *Morgan* (New Zealand), *Skelsey* (Sydney, Australia) and *Buhagiar* (Perth, Australia). Michael Buhagiar studied the shower from 1969 to 1980 and believed that two nearby radiants were active in the period February 6–8.[54] In the night of February 8/9, 1974 ($\lambda_\odot = 319.8 \pm 0.1°$), he and Skelsey noticed rates of 12–15 meteors/h.

The shower is best known for the outburst of February 8/9, 1980 at $\lambda_\odot = 320.20°$, radiating from a location close to β-Centauri (R.A. = 210.3°, Decl. = −58.5°). At that time, 169 Centaurids were plotted by three experienced WAMS observers: *Glen Blencowe*, *Neil Freckelton*, and *Craig Willoughby*. They recorded peak rates of 14, 11, and 9 meteors/h respectively in the hour from 12:10 to 13:10 UT, declining with FWHM ~0.3 h (Fig. 19.31). The mean magnitude was a bright $\langle m_v \rangle + 0.54$ ($\chi \sim 1.3$) and a high 48% of the meteors left a train. The annual α-Centaurids have $\chi = 2.3$. The large number of bright yellow meteors in the evening attracted attention of the public. Police, weather institutes, radio, and TV stations in Western Australia received reports and continued to pay attention to the event the next day. Rates declined as soon as official observations were started.[55]

The shower peak times suggest an evolution of the node −0.048°/yr, typical of a short-period Jupiter-type comet. Meteors were said to be of median velocity ($V_\infty \sim 50$ km/s). I suspect that the parent body was near Earth's orbit in 1980 (but remember the case of the June Bootids in Section 19.3). However, no parent body is known.

[53] P. Jenniskens, Meteor stream activity. I. The annual streams. *Astron. Astrophys.* **287** (1994), 990–1013; P. Jenniskens, Meteor stream activity. II. Meteor outbursts. *Astron. Astrophys.* **295** (1995), 206–235.
[54] M. Buhagiar, Southern hemisphere meteor stream list. *Western Australian Meteor Society, Internal Publication* (1980).
[55] J. Wood, *WAMS Bull.* **157** (1980) (see also: *Radiant, J. DMS* **2** (1980), 135).

19.8 Other Jupiter-family comets

A number of other Jupiter-family comets have been investigated for possible meteor shower activity. The remaining section of this chapter discusses some of these investigations and their results.

103P/Hartley 2 was discovered on March 15, 1986, by Malcolm Hartley with the 1.2 m UK Schmidt telescope at Siding Spring, Australia. Based on the preliminary orbit, *Robert McNaught* immediately realized that two returns later, in 1997, the Earth would pass close to the comet itself, making it possible that some meteor activity might occur.[56] How favorable that encounter was going to be became known only after successive observations during the return of 1991 and the recovery in 1997. The comet passed Jupiter on April 29, 1971, at only 0.085 AU, followed by two shallower encounters in November 2, 1982 (0.325 AU) and December 19, 1993 (0.374 AU). Those brought the comet in towards the Sun and near to Earth's orbit.

On November 2, 1997, the Earth was about to pass 0.0395 AU from the orbit of comet 103P/Hartley 2 and, following McNaught, I issued a warning for possible meteor activity.[57] The perihelion distance had increased from 0.95 AU in 1991 to 1.03 AU during the 1997 return. The minimum distance would be at 00.9 UT, the nodal passage at 07.1 UT on November 2, and the Earth was going to lead the comet in passing that point by only 49.0 d. The meteors should have radiated from β-Cygni (R.A. $= 297.6°$, Decl. $= +29.5°$) at 12.1 km/s, but none were seen.

Esko Lyytinen investigated dust trail encounters from comet 103P/Hartley 2. It was found that Earth passed somewhat in front of the main dust in 1997. Esko found that the trails from 103P/Hartley 2 wave in and out of Earth's orbit, but with no good encounters in the next 50 years. Only very few trails have been studied so far. The October 28, 2010 return puts the perihelion at 1.059 AU, for a minimum separation of 0.0665 AU. First in 2062 and 2068, there may be an encounter from the 1949-dust trail (Table 6i). This is true also for the 1979-dust trail. The 1973-dust trail in 2068 appeared to have rapid changes of the miss-distance within a few days from passing the ecliptic plane. Other trails were less distorted. Jupiter-family comet **P/2001 Q2 Petriew** is in a similar orbit as 103P/Hartley 2, but its trails are deflected from the Earth's orbit more than those of Hartley 2.

Michael A'Hearn et al.[58] suspected that 103P/Hartley 2 is an interloper from the Oort cloud. Comet 103P/Hartley 2 is not large, only ~1.1 km in diameter. It emits

[56] R. H. McNaught, New shower in 1997? *WGN* **15** (1986), 49.
[57] P. Jenniskens, Good prospects for a meteor outburst from comet 103P/Hartley 2. *IAU C-22 Pro-Am Working Group* (1997).
[58] M. F. A'Hearn, R. L. Millis, D. G. Schleicher, D. J. Osip and P. V. Birch, The ensemble properties of comets: results from narrow band photometry of 85 comets, 1976–1992. *Icarus* **118** (1995), 223–270.

dust over a large (71%) area of its surface. The comet was observed with the mid-infrared ISO satellite on January, 1998[59], close to its perihelion (at 1.04 AU from the Sun and 0.82 AU from the Earth). At that time, the comet emitted up to 20% CO_2 ice in addition to water. The water ortho-to-para ratio implied that the ices had been formed at temperatures of about 35 K. Jupiter-family comet 103P/Hartley 2 showed the same crystalline silicates, a high content of CO_2, and a low water ortho-to-para ratio, as comet Hale–Bopp from the Oort cloud. However Hartley 2 appeared to produce less dust relative to ice. The dust production was described well by the relation $Q_d = 100\, r^{-2}$ kg/s (r = distance from Sun, preperihelion, with little information after perihelion). Epifani *et al.*[60] found that the dust grain size distribution is on average, $\alpha = 3.2 \pm 0.2$, dominated by large grains. But the size distribution is much steeper at perihelion, increasing to $\alpha = 4.8$. These grains are ejected with a speed of about 5 m/s, but increasing to 25 m/s at perihelion.

Lubor Kresák[61] identified comet **15P/Finlay** as a likely source of meteor storms in the period 2004–2064, with meteors radiating from Ara at R.A. = 249.8°, Decl. = −44.9°, $V_g = 11.98$ km/s (2060 return, when the comet itself will pass by Earth at 0.047 AU on October 27). 15P/Finlay is a Jupiter-family comet in a low inclination orbit just outside Earth's orbit. Neslusan and coworkers searched for photographed orbits in the IAU database, expecting activity around September 24, but no orbits could be linked to this comet.[62] The comet is thought to be old and evolving towards an intermittently active stage. *Martin Beech*[63] investigated the possible meteor shower activity from this comet and found that between 1886 and 1960 no 15P/Finlay-derived meteoroids would evolve into Earth-crossing orbits. Beech found that in the years 2001 and 2008 some meteoroids can make it towards Earth's orbit, but Earth is not at the right time at that spot to cause a meteor outburst. Sergej Shanov examined future dust trail encounters with 15P/Finlay (Table 6).

Comet **46P/Wirtanen** has an orbit just outside Earth's orbit and a small inclination. It was long the target comet for ESA's ROSETTA mission, before launch delays demanded the choice of a different target. The comet is small, with a diameter of only 1.11 ± 0.08 km[64] at an axis ratio of 1:1.4, and rotates with a

[59] J. Crovisier, T. Encrenaz, E. Lellouch *et al.*, ISO spectroscopic observations of short-period comets. In *The Universe as Seen by ISO*, ed. P. Cox and M. F. Kessler, *ESA-SP* **427** (1999), 161–164.

[60] E. Epifani, L. Colangeli, M. Fulle *et al.*, ISOCAM imaging of comets 103P/Hartley 2 and 2P/Encke. *Icarus* **149** (2001), 339–350.

[61] L. Kresák, Meteor storms. In *Meteoroids and Their Parent Bodies*. (Bratislava: Astron. Inst. Slovak Acad. Sci., 1992), pp. 147–156.

[62] L. Neslusan, V. Porubčan and J. Svoren, Meteor radiants of recently discovered Earth-approaching comets. In *Meteoroids and Their Parent Bodies*, ed. J. Stohl and I. P. Williams. (Bratislava: Astron. Inst. Slovak Acad. Sci., 1992), pp. 181–184.

[63] M. Beech, S. Nikolova and J. Jones, The 'silent' world of comet 15P/Finlay. *Mon. Not. R. Astron. Soc.* **310** (1999), 168–174.

[64] D. Möhlmann, Activity and nucleus properties of 46 P/Wirtanen. *Planet Space Sci.* **47** (1999), 971–974.

period of 7.6 h.[65] The comet is unusual in that it does not have strong jets and appears to be active over a relatively large ~25% area of its surface. 46P/Wirtanen is dust poor and about 2–3 times less dusty than comet 1P/Halley. The comet has many close encounters with Jupiter. The orbital evolution is strongly affected by a close proximity to the 2:1 resonance with Jupiter. Before the encounter of August 8, 1698 (0.28 AU), the comet had q close to 1 AU and may have been responsible for meteor showers. Marco Fulle found that large grains are ejected far from the Sun.[66] The comet is currently too far from Earth's orbit for dust trail encounters. Meteor outbursts appear to be possible from orbits prior to the close encounter of August 8, 1698. However, the dust trail coherence disappears on time scales longer than 150 yr. Future activity is not excluded. Due to the slow entry speed, relatively large 0.1 g meteoroids are needed to give visible ($+5^m$) meteors.

Some individual meteoroid orbits have been measured that appear to match the orbit of **6P/d'Arrest** (Table 7). The comet spends most of its time just outside Earth's orbit, but has come very close to Earth's orbit on occasion and sometimes very close to Earth itself. This comet, 6P/d'Arrest, was once a target of NASA's CONTOUR mission, which would have passed by the comet in August of 2008, when the comet would have made a close call with Earth on August 9. The comet evolves erratically on orbits with q of either 1.17, 1.28, or 1.36 AU. It was near $q = 1.17$ in the 1678 return, during discovery in 1851 and 1857, and recently in 1970 and 1976. It came as close as $q = 1.03$ AU in recent history. The comet orbit will continue to evolve in a chaotic manner outward to 2 AU in 2200. In 1991, *Andrea Carusi* and *Giovanni B. Valsecchi* in Rome, and Kresák and Kresáková in Bratislava, independently recognized that the comet seen in 1678 by La Hire was the same object.[67] This required strong nongravitational forces being constant for a long period of time. *Syuichi Nakano*[68] has $A_1 = +0.707 \pm 0.023$, $A_2 = +0.115\,92 \pm 0.000\,55$ ($\times 10^{-8}$ AU/d^2) from observations in the period 1988–2001. The nongravitational forces are best understood if there is a thermal lag angle of about 10° and water vapor outgassing peaks 40 d after perihelion. The comet light curve also peaks 40 d after perihelion.[69] Because this comet was a target of the CONTOUR mission, *Marco Fulle*[70] studied the dust ejected and found that a total mass of meteoroids in the range 20 μm–10 cm of

[65] K. J. Meech, J. M. Bauer and O. R. Hainaut, Rotation of comet 46P/Wirtanen. *Astron. Astrophys.* **326** (1997), 1268–1276.

[66] M. Fulle, The dust environment of comet 46P/Wirtanen at perihelion: a period of decreasing activity? *Icarus* **145** (2000), 239–251.

[67] A. Carusi, G. B. Valsecchi, L. Kresák, M. Kresaková and G. Sitarski, Periodic comet d'Arrest = Comet la Hire (1678). *IAU Circ.* 5283, ed. B. G. Marsden, (Cambridge, MA: IAU Minor Planet Center, 1990).

[68] S. Nakano, *OAA Computing Section Circ. NK 800 R*, ed. S. Nakano (Dec. 22, 2001). OAA computing section circular NK 800 R.

[69] D. K. Yeomans and P. W. Chodas, An asymmetric outgassing model for cometary nongravitational accelerations. *Astron. J.* **98** (1989), 1083–1093.

[70] M. Fulle, Meteoroids from short period comets. *Astron. Astrophys.* **230** (1990), 220–226.

$8 \pm 2 \times 10^9$ kg/return, with a time-averaged size distribution index of $\alpha = 3.8 \pm 0.1$ (or $s = 2.3 \pm 0.1$).

Grains are ejected $-70 < T_p < 120$ d. The comet tends to brighten rapidly just before perihelion and then continues to be active at about the same level of activity for a long period after passing perihelion, until true anomaly 60°. During that time the active surface increases from 1 to 70 km^2 between 0° and +20°, from 70 to 140 km^2 until 60°, before rapidly decreasing again after 100° mean anomaly.[71] The comet appears to have a tenth of the dust–gas ratio of comet 1P/Halley. The comet spins fast and in a complex manner, with one period of 6.67 ± 0.03 h.[72] The nucleus is small, $D = 3.0$ km. The water production rate in 1982 (from IUE data) was 1.45×10^{28}/s at 1.29 AU.

A recent addition to the list of Jupiter-family comets is the weakly active **P/2004 CB (LINEAR)** with an orbit not far from Earth's. Some shower activity is expected, but no meteor activity has been reported yet. The radiant is at high northern latitude and it should make an observable shower. However, the comet is nearly dormant. A best fit to the light curve has $H_{10} = +17.9$ ($n = 2$), with a peak in brightness 55 d after perihelion. In April, 2004, the comet had a very compressed coma of about 13 arcsec diameter. It will come within 0.59 AU of Jupiter in 2012, leading to an Earth flyby at 0.055 AU (21.5 Lunar distances) on May 29, 2014.

The May, 2014 encounter is the most interesting for meteor activity, because the comet itself is expected to pass the node only three days later. On request, Esko Lyytinen studied this case based on an early orbit and found that Earth passes just inside a cluster of trails from this comet. Older trails can not yet be calculated reliably because the comet orbit remains uncertain. Integrating backwards, the comet orbit was found to be relatively stable, with $P = 5.02$ years. There are a few encounters with Jupiter, mostly shallow, the most recent in 1917 at 0.545 AU. Occasional close encounters with Earth can change the comet orbit and because of the unknown effects of comet outgassing, integrating to before the Earth encounter in 1582 is not meaningful.

The perihelion distance has gradually increased from a low 0.799 AU in 1698, to 0.84 AU in 1818, to 0.91 in 2004. Because the comet orbit does not change much, the trails tend to pile up and a large number of dust trails would be involved if any activity is observed. Perhaps in one of those returns the comet was more active than now.

In 2014, the very young trails pass about 0.003 AU outside the Earth's orbit (at the time of encounter) with slightly negative Δa_0 and beta values of −96 millionths for the 2004 trail and −32 millionths for the 1979-dust trail (Table 6j). These will be slow meteors ($V_g = 15.86$ km/s) from R.A. = 125°, Decl. = +78°,

[71] M. C. Festou, P. D. Feldman and M. F. A. A'Hearn, The gas production rate of periodic comet d'Arrest. In *ACM 1991*. (Houston, TX: Lunar and Planetary Institute, 1992), pp. 177–182.

[72] P. J. Gutiérrez, J. de León, L. Jorda *et al.*, New spin period determination for comet 6P/D'Arrest. *Astron. Astrophys.* **407** (2003), L37–L40.

with a peak at about 06:59 UT ($\lambda_{\odot} = 62.847°$), May 24, 2014. The miss-distance may slowly increase (to older trails) and has the value of about $\Delta r = +0.005$ AU in 1929. The 1909-dust trail, for example, has a predicted encounter with Earth for positive values of beta ($\Delta r = -0.0002$ AU) at $\lambda_{\odot} = 62.856°$. The previous trail from 1903 has $\Delta r = +0.000\,15$ AU at 62.847°. These two trails may be the most significant. Most other recent trails have a solar longitude in the range 62.847°–62.869°. Older trails ejected between AD 1693 and 1903 will pass at roughly $\Delta r = +0.001$ miss-distance between 62.817° and 62.856°. These older trails are the more likely to contribute meteors, given the lack of recent comet activity. The beta value is only +3 or +4 millionths, suggesting that large meteoroids will be observed.

We also looked at potential dust trail encounters from other Jupiter-family comets, such as 72P/Denning–Fujikawa, D/1766 G_1 Helfenzrieder, 76P/West–Kohoutek–Ikemura, 41P/Tuttle–Giacobini–Kresák, and 45P/Honda–Mrkos–Pajdusakova. Results are listed in Table 6.

Part IV

Young streams from comet fragmentation

20
Quadrantids

The mystery of the Quadrantid shower (Fig. 20.1) has always been its source. The Quadrantids have a peak ZHR $\sim$ 130/h, which is the highest of all known annual showers.[1] Unfortunately, the shower is only 8.5 h FWHM wide, and very difficult to observe because of frequent bad weather in early January and a radiant that is in under-culmination at midnight.

Even after the shower was first recognized in January of 1835 (Chapter 1), it took an 1862 report from a lady in Connecticut, stating that she had observed an unusually large number of shooting stars early in the morning of January 2, to first get Edward C. Herrick and then other observers excited.[2] Among those was *Alexander S. Hershell*, who in 1864 gave the "Shooting Stars of January" their modern name, finding a radiant point at "c Quadrantis Muralis" in the now defunct constellation *Quadrans Muralis*.[3] In the non-English speaking world, the shower is better known by the name of *Bootids*. In a modern star atlas, the radiant is at the corner of the constellations Bootes, Hercules and Draco.

The search for its parent first took a clue from the dramatic and rapid evolution of the orbit, discovered in the early modeling by *Salah E. Hamid* and *Mary N. Youssef*.[4] This 1963 study added the effect of all planetary perturbations on test particles, six in total, over many orbits. Results were confirmed later by others using more modern computering tools.[5] They found that the orbit had rotated in a nutation cycle from a low inclination of $i \sim 13^\circ$ and low perihelion distance $q \sim 0.1$ AU some 1500 years ago, to its current high $i \sim 72^\circ$ and $q = 0.78$ AU (Fig. 20.2). The orbit will continue to evolve to a peak value of $i \sim 76^\circ$ and $q \sim 1.0$ AU about 1000 years from now, before decreasing again. These dramatic changes occur in a periodic manner with the rate of rotation

[1] D. W. Hughes and I. W. Taylor, Observations of overdense Quadrantid radio meteors and the variation of the position of stream maximum with meteor magnitude. *MNRAS* **181** (1977), 517–526.

[2] D. W. Olson, The first American observations of the Quandrantid shower. *AMS Meteor News*, No. 89 (1990).

[3] J. Sauval, Quetelet and the discovery of the first meteor showers. *WGN, J. Int. Meteor Org.* **25** (1997), 21–33; I. W. Fisher, The Quadrantid meteor history to 1927. *Harvard College Observatory Circ.* **346** (1930), 1–11.

[4] S. E. Hamid and M. N. Youssef, A short note on the origin and age of the Quadrantids. *Smith. Contr. Astroph.* **7** (1963), 309–311.

[5] I. P. Williams, C. D. Murray and D. W. Hughes, The long-term orbital evolution of the Quadrantid meteor stream. *MNRAS* **189** (1979), 483–492; D. W. Hughes, I. P. Williams and C. D. Murray, The orbital evolution of the Quadrantid meteor stream between AD 1830 and 2030. *MNRAS* **189** (1979), 493–500; D. W. Hughes, I. P. Williams and C. D. Murray, The effect of orbital evolution on the influx of Quadrantid meteoroids. *MNRAS* **190** (1980), 733–741.

Fig. 20.1 A -3^m Quadrantid at 01:50:28 UT and a $+0^m$ Quadrantid at 01:56:50 UT on January 4, 1995, in a photo by Casper ter Kuile of the Dutch Meteor Society from Biddinghuizen, the Netherlands.

depending strongly on the semimajor axis. Wu and Williams calculated that a full cycle would take on average 4000 yr for a Quadrantid orbit with $a = 2.98$ AU, close to the real value, but 7000 years if $a = 2.81$ AU, and a full rotation would take 11 000 years if $a = 2.34$, even further away from Jupiter's orbit.

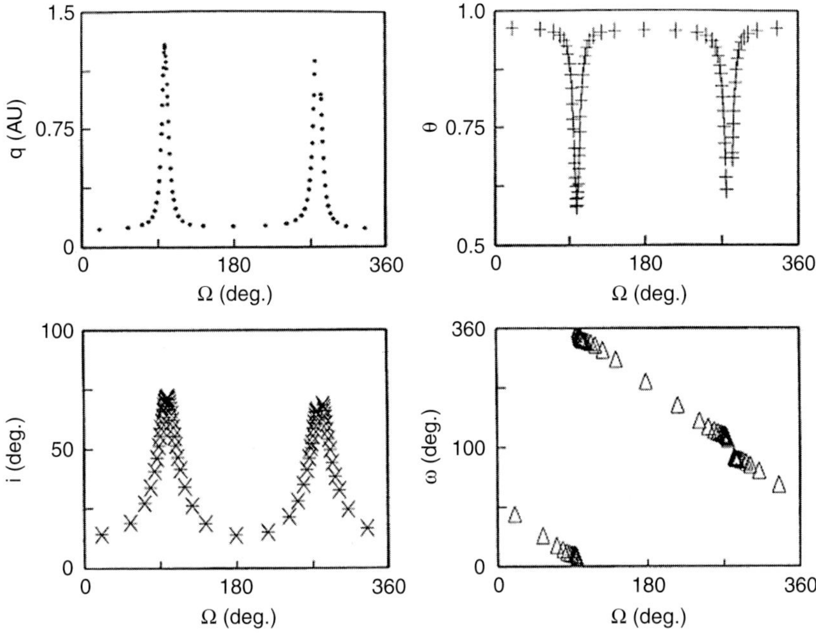

Fig. 20.2 Variations of orbital elements of the Quadrantids in a nutation cycle. Graph by Wu and Williams.[6]

Hamid and Youssef traced the orbits of six photographed Quadrantids back to a common state ~3000 years ago. They proposed that a comet was captured by Jupiter about 4000 years ago, shedding meteoroids, some of which escaped close encounters with Jupiter to stay in a narrow stream.

In 1990, *Bruce McIntosh*[7] noticed that 1500 years ago, the orbit was not unlike that of the then recently discovered comet *96P/1986* J_2 *(Machholz 1)*. At that time, comet Machholz had a perihelion closer to the Sun and a much smaller inclination. Today, Machholz has a high inclination (60.2°) and the shortest perihelion distance on record for a Jupiter-family comet that is not dormant ($q = 0.124$ AU). McIntosh proposed that the meteoroids were ejected long ago and had evolved into a different phase of the nutation cycle, further along the cycle than Machholz on account of a wider orbital period. If true, there would need to be a mechanism to bring the meteoroids back together in order to create the very narrow shower that is now the Quadrantids (Fig. 20.3).

In recent years, this idea was carried forth mostly by *Claude Froeschlé*, *Robert Gonczi*, and *Hans Rickman*, who assumed ejection occurred when comet Machholz

[6] Z. Wu and I. P. Williams, On the Quadrantid meteoroid stream complex. *MNRAS* **259** (1992), 617–628.
[7] B. A. McIntosh, Comet P/Machholtz and the Quadrantid meteor stream. *Icarus* **86** (1990), 299–304.

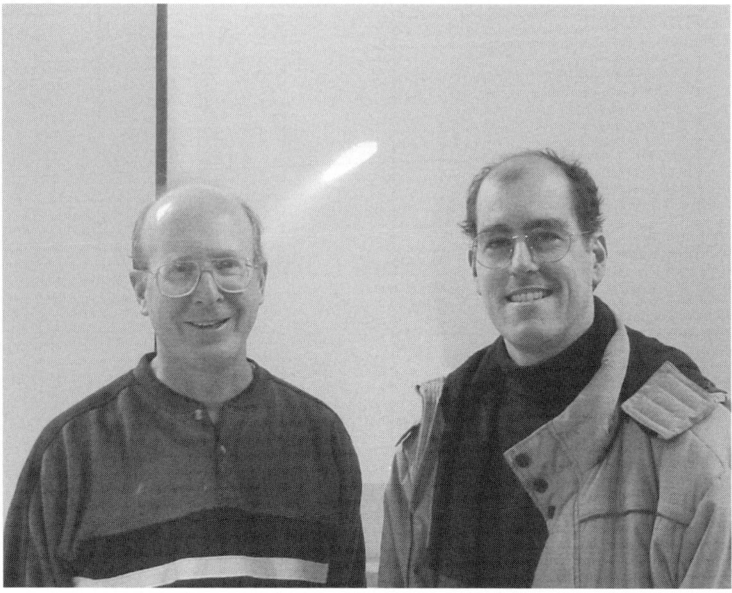

Fig. 20.3 Donald E. Machholz made his discovery of 96P/Machholz in 1986 from an observing site on Loma Prieta mountain, just south of San José, which later became notorious as the epicenter of the Loma Prieta earthquake that shook the San Francisco Bay Area in 1989. I met Don (left in photo) during one of his public seminars at the *San Jose Astronomical Association*. On the screen in the background is the comet seen by SOHO. Photo: Mike Koop.

was last in a low-q orbit. They showed that ejection could have occurred as recently as 1500 years ago.[8] They also found that the Quadrantid meteoroids might be moving in chaotic orbits, a phenomenon associated with overlapping mean-motion resonances. The idea of chaotic motion appealed to the imagination of frustrated theoreticians,[9] but is incorrect as an explanation for the Quadrantid shower.

Based on the orientation of the orbit, *Ichiro Hasegawa*[10] proposed that comet C/1490 Y_1 (1491 I) might have been the parent. When it was observed in 1491, the comet passed close to Earth. It had a perihelion on January 8.9, close to when Earth was at the Quadrantid node. There is no evidence that this was a short-period comet. To explain the apparent loss of the comet from the stream, Wu and Williams[11] proposed that it might have been lost from its observed orbit about 150 years later in a close encounter with Jupiter, not unlike the evolution of comet D/1770 L_1

[8] C. Froeschlé, R. Gonczi and H. Rickman, New results on the connection between comet P/Machholz and the Quadrantid meteor streams: Poynting–Robertson drag and chaotic motion. In: *Meteoroids and their parent bodies*. J. Stohl, I. P. Williams (eds.), (1992), pp. 169–172; R. Gonczi, H. Rickman and C. Froeschlé, The connection between comet P/Machholz and the Quadrantid meteor stream. *MNRAS* **254** (1992), 627–634.

[9] I. P. Williams, The dynamics of meteoroid streams. In: *Meteoroids and their parent bodies*. J. Stohl, I. P. Williams (eds.), Bratislava, (1992), pp. 31–40.

[10] I. Hasegawa, Orbits of ancient and medieval comets. *PASJ* **31** (1979), 257–270. (erratum 31, 829).

[11] I. P. Williams and Z. D. Wu, The Quadrantids meteoroid stream and comet 1491 I. *MNRAS* **264** (1993), 659–664.

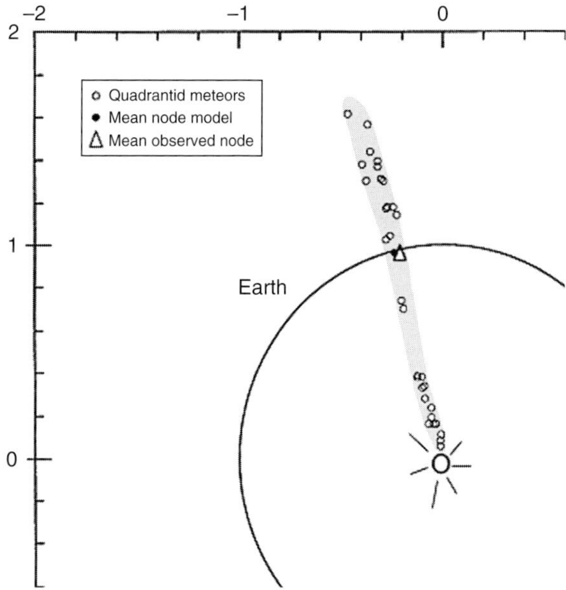

Fig. 20.4 Model of the Quadrantid meteoroid stream at the descending node in a calculation by Williams and Wu.

(Lexell) in Chapter 8. This would demand that most of the meteoroids in the shower escaped that close encounter and continued to evolve along the old orbit.

Because the orbital period is unknown, there is no evidence that Jupiter ever came close enough to C/1490 Y_1 for it to be ejected. Wu and Williams went on to calculate the orbital evolution of the meteoroids and found a wide distribution of nodes in the ecliptic plane, creating a very narrow (0.18°) shower in the Earth's path, but highly extended in heliocentric direction (Fig. 20.4). Jupiter passed close to the stream at aphelion and would disperse some of the meteoroids in each orbit. Over longer periods of time, this increased the stream width by about 0.7° in 2000 yr. From the measured width (taken to be 1.89°), Wu and Williams derived an age of 5400 years.

In 1998, Williams and Collander-Brown concluded in that same vein that asteroid 5496 (1973 NA) is a more likely candidate, even more likely than comet C/1490 Y_1.[12] This minor planet is in an orbit similar to 96P/Machholz (but with a very different Π) and was thought to be another fragment from a common body at a different stage in the nutation cycle. Other comets claimed to be part of the Machholz Complex are: *5D/Brorsen, D/1892 T_1 (Barnard 3), 23P/Brorsen–Metcalf, D/1783 W_1 (Pigott), C/1939 B_1 (Kozik–Peltier), 8P/Tuttle, 38P/Stephan–Oterma, and 12P/Pons–Brooks.* Paul Wiegert (University of Western Ontario) recently proposed a link with *C/1860 D_1 (Liais)*. None of these are a particular good match with the current Quadrantid shower.

[12] I. P. Williams and S. J. Collander-Brown, The parent of the Quadrantid meteoroid stream. *MNRAS* **294** (1998), 127–138.

20.1 Dispelling the Quadrantid myths

Because the shower is so difficult to observe, there are several misunderstandings that misguided early efforts in understanding the nature of this shower. In a variety of arguments, they hold that the shower must have significantly evolved since formation.[13]

These myths include earlier findings that the faint meteors observed by radar have a different activity curve than brighter meteors observed visually or by photography. This dates back to the early 1960 paper by Kashcheev and Lebedinets, presenting meteor echo observations at Kharkov in January 1959.[14] It was claimed that the fainter meteors precede the brighter ones in a progressive change of node. More careful analysis of radar and visual observations has since demonstrated that there is no such difference. The peak time of the shower is independent of meteoroid mass, spanning the range +6 to -2^{m}.[15]

The meteor stream activity curve[16] is also not asymmetric, as claimed before, but well represented by a symmetric Lorentz curve with full-width-at-half maximum $W = 0.35°$ and a peak rate of ZHR = 130/h at $\lambda_{\odot} = 283.28 \pm 0.01°$ (year 2000). Underlying that main peak, and significantly displaced is a broader background component more typical of annual streams with an activity curve having an exponent $W = 3.2°$, peak rate of 8/h, and a maximum at $\lambda_{\odot} = 282.2 \pm 0.2°$.

The magnitude distribution index χ does not change across the main peak of the shower, as many have claimed.[17] Instead, the observed change of χ along the Earth's path (Fig. 20.5) is understood if the background component is rich in faint meteors with $\chi = 3.5$, while the superposed main peak is rich in bright meteors with $\chi = 2.25$.

It is true that the node of the comet orbit has evolved over the past century, even though early determinations of the peak time are very uncertain. Since it was discovered in 1835, the node of the shower has gradually changed from $\lambda_{\odot} = 283.75°$ in 1863 to 283.20° now (Fig. 20.6).[18]

It is also true that the shower has become more active over the years,[19] albeit not as rapidly as often claimed. Best reported rates for a single visual observer have always

[13] D. W. Hughes, I. P. Williams and K. Fox, The mass segregation and nodal retrogression of the Quadrantid meteor stream. *MNRAS* **195** (1981), 625–637.

[14] B. L. Kashcheev and V. N. Lebedinets, The structure of the Quadrantid meteor stream. *Sov. Astron.* **4** (1960), 114–117.

[15] B. A. McIntosh and M. Simek, Quadrantid meteor shower – A quarter-century of radar observations. *BAC* **35** (1984), 14–28; P. Jenniskens, H. Betlem, M. de Lignie, M. Langbroek and M. van Vliet, Meteor stream activity. V. The Quadrantids, a very young stream. *Astron. Astrophys.* **327** (1997), 1242–1252; P. Brown, W. K. Hocking, J. Jones and J. Rendtel, Observations of the Geminids and Quadrantids using a stratosphere-troposphere radar. *MNRAS* **295** (1998), 847–859.

[16] P. Jenniskens, Meteor stream activity. I. The annual streams. *Astron. Astrophys.* **287** (1994), 990–1013; J. Rendtel, R. Koschack and R. Arlt, The 1992 Quadrantid Meteor Shower. *WGN, the Journal of the IMO* **21** (1993), 97–109; B. A. McIntosh and M. Simek, Quadrantid meteor shower – A quarter-century of radar observations. *Bull. Astron. Inst. Czechosl.* **35** (1984), 14–28.

[17] L. Bellot Rubio, Dependence of the population index on the radiant zenithal distance. *WGN, J. of the IMO* **22** (1994), 13–26.

[18] D. W. Hughes, I. P. Williams and C. D. Murray, The orbital evolution of the Quadrantid meteor stream between AD 1830 and 2030. *MNRAS* **189** (1979), 493–500.

[19] C. D. Murray, I. P. Williams and D. W. Hughes, The effect of orbital evolution on the influx of Quadrantid meteoroids. *MNRAS* **190** (1980), 733–741.

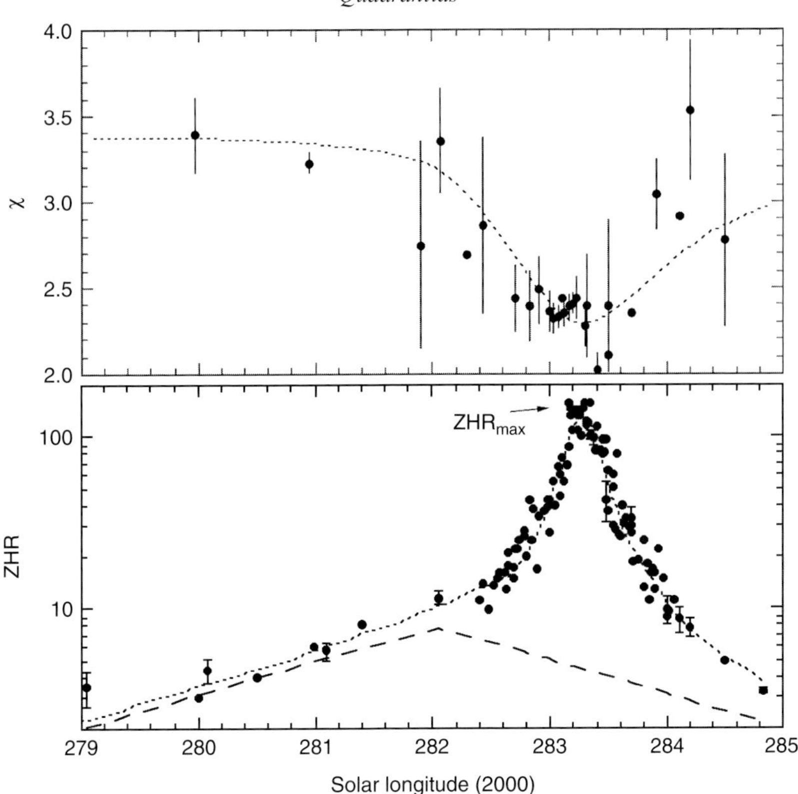

Fig. 20.5 The 1997 Quadrantid activity curve and magnitude distribution index. Two dust components are recognized.

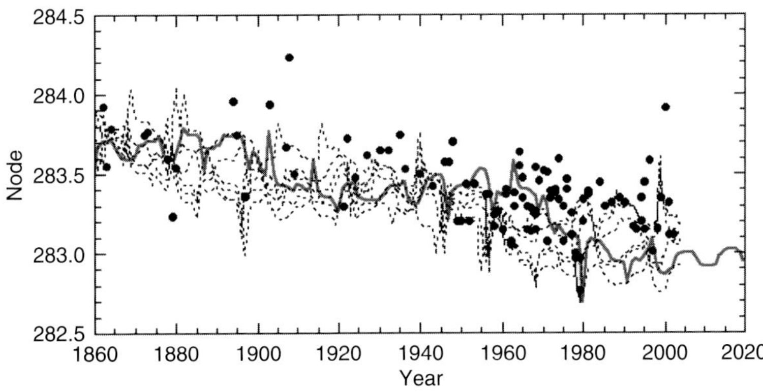

Fig. 20.6 Changes of the peak time in the past century. The dots are observations,[20] the solid line is the evolution of 2003 EH_1 (see below), the dashed lines represent model meteoroids.

[20] B. A. McIntosh and M. Simek, Quadrantid meteor shower – A quarter-century of radar observations. *Bull. Astron. Inst. Czechoslov* **35** (1984), 14–28; R. A. MacKenzie, *Solar System Debris*. Brittish Meteor Society, Dover, UK, (1980), p. 6; J. Rendtel, R. Koschack and R. Arlt, The 1992 Quadrantid Meteor Shower. *WGN, the Journal of IMO* **21** (1993), 97–109; P. Jenniskens, Bootiden 1984–1985, *Radiant, the Journal of DMS* **7** (1985), 118–122.

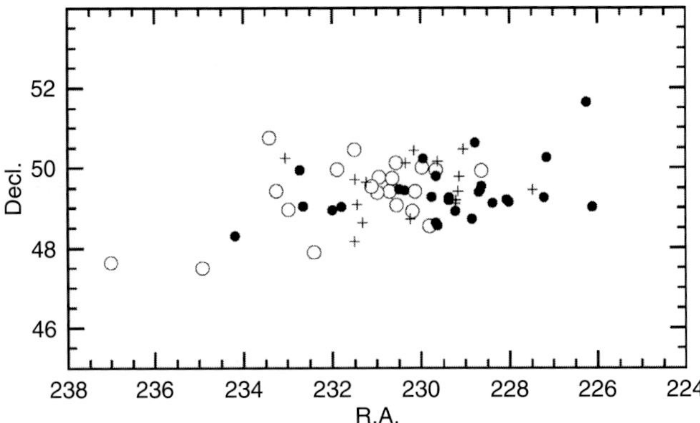

Fig. 20.9 Quadrantid radiant distribution, as measured by photographic means. Different symbols indicate meteors with different bins of entry speed.[26]

As reported before, the radiant distribution was wide, but we found that for each small range in entry speed, the radiant distribution was actually rather narrow. Meteoroids of different speeds arrive from a slightly different right ascension (Fig. 20.9). Jupiter passes close to the stream at aphelion and will always disperse some of the meteoroids, leading to spreading over time. From the dispersion of orbits calculated by Williams & Wu (Fig. 20.4), I concluded that the stream could be no older than ~500 yr.[27]

The aphelion of the photographed orbits were found clustered just inside the orbit of Jupiter around $a = 3.14$ AU or $P = 5.6$ yr (Figs. 20.8 and 20.10). The inclination scattered around a steep $i = 71.2°$ and $72.8°$. Because the orbit did not intersect Jupiter's orbit, chances for a very close encounter (<0.02 AU) are small, which ruled out a scenario where comet C/1490 Y_1 could have been lost in a 1650 AD close encounter with Jupiter.

From this, I suspected that there still was an extinct comet nucleus to be discovered among the Quadrantid meteoroids, hidden by posing as an asteroid. The mean Quadrantid orbit provided an approximate orbit with the caveat that, unlike the comet, the Quadrantid meteoroids have to intersect Earth's orbit in order to be seen as meteors. The total mass of dust in the Quadrantid stream was in excess of 1.0×10^{12} kg (for masses of 10^{-6}–10^{3} g) – I later found this closer to 10^{13} kg, see below –, and if that mass represented the mass of the remaining nucleus, then this object should have a diameter $D > 1.6$ km, a significant size (if the comet density is 0.5 g/cm^3).

[26] P. Jenniskens, H. Betlem, M. de Lignie, M. Langbroek and M. van Vliet, Meteor stream activity. V. The Quadrantids, a very young stream. *Astron Astrophys.* **327** (1997), 1242–1252.
[27] *Ibid.*

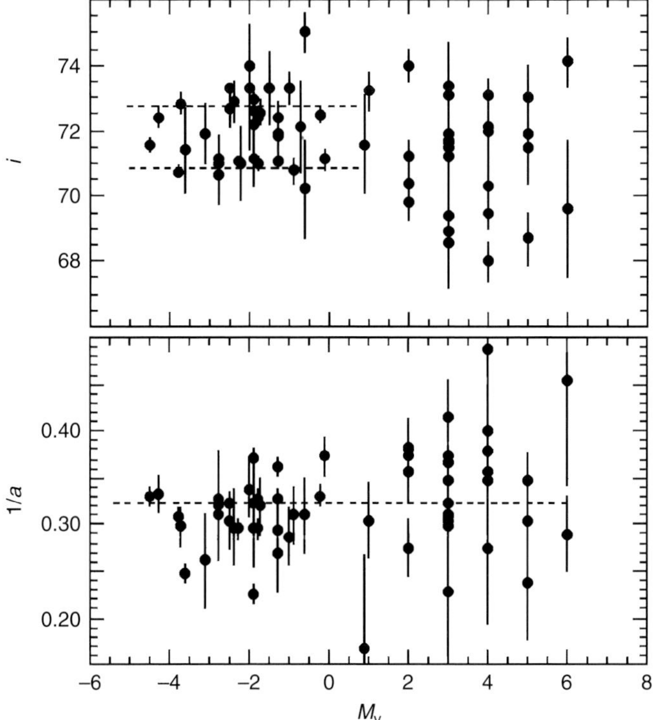

Fig. 20.10 The semimajor axis and inclination for meteors of different brightnesses.[28]

Active about 500 years ago, this now extinct comet nucleus was perhaps observed as comet C/1490 Y_1.

The young age is consistent with the relatively low mean density for the particles calculated by Verniani:[29] 0.17 g/cm^3. Higher densities were derived by Pulat Babadzhanov after taking into account the continuous shedding of small fragments in a process called quasi-continuous fragmentation, but he also arrived at a low 1.4 g/cm^3 compared to 1.2 g/cm^3 for the Perseids of 109P/Swift–Tuttle.[30] Clearly, Quadrantids are fragile cometary grains that shed fragments easily.

These results did not permit dedicated searches, because the position of the comet in its orbit remained unknown. The comet was possibly located where the dust density along the orbit was highest, but the peak intensity of the shower has not been measured very well. Based on *Jürgen Rendtel's* discovery that the stream was most active in 1965, 1970, 1987, and 1992,[31] my best guess at the time was the comet might

[28] *Ibid.*
[29] F. Verniani, Meteor masses and luminosity. *Smith. Cont. Astrophys.* **10** (1967), 181–195.
[30] P. B. Babadzhanov, Formation of twin meteor showers. *Asteroids Comets Meteors III*. Uppsala, (1989), pp. 497–503.
[31] J. Rendtel, R. Koschack and R. Arlt, The 1992 Quadrantid meteor shower. *WGN, Journal of the IMO* **21** (1993), 97–109.

return to perihelion in late 2002 or early 2003 (interpolation of the dates would give February 28, 2003).

20.2 The long-lost comet

It was a matter of time before the comet passed through one of the survey fields of the automatic near-Earth asteroid detection programs. By the time of writing this chapter, in October of 2003, about half of all near-Earth objects larger than 1 km in diameter had been discovered and a database of minor planet orbits was maintained by NASA J.P.L.. I decided to check one last time for the long sought parent, and there it was: 2003 EH_1.[32]

Minor planet designations: the minor planet now associated with the Quadrantid meteoroid stream got the preliminary object code 2003 EH_1. "EH_1" refers to the time in the year when the object was discovered. Each successive half month in the year carries a letter A–Y in the alphabet, with "I" omitted. The second letter indicates the order of minor planet within the half-month during which the discovery was made, where A = 1st, B = 2nd, C = 3rd, D = 4th, E = 5th, F = 6th, G = 7th, H = 8th, etc. Again, the "I" is omitted. After "Z", the next 25 entries carry a "1" subscript, the following 25 entries a "2", etc. Hence, the E stands for the half-month period March 1–15, while the "H_1" stands for the 33rd minor planet discovered that period. The designation of minor planets immediately recognized as comets is somewhat similar. A letter indicates the status as a comet "C/," a periodic comet "P/," or a periodic comet that has now gone extinct "D/." This is followed by the year of discovery and the first half-month period of discovery. A numeral then gives the order of discovery in that half month. For example, 1P/1682 Q_1 = periodic comet Halley (number 1 of all numbered periodic comets), which was the first comet discovered in AD 1682 in the period Q."

The minor planet had been discovered on March 6, 2003 by the automated *Lowell Observatory Near-Earth Object Survey* – LONEOS telescope (observer B. A. Skiff). The initial orbit was unlike that of any meteoroid stream, but in the next 48 d other observers measured the position of the minor planet on the sky and from that determined a high-inclination ($i = 70.8°$) orbit with an aphelion at Jupiter (Fig. 20.11).[33]

Now the orbit had been refined, I found that it agreed well with my predicted orbit of the Quadrantid parent. The aphelion of 2003 EH_1 was precisely at the peak of the meteoroid distribution. The orientation of the orbit was close to that expected,

[32] P. Jenniskens (2003) *IAU Circular* 8252 (December 08, 2003) B. G. Marsden, ed.

[33] B. G. Marsden, ed. (2003) EH_1. *MPEC 2003-E27* (March 07, 2003). Minor Planet Center, Center for Astrophysics, Harvard College, MA.

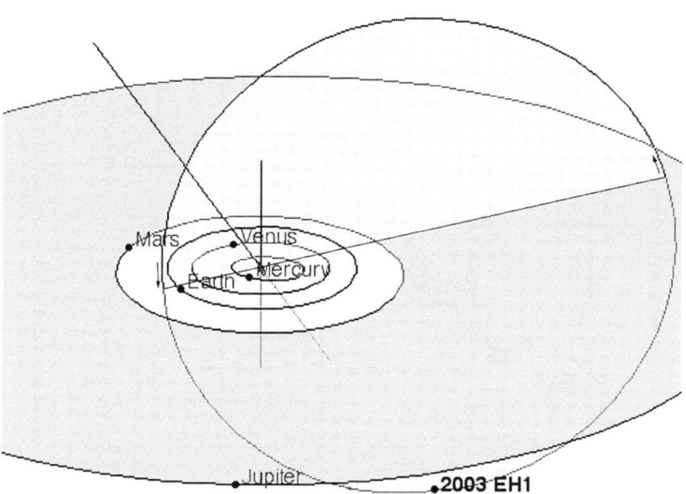

Fig. 20.11 The orbit of 2003 EH$_1$, passing just outside Earth's orbit. Position on January 4, 2004.

with no significant discrepancy in the argument of perihelion and inclination, and only a slight offset in the rapidly evolving node. Indeed, the theoretical radiant and speed for a shower from 2003 EH$_1$ (R. A. = 229.9°, Decl. = +49.6°, V_g = 40.2 km/s at $\lambda_\odot$ = 282.9° – J2000) falls in the middle of those measured for the Quadrantids.

This minor planet escaped scrutiny for many months because it was not listed as a near-Earth object. The minimum distance between the comet and Earth's orbit (0.213 AU) is larger than the limit to qualify as a "near-Earth object" (Chapter 10) and larger than the dispersion of many annual showers (<0.04 AU). However, the orbit evolves rapidly, due to frequent encounters with Jupiter, which fortunately are brief and tend to be shallow due to the high inclination. By integrating the orbit back in time using the JPL/Horizons software, I was able to show that 2003 EH$_1$ had evolved in the recent past from an orbit with a much smaller perihelion distance and a position right among the meteoroids (Fig. 20.12).

In order to investigate the general distribution of orbital elements for a stream created from object 2003 EH$_1$, I integrated the orbit of the comet back to AD 1600, and then calculated the starting orbit of a number of meteoroids released from the comet at that time with slightly higher semimajor axis $\Delta a = +0.0000$ to $+0.0124$ AU (and adjusted eccentricity). Such orbits represent meteoroids of $\beta = 1 \times 10^{-4}$ ejected in the forward direction of motion at perihelion with ejection speeds of –2.0 to +10.7 m/s, as in recent Leonid storm prediction models.

The resulting orbits show a progressive scatter of the perihelion distance as a function of time since ejection (Fig. 20.12), as found by authors in the past, but overall followed the evolution of 2003 EH$_1$, as required for this object to be associated with the stream. The dispersion relative to the current orbit of 2003 EH$_1$ accounts in sign and order of magnitude for the observed differences between 2003 EH$_1$ and the

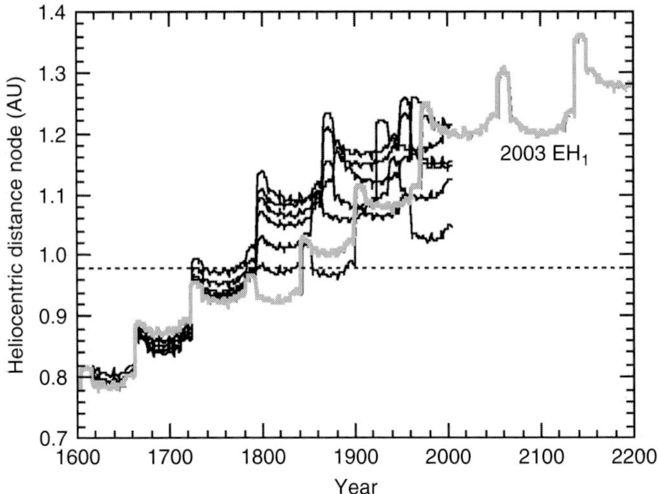

Fig. 20.12 The evolution of 2003 EH$_1$ and meteoroids ejected in 1600 over time.

Quadrantid shower at the present time. In particular, the perihelion distance stretches shortward of the present position of the comet. Williams and Wu[34] documented this spreading previously (Fig. 20.4). By calculating the dispersion since 1600, and comparing with the observed dispersion from our photographic observations, I estimated the time of release of the particles within a few hundred years prior to 1600.

Over time, the stream shifted back and forth from inside to outside Earth's orbit. By plotting the reported peak rate versus the heliocentric distance of the node of 2003 EH$_1$ over time (Fig. 20.13), the dust density is traced in a direction along the nodes away from the Sun. My calculations do not demonstrate that the peak should be about 0.213 AU inward from 2003 EH$_1$ at the present time, but a few choice encounters with Jupiter or Earth can put the comet that far from the stream center.

Together with Fig. 20.5, this defines the distribution of dust in three dimensions. From that, I find a mass of about 1×10^{13} kg for grains in the range 10^{-6}–1000 g (plus the same amount for the "annual" background component in Fig. 20.5). This is 10–100 times more than earlier estimates by David Hughes, Neil McBride,[35] and by myself because we now know how far the stream is spread out in perihelion distance.

The mass in the stream is significantly more dust than lost from a typical Jupiter-family comet in a single return ($\sim 10^{10}$ kg), and implies either a deposition for a period of $\sim$1000 years, or debris from a recent comet breakup. The long-term deposition of dust in a rapidly evolving orbit is not consistent with the narrow radiant dispersion seen today. Hence, some form of breakup is implied a few hundred years prior to AD

[34] I. P. Williams and Z. D. Wu, The Quadrantid Meteoroid Stream and Comet 1491I. *MNRAS* **264** (1993), 659–664.

[35] D. W. Hughes and N. McBride, The mass of meteoroid streams. *MNRAS* **240** (1989), 73–79; D. W. Hughes, The mass distribution of comets and meteoroid streams and the shower/sporadic ratio in the incident visual meteoroid flux. *MNRAS* **245** (1990), 198–203.

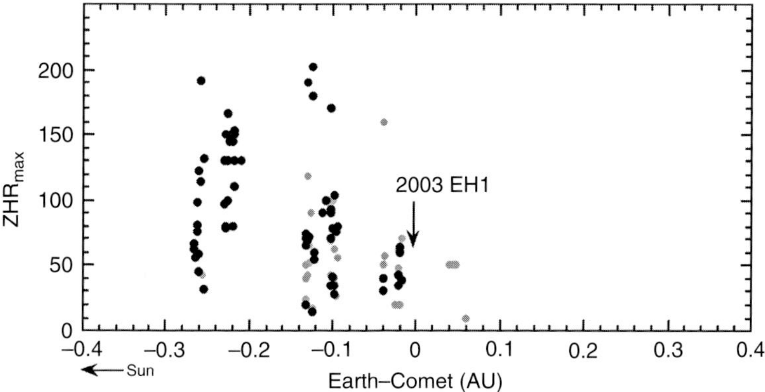

Fig. 20.13 Reported peak Quadrantid rates, as a function of the heliocentric distance of the node of 2003 EH_1. Most dust is located inside the present orbit of 2003 EH_1.

1600. 2003 EH_1 could be a remnant representing less than or equal to 38% of the original mass. Its diameter is $D \sim 2.9$ km. There could be more such fragments.

The node of 2003 EH_1 also shifts as observed for the Quadrantid shower (Fig. 20.6). Only if the meteoroids librate about the 2:1 mean-motion resonance with Jupiter does the stream as a whole avoid such close encounters and maintain its narrow structure. In that case, however, the annual shift of the node reverses sign and becomes positive. The observed nodal displacement of the Quadrantid shower is negative, identical to that of 2003 EH_1 (Fig. 20.6). This implies that the meteoroids were ejected with small enough velocities to not have had the time and the energy to get trapped in the 2:1 resonance.

The Quadrantid meteoroids are cometary in nature, given that they appear to be fragile with numerous flares from the sudden release of small fragments and a relatively high penetration depth in Earth's atmosphere (Fig. 20.14). The meteors end at altitudes similar to those of the Perseids (from 109P/Swift–Tuttle) and the Lyrids (from C/1861 G_1 Thatcher) and do not penetrate as deeply as the higher density Geminid meteoroids, cometary dust that has been sintered in a low-q orbit and is thought to be more representative of asteroidal dust. Hence, 2003 EH_1 is a comet.

20.3 Association with Comet C/1490 Y_1?

Is 2003 EH_1 the same object as that seen as a bright comet C/1490 Y_1 from China, Korea and Japan between December 31.5, 1490 and February 12.5, 1491? According to Ichiro Hasegawa,[36] comet C/1490 Y_1 passed perihelion on January 8, 1491, when Earth was near the node. The Chinese records relate how on December 31.5 UT (1490), a *hui* (comet) appeared at the south of *Thien-Chin* with its tail pointing NE. It

[36] I. Hasegawa, Orbits of ancient and Medieval comets. *Publ. Astron. Soc. Japan* **31** (1979), 257–270.

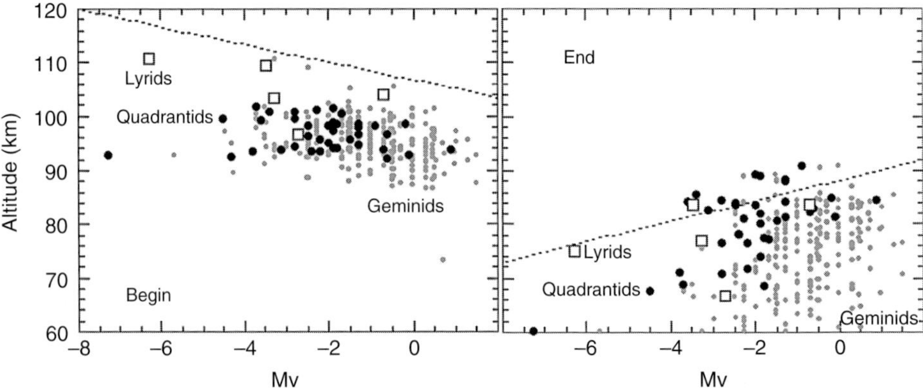

Fig. 20.14 Beginning and end heights of Quadrantids ($V_\infty = 43$ km/s), (dark dots) as a function of the meteor magnitude at 100 km distance. The results are compared to Lyrids ($V_\infty = 49$ km/s), Geminids (36 km/s), and Perseids (61 km/s, dashed line).

trespassed against *Jen-Hsing* and passed *Chhu-Chiu*. On January 10, 1491, it entered the 13th lunar mansion *Ying-Shih*. On January 22 it trespassed against *Thien-Tshang* and was seen last on January 30, when it had moved below *Thien-Tshang* and gradually faced *Tung-Pi* (Fig. 20.15).[37] In Japan,[38] the comet was first seen on January 4.4, 1491, "at the west among the *Ying-Shih*". The comet was white and had a 5° tail. It went out of sight on February 14. This information about the position of the comet on the sky traces no better than a parabolic orbit. Iwan William and Zidian Wu[39] first demonstrated that some backward integrated Quadrantids have orbital elements consistent with C/1490 Y_1 if that comet had an eccentricity of 0.77, rather than 1. Remember that Williams and Wu continued to suggest that a close encounter with Jupiter in 1650 ejected this bright comet into a very different orbit and was now lost (leaving the Quadrantid shower in place). They estimated the age of the shower at 5400 yr based on meteoroid orbits that had a large observational error.[40]

Given that AD 1491 is about the right timeframe for the origin of the Quadrantid shower, I looked into a possible link with 2003 EH_1. Although the brightness and nuclear diameter of a comet are not well related, the absolute magnitude of the comet of $H_{10} = +5.4$ implies a nucleus of about 5.8 km diameter (Eq. 6.1),[41] or a mass of about 5×10^{13} kg. This is eight times that of 2003 EH_1, and 2–5 times the mass present in the Quadrantid shower.

[37] *Ming shih*, p. 74; *Hsü Wen hsien t'ung k'ao* (18th century), (1739), p. 207; See also: G. W. Kronk (1999) *Cometography, A catalog of comets Vol. 1: ancient – 1799*. Cambridge University Press, pp. 290–291.

[38] *Nihon Temmon Shiryo* (1935), p. 586.

[39] I. P. Williams and Z. D. Wu, The Quadrantid meteoroid stream and comet 1491I. *MNRAS* **264** (1993), 659–664.

[40] Z. Wu and I. P. Williams, On the Quadrantid meteoroid stream complex. *MNRAS* **259** (1992), 617–628.

[41] D. W. Hughes (1989) Cometary absolute magnitudes, their significance and distribution. In: *Asteroids, Comets, Meteors III*. Editors C.-I. Lagerkvist, H. Rickman, B. A. Lindblad and M. Lindgren, Proc. Astron. Obs. Uppsala University, June 12–16, 1989, pp. 327–342; D. W. Hughes, The magnitude distribution and evolution of short-period comets. *MNRAS* **336** (2002), 363–372.

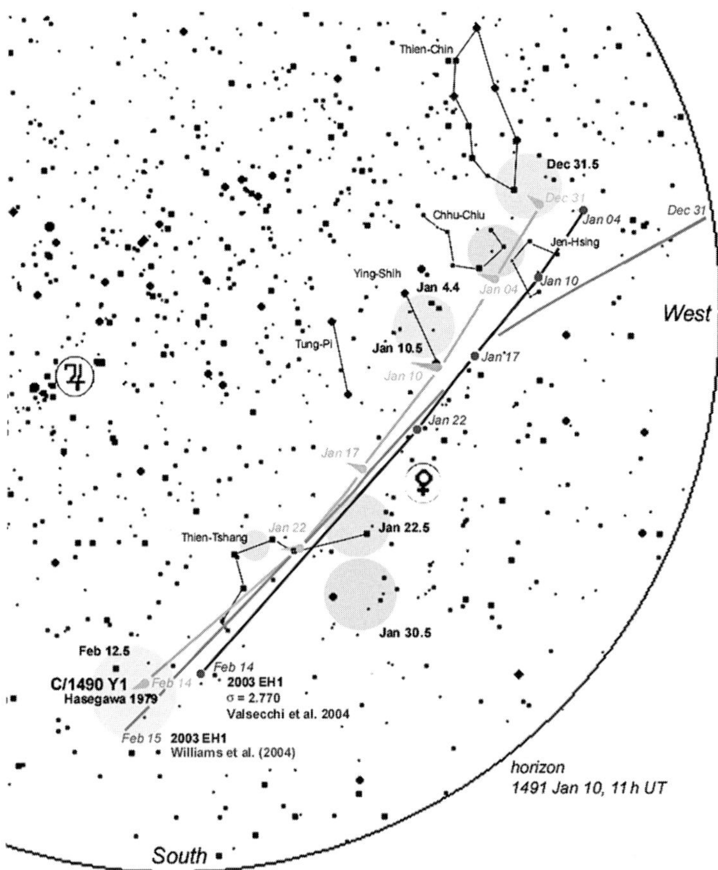

Fig. 20.15 The position of the comet C/1490 Y_1, as reported in Chinese, Japanese, and Korean observations. The best orbit by Hasegawa, and the calculated orbits of 2003 EH_1 in 1491 by Williams *et al.* (2004) and Valsecchi *et al.* are shown as well.

Efforts to find a common orbit between 2003 EH_1 and C1490 Y_1 are complicated by close encounters with Jupiter and the Earth, which can change the result dramatically for very small differences in the initial orbit. By integrating 2003 EH_1-like orbits back to 1600 and searching for perihelion times that might agree with a past perihelion in January of 1491, I found that a common orbit may exist. However, orbits for matching perihelion times tend to put the path in 1491 lower in the sky by having q and i too small.

When submitting the identification of 2003 EH_1 as the Quadrantid parent, I asked Brian Marsden of the Minor Planet Center to look into this possible link further. After various backward integrations to 1491, he arrived at the same result. Most of the potential solutions yielded $0.5 < q < 0.6$ AU, and this is probably too small to fit the data used by Hasegawa. That said, values in the more acceptable range $0.65 < q < 0.75$ AU were possible, certainly with the help of a close approach to the Earth or the presence of nongravitational forces.

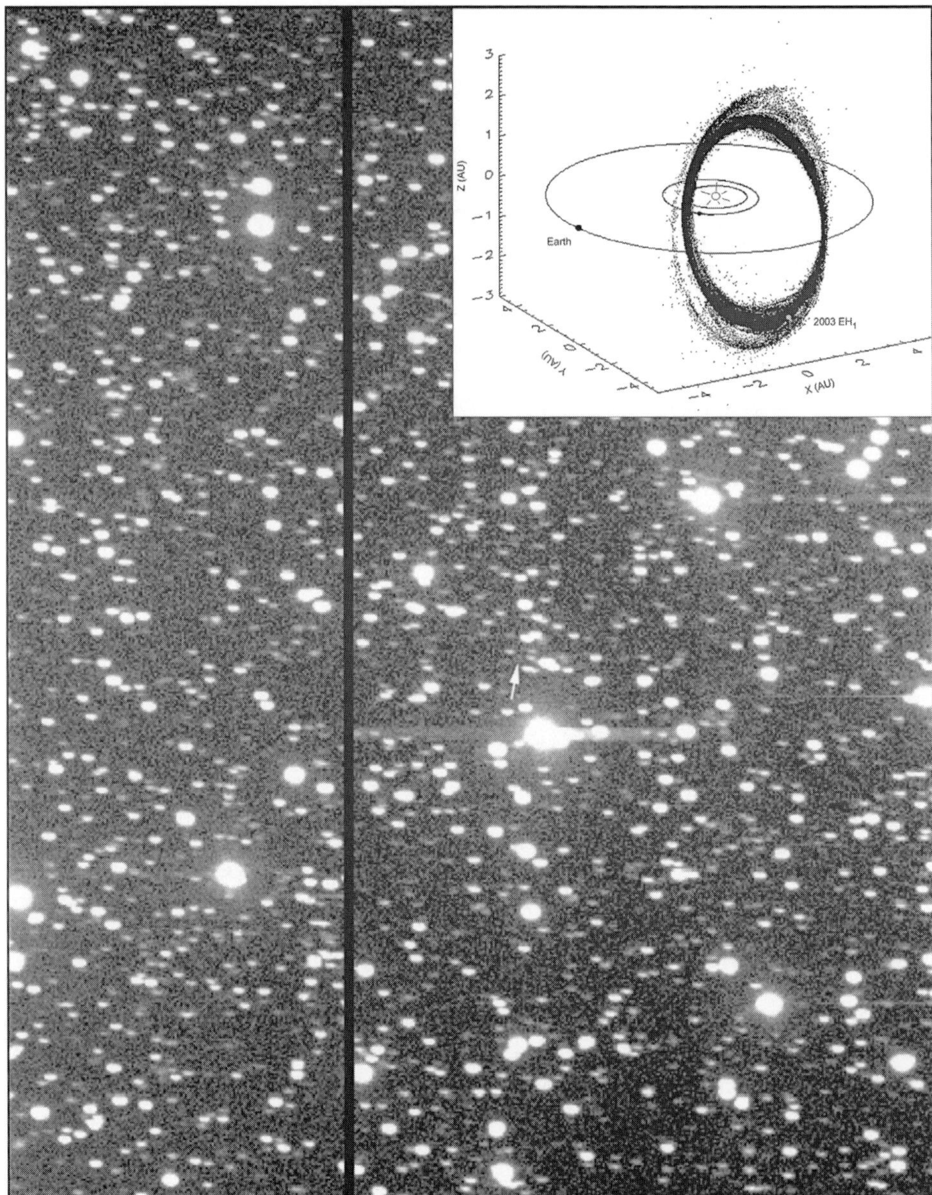

Fig. 20.16 The Quadrantid parent object 2003 EH$_1$ in the constellation of Triangulum Australe on December 24, 2003 (08:05 UT). The minor planet was only about magnitude +23 when it was recovered at the ESO *New Technology Telescope* by Emmanuel Jehin, Malvina Billeres, and the author. The inset shows a 3D model of the Quadrantid stream. Calculations by Jérémie Vaubaillon.

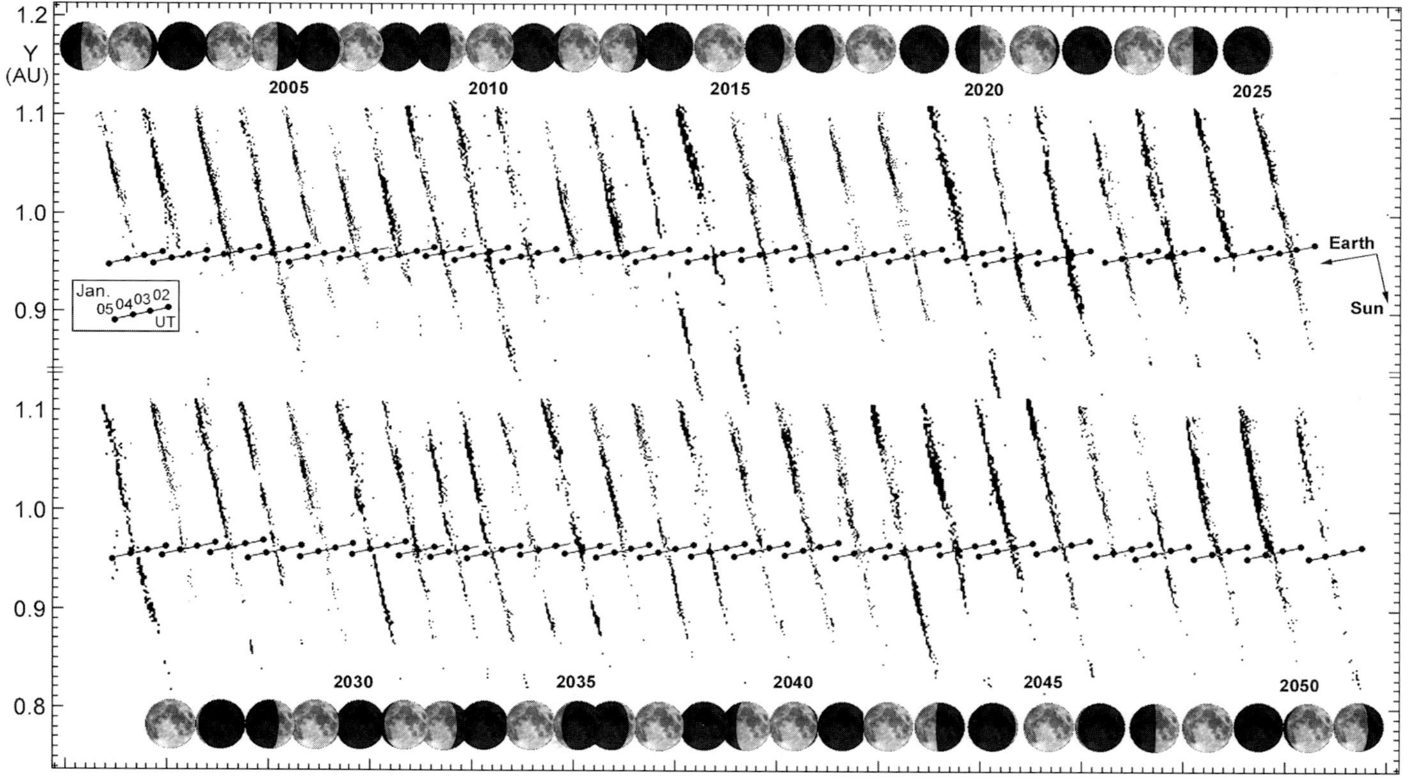

Fig. 20.17 The expected variation of dust along the nodes of the stream in the near future, calculated from ejection in the return of 2003 EH_1 in 1491. Calculations by Vaubaillon.

A more precise orbit for 2003 EH_1 was needed. I asked *Emmanuel Jehin* and *Malvina Billeres* of the European Southern Observatory to try to detect the $+22^m$ object during the one hour before dawn when the position would be high enough in the sky from locations in Chile. On December 24, 2003, the minor planet was found moving between the stars of the southern milky way (Fig. 20.16).

Follow-up observations were made in January of 2004. The resulting orbit for 2003 EH_1 was considerably more precise and decreased the range of acceptable solutions significantly. At the Meteoroids 2004 meeting in London, Ontario, Jerémié and I sat down with *Giovanni B. Valsecchi* of the Istituto di Astrofisica Spaziale e Fisica Cosmica in Rome, Italy, to study the link using Orbfit software developed by *Andrea Milani* and collaborators and designed to accomplish such dynamic associations. Giovanni found a handful of solutions within a few sigma of the current best orbit of 2003 EH_1 that have the correct perihelion time. The best solutions, with $\sigma = 2.769$ and 2.770, have perihelion times in January 18–20, 1492, slightly later than the observed orbit. Otherwise there is good agreement (Fig. 20.15). Permitting a small amount of nongravitational force existed after the breakup of the comet, this result links 2003 EH_1 to C/1490 Y_1, but not in a unique orbit.

20.4 Quadrantids in the future

Into the twenty-third century, the shower will continue to be strong, at about the same peak level as today, because 2003 EH_1 (and presumably the Quadrantid meteoroids) remain at a similar distance from the Sun. After that, the perihelion distance will decrease again.

Interestingly, the close encounters with Jupiter can lead to significant variations of peak activity from year to year and even noticeable differences in the orbital elements. One observed example may be the set of six orbits calculated by the Nippon Meteor Society in 1987. They found an average radiant slightly offset from the mean reported earlier at R.A. $= 227.7 \pm 0.5°$, Decl. $= 50.2 \pm 0.7°$ with $V_g = 41.4 \pm 0.5$ km/s. That resulted in $q = 0.982 \pm 0.001$ AU. *Katsuhito Ohtsuka* and coworkers argued that Jupiter had perturbed the observed part of the stream in August 23.7, 1984, when it passed the meteoroid stream at a mere 0.21 AU.[42]

Based on the matched orbit for 2003 EH_1 and ejection in 1491, Jérémie and I developed a preliminary stream model (Fig. 20.16) and studied the distribution of meteoroids at the node on January 3 (Fig. 20.17). Notice the strong annual variations in dust density in Earth's path. Such a model will need to be compared to observations to prove that this is correct. For various reasons the actual year-to-year rate variation may be less extreme than shown.

[42] K. Ohtsuka, M. Yoshikawa and J.-I. Watanabe, Orbital evolution of the 1987 Quadrantids. In: *Meteoroids and their parent bodies*. J. Stohl, I. P. Williams (eds.), (1992), pp. 73–76.

21

Broken comets

Whipple's ejection model *finally did away with old fashioned ideas about meteor showers originating from the breakup of comets.*[1] Or so it seemed. For many years, asteroid-looking minor planets that might be nonactive comet fragments were linked to meteoroid streams, but always leaving much discrepancy between the orbit of meteoroids and the minor planet. The one exception, when *Fred Whipple* noticed that 1983 TB was moving among the Geminids,[2] was in doubt because the minor planet looked more like an asteroid than an extinct comet nucleus. Still, with only eight objects discovered so far having $q = 0.09–0.17$ AU and $i = 19–28°$, I estimate the probability of this good a match at perhaps one in 2 million, depending on the actual distribution of minor bodies.

That has now changed. While writing this book, I discovered that minor planet 2003 EH_1 was among the Quadrantids (Chapter 20).[3] With only four high inclination prograde ($i = 60°–90°$) moving objects with $a = 2.5–3.5$ AU discovered to date, the probability of finding an object with this good a match to the Quadrantid stream is also only about 1 : 2 000 000. The narrow range of orbital elements implied a young ~500 yr age for this massive stream, tracing its history back to the sighting of comet C/1490 Y_1, at which time the stream may have been created during a breakup.

A year later, I found a second minor planet 2003 WY_{25} in the orbit of the Phoenicid meteoroid stream, which had an outburst in 1956.[4] Even though there are many more low-inclination objects discovered to date with an aphelion near Jupiter, this object matched the stream with a probability of better than 1 in about 4000. 2003 WY_{25} appears to be a ~400 m fragment of the original ~2.2 km sized nucleus of D/Banpain (see later in this chapter).

During the course of writing this book, I also identified 2004 TG_{10} (Taurids) and 2002 EX_{12} (α-Capricornids), and others as probably associated with known streams (Table 7). These parent comets and their meteoroid streams are the topic of the next chapters.

[1] I. P. Williams, Meteoroid streams: successes and problems. *WGN* **32** (2004), 11–20.
[2] F. L. Whipple, 1983 TB and the Geminid meteors. *IAU Circ.* 3881, ed. B. G. Marsden (Cambridge, MA: IAU Minor Planet Center, 1983).
[3] P. Jenniskens, 2003 EH_1 is the Quadrantid shower parent comet. *Astron. J.* **127** (2004), 3018–3022.
[4] S. Foglia, M. Micheli, P. Jenniskens and M. Marsden, Comet D/1819 W1 (Blanpain) and 2003 WY25. *IAU Circ.* 8485, ed. D. W. E. Green (Cambridge, MA: IAU Minor Planet Center, 2005).

21.1 The breakup of comets

What type of breakup could be implicated? Comets have been observed to fall apart in many different ways, from the spill-off of small boulders (e.g., C/1996 B_2 (Hyakutake)), to major fragments separating from the nucleus (73P), to catastrophic disruption and complete disintegration of the nucleus (C/1999 S_4).[5] Comets are weakly consolidated structures with binding energies only of order 0.001–0.1 J/g, as expected for a porous weakly bonded conglomerate of icy grains.[6] Possible causes include collisions with asteroids and meteoroids, thermal stresses from internal gas pressure (common for long-period comets), tidal stresses in close encounters with Jupiter and the Sun,[7] and stresses caused by comet spin-up.[8] The mechanism of separation can make an important difference in the outcome.

Comets should show the scars of such breakups, but those are not immediately obvious in the images of comets visited (Fig. 2.2). Probable scars are the large pits on comets 81P/wild 2 (center and bottom right) and 9P/Tempel 1 (top center).

Collision experiments of impact-generated breakup of icy and icy–mineral targets[9] have shown that ejection speeds can be comparable to the relative impact speed. Collisions with interplanetary boulders of $\sim 10^9$ kg ($\sim$50–100 m in size) occur typically at relative speeds as high as $\sim$41 km/s for retrograde moving comets, providing sufficient kinetic energy to break the nucleus of 1P/Halley. Larger collisions might heat the comet so much that it would explode under the pressure of the sublimating water.

For Jupiter-family comets, spin-up appears to be the only mechanism that can explain why disruptions occur at all distances from the Sun. In that case, the velocity difference of the fragments reflects the rotation rate and position of the pieces just before breakup. The lack of comet nuclei with rotation periods less than 5 h suggests that breakup typically occurs for a 5 h spinning period. For a 2 km sized comet, this means relative speeds of no more than 35 cm/s between fragments and the remaining bulk. Observed values, with few exceptions, are in the range 10 cm/s–2 m/s. For such nontidally split comets it is often found that the companions trail behind the *primary component*, eventually spreading along the orbit.[10] This is thought to be due to the secondary nuclei being smaller and more accelerated by the outgassing of water vapor once these fragments approach the Sun. Those that are accelerated most strongly also

[5] J. C. Solem, Density and size of comet Shoemaker–Levy 9 deduced from a tidal breakup model. *Nature* **370** (1995), 349–351; H. Boehnhardt, Comet splitting – observations and model scenarios. *Earth, Moon Planets* **89** (2002), 91–115.

[6] J. M. Greenberg, H. Mizutani and T. Yamamoto, A new derivation of the tensile strength of cometary nuclei: application to comet Shoemaker–Levy 9. *Astron. Astrophys* **295** (1995), L35–L38.

[7] B. J. R. Davidsson, Tidal splitting and rotational breakup of solid spheres. *Icarus* **142** (1999), 525–535.

[8] P. R. Weissman, Physical loss of long-period comets. *Astron. Astrophys.* **85** (1980), 191–196.

[9] I. Giblin, D. R. Davis and E. V. Ryan, On the collisional disruption of porous icy targets simulating Kuiper belt objects. *Icarus* **171** (2004), 487–505; J. Leliwa-Kopystynski, Impact breakup of cometary nuclei – conclusions from impact experiments. *Earth, Moon Planets* **90** (2002), 283–291.

[10] Z. Sekanina, The problem of split comets revisited. *Astron. Astrophys.* **318** (1997), L5–L8.

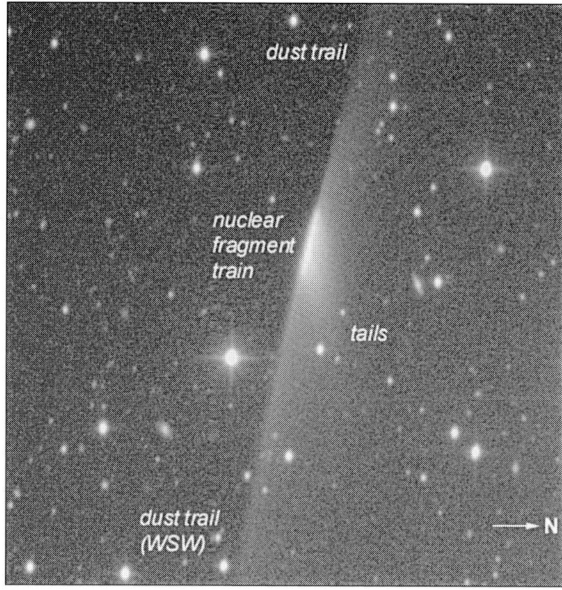

Fig. 21.1 Tidally disrupted comet D/1993 F_2 (Shoemaker–Levy 9) on March 30, 1993. According to Zdenek Sekanina, the dust trails are cm sized meteoroids and the nuclear fragment train contains greater than meter to kilometer sized fragments. Photo: J. V. Scotti, Spacewatch, LPL, University of Arizona.

tend to have less endurance. Water ice does not necessarily evaporate if the disruption occurs far enough from the Sun and water vapor drag does not need to play a role in dispersing the dust.

In tidally induced disruptions (Fig. 21.1), the fragments separate according to the separation velocities (few m/s) induced by the tidal forces. Known tidally disrupted comets were influenced by either Jupiter (e.g., D/Shoemaker–Levy 9)[11] or the Sun (sungrazers such as C/1965 S_1 Ikeya–Seki). The primary component is never the leading nucleus.

Finally, thermal stresses occur when the nucleus warms upon approach to the Sun and the resulting vapor from sublimated ices has difficulty escaping. Although this is an obvious mechanism, breakups do not occur preferentially at perihelion. Small fragments, accompanied by large dust grains, have been observed to leave the parent nucleus in cometary outbursts, perhaps driven by water vapor drag, but with low relative velocities of order 0.1–1 m/s.[12]

Larger fragments are also shed in unknown ways. A piece broke off long-period comet *C/1996 B_2 Hyakutake* on March 21.4, followed by a period of high H_2O and

[11] Z. Sekanina, P.W. Chodas and D.K. Yeomans, Tidal disruption and the appearance of periodic comet Shoemaker–Levy 9. *Astron. Astrophys.* **289** (1994), 607–636.

[12] Z. Sekanina, *Cometary Exploration*, ed. T. Gombosi. (Budapest: Hugh. Acad. Sci., 1982), p. 251.

CO gas production rates when passing Earth on March 25, 1996. Bright condensations were seen to slowly recede from the nucleus during March 22–31. Fragments were measured to be 5.4, 6, 9, 12.6, 10.8, 34.2, and 96.6 m in size, much bigger than expected from Whipple's water vapor drag theory, gradually falling behind the comet because of nongravitational forces from the preferred outgassing on the sunward side.[13]

The fresh fragments of comet D/1993 F_2 (Shoemaker–Levy 9) produced dust vigorously at a rate of 6–22 kg/s at the distance of Jupiter according to Hahn and Rettig. Millimeter-sized grains were ejected with speeds of only ~0.6 m/s, with a relatively flat size distribution, $\chi \sim 1.49 \pm 0.05$.[14] Sekanina had found earlier that the tails in Fig. 21.1 were caused by dust ejected long before the picture was taken, suggesting the comet fragments were not very active producers of dust (< 0.2 kg/s). According to him, even smaller grains were ejected with speeds <0.4 m/s, comparable to the separation speed of the secondary fragmentation. Secondary fragmentation resulted in relative velocities of up to 1.7 m/s (fragments Q_1 and Q_2 splitting at 0.4 m/s), while particle–particle collisions resulted in a velocity dispersion with a tail up to 7 m/s, in order to explain the length of the dust trails ahead of and behind the nuclear train.[15] Those dust trails formed in orbit around Jupiter not unlike the dust trails that form in orbit around the Sun. The wide fan of dust is evidence of continued fragmentation into finer dust grains, down to a few tens of microns, now widely dispersed by radiation pressure.

Breakup generates dust trails in much the same way as Whipple-type ejection, but the dust generated may differ in size frequency or relative speed from normal comet dust ejection. Some of the meteor showers associated with breakup (Andromedids, δ-Aquariids) have a very high $\chi \sim 3.0$. In those showers most of the mass is in the small meteoroids, not in boulders. Such a distribution with $\chi > 2.51$ goes beyond having equal mass per log mass interval from collisions, but instead has a lack of large grains, for example by very efficient fragmentation from the release of entrapped ices.

21.2 The Andromedids

Nontidally split comets include 3D/Biela and 73P/Schwasmmann–Wachmann 3. Comet 3D/Biela broke into at least two fragments in 1842 or 1843 that were seen as individual comets in the returns of 1846 and 1852. The larger fragment continued to shed smaller pieces. The comets were not seen again in the next returns and it was believed that the comet had continued to break apart. When this was followed by the

[13] E. Desvoivres, J. Klinger, A. C. Levasseur-Regourel *et al.*, Comet C/1996 B2 Hyakutake: observations, interpretation and modeling of the dynamics of fragments of cometary nuclei. *Mon. Not. R. Astron. Soc.* **303** (1999), 826–834; E. Desvoivres *et al.*, Modeling the dynamics of cometary fragments: application to comet C/1996 B_2 Hyakutake. *Icarus* **144** (2000), 172–181.

[14] J. M. Hahn and T. W. Rettig, Comet Shoemaker–Levy 9 dust size and velocity distributions. *Icarus* **146** (2000), 501–513.

[15] Z. Sekanina, P. W. Chodas and D. K. Yeomans, Tidal disruption and the appearance of periodic comet Shoemaker–Levy 9. *Astron. Astrophys.* **289** (1994), 607–636.

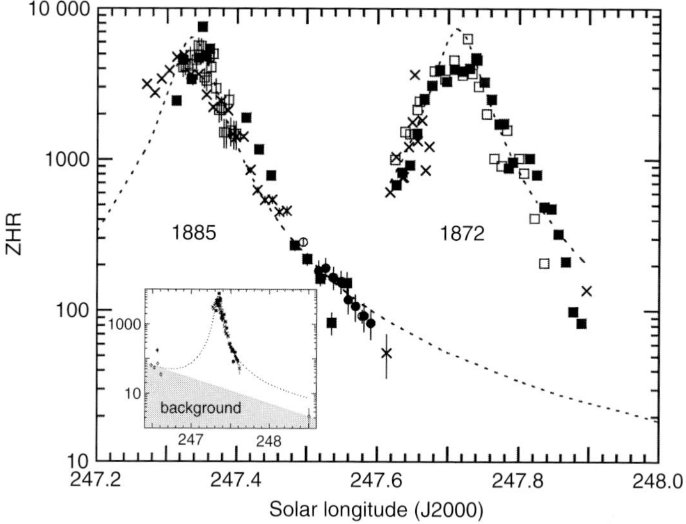

Fig. 21.2 Activity of the Andromedid shower in 1872 and 1885 from visual counts.[16]

subsequent spectacular meteor storms on November 27, 1872, and again in 1885 (Fig. 21.2), it was assumed that Earth had encountered the debris of a broken comet. Did we?

In China, local gazeteers wrote about the evening of November 27, 1885 (Dynasty *Ch'ing*, Reign: *Kuang-hsu*, Year 11, Month 10, Day 21): *At the hour* hsu *(7–9 p.m.) shooting stars fell like rain toward the west, At the second* geng *(~9 p.m.), numerous stars moved one after another, down and across like weaving. They stopped at the fourth geng (~3 a.m.), Stars fell like rain from twilight till next dawn, Stars moved like weaving, it didn't stop in the evening*, and *At night numerous stars fought in the northwest. They passed and glittered, their emissions were unusual.* And: *At night, the whole sky filled with shooting-stars flying from the north to the southeast. They tinkled and didn't cease through the night.*[17]

In the evening of December 6, 1798, student *Heinrich W. Brandes* at Göttingen University in Germany rode on top of a mail coach on his way to Bremen, when he noticed a high rate of unusually slow meteors. In the previous months, he had just completed an experiment with his fellow student Benzenberg to measure the height of meteors. Glancing in the sky he was struck by the unusual display: "I first noticed them soon after the close of evening twilight, and having no other business, I kept count of

[16] P. Jenniskens, Meteor stream activity II. Meteor outbursts. *Astron. Astrophys.* **295** (1995), 206–235.
[17] N. Nogami, Chinese local records of the 1862 Perseids and the 1885 Andromedids. *Earth, Moon Planets* **68** (1995), 435–441.

the number which appeared in the small segment of the heavens which I could, with convenience, survey from my seat." While the horses traversed the countryside, Brandes endured the shaking of the coach and only noticed a decrease in rates after 4 h, with the display ending 2 h prior to midnight. His total count was 400, in 4 h, but he suspected that many thousands of meteors must have been visible in directions that he could not observe while sitting upright.[18]

The early efforts to explain the Andromedid showers were mostly concerned with bridging the gap between the comet orbit and Earth. Several authors suggested that the showers were due to different fragments of the comet that had been ejected, with some speed, in earlier years. *Hubert A. Newton* pointed out that the nodes of 1798, 1838, and 1846 ($\Omega \sim 257.7$) were that of the comet orbit in 1772, while the node of the 1867, 1872, and 1885 returns ($\Omega \sim 247.7$) corresponded to the orbit in 1846. Hungarian astronomer *Léopold Shulhof* proposed that the Andromedids of 1798 and 1838 were from a fragment detached in 1772 that had gradually moved ahead of the comet and had gained 4 months on the comet itself by 1798 and 7 months by 1838.[19] In the same way, another fragment detached from the comet after 1840 was thought to be responsible for the 1872 and 1892 showers, while another caused the 1885 storm. As a variation on that theme, Brédikhine identified the 1877, 1885, and 1892 showers as due to "nuclear ejections" from 1846 with high speeds of 292, 342, and 279 m/s.

In 1937, it was suggested by *Jean Bosler* and *Henri Roure*[20] that Biela was split during the passage by Earth in 1832, not because of tidal forces, but because Biela passed through the Leonid stream. The impact speed of the meteoroids would have been 68.5 km/s, causing even a small meteoroid to have much kinetic energy. Although it can be shown that the distance was small, it has not yet been proven that a dust trail was in the path of the comet, at that time. Even if a dust trail was in the path of the comet, the chance of impact was small. Babadzhanov and colleagues found that a 15 g Leonid meteoroid would hit only once every hundred passages of the comet.

These theories did not consider the planetary perturbations on the individual meteoroids. In a paper written in Russian, *E. A. Reznikov*[21] identified the dust trails responsible for the 1872 storm as those of 1839 and 1846, and those responsible for the 1885 storm as ejecta from 1846 and 1852, leaving open the possibility that normal water vapor drag ejection was responsible.

Jérémie Vaubaillon and I investigated these dust trails, by calculating the evolution of all the dust trails ejected between AD 1751 and 1852. We confirmed the

18 J. F. Benzenberg and H. W. Brandes, *Sternschnuppen* (Hamburg: Friedrich Perthes, 1800), p. 80.
19 L. Schulhof, *Les Étoiles Filantes* (1894), p. 57.
20 J. Bosler and H. Roere, La cométe de Biéla et l'essaim des Léonides. *J. Observateurs* **20** (1937), 105–110.
21 E. A. Reznikov, The origin of the 1872 and 1855 Bielid meteor showers. In *Meteornoe Veshchestvo v Mezhplanetnom Prostranstve* (*Meteoric Material in Interplanetary Space*) (Moscow: Kazan, 1982), pp. 151–152.

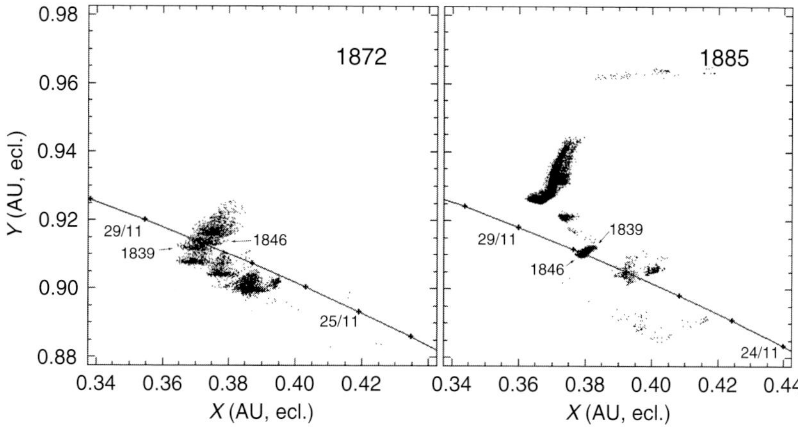

Fig. 21.3 Location of key dust trails of comet 3D/Biela in 1872 and 1885.

identification by Reznikov (Fig. 21.3). We were also able to identify some of the later Andromedid showers as a crossing of the 1846- or 1852-dust trails.

We find that all meteor outbursts can be identified with crossings of either the 1846 or 1852 ejecta when the comet was breaking up. There are no strong showers from older dust trails but few trails were in Earth's path (Table 6a). The 1832 dust trail, created just before breakup, never intersected Earth's orbit. At first glance, this confirms earlier suspicions, often quoted in the literature, that the breakup was responsible for the observed Andromedid storms. Unfortunately, this argument is weakened by the fact that few of the older trail crossings were as favorable as those of 1885 and 1872.

From the amount of dust alone, it is not obvious that the stream was created in a breakup. Based on the storm profiles, the total mass of particles in the range 1 millionth to 100 g is about 30 billion kg. From the nongravitational changes in the orbit of the comet, some of the largest on record, *Fred Whipple* calculated the loss rate of all matter to be $\Delta M/M = 0.0047$ per orbit, which translates to about 48 billion kg/orbit for his adopted upper limit to the size of 3D/Biela (3.4 km, $\rho_c = 0.5$ g/cm^3)[22] and 19.2 billion kg/orbit for the size of 2.52 km adopted by Hughes.[23] My own diameter estimate is in the middle, $D = 3.0$ km, giving a normal mass loss of 33 billion kg/orbit. In good agreement.

Compare that to the total amount of dust generated in a single return of comet 55P/Tempel–Tuttle, by Whipple-type ejection, which is about 13 billion kg. For comet

[22] F. L. Whipple, A comet model. II. Physical relations for comets and meteors. *Astrophys. J.* **113** (1951), 464–474.

[23] P. B. Babadzhanov, Z. Wu, I. P. Williams and D. W. Hughes, The Leonids, comet Biela and Biela's associated meteoroid stream. *Mon. Not. R. Astron. Soc.* **253** (1991), 69–74.

21P/Giacobini–Zinner, it is about 6 billion kg, all of the same order of magnitude. Indeed, the peak rate during the Andromedid storms was similar to that of the 1933 and 1946 Draconid storms of equally slow meteors (ZHR ~ 10 000/h). This implies that whatever was responsible for the breakup did not produce copious amounts of dust.

In contrast, the amount of dust generated in the recent breakup of long-period comet C/1994 S_4 was estimated at 400 billion kg (Chapter 6). In that case, 200 times more mass might have been deposited in dust than that which was left behind in large fragments.

21.3 Where is the Andromedid stream now?

The 1846 and 1852 dust trails would have continued to evolve under the influence of planetary perturbations. This evolution was investigated by Makhmudov,[24] Katasev, and Kulikova,[25] and more recently by Babadzhanov and collaborators. It was found that perturbations moved the node outside Earth's orbit after the return of 1892 (Fig. 21.4).

The gradual shift of the node gave Denning the wishful vision of having the two mythical showers on the sky at the same time: slow yellowish Andromedids and superfast greenish Leonids. That was not to be.[26] In 1899, rates increased to only 90/h in Vienna, ten days after the Leonids. The last to witness the Earth's passage through a dust trail of now "D" comet 3D/Biela was experienced AMS observer *R. M. Dole* of Cape Elizabeth, Maine.[27] In the 1920s and 1930s Dole published regularly, in *Popular Astronomy*, some papers with Charlies Olivier, including one on the Andromedid observations of 1923.[28] In the evening of November 15, 1940, when Europe was occupied by war, Dole observed a peak of faint Andromedid meteors of 30/h. In research for this book, Jérémie and I discovered the cause: dust from the last observed orbit of comet Biela in 1852 was in Earth's path that year.

A hundred years after the breakup of the comet and nearly 12 yr after the Andromedids were last spotted, *Gerald Hawkins*, *Richard Southworth*, and *Francis Steinon* identified a weak level of Andromedid activity among photographed meteors captured by Harvard Super-Schmidt cameras in 1952 and 1953. By that time, the center of the stream had moved to about November 15. It evolved into October at the turn of the millennium. At the same time, the center of the stream has moved away from Earth's orbit.[29]

[24] N. Makhmudov, Evolution of the orbits of the Andromedids meteoroid shower. *Akademiia Nauk Tadzhikskoi SSR, Dokl.* **25** (1982), 527–529 (in Russian).

[25] L. A. Katasev and N. V. Kulikova, Physical and mathematical modeling of the formation and evolution of meteor streams. *Solar Syst. Res.* **14** (1981), 133–139, 179–183.

[26] W. F. Denning, The meteors from Biela's comet. *Mon. Not. R. Astron. Soc.* **65** (1905), 851–855.

[27] J. P. M. Prentice, *J. Brit. Astron. Assoc.* **51** (1941), 92.

[28] R. M. Dole, The Andromedids, 1923. *Popular Astron.* **32** (1924), 389.

[29] G. S. Hawkins, R. B. Southworth and F. Steinon, Recovery of the Andromedids. *Astron. J.* **64** (1959), 183–188.

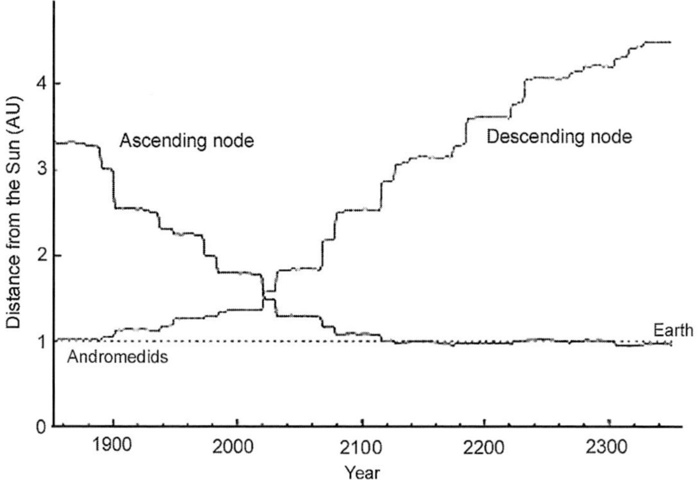

Fig. 21.4 The heliocentric distance of the nodes of the Andromedid stream over time. Calculations by Babazhanov, Wu, Williams, and Hughes (1991).[30]

That separation will continue to increase. Babadzhanov, Wu, Williams, and Hughes investigated the future evolution of comet Biela (Fig. 21.4), to find that the stream will start to intersect the Earth's orbit again only in 2120, producing a new meteor shower at the ascending, rather than the descending node.[31] If a comet fragment survives, it may break again and history may repeat itself.

21.4 No longer lost comet D/1819 W_1 (Blanpain) and the Phoenicids

Comet *D/1819 W_1 (Blanpain)* was discovered on November 28, 1819, shortly after it was strongly perturbed by Jupiter in a $P = 5.08 \pm 0.5$ yr orbit. The comet was observed until January 25, but then lost, leaving an uncertainty in the orbital period of several months. Based on the nominal orbit, the node should have evolved at a rate of –0.033°/yr from 79.80° in 1819 to 70.87° in 2003 and will continue to change to 69.56° in 2060.[32] At the same time the perihelion distance increased from $q = 0.892$ in 1819 to 1.020 AU in 2039 and then back to $q = 0.961$ AU in 2060.

In January, 2005, I used this expected current orbit of comet Blanpain to search the database of recently discovered minor planets. I came across 2003 WY_{25}, which is a tiny object with $H_{10} = +21.10^m$ (400 m diameter if albedo = 0.04), discovered by the Catalina Sky Survey on November 22, 2003.[33] It passed Earth at 0.025 AU on

[30] P. B. Babadzhanov, Z. Wu, I. P. Williams and D. W. Hughes, The Leonids, comet Biela and Biela's associated meteoroid stream. *Mon. Not. R. Astron. Soc.* **253** (1991), 69–74.

[31] *Ibid.*

[32] N. A. Belyaev, L. Kresák, E. M. Pittrich and A. N. Pushkarev, *Catalogue of Short Period Comets* (Bratislava: Astron. Inst. Slovak Acad. Sci., 1986).

[33] J. Ticha, M. Tichy, M. Kocer *et al.*, 2003 WY_{25}. *Minor Planet Electronic Circ. 2003-W41* (2003).

Object:	T_P	a (AU)	q (AU)	i (°)	ω (°)	Ω (°)	Π (°)
D/1819 W_1	1819 Nov. 20.85010	2.9570	0.8924	9.1045	350.2672	79.8038	70.071
2003 WY_{25} (a)	1819 Sep. 18.40000	3.1126	0.9181	9.2973	350.4420	79.8659	70.308
2003 WY_{25} (b)	1819 Nov. 20.27	2.9929	0.8893	9.23	349.65	80.02	69.67
2003 WY_{25} (c)	1819 Nov. 19.4674	2.9928	0.8893	9.2129	349.6602	80.0124	69.673

Fig. 21.5 Orbital elements at epoch 1819 (for the ecliptic and mean equinox of J2000.0).

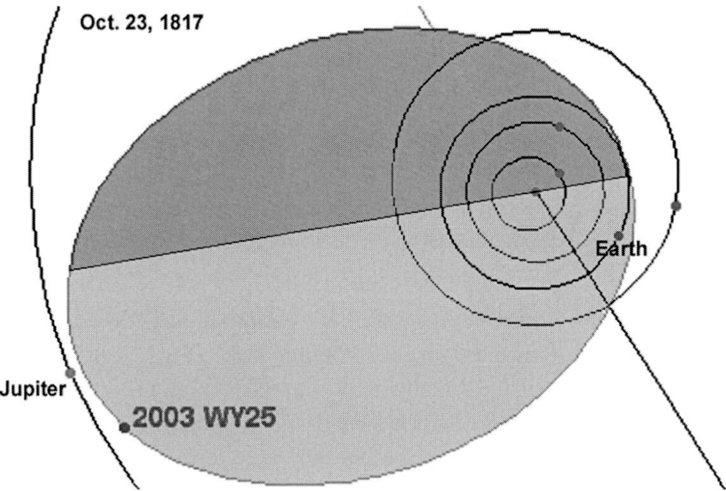

Fig. 21.6 The orbit of 2003 WY_{25}.

December 11, 2003. I integrated its present orbit back to the epoch when Blanpain was observed and found it to agree within 0.2° (Fig. 21.5, orbit a). When I informed *Brian Marsden*, he too calculated a common orbit, adjusting T by −4.28 d, and found orbit (b), which satisfactorily represents 10 of the 13 observations of the comet Blanpain made at Paris, Bologna, and Milan during December 14, 1819–January 15, 1820 within 90 arcsec. Later, *Giovanni Valsecchi* of the *Istituto di Astrofisica Spaziale e Fisica Cosmica* in Rome used *Orbfit 2.3* at my request to search for a common solution and derived orbit (c) corresponding to a $\sigma = +0.89$, in good agreement, even *without* adjusting T.

Given the uncertainty in the orbit of both Blanpain and 2003 WY_{25}, the result is very satisfying. The orbit has a node close to Jupiter's orbit and another at Earth's orbit (Fig. 21.6). Close encounters with Jupiter occurred in 1818, 1877, 1836, and 1995. The orbital period of 2003 WY_{25} in 1819 was $P = 5.178$ yr. The scenario emerging is that comet Blanpain fell apart in or shortly before 1819, and was seen as a comet that year, but none of the fragments was large enough to cause a bright comet in later years. At least one ~400 m sized fragment survived until 2003.

The announcement was made on IAU Circular 8485 (February 13, 2005).[34] Italian amateur astronomer Marco Micheli had noticed the similarity in orbits about a year earlier, before the new 2003–2004 observations, when the discrepancy was still 17°. Now, the disagreement had decreased to only 0.2°, perhaps as little as 0.01°. According to Marsden: *So, whether 2003 WY_{25} is D/1819 W_1 or a fragment of it (which maybe separated even before 1819), I think we can indeed say that a close connection has been proven.*

David Jewitt[35] of the University of Hawai'i, known for his discovery of the Kuiper Belt Objects, subsequently reported that 160 m sized 2003 WY_{25} was in fact the smallest known active cometary nucleus yet observed. On an image he had taken when the comet was 1.637 AU from the Sun back on March 20, 2004, it had a distinct tail. At the time, the comet produced dust at a rate of a mere 0.01 kg/s. At this rate, it would take an implaucible long time (300 000 years) to supply enough dust to account for the Phoenicid stream. The cometary nature of 2003 WY_{25} has now been established both from the presence of a meteoroid stream and from ongoing activity. Just as with the Quadrantids, the Phoenicids must have been created during a fragmentation event that also left 2003 WY_{25} in its wake.

Phoenicids

When Blanpain broke up, a meteoroid stream was created that was seen on Earth a century and a half later in 1956. At its postperihelion arc, the stream caused the *December Phoenicid* shower. The link between the shower and comet Blanpain was recognized by *Harold B. Ridley* in 1957, later director of the *BAA Meteor Section*, although he was cautious about this suggested link.[36] The meteors are slow, with a mean magnitude of $+2.8 \pm 0.4$ ($\chi = 2.9 \pm 0.7$), and the brighter -1 to -3^m meteors leave distinct wakes. The Phoenicid shower was discovered by *V. Williams* in Sydney, Australia, on December 3, 1887, when he observed nearly 1 meteor/min for a brief period of time ($\lambda_\odot = 252.6 \pm 0.1°$) from an apparent radiant R.A. = 25°, Decl. = −54°. Correction for zenith attraction could change the declination to −58°. The meteors were of medium brightness and displayed long yellow streaks. It was seen again by *Captain Murray* on December 5, 1938 ($\lambda_\odot = 253.6 \pm 0.1°$), who noticed a large number of meteors emanating from near the star Achernar (α-Eridani).

The next reported outburst in 1956 (Fig. 21.7) was seen by many observers in both Australia and South Africa at about the same rate (ZHR ~ 50/h),[37] suggesting a broad

34 S. Foglia, M. Micheli, P. Jenniskens and B. G. Marsden, Comet D/1819 W1 (Blanpain) and 2003 WY25. *IAU Circ.* 8485 (Feb. 13), ed. D. W. Green (2005).

35 D. Jewitt, Comet D/1819 W_1 (Blanpain): not dead yet. *Astron. J.* **131** (2006) in press.

36 H. B. Ridley, *Circ. Brit. Astron. Assoc.*, No. 382 (Feb. 07) (1957); C. A. Shain, A remarkable southern meteor shower. *Observatory* **77** (1957), 27–28; H. B. Ridley, The Phoenicid meteor shower of 1956 December 5. *Mon. Not. Astron. Soc. Southern Africa* **22** (1963), 42–49; H. B. Ridley, The Phoenicid meteor shower of 1956 December 5. *J. Brit. Astron. Assoc.* **72** (1962), 266–272; M. Huruhata and J. Nakamura, Meteoric shower observed on December 5, 1956 in the Indian Ocean. *Tokyo Astron. Bull.* (2) **99** (1958), 1053–1054.

37 P. Jenniskens, Meteor stream activity II. Meteor outbursts. *Astron. Astrophys.* **295** (1995), 206–235; A. McBeath, A re-evaluation of the Phoenicid outburst in December 1956. *WGN* **31** (2003), 148–152.

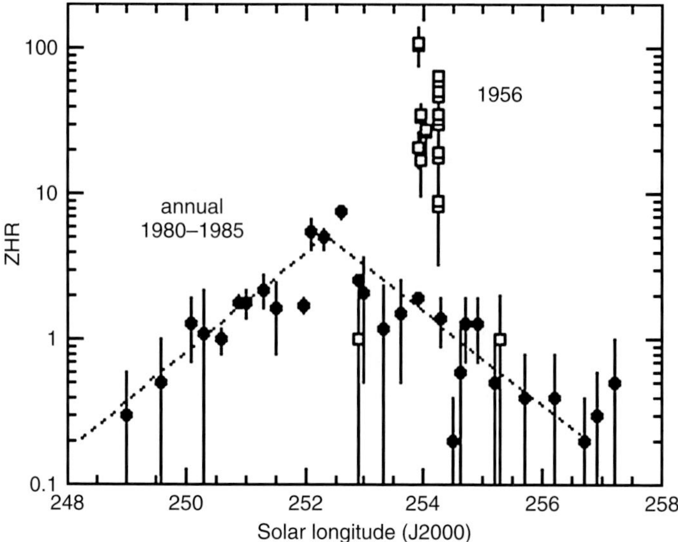

Fig. 21.7 The 1956 December Phoenicid outburst (open symbols), associated with D/1819 W_1 Blanpain. Annual Phoenicids measured by NAPO-MS in 1980–1985.

maximum around $\lambda_\odot = 254.14 \pm 0.15°$. Visual observers described the meteors as slow, with a wide range in brightness, long trails on the sky, and mostly yellow, orange, or red of color. Some left trains of several seconds duration. At Adelaide, *Alan A. Weiss*[38] detected bright radio meteors from a radiant at R.A. $= 15 \pm 2°$ and Decl. $= -58 \pm 3°$, in a brief observing interval about 6 h prior to the visual observations. This is one of the more memorable results from the radar meteor shower studies at Adelaide, led by *Graham Elford* since 1950.[39]

The Phoenicid shower was observed semiregularly since that time, and was observed to peak at ZHR ~ 20/h by *M. J. Buhagiar* on December 4/5 1972 from R.A. = 26°, Decl. = −57° near Achernar, but it is not clear if a sharp maximum was seen ($\lambda_\odot = 252.9 \pm 0.3°$). A low level of more extended activity was observed for a period December 1–8.[40] The planetary perturbations with Jupiter caused the orbital node to shift to smaller values, as a result of which the annual shower now peaks earlier than the 1956 Phoenicid outburst. A weak level of extended activity was measured by WAMS observers in 1980, stretching from November 30 to December 9 (Fig. 21.7). These meteors were faint, with $\langle m \rangle = 3.14$ and $\chi = 2.9$.

Fig. 21.8 shows the 1819 dust trail from the nominal orbit of Blanpain. We find that meteoroids are in Earth's path in December, 1951 and December, 1956, but not at the

[38] A. A. Weiss, The 1956 Phoenicid meteor shower. *Austr. J. Phys.* **11** (1958), 113–117; A. A. Weiss, Radio-echo observations of southern hemisphere meteor shower activity from 1956 December to 1958 August. *Mon. Not. R. Astron. Soc.* **120** (1960), 387–403; Annonymous, The Phoenicids. *Mon. Not. Astron. Soc. Southern Africa* **16** (1957), 2.

[39] P. Brown and M. Currie, Dr. W. G. Elford. *WGN* **21** (1993), 39–44.

[40] J. Wood, The Phoenicid meteor stream. *WAMS Bull.* **157** (1980), 1–3; J. Wood, De Phoenicidenzwerm. *WGN* **9** (1981), 9–12; J. Wood, Australia Phoenicids 1985. *WGN* **14** (1986), 186.

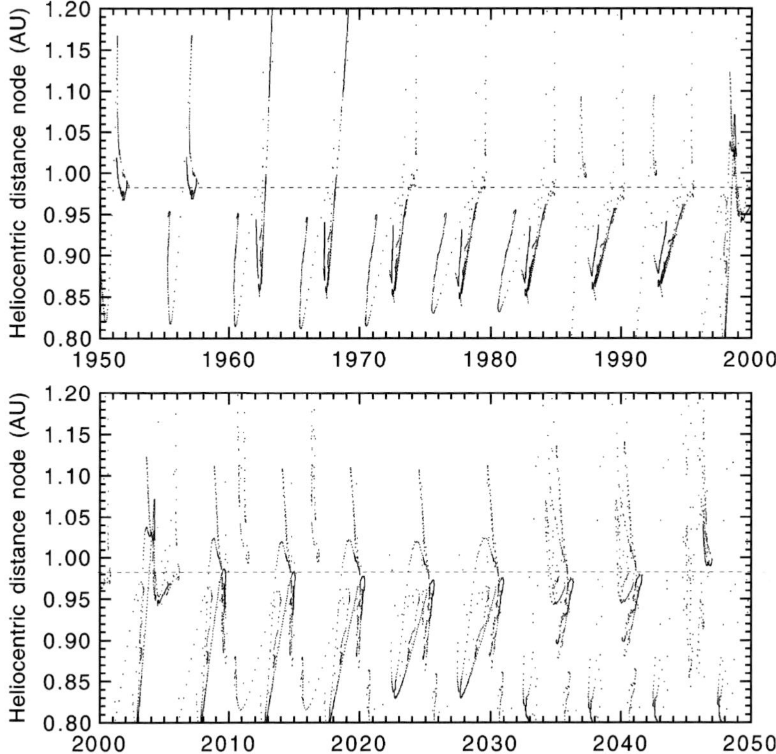

Fig. 21.8 The 1819-dust trail of comet C/1819 W_1 (Blanpain) over the beta-range of $\beta = 0.002$–0.003, bringing particles back at each return of the comet. Calculations by Esko Lyytinen.

right time when Earth is at the node in other years. The observed radiant position is close to the theoretical position (R.A. = 3.7°, Decl. = −42.1°, $V_g = 10.10$ km/s), with a predicted peak at solar longitude 254.13° on December 6/7, 1956, in good agreement with the observed outburst. Perhaps some meteors may be seen in December, 2005, but conditions are less favorable than in 1956. There are no other encounters with dense parts of this trail, at least until 2050.

μ-Pegasids

Five meteors with nearly identical radiant and speed were reportedly photographed in a two hour span between 04:15 and 06:15 UT on November 12, 1952,[41] during the Harvard Photographic Meteor Survey. Routine year-round observations were made with the two Baker Super-Schmidt meteor cameras at Doña Ana and Soledad

[41] R. E. McCrosky and A. Posen, New photographic meteor showers. *Astron. J.* **64** (1959), 25–27.

Fig. 21.9 The μ-Pegasid of 04:33:20 UT HV5370 (#1) with the original notes by the reducers.

Canyon[42] in New Mexico in the period from February, 1952 until July, 1954. Data reduction was a gigantic task and only 413 orbits have been reduced precisely.[43] μ-Pegasid HV5370 is one of those precise orbits (Fig. 21.9). The four remaining μ-Pegasid orbits were derived by a less precise graphical method, part of 2529 reduced meteor orbits by *Richard E. McCrosky* and *Anette Posen*.[44] The orbits are identical enough, and the event of brief enough duration, to suggest that Earth crossed a dust trail. Nevertheless, I was puzzled by the relatively large dispersion of the radiant ($\pm 3.5°$ in R.A., $\pm 1.5°$ in Decl.) and atmospheric speed (14.8–17.6 km/s). The beginning and end heights were also in a wide range. I suspected some problem with the data reduction and *Hans Betlem* and I decided to investigate.

On request, Dick McCrosky made available to us the original curved plates and several flat plate reproductions that were prepared from the curved Baker Super-Schmidt plates. These plates cover one time interval, and should contain four of the five μ-Pegasids. The plates at both stations were exposed from 04:33 to 04:45 UT on November 12, 1952. We found that the Doña Ana plate has three meteors radiating from near β-Pegasis. The Soledad Canyon plate has two such meteors, only one of which is multi-station (HV5370). An observer "EDM" at Doña Ana

[42] Doña Ana is at: 106° 47′ 58.50″W, +32° 30′ 21.94″N, 1412.3 m, while Soledad Canyon is at 106° 36′ 42.32″W, + 32° 18′ 13.61″N, 1567.4 m.

[43] L. G. Jacchia and F. L. Whipple, Precision Orbits of 413 Photographic Meteors. *Smithsonian Contrib. Astrophys.* **4** (1961), 97–129.

[44] R. E. McCrosky and A. Posen, Orbital elements of photographic meteors. *Smithsonian Contrib. Astrophys.* **4** (1961), 15–84.

wrote down *04:33:20 UT: a white – 1 magnitude meteor that moved with moderate speed a track of 6 degree long at a location 2 degree north of Jupiter and in a direction coming from the west.* The meteor was also noticed by an observer "IDD" at Soledad Canyon at 04:33 UT: *magnitude +3, yellow, medium fast, 12° long.* We reduced this complete dataset for HV5370 with FIRBAL[45] and derived an orbit in good agreement with the Harvard result (Table 7). The low beginning height is confirmed. The zenith angle was good: cos $(z_r) = 0.88$, as was the convergence angle: 24.2°.

We also investigated the original curved Baker Super-Schmidt plates. The plate was photographed by putting a light above the curved plate and photographing it with a macro lens from below. The third-order calculations of the Turner software resulted in the same accuracy as for our small camera work: 30 arcsec. There are no problems with projection. 30 stars were measured. The good agreement with the Harvard precise reduction is validation of our reduction method. We notice that the Baker Super-Schmidt cameras do not produce more accurate results than the current Dutch Meteor Society small-camera network. The larger field of view (60° × 20°) and poorer image quality counterbalance the gain from the larger film format.

The other meteors are not μ-Pegasids. Only one other meteor was seen by the visual observers: at 4:42:16 UT, which was white and medium fast, had a burst, was 3° long and came from the east (EDM). We were able to reproduce the solution for HV5367 by combining two unrelated trajectories with an assumed time, adjusted to match beginning and end heights (4:39:20 UT). For that incorrect solution, beginning and end heights are now much higher (88.7 and 77.8 km, respectively), the convergence angle much larger (72.8°) and the radiant differs by 7° in R. A. and 0.3° in declination.

Assuming that no other plates were involved, we conclude that three of the four μ-Pegasids in the sample by McCrosky and Posen are not multi-station, but a combination of unrelated meteor trails. The reality of the outburst is very much in doubt. Blanpain was implicated as the parent comet by McCrosky and Posen,[46] but that is now perhaps a moot point. The shape of the orbit is the same, but the node and argument of perihelion need to be rotated to intersect Earth's orbit. Esko and I checked to see if that was possible, but the dust trails of Blanpain do not seem to evolve in the necessary manner.

21.5 Comet 73P/Schwassmann–Wachmann: τ-Herculids

Comet 73P/ Schwassmann–Wachmann is perhaps the modern day 3D/Biela. The comet broke in the return of 1995, increasing by six magnitudes in brightness and leaving a trail

[45] H. Betlem, C. R. Ter Kuile, M. de Lignie *et al.*, Precisely reduced meteor orbits of the Dutch Meteor Society – photographic Survey (1981–1993). *Astron. Astrophys. Suppl. Ser.* **128** (1998), 179–185.

[46] R. E. McCrosky and A. Posen, New photographic meteor showers. *Astron. J.* **64** (1959), 25–27.

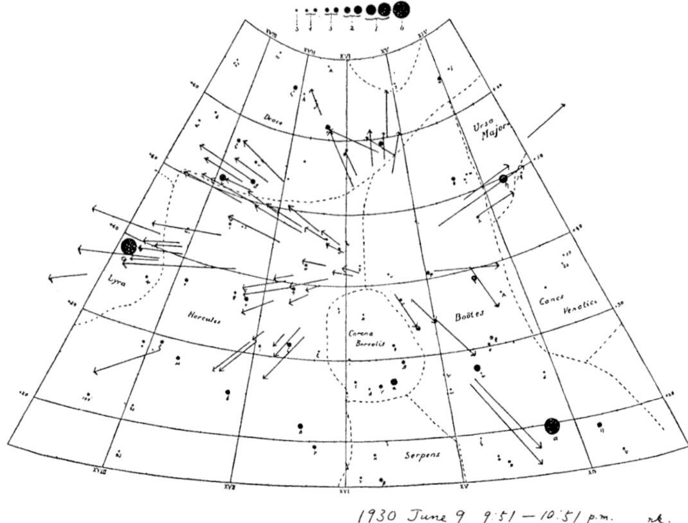

Fig. 21.10 Plots of τ-Herculids by Kaname Nakamura (Kwasan Observatory, Japan) in the night of June 8 (12:51–13:51 UT), 1930. From: *Mon. Not. R. Astron. Soc.*[47]

of fragments. *Hartwig Lüthen*, *Rainer Arlt*, and *Michael Jäger*[48] were the first to point out that the dust trail from this breakup will be encountered by Earth in 2022 on May 31.205 UT at a 'miss-distance' of only 0.0004 AU! Get ready for a nice storm. The moon will be new at that time, and the expected storm is well positioned for the USA (Table 6g).

After receiving the telegram of the discovery of the comet in May of 1930, observers at Kwasan Observatory (Kyoto, Japan) calculated the theoretical radiant and started watching the skies on May 21 in search of meteors. On May 24, they noticed a stationary meteor at R.A. = 230°, Decl. = +48°. On May 25, several more meteors were observed to radiate from a position less than 1° away from this.[49] After a period of no observations, further such τ-Herculids were observed on June 3 from (232°, +46°). Now, newspapers around the world reported that a strong shower might occur in early June. Weak activity continued on June 6 and 7. Bright Moonlight interfered at this time.

It was only well-known observer *Kaname Nakamura* of Kwasan Observatory who reported seeing 59 faint meteors in the time span of one hour in the evening of June 9 (Fig. 21.10). Only 12 were as bright or brighter than magnitude +4. The radiant was found at R.A. = 236°, Decl. = +41.5°. In the next evening, 36 more meteors were seen in a 30 min interval, only three of +4 and brighter. Observers at other locations in the world did not see anything unusual. As Nakamura recalls: *It is true that the present display was*

[47] K. Nakamura, On the observation of faint meteors, as experienced in the case of those from the orbit of comet Schwassmann–Wachmann, 1930 d. *Mon. Not. R. Astron. Soc.* **91** (1930), 204–209.

[48] H. Lüthen, R. Arlt and M. Jäger, The disintegrating comet 73P/Schwassmann–Wachmann 3 and its meteors. *WGN* **29** (2001), 15–28.

[49] K. Nakamura, On the observation of faint meteors, as experienced in the case of those from the orbit of comet Schwassmann–Wachmann, 1930 d. *Mon. Not. R. Astron. Soc.* **91** (1930), 204–209.

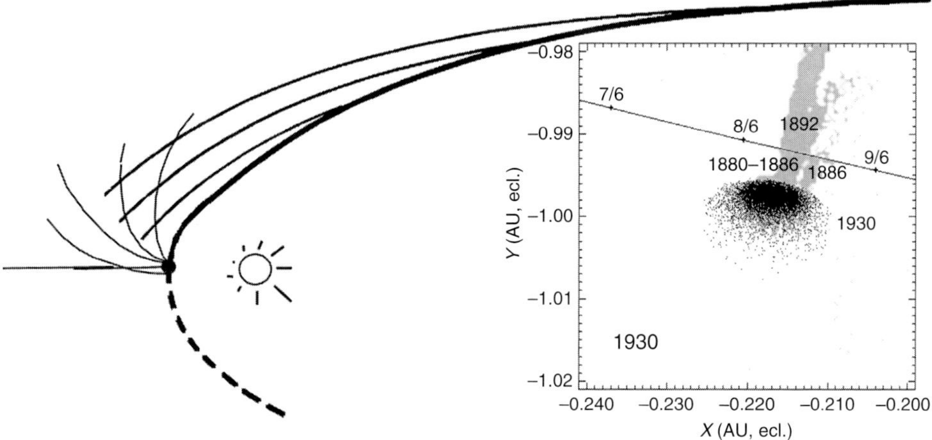

Fig. 21.11 73P/SW-3 dust ejected in 1930 was located just outside Earth's orbit at the time of the reported 1930 outburst. This is a zero-revolution trail, or better the large-particle comet tail. Left: The orbit of small meteoroids released on approach. The gray shaded area shows the position of older dust trails as identified by Jérémie Vaubaillon and Paul Wiegert.[50]

poor for the world-wide expectations suggested by the daily press. After that, rates appeared to have decreased. Weak activity continued to be seen by an independent observer in Japan on June 12 and 13 (237.5°, +41°), and the last τ-Herculid was seen on June 19.

This observation has not been confirmed. Nakamura used an unusual observing technique on the nights of June 9 and 10, in which he focused intensely on faint meteors close to the radiant and plotted all impressions of meteors. This certainly changed his detection probability function and it is hard to estimate a zenith hourly rate from the reported counts.

The shower was not so rich in faint meteors because of dust ejected during the 1930 return itself. Perhaps someday we may see such an exceptional shower from a zero-revolution trail (Fig. 21.11). We now know that the zero-revolution dust trail was relatively far from Earth's orbit at that time. Fig. 21.11 shows the position of the meteoroids calculated by Vaubaillon. The large beta particles should have been about 0.005 AU outside Earth's orbit at that time. Instead older trails are responsible. *Mikiya Sato* found a segment of the 1925 dust trail at the correct solar longitude (Table 6g). Recently, Vaubaillon and Wiegert identified also dust from 1892 and 1880–1886 in Earth path at the right time.[51]

In 1941, there was one encounter with the 1930-dust trail at a relatively large miss-distance (first recognized by *Hartwig Lüthen*). In later years, dispersed low level annual activity has been identified among photographic orbits, with a radiant at

[50] P. A. Wiegert, P. G. Brown, J. Vaubaillon and H. Schijns, The tau-Herculid meteor shower and Comet 73P/Schwassmann–Wachmann 3. *Mon. Not. R. Astron. Soc.* **361** (2005), 638–644.
[51] *Ibid.*

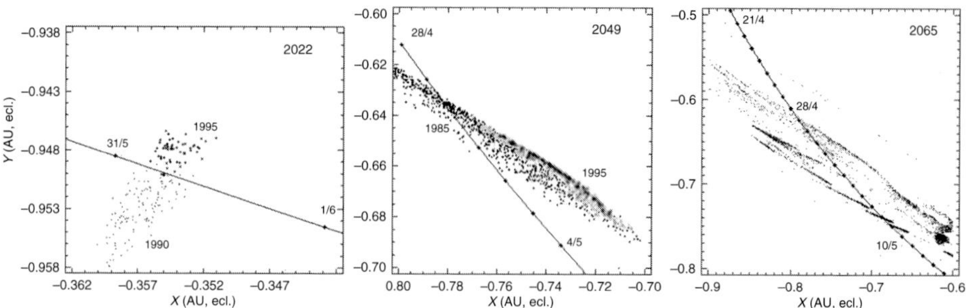

Fig. 21.12 Left: the 1990- and 1995-dust trails of comet 73P/Schwassmann–Wachmann 3 in the year 2022. Middle: the 1985-dust trail of 73P/SW-3 as encountered in 2049. Right: The dust trail of 73P/ Schwassmann–Wachmann 3 in 2065.

(228°, +39°) on June 4, moving at +0.9° and −0.1°/d, $V_g = 15$ km/s, over the period May 19–June 14. *Richard B. Southworth* and *Gerald S. Hawkins*[52] identified two meteors in 1963 with good similarity to the orbit of 73P/Schwassmann–Wachmann, while *Bertil Lindblad* found 14 associated meteors in the Harvard Meteor Project.[53] Among annual showers possibly associated with 73P/S-W 3 are the *May Bootids* (not included in Table 7). This possible shower was observed on May 30, 1970, in Japan, and observed again on May 23, 1998 (R.A. = 214°, Decl. = +18°), and at a rate of ZHR = 2/h on May 23, 2000 (R.A. = 213°, Decl. = +17°).

Sato has predicted returns in 2006 and 2008. Lüthen, Arlt and Jäger[54] pointed out that especially the 2022 return is exciting to look forward to, because debris from the 1995 breakup will be very close to Earth's path (Fig. 21.12). Jérémie and I find that the 1995 debris is just outside of Earth's orbit. In case it does pass into Earth's path, a storm of slow meteors is expected on May 31st (evening of May 30th) that year.

In 2049 and 2065, we will encounter an accumulation of dust from a number of returns (Fig. 21.12). Especially in 2049, rates may go up significantly, even though the dust trails that year are some of the most distorted on record, positioned at a very shallow angle to Earth's orbit. A similar distribution of dust is expected in the return of 2065.

At that time, trails are also dispersed considerably in node, with shower activity possible for nearly two weeks. The radiant position of the dust trails will scatter over a wide region of the sky in the Bootes/Canes Venatici constellations, as pointed out by Lüthen.

Finally, *Lubor Kresák*[55] remarked that comet Schwassmann–Wachmann 3 could produce meteor showers at the opposite node from a radiant R.A. = 200°, Decl. = +10°

[52] R. B. Southworth and G. S. Hawkins, Statistics of meteor streams. *Smithsonian Contrib. Astrophys.* **7** (1963), 261.

[53] B. A. Lindblad, A stream search among 865 precise photogrphic meteor orbits. *Smithsonian. Contrib. Astrophys.* **12** (1971), 1–13.

[54] H. Lüthen, R. Arlt and M. Jäger, The disintegrating comet 73P/Schwassmann–Wachmann 3 and its meteors. *WGN* **29** (2001), 15–28.

[55] L. Kresák, Meteor storms. In *Meteoroids and Their Parent Bodies*. (Bratislava: Astron. Inst. Slovak Acad. Sci., 1992), pp. 147–156.

(my radiant) near ε-Virginis with a peak on May 8 ($\lambda_\odot = 48$, $V_g = 13.6$ km/s) in the period 2025–2072, especially in the years 2033 and 2049. I find a miss-distance of 0.008 AU between the comet orbit and the shower. This needs further study.

21.6 2004 HW and the Corvids

One final meteor outburst discussed here is that seen by Cuno Hoffmeister during his southern hemisphere campaign. He recorded 109 slow meteors in the period June 25–July 2, 1937. The outburst was broad, with a maximum on June 27 ($\lambda_\odot = 96.3°$), and $\chi = 1.9$. No activity was observed in later years.[56] The radiant was a diffuse ~15°. If the inclination had been 90° instead of 2.5°, then the shower would have lasted 7 h rather than 6 d. Hence, this is a possible crossing of a recent dust trail.

The outburst has sparked some debate as to whether an event recorded in the chronicles of *Gervase of Canterbury* on June 25, 1178, which some believe pertains to the creation of the Giordano Bruno impact crater on the Moon, could have created this shower. From that hypothesis, *J. B. Hartung* predicted returns of the Corvids in 2003 and 2006, assuming that the meteoroids move in a solar orbit with an origin at the moon in 1178.[57] This hypothesis was rejected by *Al Harris* on the contention that it is implausible that a clump of ejecta could be launched from the Moon into a heliocentric orbit with a low enough dispersion in velocity among separate pieces that it would produce a meteor shower in just one year and not others.[58]

Hoffmeister noticed "a rather striking resemblance" to the orbit of *11P/Tempel–Swift–LINEAR*, but the argument of perihelion is off (Fig. 21.13). Gary Kronk pointed out that there is better agreement with the intermittently active comet *107P/Wilson–Harrington*. Indeed, there may have been a shower associated

	Epoch	a	q	i	ω	Node	Π	R.A.	Decl.	V_g
Corvids	(1937)	(2.5)	0.93	2.5	38.8	275.7	283.3	192.5	−19.5	9.2
2004 HW	(1937)	2.697	0.983	0.92	46.83	234.050	280.88	202.4	−12.8	9.95
11P	(1938)	3.369	1.488	13.27	161.97	242.186	44.16	-.-	-.-	-.-
107P	(1936)	2.643	1.005	2.81	81.00	280.183	361.2	279.1	−26.9	8.6

Fig. 21.13 Comparison of the orbit of the Corvid shower and proposed parent bodies.

56 C. Hoffmeister, *Meteorströme* (Leipzig: Barth, 1948).
57 J. B. Hartung, Corvid meteoroids are ejecta from the Giordano Bruno impact. *J. Geophys. Res.* **98** (1993), 9141–9144.
58 A. W. Harris, Corvid meteoroids are not ejecta from the Giordano Bruno impact. *J. Geophys. Res.* **98** (1993), 9145–9149.

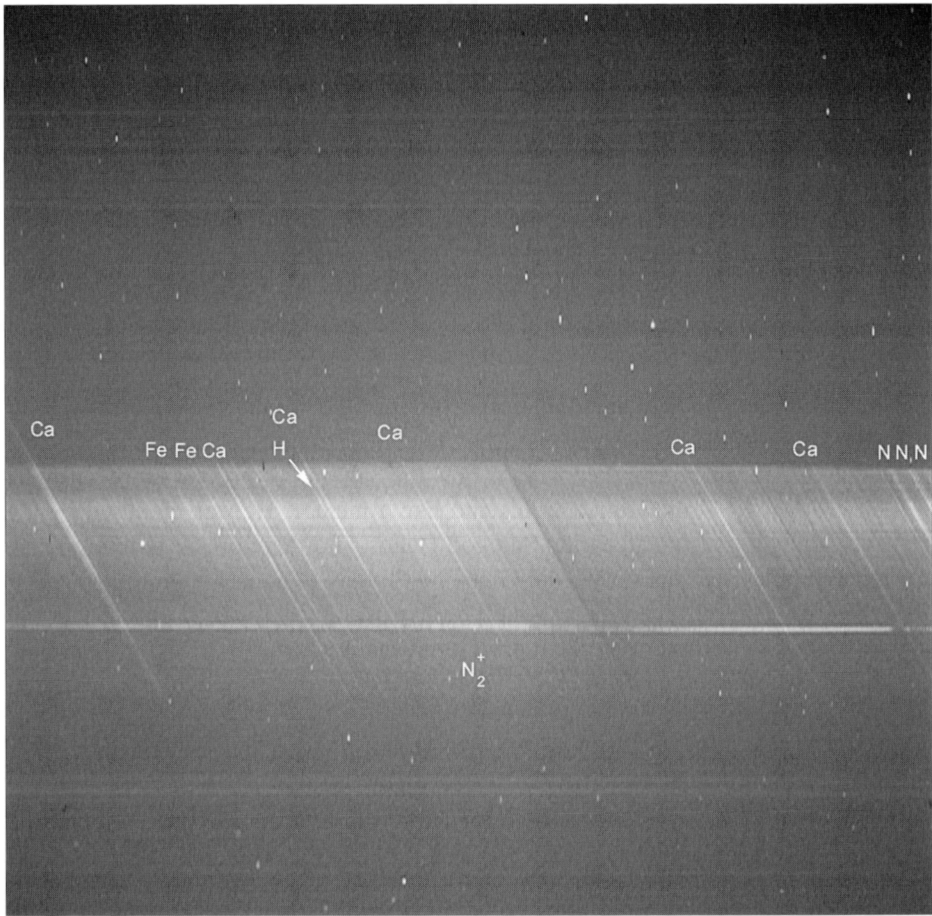

Fig. 22.1 Spectrum of the Geminid meteor of 08:41:27 UT, 2001 December 15, from Fremont Peak Observatory by Peter Jenniskens. The meteor was outside the field of view towards the left, and moved from top to bottom. Each line in the band is an image of the meteor trail emitted by the atoms identified in a specific color of light (left to right: orange to red). The sharp edge at the top of the band marks the moment the shutter was opened.

Phaethon was long an active comet. Instead, Phaethon and the Geminid stream are the likely product of a breakup, much like that which caused the Quadrantids.

22.1 The Sextantids

One clue to their cometary origin is that Phaeton is not alone. The daytime Sextantid shower, first detected at Adelaide by *A. A. Weiss* in 1957, has an orbit very similar to the Geminids. The activity profile of Fig. 22.2 was derived from the AMOR

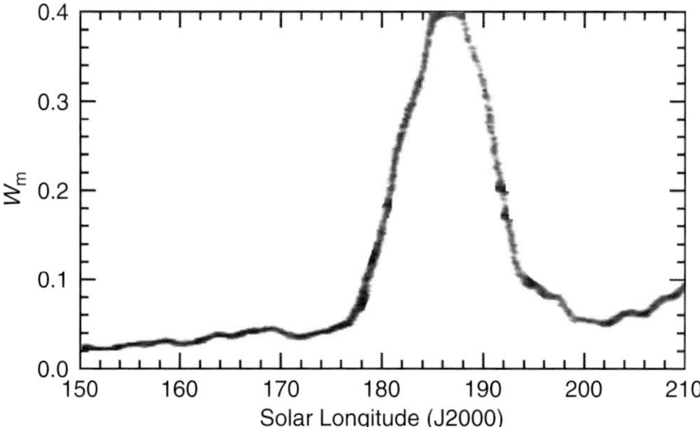

Fig. 22.2 Rate of observed Sextantids in the AMOR radar orbit database (1995–1998). Search in the helion region with a 3° wavelet, a 6° sliding solar longitude window, and $V_g = 30 \pm 10$ km/s.

radar orbit survey.[3] The shower is about one-third the intensity of the Geminids. The Sextantids peak at $\lambda_\odot = 188.35 \pm 0.10°$ with a full-width-at-half maximum of $2.0 \pm 0.2°$.

Initially, *Carl Nilsson* at Adelaide thought that the stream was the daytime manifestation of the Geminids, if the stream were wide enough.[4] However, the Geminids currently have the other node inside Mercury's orbit. *Katsuhiro Ohtsuka et al.* suggested instead that the Sextantids are at a more progressive stage of orbital evolution, with meteoroids spread between the two showers in a continuous structure.[5]

The agreement of 3200 Phaethon and Geminids is so good, that the Sextantid shower can *not* originate from 3200 Phaethon. Both showers appear to be the product of a recent fragmentation of separate minor bodies that themselves are siblings from an earlier breakup. I suspected at least one fragment would be found in a Sextantid-like orbit. Recently Ohtsuka identified ~2-km sized 2005 UD as a probable parent, now further evolved along the nutation cycle.[6] And there may be more. The distribution of inclinations of the 24 minor-bodies with $q < 0.20$ AU, $a < 3$ AU, is sharply peaked at $i = 23°$ (eight have 22°–24°) with two three-member clusters at $a \sim 1.27$ AU (Phaethon, 2001 KR_1, 2004 UL) and 1.07 AU (Icarus, Talos, 2003 BA_{21}). A large number have $\omega \sim 212°$. How the parent initially got into such a short-period orbit may have involved close encounters with Earth and Venus, and possibly also nongravitational

[3] D. P. Galligan and W. J. Baggaley, Wavelet enhancement for detecting shower structure in radar meteoroid orbit data. In *Dust in the Solar System and Other Planetary Systems*, ed. S. F. Green, I. P. Williams, J. A. M. McDonnell and N. McBride. (Amsterdam: Pergamon, 2002), p. 48–60.

[4] C. S. Nilsson, The Sextanid meteor stream. *Austral. J. Phys.* **17** (1964), 158–160.

[5] K. Ohtsuka, C. Shimoda, M. Yoshikawa and J.-I. Watanabe, Activity profile of the Sextantid meteor shower. *Earth, Moon Planets* **77** (1997), 83–91.

[6] K. Ohtsuka 2005 UD and the Daytime Sextantids. CBET 283 (2005), Minor Planet Center, Cambridge, MA (also Yamamoto Circular 2493, p. 2 (November 14, 2005), S. Maleano ed., Oriental Astronomical Association.

effects.[7] Could some of these bodies be Phaethon's sisters? According to Ohtsuka, 2005 UD will be in a very similar orbit to 3200 Phaethon now in AD 6590.

22.2 The Geminid shower

The Geminid shower was only discovered in December 1861 by *Robert Philips Greg* in Manchester, England[8] and independently by *B. V. Marsh* and Professor *Alex C. Twining* in the United States the following year.[9] It has since been seen every year. The late discovery may have had something to do with the bad weather in December, because Greg wrote in 1872 that this was *an important meteoric shower*. There is good evidence[10], in fact, that the Geminids were quite active as early as 1833, when ten meteors (sporadics too?) were seen simultaneously, and a similar phenomenon was observed in 1836 on December 11.[11] In December 12, 1860, (8:20–12:00), 180 meteors were seen in total, with 60 in the hour at 11:30 UT. In 1864, on December 12, meteors were seen falling at a rate of one per minute. These were *blue and white, of momentary duration, and the majority without trains.* – a typical description of a modern Geminid display. In 1868 the meteors were also described as *blue and white, of momentary duration, and the majority without trains.*

That said, there is ample evidence for a progressive increase in activity. There are no historic Geminid accounts (Table 1), while reported peak hourly rates have gradually climbed since discovery (Fig. 22.3). British visual observers reported a fairly constant hourly rate between 20 and 60 in the first half of the twentieth century.[12] Over the past quarter century (1970–2003) the ZHR = 80/h consistently.[13] Lately, IMO puts the ZHR around 120/h. It is unclear how much of that is a change in reduction techniques, with careful reductions attempted only in recent years.

There has not been a significant change to the activity profile over the period 1958–2003. As late as 1920 it was unclear when the exact peak of the shower was, and records have been fragmentary since.[14] More recent observations show a nearly constant node over time (Fig. 22.4).[15] That is unexpected, because all models predict a rapid regression with time as shown by the dashed lines calculated by Williams and Wu.

[7] A. K. Terent'eva and O. A. Baiuzh, On the possible cometary origin of Geminid type meteor streams. *Bull. Astron. Inst. Czech.* **42** (1991), 377–378.

[8] R. P. Greg, E. W. Brayley, A. S. Herschel and J. Glaisher, *Report on Observations of Luminous Meteors, 1862–63* (London, 1864); R. P. Greg, A general comparative table of radiant – positions and duration of meteor-showers. *Mon. Not. R. Astron. Soc.* **32** (1872), 345–354.

[9] A. King, An ephemeris of the Geminid radiant-point. *Mon. Not. R. Astron. Soc.* **86** (1926), 638–641.

[10] W. E. Besley, The Geminid meteor-shower. *Observatory* **23** (1900), 366–370.

[11] K. Fox, I. P. Williams and D. W. Hughes, The evolution of the orbit of the Geminid meteor stream. *Mon. Not. R. Astron. Soc.* **200** (1982), 313–324.

[12] A. C. B. Lovell, *Meteor Astronomy* (Oxford: Clarendon Press, 1954), Ch. XV.

[13] K. Izumi, The long-term variation of the Geminids meteor shower. *At the 23rd General Assembly of the IAU*, Joint Discussion 1, 21 August 1997, Kyoto, Japan (Cambridge MA: IAU, 1997).

[14] W. F. Denning, The meteoric showers of January and December (Quadrantids and Geminids). *Mon. Not. R. Astron. Soc.* **84** (1924), 178–179.

[15] K. Fox, I. P. Williams and D. W. Hughes, The rate profile of the Geminid meteor shower. *Mon. Not. R. Astron. Soc.* **205** (1983), 1155–1169.

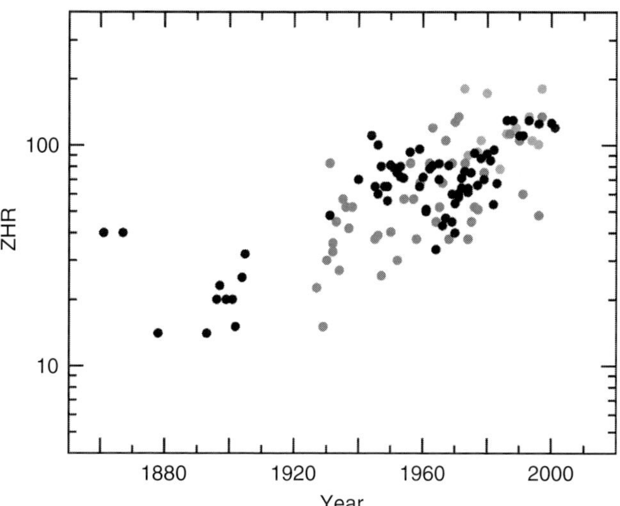

Fig. 22.3 The change of the reported Geminid peak hourly rates over time.[16]

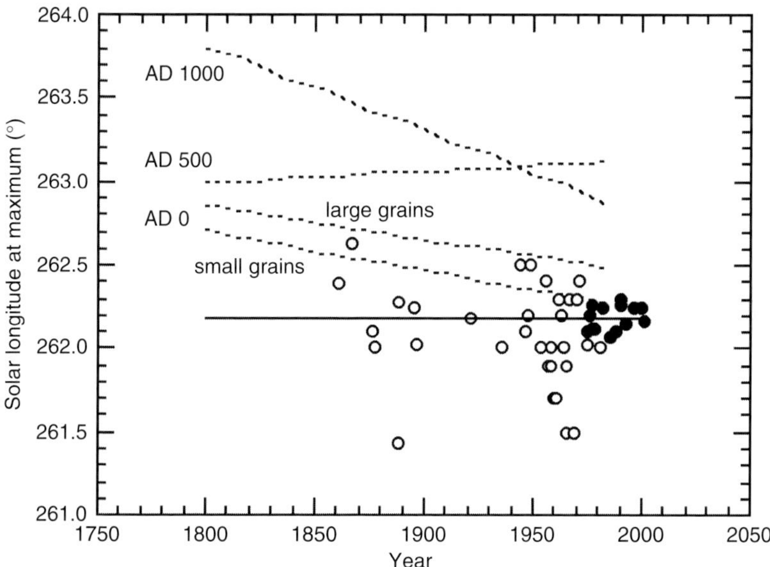

Fig. 22.4 The change of node over time.[17] Lines are models by Williams and Wu.[18]

[16] Compilation of counts from various sources, a.o.: Izumi K., The long-term variation of the Geminids meteor shower. *Meteoroids 2004* (abstract) (2004); M. Simek, Longitudinal structure of the Geminid stream. *Bull. Astron. Inst. Czechoslov.* **27** (1976), 168–173.

[17] P. Jenniskens, Winter 1991: Geminiden, Monocerotiden and snelle meteoren uit de Leeuw. *Radiant, J. DMS* **14** (1991), 28–33.

[18] I. P. Williams and Z. Wu, The Geminid meteor stream and asteroid 3200 Phaethon. *Mon. Not. R. Astron. Soc.* **262** (1993), 231–248.

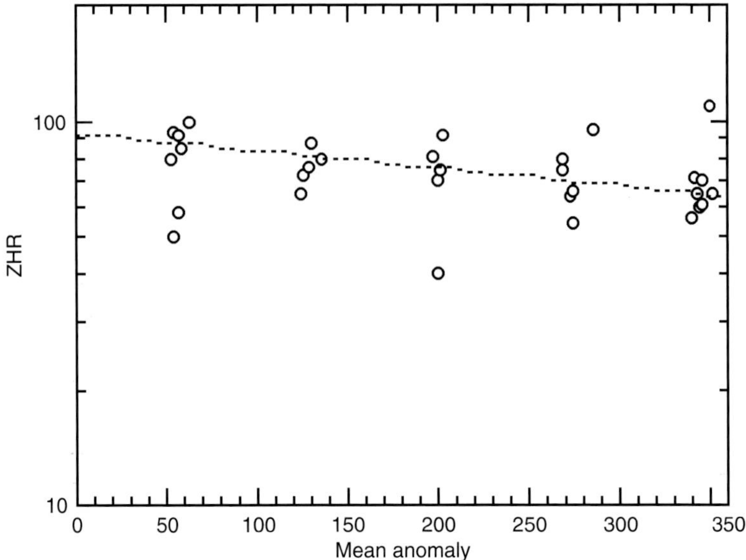

Fig. 22.5 The peak rate (1940–1980) as a function of position along the orbit, if $P = 1.667$ AU (the dust is in the 3:5 mean motion resonance with Earth).

Bruce Mcintosh[19] in Canada and *Milos Simek*[20] at Ondrejov investigated the variation around the orbit of the shower for a range of adopted mean orbital periods and found the largest variation in radar meteor flux with a 2:1 variation of peak flux for an orbital period of about 1.7 yr. Fig. 22.5 is a recreation of this effort, based on visual and radar observations of the shower and shows the result for $P = 1.667$ yr. Kiyoshi Izumi recently found, from visual observations, that the Geminids have become gradually brighter on average between 1971 and 1996.

22.3 The Geminid shower activity and mass

Figure 22.6 pictures the size and orientation of the orbit of 3200 Phaethon. Unlike many other streams, we encounter the Geminid shower at a relatively high true anomaly, far from perihelion. Several attributes of the shower are directly related to that.

The peak is on December 14 and lasts about a day, with some activity visible for at least 22 d. It is the only major shower that is visible from both the northern and the southern hemispheres. That said, there is much confusion about the properties of the activity of the shower and variation of meteor mean magnitude in the Earth's path. When all meteors are considered on a given date, it is clear that the stream profile is

[19] B. A. McIntosh, Geminid meteor shower: further note on the distribution of particles around the orbit. *Bull. Astron. Inst. Czechoslov.* **25** (1974), 362–365.

[20] M. Simek, Longitudinal structure of the Geminid stream. *Bull. Astron. Inst. Czechoslov.* **27** (1976), 168–173.

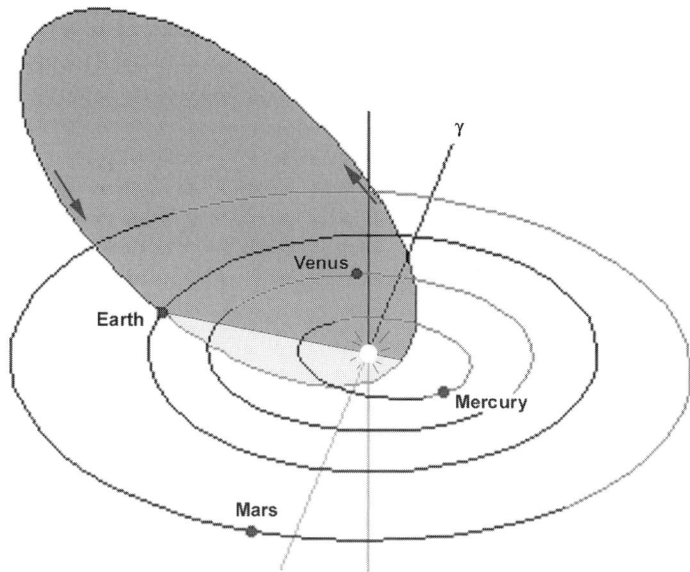

Fig. 22.6 The orbit of 3200 Phaethon and position of the planets on December 13, 2000.

asymmetric,[21] dominated by fainter meteors early on and with more bright meteors at the end (Fig. 22.7).[22] This causes the position of the center of the profile to change with meteoroid mass.[23]

Instead of a smooth transition of parameters, most (but not all) features can be understood if the data are the sum of a narrow main peak and a broader background component. The radar observations by Jones and Morton, for example, appear to mainly pick up that broader background component, with only a small contribution from the main peak (gray line in Fig. 22.7). This instrument has a limiting magnitude of +9, and a mean magnitude of about +7. The effect is also noticed when comparing the activity curves of short and long duration radar echoes.[24]

In contrast, the radar activity curves of Fox, Williams, and Hughes[25] are dominated by the narrow component for meteors brighter than magnitude +5. Visual observers[26], are also mostly impressed by the narrow main peak, which has $s \sim 1.75$, where

[21] A. R. Webster, T. R. Kaiser and L. M. G. Poole, Radio-echo observations of the major night-time meteor streams. II, Geminids. *Mon. Not. R. Astron. Soc.* **133** (1966), 309–319; G. Spalding, The Geminid meteor stream in 1980. *J. Brit. Astron. Soc.* **92** (1982), 227–233.

[22] M. Simek, Some characteristics of the Geminid meteor shower. *Bull. Astron. Inst. Czechoslov.* **29** (1978), 331–340.

[23] B. A. McIntosh and M. Simek, Geminid meteor stream – structure from 20 years of radar observations. *Bull. Astron. Inst. Czechoslov.* **51** (1980), 39–50; G. H. Spalding, The time of Geminid maximum as a function of visual meteor magnitude. *J. Brit. Astron. Assoc.* **94** (1984), 109–112.

[24] M. Simek, P. Pecina, R. P. Chebotarev, S. O. Isamutdinov and V. Znojil, Geminid meteor shower as observed on the long base. *Bull. Astron. Inst. Czechoslov.* **33** (1982), 349–358.

[25] K. Fox, I. P. Williams, and D. W. Hughes, The rate profile of the Geminid meteor shower. *Mon. Not. R. Astron. Soc.* **200** (1983), 313–324.

[26] J. Zvolankova, Activity of the Geminid meteoric shower in the years 1944–1974. *Contrib. Astron. Obs. Skalnate Pleso* **14** (1986), 111–120.

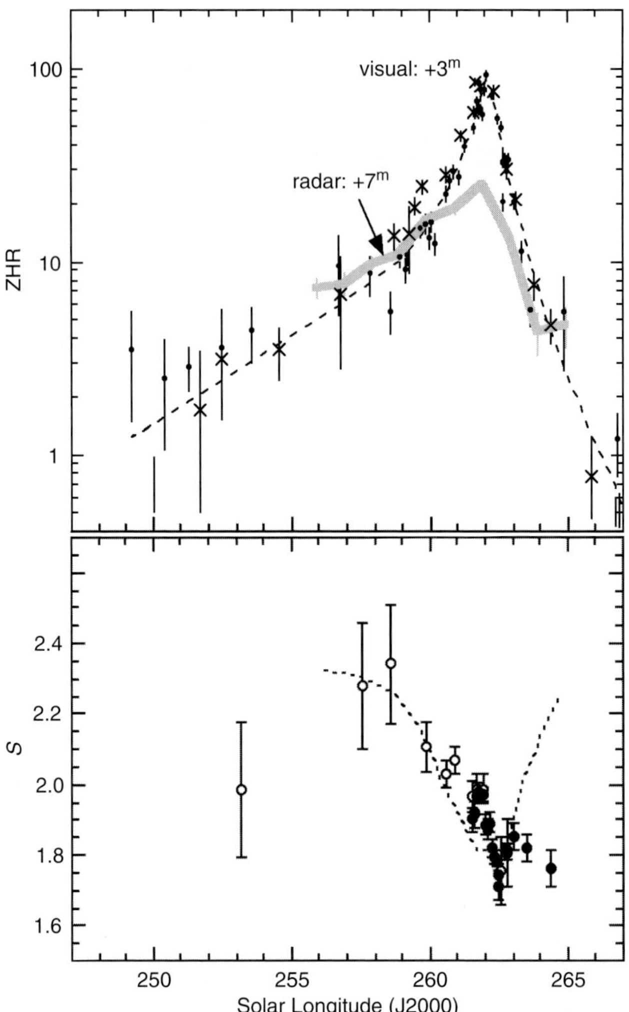

Fig. 22.7 The Geminid zenith hourly rate curve and mass distribution index (*s*) from DMS and NAPO-MS observations.[27] The gray line shows radar observations by Jones and Morton[28], scaled down to match the background component in the ZHR profile. The dark points in the s-curve are the IMO 1988–1997 averages.[29]

[27] P. Jenniskens, Meteor stream activity I. The annual streams. *Astron. Astrophys.* **287** (1994), 990–1013.

[28] J. Jones and J. D. Morton, High-resolution radar studies of the Geminid meteor shower. *Mon. Not. R. Astron. Soc.* **200** (1982), 281–291.

[29] J. Rendtel and P. Brown, Visual observations of the Geminids 1988–1997. In *Meteoroids 1998*, ed. W. J. Baggaley and V. Porubčan. (Bratislava: Astron. Inst. Slovak Acad. Sci., 1999), pp. 243–246.

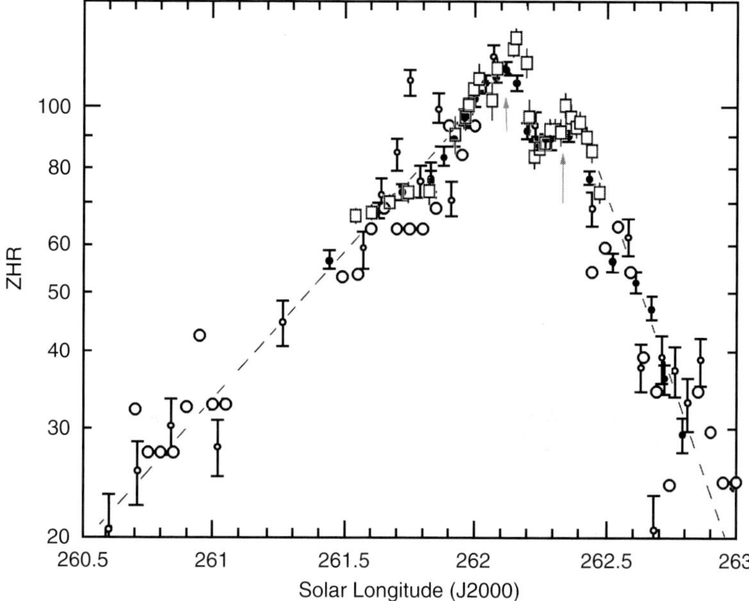

Fig. 22.8 Detail of the Geminid peak. ○ = DMS-data: Geminids 1990[30] and the DMS average; solid symbols: the IMO average over the period 1988–1997, □: detail at peak.[31] The background component has been subtracted.

$s = 1 + 2.5 \log(\chi)$, compared to $s = 2.3$ for the background component. The main peak is dominant in photographic observations, as is clearly seen, for example in the rate curve of Evans and Bone.[32]

The main peak does not have a constant mass distribution index across the profile. The value is lower at the peak of the visual curve, and increases towards higher solar longitude (Fig. 22.7). Even after subtraction of the assymmetric background component, the shape of the main peak too shows slight assymetry, with $B = 0.5/°$ for the ascending branch and $B = 0.8/°$ for the descending branch (with a peak at $\lambda_\odot = 262.12°$ in 1998–1997). Finally, there is a double maximum at the peak of the narrow component (arrows in Fig. 22.8). Sometimes, the background component, responsible for a secondary peak around $\lambda_\odot = 260.6°$ for underdense radar meteors, is confused with this double maximum.[33]

[30] P. Jenniskens, Winter 1991: Geminiden, Monocerotiden and snelle meteoren uit de Leeuw. *Radiant, J. DMS* **14** (1991), 28–33.

[31] J. Rendtel and P. Brown, Visual observations of the Geminids 1988–1997. In *Meteoroids 1998*, ed. W. J. Baggaley and V. Porubčan. (Bratislava: Astron. Inst. Slovak Acad. Sci., 1999), pp. 243–246; J. Rendtel, Almost 50 years of visual Geminid observations. *WGN* **32** (2004), 57–59.

[32] S. J. Evans and N. M. Bone, Photographic and visual observations of the Geminid meteor shower in 1991. *J. Brit. Astron. Assoc.* **103** (1993), 300–304.

[33] O. I. Belkovich, M. G. Ishmukhametova and N. I. Suleymanov, Comparative analysis of meteor shower observations processed by three different methods. *ESA SP* **495** (2001), 91–93.

The background component has a FWHM of 98 ± 5 h ($B = 0.07/°$) centered on $\lambda_\odot \sim 259.0°$. This component appears to be asymmetric with a shallower slope to shorter solar longitudes and peaks at $\lambda_\odot \sim 260.6°$. The background component also appears to have a larger dispersion in the radiant position (not unexpected because of the larger dispersion in node), as a result of which the mean radiant measured by Jones and Morton is less diffuse later in solar longitude, starting out at a large 1.3° at $\lambda_\odot = 255°$ and decreasing to 0.82° at $\lambda_\odot = 263°$. The meteoroids in the background may be in a shorter orbit, with an orbital period of about 1.51 yr,[34] but photographic orbits have $P \sim 1.62$ yr[35]. 3200 Phaethon has a current orbital period of about 1.617 yr, in close agreement.

Hughes and McBride[36] estimated the total mass of the Geminid shower at 1.6×10^{13} kg. It is generally assumed that the Geminid stream is formed by the persistent ejection of particles from a single body over a significant time interval. Such a large mass would need an extended period of time, about 1000 yr, of normal comet dust ejection.

22.4 Physical properties of Geminid meteoroids

Geophysicist *George Wetherill* of the Carnegie Institute of Washington, for years active in the *American Meteor Society*, pointed out the fact that frequent exposure to the intense solar radiation near perihelion (0.14 AU), which heats the Geminids to ~800 K and may lead to a loss of volatiles and produce materials more similar to the crust of a comet than to its interior.[37] Perhaps even more similar to asteroidal stony matter. *Bo Gustafson*, an expert on light scattering by dust, proposed that Geminids may actually represent the product of a cometary crust.[38]

Geminids are more fragile than the meteorites from asteroids. *Ian Halliday*[39] obtained results for twelve bright Geminid fireballs from the MORP network (with somewhat uncertain appearance times) and found that the Geminids decelerated from entry velocities of 36 km/s to a lower limit near 15 km/s and disappeared at heights above 38 km. No fragments larger than dust particles survived the descent. Asteroidal meteoroids of the same mass penetrate considerably deeper than the Geminids.

On the other hand, Geminids are more cohesive than other cometary shower fireballs. They are more strongly decelerated, but penetrate relatively deep. Jacchia *et al.*[40] derived a normal *fragmentation index* $\chi_f = 0.21 \pm 0.05$, compared to 0.28 ± 0.10 for the

[34] J. Jones, On the period of the Geminid meteor stream. *Mon. Not. R. Astron. Soc.* **183** (1978), 539–546.

[35] H. Betlem, C. R. Ter Kuile, M. de Lignie *et al.*, Precision meteor orbits obtained by the Dutch Meteor Society – photographic meteor survey (1981–1993). *Astron. Astrophys. Suppl. Ser.* **128** (1988), 179–185.

[36] D. W. Hughes and N. McBride, The mass of meteoroid streams. *Mon. Not. R. Astron. Soc.* **240** (1989), 73–79.

[37] G. W. Wetherill, Sun-approaching bodies: neither cometary nor asteroidal? *Meteoritics* **21** (1986), 537–538.

[38] B. Å. S. Gustafson, Are the Geminids high density porous flakes from a surface crust on Phaethon? In *Asteroids Comets Meteors III* (Uppsala: Astronomical Observatory 1989), pp. 523–526.

[39] I. Halliday, Geminid fireballs and the peculiar asteroid 3200 Phaethon. *Icarus* **76** (1988), 279–294.

[40] L. G. Jacchia, F. Verniani and R. E. Briggs, An analysis of the trajectories of 413 precisely reduced photographic meteors. *Smithsonian Contrib. Astrophys.* **10** (1967), 1–139.

cometary Perseids and a normal ablation coefficient $\sigma = 0.059\ s^2/km^2$ (compared to $\sigma = 0.055 \pm 0.08\ s^2/km^2$ for the Perseids). Halliday derived particle densities of between 0.7 and 1.3 g/cm^3 (best around 1.0), and a luminous efficiency of 2–6% (most likely 3.4%). This is in good agreement with the mean value of 1.14 g/cm^3 derived by Verniani[41] in his classical study of the Super-Schmidt meteor photographs.

The results of these calculations depend on the treatment of fragmentation. *Pulat Babazhanov* has $\rho = 1.6$ g/cm^3, after taking into account the effects of progressive shedding of small fragments in a process called quasi-continuous fragmentation. Again, his value for the Geminids does not differ significantly from Perseids (1.2 g/cm^3) or Ursids (1.6 g/cm^3), suggesting a cometary origin. These densities are comparable to those of primitive carbonaceous chondrites (CI ~ 1.6 g/cm^3, 35% porosity; CM ~ 2.2 g/cm^3, 23% porosity) that consist mostly of less consolidated matrix material.[42] In contrast, ordinary chondrites have densities of order 3.1–3.6 g/cm^3 and porosities of only 0–10%.

This is not the complete story. *Zdenek Ceplecha* showed that often a better fit to the lightcurve and deceleration was obtained if the meteoroid was taken to break up at choice points along the trajectory. In between those breaks, the meteoroid behaves as above. This *gross-fragmentation* model does not always put the breaks during flares. Bright Geminids broke at altitudes in the range between 81 and 56 km. Most significantly, the computed density of the meteoroids changed dramatically, now raising the values to 3.2 ± 0.4 g/cm^3, the same as stony meteorites.[43] It is not clear if this approach is better.

Ceplecha derived an ablation coefficient $\sigma = 0.01\ s^2/km^2$, give or take a factor of two, with or without this "gross-fragmentation" approach. Although a higher density for the resulting fragments follows, relatively small energies are sufficient to break the meteoroids apart. Martin Beech[44] found that 100 J/g was sufficient, compared to 300 J/g for Perseid meteoroids. This, again, points to a cometary origin for the Geminids.

22.5 Models of Geminid stream evolution

The Geminids, more than other streams, have attracted attention from meteoroid stream modelers, because interactions with Jupiter are weak and the secular perturbations can be calculated by assuming that Jupiter is spread out around its orbit proportional to the amount of time the planet is at a given position. From this, *Miroslav Plavec* first modeled the stream evolution and correctly noticed the main features of the orbital evolution: a secular nutation cycle. He placed the node 0.134 AU

[41] F. Verniani, Meteor masses and luminosity. *Smithsonian Contrib. Astrophys.* **10** (1967), 181–195.

[42] F. J. M. Rietmeijer and J. A. Nuth, Collected extraterrestrial materials: constraints on meteor and fireball compositions. *Earth, Moon Planets* **82–83** (2000), 325–350.

[43] Z. Ceplecha and J. Borovicka, Meteors. In *Interrelations between Physics and Dynamics for Minor Bodies in the Solar System*, ed. D. Benest and C. Froeschlè. (Gif-sur-Yvette: Editions Frontiéres, 1992), pp. 309–367.

[44] M. Beech, The structure of meteoroids. *Mon. Not. R. Astron. Soc.* **211** (1984), 617–620.

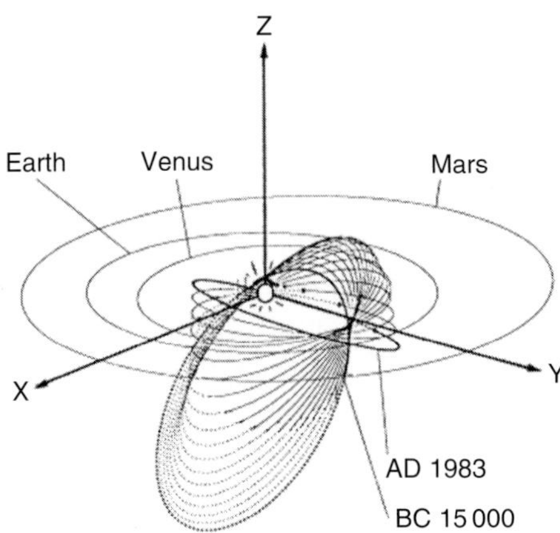

Fig. 22.9 Change of the orbit of Phaethon at regular intervals of 1000 yr over the period 15 000 BC to AD 1983. Crosses mark the ascending node. Calculations and graph by Williams & Wu.[45]

inside Earth's orbit in AD 1700, and 0.107 AU outside Earth's orbit in AD 2100. This shift of the node would account for the increase in activity over the years, and the shower is expected to decrease when the node moves beyond Earth's orbit.

In the early eighties, the evolution of the stream was studied by *Ken Fox*, *Iwan P. Williams*, and *David W. Hughes*. After that, improved models were made by Jones and Hawkes, and Williams and Wu. It was found that the long range perturbations rotate the argument of perihelion around the comet orbit in about 20 000 yr, creating a periodic change in inclination, perihelion distance, eccentricity, and the ascending node (Fig. 22.9 and 22.10).

About 3500 years ago, the ascending node was at Earth's orbit and a shower could have been seen. Around AD 0, the perihelion was at a minimum of 0.12 AU, the inclination at a minimum of 12° and the eccentricity at a maximum of 0.91. The other extreme position in rotation, a maximum $q = 0.25$ AU, maximum inclination of 45°, and minimum eccentricity of 0.81 was about 10 000 BC.

Fig. 22.11 shows how the cross section of the stream evolved and how the stream moved by Earth's orbit over the years. The shower quickly evolved to spread out perpendicular to Earth's orbit. This implies that the stream is more massive than suggested by the distribution of dust in Earth's path alone.

I find that by shifting the scale "Time since stream formed" in the figure of Jones and Hawkes to start at AD 1030 (±10 yr), the calculated variation of rates agrees well with the observed variation (•, Fig. 22.12).[46] The rate rapidly increased about 1000 years after formation. The mean brightness of the meteors gradually increased and is expected to peak when the shower rate is highest.

[45] I. P. Williams and Z. Wu, The Geminid meteor stream and asteroid 3200 Phaethon. *Mon. Not. R. Astron. Soc.* **262** (1993), 231–248.

[46] J. Jones and R. L. Hawkes, The structure of the Geminid meteor stream – II. The combined action of the cometary ejection process and gravitational perturbations. *Mon. Not. R. Astron. Soc.* **223** (1986), 479–486.

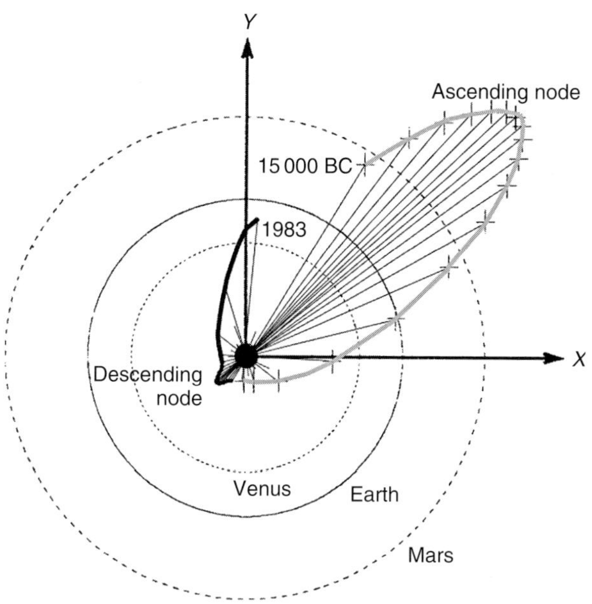

Fig. 22.10 Change of the orbit of minor planet 3200 Phaethon at regular intervals of 1000 yr over the period 15 000 BC until AD 1983. Graph after Williams and Wu.

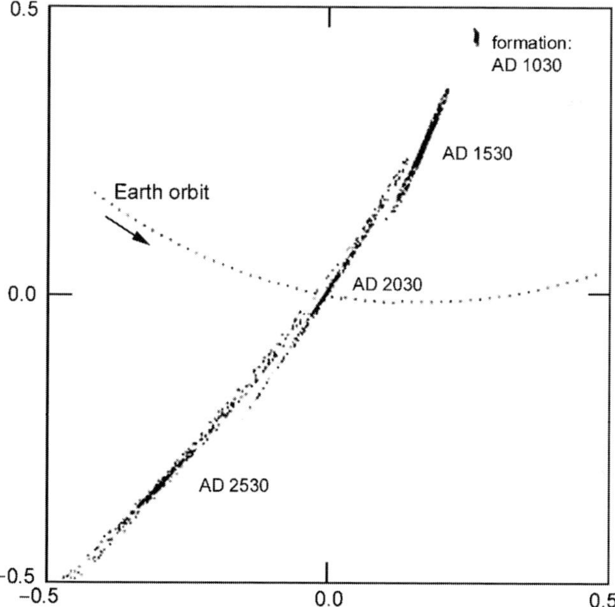

Fig. 22.11 Evolution of the Geminid stream: nodes at formation (a), after 500 yr (b), after 1000 yr (c), and after 1500 yr (d), in calculations by Jones and Hawkes.

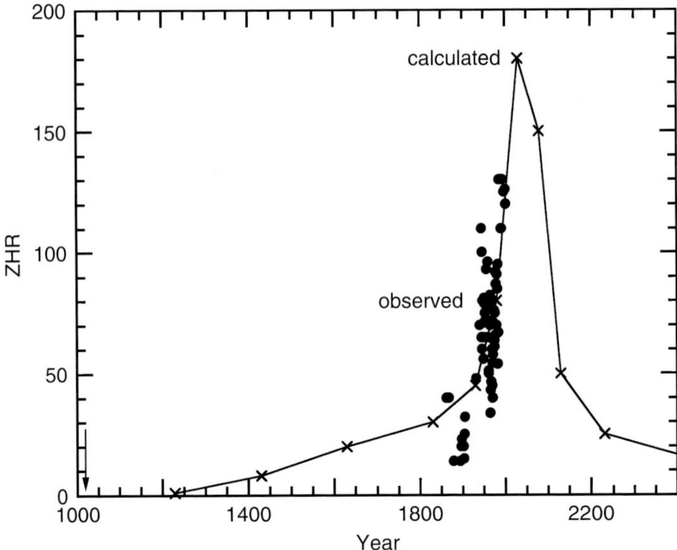

Fig. 22.12 Expected activity of the Geminid shower as a function of year after formation, calculations by Jones and Hawkes.

The model also predicts that the node is rapidly shifting backward in time, one day earlier every 60 years. Observations show a nearly constant node (Fig. 22.5). To explain this peculiar feature, Fox, Williams, and Hughes[47] pointed out that the intersection points of Geminids with the ecliptic plane are at an angle with the normal line to the Earth's orbit and, as a result, the gradual outward movement of the stream shifts the mean node of shower meteoroids forward. This forward shift is counteracted by the backward precession of the node from planetary perturbations.

The reason for the skewed distribution in the model is because each profile is a superposition of dust ejected at different parts in the orbit, each having an elliptical distribution of nodes (Fig. 22.13).[48] In this model, the position of dust in the stream activity profile is a function of the mean anomaly of ejection. Jones and Hawkes[49] found a nodal change $\Delta\text{node}/\Delta t = -0.0025^\circ/\text{yr}$ for the shower as opposed to $-0.152^\circ/\text{yr}$ for the stream as a whole.

More recently, Williams and Wu[50] calculated a more elongated distribution of nodes. In their model, the node stays constant because the precession rate and ejection speeds are mass-dependent, with implications for the expected variation of meteor

[47] K. Fox, I. P. Williams and D. W. Hughes, The evolution of the orbit of the Geminid meteor stream. *Mon. Not. R. Astron. Soc.* **199** (1982), 313–324.

[48] K. Fox, I. P. Williams and D. W. Hughes, The rate profile of the Geminid meteor shower. *Mon. Not. R. Astron. Soc.* **205** (1983), 1155–1169.

[49] J. Jones and R. L. Hawkes, The structure of the Geminid meteor stream. II – The combined action of the cometary ejection process and gravitational perturbations. *Mon. Not. R. Astron. Soc.* **223** (1986), 479–486.

[50] I. P. Williams and Z. Wu, The Geminid meteor stream and asteroid 3200 Phaethon. *Mon. Not. R. Astron. Soc.* **262** (1993), 231–248.

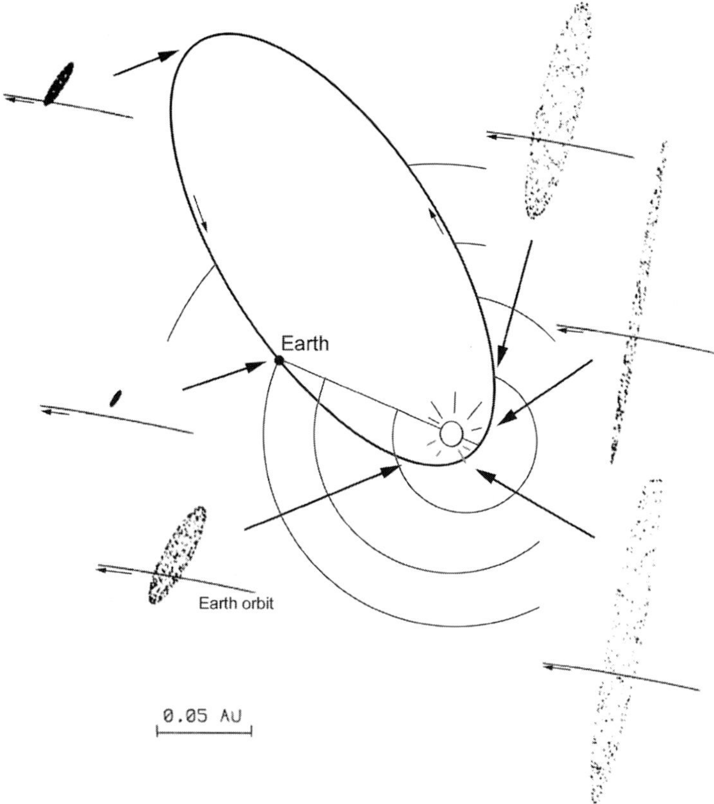

Fig. 22.13 Stream cross section resulting at Earth's orbit from different points of ejection along the orbit – from: Fox, Williams, and Hunt (1993).[51]

brightness in Earth's path and over time. The Geminids should have become steadily brighter, building up to the highest rates later this century.

A large, still outstanding problem with these early model calculations is how to explain the large width of the shower. *Kramer* and *Shestaka*[52] integrated orbits backward and concluded that most were ejected from the parent body in the last 1000 years, but with speeds of ~5 km/s if ejection occurred at perihelion and ~0.5–1 km/s if ejection occurred at aphelion. These are much higher speeds than expected.

If the observed dispersion was due to ejection, then more reasonable models give a stream width of about 0.18° (4.4 h),[53] representative of 10^{-4} g particles. Only if

[51] K. Fox, I. P. Williams and J. Hunt, The past and future of 1983 TB and its relationship to the Geminid meteor stream. In *Dynamics of Comets: Their Origin and Evolution*, ed. A. Carusi and G. B. Valsecchi. (Dordrecht: Reidel, 1985), pp. 143–148.

[52] E. N. Kramer and I. S. Shestaka, The age, origin and evolution of Geminids' meteor swarm. *Kinematika i Fizika Nebesnykh Tel* **2** (1986), 81–86 (in Russian).

[53] K. Fox, I. P. Williams and J. Hunt, The past and future of 1983 TB and its relationship to the Geminid meteor stream. In *Dynamics of Comets: Their Origin and Evolution*, ed. A. Carusi and G. B. Valsecchi. (Dordrecht: Reidel, 1985), pp. 143–148.

ejection occurs at one mean anomaly at ~50° or 315° would the width be about 1.2° (28 h).

Because the perturbations on the stream by Jupiter are relatively gentle and quasi-random, the width of the stream varies with time as in a random-walk migration: $W \sim t^{1/2}$. Jones and Wheaton[54] found that the dispersive effects of planetary perturbations by Jupiter are extremely small, leading to only a root-mean-square width of the Geminid stream at 1 AU of 0.017° (0.4 h) after 1000 yr.[55] The dispersion in the plane of the orbit is much more rapid, at rms = 0.084°/1000 yr. Close encounters with the terrestrial planets add to the dispersion of the shower.

Similarly, Williams and Wu have a scatter of 0.7° in 2000 years. Hence, the observed width of 1.89° implies an age of ~5400 yr. Other such early estimates of the Geminid shower age were in the range 5000–10 000 yr.

Jim Jones also found a double-peaked structure in the stream. Even with parts of the rings catching up on each other to form a Filament-like structure, this still had the effect of creating dust grains distributed on the outside of a hollow tube. Jones found a gradual dispersion and widening of this structure with time, with the peak maxima increasing by about 1.3° in 5000 yr. From a tentative detection of such a hollow structure in the 1983–1985 ZHR curves, I derived a stream age of about 1180 ± 100 yr for the central structure, and about 2000 yr for the profile wings.[56] I had the maximum activity at solar longitudes 262.01 ± 0.02° and 262.34 ± 0.01°. More recently, *Jürgen Rendtel*[57] derived a mean curve for the 1988–1997 Geminids observed by IMO members with peaks of the same strength and ratio (ZHR ~ 140 and 110/h minimum to 90/h), but now better defined. The two peaks are at 262.12 ± 0.02° and 262.33 ± 0.02° (center of the bumps in Fig. 22.5), from which I derive the epoch for the breakup of the parent body and formation of the Geminid stream of: AD 1240 ± 140 yr, which is not very different from the AD 1030 ± 10 yr from the increase of Geminid rates.

Finally, Galina Ryabova of Tomsk State University[58] showed that particles ejected preperihelion (+/●, Fig. 22.14) tended to be concentrated more in the center of the stream at the descending node than those ejected postperihelion (●/○). After 2000 years, this structure evolved into a bimodal distribution, with pre- and postperihelion grains forming two separate dust sheets. The reason for the separation is that the meteoroids ejected preperihelion move in slightly different orbits than those ejected postperihelion. That difference enhanced over time. The distance between the two sheets is 1.3° in solar longitude for particles of meteoroid mass 3×10^{-3} g, and 1.8° for particles of 3×10^{-4} g after 2000 years. This puts the age at around AD 1460. This is

[54] J. Jones and K. R. Wheaton, The dispersion of the Geminid stream by planetary perturbations. *Observatory* **105** (1985), 34–36.

[55] J. Jones, The structure of the Geminid meteor stream – I. The effect of planetary perturbations. *Mon. Not. R. Astron. Soc.* **217** (1985), 523–532.

[56] P. Jenniskens (1986), De structuur van het Geminiden maximum. *Radiant, J. DMS* **8** (1986), 58–59.

[57] J. Rendtel, Almost 50 years of visual Geminid observations. *WGN* **32** (2004), 57–59.

[58] G. O. Ryabova, The Geminid meteor stream activity profile. Solar Syst. Res. **35** (2001), 151–157; G. O. Ryabova, On the bimodality of the Geminid meteor shower. *WGN* **17** (1989), 240–241; G. O. Ryabova, Mathematical model of the Geminid meteor stream formation. *ESA SP* **495** (2001), 77–81.

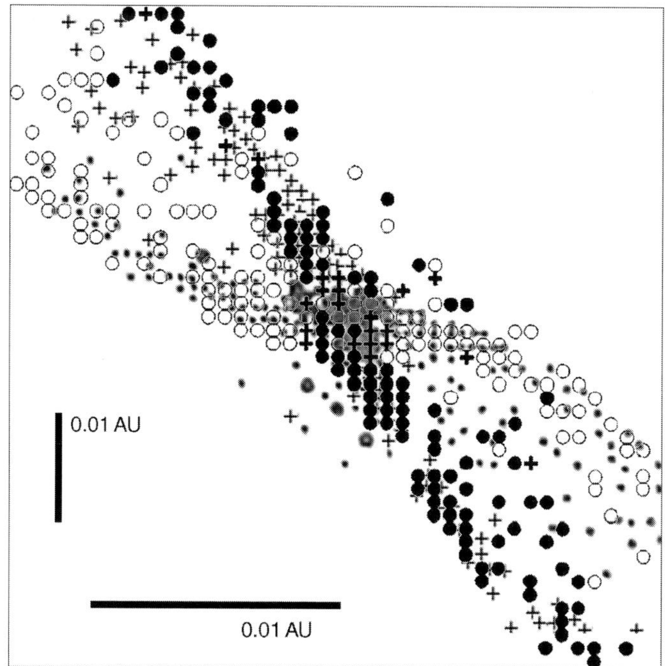

Fig. 22.14 Descending node of pre (+, •) and postperihelion (., ○) ejected particles of mass 0.000 214 g after the stream has evolved for 2000 yr, in a model by Galina Ryabova.

only possible if there was prolonged ejection from the fragments, a discrete deposition of matter would only develop one of the two structures.

22.6 Tracing back the meteoroids to the time of origin

University of Florida at Gainesville astronomers *Bo Å. S. Gustafson*[59] and *Lars G. Adolfsson* traced back the orbits of individual Geminid meteoroids and parent 3200 Phaethon in search of a point in time when meteoroid and comet where at the same position in the orbit (Fig. 22.15 and 22.16). If the Geminid stream was generated in cometary activity, most particles would have been released near perihelion, while if Phaethon was an asteroid and a collision was responsible, it is likely to have occurred near aphelion in the asteroid belt.

Gustafson used orbits derived from the *Harvard Meteor Program* in New Mexico (meteors of about magnitude +2) and the *Meteorite Orbit and Recovery Program* in Canada decades ago (fireballs of about magnitude −6) and concluded that most meteoroids were released at perihelion under conditions typical for comet activity.

[59] B. Å. S. Gustafson, Geminid meteoroids traced to cometary activity on Phaethon. *Astron. Astrophys.* **225** (1989), 533–540.

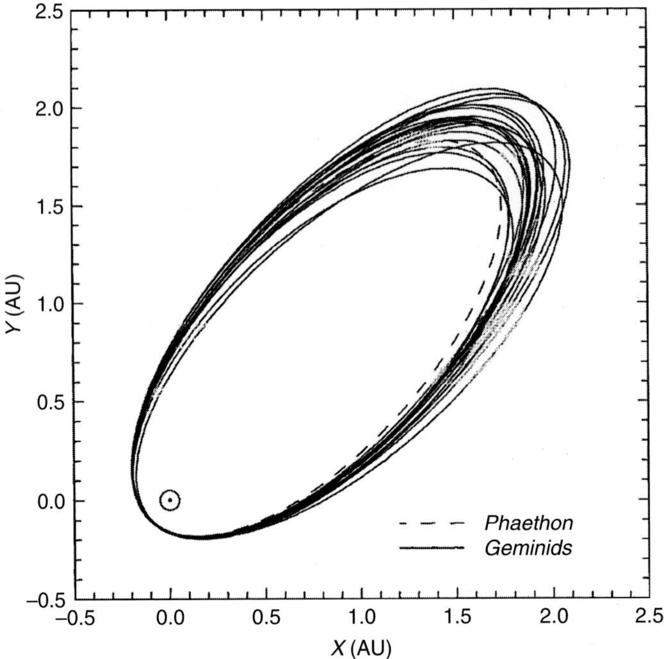

Fig. 22.15 Geminid orbits have a small range in perihelion, and a larger range in aphelion. Calculations by Bo Å. S. Gustafson, University of Florida, Gainesville.[60]

The meteoroids were mostly ejected towards the Sun with ejection velocities of order 40 m/s (at $q = 0.14$ AU). Most small particles produced the best result for ejection epoch about 2300–1700 years ago (~300 BC), and were ejected over a period of less than a few hundred years. Larger meteoroids, however, appeared to be emitted in all directions and ejected over a 3000 year period, as recently as AD 1600.[61]

One problem was that the orbits in hand did not provide both accurate orbits and deceleration information that could help decide the beta value. Instead, beta was assumed to be proportional to meteor brightness. Better orbits were needed.

22.7 The Geminid experience

It was the French novelist *Michel Tournier* ("Les météores") who saw in meteors a symbol of the dynamic adventurous life. In the winter of 1990, we organized our first observing adventure abroad in an effort to provide Gustafson with better Geminid meteoroid orbits. The Moon was new and the shower maximum was

[60] *Ibid.*

[61] L. G. Adolfsson and B. Å. S. Gustafson, Dynamics and probabilistic relation between meteoroids and their parent bodies. *ASP Conf. Ser.* **104** (1996), 133–136; see also: G. O. Ryabova, Age of the Geminid meteor stream (review). *Solar Syst. Res.* **33** (1999), 224–238.

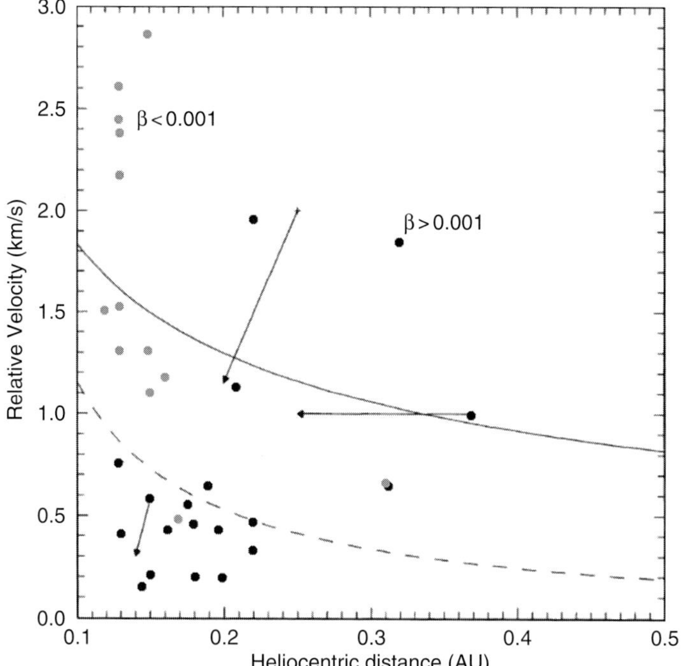

Fig. 22.16 Point of ejection and ejection speed for Geminid meteoroids traced back to their most likely point of ejection. The two shades of gray denote particles of different size. The dashed line is for 0.1 cm diameter grains ejected from a 5 km comet using Whipple ejection speeds, while the solid line shows the speed of very small grains in the tail of a comet. From: Gustafson (1989).[62]

expected to fall in the early morning of December 14, when the radiant position was high in the sky.

Until that time, my best Geminids were seen seven years earlier, when I had teamed up with my mentor and visual observing partner Rudolf Veltman in the town of Oegstgeest. The night fell in the middle of an exam period, and when the freezing cold was too much to bear, I escaped inside the house to read another chemistry chapter. That night was supposed to be the one after the main peak, but we were showered by bright Geminids.

In the fall of 1990, a cabin was rented in the village of Le Thouron in the Haute Provence of southern France for a period of one week. Our plan was for DMS photographer *Casper ter Kuile* to join a group of visual observers from Belgium in the town of Lardiers some 55 km to the east. Photographers *Marc de Lignie* and I would stay in Le Thouron, where we would get visual support from four observers

[62] B. Å. S. Gustafson, Geminid meteoroids traced to cometary activity on Phaethon. *Astron. Astrophys.* **225** (1989), 533–540.

of group "Loosdrecht." At the night of the maximum, I would create a third photographic station at a location further south to form an equal sided triangle. If everything worked out, we hoped to photograph a few good bright Geminids.

It was under the threat of an approaching winter storm that Marc and I drove south from the Netherlands to France, following Casper's tail lights. When we finally drove out of the snowy French Alps and into the Haute Provence, a line of navy blue gradually rose from the horizon in front of us and became a nice transparent canopy. Under the artificial lights of a shooting Christmas star we separated. Marc and I went left in the direction of Le Thouron, while Casper went right in the direction of Lardiers.

Our site in Le Thouron was reached by driving over gradually narrowing roads deeper into the mountains. When the road suddenly opened into a wide valley guarded by two tall poplar trees, we felt like Frodo entering the land of Gondor. The surrounding hill tops were bathed in the orange light of sunset. The road followed a small river upstream to the town of Tartonne. Above Tartonne was a mountain with the idyllic village of Le Thouron on its slopes. To our relief, the rental *gîtes* did have a back yard with a sufficiently free field of view to operate our cameras. Some hills prevented us from seeing the horizon, but also shielded us from disturbing artificial lights.

Our visual support arrived later that evening. Group *Loosdrecht*, led by Paul van der Veen, drove in straight from the hell of northern Europe, exuberant that they had just managed to stay ahead of the snow storms. That night it started snowing in Le Thouron and it continued during the day. The roads were cleared in the afternoon, but our car was sliding more than driving down the hill on our way to Tartonne. The television in the local pub showed images of people stranded in a blizzard on both the French and German highways. Telephone lines were dead. We rested and spent the next day enjoying the fresh snow in the mountains.

Just when we thought that our Geminid campaign was doomed, the clouds parted. The next four nights had clear star sprinkled skies. Already in the first night of December 11/12, Geminids were seen at a rate of ZHR = 16/h. Group *Loosdrecht* found an observing site on the Col du Défense, somewhat higher on the mountain, where they had a clear view all around. Hiding behind a low windscreen did not fully temper the cold −10 °C wind and ever so often snow would blow in their faces. Marc had spent the afternoon clearing a square spot in the snow near the gîtes, where cameras were now recording the bright meteors. We were hoping that Casper was recording the same meteors in Lardiers.

During the second night, the Earth had penetrated deeper into the Geminid stream and the rate had increased to ZHR = 38/h. In front of the cabin, Marc battled with the French utility system. Snow and wind caused periodic blackouts, during which heaters and rotating shutters did not work. Those shutters interrupt the meteor image 25 times per second. This makes it possible to follow the meteor down its track and measure the speed. Marc caught the Taurid fireball that I observed through the cloth of a windscreen.

In the night of December 13, Earth passed through the center of the Geminid stream. Early in the evening, I drove slowly down the slippery road to Tartonne, on my way to a spot on the map that was 80 km south near the town of Riez. There, for one night, a temporary station would complete an equal sided triangle with Lardiers and Le Thouron for stereoscopic observations. I was alone and a little worried. A -1 Geminid sailed to the horizon, then another. When I finally arrived in Riez, I found a big antenna complex full of bright red lights at the exact spot that I had chosen on the map. Further south, up on the plateau in the neighborhood of Quinson, I discovered an unpaved road that led to the edge of a forest. Nearby hills covered a few degrees above the horizon, shielding artificial lights. The forest offered some protection from the wind and added to the aroma of lavender in the fields. A perfect place.

In the dark, I set up an observing site on the side of that road. The temporary station would consist of one platform of six regular 35 mm photographic cameras with standard f1.8/50 mm lenses and antidew heating ribbons plus a rotating shutter operated by a car battery. While I took the camera platform out of the car, I saw five Geminids within three seconds! That flurry was followed by a lull in activity, because the mean rate was only tens per hour. But boy, the adrenalin! Seconds later, all cameras were open and I was down on a reclining lawn chair, huddled in warm blankets, watching the sky intently. It was 19:43:20 UT when I officially recorded my first meteor.

The Geminids were everywhere, that night (Fig. 22.17). The number of meteors was frightening. Hiding in the car for a quiet cup of coffee was impossible. The front window still exposed some 20° of restless sky and each bright meteor needed to be recorded. The Geminids are quite unusual meteors. They tend to be bluish because of a relative lack of air plasma emissions. Geminids have no flares and almost no persistent trains, which is very unusual. The meteors gradually light up and elegantly disappear, all within a fraction of a second.

I saw 648 Geminids in a period of just over six hours that night. 150 meteors were captured on film. Fortunately, there was no snow at my site and operating the cameras was a delight. Later that night, a small Moon rose above the horizon and twilight came abruptly. After the last stars had disappeared in the glow of dawn, I put the camera battery on the front seat of the car, and locked the doors for a short nap in the back.

In all the uncertainty of our life, it is good to know that the Geminids come back every year, although not always under the same circumstances. When analyzing the visual counts of Geminids, we found that the mean ZHR was 77.8 ± 1.3/h, a little shy of the peak which occurred later that morning. The activity changed smoothly during the night.

In the final night, December 14/15, rates had fallen back to ZHR = 31 Geminids per hour. Because I was still tired from my previous excursion, I decided to leave Le Thouron for the relatively nearby town of Barrème, a place as cold as

Fig. 22.17 A–2 magnitude Geminid from de Bilt on 1994 December 14 at 05:18:14 UT. Photo: Casper ter Kuile, DMS.

its name sounds. I was reminded that a raincoat is essential protection against wind. Just as that other postmaximum Geminid night in 1983, the Geminids were bright and glorious.

One hundred Geminid orbits were measured that season. We have yet to track the measured orbits back in time. Let us first check if the semimajor axis changes as a function of meteoroid mass, as in the models by Ryabova and others. This is expected

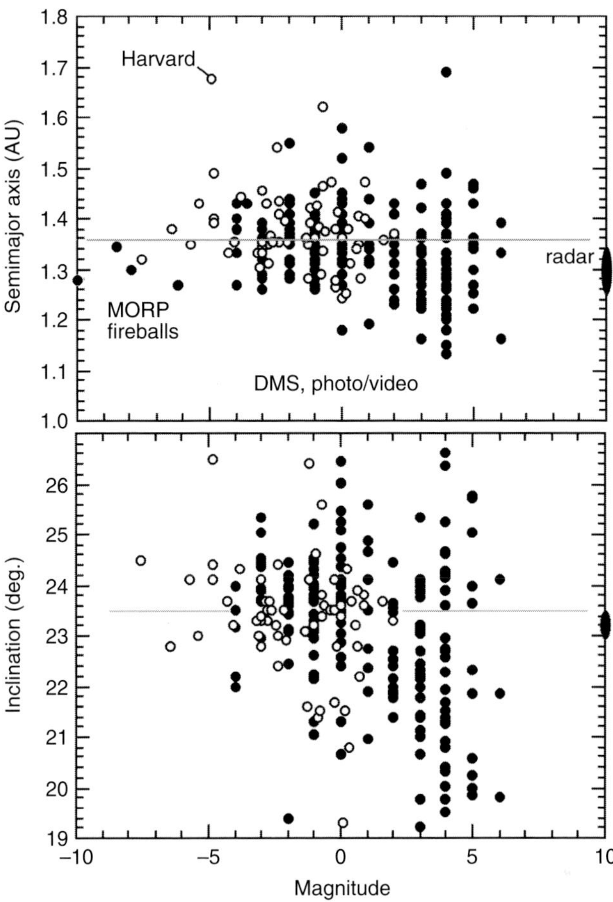

Fig. 22.18 Change of semimajor axis and inclination with meteoroid mass (meteor magnitude).

for normal ejection at perihelion and if the radiation pressure puts the smaller grains in wider orbits than the larger grains.

The answer is: no! The measurement error in the semimajor axis (a) is less than the observed spread. The most precise orbits (quality factor $Q > 5$) have a mean formal error ± 0.014 AU in $1/a$, and the 1σ dispersion is ± 0.034 AU (Fig. 22.18). DMS and Harvard Meteor Program data agree well, and have the same semimajor axis measured from radar observations, a range that covers -10 to $+10^{\text{m}}$ and masses from about 0.5 kg to 1e^{-6} g.

The inclination is also nearly independent of mass in the photographic data sets, but there is a population of smaller meteoroids that have lower inclination. Radar orbits from the Harvard Meteor Project have the same inclination, but the Adelaide radar has a lower 16°–18° (Table 7). It is possible that this is the background component.

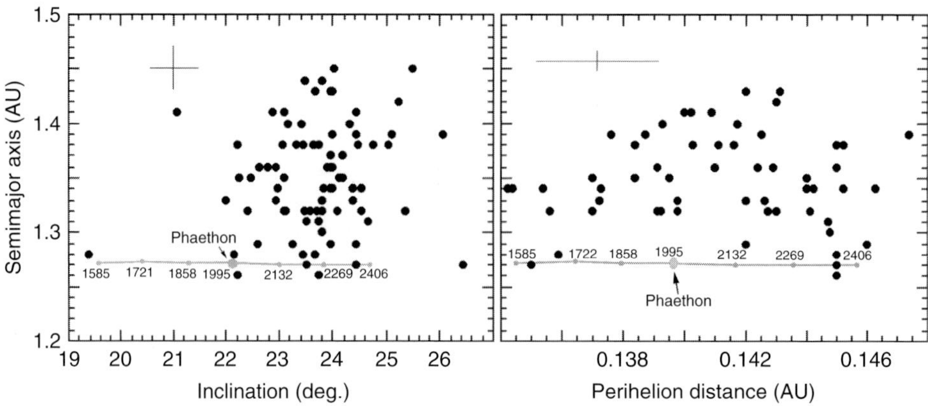

Fig. 22.19 Orbital elements of photographed Geminids in 1990, 1991, and 1996 from DMS data (quality parameter 6 and higher). The past and future orbit of 3200 Phaethon is also plotted.

The meteoroids have significantly wider orbits than Phaethon (Fig. 22.19). There is no concentration at a specific value. We find no evidence that the meteoroids are trapped in an orbital resonance with the terrestrial planets (Phaethon is suspected to have librated about the 3:7 mean-motion resonance with Venus prior to 1891), but meteoroids scatter widely and most are in between the 7:1 and 8:1 mean-motion resonances with Jupiter.

The eye is drawn to a semicircle of solutions that has Phaethon just off-center. The distribution of inclination, too, seems to have a hole at the center (Fig. 22.19). This could be a selection effect of Earth sampling only certain meteoroids, but only if the stream is highly stratified. That is more consistent with a single breakup some time ago than with an extended period of Whipple-type ejection.

All angular elements of the Geminids are off set from Phaethon (Fig. 22.20), but the inclination, argument of perihelion, and ascending node are centered at the value that Phaethon is expected to have in about AD 2200. Hence, the shower appears to have moved ahead of the parent body, which is now located slightly outside those meteoroids that intersect Earth's orbit. The larger meteoroids are found at the longer nodes, suggesting that the larger meteoroids evolved more slowly than the smaller meteoroids relative to 3200 Phaethon. Of course, Phaethon in this diagram is the only orbit that is not intersecting with Earth's orbit.

The length of perihelion of Phaethon is about 1° smaller than that of most of the Geminids, but Phaethon matches the trend of length of perihelion and orbital period (Fig. 22.21). This is not an evolution effect, because the argument of perihelion and orbital period are expected to be nearly constant over time. This trend implies that the meteoroids were put in the longer orbits from the start, those most different from Phaethon had also the greatest difference in the length of perihelion. The observed trend is evidence of a discrete breakup event, which must have occurred on the outward leg near perihelion in order to get a sufficient range in semimajor axes and a decrease in the length of perihelion for the longer orbits.

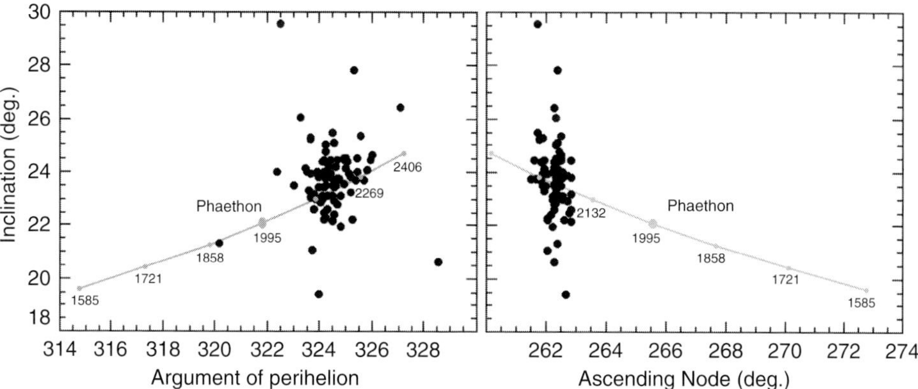

Fig. 22.20 Distribution of inclination as a function of argument of perihelion and node.

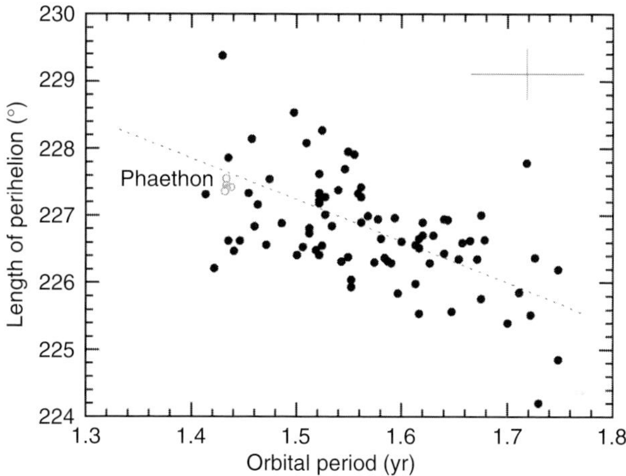

Fig. 22.21 What should have been constant: orbital period and length of perihelion.

Most current models of the Geminid stream consider 3200 Phaethon to have gradually released the Geminid dust over many orbits. Because the effects of ejection at certain positions in the orbit are slow, we can use the model by Fox, Williams, and Hunt (Fig. 22.13) to infer how the stream profile would look for a discrete breakup. The distribution of nodes most consistent with observations is a breakup around true anomaly $v = +45\,^{\circ}$, close to and just after passing perihelion where thermal stresses are largest. That would create the elongated stream profile observed. The distribution of large particles shows the same hollow distribution of inclinations centered on 23.6 °. The dispersion in perihelion distances appears to be slightly smaller than observed, but otherwise there is good agreement between the model distributions and those observed.

By comparing the new orbits with the precisely reduced orbits obtained in the 1951–1952 Harvard Meteor Project, the change in the orbital elements over time is measured. The results are given in Table 7 and show a progression of the inclination at a rate of about that predicted, but a less rapid increase of perihelion and the constant node noticed before. Selection effects are important here, because for a Geminid meteoroid to hit the Earth: $q \sim 1 - e\cos(\omega)$ for given e. If ω is advancing, then perihelion distances that are smaller will be sampled. The node may decrease less fast than expected if meteoroids in longer orbital periods are observed progressively.

That leaves the question of how the background of fainter meteoroids came to be, earlier attributed to outgassing effects.[63] Jones and Morton[64] invoked planetary perturbations, such as with Earth, which could make the particles jump into a lower energy (smaller semimajor axis) orbits with a small transition probability. Perhaps the smaller meteoroids met Earth earlier in their orbital evolution and are now further dispersed accordingly.

22.8 Geminids and 3200 Phaethon in the future

If my matching of the model results of Jones and Hawkes to the observed Geminid rates in recent years (by shifting the zero point of the time scale) is correct, then the shower will continue to increase and peak around ZHR = 190 in AD 2050 (Fig. 22.12).

Iwan Williams[65] calculated that 250 years from now, comet Phaethon itself will come very close to Earth's orbit, but is not expected to make a giant Geminid, so to speak. During one particular close encounter at that time, Phaethon will make naked-eye visiblity. After that occasion, Earth will have changed the orbit radically and this will prohibit further close encounters for some time.

63 D. W. Hughes, I. P. Williams and K. Fox, The mass segregation and nodal retrogression of the Quadrantid meteor stream. *Mon. Not. R. Astron. Soc.* **195** (1981), 625–637.

64 J. Jones and J. D. Morton, High-resolution radar studies of the Geminid meteor shower. *Mon. Not. R. Astron. Soc.* **200** (1982), 281–291.

65 K. Davies, Is 3200 Phaethon a dead comet? *Sky Telescope* October (1985), 317–318.

23

The sunskirting Arietids and δ-Aquariids

The most obviously broken comets, with a long slew of fragments to prove it, are the Sungrazers and Sunskirters. They are in similar orbits, which pass close to the Sun on each orbit. Comet C/1965 S_1 (Ikeya–Seki) is a *Sungrazer*. The comet returned after 877 years, had a perihelion at 0.008 AU (1.2 million km), brightened in the glare of the Sun to -10^m, and became visible in broad daylight on October 21, 1965.[1] In recent years, hundreds of such comets, albeit smaller, have been found by the space-borne *SOLWIND*, *Solar Maximum Mission* (SMM), and *Solar and Heliospheric Observatory* (SOHO) satellites. SOHO routinely takes images of the Sun's corona by blocking out the bright solar disk (Fig. 23.1). The comets that travel through the field of view are visible because of their proximity to the Sun, from high activity and an efficient forward scattering of the light. At their closest point to the Sun, they move at around 600 km/s, and few survive this encounter.

About 90% of all sungrazer comets detected by SOHO have a retrograde $i = 144.0°$ low, perihelion distance $q \sim 0.0055$ AU, orbit and form a group, the *Kreutz Sungrazing group*. These comets pass within $q = 0.0047$–0.0095 AU from the Sun, but only the large ones with $q > 0.005$ AU survive. These are small comets, with typical diameters $D \sim 16$–130 m, which brighten dramatically on approach to the Sun. There are many, 464 and counting.

Although sungrazers are a dynamic end-state of comet evolution,[2] the (Heinrich Carl Friedrich) *Kreutz Sungrazing Group*[3] on 600–1100 yr orbits is thought to have originated from one large 120 km sized retrograde moving Sungrazer parent comet,

[1] B. G. Marsden, The Sungrazing Comet Group. *Astron. J.* **72** (1967), 1170–1183.

[2] M. E. Bailey, J. E. Chambers and G. Hahn, Origin of sungrazers – a frequent cometary end-state. *Astron. Astrophys.* **257** (1992), 315–322.

[3] D. Kirkwood, On the great southern comet of 1880. *Observatory* **3** (1880), 590–592; H. C. F. Kreutz, Untersuchungen über das Comentesystem 1843 I, 1880 I und 1882 II. *Publ. Sternwarte Kiel*, No. 3, ed. C. Schaidt and C. F. Mohr (1888); H. C. F. Kreutz, Untersuchungen über das System der Cometen 1843 I, 1880 I und 1882 II Theil II. *Publ. Sternwarte Kiel*, No. 6 (Kiel: Schaidt, 1891), 67pp; L. Kresák, On the reality and genetic association of comet groups and pairs. *Bull. Astron. Inst. Czech.* **33** (1982), 150–160; D. A. Biesecker, P. Lamy, O. C. St. Cyr, A. Llebaria and R. A. Howard, Sungrazing comets discovered with the SOHO/LASCO coronagraphs 1996–1998. *Icarus* **157** (2002), 323–348; Z. Sekanina, Runaway fragmentation of sungrazing comets observed with the solar and heliospheric observatory. *Astrophys. J.* **576** (2002), 1085–1089; Z. Sekanina and P. W. Chodas, Fragmentation origin of major sungrazing comets C/1970 K1, C/1880 C1, and C/1843 D1. *Astrophys. J.* **581** (2002), 1389–1398; R. Strom, *Astron. Astrophys.* **387** (2002), L17–L20.

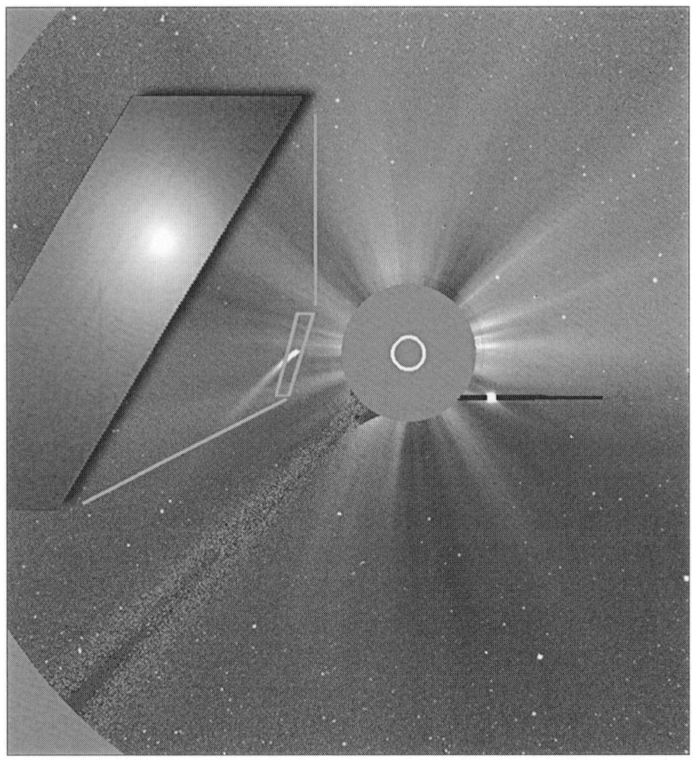

Fig. 23.1 Comet 96P/Machholz seen towards the Sun at $\sim -2^{\mathrm{m}}$ at 02:42 UT on January 8, 2002. Photo: NASA/SOHO.

which appeared between 20 000 and 10 000 yr ago, according to *Brian Marsden*.[4] This original comet had to be large with $H_{10} = -5.0^{\mathrm{m}}$. The orbital period would have been about 1000 years and 10–20 orbits ago, this comet fell apart from the violent loss of gas, or from tidal stresses. The breakup occurred in a few steps, with one piece holding together until 371 BC, when it split into at least three pieces. The Greek historian Ephorus (c. 400–330 BC) mentioned a bright comet that *split into two stars as it departed* at the time of the destruction by earthquake of the Peloponnesian cities Helike and Boura (the winter of 373–372 BC), which just might have marked that occasion.

One of the three fragments had a period of around 350 yr and the other a period of about twice this length. The shorter-period object returned in the first, fourth, eighth and eleventh centuries and may be identical to the comet of 1487. A breakup during the eleventh century return gave rise to the so-called "subgroup I" of the Kreutz

[4] B. G. Marsden, The Sungrazing Comet Group. *Astron. J.* **72** (1967), 1170–1183; B. G. Marsden, The Sungrazing Comet Group. II. *Astron. J.* **98** (1989), 2306–2321.

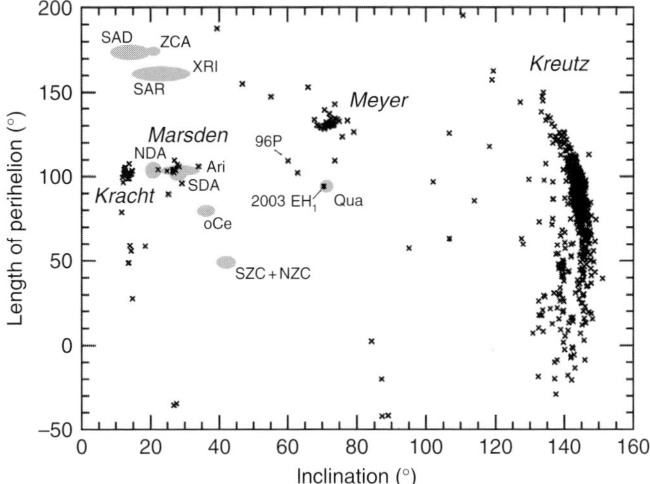

Fig. 23.2 Sungrazer groups with individual comet fragments shown as crosses (observations until October 2004). Meteor showers with $q < 0.2$ AU are shown as gray areas. Also shown are 96P/Machholz and the Quadrantids with 2003 EH_1.

Sungrazers. This group includes C/1843 D_1 (The Great March Comet), comet C/1963 R_1 (Pereyra), the objects detected by SOLWIND and SMM, and some 67% of those discovered by SOHO. Fragmentation continued after that. Marsden found that 1843 D_1 and 1963 R_1 broke apart in the return of AD 1487.[5]

The longer period piece returned in the fourth century and held together until AD 1106, when it too split into at least three pieces. This breakup gave rise to the subgroup II Kreutz Sungrazer comets. The group includes C/1882 R_1 (The Great September Comet), C/1965 S_1 (Ikeya–Seki), and about 19% of the objects discovered by SOHO.[6]

Since that time, three other groups have been discovered in prograde moving orbits, called the *Sunskirting comets* by Sekanina and Chodas (Fig. 23.2).[7] It is not at all clear that they are related to the Kreutz Sungrazers. Their perihelion distance is larger and two groups form a complex with comet 96P/Machholz 1, the *Machholz complex*. They are called the Marsden, Meyer, and Kracht groups. The discovery of the Marsden and Meyer groups was announced on February 18, 2002.[8] The (Brian) *Marsden group* has inclinations around 26° and $q \sim 0.05$ AU. The (Maik) *Meyer group* hovers around

[5] B. G. Marsden, The sungrazing comets revisited. In *Asteroids Comets Meteors III*. (Uppsala: Astronomical Observatory 1989), pp. 393–400.

[6] Z. Sekanina, Statistical investigation and modeling of sungrazing comets discovered with the solar and heliospheric observatory. *Astrophys. J.* **566** (2002), 577–598.

[7] D. W. E. Green, *IAU Circ.* 7832, ed. B. G. Marsden (2002) *MPEC 2002-C28 + MPEC 2002-E18 + MPEC 2002-E25 + MPEC 2002-F03 + MPEC 2002-F43 + MPEC 2002-M47 + MPEC 2002-O35*; M. Meyer, New groups of near-sun comets. *Int. Comet Q.*, July (2003), 115–122; Z. Sekanina and P. W. Chodas, Origin of the Marsden and Kracht groups of sunskirting comets. I. Association with comet 96P/Machholz and its interplanetary complex. *Astrophys. J.* suppl. ser. 161 (2005), 551–586

[8] D. W. E. Green, Non-Kreutz near-Sun groups. *IAU Circ.* 7832 (2002).

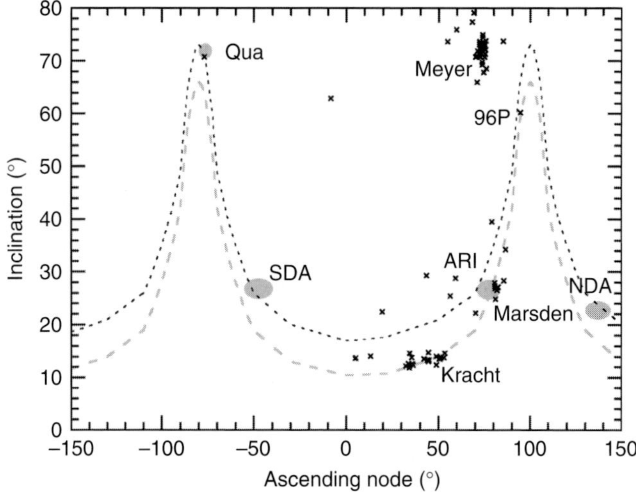

Fig. 23.3 The nutation cycle of Sungrazer comets (x) and their streams (gray). Dotted line from Fig. 20.2. The dashed line shows another possible solution.

$i = 72°$ and $q = 0.04$ AU. The most recent addition is a rich group of comets called the (Rainer) *Kracht group*, first reported on August 10, 2002. Like the Marsden group, the comets have $q \sim 0.047$ AU, but the inclination is smaller at $i = 13.4°$. Because of their larger perihelion distance, they are expected to have broken for reasons other than the tidal force from the Sun's gravity.

The Marsden and Kracht groups are related to each other (Fig. 23.3), and probably also to comet 96P/Machholz (Fig. 23.1). Machholz is now in a much steeper orbit with a higher perihelion distance, but has nearly the same length of perihelion. They may be siblings from the same progenitor body, evolved in a secular nutation cycle from a high inclination, high q orbit to a low inclination, low q orbit at different rates (Fig. 20.2).[9] The evolution is typically such that the objects spend a lot of time at the high inclination end and little time at small inclinations of the cycle. Moreover, q is low for a wide range of small inclinations. The evolution from a 96P-type orbit to that of the Kracht group may take only ~360 yr, even though the whole cycle takes about 4000 yr.

96P/Machholz has the fifth shortest known orbital period among comets of $P \sim 5.23$ yr, just below the 2:1 resonance with Jupiter, which has kept the comet orbiting on the inner side of the resonance boundary for at least 4000 yr, where its mean motion oscillates around the 9:4 resonance, keeping it >0.5 AU from Jupiter at all times.[10] It travels closer to the Sun than any known active comet with $P < 150$ yr.

[9] K. Ohtsuka, S. Nakano and M. Yoshikawa, On the association among periodic comet 96P/Machholz, Arietids, the Marsden comet group, and the Kracht comet group. *Publ. Astron. Soc. Japan* **55** (2003), 321–324.

[10] D. W. E. Green, H. Rickman, A. C. Porter and K. J. Meech, The strange periodic comet Machholz. *Science* **247** (1990), 1063–1067.

The comet measures 4–5 km in size, based on activity near aphelion, and was observed to have a sudden increase in activity shortly after passing perihelion in 1986. Sekanina believed this to be caused by a source region within 15° of the rotational north pole, emerging from the comet winter night when it passed perihelion. The outgassing area was estimated at only $0.5\,km^2$.

Until recently, the Marsden and Kracht groups were suspected to be of short orbital period like Machholz, but only parabolic orbits could be calculated from the brief interval in which the comets were observed. One of the best observed objects, C/1999 J_6, suggested a solution $P = 2–5$ yr. Then, on December 14, 2004, Brian Marsden announced the discovery of yet another such Sunskirter, C/2004 V_9, that appeared to be the return of comet C/1999 J_6, which implied $P = 5.49$ yr, $q = 0.0491$ AU, $i = 26.58°$, $\omega = 22.32°$, and node = 81.680° (epoch November 11, 2004), with a perihelion in early May.[11]

This comet had passed only 0.0091 AU from the Moon and 0.0087 AU from the Earth on June 12.22, 1999 and June 12.31 UT, respectively. Only the Marsden group members pass very close to Earth's orbit in this manner, if their perihelion happens to be on May 11–12. If they survive perihelion, then they will continue to pass close to Earth itself on June 10–11 and are visible from the northern hemisphere. This particular comet was not detected, probably because it was only between $+19$ and $+21^m$, according to Sekanina and Chodas.

23.1 The Machholz complex showers

David Seargent of the Australian Comet Section first remarked on the similarity of the Marsden group comets to that of the *Daytime Arietid* meteoroid stream (Fig. 23.4).[12] The agreement is quite good, with a discrepancy only in the perihelion distance, which is about twice as large as that of the Marsden group comets, while the orbital period may be smaller.[13] The Earth samples only the outskirts of the meteoroid stream.

The Arietids are a strong shower with a colorful history, mostly hidden from view in the daytime. Fortunately, such daytime showers can be heard on radio. When meteoroids hit the atmosphere, they ionize the air to create a long antenna of ionized gas that can re-emit radio signals (so called *under-dense echoes*) or reflect them back (*over-dense echoes*). Since the late 1940s, astronomers have bounced back radar pulses to study these showers. Amateur astronomers may be more familiar with the "pings" and "bursts," respectively, of AM/FM waves reflected over the horizon from distant radio and TV stations. Fig. 23.5 shows meteor counts obtained in this manner during Spring and Summer by amateur astronomer *Ilkka Yrjölä* of Finland. He records the carrier wave of stations that can not be heard directly at frequencies where there are no local

[11] B. G. Marsden, COMETS C/2004 V9, 2004 V10 (SOHO). *MPEC 2004-X73* (Dec. 14, 2004).

[12] R. Kracht, M. Meyer, D. Hammer, B. G. Marsden and D. A. J. Seargent, Comets C/1999 173, 2002 E1 (SOHO). *MPEC 2002-E18* (2002).

[13] B. G. Marsden, Comets C/1998 A2, 1998 A3, 2000 B8 (SOHO). *MPEC 2002-E25* (2002).

	a (AU)	q (AU)	i (°)	ω (°)	Ω (°)	Π (°)
Marsden group	3.27	0.048	26.8	23.24	81.46	104.7
Daytime Arietids	1.56	0.090	26.5	27.7	76.7	104.4
δ-Aquariids North	2.54	0.071	23.0	332.6	139.0	111.6
δ-Aquariids South	3.11	0.087	26.4	148.9	312.2	101.1

Fig. 23.4 Comparison of orbital elements. Angles in J2000.

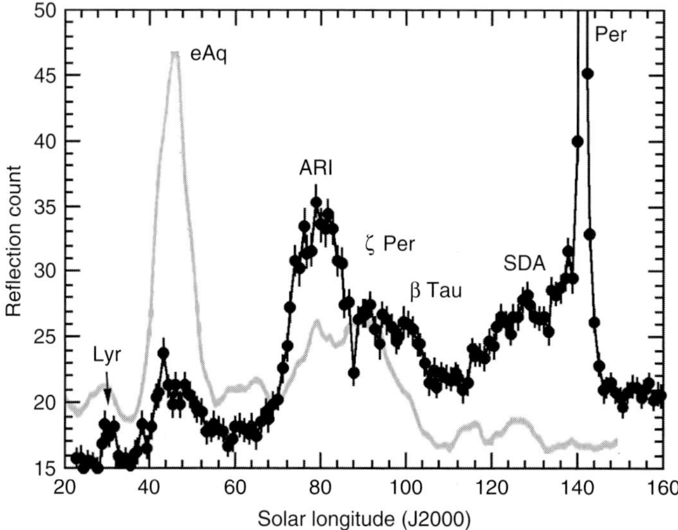

Fig. 23.5 The daily rate of forward meteor scatter counts from observations by Ilkka Yrjölä, Finland (+61 N, ●) and AMOR counts (−43 S, gray line) of apex source meteor orbits.

stations. For best results, that station should be strong (100 kW) and far away (700–1200 km). If a close transmitter is used, then fainter meteors are sampled and the echo-height ceiling becomes more important, which does not condone picking up fast meteor showers.

The Daytime Arietids (ARI) and other daytime showers were discovered at Jodrell Bank, UK, in 1947. Their discovery is narrated in, and was a big motivation for, *Bernard Lovell*'s influential book *Meteor Astronomy*.[14] The discovery was made immediately after the wartime radar systems had been adapted for meteor studies. During the war years, meteor reflections were a nuisance that interfered with their military use. The showers were first detected by *James Stanley Hey* and *G. S. Stewart*

[14] A. C. B. Lovell, *Meteor Astronomy* (Oxford: Clarendon Press, 1954), 463pp.

during the summer of 1945.[15] But their true significance was only apparent, when in the winter of 1946 the Geminids were observed and it was found that radar rates were in proportion to the rate of visible meteors.

What then happened is best told in the words of Lovell, who wrote: *The results obtained with this apparatus in daylight were not exceptional until May. Then, during the investigation of the eta-Aquariid shower on 1947 May 1, it was found that the well-known shower with its radiant near eta-Aquarii was not an isolated event, but merely the beginning of a great belt of meteoric activity extending towards the Sun, observable only in daylight. Initially, the main radiant was in Pisces, but the phenomena developed with great rapidity, and by the end of June at least seven centers of considerable activity had been delineated, extending in the ecliptic up to right ascension 90 degrees. The day-time activity continued throughout July and August, and comparison with the known major showers indicated that it was without precedent in extent and duration.*

For the connoisseurs, some meteors from the Daytime Arietid shower are visible by the naked eye from southern latitudes, albeit with a very low radiant position in the early morning hours in the first two weeks of June, with typical apparent rates of only 1–2/h.[16] Since the radiant lies 32° west of the Sun, their visibility is limited to the hour or two before dawn.

The most active center of radiation was in Aries, with a peak on June 8.[17] Most observations appear to show that there are two centers of radiation, at slightly different declinations. *Mary Almond* first found an increase of inclination and perihelion with position in the shower. This was disputed in later work,[18] but confirmed more recently. The latest radar measurements of $+8^{m}$ meteors ($M \sim 10^{-5}$g) by *Margaret Campbell-Brown*[19] showed that the semimajor axis decreased from $a = 2.0$ to 1.3 AU and q changed from $q = 0.115$ to 0.070 AU with increasing node (−0.0020 AU/°), while e remained constant at about $e = 0.945$. At the same time, the inclination changed from $i = 25°$ to 35°, when the node changed from 69° to 91° (+0.45°/°). This shows that the stream is highly stratified and we sample different parts of the stream at different times.

Interestingly, the Daytime Arietids that are observed at the shortest solar longitudes have semimajor axes and inclinations most like those of the Marsden group, but have

[15] J. S. Hey and G. S. Stewart, Radar observations of meteors. *Proc. Phys. Soc.* **59** (1947), 858–883; J. S. Hey, S. J. Parsons and G. S. Stewart, Radar observations of the Giacobinids meteor shower, 1946. *Mon. Not. R. Astron. Soc.* **107** (1947), 176–183.

[16] J. Wood, Australia, alpha Crucids, alpha Centaurids, gamma Normid, Lyrids, eta Aquariids and 1983 Arietids. *WGN IMO* **12** (1984), 7.

[17] J. A. Clegg, V. A. Hughes and A. C. B. Lovell, The daylight meteor streams of 1947 May–August. *Mon. Not. R. Astron. Soc.* **107** (1947), 369–378; A. Aspinall, J. A. Clegg and A. C. B. Lovell, The daytime meteor streams of 1948. Part I. Measurement of the activity and radiant positions. *Mon. Not. R. Astron. Soc.* **109** (1949), 352; J. G. Davies and J. S. Greenhow, The summer daytime meteor streams of 1949 and 1950. II. Measurement of the velocities. *Mon. Not. R. Astron. Soc.* **111** (1951), 26.

[18] G. Gartrell and W. G. Elford, Southern hemisphere meteor stream determinations. *Austr. J. Phys.* **28** (1975), 591–620.

[19] M. Campbell-Brown, Radar observations of the Arietids. *Mon. Not. R. Astron. Soc.* **352** (2004), 1421–1425.

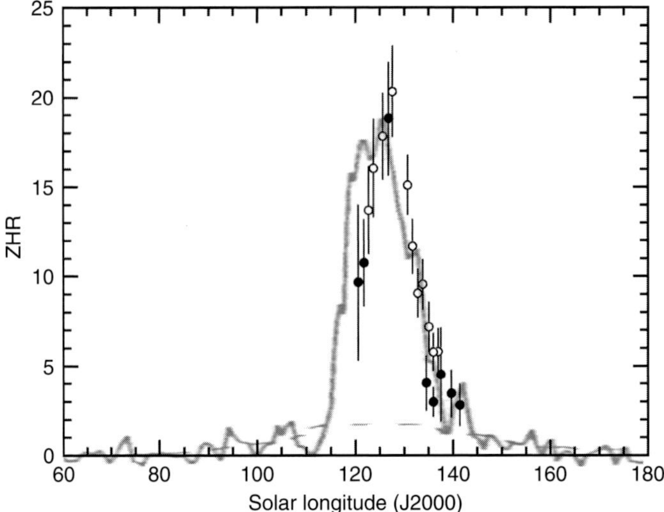

Fig. 23.6 The activity curve of the Southern δ-Aquariids from observations by the Dutch Meteor Society (●) and NAPO-MS (○),[20] and compared with scaled Southern δ-Aquariid rates measured by the AMOR radar (solid line, in arbitrary units).[21]

also the highest (most discrepant) q. Only the meteoroids with highest perihelion distance hit Earth.

The δ-Aquariids

The Marsden group has elongated elliptical orbits that pass Earth's orbit at the preperihelion arc in July, but now above the ecliptic plane. The active Southern δ-Aquariid shower is encountered at that other node with a peak on July 28 (Fig. 23.6).[22] Although the shape of the orbit and the length of perihelion are the same, the δ-Aquariids have a perihelion above the ecliptic plane, while the Arietids have a perihelion below the plane. The showers represent different parts of the nutation cycle (Fig. 23.3). Again, the perihelion distance of the meteoroids is larger than that of the Marsden group Sungrazers but similar to the Arietids.

Four streams are active in the same timeframe and have radiants in the same general direction on the sky. Those are the Southern δ-Aquariids (SDA), the Northern δ-Aquariids (NDA), the Southern ι-Aquariids (SIA) and the Northern ι-Aquariids (NIA).

[20] P. Jenniskens, Meteor stream activity. I. The annual streams. *Astron. Astrophys.* **287** (1994), 990–1013.

[21] D. P. Galligan and W. J. Baggaley, Wavelet enhancement for detecting shower structure in radar meteoroid orbit data. In *Dust in the Solar System and Other Planetary Systems*, ed. S. F. Green, I. P. Williams, J. A. M. McDonnell and N. McBride. (Amsterdam: Pergamon, 2002), pp. 42–47.

[22] M. Almond, The summer daytime meteor streams of 1949 and 1950. III. Computation of the orbits. *Mon. Not. R. Astron. Soc.* **111** (1951), 37–44.

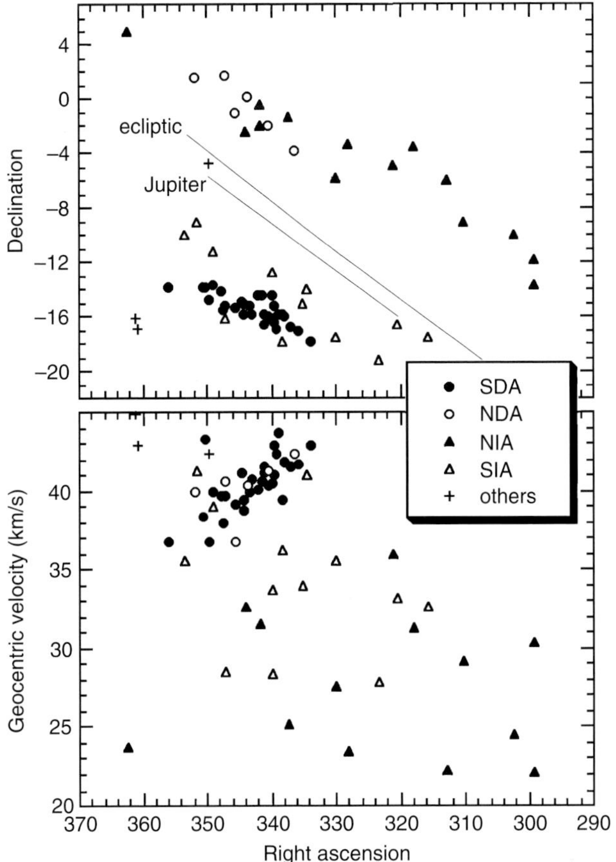

Fig. 23.7 The radiant position of Aquariid showers in July and August from meteor orbits in the IAU, DMS, and MORP databases.

First detected as a minor shower as early as 1849 by *I. F. Julius Schmidt*,[23] the Southern δ-Aquariids were observed frequently by Denning in 1869–1898.[24] Peak activity appears to have held fairly constant at peak ZHR $= 18 \pm 4$ over the past century.[25] There were no past meteor storms, although others have suggested that the showers caused the AD July 10.7, 784 and July 15.8, 714 showers, as well as fireballs seen in the period AD 1022–1095 (Table 1).

The Southern δ-Aquariids stand out well above the sporadic background in late July and are easily recognized in photographic orbit surveys (Fig. 23.7). The

[23] I. S. Astapovic and A. K. Terenteva, Fireball radiants of the 1–15 centuries. In *Physics and Dynamics of Meteors*, ed. L. Kresák and P. M. Millman, *IAU Symp.* 33, (Dordrecht: Reidel, 1968), pp. 308–319.

[24] W. F. Denning, General catalogue of the radiant points of meteoric showers and of fireballs and shooting stars observed at more than one station. *Mem. R. Astron. Assoc.* **53** (1899), 202–292.

[25] R. A. MacKenzie, *Solar System Debris* (Dover: British Meteor Society, 1981), p. 13.

shower has been the target of photographic meteor surveys,[26] and has also been studied by radar[27] and video observations.[28] It is especially well seen from southern hemisphere locations,[29] but was first investigated in the southern hemisphere only in the period 1926–1933.[30] The δ-Aquariid shower is the dominant shower in the AMOR orbit radar data.[31] Even Arietids and Geminids are not detected this well.

Interestingly, visual observers with a limiting magnitude of $+6.5^m$ observe the same activity profile as the AMOR radar with a limiting magnitude of $+14^m$ (Fig. 23.6). Whatever is responsible for the nodal dispersion in the shower, is not strongly mass-dependent.

The Southern δ-Aquariids have a relatively narrow dispersion in declination and a distinct increase in speed as a function of right ascension (bottom, Fig. 23.7). There is no change of the radiant position with meteoroid mass, as suggested by Sarma and Jones.[32] The magnitude distribution index is a high $\chi = 3.3$ ($s = 2.30$). Znojil found that the maximum of the bright meteors preceded the faint ones,[33] but that is not confirmed by the AMOR results.

There is a nonrandom dispersion of orbital elements, with the eye recognizing a hollow distribution in the diagram of inclination versus perihelion distance (Fig. 23.8). This, again, suggests that Earth samples the meteoroids from a stratified assembly. The perihelion distance increases with increasing node by $+0.0041$ AU/° (compared to -0.0020 AU/° for the Daytime Arietid shower) and eccentricity decreases (-0.0017/°) for a constant semimajor axis. The argument of perihelion increases by $+0.29°$/°. The inclination decreases by $-0.25°$/° (compared to $+0.45°$/° for the Arietids), but is less correlated. I conclude that the breakup that led to the Southern δ-Aquariids happened longer ago than that which caused the Daytime Arietids.

[26] F. W. Wright, L. G. Jacchia and F. L. Whipple, Photographic delta Aquariid meteors and evidence for the northern iota Aquariids. *Harvard Reprint Ser. II* **108** (1957), 225–233; F. L. Whipple, Photographic iota-Aquariid evidence for northern delta Aquariids. *Astron. J.* **62** (1957), 225–233.

[27] D. W. R. McKinley, Radio determination of the velocity and radiant of the delta Aquariid meteors. *Astrophys. J.* **119** (1954), 519–530.

[28] T. J. Jopek and Cl. Froeschlé, A stream search among 502 TV meteor orbits. An objective approach. *Astron. Astrophys.* **320** (1997), 631–641.

[29] R. A. McIntosh, Ephemeris of the radiant-point of the delta Aquariid meteor stream. *Mon. Not. R. Astron. Soc.* **94** (1934), 583–587; R. A. McIntosh, The delta Aquariids and contemporary meteor showers. *Observatory* **53** (1930), 235–236.

[30] R. A. McIntosh, The delta Aquariids and contemporary meteor showers. *Observatory* **53** (1930), 235–236; R. A. McIntosh, Ephemeris of the radiant point of the delta Aquariid meteor stream. *Mon. Not. R. Astron. Soc.* **94** (1934), 583–587.

[31] D. P. Galligan, Radar meteoroid orbit stream searches using cluster analysis. *Mon. Not. R. Astron. Soc.* **340** (2003), 899–907; D. P. Galligan, A direct search for significant meteoroid stream presence within an orbital data set. *Mon. Not. R. Astron. Soc.* **340** (2003), 893–898.

[32] T. Sarma and J. Jones, Television observations of the delta-Aquariid shower. In *Solid Particles in the Solar System. IAU Symp.* 90 (August 1980), p. 167.

[33] V. Znojil, Occurrence of minor particles in summer meteor streams of the Northern hemisphere. *Bull. Astron. Inst. Czech.* **33** (1982), 201–210.

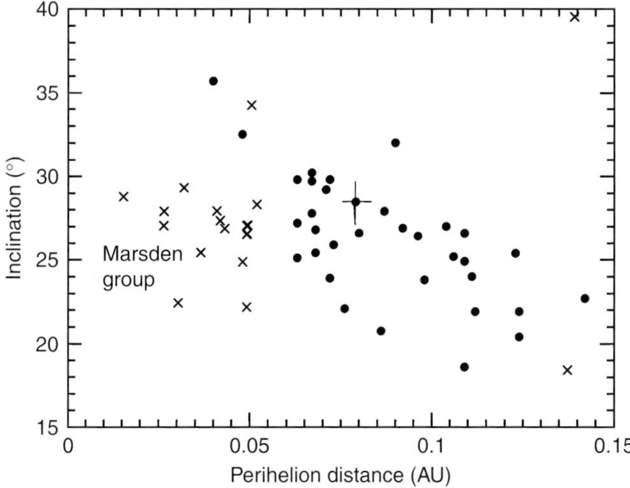

Fig. 23.8 Southern δ-Aquariids: change of inclination with perihelion distance.

23.2 Origin and evolution of the Machholz complex showers

Based on the orbital evolution of the Quadrantid shower (Fig. 23.9),[34] Hamid and Whipple first proposed that the Arietid and the δ-Aquariid stream had a common origin with the Quadrantids, having evolved over at least one cycle of ~4000 years.[35] Based on the secular perturbations, *Pulat Babadzhanov*[36] recognized five other nodal crossings (including one identified as the Ursids, now proven to be from 8P/Tuttle, Chapter 16).[37]

Since, I have shown that the Quadrantid shower is not old enough (<500 yr) for all meteoroids to have originated from Whipple-type ejection that long ago.[38] Before reading the recent paper by Sekanina and Chodas (see below), I concluded from the diagrams by Williams *et al.* (Fig. 23.9)[39] that the original parent body broke up about 1800 years ago (around AD 200), when in a high inclination orbit (which last longest), after which individual fragments evolved at different rates along the nutation cycle, only then to break and create the different meteoroid streams at different times in a

[34] S. E. Hamid and F. L. Whipple, Common origin between the Quadrantids and the delta Aquariids streams. *Astron. J.* **68** (1963), 537; A. F. Zausaev and A. N. Pushkarev, Orbit evolution of the delta-Aquariids meteoroid shower over 4000 years (1950 AD to 2050 BC). *Akad. Nauk Tadzhik. SSR, Dokl.* **31** (1988), 787–790.

[35] P. B. Babadzhanov and A. F. Zausaev, Orbital evolution of the Quadrantid, delta Aquariid, and alpha Capricornid meteor showers. *Akad. Nauk Tadzhik. SSR, Inst. Astrofiz., Biull.*, No. 65, 40–50 (1975). In Russian.

[36] P. B. Babadzhanov, Yu. V. Obrubov and A. N. Pushkarev, Evolution of the Quadrantid meteoroid swarm. *Solar Syst. Res.* **25** (1991), 63–70; B. P. Babadzhanov and V. Yu. Obrubov, Dynamics and relationship between interplanetary bodies: comet P/Machholz and its meteor showers. In *Meteoroids and Their Parent Bodies*, ed. J. Stohl and I. P. Williams. (Bratislava: Astron. Inst. Slovak Acad. Sci., 1993), p. 49.

[37] P. Jenniskens, E. Lyytinen, M. C. de Lignie *et al.*, Dust trails of 8P/Tuttle and the unusual outbursts of the Ursid shower. *Icarus* **159** (2002), 197–209.

[38] P. Jenniskens, H. Betlem, M. de Lignie, M. Langbroek and M. van Vliet, Meteor stream activity. V. The Quadrantids, a very young stream. *Astron. Astrophys.* **327** (1997), 1242–1252.

[39] I. P. Williams, C. D. Murray and D. W. Hughes, The long-term evolution of the Quadrantid meteor stream. *Mon. Not. R. Astron. Soc.* **189** (1979), 483–492.

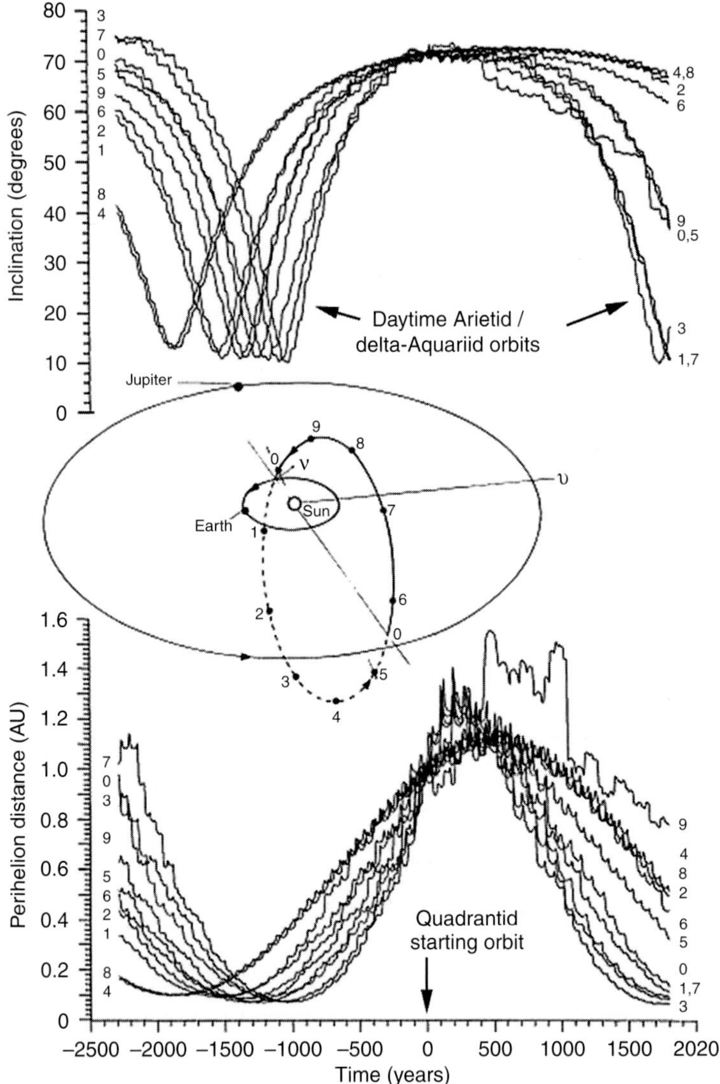

Fig. 23.9 Orbital evolution of meteoroids on an initial Quadrantid-like high-inclination orbit. Graphs from Williams *et al.* (1979).[40]

progressive pattern of fragmentation. This would explain the isolated nature of the Marsden and Kracht groups and the different dispersions of the streams encountered by Earth. The evolution from an Arietid-like orbit to that of the δ-Aquariids is fast, about 400 years according to these graphs, and the dispersion could have been accomplished in the same amount of time.

40 *Ibid.*

Recently, Sekanina and Chodas[41] studied the evolution of the Machholz complex by statistically examining all the possible evolution scenarios of fragments released from a progenitor body, assuming that 96P/Machholz was the largest remaining fragment. They identified the mechanism that caused the evolution along the cycle at different rates. They found that the Marsden and Kracht groups (and the Southern δ-Aquariids, Fig. 23.3) can be explained by a breakup most likely some time around AD 710, 530, 340, or 170, but certainly before AD 950, followed by a close encounter with Jupiter in AD 1059. Depending on the severity of that specific encounter, a series of subsequent close encounters of these fragments accelerated the evolution along the cycle differently for different fragments. In later years, the comets continued to evolve by progressive bulk fragmentation and erosive sublimation, each fragment losing $\sim$6 m of their surface area in each return. All groups and individual bodies continued to evolve along the cycle, with the δ-Aquariids having hit the lowest inclination in AD $\sim$1820, while the Kracht group will follow in 2020, and the Marsden group in 2110. 96P/Machholz will not reach this phase until AD $\sim$2450.

23.3 Earlier progenitors

The progenitor of comet 2003 EH_1, associated with the Quadrantid shower (Qua), could be part of the same group (Fig. 23.3), but only if one fragment had an AD 1059 encounter with Jupiter as close as 0.02 AU or less. Sekanina and Chodas pointed out that likely progenitor bodies for the Machholz complex did not evolve into Quadrantid-like orbits, nor orbits similar to those of the other suspected high-inclination meteoroid streams (Ursids, Puppids-Velids). Also, progenitor bodies did not include low-inclination orbits such as those of Babadzhanov's α-Cetids and N ι-Aquariids (their "Northern δ-Aquariids"). Perhaps such showers were created from fragments that evolved from an earlier progenitor (or progenitors) rather than those responsible for the Machholz complex. This has not yet been studied.

The Northern *δ-Aquariids* have the same speed, and speed-dependence as the southern branch, and are found symmetrically relative to Jupiter's orbital plane (Fig. 23.7). They are active over the same time interval as the southern branch (Fig. 23.10). However, their dispersion (an equivalent cross section of 13.8°) is larger than that of the southern branch (9.5°) and larger than the Daytime Arietids (8.7°). Their nodes coincide with a prior cycle, representing the same phase in the cycle as the Southern δ-Aquariids. (Fig. 23.3). It is not clear at present how they would fit into the fragmentation scenario, other than by extending the fragmentation scenario futher back in time.

There is evidence of other clusters of comets (i.e. the Meyer group in Fig. 23.2) and of other meteoroid streams with low perihelion distances that have different longitudes of perihelion (gray markings in Fig. 23.2). One of those, the *o-Cetids*, was

[41] Z. Sekanina and P. W. Chodas, Origin of the Marsden and Kracht groups of sunskirting comets. I. Association with comet 96P/Machholz and its interplanetary complex. *Astrophys. J.* Supp. Ser. 161 (2005), 551–586.

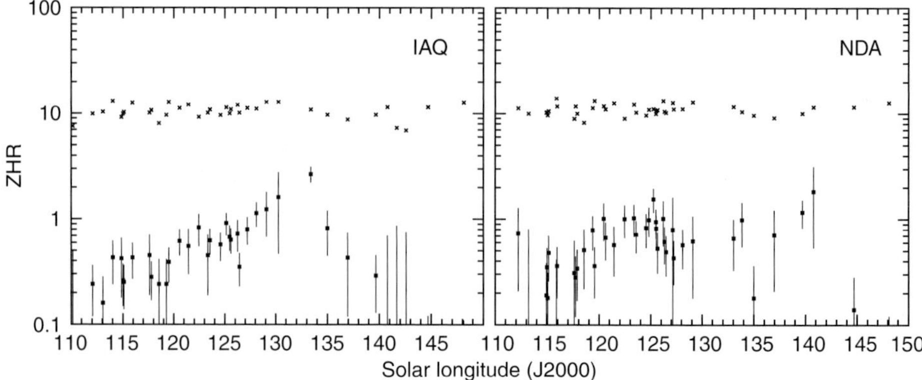

Fig. 23.10 Shower activity curves from visual observations for southern ι-Aquariids and northern δ-Aquariids. Crosses mark sporadic hourly rates. Data: Jeff Wood NAPO-MS.[42]

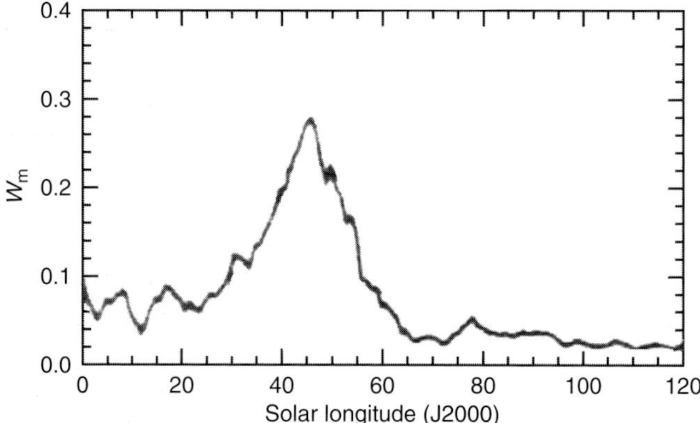

Fig. 23.11 Rate of *o*-Cetids detected in AMOR radar data using a 3° wavelet, a 6° sliding $\lambda_\odot$ window and V_g of 30 ± 10 km/s.

recognized in the radar surveys by *A. Aspinall* and *Gerald S. Hawkins* of the Jodrell Bank radio observatory in the summer of 1950. The stream is active from May 13 to 23. Its perihelion is close to that of the Arietid shower. It appears that earlier Daytime "Piscids" were in fact this shower. Because the argument of perihelion is 211° and the inclination is a high 34°, the only encounter with Earth's orbit is in May. The stream was strong in AMOR meteor radar detections (Fig. 23.11),[43] and demonstrates a very broad activity curve. The database contains more *o*-Cetids than Arietids or Geminids. This suggests that the shower is rich in small meteoroids, just like the δ-Aquariids. The

[42] P. Jenniskens, Meteor stream activity I. The annual streams. *Astron. Astrophys* **287** (1994), 990–1013.

[43] D. P. Galligan and W. J. Baggaley, Wavelet enhancement for detecting shower structure in radar meteoroid orbit data. II. Application to the AMOR data set. In *Dust in the Solar System and other Planetary Systems*, ed. S. F. Green, I. P. Williams, J. A. M. McDonnell and N. McBride. (Amsterdam: Pergamon, 2002), pp. 48–60.

length of perihelion for this stream $\Pi \sim 80^\circ$ is smaller than that of the Marsden group. This stream is not part of the same nutation cycle.

The *ι-Aquariids*, both South and North (Table 7), have the same length of perihelion, $\Pi = 84^\circ$ and 83° respectively, but a smaller inclination ($\sim 7^\circ$). Both have a relatively low perihelion distance, albeit not as small as that of the o-Cetids. Only one Sungrazer has been found with a similar longitude of perihelion, C/2004 E_2 (SOHO), but it has perhaps too small a perihelion distance to be related.

The *Southern ι-Aquariids* have radiants close to the Southern δ-Aquariids, but can be discriminated by their lower speed. This shower is more diffuse even than the Northern δ-Aquariids. The Northern ι-Aquariids have a low inclination and a wide range of perihelion distances. The Northern ι-Aquariids are clearly more evolved than the southern ι-Aquariids and are not distributed symmetrically relative to the orbital plane of Jupiter. The relationship between these streams, and to that of the Machholz group if any, is unclear at present.

23.4 Space weathering

If all Machholz-complex streams originated from the same progenitor (presumably having a homogenous meteoroid composition and morphology), but were released into the interplanetary medium by the breakup of its fragments at different times, then these streams may be used to study the effect of space weathering on the meteoroids, not in the least because the metoroids are heated to $\sim$1000 K at $q = 0.09$ AU.[44]

Jacchia *et al.*[45] found that both δ-Aquariids and Quadrantids are efficiently decelerated in the Earth's atmosphere. Both showers have a high fragmentation index ($\chi_f = 0.52 \pm 0.08$, SDA, and 0.44 ± 0.11 for the QUA), showing that the breakup of the meteoroid quickly accelerates when the meteoroid penetrates deeper in the Earth's atmosphere. Not as efficient as Capricornids, but better than Perseids and Taurids. Both show abundant wake, perhaps also indicative of small fragments. Hence, the sintering of the grains is skin deep.

The breakup sequence QUA (recent, high q), SDA (older, low q), SIA (even older, low q), NIA (oldest, low q) is reflected in the particle densities calculated by Verniani:[46] $\rho = 0.17$ g/cm^3 for the Quadrantids, 0.27 g/cm^3 for the δ-Aquariids, 0.32 g/cm^3 for the Southern ι-Aquariids, and 0.63 g/cm^3 for the Northern ι-Aquariids. No density estimates are available for the Arietids, but they were found to peak at elevations of about 92 km with a narrow distribution of heights, whereas Quadrantids of a similar size peaked at $\sim$95 km, suggesting the latter are slightly more fragile. But both are relatively young streams.[47]

44 B.-A. Lindblad, The peculiar orbit of the delta Aquariid meteors. *Observatory* **73** (1953), 157–159.
45 L. G. Jacchia, F. Verniani and R. E. Briggs, An analysis of the atmospheric trajectories of 413 precisely reduced photographic meteors. *Smithsonian Contrib. Astrophys* **10** (1967), 1–139.
46 F. Verniani, Meteor masses and luminosity. *Smithsonian Contrib. Astrophys.* **10** (1967), 181–195.
47 D. W. R. McKinley, *Meteor Science and Engineering* (New York: McGraw-Hill, 1961).

24

α-Capricornids and κ-Cygnids

The *α-Capricornids* and *κ-Cygnids* are considered to be old streams that are just about to disappear into the sporadic background. They are active in July and August as a sidekick to the summer Perseids. Both streams are very dispersed, stretching from July into early September, have a large radiant area, reportedly with multiple centers, and are detected just above the sporadic background. No parent bodies are known, although many have been proposed that would have evolved far from the current stream.

I will now challenge that perception, and argue that these are relatively young streams created in the breakup of a parent body not long ago, much like the Sunskirter streams. My arguments are that the meteoroid orbits are not randomly dispersed, the fireballs look fresh, and a spectacular outburst of κ-Cygnids was observed in 1993.

24.1 Stream structure

Activity from the constellation of Capricorn was first mentioned in the 1871 account by *N. de Konkoly* in Hungary, who plotted six meteors from a radiant at R.A. = 307°, Decl. = −4° during July 28–29.[1] Later that century, Denning[2] knew the Capricornids as an annual shower of slow and often bright meteors from the antihelion source direction. Visual observers continued to map the radiant and its daily movement,[3] but the ecliptic low-inclination stream was established only with the publication of 12 multistation photographic orbits from the Harvard Meteor Survey in 1956 (Fig. 24.1).[4] Several of the Capricornids detected between July 16 and August 1 had similar orbits ($i \sim 8°$), and this stream was called the *α-Capricornids*.

A second more diverse group in the August 8–22 timeframe (August Capricornids) had lower inclinations, while a third group with smaller semimajor axis was found active during the same timeframe as the α-Capricornids. In the latest analysis of

[1] G. Kronk, *Meteor Showers. A Descriptive Catalogue* (Hillside, NJ: Enslow, 1987).
[2] W. F. Denning, Meteor-shower in Capricornus. *Observatory* **16** (1893), 356–357.
[3] R. A. McIntosh, An index to southern meteor showers. *Mon. Not. R. Astron. Soc.* **95** (1935), 709–718.
[4] F. W. Wright, L. G. Jacchia and F. L. Whipple, Photographic α-Capricornid meteors. *Astron. J.* **61** (1956), 61–69.

Fig. 24.1 The characteristically irregular lightcurves of the 00:15:58 UT -4^{m} α-Capricornid of July 31, 1981, photographed in the constellation Camelopardalis by Klaas Jobse at Oostkapelle and a -2^{m} α-Capricornid (inset) photographed by Casper ter Kuile from Biddinghuizen on August 4, 1995 at 23:16:27 UT. Photo: DMS.

meteoroid orbits by Hasegawa,[5] as many as ten streams are distinguished, tentatively linked to seven potential parent bodies.

These latter streams are more spurious. The *α-Capricornids*, on the other hand, can be recognized by visual observers from early July until mid August (Fig. 24.2).

They are also evident in the 2003 *IAU Photographic Meteor Orbit Database* for July and August (Fig. 24.3). Black dots in the lower half of the diagram identify all meteors that are called "α-Capricornids" (CAP). Some are clearly misidentified, but others are dispersed along a series of narrow strings that distinguish these meteors from the sporadic background.

The α-Capricornids are active from a radiant at: R.A. $= 306.2 + 0.538°\ (\lambda_{\odot} - 128.0°)$ and Decl. $= -8.4 + 0.245°\ (\lambda_{\odot} - 128.0°)$ in the period July 19 until August 18 ($\lambda_{\odot} = 116° - 146°$). The other showers scatter much wider than the α-Capricornids.

The radiant positions are distributed along narrow strings and when plotting the inclination of the meteoroids versus their longitude of perihelion (Fig. 24.4), I recognize at least three. Those could be the product of three fragments. Because both nodes are near Earth's orbit (the argument of perihelion is close to 270°), there are no close encounters with Jupiter that could create such a structure.

There is a remarkable stratification in the stream. Although the stream is dispersed by 30° in solar longitude in Earth's path, the node is much less dispersed at the other

5 I. Hasegawa, Parent objects of α-Capricornid meteor stream. *ESA SP* **495** (2001), 55–62.

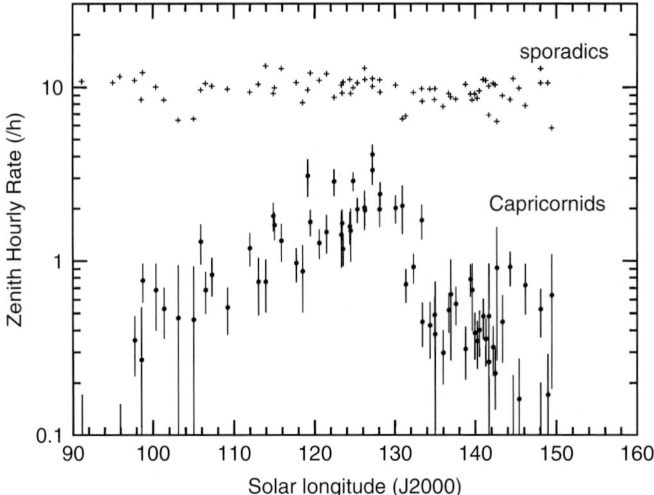

Fig. 24.2 A meteor shower activity curve derived from DMS and NAPO-MS counts of Capricornid meteors.

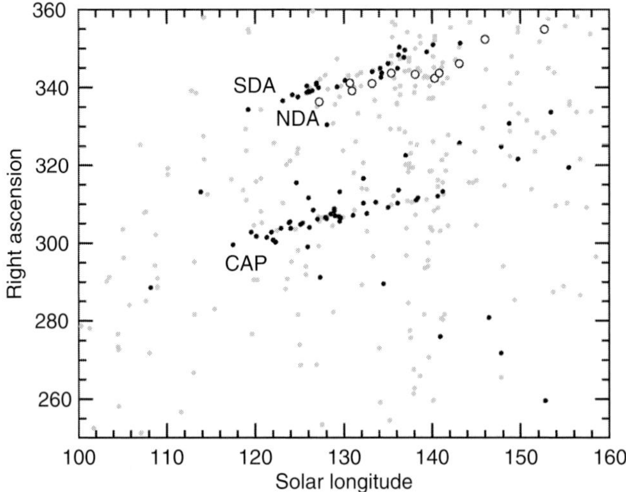

Fig. 24.3 Radiant drift for the ecliptic summertime meteors of July and August ($-30^\circ <$ Decl. $< +30^\circ$). Black dots and open circles are stream members as identified in the 2003 IAU photographic database.

side of the orbit. Plotted on the ecliptic plane, the other node forms a band at a steep angle to, and starting just at, Earth's orbit (Fig. 24.5).

The twin shower of this stream is a daytime shower, and would be located near θ-Sagittarii at R.A. $= 306^\circ$, Decl. $= -30^\circ$, and $V_g = 22$ km/s on January 20 ($\lambda_\odot = 299^\circ$).

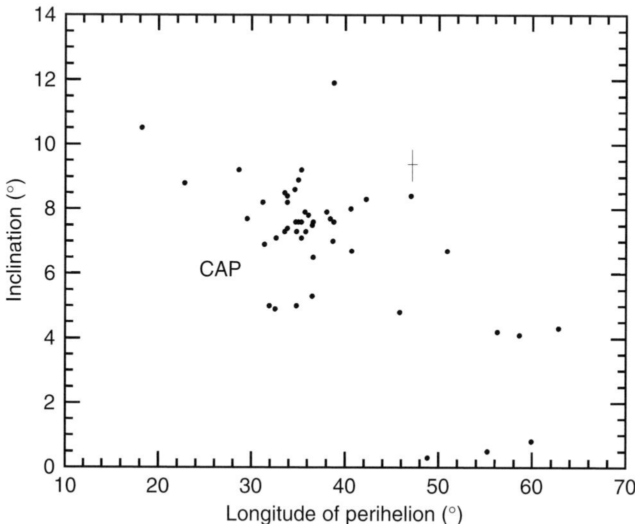

Fig. 24.4 Longitude of perihelion and inclination for all meteors labeled #1 (= α-Capricornids) in the 2003 IAU database.

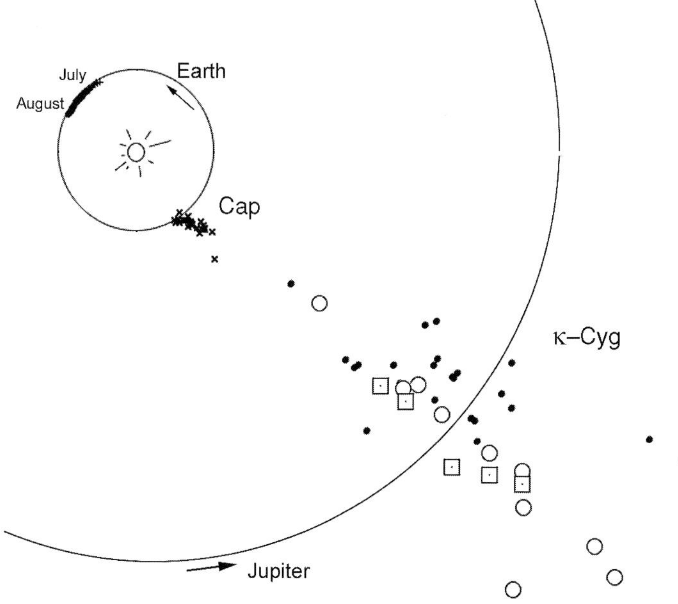

Fig. 24.5 The position of the meteoroid nodes in the ecliptic plane for α-Capricornids and κ-Cygnids.

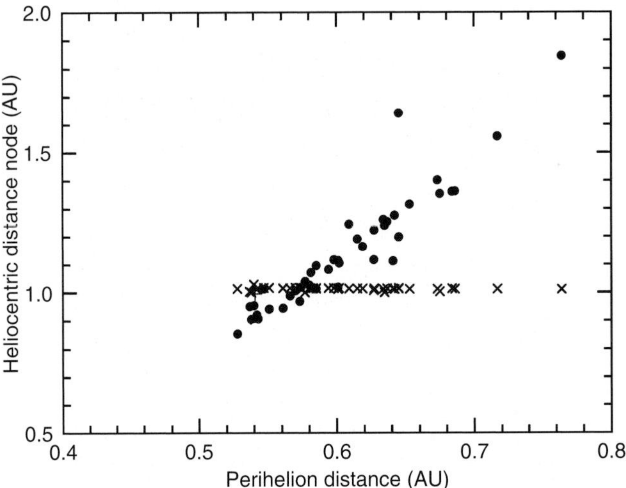

Fig. 24.6 The heliocentric distance of the ascending and descending nodes of photographed α-Capricornids. Crosses mark the node at Earth's orbit, where the meteoroids were observed.

The orbit with both nodes in Earth's path ($\omega = 270°$) has $q = 0.572$ AU, $i = 7.5°$, and $\Pi = 36°$. Unlike the Capricornids, this twin-shower should only be a few days wide.

A much more widely dispersed stream is known to be active at that time from radar observations, called the *Daytime Capricornids–Sagittariids* (Table 7), but it has a higher V_g and higher declination. Hence, this is not the α-Capricornid twin shower.

The rotation of the orbit is also manifested in a change of the perihelion distance with length of perihelion. There is a progressive increase of the perihelion distance with solar longitude and a progressive decrease of the argument of perihelion, not unlike the Arietid and δ-Aquariid showers. The effect is that the position of the opposite node from where the meteors are observed has a striking correlation with perihelion distance (Fig. 24.6).

This stream is a young shower. Given the absence of an active parent comet and two nodes near Earth's orbit, this stream was likely to have been created in a comet breakup.

The κ-Cygnids

There is another shower rich in bright meteors with spectacular flares. It is active in late August and is called the *κ-Cygnids*. The κ-Cygnids are a well-defined shower with a long activity period from August 3 to 31, peaking around August 18. Each season has some nice κ-Cygnid fireballs and, just like the Capricornids, they show more than the usual share of flares (Fig. 24.7).

This shower also appears to have been first noticed by *N. de Konkoly* of the O'-Gyalla Observatory, Hungary, during a Perseid watch in 1874. He observed

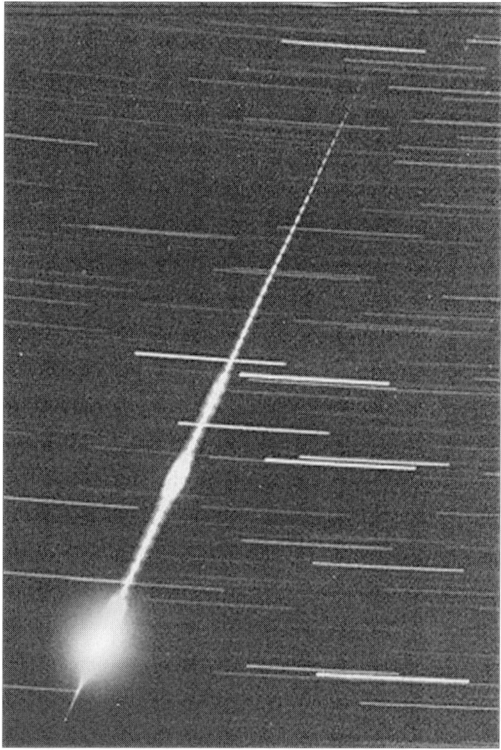

Fig. 24.7 A κ-Cygnid photographed in the constellation Delphinus during the 1993 outburst at 00:34:53 UT, August 12. Photographed from Rognes, by Robert Haas (DMS).

seven meteors from R.A. = 292°, Decl. = +50°.[6] William F. Denning gave the shower its name by noting in September 1893, how he was *struck with the frequency and brightness of meteors from a contemporary radiant on the N.W. limits of Cygnus near the star κ.*[7] Denning also published a note in 1907,[8] where he reported seeing five meteors in 40 min (on August 15, 1907), radiating from R.A. = 288°, Decl. = +61°. One bright member showed curious fluctuations in its light, suggesting these were also κ-Cygnids. He pointed out that the shower was not always active at the same level, but had been strong in August 21–25, 1879, when 56 out of 225 meteors belonged to this shower.

The shower was established in 1954 when Fred Whipple[9] published five orbits from multistation photographed meteors in the period August 9–22, with an average radiant near R.A. = 291.8°, Decl. = +53.1°. Whipple noticed the relatively large

[6] G. W. Kronk, *Meteor Showers, a Descriptive Catalogue* (Hillside, NJ: Enslow, 1987).
[7] W. F. Denning, The August meteors, 1893. *Observatory* **16** (1893), 317–319.
[8] W. F. Denning, Note on a meteoric shower (The August Draconids). *Mon. Not. R. Astron. Soc.* **67** (1907), 566.
[9] F. L. Whipple, Photographic meteor orbits and their distribution in space. *Aston. J.*, **59** (1954) 201–217; V. Letfus, The orbits of the Virginids and κ Cygnids. *Bull. Astron. Inst. Czechoslov.* **6** (1955), 143–145.

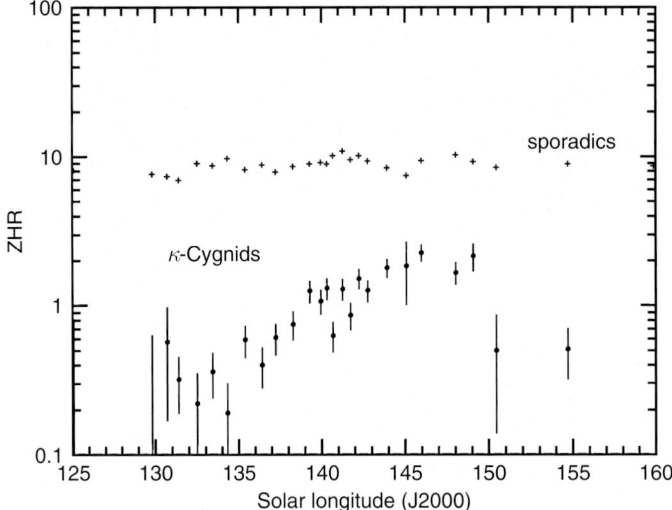

Fig. 24.8 κ-Cygnid ZHR curve.

orbital period (7–8 yr) and from the dispersion in node (Fig. 24.8) concluded that there had to be a large amount of mass in this shower, adding: *The long duration . . . suggests remnants of a large comet.*

Jacchia *et al.*[10] found that both the α-Capricornids and κ-Cygnids are more efficiently slowed down than Taurids and Perseids, but not as efficiently as Draconids. The progressive fragmentation index for the Capricornids $\chi = 0.38 \pm 0.07$ is relatively high, suggesting a rapid progression of breakup when the meteoroids penetrate deeper into the atmosphere. They also appear to be efficient in transferring kinetic energy into mass loss.

Capricornids and κ-Cygnids show wakes from shedding small fragments. Verniani[11] derived a low median particle density of 0.16 g/cm^3 for the α-Capricornids and 0.17 g/cm^3 for the κ-Cygnids, respectively, compared to 0.32 g/cm^3 for the Perseids. Taking into account quasi-continuous fragmentation by shedding of small fragments, systematically higher values are found and Babadzhanov measured a density of 1.0 g/cm^3, less than the Perseid mean of 1.2 g/cm^3.[12]

1993 κ-Cygnid outburst

In 1993, a large number of bright κ-Cygnid fireballs were recorded by members of the Dutch Meteor Society in southern France in the night of the famous 1993 Perseid

[10] L. G. Jacchia, F. Verniani and R. E. Briggs, An analysis of the atmospheric trajectories of 413 precisely reduced photographic meteors. *Smithsonian Contrib. Astrophys.* **10** (1967), 1–139.

[11] F. Verniani, Meteror masses and luminosity. *Smithsonian Contrib. Astrophys.* **10** (1967), 181–195.

[12] P. B. Babadzhanov, Formation of twin meteor showers. In *Asteroids Comets Meteors III* (Uppsala Astronomical Observatory, 1989) pp. 497–503.

Fig. 24.9 This κ-Cygnid in the Summer Triangle, photographed by the author from California at 07:51:41 UT on August 12, 1993, suggests that the outburst was still ongoing at that time.[13]

outburst. At times, the κ-Cygnids had the observers more in awe than the Perseids, with multistation meteors from one night including a − 10, four (!) − 8, a − 6, a − 4 and two + 0^m κ-Cygnids! There was no strong increase of rates among fainter κ-Cygnids. As before, the meteors erupted spectacularly in many flares (Fig. 24.7). While observing in California, I did see some of these as well (my second outburst!), and photographed one particularly bright κ-Cygnid at 07:51:41 UT, August 12 (Fig. 24.9).[14] I observed the persistent train through binoculars, curling up like a thread of wool on a damp day.

This event, by all accounts, is a meteor outburst. Nine out of 10 of those meteors radiated from a region 3.4° diameter and clearly resolved, each position measured to ±0.07° accuracy. The speed of the 1993 meteors $V_g = 23.6 \pm 1.8$ km/s was slightly higher than that of earlier κ-Cygnids, and resolved (measurement error = ±0.24 km/s).

[13] P. Jenniskens, The other side .... *Radiant, J. DMS* **15** (1993), 117–118.
[14] *Ibid.*

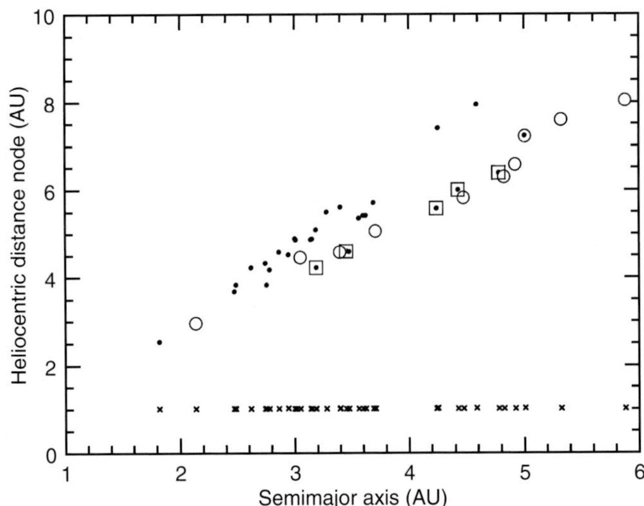

Fig. 24.10 The heliocentric distance of the nodes of *κ*-Cygnids. The open circles are all DMS orbits measured in 1993, the circles with a dot are those from 1994, and the squares with a dot are 1964 data in IAU database.

The error margins here imply a real spread of orbital elements with a significant range in the semimajor axis.

A striking result is found when plotting the semimajor axis versus the heliocentric distance of the nodes (Fig. 24.10). There is a narrow correlation, but with two different components intersecting at about $a = 2$ AU. One structure was only seen in 1993 (with a straggler photographed in 1994) and in 1964 (HV41623, 41701, 41782, and 41872). Hence, it appears that the stream responsible for the 1993/4 outburst was also detected in 1964, but not in 1963 and 1965, and not in 1996. It was also not seen in 1952, 1953, 1956, 1957, 1958, 1975, 1977, 1983, and 1989, when all photographed *κ*-Cygnids conformed to the other structure.

The dispersion is not random in other ways also. The graph of inclination versus longitude of perihelion (Fig. 24.11) has the older κ-Cygnids (o) distributed in a circle.

While many of the normal *κ*-Cygnids have a node near Jupiter's orbit, they are not clustered there, but have a distinct range of semimajor axes and orbital periods. The Tisserand parameter with respect to Jupiter is not preserved (Fig. 24.12), but stretches along the whole domain of Jupiter-family comets, perhaps even into the asteroid domain with $T_J < 3$. Interestingly, the 1993 outburst meteors are mostly clustered at $T_J = 2.1$, close to the domain of Halley-type comets ($T_J < 2$). This implies that the perturbations on the Cygnid meteoroids are not by close encounters with Jupiter at the node alone. Because of the large range in aphelion distance and orbital periods among the outburst meteors and those of other years, this outburst can not be due to a clump of dust in the 1:2 mean-motion resonance.

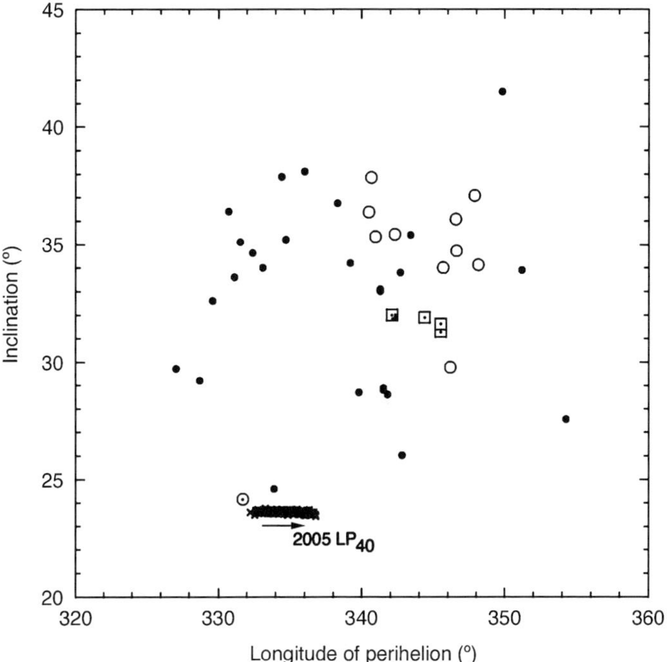

Fig. 24.11 Orbital elements of κ-Cygnids in a diagram showing inclination versus the longitude of perihelion. Symbols as in Fig. 24.10.

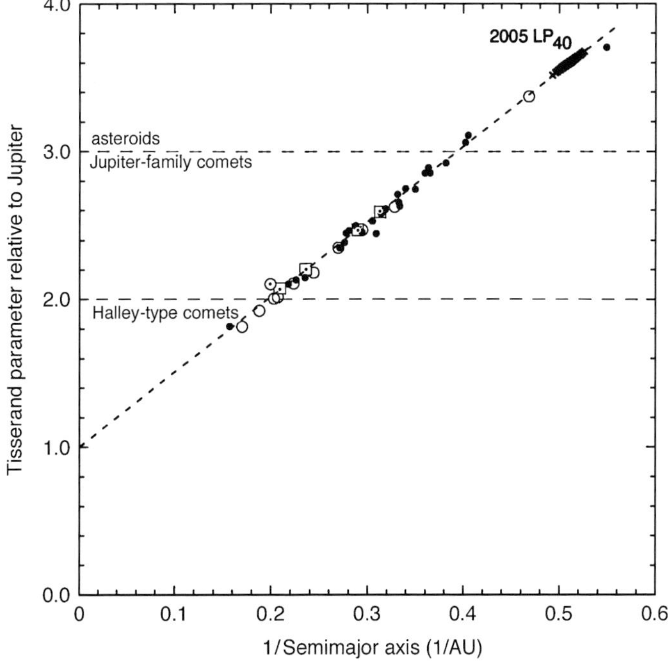

Fig. 24.12 Tisserand parameter relative to Jupiter for κ-Cygnids. Symbols as in Fig. 24.10.

Could it be that the α-Capricornids are a relatively young structure, less than one nutation cycle old and now observed near the low-i part of the cycle, while the κ-Cygnids are older, at least two cycles old and currently observed at the high-i part?

24.2 Parents bodies

As soon as α-Capricornid orbits became available, astronomers tried to identify the parent of this fascinating meteoroid stream. A large number of candidates have been proposed (Fig. 24.13), none of which fits the bill. The first candidate, comet *C/1457* L_2, was proposed by Kramer in 1953.[15] A year later, Bernhard, Bennett, and Rice[16] proposed that comet *72P/Denning–Fujikawa* was responsible (following a speculation by Denning back in 1920).[17] In 1956, Wright, Jacchia, and Whipple[18] suggested that comet *45P/Honda–Mrkos–Pajdusáková* may be the parent comet for the Capricornids seen in August. Recently, Nesulsan[19] argued that *14P/Wolf*, which was at Earth's orbit

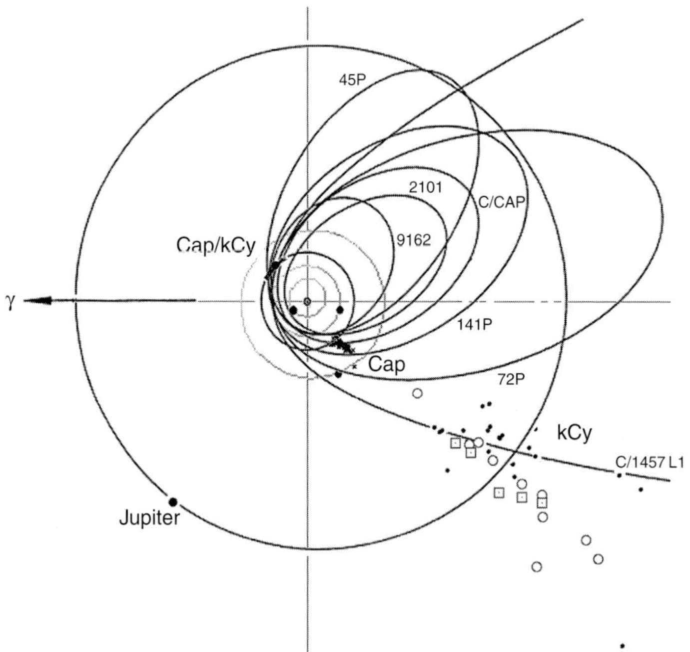

Fig. 24.13 Proposed Capricornid parents and the position of meteoroid nodes.

15 E. N. Kramer, (In: G. W. Kronk (1988) *Meteor Showers, a Descriptive Catalog* (Hillside, NJ: Enslow, 1988)

16 H.-J. Bernhard, D. A. Bennett and H.-S. Rice, Handbook of the heavens. A simple introduction to the study of the stars. *Ciel Terre* **53** (1954), 31.

17 W. F. Denning, Comet 1881 V and the shower of July–August alpha Capricornids. *Observatory* **43** (1920), 262–263.

18 F. W. Wright, L. G. Jacchia and F. L. Whipple, Photographic α-Capricornid meteors. *Astron. J.* **61** (1956), 61–69.

19 L. Neslusan, Comets 14P/Wolf and D/1892 T_1 as parent bodies of a common, α-Capricornids related, meteor stream. *Astron. Astrophys.* **351** (1999), 752–758.

only between dramatic encounters with Jupiter in 1875–1922, was a likely parent, but the agreement is less good than that for 45P.[20]

In 1973, Sekanina[21] confirmed that 45P/Honda–Mrkos–Pajdusáková was likely to be capable of producing Capricornids, but in his 1976 study of radar orbits he proposed the Apollo asteroid (2101) Adonis as a more likely alternative.[22] In 2003, Babadzhanov again implicated asteroid (2101) Adonis in producing Zekanina's *σ-Capricornids*, *χ-Sagittariids*, *Daytime χ-Capricornids* and *Capricornids–Sagittariids* showers, and pointed out that the 50 m near-Earth asteroid 1995 CS may be a fragment of the larger Adonis and perhaps responsible for the *χ-Capricornid* shower.[23] Other recently proposed minor planets include (9162) 1987 OA.[24] None of these is an obvious match.

Comet **45P/Honda–Mrkos–Pajdusáková** is now a small object with a diameter of only about 0.68 km (assuming 4% albedo), and an axis ratio of at least 1.3.[25] The nucleus has a reddish color at ultraviolet wavelengths, but a nearly neutral gray color in the near-infrared. About 11% of the surface is estimated to be active, with a dust production rate of 1.0 kg/s, or 86 metric tons/d. Sitarski[26] concluded that the comet motion was controlled by a precessing rotating nucleus with a lag angle of $12.16 \pm 0.85°$, an equatorial obliquity of $84.44 \pm 0.56°$ and a cometo-centric solar longitude at perihelion of $161.62 \pm 0.80°$, a precession factor of $0.421 \pm 0.003 \times 10^6$, an oblateness of the nucleus of 0.437 ± 0.014, and an asymmetric activity with respect to perihelion of -13.65 ± 0.43 d. Because of the small size, the comet motion is strongly affected by the nongravitational forces from the ejection of the grains.[27] There are frequent close encounters with Jupiter at aphelion.[28] The current small perihelion distance is unusual, but will persist until into the twenty-second century. Showers from the current orbit were first possible following a close encounter in December 2, 1876 (0.050 AU). In recent years, the most conspicuous change followed a close encounter with Jupiter in 1983 (to 0.111 AU). The orbit inclination is small and sometimes the plane tilts the opposite way, reversing the ascending and descending nodes at the beginning of the twenty-third century. The line of apsides, on the other hand, remains rather

[20] N. V. Kulikova, The genetic connection between the α-Capricornid meteoroid shower and comets. *Astronom. Vestnik* **21** (1987), 242–250 (in Russian).

[21] Z. Sekanina, Statistical model of meteor streams. III. Stream search among 19303 radio meteors. *Icarus* **18** (1973), 253–284.

[22] Z. Sekanina, Statistical model of meteor streams. IV – A study of radio streams from the synoptic year. *Icarus* **27** (1976), 265–321.

[23] P. B. Babadzhanov, Meteor showers associated with the near-Earth asteroid (2101) Adonis. *Astron. Astrophys.* **397** (2003), 319–323.

[24] I. Hasegawa, Parent objects of a Capricornid meteor stream. *ESA SP* **495** (2001), 55–62.

[25] P. L. Lamy, I. Toth, M. F. A'Hearn and H. A. Weaver, Hubble Space Telescope observations of the nucleus of comet 45P/Honda–Mrkos–Pajdusáková and its inner coma. *Icarus* **140** (1999), 424–438.

[26] G. Sitarski, Motion of comet 45P/Honda–Mrkos–Pajdusáková. *Acta Astronom.* **45** (1995), 763–770.

[27] B. A. Marsden and Z. Sekanina, Comets and nongravitational forces. IV. *Astron. J.* **76** (1971), 1135–1151.

[28] A. Carusi, L. Kresák and G. B. Valsecchi, *Electronic Atlas of Dynamical Evolutions of Short-Period Comets* (Rome: IAS Computing Centre, 1998). (Website, http://www.rm.iasf.cnr.it/ias-home/comet/catalog.html).

stable. It is therefore striking that the angle of perihelion ($\omega + \Omega$) for the comet is 20° less than that for the α-Capricornids. *Bo Gustafson* and *Lars Adolfsson*[29] calculated the past orbit of the comet and four photographed α-Capricornids. They confirmed that all four had intersections with the comet at some point in the past, but the relative velocities needed to eject the meteoroids from the comet orbit at that time into their observed orbits would be over 6 times the estimated gas velocity. This makes an association unlikely. Matching angles of perihelion can be found with Capricornids photographed in late August (15–21). Neslusan and coworkers called these the *λ-Capricornids.*[30] This grouping is much less convincing, but included in Table 7 as the August (δ-)*Capricornids.*

To find the parent body of the α-Capricornids, it is possible to search along the nutation cycle of the backwards integrated orbit. Fig. 24.14 shows the back integrated orbit for the nominal semimajor axis of $a = 2.618$ AU and for a slightly higher value of $a = 2.70$ AU. One of the nodes was near Jupiter's orbit only ~1000 yr ago, while the other node tends to be near ~0.68 AU when the orbit is in a high inclination. With $a = 2.70$ AU the highest inclination was around AD 1140 at about 21.9°, while a further increase of the semimajor axis to 2.8 AU did not give a further increase in the maximum inclination.

The nodes are near Earth's orbit for only a brief moment. Hence, the presence of the shower since the late nineteenth century implies that the meteoroids are distributed along different parts of the cycle. Indeed, I find that the semimajor axis of α-Capricornids increased from a median value $\langle a \rangle = 2.36 \pm 0.23$ AU in 1955 to $\langle a \rangle = 2.61 \pm 0.23$ AU 1982. The observed high level of stratification in the orbits implies a young age and, in all likelihood, a discrete moment in time when the stream was created during the breakup of a comet. The large dispersion in semimajor axes suggests that breakup occurred about one cycle ago, when the inclinations were low, perhaps around AD 10.

The Capricornid parent body, or other remaining fragments, would be expected along the nutation cycle, with a similar longitude of perihelion ~35°, a perihelion distance between 0.55 and 0.75 AU and a semimajor axis around 2.3–3.0 AU. The longitude has evolved slowly from 16° in AD 500 to 35° now.

This allows only comet 141P/Machholz 2 and one minor body (2002 EX_{12})[31] to be potential candidate parent bodies (Fig. 24.15). 2002 EX_{12} lags the Capricornids by about 300 years in the cycle. 141P/Machholz 2, with a larger semimajor axis, would be about 2000 years behind.

29 L. G. Adolfsson and B. Å. S. Gustafson, Dynamics and probabilistic relation between meteoroids and their parent bodies. *ASP Conf. Ser.* **104** (1996), 133–136.

30 L. Neslusan, V. Porubčan and J. Svoren, Meteor radiants of recently discovered Earth-approaching comets. In *Meteoroids and Their Parent Bodies.* (Bratislava: Astron. Inst. Slovak Acad. Sci., 1992), pp. 181–184.

31 J. Ticha, M. Tichy, M. Kocer *et al.*, 2002 EX12. *Minor Planet Electronic Circ., 2002–F30* (2002).

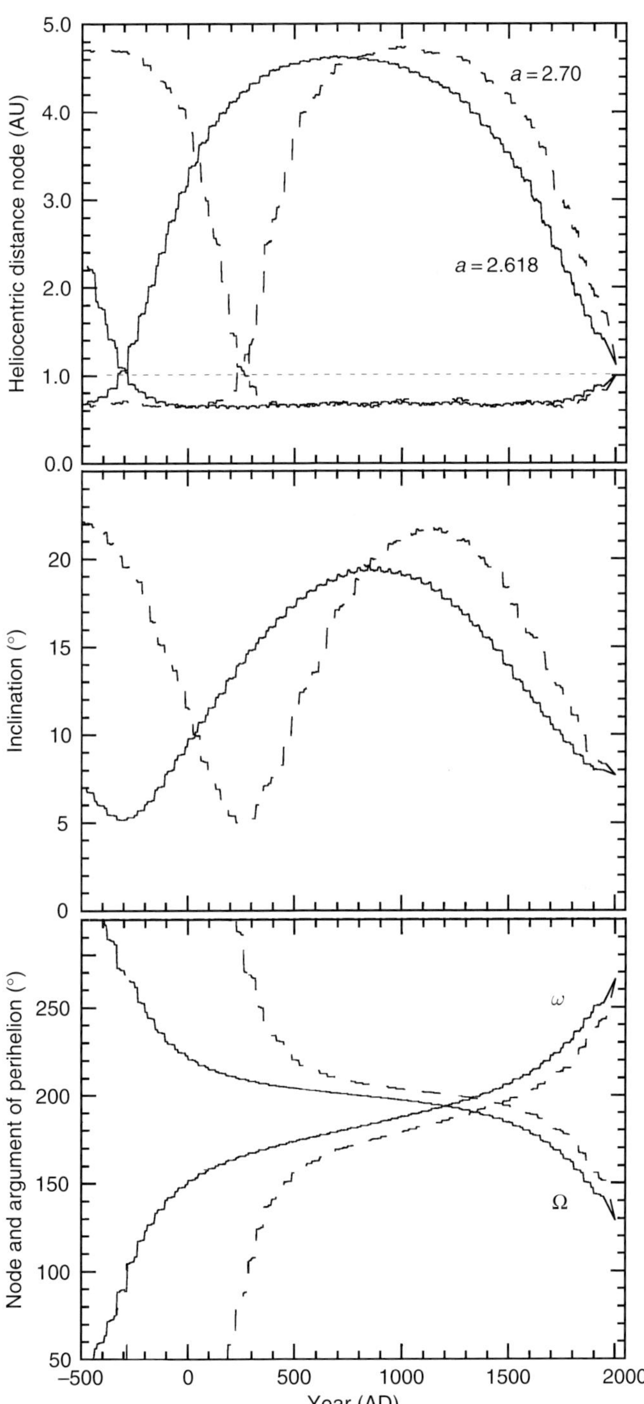

Fig. 24.14 Evolution of Capricornid orbits (J2000). The nominal orbit (solid line) is for $a = 2.618$ AU, $q = 0.602$ AU, $i = 7.68°$, $\omega = 266.67°$, node $= 128.9°$ at epoch 2000. The dashed line is for an orbit with $a = 2.70$ AU. Calculations by Esko Lyytinen.

Object	Epoch	a (AU)	q (AU)	i (°)	ω (°)	Ω (°)	Π (°)
141P/Machholz 2-A	1999	3.009	0.749	12.811	149.302	246.136	35.438
2002 EX$_{12}$	2005	2.603	0.606	11.317	217.921	176.258	34.179
α-Capricornids	1982	2.618	0.602	7.68	266.67	128.9	35.570

Fig. 24.15 Orbital element comparison of proposed Capricornid parents (J2000).

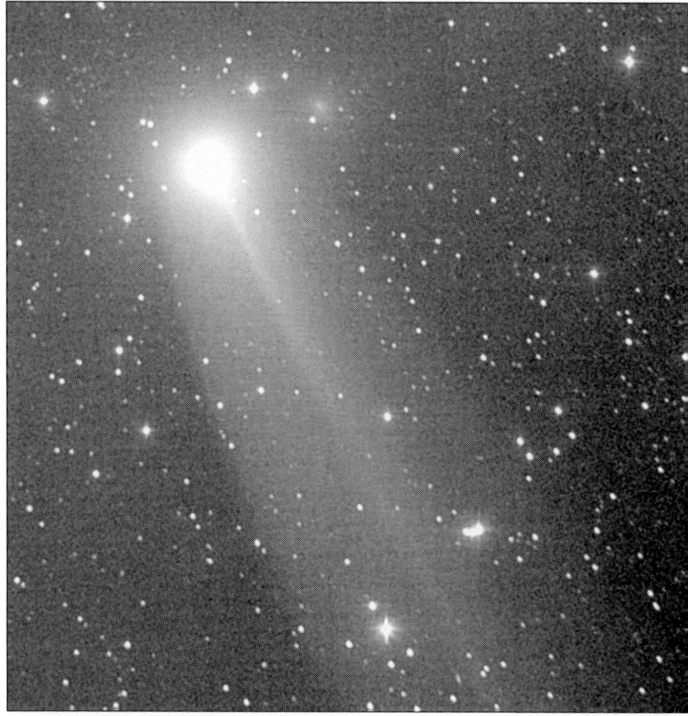

Fig. 24.16 141P/Machholz 2 with fragment on September 9, 1994, courtesy of J. V. Scotti (Spacewatch, LPL, University of Arizona).

Asher and Steel looked into the likelihood of meteors from comet **141P/Machholz 2** (Fig. 24.16).[32] The comet was discovered by Don Machholz on August 13, 1994, the night after the Perseid outburst of 1994, when California Meteor Society observers chased their first meteor outburst. It is currently in a fairly stable orbit due to its proximity to the 9:4 mean-motion resonance with Jupiter and has split up into a number of fragments over recent years.

[32] D. J. Asher and D. I. Steel, The orbital evolution of P/Machholz 2 and its debris. *Mon. Not. R. Astron. Soc.* **280** (1996), 1201–1209.

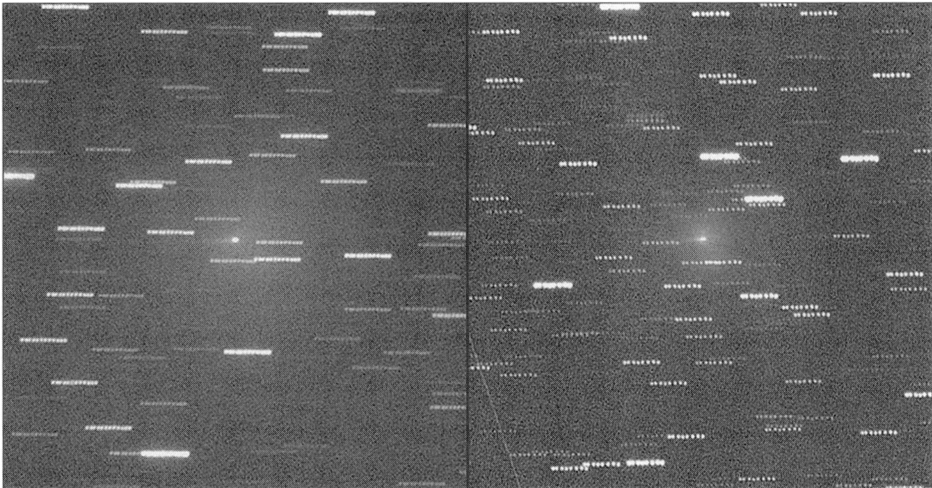

Fig. 24.17 Intermittently active comet 169P/NEAT (2002 EX_{12}), parent of the α-Capricornids, outbursts on August 5 (21:45 UT) and August 12 (20:50 UT) 2005. Photo by Michael Jäger and Gerald Rhemann.

Earth reached the descending node 55 d after the comet on November 28.6, 1994, but at a distance of 0.186 AU at the outside of the orbit of the comet. That distance is too large for high meteor activity. The most recent approach of cometary and terrestrial orbits occurred in two epochs in the eighteenth and nineteenth centuries, but for various reasons outbursts were unlikely. The next epochs in which such intersections occur are over a millennium away, although some meteoroids ejected with high speeds could meet the Earth within a few centuries.

2002 EX_{12} was proposed as a possible parent of the Capricornid shower on dynamical arguments by *Paul Wiegert* of the University of Western Ontario, Canada, at the Meteoroids 2004 meeting in London, Ontario. His results have not yet been published. I now confirm that this is indeed a probable association. With $H_{10} = +16.10$, this is a large extinct comet nucleus, about 4.0 km in diameter if the albedo is 0.04. The JPL orbit database even gives $H_N = +15.44$ (D = 5.4 km). Not much more is known at the present time. It is perhaps possible that 2002 EX_{12} and 141P/Machholz 2 were split long ago.

Just before going to press, it came to my attention that *Brian Warner* at the Palmer Divide Observatory, Colorado, had discovered a straight 90 arcsec long tail in images of 2002 EX_{12} taken on the 28th and 29th July 2005. Independently, *Alan Fitzsimmons* using the Faulkes Telescope North at Haleakala also reported a 30 arcsec long tail on images taken on the 29th (Fig. 24.17). The object was renamed 169P/NEAT and did not show such comet activity in images taken on July 12 by Alan Cahill and back in May. At the time, the comet was 1.1 AU from the Sun, due at perihelion (0.6 AU) on September 17. It is in an eccentric orbit with an orbital period of 4.2 yr. It is now very likely that the α-Capricornid parent has been identified. Or at least one of a number of fragments.

The κ-Cygnid parent

Like the Quadrantids, the κ-Cygnids have a node close to Jupiter's orbit and a high inclination. However, the mean orbit in Table 7 with $a = 3.19$ AU ($Q = 6.36$ AU) librates about the 1:2 mean-motion resonance with Jupiter ($a_c \sim 3.28$ AU), with q, a, and e oscillating with a period of ~230 yr. This creates a large natural dispersion of semimajor axes, but not big enough to explain the large range of values for the κ-Cygnids. This oscillation is due to a nearby mean-motion resonance and is unique for the semimajor axis of the the mean orbit. The unique feature of the κ-Cygnids is their range in semimajor axes.

The obvious approach to finding the parent body is to search for objects with the semi-major axis and the position of the node along the lines of Fig. 24.10. The longitude of perihelion should be in the range 326°–355° and the inclination in the range 23°–43° (Fig. 24.11). Then try and understand how the stream could have dispersed in the manner observed from a fragmentation event in the recent past.

I find that *2005 LP*$_{40}$, with $a = 1.964$ AU and a heliocentric distance of the node of 2.244 AU, is close to the intersection of the lines in Fig. 24.10. The low 23.6° inclination is perhaps within the suggested evolution of the meteoroids (Fig. 24.11). The perihelion distance of 0.890 AU is lower than expected, although this could be a selection effect of Earth sampling only that part of the stream with q close to 0.987 AU. The orbit of 2005 LP$_{40}$ from AD 1600 until 2200 is shown by "x" symbols in Figs. 24.10 and 24.11. It's node regresses at −0.00571°/yr, the longitude of perihelion increases at +0.00606°/yr, while the inclination and semi-major axis are constant.

The unanswered question remains as to how the Tisserand parameter could have evolved as much as it has. 2005 LP$_{40}$ is not the source of the meteoroids, given that the younger-looking 1993 Cygnids have a lower and more comet-like Tisserand parameter. An ongoing cascade of fragmentation events is implicated.

25

The Taurid complex

It has been proposed in various papers by *Victor S. Clube*, *Bill Napier*, *David Asher*, and Duncan Steel[1] that only about 20 000–30 000 yr ago, a giant (~42 km sized) comet, in an orbit passing not far from Earth, broke up and left in its wake a massive meteoroid stream now known as the Taurid shower, as well as a large number of extinct comet nuclei, all on a potential collision course with Earth.

Long before that time, this large comet was supposedly captured by Jupiter and evolved into the unusually short asteroid-like orbit of comet 2P/Encke. The orbital period of Encke is only $P = 3.3$ yr. For a Jupiter-family comet to arrive in such an orbit, nongravitational forces working over a long time are needed (this is difficult for a large comet). In this hypothesis, comet 2P/Encke is just a small remnant of the progenitor parent body of the Taurid meteoroids, following Whipple's original proposal that 2P/Encke was the main source of the dispersed and massive Taurids and was the main contributor to the zodiacal cloud.[2]

Clube *et al.* went on to argue that so much dust and fragments were suddenly released that it could have had an effect on the Earth's climate from dust loading of our atmosphere and from giant impacts on our planet.[3] Whether or not this hypothesis is correct can be tested from the nature of the sibling minor planets. Are they indeed extinct comet nuclei, and can they be dynamically linked?

25.1 The old age of the Taurid shower

The Taurid shower is active over a period of ~80° in solar longitude, with a peak in the first week of November. The shower is known for its spectacular fireballs (Fig. 25.1)

[1] S. V. M. Clube and W. M. Napier, *The Cosmic Serpent* (London: Faber and Faber, 1982), p. 299; S. V. M. Clube and W. M. Napier, The microstructure of terrestrial catastrophism. *Mon. Not. R. Astron. Soc.* **211** (1984), 953–968; S. V. M. Clube and W. M. Napier, The cometary breakup hypothesis re-examined – a reply. *Mon. Not. R. Astron. Soc.* **225** (1987), 55P–58P. D. J. Asher, S. V. M. Clube and D. I. Steel, Asteroids in the Taurid Complex. *Mon. Not. R. Astron. Soc.* **264** (1993), 93–105.

[2] F. L. Whipple, Photographic meteor studies. III. The Taurid shower. *Proc. Am. Phil. Soc.* **83** (1940), 711–745; F. L. Whipple, On maintaining this meteoric complex. *SAO Special Report*, **239** (1967), 1–45.

[3] S. V. M. Clube and W. M. Napier, *The Cosmic Winter* (Oxford: Basil Blackwell, 1990), 307 pp. S. V. M. Clube, Large scale perturbations of the terrestrial atmosphere – some considerations relating to the role of cometary trails and disintegrating meteoroids. *Adv. Space Res.* **11** (1991), 63–66.

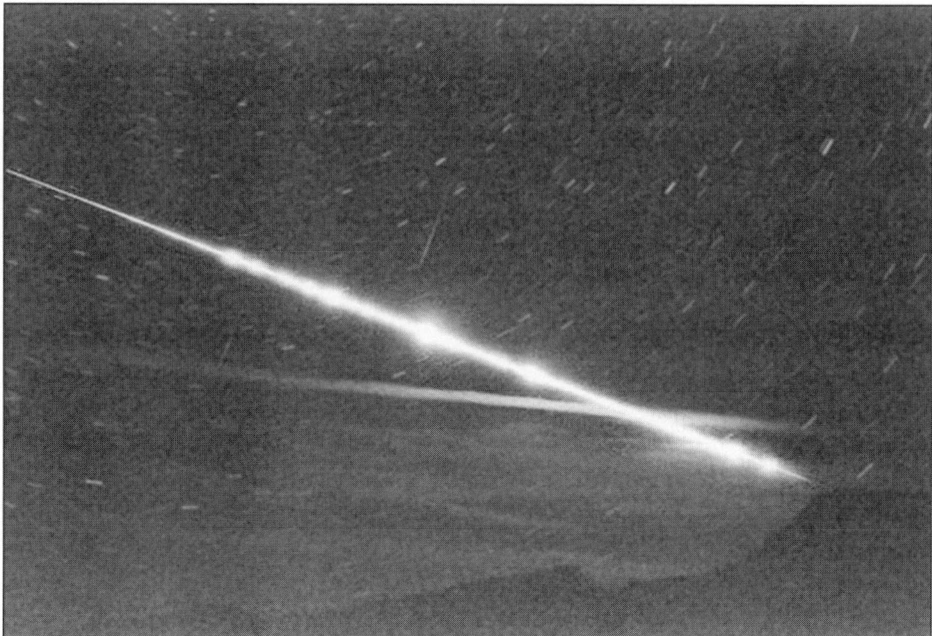

Fig. 25.1 At the tail of the Leonid storm, this November 18, 2001, 12:52:42 UT Taurid fireball flashed multiple metallic colors, chiefly green and blue in a direction NNE from Kingsman, Arizona. This 43 s exposure is by Wil Milan from Phoenix, using a Finger Lakes Instruments IMG6303 scientific CCD camera and a Nikon 28 mm f/4.0 wide-angle lens. A video of the same fireball was the highlight of the 2001 Leonid MAC mission. The foreground glow might be the exhaust trail of the FISTA research aircraft. Photo: Wil Milan.

that frequently light up the Autumn skies, and have been doing so at least since the middle ages. Its percentage of fireballs is not different from that of the Perseid shower,[4] but the stream is wide (Fig. 25.2). It is so dispersed that different parts have been called different names: the *Piscids* in September, the *Arietids* in October, the *Taurids* in November, and the *χ-Orionids* in December (Table 7), possibly reflecting real concentrations of dust. Over its full length, the stream has northern and southern branches. In addition there are two daytime showers at the preperihelion arc: the southern branch in June is called the *β-Taurids* (in May: *S. May Arietids*), while the northern branch is called the *ζ-Perseids* (*N. May Arietids*).[5]

The presence of northern and southern branches has always been interpreted as a sign of old age. It takes at least ~5500 yr for the argument of perihelion to rotate 360° by secular perturbations in a nutation cycle to create showers at both ascending and descending nodes with northern and southern branches (Fig. 9.1). Fig. 25.3 shows the

[4] A. McBeath, On the occurrence of bright Taurids. *WGN* **27** (1999), 53–56.

[5] A. Aspinall, J. A. Clegg and A. C. B. Lovell, The daytime meteor streams of 1948. Part I. Measurement of the activity and radiant positions. *Mon. Not. R. Astron. Soc.* **109** (1949), 352–358.

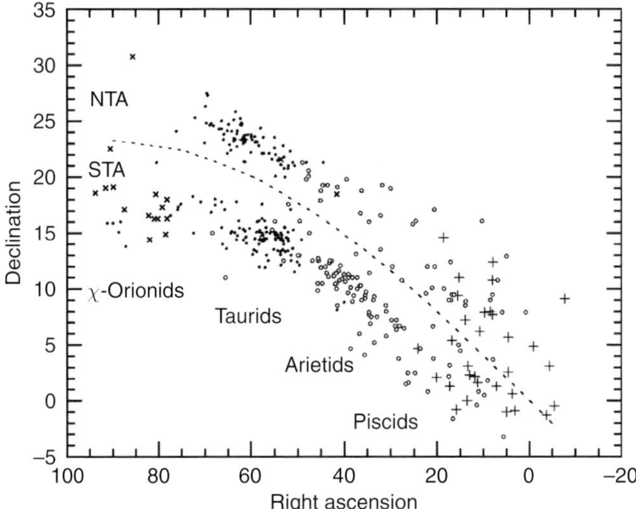

Fig. 25.2 Radiant positions of photographed Taurids in September (+), October (o), November (●), and December (×) from the IAU database, not corrected for daily motion. The dashed line is the ecliptic plane.

orbit of some of the minor planets associated with the Taurid Complex, which span the full range of the rotating ω. All orbits have the same length of perihelion, meaning that the orbit is oriented in the same manner. The gray part of the orbit is above the ecliptic plane. The corresponding orbit of each of the four showers is marked. With four permitted values of ω (see Appendix E), this means that a meteoroid stream ejected by a single comet can create four showers, if the meteoroids have enough time to disperse along the full cycle.[6]

In this scenario, the meteoroids in each stream sample a subset of the whole stream, a narrow range in inclination and epoch, since ejection. The meteoroids of the Northern and the Southern Taurids cluster around $i = 3.1°$ ($q = 0.41$ AU) and $5.7°$ ($q = 0.33$ AU), respectively.

From the first photographed Taurid orbits, Fred Whipple[7] identified comet 2P/Encke as a likely parent. However unlike most of the Taurids, 2P/Encke has a node near Mercury's orbit, a relatively steep orbit (12°), and passes far from Earth's orbit.

The duality of the North and South branches of the Taurid showers followed from the first analysis of precisely reduced photographic orbits. The most recent study, using all 4581 photographic orbits in the IAU database by Vladimin Porubčan and Leos Kornoš[8] identified 80 northern Taurids and 144 southern Taurids in the period September 16–December 29 assuming a D-criterion of $D < 0.25$ (Chapter 26). During

[6] J. D. Drummond, Theoretical twin meteor showers. *Icarus* **49** (1982), 135–142.
[7] F. L. Whipple, Photographic meteor studies III. The Taurid shower. *Proc. Am. Phil Soc.* **83** (1940), 711–745.
[8] V. Porubčan and L. Kornoš, The Taurid meteor shower. *ESA-SP* **500** (2002), 177–180.

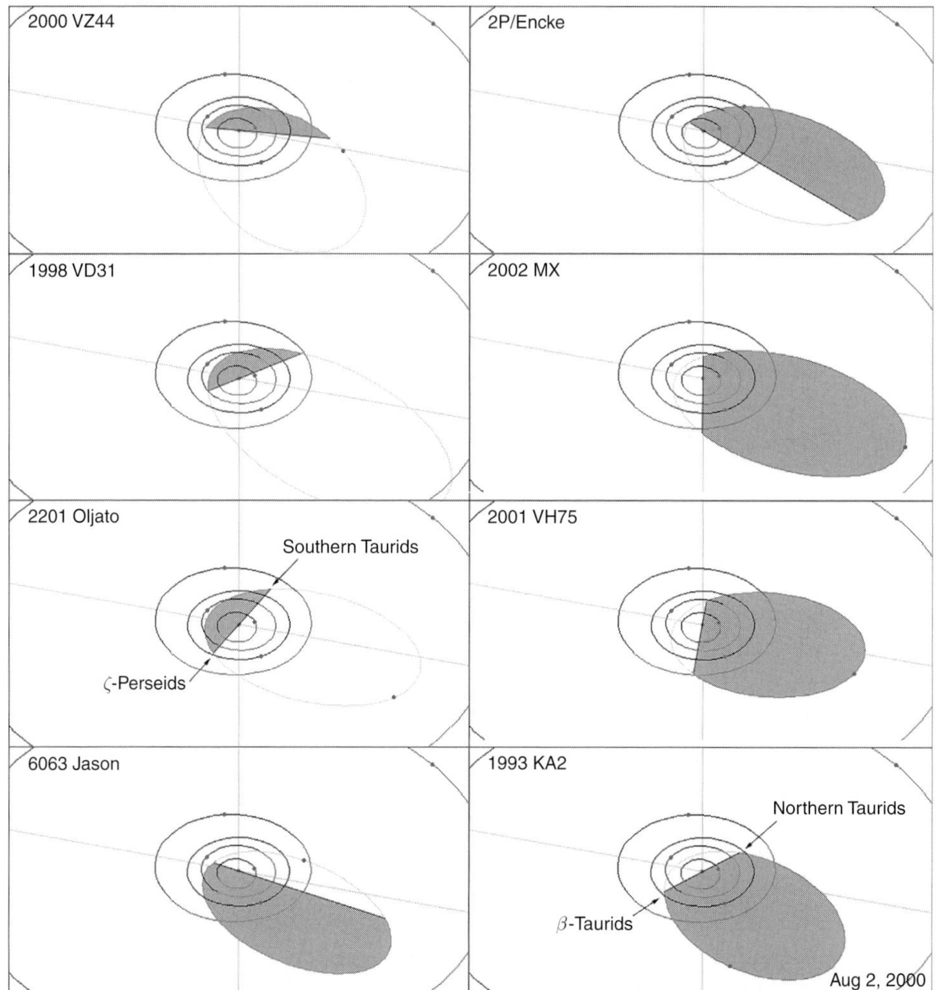

Fig. 25.3 Secular nutation cycle in orbit diagrams of some of the minor planets associated with the Taurid complex. The group has a constant longitude of perihelion and a rotating argument of perihelion (Fig. 9.1). When the nodes are near Earth's path, showers are observed.

that timeframe, the orbital elements gradually changed from a low value of $q \sim 0.25$ to a higher value of $q \sim 0.58$ AU, and from a low longitude of perihelion (138°, $\lambda_\odot = 190°$) to a high longitude of perihelion (183°, $\lambda_\odot = 277°$).

If both branches were created by one comet, this would put the age of the Taurid shower beyond ~5500 yr, the duration of one full rotation of ω because all meteoroids move at some rate along the cycle. This lower limit is essentially the first age estimate of the Taurid shower by Whipple in 1940, who estimated the age from the observed range of nodes and argument of perihelion compared to the calculated changes by Jupiter's perturbing the comet at aphelion. Whipple's biggest problem was on deciding what the aphelion distance of the Taurids was, because the rates of change in Ω and ω depend strongly on aphelion distance. From the measured mean aphelion distance from only five multistation photographic orbits measured in 1937 and 1938 at the northern stations of the Harvard Observatory, he derived an age of at least 5000 yr.

Whipple obtained another estimate, 14 000 yr, by comparing the dispersion in $\Pi = \Omega + \omega$ (which grows slowly with time) of four Taurid meteors to the rate at which these elements had been observed to change in perturbations of 2P/Encke.

The first effort to model the evolution of individual Taurid meteoroids was that by *Fred Whipple* and *Salah El-Din Hamid*,[9] published in 1952. They found two episodes of ejection (about 4700 and 1400 years ago) when the orbits were most similar, but ejection speeds had to be high. Moreover, a parent body other than comet 2P/Encke was involved.

Until that time, calculating planetary perturbations had not been an efficient process, with hundreds, if not thousands, of calculations needed to work out the perturbation of each planet in small steps of time along the comet orbit. Mathematical tools were developed to consider only the effect over a whole orbit, saving enormous amounts of time. While gravitational perturbations depend on the exact distance of the meteoroid from each of the planets, over many orbits these perturbations tend to average out. What mattered were the *secular perturbations*, those that continue to act in one direction without limit, in contrast to the short periodic perturbations (Chapter 9.1). Unfortunately, the secular perturbation methods available initially only worked well on near-circular orbits.

In 1947, Dirk Brouwer[10] published a method of calculating secular perturbations that worked well even for orbits of high eccentricity and Whipple and Hamid were the first to apply this to the evolution of meteoroid orbits.[11] This method ignored close encounters with the planets, because it does not care where the meteoroid is along its orbit. For 2P/Encke that was not a problem.

While the calculations explain that there is a range in node and argument of perihelion in the complex, Earth should encounter only a narrow range in node. If a larger range is observed, there has to be some change in other elements as well. The model also predicts that the inclination should vary with the argument of perihelion, which is not observed.

The planetary perturbations do disperse the perihelion distance and semimajor axis over time, but not rapidly enough unless high ejection speeds create significantly different initial conditions (Fig. 25.4).[12] Hamid and Whipple noticed the large spread in perihelion distances, and found that it could not be explained from ejection at perihelion (which tends to keep q the same as that of the comet). Hence, they suggested that the stream was created at aphelion from a collision with an asteroid.

[9] F. L. Whipple and S. E. Hamid, On the origin of the Taurid meteor streams. *Helwan Obs. Bull.* No. 41, Cairo (1952).

[10] D. Brouwer, Secular variations of the elements of Encke's comet. *Astron. J.* **52** (1947), 190–198.

[11] At the same time, M. Plavec was using another method, the Gauss–Halphen–Goryachev method, to investigate the age of streams, i.e. M. Plavec, *Nature* **165** (1950), 362.

[12] D. I. Steel and D. J. Asher, The orbital dispersion of the macroscopic Taurid objects. *Mon. Not. R. Astron. Soc.* **280** (1996), 806–822.

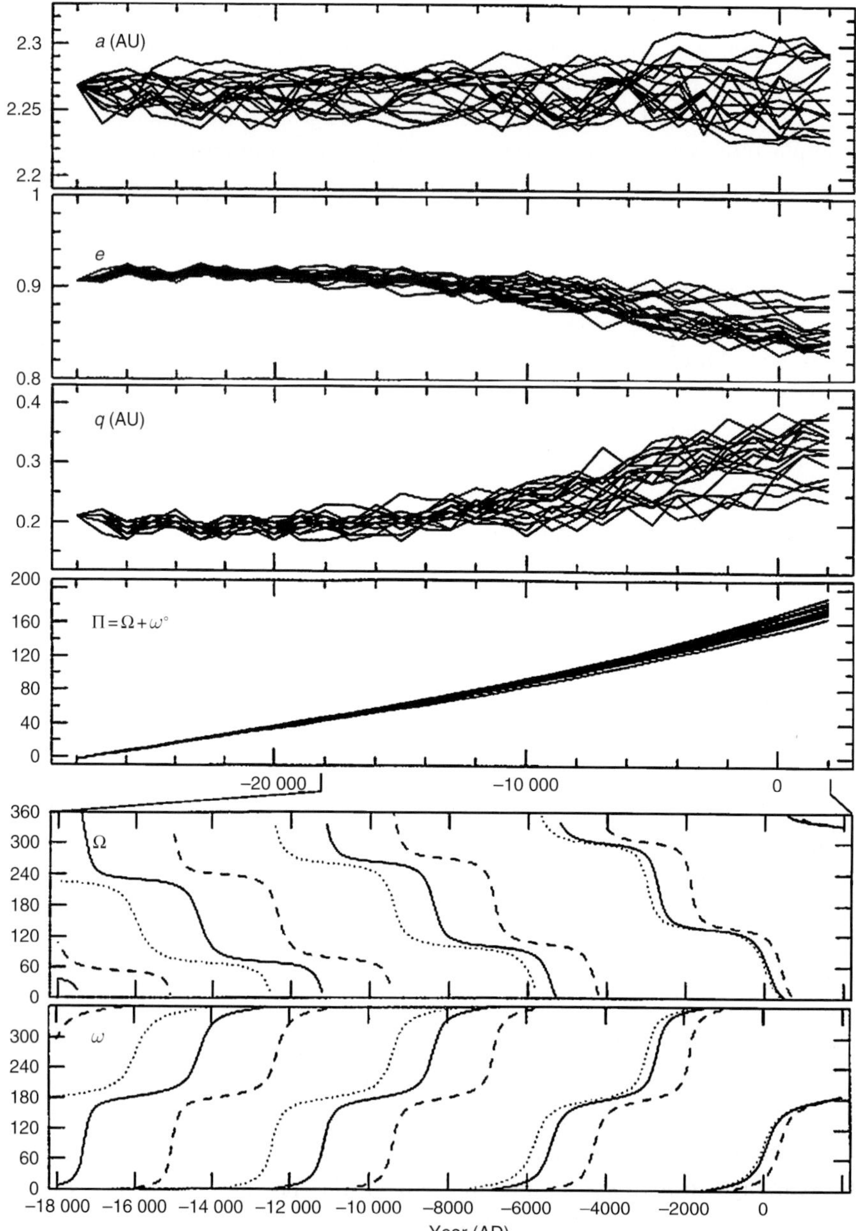

Fig. 25.4 The dispersion of particles released from the orbit of 2P/Encke in a backward integration by Steel and Asher (1996).[13]

[13] *Ibid.*

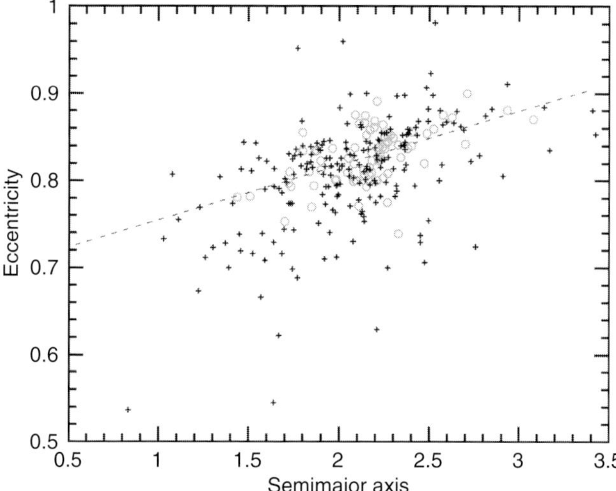

Fig. 25.5 **An *a* versus *e* diagram** from photographed northern (o) and southern (+) Taurids in the IAU database.

George Wetherill[14] argued that such collisions with asteroids are rare and unlikely. Then, *E. M. Pittich*[15] pointed out that a comet has a 10–50% chance of splitting at least once in a period of 200 yr, with typical fragments separating by less than 50 cm/s. Comets spend most of their time at aphelion. Hence, a breakup could also create showers at aphelion.

With more Taurids photographed, it became evident that breakup near aphelion could not explain the observed dispersion. The dispersion in semimajor axes and eccentricity is large (Fig. 25.5). Simulations by *Duncan Steel* and coworkers showed that breakup at aphelion tends to produce a dispersion of semimajor axes and eccentricity that slopes in the wrong direction. In that case, there is a narrow range of semimajor axes and no correlation with eccentricity, while meteoroids ejected at perihelion tend to have q constant and increasing a with increasing e. In reality, both aphelion and perihelion distance have a range of values.

Ejection of dust over time, combined with a changing orbit of 2P/Encke would disperse the meteoroid orbits both in a and q, and in node (Fig. 25.6). Steel and Asher showed that comet 2P/Encke has evolved from a much lower $q < 0.2$ AU less than 20 000 years ago to its present value of $q = 0.34$ AU. The reason for this is that the longitude of aphelion has gradually approached the longitude of perihelion of Jupiter, which had the effect of reducing the eccentricity of the comet, whilst maintaining its

[14] G. W. Wetherill, Collisions in the asteroid belt. *J. Geophys. Res.* **72** (1967), 2429–2444.

[15] E. M. Pittich, Splitting and sudden outbursts of comets as indicators of nongravitational effects. In *The Motion, Evolution of Orbits, and Origin of Comets*, ed. (Dordrecht: Reidel, 1972) G. A. Chebotarev, E. I. Kazimirchak-Polonskaia and B. G. Marsden, *IAU Symp.* 45, pp. 283–286.

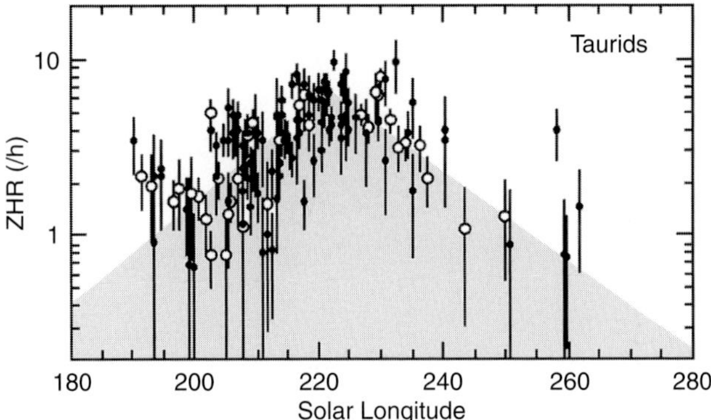

Fig. 25.6 The zenith hourly rate of the Taurid shower. Dark symbols are measurements from observers of the Dutch Meteor Society, while open symbols were obtained by the NAPO – Meteor Section.

semimajor axis.[16] This effect alone could have created a range in perihelion distances with ejection of meteoroids over time.

Steel and Asher continued to argue that a model that produced meteoroids near perihelion over a period of 10 000 years comes close to explaining the observed trends of a and e.[17] The models do, however, tend to produce less scatter in the observations than reported. A very high ejection speed of ~2 km/s would be needed to explain the observed dispersion in semimajor axes, which is not consistent with the speed of dust released from Encke.

Asher and Steel integrated the orbit of comet 2P/Encke back in time, to find that around AD 210 and 80 BC the orbit intersected with Earth's orbit. At those times, meteor outbursts may have occurred. The ascending node of the meteoroids was at about 60° at the time and the descending node at about 240°. The result is uncertain because the nongravitational forces may have varied over time. Moreover, it is not known if 2P/Encke was active at the time. No historic meteor storm records match these dates (Table 1).

25.2 No Taurid complex of minor planets?

Apart from 2P/Encke, at least 22 other minor planets have been associated with this stream.[18] The full list[19] as of 2003 is given in Fig. 25.7, with annotations gathered from

[16] D. I. Steel, D. J. Asher and S. V. M. Clube, The structure and evolution of the Taurid complex. *Mon. Not. R. Astron. Soc.* **251** (1991), 632–648.

[17] P. B. Babadzhanov, Y. V. Obrubov and N. Makhmudov, Meteor streams of comet Encke. *Solar Syst. Res.* **24** (1990), 12–19.

[18] S. V. M. Clube and W. M. Napier, The microstructure of terrestrial catastrophism. *Mon. Not. R. Astron. Soc.* **211** (1984), 953–968; D. Olsson-Steel, Asteroid 5025 P-L, Comet 1967 II Rudnicki, and the Taurid meteoroid complex. *Observatory* **107** (1987), 157–160; J. Stohl and V. Porubčan, *Publ. Astr. Inst. Czechoslov. Acad. Sci.* **67** (1987), 163–166; D. J. Asher, S. V. M. Clube and D. I. Steel, Asteroids in the Taurid Complex. *Mon. Not. R. Astron. Soc.* **264** (1993), 93–105.

[19] D. Steel and D. Asher, The orbital dispersion of the macroscopic Taurid objects. *Mon. Not. R. Astron. Soc.* **280** (1996), 860–822.

Object	D (km)	Epoch	a	q	i	Node	Π	D_{SH}	Type	Albedo
2P/Encke	4.6	(2003)	2.217	0.339	11.77	334.59	161.09	0.22	Comet	
5025 P-L	4.5	(1961)	4.287	0.435	6.29	356.84	149.76	0.21	-.-	-.-
1937 UB	2.1	(2003)	1.655	0.622	6.07	34.52	126.92	0.38	S	-.-
1947 XC	2.1	(2005)	2.172	0.623	2.52	76.62	172.84	0.35	E, Sq	0.24
1959 LM	4.5	(2003)	1.982	0.721	6.75	295.68	171.08	0.44	Sq	mh
1979 XB	1.3	(1979)	2.372	0.646	25.25	86.02	161.37	0.50	-.-	-.-
1982 TA	1.8	(2003)	2.300	0.525	12.22	9.99	129.44	0.27	S, Sq	0.33
1984 KB	1.4	(2004)	2.218	0.525	4.85	169.86	146.45	0.22	S	0.16
1987 KF	2.5	(2004)	1.836	0.590	11.85	108.17	123.74	0.40	O	mh
1988 VP_4	4.7	(2002)	2.263	0.784	11.44	281.80	138.76	0.52	-.-	-.-
1990 HA	1.5	(2002)	2.570	0.697	3.91	184.77	133.09	0.40	-.-	m
1991 TB_2	2.7	(1991)	2.077	0.432	7.90	291.78	131.21	0.24	-.-	-.-
1991 BA	0.01	(1991)	2.188	0.715	1.94	118.88	189.57	0.55	-.-	-.-
1991 GO	0.73	(2004)	1.927	0.668	9.56	24.73	113.62	0.49	-.-	-.-
1991 VL	5.0	(2004)	1.833	0.416	9.13	310.33	177.18	0.26	O, Sk	mh
1993 KA_2	0.01	(1993)	2.227	0.501	3.19	239.63	140.87	0.18	-.-	-.-
1994 AH_2	2.2	(2002)	2.534	0.738	9.58	164.22	189.22	0.60	O	m
1995 FF	0.03	(2005)	2.323	0.680	0.59	175.56	111.82	0.52	-.-	-.-
1997 US_2	0.73	(2004)	1.674	0.568	3.17	66.26	166.15	0.30	-.-	-.-
1999 VK_{12}	0.12	(1999)	2.254	0.502	9.54	48.96	151.68	0.22	-.-	-.-
2000 VZ_{44}	0.42	(2000)	2.076	0.539	5.32	8.70	137.15	0.24	-.-	-.-
2001 VH_{75}	1.8	(2005)	2.104	0.548	10.70	277.42	160.41	0.26	-.-	-.-
2002 MX	0.30	(2002)	2.504	0.511	1.96	284.41	161.98	0.17	-.-	-.-

Fig. 25.7 A list of proposed minor bodies of the Taurid Complex by various authors up to 2003. The *D*-criterion is given relative to the meteoroid stream (Southern Taurids).

the EARN (European Asteroid Research Node) database of physical properties of asteroids. All these objects have a *D*-criterion (*a*, *e*, *i* only) less than about 0.2 and a theoretical radiant similar to that of one of the Taurid group streams. They were selected among all minor planets based on having a semimajor axis in the range 1.8–2.6 AU, eccentricities in the range 0.64–0.85, a low inclination ($i < 12°$), and longitudes of perihelion in the range $100 < \Pi < 190$. The orbits pass relatively close to Earth's orbit, mainly because of a low inclination.

These objects come in all shapes and sizes. The 4.5 km (for albedo = 0.04) *5025 P-L* (Palomar–Leiden), for example, was discovered by the Leiden University astronomers *Cees J. van Houten* and *Ingrid van Houten-Groeneveld* from plates taken at Palomar Mountain on October 22, 1960. The object is reportedly lost. And at the other extreme is the 7 m sized *1993 KA*$_2$, which had a close encounter with Earth in 1993 (0.0010 AU, 150 000 km) on May 20.

Remarkably, all candidates have a relatively high perihelion distance $q > 0.4$ AU, larger than the mean orbit of the Taurids and that of comet 2P/Encke, those with $D_{SH} < 0.1$ have $q \sim 0.5$ AU. This "Taurid Complex" is found right among the population of Apollo asteroids, which does not exclude that many, if not all, of the proposed members are in fact asteroids. If so, they are not responsible for the Taurid meteors, which are cometary in nature.

Candidate extinct comet nuclei according to *Bill Botke*'s criteria are given in bold face. There is only one, and this object 5025 P-L has an aphelion very different from the Taurids. The proposed members of this *Taurid Complex* show a broad dispersion of orbits in different stages of rotation of the argument of perihelion. They are in fact so dispersed that a single breakup can not explain the distribution in orbits.

The big question is whether or not *any* of these minor planets are related to 2P/Encke or the meteoroid stream. The candidates for *extinct comet nuclei* must be dark (albedo of ~0.03–0.06), of a "primitive-class" spectral taxonomic type (most likely D, X, T, C, B, or F), with a long spin period > a few hours, be of a significant size (>500 m), show intermittent activity, or be undisputably linked to a meteoroid stream or dust trail.

Until now, only the last issue was addressed, showing that none has $D_{SH} < 0.06$, if Ω and Π are included (Chapter 26), a criterion that defines the similarity of the orbits. Most turn out to be asteroids, mostly of S and O type.[20] From the original 1992 list proposed by Asher and Clube,[21] *all but one* are asteroidal, with that one still in doubt due to a lack of data.

Even if the giant comet hypothesis is later proven to be correct, the hypothesis was not based on viable objects. This implies that the original comet responsible for the Taurids was much smaller than claimed.

The fact that asteroids are found among the Taurid stream shows that there is an important dynamical pathway from asteroids in the main belt to a comet orbit such as that of 2P/Encke. This, in fact, is the route along which many small asteroid fragments traveled to Earth before being recovered as meteorites.[22]

25.3 A young age for the Taurid shower?

There are signs that part of the Taurid stream is young. The meteoroids have not sintered much from having been released long ago into the interplanetary medium; returning every ~3.5 yr to the orbit of Mercury! Instead, some meteoroids were relatively recently ejected. Taurid fireballs exhibit all the key features of fragile

[20] S. J. Bus and R. P. Binzel, Phase II of the small main-belt asteroid spectrographic survey: a feature-based taxonomy. *Icarus* **158** (2002), 106–145.

[21] D. J. Asher and S. V. M. Clube, The Taurid complex asteroids. In *Meteoroids and Their Parent Bodies*, ed. J. Stohl and I. P. Williams. (Bratislava: Astron. Inst. Slovak Acad. Sci., 1992), pp. 93–96.

[22] G. B. Valsecchi, A. Morbidelli, R. Gonczi *et al.*, The dynamics of objects in orbits resembling that of P/Encke. *Icarus* **118** (1995), 169–180.

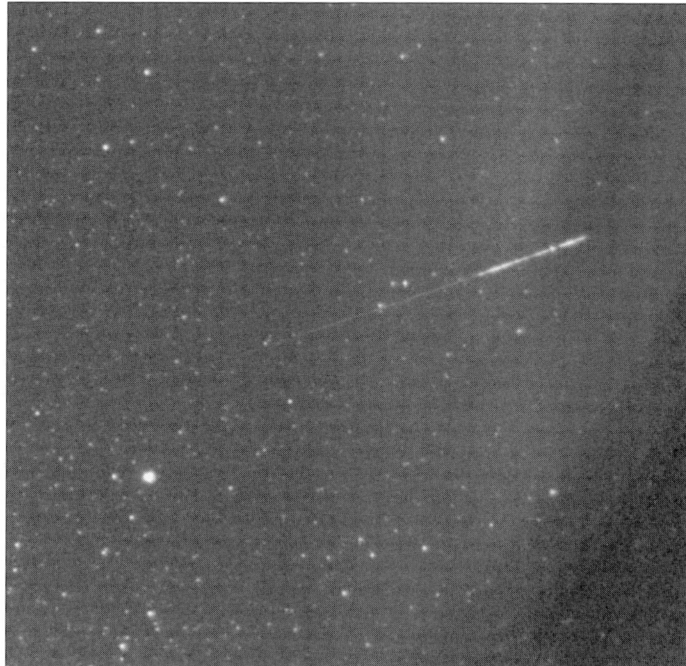

Fig. 25.8 Normal southern Taurid meteor from the night October 26/27, 1995, (18:03:59 UT) at Ondrejov Observatory by Vlastimil Marx. This meteor has a faint companion. Photo courtesy: Pavel Spurný, Ondrejov Observatory.

cometary grains. They enter at ~28 km/s, but leave nonetheless persistent trains. The Taurid fireball observed during the 2001 Leonid MAC mission (Fig. 25.1) had a train that was still bright when it moved out of the field of view 2 min later. Taurid light curves are irregular, with characteristic flares in the middle of the track, plus a range of smaller flares at various other places, especially towards the end (Figs. 25.8 and 25.9). They are decelerated as efficiently as Perseid meteoroids, more so than Geminids, but not as much as α-Capricornids and κ-Cygnids.[23]

Jacchia *et al.*[24] derived a low *progressive fragmentation index* $\chi = 0.04 \pm 0.05$, STA, and 0.03 ± 0.07, NTA (compared to 0.28 ± 0.10 for the Perseids), suggesting that the fragmentation occurs in spurs and does not rapidly accelerate when the meteoroids penetrate deeper into the atmosphere. Verniani[25] derived a mean particle density of 0.25 g/cm^3 for STA and 0.27 g/cm^3 for NTA, respectively, compared to 0.32 g/cm^3 for the Perseids. The biggest and brightest fireballs have been heard to produce a deep

[23] I. Halliday, Geminid fireballs and the peculiar asteroid 3200 Phaethon. *Icarus* **76** (1988), 279–294.

[24] L. G. Jacchia, F. Verniani and R. E. Briggs, An analysis of the atmospheric trajectories of 413 precisely reduced photographic meteors. *Smithsonian Contrib. Astrophys.* **10** (1967), 1–139.

[25] F. Verniani, Meteor masses and luminosity. *Smithsonian Contrib. Astrophys.* **10** (1967), 181–195.

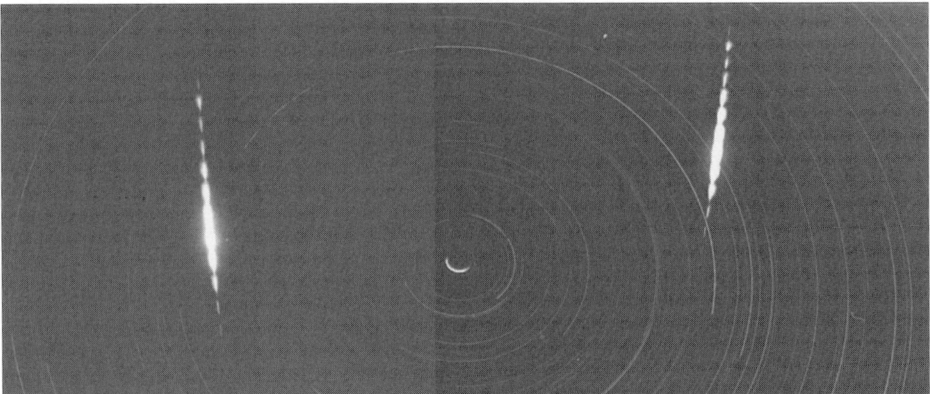

Fig. 25.9 Taurid outburst fireball EN27–281095 as seen from stations EN14 (left, Cervena hora – operator Pavel Barinka) and EN16 (right, Lysa hora – operator Josef Krsnak). Photo courtesy: Pavel Spurný, Ondrejov Observatory.

rumble a minute or so after the meteor because of a sonic boom penetrating to the ground.[26]

The northern Taurids are strong in the video observations, while almost all bright fireballs originate from the southern Taurid branch. This, again, is not consistent with both branches originating from the same stream having rotated the line of nodes a full circle.

Here, *Pavel Spurný* of Ondrejov Observatory and I would like to add another little known fact: the Taurid shower has occasional outbursts! Between October 22 and 31 of 1995 ($\lambda_\odot = 208.3–218.0°$), all-sky cameras of the *European Network* equipped with Zeiss-Distagon fish-eye objectives ($f/3.5$, $f = 30$ mm) photographed 14 fireballs brighter than magnitude −6 over central Europe.[27] This rate was several times higher than is usual for the European Network at that time of year.

Five of these fireballs had very similar orbits, not unlike those of Taurids, radiating from R.A. $= 50.48 + 0.631° (\lambda_\odot - 214.0°)$, Decl. $= +15.43 + 0.235° (\lambda_\odot - 214.0°)$, near the southern Taurid radiant (Fig. 25.10), but with higher initial velocities of $V_g = 33.1 \pm 0.3$ km/s. The five orbits have a higher semimajor axis scattering around $a = 2.530 + 0.025 (\lambda_\odot - 214.0°)$ AU (rather than $a = 1.93$ AU), and a *smaller* perihelion distance scattering around $q = 0.2423 + 0.0026 (\lambda_\odot - 214.0°)$ AU (rather than q = 0.38 AU). The argument of perihelion scatters around $\omega = 126.35°–0.415° (\lambda_\odot - 214.0°)$. Only the inclination appears to correlate well with the node: $i = 6.199°–0.167° (\lambda_\odot - 214.0°)$. Current photographic databases contain many Taurid orbits, but no orbits derived from a radiant, speed, and time in agreement with the observed range during this outburst. Unlike other Taurids, these meteoroids

[26] P. Jenniskens, C. ter Kuile and M. de Lignie, Geminiden 1990 in Zuid Frankrijk. *Radiant, J. DMS* **13** (1990), 8–19.

[27] P. Spurný, High fireball activity of the new subsystem of the Taurid complex. *ACM 1996*, Berlin, Poster presentation (1996).

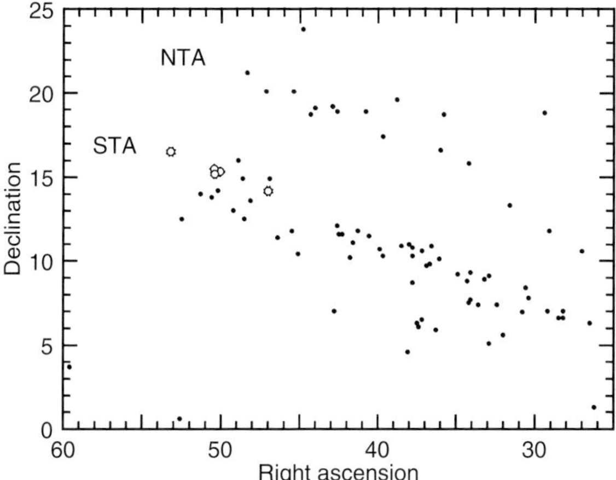

Fig. 25.10 Outburst radiants (○) among October Northern and Southern Taurids (●) in IAU Photographic Database.

Fig. 25.11 The brightest member (-17^m) EN221095 of the 1995 Taurid outburst over Poland. Photo: Pavel Spurný, Ondrejov Observatory.

had an aphelion of $Q = 4.9 \pm 0.3$ AU, close to the orbit of Jupiter ($r \sim 5.2$ AU), which means that these orbits are less decoupled than comet Encke.

The fireballs reached absolute magnitudes in the range −5.8 to −17.1 (Fig. 25.11), corresponding to initial masses of 900 kg, 65 kg, 420 g, 190 g, and 30 g, respectively (estimates by Spurný). This means that this shower may have included meter-sized bodies. These were clearly cometary meteoroids, with very irregular lightcurves (not uncommon among the Taurids). The brightest members (900 kg, 65 kg) were classified as type IIIb (fragile, cometary) in the scheme of Ceplecha. The less bright members (190 g, 30 g) were of type II, suggesting less fragile material. The fireballs penetrated to about 58 km altitude, the brightest to 52.8 km.

The measured pre-atmospheric entry speed is higher for meteors with brighter absolute magnitude, and more so with increasing uncertainty. This points to a measurement error in translating measured speeds to initial pre-atmospheric velocity speeds, for example, when decelerations are over-estimated slightly more for the brighter members in the group. Thus, the weak correlation of q and mass could be due to measurement error.

This observation identifies for the first time a meteor outburst associated with the Taurid shower, and by implication the Earth crossing a relatively young dust trail. This trail, however, appears to be dominated by large meteoroids, so large in fact that their origin is more likely to be from a comet breakup than Whipple-type comet ejection from water vapor drag when the comet is near perihelion.

There may have been other such outbursts in the past. On November 22, 1990, *Anna Petrenko*[28], an experienced meteor observer, first noticed a bright -4^{m} Taurid near the Moon in the early morning twilight (15:00 UT). She kept watching in the next half hour and saw five more fireballs from the same radiant. Part of the sky was obstructed by houses. One fireball passed near α-UMi and peaked at magnitude −6 to −7, while the others were of magnitudes −3 to −5. All meteors were bright yellow, with a reddish hue, and were of medium speed. Flares were seen at the end and there were short wakes. The radiant was determined to be between R.A. = 10°–90°, Decl. = +15° to +30°, with the Northern Taurid radiant being at about R.A. = 64° and Decl. = +24° at that time.

In Japan, *Yoshihiko Shigeno* and *Hiroyuki Shioi* observed strong activity of the southern δ-Piscids on September 8, 2001 and eight meteoroid orbits were calculated from intensified TV camera observations. At the same time, the visual zenith hourly rate was low (~0.5/h).[29] It is unknown whether or not this high rate of faint meteors is unusual. The Piscid shower is thought to be an extension of the northern Taurid radiant.

A possible outburst of mostly bright meteors (long duration echoes) was detected by radio forward meteor scatter techniques in Japan between 05:00 and 05:30 UT (14:00–14:30 JST) on July 6, 2001. *Hiroshi Ogawa* of Japan first noticed a spike of long-duration echoes while gathering forward meteor scatter observations for the Nippon Meteor Society's *International Project for Radio Meteor Observation. Brian Fuller* (Australia) reported that SkiYMet meteor radars had determined a center of activity at that time at R.A. = 92°, Decl. = +25°, but it is not clear if that center was responsible for the bright daytime meteors. Other radio-MS observers in Australia did not confirm this event. Finally, *David Asher*[30] of Armagh Observatory and *Kiyoshi Izumi* of the Nippon Meteor Society have pointed out that the Taurid showers are more active than usual in a periodic cycle of 3.39 yr, presumably because of dust

[28] A. Grishchenyuk, V. Martynenko and A. Petrenko, The observation of a fireball cluster probably belonging to the Taurid family. *WGN* **19** (1991), 30–31.
[29] Y. Shigeno and H. Shioi, Outburst of faint Piscids in 2001. *WGN* **30** (2002), 56–58.
[30] D. J. Asher, K. Izumi, Meteor observations in Japan: new implications for a Taurid meteoroid swarm. *Mon. Not. R. Astron. Soc.* **297** (1998), 23–27.

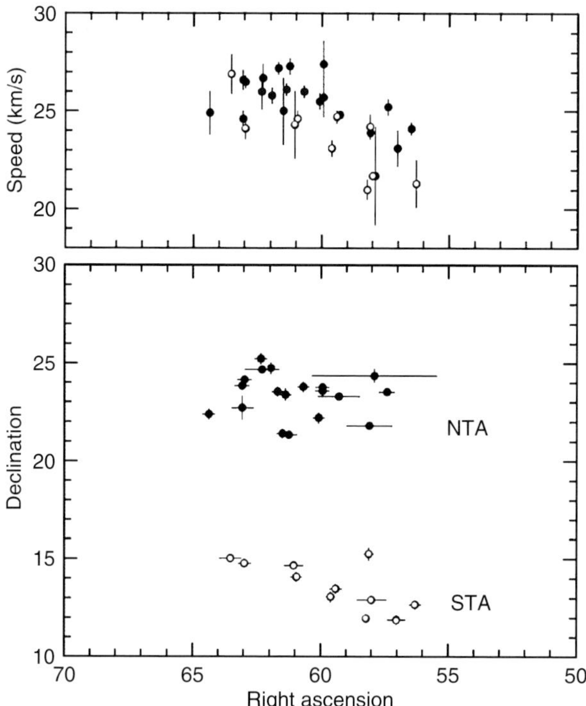

Fig. 25.12 Position of the Taurid radiant on November 17–19, 1998/1999, as derived from multistation video observations by Marc de Lignie (Dutch Meteor Society) and Masayoshi Ueda (Nippon Meteor Society), and coworkers.

trapped in the 7:2 mean motion resonance. If so, other good years would be 2005, 2008, 2012, 2015, 2022, 2025, 2032, 2039, 2042, 2049, and 2052. This needs confirmation.

25.4 How old is "young"?

There is also significant nonrandom structure in the annual Taurid radiant. It is difficult to get a good picture of the Taurid radiant from photographic observations, because the derived radiant structure is sensitive to the adopted value for the radiant drift. Both branches show a narrow range in declination, and a very smooth variation of right ascension with node. This makes it possible to recognize the Taurids from the sporadic background, but only clearly between solar longitude 205°–248° for NTA and 195°–245° for the STA. The mean change of the radiant with node is ΔR.A./$\Delta\lambda_{\odot} = +0.615°$ and $+0.825°$, respectively, and ΔDecl./$\Delta\lambda_{\odot} = +0.073°$ and $+0.239°$. If so corrected, the radiant of both streams look patchy. After correcting for daily motion, Porubčan and Kornoš found the same dispersion in R.A. (0.80° versus 0.73°

Object	D (km)	Epoch	a (AU)	q (AU)	$i°$	$\omega°$	Node	$\Pi°$	D_{SH}
2004 TG$_{10}$	0.88	(1952)	2.227	0.302	3.09	290.80	231.22	162.02	0.06
* Northern Taurids	Photo	(1952)	2.12	0.350	3.1	294.9	226.2	161.1	-.-
* June β-Taurids	Radar	(1952)	2.2	0.34	6.0	224.0	278.1	142.1	0.29
2P/Encke	4.6	(1951)	2.216	0.338	12.38	185.21	334.74	159.95	0.22
* Southern Taurids	Photo	(1955)	2.07	0.352	5.4	115.4	37.3	152.70	-.-
* June ζ-Perseids	Radar	(1952)	2.33	0.35	8.0	61.0	77.0	138.00	0.23

Fig. 25.13 Comparison of orbital elements. D_{SH}-criterion is relative to NTA and STA, respectively.

for NTA and STA) and in Decl. (0.16° versus 0.18° for NTA and STA, respectively). This corresponds to an inclination dispersion of 1.1° for both.[31]

Video observations provide more meteors and permit the viewing of the radiant structure on a given date. From observations by the Dutch Meteor Society during the Leonid season (Fig. 25.12), reduced by Marc de Lignie, I find that the radiant is elongated in right ascension, and significantly dispersed in declination, but with velocity changing with position in the radiant. For a single day, the graph of π versus q shows a narrow correlation, with q decreasing from $q = 0.55$ to 0.35 AU for increasing π from 147° to 170° (for Nov. 16–18). The Northern and Southern branches have a range in π, with $\langle\pi\rangle = 162.9°$ for NTA and $\langle\pi\rangle = 157.2°$ for STA. Such a stratified radiant can not result from a random accumulation of planetary perturbations over a long life, but could arise from the relatively recent breakup of dispersed fragments.

25.5 Progressive comet fragmentation

While writing this chapter, I noticed that minor planet **2004 TG$_{10}$**, discovered in the *Spacewatch Near-Earth Object Search Program*, matches the particular branch of the Northern Taurid stream well. It has the low perihelion distance of the Taurid stream, a perfect match to the semimajor axis of the meteoroids, and matches the length of perihelion (Fig. 25.13). This is probably the first comet nucleus fragment of a series that are to be found among the Taurid meteoroid stream. It is a better match ($D_{SH} = 0.058$) than 2P/Encke for any of the Taurid branches ($D_{SH} = 0.22$).

Does this finally validate the Giant Comet hypothesis of Clube and Napier? The fragment is about 0.88 km in size (assuming an albedo of 0.04 and density 1 g/cm^3). If there are 20 other cometesimals like 2004 TG$_{10}$, and if I include the dust in the Northern Taurid stream, then the parent body would have measured about $D = 5.5$ km, about the same size as comet 2P/Encke. Both objects could be part of

[31] V. Porubčan and L. Kornoš, The Taurid meteor shower. *ESA SP* **500** (2002), 177–180.

an earlier breakup. If, at best, 20 such objects existed, then this initial "giant" comet measured a mere 15 km in diameter, the size of comet Halley. No giant comet. Nevertheless, a few hundred sub-km sized fragments could still exist among the stream, which are a potential impact hazard for Earth.

The dispersion in length of perihelion of the Northern Taurids ($\sigma = 7.7°$), would imply a breakup age about 20 000 years ago (Fig. 25.4). However, it is possible that the fragments such as 2004 TG_{10} continue to break apart and the overall distribution of northern Taurids reflects the current distribution of fragments. In that case, the original breakup was 20 000 years ago (Table 8), but most of the current Taurids are much younger. Much of the older meteoroids have presumably been lost by collisions with other dust grains contributing to the zodiacal cloud (Chapter 31).

Part V

Old streams and sporadic meteoroids

26
Annual showers

Meteor showers are called *annual* when they return year after year at much the same intensity.[1] In that case, we can count on predictable levels of activity from year to year. This implies that the distribution is sufficiently dispersed for meteor activity to not be greatly influenced by trail motions relative to Earth's orbit from the cyclic motions of planets, by the position of the dust along the orbit of the comet, or by resonances. All known Halley-type showers and long-period comet showers have a dust component that returns annually, typically $W = 3^\circ$ wide or having an increase in rates by a factor of ten per $\sim 5^\circ$ of solar longitude ($B = 0.20$), depending on the encounter conditions. These annual shower activity profiles typically have a broader background component with a shallower tail (usually) towards shorter solar longitude ($W \sim 10^\circ$–20°). These annual showers tend to contain fainter meteors on average than the outbursts ($\chi = 2.5$ versus 1.8). The annual shower rate does not change much (<20%) when the comet returns to perihelion. The off-season Leonid shower (Fig. 26.1), for example, is a mild shower with a peak rate of about ZHR $\sim$ 13/h and a width $W = 3.0 \pm 0.6^\circ$.

The $\sim$230 most reliable annual showers are included in Table 7. *David Hughes* used comet orbit statistics to guess how many annual showers may exist at different levels of activity. He found that 12 showers have $2.8 < \mathrm{ZHR} < 8$/h, 53 showers should have $0.8 < \mathrm{ZHR} < 2.8$/h, and another 95 should have $0.28 < \mathrm{ZHR} < 0.8$/h.[2] The recorded number is 26, 80, and 166, respectively ($s = 0.8$). *Alexandra K. Terentjeva*[3] found some 249 minor showers from a list of 3700 photographic orbits and 200 radiant observations by visual observers. On a typical nonshower night, an estimated 20% of observed visual meteors belonged to minor showers, while the remaining 80% were considered to be "sporadic."

It is possible to calculate the expected *apparent* activity of annual showers, seen from different locations on the globe for any time in the night (Fig. 26.2). Those are

[1] P. Jenniskens, Meteor stream activity I. The annual streams. *Astron. Astrophys.* **287** (1994), 990–1013.

[2] D. W. Hughes, The mass distribution of comets and meteoroid streams and the shower/sporadic ratio in the incident visual meteoroid flux. *Mon. Not. R. Astron. Soc.* **245** (1990), 198–203.

[3] A. K. Terentjeva, Investigation of minor meteor streams. In *Physics and Dynamics of Meteors*, ed. L. Kresák and P. M. Millman. (Dordrecht: Reidel, 1968), pp. 408–427.

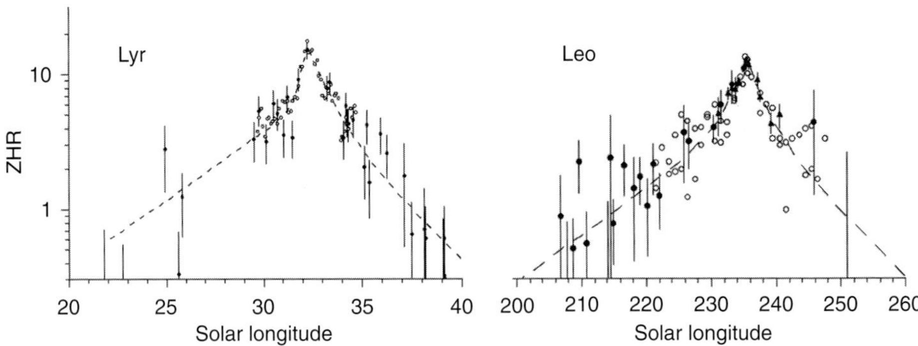

Fig. 26.1 Annual shower profiles of the long-period Lyrid and Halley-type Leonid showers.[4]

simply the measured zenith hourly rate profiles (such as Fig. 26.1), corrected for the geometric dilution due to the radiant elevation.

Individual lines in Fig. 26.2 are the hourly rates of meteors for different times in the night, in steps of 1 h with the dashed line at local midnight. The lowest line is that of early evening, the highest are rates just before dawn. The actual observed rates may be less (or more) than this, depending on the sky conditions, disturbing lights from artificial sources, or the Moon, the presence of obstruction in the field of view, viewing direction, and observer perception.

26.1 The evolution towards annual showers

The formation of an annual stream component in Halley-type showers is rapid Fig. 26.3 shows the dispersion of dust ejected from 55P/Tempel–Tuttle between about 10 and 20 revolutions ago as seen in 1993. Instead of a clustering of trails in a region about 1 d wide as typical for a Filament, we find that perturbations by Jupiter have already scattered the position of the trails over the wide region typical of the annual Leonid shower, from November 10 until December 1. These are suspected to be the meteoroids that are not trapped in mean-motion resonances and that felt the wrath of Jupiter in full force.

Leonid meteoroids disperse more rapidly along the Earth's path than they do perpendicular to it, in part due to their low inclination. That raises the question of whether the dust in annual showers is dispersed widely enough to not cause variations in peak rates by the change from barycentric to heliocentric orbits.

Indeed, *Audrius Dubietis* and *Rainer Arlt*[5] derived a particularly nice activity curve for the annual Lyrid shower observations gathered by the *International Meteor*

[4] P. Jenniskens, Meteor stream activity. III. Measurement of the first in a new series of Leonid outburst. *Meteoritics Planet. Sci.* **31** (1996), 177–184.
[5] A. Dubietis and R. Arlt, Thirteen years of Lyrids from 1988 to 2000. *WGN* **29** (2000), 119–133.

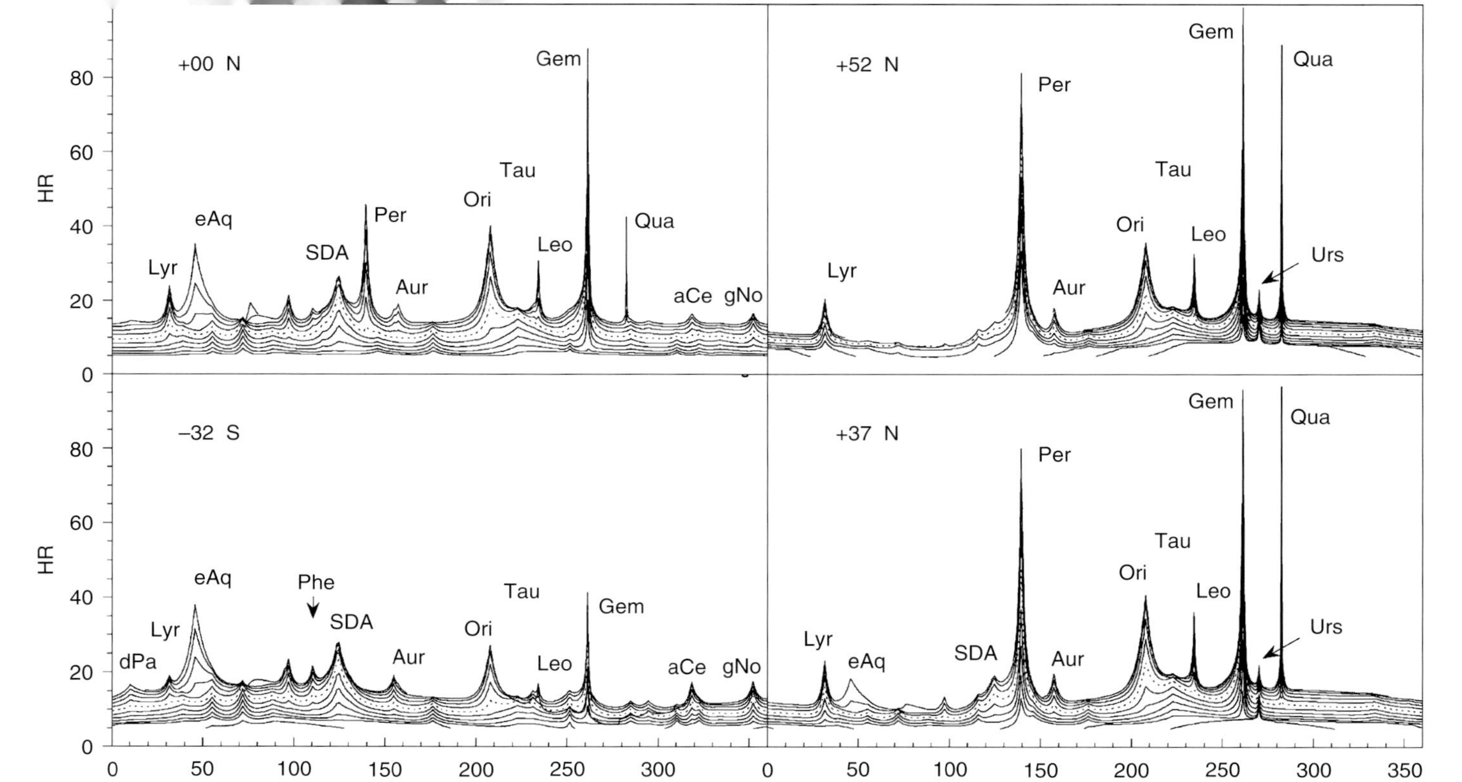

Fig. 26.2 The apparent hourly rate of meteors throughout the year for a typical visual observer under dark and clear skies from locations at latitudes +52 North, +37 North, the equator, and −32 South. Annual streams and sporadic meteors only. Choose the graph for latitude nearest to your observing site, choose the solar longitude (J2000) of the date of observation, then read off the expected rate of meteors (per hour) along the lines. The dashed line is for 0 h local time, other lines are for steps of 1 h, increasing from early evening (bottom) to morning dawn (top).[6]

[6] P. Jenniskens, Meteor stream activity I. The annual streams. *Astron. Astrophys.* **287** (1994), 990–1013.

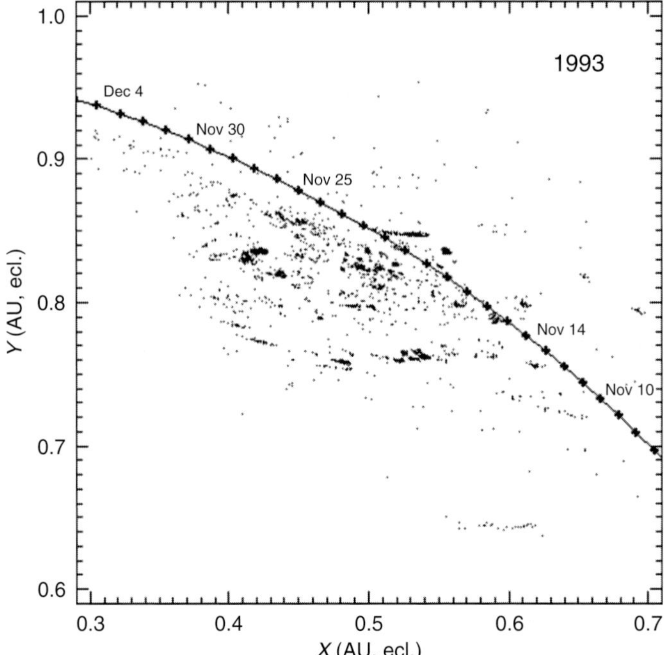

Fig. 26.3 The dispersion of the 10–20 revolution old dust trails of 55P/Tempel–Tuttle as seen in 1993. Younger trails are included in Fig. 14.11. Calculations by Jérémie Vaubaillon.

Organization over the period 1988–2000 and, fitting the profile of each year with a set of exponential curves, they found that the peak rate varied each year between about ZHR = 14.0 and 24.3/h. More interestingly, the peak rate appeared to correlate with the time of the maximum. This suggests to me that this annual shower also follows the Sun's reflex motion. In that case, I find that the observed variation in rates is only large enough if the annual Lyrid shower rates fall off towards the Sun a factor of 3 faster than they do in the Earth's path, much as the Leonid dispersion in Fig. 26.3.

In contrast, the Perseids spread more widely, perpendicular to their orbit, due to a higher inclination (Fig. 26.4). The intrinsic dispersions in the stream (after correction for measurement errors)[7] can be used to estimate the age of the shower. In 1972, *Boris Iul'evich Levin et al.*[8], using Sherbaum's program of numerical integration based on Cowell's method,[9] were the first to show that planetary perturbations tend to broaden meteoroid streams over time. From the calculated increase in the Perseid shower radiant dispersion of: FWHM $= 4.74 \pm 0.84 \times 10^{-3}$/yr, *Peter Brown* and *Jim Jones*

[7] H. Betlem, C. R. Ter Kuile, M. de Lignie *et al.*, Precision meteor orbits obtained by the Dutch Meteor Society – photographic meteor survey (1981–1993). *Astron. Astrophys. Suppl. Ser.* **128** (1998), 179–185.

[8] B. Y. Levin, A. N. Simonenko and L. M. Sherbaum, Deformation of a meteor stream caused by an approach to Jupiter. In *The Motion, Evolution of Orbits and Origin of Comets*, ed. G. A. Chebotarev *et al.*, *IAU Symp.* 45. (Dordrecht: Reidel, 1972), pp. 454–461.

[9] L. M. Sherbaum, *Vestun. Kiev Un-ta. Ser. Astron.* **12** (1970), 42 (in Russian).

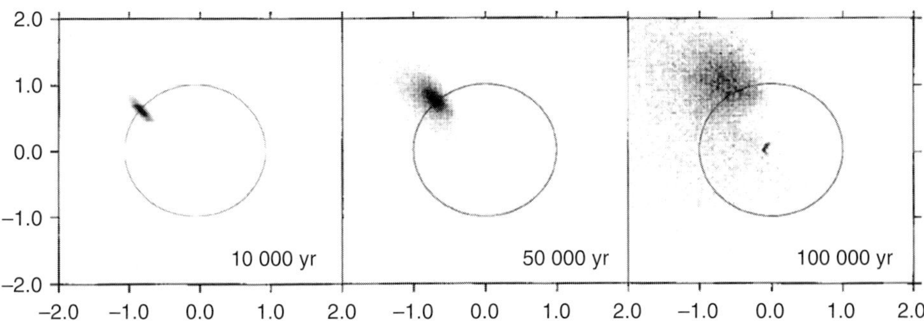

Fig. 26.4 Evolution of the Perseid shower for long periods of time. The circle is the Earth's orbit. The distribution of nodes of the meteoroid orbits is shown in calculations by Peter Brown.[10]

derived an age of about 15 000–20 000 years for the central portion of the annual Perseid stream. That amounts to roughly 140 orbits of the comet 109P/Swift–Tuttle. On a human time scale, that is about as far back in time as when the first settlers populated the Americas.

According to Brown and Jones, the fate of the meteoroids over long periods of time is mostly determined by the planetary perturbations from Jupiter, and to a smaller degree from Saturn. The terrestrial planets add only 5–10% to the dispersion of the orbits.

A Perseid meteoroid is expected to survive for several hundred thousand years before being removed from the stream. The main loss mechanism is not collisions with Earth, but ejection into hyperbolic orbits by Jupiter (and Saturn), sometimes via a Sungrazer state (see the meteors near the Sun in the rightmost part of Fig. 26.4). It requires 40 000–80 000 years for a significant fraction (>0.1%) of orbits to be ejected and after 100 000 years only 1–35% of meteoroids will have been lost.

26.2 Recognizing minor showers in the sporadic background

At what point are the meteoroids fully dissipated into the sporadic background? Interestingly enough, that may be far beyond the point in time when the shower can no longer be recognized by a visual observer. Even when rates are low, the meteoroids may still be recognized as originating from the same source if the orbit can be measured, because the planetary perturbations do not change the Tisserand invariant much.

The evolution of the dust of Jupiter-family comets is very fast due to Jupiter's perturbations, as we saw earlier. For example, projected onto the ecliptic plane, meteoroids ejected by comet 26P/Grigg–Skjellerup are difficult to recognize as a

[10] P. Brown and J. Jones, Simulation of the formation and evolution of the Perseid meteoroid stream. *Icarus* **133** (1998), 36–68.

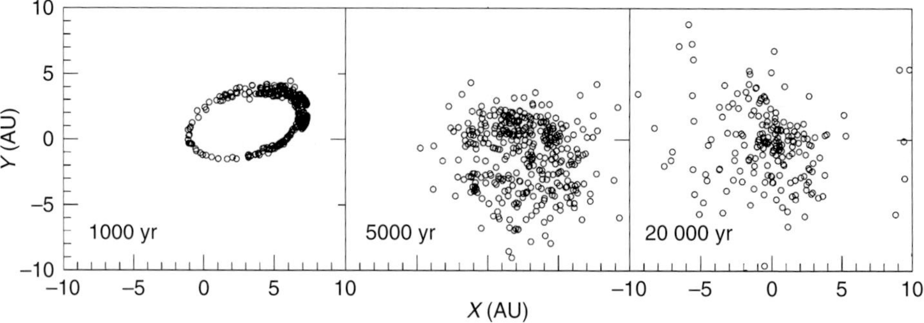

Fig. 26.5 The orbital evolution of tiny 20 μm dust grains ejected by comet 26P/Grigg–Skjellerup at perihelion in 1992 after 1000, 5000, and 20 000 yr.[11]

stream after 5000 years, as shown in calculations by *Gabriele Cremonese et al.* (Fig. 26.5). However, the orbital elements of the evolved stream have a semimajor axis and eccentricity that follow the lines of a constant Tisserand parameter with respect to Jupiter (Fig. 26.6).

Radiation forces such as the Poynting–Robertson drag (see Chapter 31) and perturbations by other planets do not conserve the Tisserand invariant relative to Jupiter and add to the dispersion over long timescales. Particles of only 20 μm in size are strongly affected by these radiation forces, even after 20 000 years, but the larger grains that cause meteor showers are not.

Hence, as long as a memory of the original orbit is retained, meteoroid streams may still be recognized in the population of the sporadic meteoroids, even when their radiant is diffuse. The search for meteoroid streams solely among the 19 698 orbits from the Harvard Radio Meteor Project at Havana, Illinois (operated by the Smithsonian Astrophysical Observatory) resulted in 275 "streams."[12]

For such work to be meaningful, a criterion had to be used that identifies a meteoroid orbit as part of a larger complex. The first such criterion, and still the one most commonly used, is that by Southworth and Hawkins,[13] called the *D-criterion* (for a, e, i only) or the D_{SH}-*criterion* if node and Π are considered too (Appendix F). The orbit is more similar to the proposed mean of the stream, and the stream association is more likely, if the D-criterion is smaller. For the major streams $\langle D_{SH} \rangle$ is about 0.06, for the minor streams $\langle D_{SH} \rangle = 0.17$. Usually D_{SH} is taken to be about <0.20 if associated with the stream. *Bertil-Anders Lindblad*[14] concluded that a good working rule was to put the limiting value at $D_{SH} < 0.80\ N^{-0.25}$, where N is the number of orbits in the

[11] G. Cremonese, M. Fulle, F. Marzari and V. Vanzani, Orbital evolution of meteoroids from short period comets. *Astron. Astrophys.* **324** (1997), 770–777.
[12] Z. Sekanina, Statistical model of meteor streams. IV. A study of radio streams from the synoptic year. *Icarus* **27** (1976), 265–321.
[13] R. B. Southworth and G. S. Hawkins, Statistics of meteor streams. *Smithsonian Contrib. Astrophys.* **7** (1963), 261–285.
[14] B.-A. Lindblad, 2. A computerized stream search among 2401 photographic meteor orbits. *Smithsonian Contrib. Astrophys.* **12** (1971), 14–24.

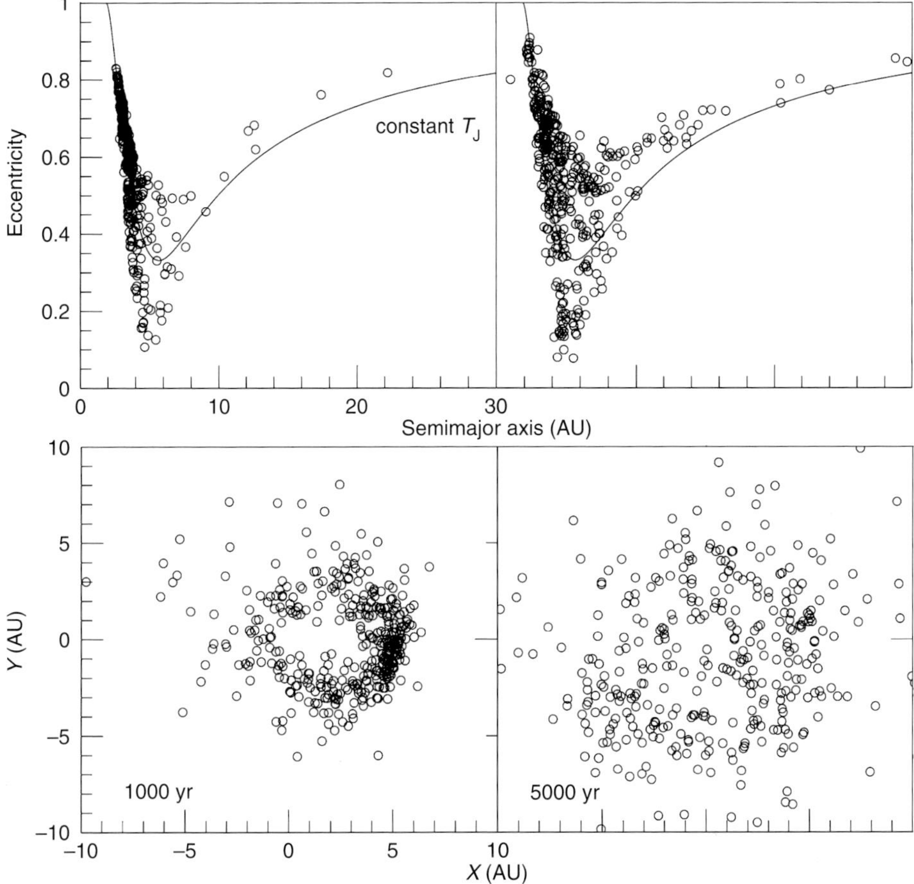

Fig. 26.6 The change in orbital elements by planetary perturbations follows approximately the Tisserand invariant (solid line). Shown is the distribution of 200 μm π-Puppids after 1000 and 5000 years. Calculations by Gabriele Cremonese *et al.*

sample. Some care must be taken. On a sample of 2400 photographic orbits, Lindblad found that 50% of the streams with only two or three members were chance associations of sporadic meteors for $D_{SH} < 0.115$.

In applying the D_{SH}-criterion, it makes a difference if one compares the meteoroids to a stream average (e.g., Table 7) or if one only demands that each meteoroid orbit is associated with at least one other member of the stream. Only the latter method of serial association makes it possible to detect dispersed streams such as the Taurids and Capricornids. Hence, the serial association (or progressive search) is often preferred over the direct association.

The main assumption is that the distribution of longitude of perihelion is random. More evolved criteria consider the fact that the direction of the perihelion is often constant, but these criteria miss dynamical evolutions that change the longitude of perihelion over time.[15]

Visual observers have traditionally tried to discover annual meteor showers by plotting the meteors on (often bad) gnomonic star charts and searching for radiants. This has resulted in long lists, the most extensive is the *British Meteor Society Radiant Catalogue* (containing some 1200).[16] No effort was made to check whether or not different names pertain to the same stream. Others have focused on the most active showers, using a working list of meteor showers in the footsteps of the 1973 publication by *Alan Cooke*.[17]

Table 7 is a new working list, where I tried to consolidate all meteor showers detected in orbit surveys, showers detected by visual observers as meteor outbursts, and some visually detected showers on the southern hemisphere that lack orbit survey underpinning. I have restricted myself to those ecliptic streams that are confirmed by others, except for some helion sources, for which few data are available. By providing the current evidence for each stream, my hope is that this list will be a useful starting point, from which to investigate further.

26.3 What fraction of all meteors are in showers?

The question has been raised as to what percentage of all meteoroids belong to recognized streams? There should be a higher fraction among brighter meteors on account of sporadic meteors having a steeper magnitude distribution index (χ). Visual observers often estimate $\chi = 2.5$–3.4, but only after including minor showers in the sample. Radar observations of fainter meteors consistently have $\chi = 4.3$.

The answers vary widely with the association criteria used and the brightness of the meteors (Fig. 26.7). Most surprising are the high percentages reported for radar observations. Tadeusz J. Jopek *et al.*[18] found 28% of 3675 radar orbits in the IAU database to be part of meteoroid streams. The *Harvard Radar Survey*[19] recorded meteors down to magnitude +12, with most meteors probably in the range +6 to +10. 16% of all meteors were part of the 275 meteoroid streams identified by Sekanina. Southworth and Sekanina[20] estimated that about 30% of 12 600 orbits belonged to these streams. 40.4% of a sample of 1667 meteor orbits in the *Adelaide Radar Survey* could be assigned

[15] L. Neslusan and P. G. Welch, Comparison among the Keplerian-orbit-diversity criteria in major-meteor-shower separation. In *Proc. Meteoroids 2001 Conf.*, 6–10 August 2001, Kiruna, Sweden, ed. B. Warmbein, *ESA SP* 495. (Noordwijk: ESA Publications Division, 2001), pp. 113–118.

[16] R. A. MacKenzie, *British Meteor Society Radiant Catalogue* (Dover: British Meteor Society, 1981).

[17] A. F. Cook, A working list of meteor streams. In *Evolutionary and Physical Properties of Meteoroids, Proc. Coll.* 13, Albany, NY, 14–17 June 1971, ed. C. L. Hemenway, P. M. Millman and A. F. Cook, *NASA SP* 319 (1973), p. 183.

[18] T. J. Jopek, G. B. Valsecchi and Cl. Froeschlé, Meteor stream identification: a new approach. Application to 3675 radio meteors. In *Meteoroids 1998* (Bratislava: Astron. Inst. Slovak Acad. Sci., 1999), pp. 307–310.

[19] Z. Sekanina and R. B. Southworth, Physical and dynamical studies of meteors. Meteor-fragmentation and stream-distribution studies. Final Report. (Cambridge, MA: Smithsonian Astrophysical Observatory, 1975); Z. Sekanina, Statistical model of meteor streams. III. Stream search among 19303 radio meteors. *Icarus* **18** (1973), 253–284.

[20] R. B. Southwort and Z. Sekanina, *NASA Contractor Rep. CR-2316* (1973).

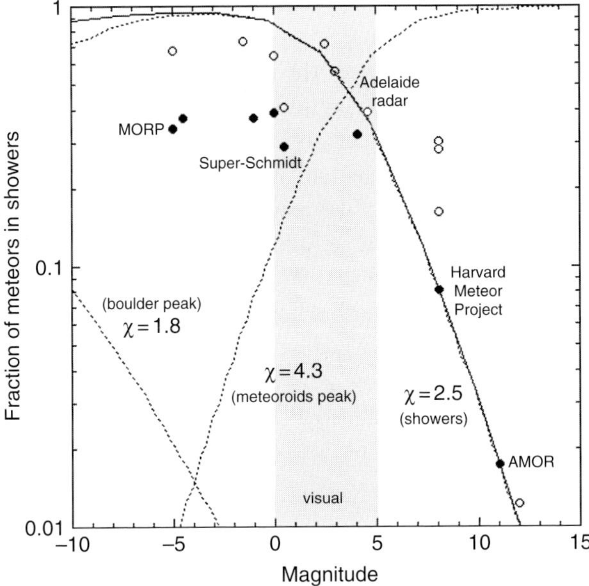

Fig. 26.7 Percentage of meteors that belong to meteor showers. The solid line shows the expected pattern from the observed number density of meteors in the simplified case that all sporadic meteors have $\chi = 4.3$, all streams have $\chi = 2.5$, and a fireball component has $\chi = 1.8$ (of which 30% are assumed to be in showers). Individual contributions of each component to the total are given by dashed lines.

to at least one other orbit, and 29.8% with two or more.[21] Nillson's survey of 2200 radio meteors found a percentage of ~33% (allowing for progressive searches). The Adelaide radar survey considered brighter meteors, mostly in the range +2 to +6 (with a limiting magnitude of +8). These radar observations were performed with a combined multi-station CW and pulse radar system operating at 27 MHz.

Such high percentages are not expected if the sporadic meteors have a magnitude distribution index $\chi = 4.3$ and streams have $\chi = 2.5$. In that case, only 2% of meteors at $+11^{\mathrm{m}}$ should belong to showers, rising to only 10% at $+5^{\mathrm{m}}$. Indeed, I estimate from the published AMOR radar results (typical meteor magnitude of +11) that the percentage of showers is only about 1.7%. If so, at least half of the meteor showers reported in the Harvard and Adelaide meteor radar surveys are real (Fig. 26.7).

Visual plotting of meteor trails mostly concerns meteors in the range +0 to +5 (gray area Fig. 26.7). About 50–80% of all meteors are expected to be in showers, a higher percentage than estimated by Hughes, but consistent with the high percentages of 56% and 72% reported in the months of October and the time period of

[21] G. Gartrell and W. G. Elford, Southern Hemisphere meteor stream determinations. *Austr. J. Phys.* **28** (1975), 591–620.

November–January, respectively.[22] A similar magnitude range is covered by results from multistation video observations.[23]

Among the photographed meteors of the Harvard Meteor Survey,[24] mostly covering meteors of +2 to -4^{m} (limiting magnitude +4) Southworth and Hawkins found 41% associated in groups (12% perhaps by chance). Bertil Lindblad found that 37% of a sample of 2401 photographic meteor orbits were associated into streams after rejecting chance associations among two- and three-member groups. That may be too low an estimate. Jacchia *et al.* were able to identify 65% of all orbits with one shower or the other. Babadzhanov[25] found that 73% of 185 doubly photographed meteors had dynamic properties consistent with a cometary origin.

Among the brightest fireballs analyzed is a relatively large proportion of asteroidal objects. These fireballs have a shallow $\chi \sim 1.8$, stretching all the way to the regime of minor planets. About 37% of all MORP fireballs can be associated with meteor showers.[26] Terentjeva[27] found that 68% of 554 fireballs were part of a shower, but half of those represented groupings of what may have been meteorite dropping fireballs and have only two members (Chapter 30).

Hence, among visual and photographic meteors, a large percentage belongs to (minor) meteor showers.

[22] M. S. Rao and A. G. Murthy, Minor meteor showers of October observed over Waltair during 1962–1971. *Austr. J. Phys.* **27**, 679–686; M. S. Rao, P. V. S. Rama Rao and P. Ramesh, Minor meteor showers of November, December, and January. *Austr. J. Phys.* **22** (1969), 767–774.

[23] T. J. Jopek and Cl. Froeschlé, A stream search among 502 TV meteor orbits. An objective approach. *Astron. Astrophys.* **320** (1997), 631–641.

[24] L. G. Jacchia and F. L. Whipple, Precise orbits of 413 photographic meteors. *Smithsonian Contrib. Astrophys.* **4** (1961), 97–129.

[25] P. Babadzhanov, Orbital elements of photographic meteors. *Smithsonian Contrib. Astrophys.* **7** (1963), 287–288.

[26] I. Halliday, A. A. Griffin and A. T. Blackwell, Detailed data for 259 fireballs from the Canadian camera network and inferences concerning the influx of large meteoroids. *Meteoritics Planet. Sci.* **31** (1996), 185–217.

[27] A. K. Terentjeva, Fireball streams. *WGN* **17** (1989), 242–245.

27

Dispersion from gradually evolving parent body orbits

Meteoroid streams will be dispersed by planetary perturbations, because the perturbations are different for meteoroids on slightly different orbits. Earlier, I gave the examples of Quadrantids, Geminids, Daytime Arietids, δ-Aquariids, α-Capricornids, and κ-Cygnids. While these are relatively young streams, the long-range perturbations of Jupiter have dispersed the meteoroids far enough to be seen annually. Their shower activity is not stable over tens of years (Chapter 1), however, possibly not even over one year (Chapter 24). Think of these as the broken-comet showers.

Halley-type showers also evolve in an oscillating manner, because the planets (mostly Jupiter) exert a force on the orbit of the meteoroids. Again, the slow changes (of the longitude of perihelion, for example) are called precession (Fig. 27.1), while the faster periodic terms are called nutation. While most short-period broken-comet showers appear to have evolved along less than a single nutation cycle before being dispersed beyond recognition, the Halley-type showers could, in principle, evolve over multiple nutation cycles, separated in space by the effect of precession. This chapter investigates the claims to that effect.

27.1 The tail of the Perseid shower

William F. Denning of Bristol was the first to discover that the Perseid shower stretched from the beginning of July until the third week of August, way beyond the core of the stream in mid August, and noticed that the stream had a more gradual increase of rates leading up to the peak than the subsequent decrease after the peak (Fig. 27.2).[1]

A possible reason for this tail was first identified by Richard Southworth,[2] but recent modeling has quantified the effect more clearly. Jupiter's gravitational force on the orbiting comet or meteoroid will tend to rotate the orbit so that the direction of perihelion and Π gradually shifts over time (Fig 27.1). This is called *precession*[3] if Π moves in prograde manner and *regression* if Pi moves against the motion of the

[1] P. Jenniskens, Meteor stream activity I. The annual streams. *Astron. Astrophys.* **287** (1994), 990–1013.
[2] R. B. Southworth, Planetary perturbations and the Perseid meteor stream. *Astron. J.* **70** (1961), 295–296.
[3] K. Fox, I. P. Williams and D. W. Hughes, The evolution of the orbit of the Geminid meteor stream. *Mon. Not. R. Astron. Soc.* **200** (1982), 313–324.

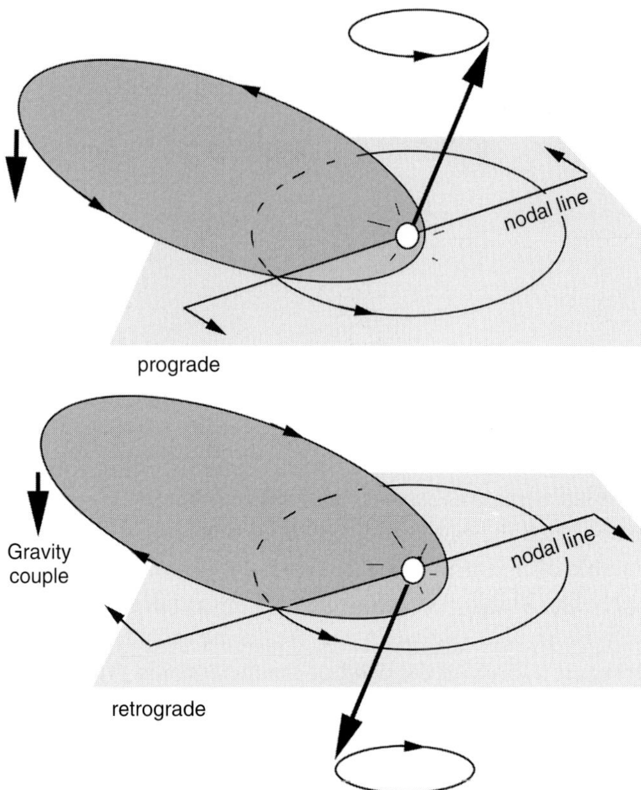

Fig. 27.1 Principle of precession (top) and regression (bottom) as the planet exerts a force on the orbit, the orbit will start to rotate in a manner that changes Π. After: Fox, Williams, and Hughes.[4]

planets. Because the comet orbit librates about the 1:11 mean motion resonance, there is no nutation that overwhelms the change of Π.[5] The rate of precession is expected to be a function of the orbital period and is weaker for high inclinations.

The precession of the Perseid orbit is very slow due to its high inclination ($i = 113.4°$). The node has shifted only 0.22° over the course of a thousand years. This compares to a typical change from planetary perturbations of $\Delta\Omega = 0.3°$/orbit (or about 50° per thousand years) for Jupiter-family comets. *Nathan Harris*, *Kevin Yau*, and *David Hughes* explained the extended tail of the Perseids by noting that meteoroids that are put in longer orbits than the parent comet tend to precess more slowly. They believed that in this way the meteoroids trace the orbital evolution of the comet, leaving meteoroids in wider orbits at the old node, and found that most of the dust is located just outside Earth's orbit, creating a tail in the annual shower profile (Fig. 27.3). To do so, however, they needed high (a few 100 m/s) ejection speeds.

[4] *Ibid.*
[5] J. E. Chambers, The long term dynamical evolution of comet Swift-Tuttle. *Icarus* **114** (1995), 372–386.

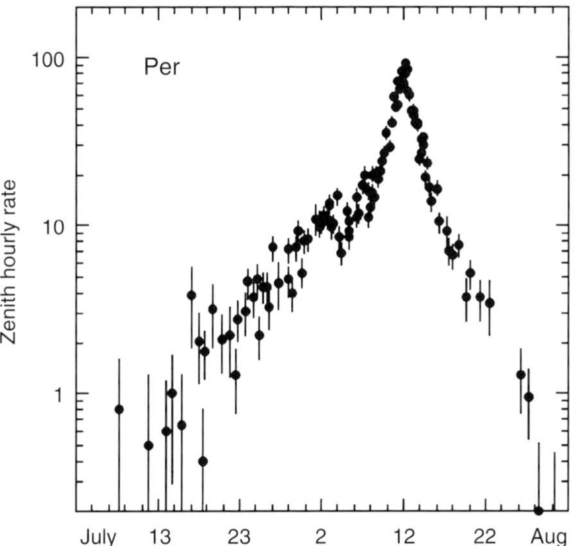

Fig. 27.2 Background to the Perseid shower activity curve as traced to early July from observations by the Dutch Meteor Society (1981–1992).

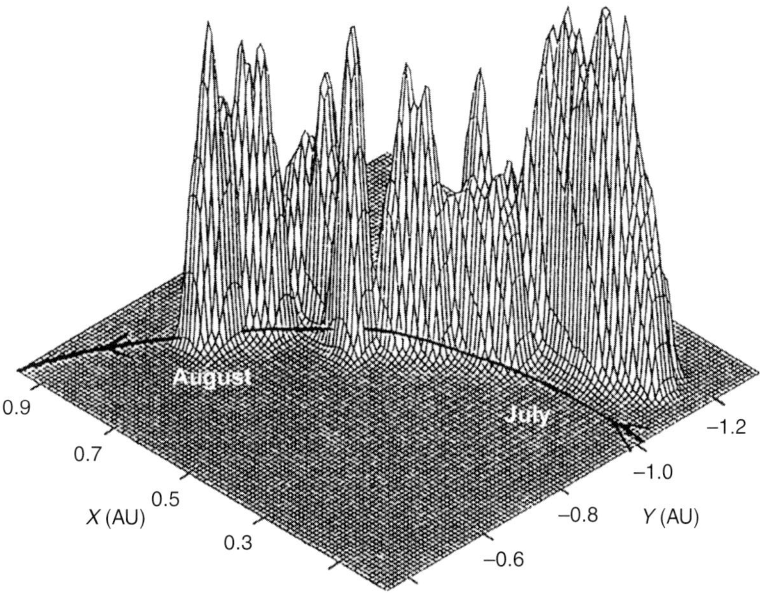

Fig. 27.3 The full extent of the Perseid shower as calculated by Nathan Harris, Kevin Yau, and David Hughes. Earth moves from right to left. Most of the dust is located just outside Earth's orbit, creating a tail in the annual shower profile. Photo: David Hughes, *MNRAS*.[6]

[6] N. W. Harris, K. K. C. Yau and D. W. Hughes, The true extent of the nodal distribution of the Perseid meteoroid stream. *Mon. Not. R. Astron. Soc.* **273** (1995), 999–1015.

Harris *et al.* did not consider perturbations by the planets. In my opinion, that is a more likely mechanism to disperse meteoroids into wider orbits. Alternatively, when the meteoroids are not trapped in the mean-motion resonance of the comet, they may evolve at a different rate than the comet.

Other mechanisms that delay the orbital evolution of the meteoroids relative to that of the parent comet would not necessary change the age estimates, which are based on the rate of evolution of the parent comet. According to Harris *et al.*, the main peak is due to dust particles that were produced in the past 5000 yr. This compares to the early age estimates of 5300–27 000 yr by Southworth[7] and 15 000–20 000 yr by Brown and Jones.[8] The current mass-loss of 109P/Swift–Tuttle of about 3×10^{11} kg/orbit, and a total mass in the main peak component of 1.1×10^{13} kg[9] would also suggest about 5000 yr.

Harris *et al.* found that particles in the tail were produced as far back as 160 000 years ago, after a close encounter with Jupiter moved the perihelion distance of the parent comet sufficiently inward for water-ice evaporation. Long-term integration of the comet orbit shows no such close encounter at least as far back as 27 000 yr.[10] This age estimate compares well with the estimated loss rates of particles in the Perseid stream by Brown and Jones (Chapter 26), and is only a factor of 4 higher than the early estimate of about 40 000 yr by *Salah E. Hamid.*[11] On the other hand, if the dispersion away from the Sun is the same, then there is only 3.1 times more mass in the background component than in the main peak,[12] demanding a deposition age of only $\sim$15 000 years. There should be about 30 times more mass in the stream than observed. This is understood if most of the mass is just outside Earth's orbit and the July tail is only the edge of a massive dust stream (Fig. 27.3).

27.2 The showers from 1P/Halley

The long-term orbital evolution of comet 1P/Halley is move rapid and is claimed to be responsible for some unusual features of the Orionid and the η-Aquariid showers (Fig. 27.8). These showers are unusually broad, seen over periods of about 5 and 11 d around their maximum, and have similar intensity despite Earth passing at different distances from the current orbit of parent 1P/Halley (0.155 AU and 0.066 AU, respectively). Those minimum distances are unusually large for meteor showers associated with comets and it was not universally accepted that comet Halley was the parent of these showers until a few decades ago.[13]

7 R. B. Southworth, Dynamical evolution of the Perseids and Orionids. *Smithsonian Contrib. Astrophys.* **7** (1963), 299–303.

8 P. Brown and J. Jones, Simulation of the formation and evolution of the Perseid meteoroid stream. *Mon. Not. R. Astron. Soc.* **295** (1998), 847–859.

9 P. Jenniskens, Meteor stream activity I. The annual streams. *Astron. Astrophys.* **287** (1994), 990–1013.

10 J. E. Chambers, The long-term dynamical evolution of comet Swift–Tuttle. *Icarus* **114** (1995), 372–386.

11 S. E. Hamid Doctoral dissertation, Harvard (1950), as quoted in: A. C. B. Lovell, *Meteor Astronomy* (Oxford: Clarendon Press, 1954), p. 429.

12 P. Jenniskens, Meteor stream activity I. The annual streams. *Astron. Astrophys.* **287** (1994), 990–1013.

13 D. W. R. McKinley, *Meteor Science and Engineering* (New York: McGraw-Hill, 1961).

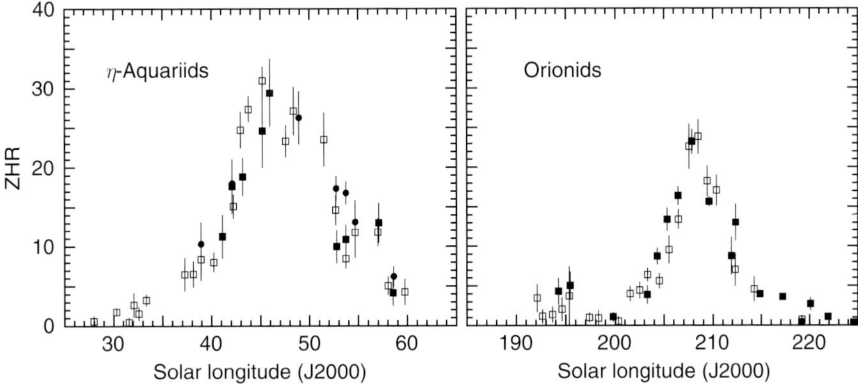

Fig. 27.8 The Halley showers: the η-Aquariids and Orionids as observed by NAPO-MS and the DMS visual observers.[14]

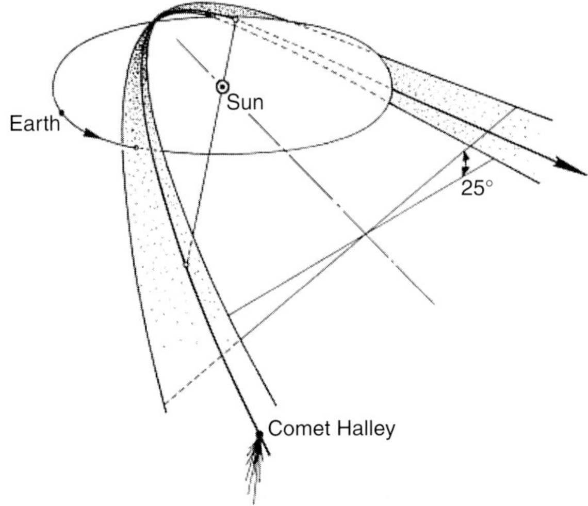

Fig. 27.9 The Shell Model of comet 1P/Halley streams showing one shell of dust. Figure by: Bruce McIntosh and Jim Jones, *MNRAS*.

The return of comet Halley in 1985/86 spurred new interest in the origin of its meteor showers. As soon as past observations of the comet and advances in computing techniques made it possible to trace the orbit of 1P/Halley back to 1404 BC,[15] it was recognized that the node of the comet orbit advanced rapidly with time, due to the gravitational perturbations by the major planets (Fig. 27.9). The meteoroid orbits now advance at a rate of about 0.19°/century (causing the radiant position to move with about 0.18°/century).[16]

[14] P. Jenniskens, Meteor stream activity I. The annual streams. *Astron. Astrophys.* **287** (1994), 990–1013.
[15] D. K. Yeomans and T. Kiang, The long-term motion of comet Halley. *Mon. Not. R. Astron. Soc.* **197** (1981), 633–646.
[16] B. A. McIntosh and J. Jones, The Halley comet meteor stream – numerical modelling of its dynamic evolution. *Mon. Not. R. Astron. Soc.* **235** (1988), 673–693.

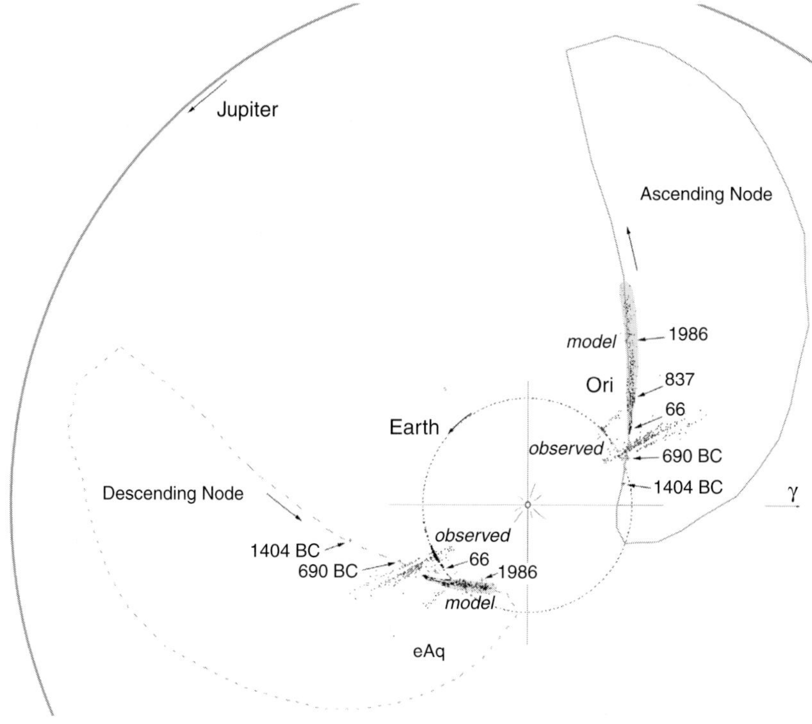

Fig. 27.10 The node of Orionid (descending) and η-Aquariid (ascending) orbits during one nutation cycle in the IAU and 2003 photographic database and DMS video database, projected on top of calculated nodes of meteoroids ejected by comet Halley since 1404 BC by McIntosh and Jones. The libration of the nodes is according to the secular variations calculated by Obrubov.

Moreover, the argument of perihelion, the inclination ($\pm 12.5°$) and the node oscillated over time in a nutation cycle.[17] The effect of rotating the nodal line is to tilt the orbit up and down over a 25° arc, sweeping out a volume that measures 0.44 AU perpendicular to the orbital plane, and 0.05 AU in the orbital plane near Earth's orbit. In this case, the variation of node and argument of perihelion are not over a full circle, but over a narrower range of values (due to a large "forced" part of the secular perturbation unique to the semi-major axis of 1P/Halley). Fig. 27.10 traces the position of the nodes over a full nutation cycle as calculated by Yu. V. Obrubov.

Bruce McIntosh of the University of Western Ontario and *Anton Hajduk* of the Astronomical Institute of the Slovak Academy of Sciences proposed a *Shell Model*[18] in which particles released in the past would evolve at different rates and, with each

[17] Y. Kozai, Secular perturbations of asteroids and comets. In *Dynamics of the Solar System*, ed. R. L. Duncombe, *IAU Symp*. 81. (Dordrecht: Reidel, 1979), pp. 231–237.

[18] B. A. McIntosh and A. Hajduk, Comet Halley meteor stream – a new model. *Mon. Not. R. Astron. Soc.* **205** (1983), 931–943; B. A. McIntosh, The structure and evolution of the comet Halley meteor stream. In *Comet Halley: Investigations, Results, Interpretations. Vol. 2: Dust, Nucleus, Evolution*, ed. J. W. Mason and E. Cliffs. (Chichester, NJ: Ellis Horwood/Prentice-Hall, 1990), p. 121.

nutation cycle of rotating ω, would form a shell that is not cylindrical but rather a flat ribbon with uniform thickness perpendicular to the orbital plane of the comet (Fig. 27.9). Earth passing such shells would result in meteor showers of the same duration, for a wide range of miss-distances. A number of (nonoverlapping) shells could accumulate over time, accounting for secondary peaks in the activity profiles.[19]

Five such secondary peaks were claimed present, adding up to 100 000 yr (about 1500 comet revolutions). McIntosh and Jones[20] subsequently confirmed that the rate of evolution was very sensitive to ejection conditions, ranging from a short of 20 000 yr to a more typical nutation period of 200 000 yr, but suggested that populating five shells could take as short as a few thousand years. They proposed that the stream began to develop after the close approach of the comet to Jupiter about 20 000 years ago.

Because the stream is a reflection of the orbital evolution of the comet, it does not tell us how old the comet is. On the other hand, the age of the comet (since first entering the inner parts of the solar system) does put an upper limit to the size of the stream. A 2 km sized crater on Halley, if caused by a collision with an interplanetary boulder, implies an age of about 6000 revolutions ($\sim$500 000 yr).[21] In comparison, Halley's expected lifetime against being ejected from the solar system is only about five times higher: 70 000 orbits, but the comet will break in a collision in half that time, after about 30 000 revolutions.

The claim of five secondary peaks in the Orionid shower activity,[22] one of the fundamentals of the early Shell model, has not weathered the scrutiny of better data. With better statistics, the structure disappears, except for one secondary peak in the η-Aquariid profile, which may be variable in intensity (Chapter 18). Moreover, meteoroids quickly show a relatively large spread in their nodes, not only along the path of the librating nodes of the comet, but also perpendicular to that. This would wash out shells in the stream profile. Finally, ejection of meteoroids at all returns of the comet would also wash out any evidence of multiple libration periods.[23]

I have also studied whether or not the shells are recognized in the radiant structure (Figs. 27.11 and 27.12). A stubborn *W. F. Denning* believed that the Orionid radiant was stationary over the three weeks of its activity, with two active centers.[24] His successor *J. P. M. Prentice*[25] found a daily motion instead, but claimed four different active radiants that peaked at different times, thus explaining Denning's incorrect perception of a stationary radiant. Recent photographic and video orbit

[19] Y. Kozai, Secular perturbations of asteroids and comets. In *Dynamics of the Solar System*, ed. R. L. Duncombe, *IAU Symp*. 81. (Dordrecht: Reidel, 1979), pp. 231–237.

[20] B. A. McIntosh and J. Jones, The Halley comet meteor stream – numerical modelling of its dynamic evolution. *Mon. Not. R. Astron. Soc.* **235** (1988), 673–693.

[21] J. A. Fernández, Collisions of comets with meteoroids. In *Comets Asteroids Meteors III* (Uppsala: Astronomical Observatory, 1989), pp. 309–312.

[22] V. Porubčan, A. Hajduk and B. A. McIntosh, Visual meteor results from the International Halley Watch. *Bull. Astron. Inst. Czechoslov.* **42** (1991), 199–204.

[23] Z. Wu and I. P. Williams, Comet P/Halley and its associated meteoroid streams. In *Meteoroids and Their Parent Bodies*, ed. J. Stohl and I. P. Williams. (Bratislava: Astron. Inst. Slovak Acad. Sci., 1992), pp. 77–80.

[24] W. F. Denning, Observations of the Orionids. *Mon. Not. R. Astron. Soc.* **73** (1913), 667–668.

[25] J. P. M. Prentice, The radiant of the Orionids. *Observatory* **46** (1923), 47–49.

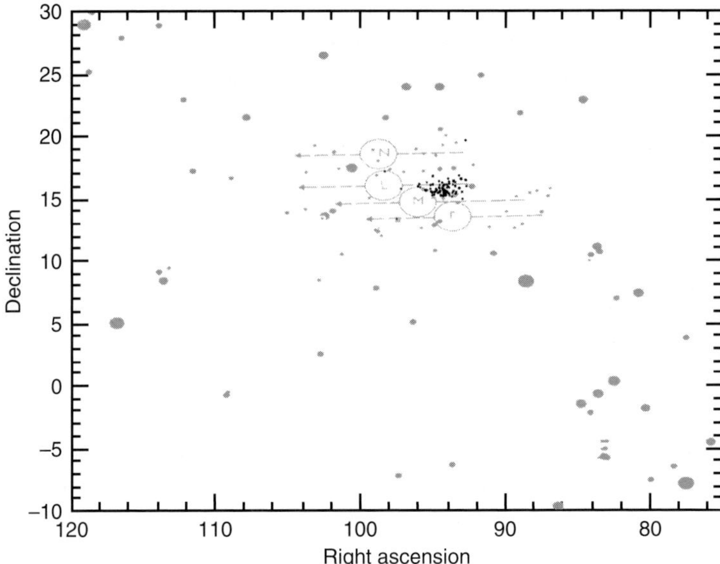

Fig. 27.11 The radiant of DMS video and photographic Orionids compared to the various stream components derived from visual observations by J. P. M. Prentice.

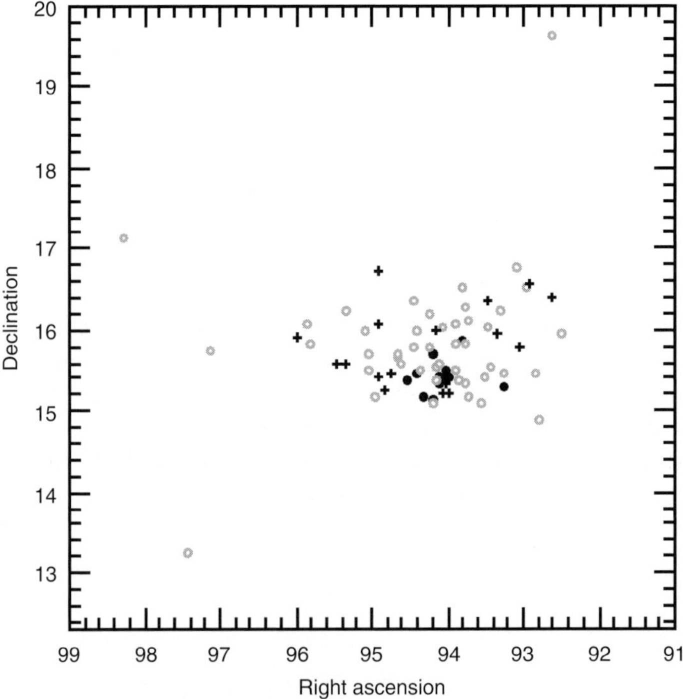

Fig. 27.12 Radiant of DMS video (○) and photographic (+) Orionids compared to the various stream components positions corrected to node = 27.0°.

measurements by the DMS (*Marc de Lignie*) and at Ondrejov (*Pavel Koten*) sample the same brightness meteors as on which the visual results were based. The verdict is not good. There is no evidence of the leading and following branches recognized by Prentice.

Fig. 27.10 shows the nodes of photographed Orionids and η-Aquariids in the 2003 IAU photographic meteor orbit database and in the DMS video database on top of the theoretical calculations by McIntosh and Jones. The opposite stream nodes on the other side of Earth's orbit trace the outline of the meteoroid stream. In the Earth's path, the stream is $W = 0.19 \pm 0.02$ AU wide. The meteoroid nodes are distributed in a band, twice that wide, centered on the path of the current libration cycle of Halley and stretching in the heliocentric direction for $W \sim 0.5$ AU.

For this dispersion of dust, the total mass in the stream would be about 21 000 billion kg.[26] Compared to the modern era dust ejection by comet 1P/Halley of about 100 billion kg/orbit, I calculated that the stream is ~210 orbits old (= 16 000–18 000 yr if the mean orbital period of Orionids is 76–85 yr). If so, then Halley has lost a layer of about 120 m of its surface since the fatal encounter with Jupiter.

Finally, *Yu. V. Obrubov*[27] found that a full nutation cycle (70 000 yr in his calculation) of ω between 45° and 135° causes the distance of the node to vary in a cyclic manner, while the node is also changing over a 90° angle, and should therefore result in four meteor streams, two more in addition to the Orionids and η-Aquariids. The radar catalog of Lebedinets *et al.* contains such September "Southern Orionids" and June "Northern η-Aquariids," if you like, but I find no evidence of these showers in other sources. Unless activity from these radiants is confirmed, I conclude the age of the Orionid and η-Aquariid showers is less than a single nutation cycle.

Wu and Williams calculated that the standard deviations in the observed and calculated streams deviated considerably, with the observed streams still being much broader. The planetary perturbations are expected to gradually increase the scatter in nodes and the argument of perihelion. From that, an age of 19 700 yr was found for the η-Aquariid shower, and 18 700 yr for the Orionid shower.

All evidence points to a scenario in which 1P/Halley only completed 1/2 or 3/4 cycle since its last (or one before last at the other node) encounter with Jupiter and the comet was captured from a long-period comet orbit at that time. From Kozai's calculations, McIntosh and Jones put this ~180 comet orbit revolutions ago, in good agreement with age estimates from the orbital element dispersion and the total mass in the stream.[28] Obrubov's results put the descending node near Jupiter's orbit in 8000 BC, while the encounter before that was with the ascending node around 12 000 BC.

[26] P. Jenniskens, Meteor stream activity. I. The annual streams. *Astron. Astrophys.* **287** (1994), 990–1013.

[27] Yu. V. Obrubov, Long-period evolution and new meteor showers of comet P/Halley. In *Meteoroids and Their Parent Bodies*, ed. J. Stohl and I. P. Williams. (Bratislava: Astron. Inst. Slovak Acad. Sci., 1992), pp. 69–72.

[28] B. A. McIntosh and J. Jones, The Halley comet meteor stream – numerical modelling of its dynamic evolution. *Mon. Not. R. Astron. Soc.* **235** (1988), 673–693; J. Maddox, Halley's comet is quite young. *Nature* **339** (1989), 95.

28
The ecliptic streams

The cardinal directions in meteor astronomy are relative to the direction of Earth's motion (the Apex) and the direction to the Sun (Helios). Instead of "North" and "South" or "East" and "West," it is common to speak of *Apex* and *Antapex*, *Antihelion*, and *Helion*. Figure 28.1 shows the distribution of approach directions,[1] on top of a typical sporadic meteoroid orbit. Most meteoroids arrive from one of three distinct directions: the Earth's direction of motion (Apex source), the direction of the Sun (Helion source), or the opposite direction (Antihelion source).

28.1 The Helion and Antihelion sources of ecliptic showers

The early-morning Apex source consists mostly of meteoroids in retrograde orbits from long-period and Halley-type comets. Because planets perturb low-inclination orbits strongly, this source is split into two components: a Northern and a Southern Apex source, 15° below and above the ecliptic plane. The Apex source meteoroids arrive mostly at between 45 and 72 km/s (peaking around 62 km/s).

The midnight Antihelion and noon Helion sources are the same strength, but daytime observations are more difficult, even for radar. The meteoroids originate from Jupiter-family comets in prograde Apollo-like orbits (Fig. 28.1), with an aphelion outside and a perihelion inside Earth's orbit. The meteoroids arrive at speeds of typically 11–45 km/s, peaking around 20 km/s from an apparent direction $\pm 70^\circ$ from the Apex (true direction $\pm 105^\circ$, Fig. 28.1). The Antihelion source is active throughout the year and the center of this diffuse region of activity moves in position on the sky by about 1° each night, forward in right ascension, passing through all the constellations of the zodiac (ecliptic plane). Associated *ecliptic showers* are called from January to December: the Cancrids, Leonids, Virginids, Librids, Scorpiids, Sagittariids, Capricornids, Aquariids, Piscids, Arietids, Taurids, and the (ρ-)Geminids.

[1] W. J. Baggaley and L. Neslusan, A model of the heliocentric orbits of a stream of Earth-impacting interstellar meteoroids. *Astron. Astrophys.* **382** (2002), 1118–1124.

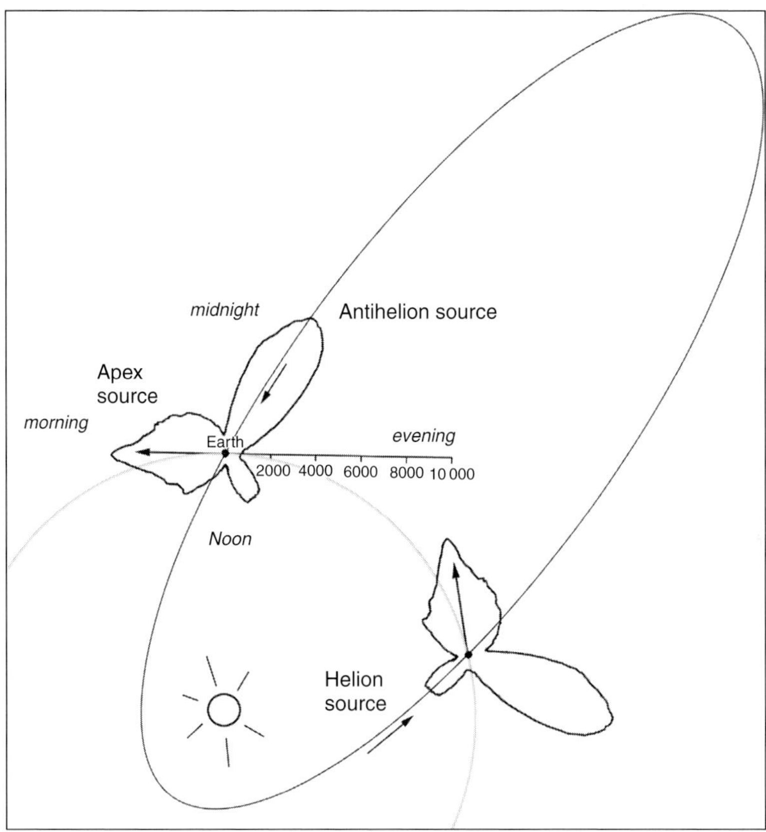

Fig. 28.1 The direction of origin of 132 996 meteors detected in 1990 and 1991 with the AMOR radar by Jack Baggaley, University of Christchurch.

The annual variation of activity from the various sources is shown in Fig. 28.2. Each plot depicts the daily rate of radio reflections in a forward meteor setup by *Ilkka Yrjölä* at Kuusankoski, Finland, representative of visual meteors.

Best rates from the (ecliptic) Antihelion source are just *after* midnight (local standard time), when we view in the direction from where the meteoroids appear to arrive. Remember to add Earth's velocity vector to the directions of the meteoroids shown in Fig. 28.1 to get the apparent radiant (Chapter 4)!

The meteoroids are concentrated near the ecliptic plane (Fig. 28.3), so the highest sporadic rates are in the fall (in the northern hemisphere), up to HR = 15/h for a visual observer under transparent moonless skies, because that is when the ecliptic culminates at the highest elevation (Fig. 26.2). The lowest rates are in the Spring, when sporadic rates drop to 5/h.

In the late afternoon the Apex source is below the horizon and the yearly variation of the Antihelion and Helion source of ecliptic showers is best recognized. The Apex

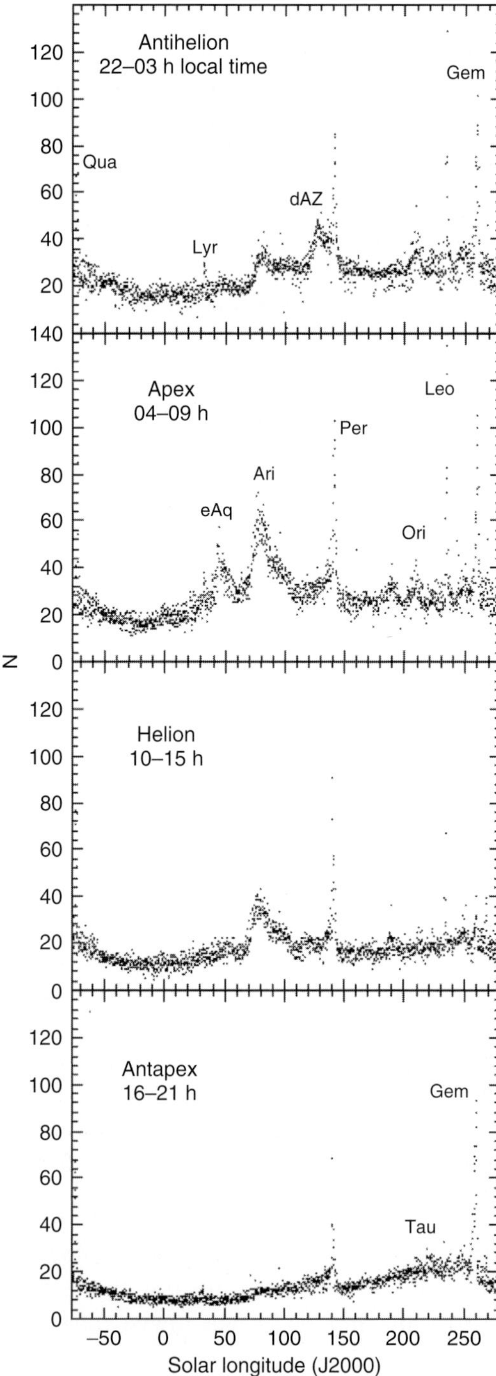

Fig. 28.2 Meteor rates from forward meteor scatter at a location in Finland (+61N). Data: Ilkka Yrjölä, Kuusankoski.

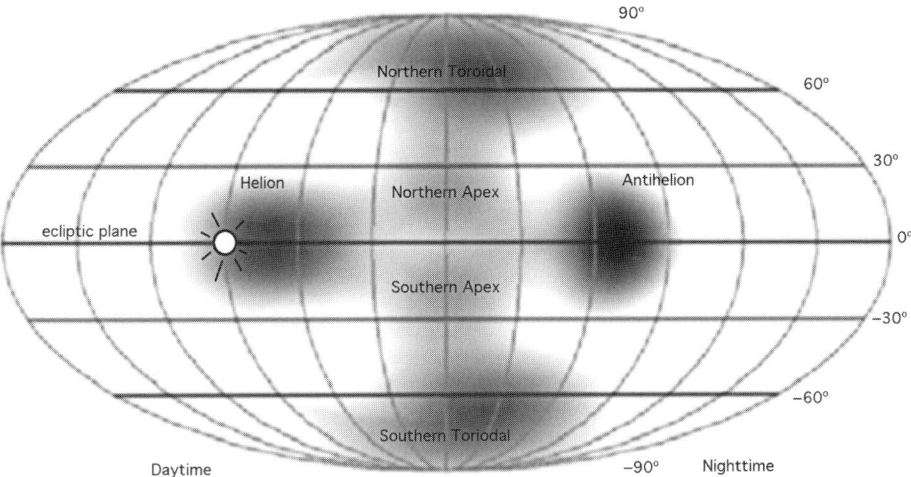

Fig. 28.3 Schematic distribution of apparent radiants on the sky. Earth Apex is at the center. The northern hemisphere part of the plot shows Canadian Meteor Orbit Radar data for April 10, 2004, results by Margaret D. Campbell-Brown and Peter Brown (University of Western Ontario).[2]

source of Halley-type showers is responsible for the yearly variation of rates in the early morning hours.

There are real peaks in fall and summer that coincide with the Taurid showers of comet 2P/Encke (see below), with the various Sunskirter streams (Daytime Arietids, δ-Aquariids), and with the Geminids.

The ecliptic showers tend to be active for many days, weeks, or even months, because of their low inclination (the Earth travels through the stream at a small angle), and Jupiter is not avoided at aphelion, causing a rapid dispersion of orbits. Streams tend to disperse their longitude of perihelion over time.

Stream searches among Antihelion (and Helion) source meteoroids in different orbit databases tend to extract similar streams, suggesting some of these ecliptic streams are as real as the Taurids, for example. The more likely showers are listed in Table 7, including a mean orbit derived from each sample of associated meteoroid orbits.

The mean orbit alone is not enough to validate a shower and much work still needs to be done. More information is in the spatial distribution of radiants, patterns of speed, distribution of nodes, and the Tisserand invariant, for example. Fig. 28.4, for example, illustrates how meteoroids associated with asteroid 4179 Toutatis do not show the expected concentration of radiants near the center and lack the dispersion of radiants along the ecliptic plane, like the Taurids matching the orbit of comet 5025 P-L. Also, meteoroids associated in this manner with asteroid 4179 Toutatis are dispersed over 58 d, without a typical activity curve.

[2] M. D. Campbell-Brown, P. Brown, The meteoroid environment: shower and sporadic meteors. *Workshop on Dust in Planetary Systems, Kauai, Hawaii from September 26–30, 2005*. Abstract (2005).

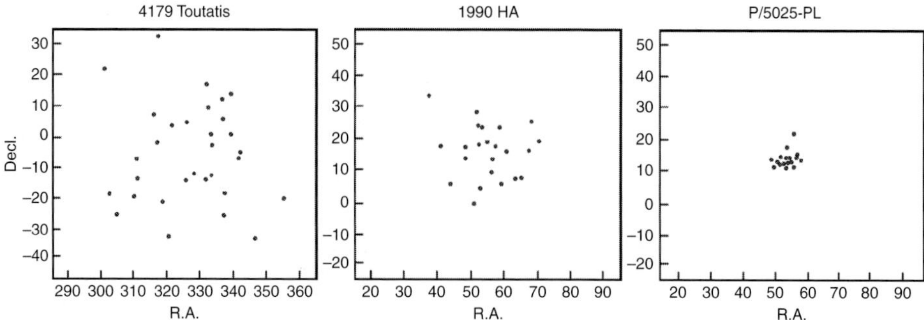

Fig. 28.4 The distribution of radiants for meteoroids associated ($D < 0.25$) with three potential parent bodies with different entry speeds, from left to right: $V_g = 11.9$, 15.9, and 28.5 km/s. Figure by Jan Stohl and Vladimir Porubčan.[3]

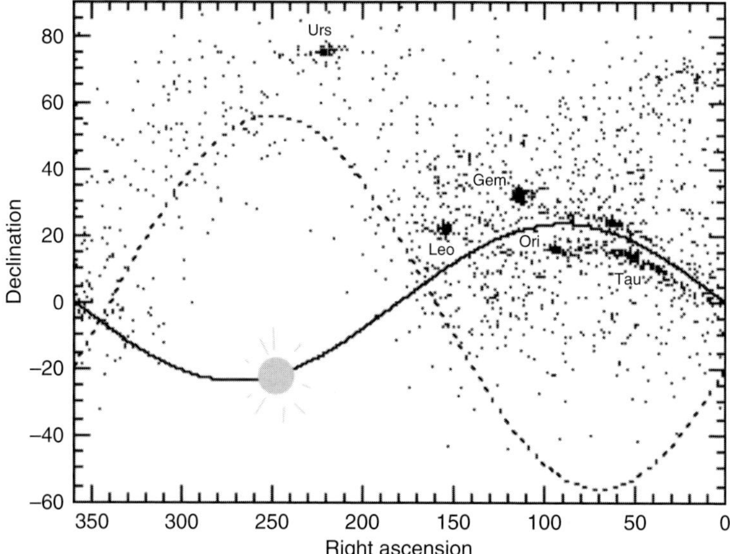

Fig. 28.5 Radiant positions of photographed video meteors in the DMS and IAU database in the months of September–December. The solid line shows the ecliptic plane, while the dashed line indicates the local horizon for a site at +37N.

28.2 Ecliptic showers in the fall

Let us first discuss the ecliptic streams in the fall, strictly the months from September to December, when the Antihelion source moves from Aries, Taurus, and Gemini, into Cancer.

[3] J. Stohl and V. Porubčan, Meteor streams of asteroidal origin. In *Meteoroids and Their Parent Bodies*, ed. J. Stohl and I. P. Williams. (Bratislava: Astron. Inst. Slovak Acad. Sci., 1992), pp. 41–47.

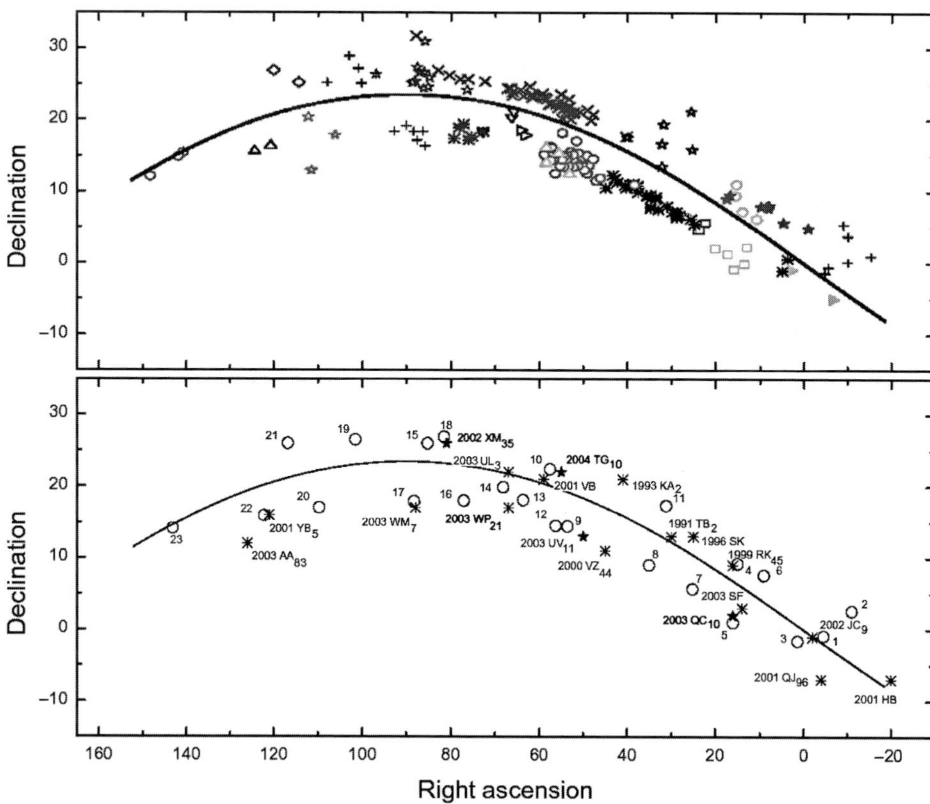

Fig. 28.6 23 possible clusters of meteoroid orbits in the Taurid Complex (○) were compared to the orbit of 19 potential parent minor bodies (∗) by Vladimír Porubčan, Leoš Kornoš, and Iwan P. Williams. Stars identify my four earlier identifications.[4]

Now, the chance of a mis-identification increases because of the abundance of low-inclination orbits among near-Earth objects. A dynamical study is needed to confirm that the deviations in the orbit between minor body and stream are not too large for them to separate in a short time by planetary perturbations. Many older Taurid shower parent bodies were dismissed in Chapter 25, but recently *Vladimír Porubčan, Leoš Kornoš*, and *Iwan P. Williams* have investigated more recently discovered candidate parent bodies. They identified 23 clusters of orbits in the Taurid Complex, each of one of which could derive from, or be associated with, a separate minor body (top Fig. 28.6). Numerous associations were investigated (bottom Fig. 28.6), but few showed the same evolution in orbital elements between shower and minor body. They identified four potential parents. They confirmed my link of **2004 TG$_{10}$** with the Northern Taurids (cluster 10, $D_{SH} = 0.05$), with a nutation period of 5000 yr. They

[4] V. Porubčan, L. Kornoš, I. P. Williams, The Taurid complex meteor showers and asteroids. Presentation at: *Asteroids, Comets, Meteors. IAU Symposium 229*. Búzios, Rio de Janeiro, Brazil. August 7–12, 2005 (2005), 20 pp.

also confirmed my identification of **2003 QC_{10}** with the N. δ-Piscids (cluster 5, $D_{SH}=0.05$), with both meteoroid stream and parent body having a similar short semi-major axis. This body also has a 5000 yr nutation period. They did not confirm my tentative link of *2003 UV_{11}* with the Southern Taurids cluster 9, the orbits being sufficiently different for minor body and stream to evolve differently. This object has subsequently been dismissed from Table 7.

In addition, they found a good match for 2002 XM_{35} with the N. χ-Orionids (cluster 15). In the same way, they found that **2003 WP_{21}** was a match with cluster 9 ($D_{SH}=0.15$, 5000 yr nutation period), with a radiant in late November from R.A. = 67.0°, Decl. = +17.2° on Nov. 25.3, close to the Southern Taurids, but with a lower speed $V_g = 24.63$ km/s. These meteors do not stand out well from the Southern Taurids and are not listed as a separate stream in Table 7.

The Taurid Complex fragments may include **2005 UR**, a match for the Northern October Arietids, which has not been investigated yet. *2201 Oljato (1947 XC)* in a shorter orbit with $a = 2.172$ AU (Fig. 25.7) is often associated with the S. χ-Orionids, but the object appears to be an asteroid (Chapter 25). *2003 UL_3* provides a match to cluster 16 (R.A. = 68.8°, Decl. = + 21.8°, $V_g = 25.45$ km/s on Nov. 25.9), but has a large miss-distance of 0.25 AU.

There are several non-Taurid ecliptic streams in the fall as well. The γ-Piscids in October may derive from minor planet 6344 P-L. The Daytime δ-Scorpiids agree well with the orbit of 2004 YD_5. Other potential identifications are less certain, with significant dicsrepancies in perihelion distance or inclination, and need further study.

28.3 The ecliptic showers in winter and spring

Comet Encke is probably a remnant from the breakup of a larger comet forming the Taurid Complex. There is one other such complex proposed, with five members, which has minor planet *(2212) Hephaistos* as its archetype. These objects have been associated with the *Virginid shower* complex in late winter and early spring. One comet, ***D/1766 G_1 (Helfenzrieder)***, has been put in this group.[5] This comet has the second shortest orbital period known to date, $P = 4.35$ yr. At $H_{10} = +6.8$, the nucleus diameter of the comet could be about $D = 3.8$ km. The comet has only been observed during one month in April–May, 1766, after *Johann Evangelist Helfenzrieder* discovered the comet at Dillingen, Germany on the evening of all fools day, April 1. Messier described the comet as being visible to the naked eye near the horizon, exhibiting a tail 4° long. Wirtz found that the comet had just survived a close passage by Jupiter at a mere 0.03 AU on November 11, 1763. It faded rapidly in the next weeks and was last seen on May 13. The comet has not been recovered. As a result, the orbital period is uncertain by at least one year. If the orbital period is 4.35 yr, then it is not fully

[5] D. Steel and D. Asher, P/Helfenzrieder (1766 II) and the Hephaistos group of Earth-crossing asteroids. *Observatory* **114** (1994), 223–226.

decoupled from Jupiter, because the orbital period is close to its 8:3 mean-motion resonance. The comet also passes close to Earth, Venus, and Mercury. This comet has one of the highest impact probabilities with asteroids and could have episodes of activity associated with asteroid impacts or hypervelocity meteoroid impacts from asteroidal dust.[6] No such event is known. The inclination of Helfenzrieder is small and passes through zero several times between AD 1600 and 2400, at which time the miss-distance decreases to improve the chance of meteor showers. The most likely scenario, however, would be that the observed shower is the remnant of a breakup during the return of 1766.

The comet has been linked to the δ-Leonids in February.[7] However, I find that the orbit is in better agreement with the η-*Virginids* in March if that shower were to extend into February. The forward projected orbit in AD 2004 would pass Earth's at 0.0198 AU on February 23, with meteors radiating from R.A. = 165.3°, Decl. = +6.1°. The δ-Leonids are indiscriminate from the Antihelion source at that time.

The Virginid Complex

During the months of January to April, from the dramatic Quadrantids to the beautiful Lyrid showers, little is going on that brings casual meteor observers on the northern hemisphere outside. This is the off-season even for many experienced meteor observers. The Antihelion source moves from the constellations of Cancer (January), Leo (February), Virgo (March), into Libra in April. Those that do fight the cold and bad weather always spot some relatively slow meteors emanating from the constellations of Leo (January–February) and Virgo (March–April) from a series of diffuse radiants, called the *Virginid Complex*.[8]

After correcting for the curved path of Earth, the radiant positions scatter as shown in Fig. 28.7. Several orbit searches have isolated clusters of Virginid shower orbits,[9] observed over a period of many weeks. The more believable groups are listed in Table 7. Most showers are very diffuse and may simply have been the result of search routines isolating meteoroids from an otherwise smooth distribution of sporadic orbits. Only the α-*Virginids* (VIR) are well recognized, and stand out when selecting a narrow speed range of 28.0–31.5 km/s.

Jacchia *et al.*[10] found the Virginids penetrated relatively deeply into the atmosphere. These meteoroids are as inefficiently slowed down as the Geminids. A particle density

[6] M. Beech and K. Gauer, Cosmic roulette: comets in the main belt asteroid region. *Earth, Moon Planets* **88** (2002), 211–221.

[7] D. Olsson-Steel, Theoretical meteor radiants of Earth-approaching asteroids and comets. *Austr. J. Astron.* **2** (1987), 21–35.

[8] D. W. Olson, An early observation of Virginid activity. *WGN* **18** (1990), 42–43.

[9] V. Letfus, The orbits of the Virginids and κ-Cygnids. *Bull. Astron. Inst. Czechoslov.* **6** (1955), 143–145.

[10] L. G. Jacchia and F. L. Whipple, Precision orbits of 413 photographic meteors. *Smithsonian Contrib. Astrophys.* **4** (1961), 97–129.

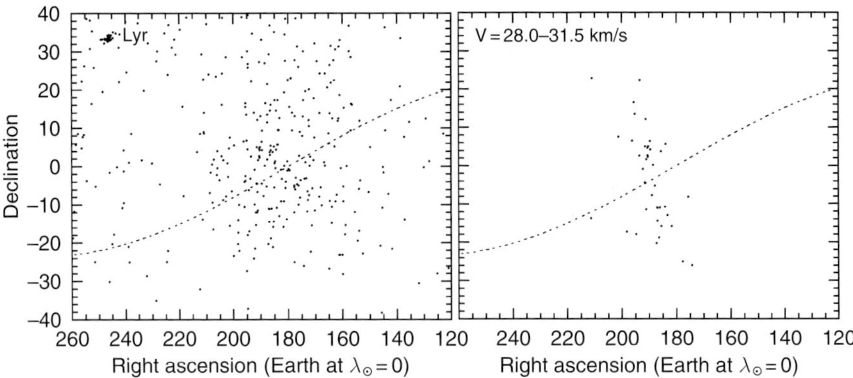

Fig. 28.7 Disbribution of all radiants and those of Virginids in the period from February 1 until April 30, derived from photographic surveys collected in the 2003 IAU database.[11] Positions are corrected for the Earth's motion at solar longitude = 0.

of 3.5 times that of sporadic meteors was derived,[12] or about 1.8 g/cm^3. The fragmentation index is similar to that of the Geminids, suggesting relatively compact material. Hoffmeister found that only 4% of Virginids leave a persistent train (or better: wake).[13] The meteors tend to be faint, with $\chi = 3.04 \pm 0.33$.[14] The meteoroids do not come as close to the Sun as the Geminids and have a longer orbital period. It is possible that space weathering is the reason for this property if these meteoroids are much older.

In 1973, *E. I. Kazimirchak-Polonskaya* and *A. K. Terent'eva*[15] published a paper examining the evolution of the Virginid Complex over the period 1860–2060 AD. With an aphelion close to Jupiter's orbit and a low inclination, there are frequent close encounters with Jupiter to within 0.32 AU. The orbits were found to evolve very rapidly, with the radiant point moving erratically in a region measuring 15° by 31°, while the node moved back and forth across a period spanning 51 d.

The following centers of radiation are still recognized. The month of January has the *ρ-Geminids* (RGE), recognized in photographic and radar surveys. On January 16/17, 1993, around 00:36 UT ($\lambda_\odot = 297.010°$), DMS observer *Koen Miskotte* witnessed a small burst of slow, fragmenting meteors radiating from a point south of Pollux (Fig. 28.8).[16] The seven observed meteors all had the remarkable appearance of a ball with a small tail. The magnitudes estimated were +1.5, +2.5, +0.5, +0.5, +3.0, +3.5, and +2.0, respectively. *Michiel van Vliet* calculated a ZHR = 12/h over the observing interval. No parent body is known. No potential extinct comet nucleus has been found yet that could be responsible.

[11] R. E. McCrosky and A. Posen, Orbital elements of photographic meteors. *Smithsonian Contrib. Astrophys.* **4** (1961), 15–84.
[12] F. Verniani, On the luminosity efficiency of meteors. *Smithsonian Contrib. Astrophys.* **8** (1965), 141–172.
[13] C. Hoffmeister, *Meteorströme* (Leipzig: Verlag Johan Barth, 1948), p. 147.
[14] R. Arlt, *Mitteilungen Meteore Arbeits Kreis Meteore* **82** (1987), 4.
[15] E. I. Kazimirchak-Polonskaya and A. K. Terent'eva, Orbital and radiant evolution of a meteor stream of the Jupiter family. *Sov. Astron.* **17** (1973), 368.
[16] M. Van Vliet, Meteorenzwerm aktief op 17 January! *Radiant, J. DMS* **15** (1993), 52–53.

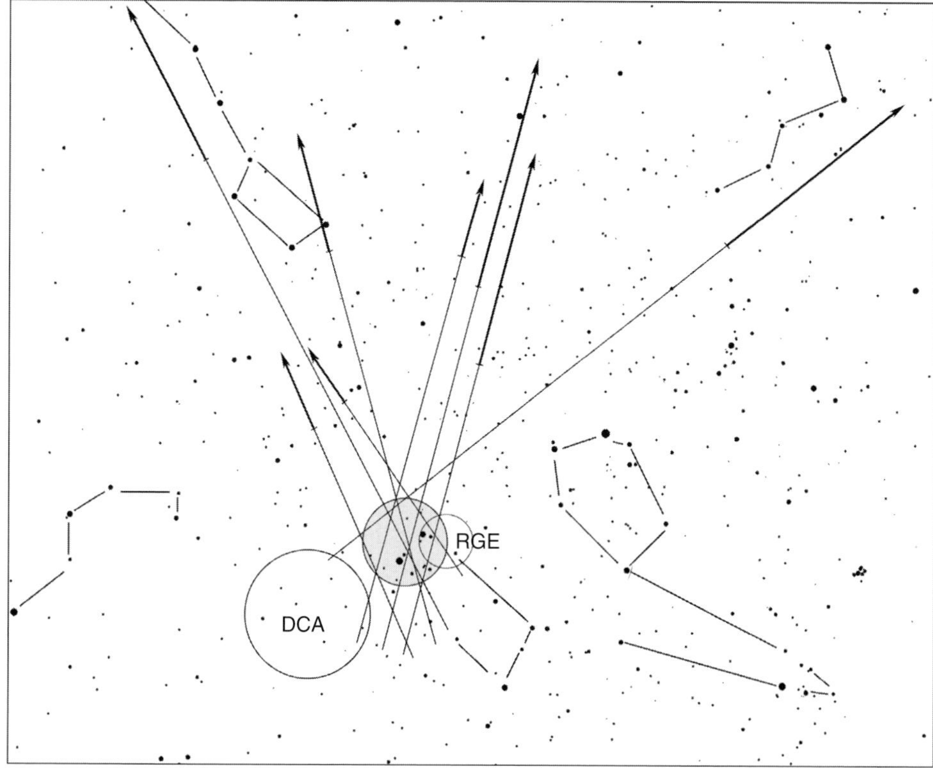

Fig. 28.8 Drawings of 1993 ρ-Geminids, by Koen Miskotte. Figure by Marco Langbroek (DMS).

The radiant of this outburst was not far from that of the annual *δ-Cancrid* shower, albeit 15° lower in right ascension and 5° higher in declination (Fig. 28.8). The δ-Cancrids are at the center of the Antihelion source. A northern branch (the most active) and a southern branch has been identified.[17] *Gary Kronk* has reported on AMS observations of this shower, suggesting rates of 2–3/h. In good agreement, I found rates around ZHR = 3/hr from a very dispersed shower. This could mark just the center of the Antihelion source activity. Potential parent bodies are *1991 AQ* (NCC) and *2001 YB*$_5$ (SCC).

The month of February is said to show activity from the *δ-Leonids* (Table 7). Again, the position is that of the Antihelion source and the radiant area is very diffuse. This stream has been associated with the asteroid *4450 Pan* (1987 SY).[18] According to the dynamical criteria of *Bill Bottke*, it is unlikely that this object is an extinct comet nucleus. A more likely parent body is **1999 RD$_{32}$**, a large $H_N = +16.32$ ($D \sim 3.2$ km) object with a 46% likelihood of being an extinct comet nucleus (Table 2). The

[17] Z. Sekanina, Statistical model of meteor streams. IV – A study of radio streams from the synoptic year. *Icarus* **27** (1976), 265–321.

[18] D. Olsson-Steel, Apollo asteroid 1987 SY and the σ-Leonid meteor shower. *WGN* **15** (1987), 179–180.

Object	Epoch	a (AU)	q (AU)	i (°)	ω (°)	Ω (°)	Π (°)
1999 RD_{32}	(1999)	2.640	0.605	6.788	310.286	299.644	249.98
4450 Pan (1987 SY)	(1987)	2.287	0.596	5.53	291.53	312.087	243.62
δ-Leonids	(1953)	2.618	0.643	6.2	259.0	338.8	237.80

Fig. 28.9 Orbital elements of possible δ-Leonid parents (J 2000).

δ-Leonid orbit in Fig. 28.9 is that by Lindblad. 1999 RD_{32} matches a pattern with increasing Π for increasing Ω.

Another often proposed candidate, comet **41P/Tuttle–Giacobini–Kresák** ($\lambda_\odot = 321°–324°$, R.A. = 127.6, Decl. = −2.9, $V_g = 16.9–10.9$ km/s) with a colorful history, is best known for an incredible cometary outburst in 1973, when the comet became nine magnitudes brighter.[19] The event was thought to coincide with a rotational breakup due to spin-up, emitting the same amount of gas as in 80 normal orbits.[20] The comet nucleus is thought to be very small, 1.4 km in size,[21] but still ice laden. In later returns, the comet continued to have episodes of brightening. The comet has many close encounters with Jupiter, but constantly remains below the 2:1 resonance. The perihelion has gradually decreased in many close encounters with Jupiter. As a result, the nodal line rotates rapidly with respect to the line of apsides; the long-period variations of eccentricity are closely tied to those of the perihelion argument and to those of inclination. The latitude of the pericenter increased substantially in the seventeenth and eighteenth centuries. The theoretical radiant has moved from a high inclination for the 1858 III orbit to a low inclination for the current one. The comet remains just outside Earth's orbit, with a minimum distance to Earth's orbit of 0.138 AU. Over time, dust may evolve into the Earth's path, but at the moment the corresponding meteor shower is not detected.

In March, photographic orbit surveys by *Fred Whipple* and coworkers first identified a stream called "Virginids," or southern Virginids, which I call the *η-Virginids*.[22] In what may have been an outburst, *Dick McCrosky* and *Annette Posen* photographed five meteors of magnitude −0.2 to +2.0 on 1953 March 12–13, that radiated from

[19] L. Kresák, The outbursts of periodic comet Tuttle–Giacobini–Kresák. *Bull. Astron. Inst. Czechoslov.* **25** (1974), 293–304.

[20] M. B. Niedner, Jr., Interplanetary gas. XXV – A solar wind and interplanetary magnetic field interpretation of cometary light outbursts. *Astrophys. J.* **241** (1980), 820–829.

[21] G. Tancredi, J. A. Fernández, H. Rickman and J. Licandro, Catalog of nuclear magnitudes of Jupiter-family comets. *Astron. Astrophys. Suppl. Ser.* **146** (2000), 73–90.

[22] B. A. Lindblad, A computerized stream search among 2401 photographic meteor orbits. *Smithsonian Contrib. Astrophys.* **12** (1971), 14–24.

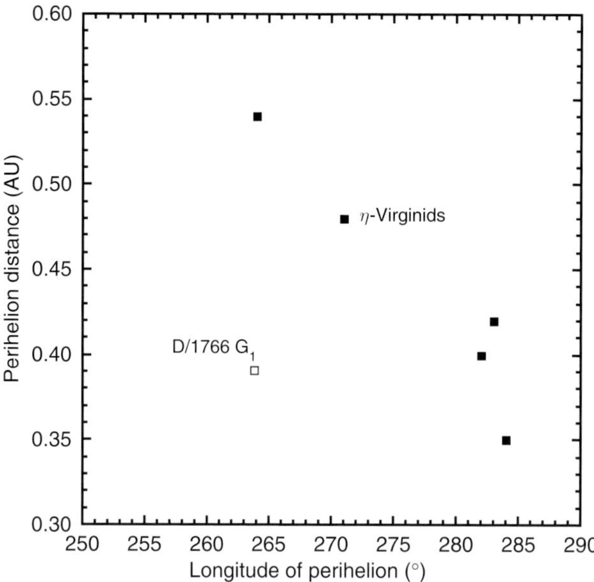

Fig. 28.10 Change of perihelion distance with inclination for outburst "Virginids" in the database of McCrosky & Posen.

R.A. = 182°, Decl. = +0°, $V_g = 26.6$ km/s. No other members were photographed in later days. The duration of the event was centered on solar longitude $\lambda_{\odot} = 358.4 \pm 0.4°$, and extended from at least 351.98° to 353.05°. Surprisingly, the orbital elements show a range of perihelion distances and longitude of perihelion (Π). However, the perihelion distance steadily decreases with Π (Fig. 28.10).

The forward integrated orbit of comet D/1766 G_1 (Helfenzrieder) in the period 1937–1956 has the same q and i, and may have been responsible for this meteor outburst. No effort has yet been made to verify this with a dynamic model. The comet does not fall on the q–Π trend of the meteoroids (Fig. 28.11).

Other Virginids arrive with a lower entry speed. Those Antihelion source meteors are generally summarized as "March Virginids" (Table 7). Minor planet *1998 SJ*$_{70}$ may be responsible for some of these meteoroids, if it is not an asteroid. I find that minor planet **1999 RM$_{45}$** may be responsible for the *ν-Hydrids* (Fig. 28.11). Minor planet **2003 YG$_{118}$** (a 2.4 km sized object) has been associated with the *N. α-Leonids* (= Sekanina's Leonids–Ursids) shower by Vladimir Porubčan and M. Gavajdova.[23] Both have a shorter semimajor axis and could be an asteroid instead. In early April, the Antihelion source meteors are called *γ-Virginids*. *2003 BD*$_{44}$ and *2002 FC* are potential parent bodies.

[23] V. Porubčan and M. Gavajdova, A search for fireball streams among photographic meteors. *Planet. Space Sci.* **42** (1994), 151–155.

Object	Epoch	a (AU)	q (AU)	i (°)	ω (°)	Ω (°)	Π (°)
1998 SH_2	(1998)	2.703	0.760	2.482	259.94	14.274	274.21
α-Virginids (L'73)	(1953)	2.349	0.753	0.7	247.5	28.9	276.4
α-Virginids (L'71)	(1953)	2.56	0.812	3.1	239.7	31.8	271.5
1999 RM_{45}	(1999)	1.681	0.598	10.89	88.46	163.114	251.57
ν-Hydrids	(1976)	1.622	0.626	12.5	88.4	163.1	251.5
2003 YG_{118}	(2004)	2.283	0.812	8.132	232.147	348.627	220.75
N. α-Leonids	(1976)	1.978	0.878	5.8	226.0	343.9	209.9

Fig. 28.11 Orbital elements of α-Virginids, from Bertil Lindblad's 1971 and 1973 publications, and potential parent 1998 SH_2. Two other associations are shown too.

In the second week of April, meteors radiate from R.A. = 208°, Decl. = −10°, V_g = 23 km/s, and are called the *μ-Virginids* if they have an entry speed of 17–23 km/s ($\Pi \sim 280°$, $q \sim 0.7$ AU), or *λ-Virginids* if they are 27–32 km/s, fast ($\Pi \sim 315°$, $q \sim 0.3$ AU).

Another shower, the α-Virginids is perhaps the strongest center of activity in the Virginid Complex. **1998 SH_2** is a possible extinct comet nucleus, originating from the Jupiter-family comets with a 33% likelihood. It is very small with $H_N = +20.81$ (Table 2). The scenario would be that this object is the remnant of a breakup that once created the stream. Other possible associated objects include *2002 GM_5* (linked to the σ-Leonids), which is possibly an asteroid given its smaller semimajor axis. In the same manner, *1996 AJ_1* is perhaps associated with the λ-Virginids, but its semimajor axis is smaller than expected.

Later in April, at the time of the Lyrid shower, meteors radiate from R.A. = 221°, Decl. = −5°, V_g = 25 km/s. McIntosh called this stream the "*Librids*". Several centers of radiation are found in radar data, with a small $q \sim 0.16$ AU (*γ-Librids*) and a higher $q \sim 0.42$ AU (*δ-Librids*). Again, perhaps because of the way the search routines work a number of sources may exist, but no parent is known.

In early May, there is still activity from radiants in Virgo. McCrosky & Posen identified 10 meteors in the period May 1–9 as belonging to a stream they call the *May "λ-Virginids.*" Lindblad found two meteors photographed on May 5 and 6, and called the stream the southern λ-Virginids. The stream is dispersed over that time interval, but appears to peak around solar longitude, $\lambda_\odot = 45°$. Most significantly, the perihelion distance is a steep function of the solar longitude, with no such strong correlation in the radiant position and speed. Significantly, these May Virginids are not an extension of the Virginids seen in early March, with no smooth correlation between solar longitude and q for both the streams combined. There is also no strong correlation of meteor altitude with meteor magnitude, these Virginids typically first being observed at 99 km. Overall, these meteors remain further from the Sun. Minor planet **1998 HJ_3** may be related.

The Hephaistos group

Could these minor bodies of the Virginid Complex be somehow related? Asher, Clube, and Steel[24] proposed that the Taurid complex included a group of five others aligned with the Apollo asteroid *(2212) Hephaistos.*[25] Unlike the remainder of the Taurid complex, their angle of perihelion falls in the range 220°–260°, with an Earth encounter in January–March (night time) and July–August (daytime). Now in an orbit about 90° rotated from 2P/Encke, Steel and Asher have proposed that the Hephaistos group is the product of an early fragmention of the Taurid complex progenitor, proposed to originate from a giant comet in a cis-Jovian orbit that had a hierarchical disintegration in the past 20 000 years.[26]

The key question, again, is whether any of these objects are extinct comet nuclei. This time the jury is still out. Of all the proposed members, *Mithra*, *1990 SM*, and *1990 TG₁* are the largest. The earliest discovered member of that group, *1978 SB* (*2212 Hephaistos*) is classified as an SG- or C-type asteroid.[27] The object measures 3.3 km if of taxonomic type SG, according to Binzel *et al.*,[28] not unlike Jupiter-family comets, and rotates slowly, with a period larger than 20 h, causing a modest lightcurve variation of about 0.1 magnitudes. Hephaistos is now too far from Earth's orbit to contribute meteors, if it was ever an active comet. The object has numerous close encounters with Mercury, as well as with other inner planets. Its perihelion is much less than that of the bulk of Virginid meteoroids. Minor planet *4486 Mithra* (1987 SB = 1974 DN_1), was discovered on September 22, 1987 by *Eric Elst* and *Vladimir Shkodrov* in Bulgaria. It was later observed by the Goldstone radar (Fig. 28.12), measuring a projected $D = 0.8$ km, with the possibility that the object was significantly larger because of the irregular shape. Indeed, the brightness of the object suggests a size of 2.3 km (if S-type) or 5.1 km (if C-type).

The radar images show a double-lobed object, more severely bifurcated than any other near-Earth asteroid imaged to date.[29] Because there was no obvious periodicity in the observed images, nonprincipal-axis rotation was suggested with slow rotation, or a line-of-sight along the rotation axis. These observations, too, could point at an extinct comet nucleus. Based on observed meteor shower activity, Mithra is not a source of meteors on Earth despite the relatively low inclination. The node is at the orbit of Venus.

[24] D. J. Asher, S. V. M. Clube and D. I. Steel, Asteroids in the Taurid Complex. *Mon. Not. R. Astron. Soc.* **264** (1993), 93–105.

[25] D. I. Steel and D. J. Asher, The orbital dispersion of the macroscopic Taurid objects. *Mon. Not. R. Astron. Soc.* **280** (1996), 806–822.

[26] D. I. Steel and D. J. Asher, P/Helfenzrieder (1766 II) and the Hephaistos group of Earth-crossing asteroids. *Observatory* **114** (1994), 223–226.

[27] D. J. Tholen, Asteroid taxonomic classifications. In *Asteroids II*, ed. R. P. Binzel *et al.* (Tuscon, AZ: University of Arizona Press, 1989), pp. 1139–1150; J. X. Luu and D. C. Jewitt, Charge-coupled device spectra of asteroids. I – Near-earth and 3:1 resonance asteroids. *Astron. J.* **99** (1990), 1985–2011.

[28] R. P. Binzel, D. Lupishko, M. di Martino *et al.*, Physical properties of near-Earth objects. In *Asteroids III*, ed. W. F. Bottke *et al.*, (Tucson, AZ: University of Arizona Press, 2002) pp. 255–271.

[29] S. J. Ostro, R. S. Hudson, L. A. M. Benner *et al.*, Radar observations of asteroid 4486 Mithra. *DPS Pasadena Meeting 2000*, 23–27 October 2000, abstracts (2000).

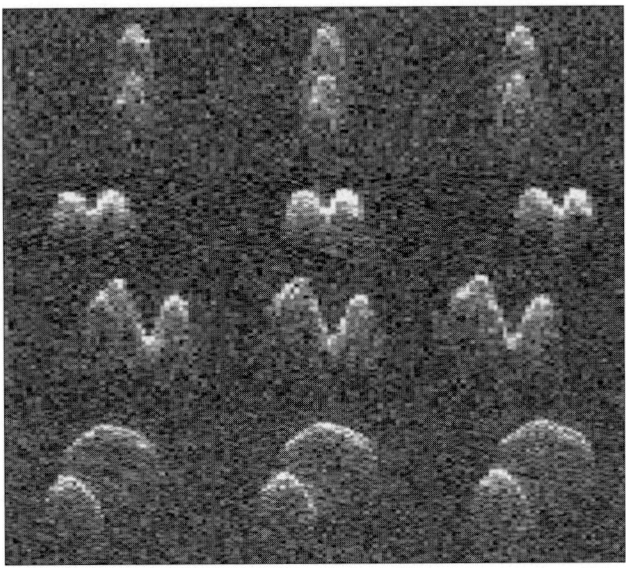

Fig. 28.12 Minor planet 4486 Mithra, a candidate extinct comet nucleus and member of the Virginid complex, as observed by the Goldstone radar. NASA/JPL radar images courtesy of Steve Ostro (see Ostro *et al.* 2000).[30]

Other members of the Hepaistos group include *1990 TG*$_1$ = *2000 YP*$_{29}$ *(30825)* and *1990 SM*. Nothing is known about the identity of these bodies. None of them have a node close to Earth's orbit.

As it stands, there may be a larger progenitor comet for the Virginid Complex, but the meteor showers identify other minor bodies than those proposed earlier.

28.4 Ecliptic showers in summer

The Scorpiid–Sagittariid Complex

The Antihelion continues to move through Scorpius and into lower Ophiuchus in May, through Sagittarius in June, into Capricorn in July, and into Aquarius in August (Fig. 28.13). During that period, ecliptic shower activity from Apollo-type orbits continues to be observed. Cuno Hoffmeister (1948) called (part of) this the *Scorpio–Sagittarius Complex*.

After plotting all meteoroid radiants derived from photographic studies[31] and included in the 2003 IAU meteor orbit database, it is clear that the Scorpiid-Sagittariid Complex does not stand out well. The main summer showers are the

[30] *Ibid.*
[31] R. E. McCrosky and A. Posen, Orbital elements of photographic meteors. *Smithsonian Contrib. Astrophys.* **4** (1961), 15–84.

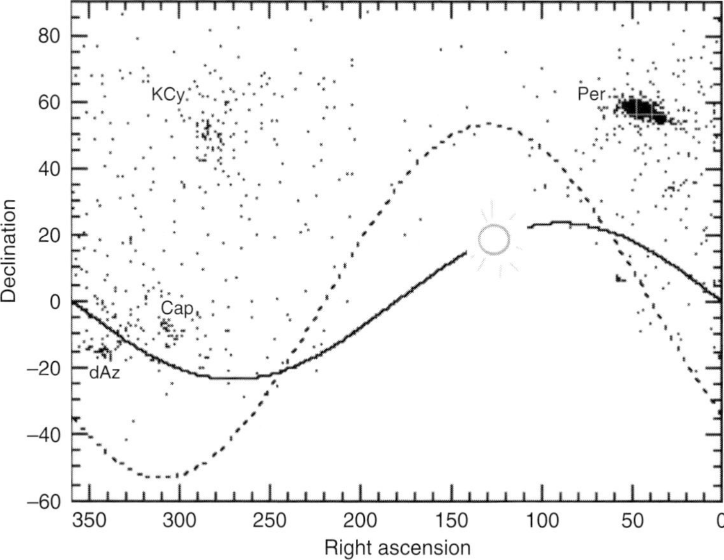

Fig. 28.13 Meteoroid orbit radiants from May–August from the IAU photographic database.

δ-Aquariids, the α-Capricornids, the Perseids and the κ-Cygnids. The comparative lack of Antihelion source meteors along the ecliptic plane is due to the fact that the ecliptic plane is at its lowest point on the northern hemipshere in July. Southern hemisphere observers have a good view of these streams, but during their winter.

The situation is different for instruments that can work more continuously and are sensitive to fainter meteoroids. In the SkyMet wind radar observations at Andoya (Fig. 28.14), there is a pronounced excess of meteoroids in the 40°–150° interval from early May until the end of August. There does not seem to be an increase in sensitivity with the position of the Sun in the sky and climate effects appear to be excluded. This summer months peak is not so obvious in the forward meteor scatter results of Global-MS-Net (Fig. 28.15), a type of instrument sensitive to brighter meteors. Perhaps, this broad distribution of dust in the summer months is a mirror-image of the broad distribution of larger meteoroids known to peak in November. The Taurid Complex is implicated, but the radiant distribution measured with the Andoya wind radar in that period of time shows no strong concentration of radiants near the expected positions.

In the month of May, the main center of activity from the Antihelion direction is called the *α-Scorpiids* (R.A. = 247°, Decl. = −24°, V_g = 32 km/s). These are relatively fast meteors that stand out from the bulk of slower Antihelion source meteors and a parent body is implicated. Among the candidate extinct comet nuclei, there is one that jumps out: **2004 BZ$_{74}$**, discovered by LINEAR (Fig. 28.16). With a theoretical radiant at R.A. = 249.2, Decl. = −34.0° (V_g = 32.2 km/s) on May 20.1 (59.4), it is a good match to the α-Scorpiid shower. According to Bottke, this ~1.2 km sized minor planet

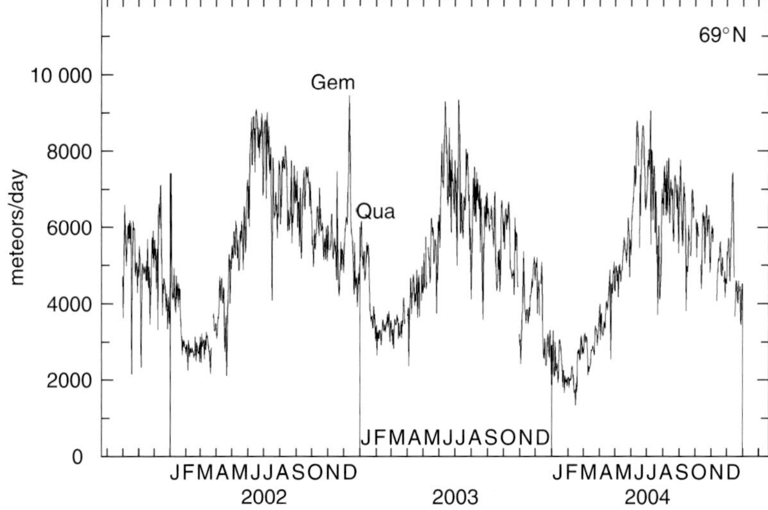

Fig. 28.14 Daily echo rate from SkiMet radar in Andoya (Norway) – the number of echoes per day measured with the Kuelungsborn all-sky meteor radar system and posted in near-real time on the internet. Note the high counts in June and July. Downward depressions are lack of data due to instrumental effects. Courtesy: Werner Singer, Leibniz-Institute of Atmospheric Physics.[32]

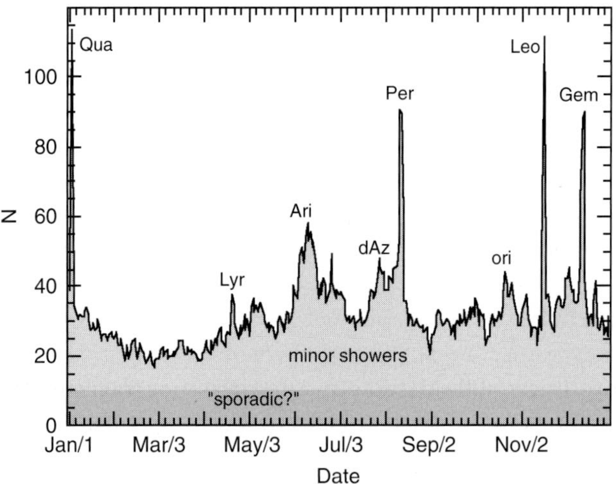

Fig. 28.15 Average radio-MS observations at 04:00 local time in 1998 from several observers participating in Global-MS-Net.

[32] D. Keuer, J. Mielich and U. von Zahn, *SKiYMET All-Sky Meteor Radar – at ALOMAR/Andoya Rocket Range* (Norway) – (website Leibniz-Institut für Atmosphärenphysik, Universität Rostock – Außenstelle Juliusruh, Kühlungsborn, Germany) (2003).

Object	Epoch	a (AU)	q (AU)	i (°)	ω (°)	Ω (°)	Π (°)
2004 BZ$_{74}$	(2004)	3.046	0.330	16.57	120.91	234.208	355.12
α-Scorpiids	(1954)	2.640	0.330	9.8	116.5	236.8	353.30

Fig. 28.16 Orbital elements of the α-Scorpiids and a possible parent (J 2000).

Object	Epoch	a	q	i	ω	Ω	Π
2001 ME$_1$	(2001)	2.654	0.357	5.78	299.87	86.891	26.76
N. σ-Sagittariids	(1954)	1.72	0.38	4.5	296.6	86.6	23.2
N. σ-Sagittariids	(1976)	1.133	0.332	8.2	309.4	92.3	41.7

Fig. 28.17 Orbital elements of σ-Sagittariids and a possible parent.

is an extinct comet nucleus with a 91% likelihood, its 5.32 yr orbit stretching to just outside Jupiter's orbit at an inclination of 16° and a node just outside Earth's orbit. The α-Scorpiids, however, include meteoroids at much lower inclinations.

In the month of June, the showers are known as the Ophiuchids or Scorpiids–Sagittariids (Table 7). Despite the lack of observations, several streams are discriminated, with much confusion caused by different authors adopting different names for similar streams. To the best of my ability, I have summarized those into the *N. and S. ω-Scorpiids* with $\Pi \sim 329°$ and $q \sim 0.68$ AU, the *N. and S. μ-Sagittariids* with $\Pi \sim 5°$ and $q \sim 0.4$–0.6 AU, and the *N. and S. σ-Sagittariids* with $\Pi \sim 20°$ and $q \sim 0.35$ AU. Potential parent bodies include *2001 ME$_1$* (NSS) and *1996 JG* (NSC).

Duncan Steel has suggested that *72P/Denning–Fujikawa* might contribute meteor activity from Sagittarius,[33] but the comet is too far outside Earth's orbit, as is comet *15P/Finlay* at the present time. Comet *Lexell* did not put down enough mass during its two returns near Earth's orbit to account for the μ-Sagittariid stream.[34] I find it more believable that the 55% likely extinct comet nucleus **2001 ME$_1$** is a source of *N. σ-Sagittariids*, or a tracer of the debris generated in a prior fragmentation (Fig. 28.17).

In the months of July and August, the showers of *δ-Aquariids* (part of the Marsden Sungrazer Complex) and the *α-Capricornids* dominate the ecliptic streams. There is plenty of additional sporadic activity not associated with these streams from radiants in Sagittarius, Capricorn, and Aquarius. Potential sources include low-inclination **2000 QS$_7$**, which would radiate from the border of Sagittarius and Capricorn in late August.

[33] D. Olsson-Steel, Theoretical meteor radiants of Earth-approaching asteroids and comets. *Austr. J. Astron.* **2** (1987), 21–35.
[34] D. Olsson-Steel, A radar orbit search for meteors from comet Lexell. *Astron. Astrophys.* **195** (1988), 338–344.

Object	Epoch	a	q	i	ω	Ω	Π	Mass (billion kg)
2004 NL_8	(1976)	2.601	0.760	4.93	265.00	163.17	68.2	7300
κ-Aquariids	(1976)	2.606	0.741	7.6	247.8	179.5	67.3	1500

Fig. 28.18 Orbital elements of κ-Aquariids and associated minor planet.

In early September, the Antihelion source is in Aquarius. A stream of κ-*Aquariids* has been recognized by different sources. Interestingly, I find that minor planet **2004 NL_8** is a very good match to the mean orbit given by Zekanina. The minor planet has a mass larger than the stream and the diameter is about 2.4 km. Without further evidence, I suspect this is another example of a stream created in the recent breakup of a comet (Fig. 28.18).

Other potentially interesting sources among extinct comet nuclei are the relatively large **1979 VA (4015 Wilson–Harrington)**, radiating from Sagittarius on September 6, and minor planet **2001 HA_4** with a radiant near β-Ceti on September 22.

29
Toroidal streams

In the early nineties, satellite impact hazard models erroneously assumed that geocentric meteor radiants were spread evenly over the sky relative to the moving Earth.[1] *Jim Jones* and *Peter Brown*[2] in Canada and *Andrew Taylor* at Adelaide[3] took stock of our knowledge of the overall radiant distribution from radar data by tackling the observing bias. They identified six principal source areas on the sky. Next to the *North and South Apex source* from Halley-type and long-period comet showers (15–28% of meteoroids), and the *Helion and Antihelion sources* of the ecliptic Jupiter-family comet streams (30–40% each), they also confirmed earlier reported *North and South "Toroidal" sources* at high ~67° latitude (Fig. 28.4). These meteoroids moved in nearly circular orbits at steep angles to the ecliptic plane. They had a very small semimajor axis $a \sim 1.0$ AU, not so elongated $e \sim 0.3$, and a prograde high inclination ($i \sim 60°$). The Toroidal meteoroids surround Earth in a volume that has the shape of a car tire, with Earth on the open inside of the torus.

Toroidal meteoroid orbits dominate the radar orbit database and are a separate source from ecliptic meteoroids only because of the competing factors of increased likelihood of detection by radar and of decreasing numbers with increasing latitude of the radiant and entry speed of the meteors. After correcting for bias, the Toroidal source all but disappeared, with only 3–6% of all meteors coming from this region. *Jim Jones* has since taken the view that the Toroidal meteoroids are merely the high-inclination tail of a continuous distribution of dust from the Helion and Antihelion sources.[4]

Some 100 out of 275 streams extracted from the Harvard Radio Meteor Project radar data by *Zdenek Sekanina* are Toroidal streams, most in the fall and the winter.[5] In Table 7, I mainly list streams that were confirmed by other techniques, and that

[1] N. Divine, Five populations of interplanetary meteoroids. *J. Geophys. Res.* **98** (1993), 17 029–17 048.

[2] J. Jones and P. Brown, Sporadic meteor radiant distributions: orbital survey results. *Mon. Not. R. Astron. Soc.* **265** (1993), 524–532; P. Brown and J. Jones, A determination of the strengths of the sporadic radio-meteor sources. *Earth, Moon Planets* **68** (1995), 223–245.

[3] A. D. Taylor, Radiant distribution of meteoroids encountering the Earth. *Adv. Space Res.* **20** (1997), 1505–1508; A. D. Taylor and N. McBride, A radiant-resolved meteoroid model. *ESA SP* **393** (1997), 375–380.

[4] J. Jones, M. Campbell and S. Nikolova, Modelling of the sporadic meteoroid sources. *ESA SP* **495** (2001), 575–580.

[5] G. S. Hawkins, Radar determination of meteor orbits. *Astron. J.* **67** (1962), 241–244; Z. Sekanina, Statistical model of meteor streams. III. Stream search among 19303 radio meteors. *Icarus* **18** (1973), 253–284.

includes most of Sekanina's Antihelion, Helion, and Apex streams, but very few of his Toroidal streams. Indeed, Sekanina's total fraction of meteors assigned to streams is higher (16%) than expected (~3%) for $+10^{m}$ meteors. It appears that the radar samples the sporadic background efficiently at these high latitudes, which accounts for the steep magnitude distribution index derived for these meteors (Chapter 32). None of Sekanina's 100 toroidal streams have been confirmed.

There is, however, a natural mechanism to create short-period, high inclination orbits, without explaining the short semimajor axis. *Mark Bailey et al.*[6] found that all known Jupiter-family comets (JFCs) have large variations in inclination from secular perturbations, but more so if their average inclination was high (and their perihelion distance was small). Most comets reside at the highest point of their nutation cycle. In the same way, most meteoroids will tend to be at the high point of their cycle, but with q near the maximum value. I would like to point out that higher inclination JFCs therefore have a larger likelihood of having q close to Earth's orbit at one point in their cycle, which would increase the rate of Antihelion source meteoroids at high latitudes. Recently captured comets will have $\omega \sim 0°$ or 180°,[7] leading to meteoroid streams with $q \sim 0.93$–1.00 AU.

So far, most minor bodies discovered in high inclination orbits approach Earth from the south, but none are likely to be the parent bodies of southern hemisphere showers.

29.1 The Puppid-Velid Complex

Southern hemisphere observers are familiar with numerous minor streams with radiants at deep southern declinations that are active from December to March,[8] called the *Puppid–Velid Complex*, first noticed by *Cuno Hoffmeister* during his South Africa expedition. From late October to early January there are a series of radiants active in the constellations of Carina (= Keel), Puppis (= Stern), Pyxis (= Compass Box), and Vela (= Sail), which used to form the ancient constellation of *Argo Navis*, the ship of the Argonauts. These are high inclination (>40°) streams. The best period to observe these showers is in early December, when rates increase to ZHR = 3–6 (Fig. 29.1). These are followed by activity from various Centaurid showers in January–March.

At present, a large number of showers are discriminated by visual observers (Fig. 29.2). I plotted the solar longitude versus the radiant position to search for a pattern and found that three groups stand out (Fig. 29.3), which I call the *Puppid-Velid I Complex* (Decl. ~−45°, $V_g = 33$–39 km/s), the more dispersed *Puppid-Velid II Complex* (range in Decl., $V_g = 31$–33 km/s), and the *Carinid Complex* (Decl. = −54 to −63°, $V_g = 33$–39 km/s). The *Centaurids* have a higher speed $V_g = 48$–58 km/s and

[6] M. E. Bailey, J. F. Chambers and G. Hahn, Origin of sungrazers: a frequent cometary end-state. *Astron. Astrophys.* **275** (1992), 315.
[7] H. F. Levison and M. J. Duncan, The long-term dynamical behavior of short-period comets. *Icarus* **108** (1994), 18–36.
[8] G. Gartrell and W. G. Elford, Southern Hemisphere meteor stream determinations. *Austr. J. Phys.* **28** (1975), 591–620.

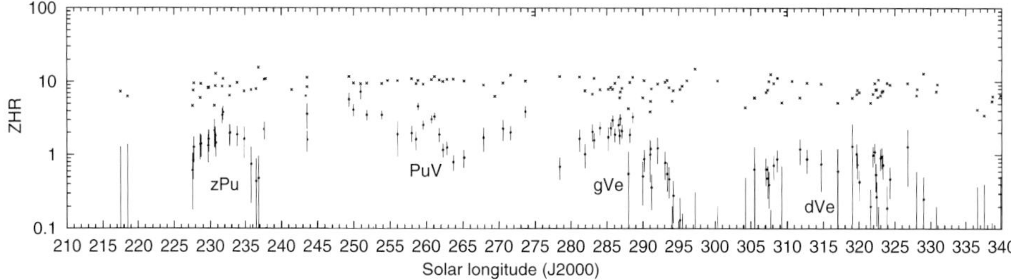

Fig. 29.1 Rates of the Puppid-Velid Complex showers from observations by NAPO-MS members (courtesy Jeff Wood). Crosses indicate the level of sporadic hourly rate.[9]

Code:	Stream:	Dates:	λ_o (°)	R.A.	Decl.	ΔR.A.	ΔDecl.	V_g (km/s)	ZHR
	Puppids-Velids I:								
ZPU	ζ-Puppids	11/02–12/20	254.7	123.0	−43.0	-.-	-.-	39	3.2 ± 0.8
PUP	Puppids-Velids I	12/01–12/15	255.0	123.0	−45.0	+0.50	+0.00	38	-.-
PVE	Puppids-Velids II	12/02–12/06	256.3	128.0	−45.0	-.-	-.-	39	4.5 ± 0.7
LVL	λ-Velids	01/18–01/26	269.7	133.0	−46.0	+0.60	−0.20	33	-.-
CVE	c-Velids		273	135	−46	-.-	-.-	36	-.-
SPU	σ-Puppids	11/27–12/12	253.7	102.0	−45.0	+0.50	+0.00	36	-.-
	Puppids-Velids II:								
DOR	α-Doradids	10/08–10/31	207.7	69.0	−56.0	-.-	-.-	-.-	-.-
COL	Columbids	11/08–12/08	244.7	86.0	−34.0	-.-	-.-	-.-	-.-
TPU	τ-Puppids	12/19–12/30	270.7	104.0	−50.0	+0.40	−0.10	31	-.-
PIP	π-Puppids	01/06–01/14	290.7	113.0	−43.0	+0.50	−0.20	33	2.4 ± 0.4
= GVE	γ-Velids	01/06–01/14	288	125	−47			33	
APY	α-Pyxids	01/14–02/03	304.7	132.0	−34.0	-.-	-.-	-.-	-.-
DVE	δ-Velids	01/22–02/21	319.7	132.0	−56.0	-.-	-.-	33	1.2 ± 0.5
	Carinid Complex:								
ECA	ε-Carinids	12/04–12/26	263.7	129.0	−58.0	-.-	-.-	39	-.-
ECR	η-Carinids	12/28–02/04	280.7	160.0	−60.0	-.-	-.-	-.-	-.-
	Centaurids I:								
ACR	α-Crucids	01/06–01/28	299.7	192.0	−63.0	+1.00	−0.30	48	-.-
ACE	α-Centaurids	01/28–02/21	319.2	210.0	−59.0	+1.18	−0.26	55	-.-
OCA	o-Centaurids	01/31–02/19	322.7	177.0	−56.0	+0.70	−0.30	51	-.-
	Centaurids II:								
BHD	β-Hydrids	01/28–02/24	320.7	187.0	−34.0	-.-	-.-	-.-	-.-
TCN	θ-Centaurids	01/23–03/12	322.7	210.0	−40.0	+1.00	−0.30	59	-.-
MVE	μ-Velids		262	155	−41	-.-	-.-	54	-.-
	Unrelated:								
GPU	γ-Puppids	09/28–10/30	202.7	109.0	−44.0	+0.60	−0.20	42	-.-
GDO	γ-Doradids	08/19–09/06	157.7	67.0	−54.0	-.-	-.-	-.-	-.-

Fig. 29.2 Showers of the Puppid-Velid Complex. Rates are from NAPO-MS observations.

[9] P. Jenniskens, Meteor stream activity. I. The annual streams. *Astron. Astrophys.* **287** (1994), 990–1013.

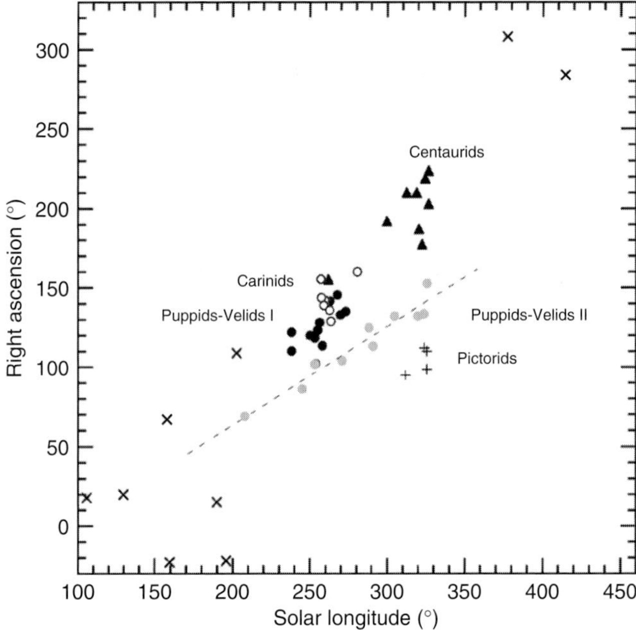

Fig. 29.3 Southern high declination shower radiants from visual and radar observations.

at least one group, the α-Centaurids, is known for meteor outbursts. Each may be the product of a comet breakup, the Puppid–Velid I Complex perhaps more recently than the Puppid-Velid II Complex.

The speeds were measured at Adelaide by radar.[10] The Puppids-Velids I (Fig. 29.3) were also observed by radar in 1956 from Christchurch, New Zealand, where *C. D. Ellyett* and *K. W. Roth* determined the radiant at R.A. = 122°, Decl. = −45° for November 17–21, and at R.A. = 120°, Decl. = −43° for November 27–December 8. Early radar measurements implied a short semimajor axis, but they suffered from underestimated speeds. More modern measurements result in orbits with a semimajor axis in excess of 2 AU and are not necessarily always decoupled from Jupiter. These are not Toroidal streams.

The Carinids and Puppids-Velids I form a low and high declination pair (having different Π), while the Adelaide radar found two such concentrations for the Centaurids (Fig. 29.4).[11] Similar two clusters were found by Sekanina for the northern hemisphere showers.[12]

[10] L. Jacchia and F. L. Whipple, Precision orbits of 413 photographic meteors. *Smithsonian Contrib. Astrophys.* **4** (1961), 97–129; R. E. McCrosky and A. Posen (1961) Orbital elements of photographic meteors. *Smithsonian Contrib. Astrophys.* **4** (1961), 15–84.

[11] G. Gartrell and W. G. Elford, Southern Hemisphere meteor stream determinations. *Austr. J. Phys.* **28** (1975), 591–620.

[12] Z. Sekanina, Statistical model of meteor streams. III. Stream search among 19303 radio meteors. *Icarus* **18** (1973), 253–284.

Object	Epoch	*a* (AU)	*q* (AU)	*i* (°)	*ω* (°)	Ω (°)	Π (°)
Carinids:	(1960)	2.94	0.91	66.6	323.7	76.9	40.60
?2000 $\mathbf{DK_{79}}$	(2000)	1.776	1.041	60.676	362.396	43.499	45.90
Puppids-Velids I:	*(1960)*	*1.3*	*0.98*	*70*	*360*	*82.7*	*82.70*
Puppids-Velids II:	*visual*	*1.71*	*0.868*	*53.7*	*49.9*	*106.0*	*155.9*
	(1960)	*2.4*	*0.93*	*62.0*	*33*	*145.7*	*178.7*
?1975 YA (2102 Tantalus)	(2003)	1.290	0.904	64.007	61.593	94.389	156.0
Centaurids I:	(1960)	2.5	0.973	105.0	344	146.7	130.7
Centaurids II:	(1960)	5.0	0.87	114.7	43	146.7	89.7

Fig. 29.4 Orbital elements of Southern Hemisphere high latitude showers (J2000).

The existence of a young and an old pair of streams such as the Puppid-Velid I and II Complexes (Fig. 29.3) hints that there are still other remaining comet fragments, now inactive as an extinct comet nucleus. I searched the asteroid database, selecting those objects that have a high inclination and a node near Earth's orbit, but found no minor planets yet that are definitely associated with the Puppids Velids.

Minor planet *2000 DK_{79}* was discovered by the LINEAR (Lincoln Laboratory ETS) near-Earth survey in New Mexico on February 26, 2000. If the albedo of this object is 0.04, as expected for an extinct comet nucleus, then the radius of the object is 4.7 km. After the discovery, the object was found to be on the 1986 Palomar Observatory plates. Its orbit does not precisely match that of the Carinid Complex, and I consider this association doubtful. Most importantly, the semimajor axis is too small. In addition, these showers have an inclination somewhat higher than 2000 DK_{79} and an ascending node that is somewhat lower (Fig. 29.4). I integrated the orbit back to 1600, using the JPL/Horizons software, only to find a change in the argument of perihelion, but with other orbital elements remaining unchanged.

It has earlier been suggested that *1975 YA* (2102 Tantalus) could be associated with the Puppid–Velid Complex. This is a Q-type asteroid of 3.3 km size with a rotation period of only 2.391 h, a potential source of meteorites. If true, this complex would be an asteroidal meteoroid stream. However, the semimajor axis of the meteoroids is higher than that of Tantalus and I consider this association to be doubtful as well.

30

Meteor showers from asteroids

Ever since the discovery of 3200 Phaethon among the Geminids in 1983, it has become popular to link meteoroid orbits to asteroids. So far, no asteroid has been convincingly identified as the source of meteor showers. 3200 Phaethon is an extinct comet nucleus.

Indeed, the general population of mm–cm sized meteoroids is not dominated by asteroidal dust, judging from the Tisserand parameter (T_J) of all photographic meteoroid orbits in the 2003 IAU database (Fig. 30.1). Most have $T_J < 3.0$, in the domain of Jupiter-family comets. Those that do not, tend to straggle the border with the asteroid domain ($T_J > 3.0$), where cometary showers such as the Taurids are found. Geminids are cometary, but orbit among asteroids with $T_J = 4.51$.

What about the association with real *asteroids*, the minor planets that were born in the inner solar system? They are expected to contain less volatile matter, be more cohesive, and have different chemical and mineralogical properties.[1] Asteroids are the source of meteorites.

How to recognize asteroidal meteor showers? For one, the group of deeply penetrating meteors are identified as asteroidal in origin (Fig. 30.2), on account of their compact structure and relative lack of fragmentation. Many published fireball orbits are deeply penetrating bodies, but only because they are the target of meteorite recovery projects. Most 10 m sized bodies are in fact cometary with the weakest known structure.[2]

Strong materials are also found among smaller meteoroids. For example, *Jiri Borovicka* found that about 5% of $\sim+2^{m}$ sporadic meteors consist of pure iron. All of these iron meteoroids have an aphelion in the asteroid belt.[3] These meteoroids may originate from a discrete source, but no stream association was found.

Another criterion used to recognize stony matter is that the asteroidal meteoroids start ablating relatively high in the atmosphere, because of a larger heat conductivity.[4]

[1] F. J. M. Rietmeijer, Interrelationships among meteoric metals, meteors, interplanetary dust, micrometeorites, and meteorites. *Meteoritics Planet. Sci.* **35** (2000), 1025–1041.

[2] Z. Ceplecha, Impacts of meteoroids larger than 1 m into the Earth's atmosphere. *Astron. Astrophys.* **286** (1994), 967–970.

[3] J. Borovicka, P. Koten, P. Spurný, J. Bocek and R. Stork, A survey of meteor spectra and orbits: evidence for three populations of Na-free meteoroids. *Icarus* **174** (2004), 15–30.

[4] Z. Ceplecha, Classification of meteor orbits. *Bull. Astron. Inst. Czechoslov.* **17** (1966), 96–98; G. W. Wetherill and D. O. ReVelle, Which fireballs are meteorites? A study of the Prairie Network photographic meteor data. *Icarus* **48** (1981), 308–328.

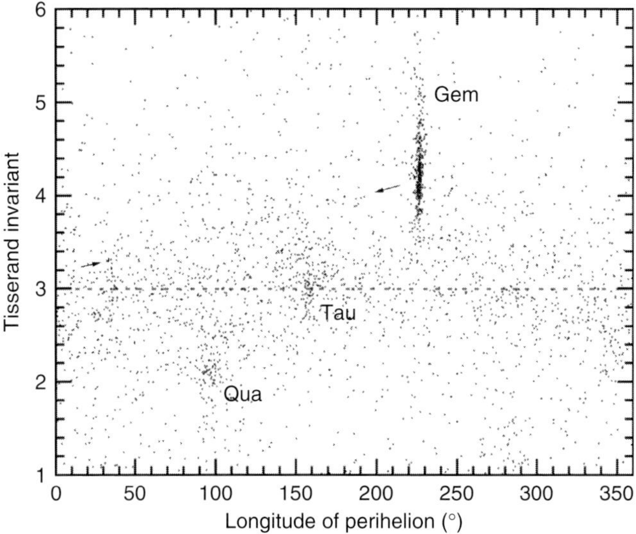

Fig. 30.1 Tisserand invariant of all photographic meteor orbits in the 2003 IAU photographic meteor orbit database. Arrows mark an unidentified cluster.

From a parameter that is based on the heat conduction in a model of meteoroid ablation, *Zdenek Ceplecha*[5] developed a classification scheme for meteors. The parameter is a function of meteor beginning height (the point where rapid ablation starts) and takes away effects of speed and the entry angle of the meteor, his *k-criterion*:

$$k = \log(\rho_a) + 5/2 \log(V_{inf}) - 0.5 \log(\cos Z_r) \tag{30.1}$$

where k and ρ_a (the air density) refer to the conditions at the beginning of the trajectory.

This led to the recognition of four classes of meteors, the same classes are recognized in the end height of the meteors. Group I is associated with ordinary chondrites of density $\rho = 3.7\,g/cm^3$ and ablation coefficient $\sigma = 0.015\,s^2/km^2$, and are mostly the cloud of points in the lower-left part of Fig. 30.2, with $H_e < 43$ km. Group II is associated with carbonaceous chondrites ($\rho = 2.0\,g/cm^3$, $\sigma = 0.035\,s^2/km^2$) and are the objects forming a scatter of points between $43 < H_e < 57$ km (perhaps, just the lower end of group III, the cometary matter). Group IIIA is thought to be relatively dense cometary matter with $\rho = 0.8\,g/cm^3$ and $\sigma = 0.15\,s^2/km^2$, while Group IIIB is very fragile cometary matter with $\rho = 0.4\,g/cm^3$ and $\sigma = 0.59\,s^2/km^2$, such as that of the Draconid meteor shower.[6] Most of these differences are a result of the fragmentation properties of the different materials.

[5] Z. Ceplecha, Classification of meteor orbits. *Bull. Astron. Inst. Czechoslov.* **17** (1966), 96–98.

[6] Z. Ceplecha, Fireballs photographed in central Europe. *Bull. Astron. Inst. Czech.* **28** (1977), 328–340; Z. Ceplecha, Meteoroid properties from photographic records of meteors and fireballs. In *Asteroids, Comets, Meteors 1993*, ed. A. Milani *et al.* (Dordrecht: Kluwer, 1993), pp. 343–356.

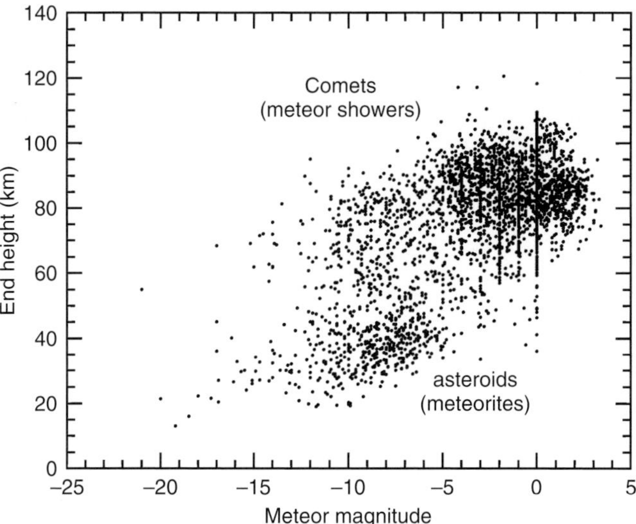

Fig. 30.2 The end height of photographed meteors in the 2003 IAU meteor orbit database. The asteroid population is overly represented among fireballs because these orbits are the harvest from meteorite recovery projects.

Measurements of density can only be made when the fragmentation properties of the materials are known. The meteoroid density of stony asteroids is higher than that of cometary matter, about 3.6 g/cm^3 (8% porosity). However, the situation is less distinct from cometary matter when considering carbonaceous meteorites, which have densities of 3.5 g/cm^3 (CV), 2.8 g/cm^3 (CM), and 2.2 g/cm^3 (CI), with a porosity of about 10% and sometimes as high as 20%. Cometary matter (devoid of ices) is believed to have about 0.1–1.5 g/cm^3, depending on the amount of porosity, which can be quite high.

30.1 Streams among the population of fireballs with low end heights

Alexandra Konstantinovna Terentjeva[7] and *Ian Halliday*[8] have studied meteoroid orbit groupings in the small database of 554 fireballs photographed in the Prairie Network and MORP projects, many of which are the deep penetrating kind. Forty percent have a Tisserand invariant >3.0. Terentjeva found 78 streams, consisting of 68% of the whole sample, most were only pairings or small groupings. Half of those streams consist of stony or metallic asteroidal bodies of Type I, capable of producing meteorites, few if none linked to the meteoroid streams detected by radar. A subset of the

[7] A. K. Terentjeva, Fireball streams. In *Asteroids Comets Meteors III.* (Uppsala: Astronomical Observatory, 1989), pp. 579–584.

[8] I. Halliday, A. T. Blackwell and A. A. Griffin, Evidence for the existence of groups of meteorite-producing asteroid fragments. *Meteoritics* **25** (1990), 93–99.

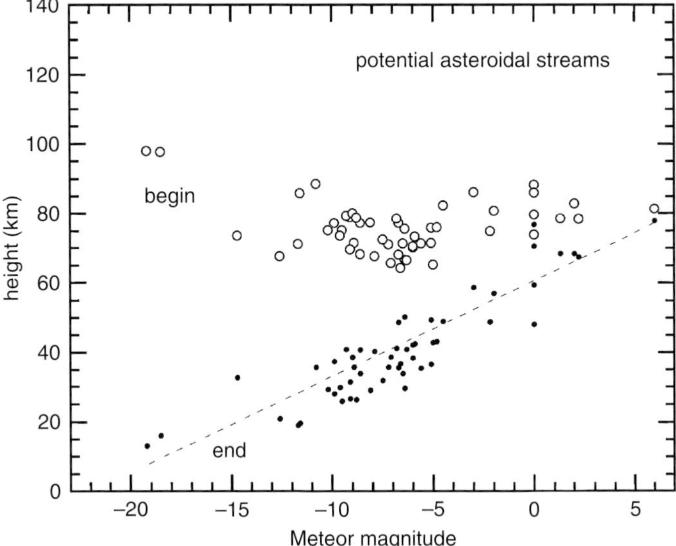

Fig. 30.3 Beginning and end heights of members of potential asteroidal meteoroid streams.

more likely asteroidal meteoroid streams are listed in Table 9. Fig. 30.3 shows how deep these meteoroids penetrate.

In efforts to identify their parent bodies, *Vladimir Porubčan* and *Jan Stohl*[9] matched the IAU database of 3500 orbits of meteors with the population of known asteroids. In this manner, they identified the following objects as potential parents: **3200 Phaeton**, 4179 Toutatis, **5025 P-L**, 1990 HA, 1991 VK, 3671 Dionysius, 1989 VB, 1986 JK, 1990 OS, 1991 BA, 1990 UA, 4660 (1982 DB), 1980 AA, 2201 Oljato, and 1988 TA. Only those marked in bold are suspected extinct comet nuclei. Similarly, Gavajdova[10] linked eight asteroids to 18 fireball streams.

On closer inspection, few of these associations are likely. Due to the search algorithms, 80% of the candidate parents are on Amor-type orbits that lie fully outside of Earth's orbit ($1.017 < q < 1.3$ AU), unlike the extinct Jupiter-family comets that are identified with meteor showers, which are on Apollo-type orbits.

30.2 A case study: Príbřam, Neuschwanstein, and Glanerbrug

Ian Halliday first postulated "the existence of *asteroidal meteoroid streams* (*resulting in asteroidal meteor showers, the name 'meteorite shower' being reserved for showers of meteorite fragments that originate from one asteroid impact*)," after finding a pairing

[9] V. Porubčan, J. Stohl and R. Vana, In *Asteroids Comets Meteors 1991*. ed. A. W. Harris and E. Bowell. (Flagstaff, AZ: LPI, 1992), p. 473.

[10] M. Gavajdova, Search for associations between fireball streams and asteroids. *Earth, Moon Planets* **68** (1995), 289–292.

among the recovered meteorite fall of Innisfree and the possible fall at Ridgedale, in 1977 and 1980.[11]

In fact, very few meteorites (only six) have been recovered, for which a precise orbit in space could be measured from photographic and video observations of the fall. The first case was *Príbřam*, a meteorite fall in the former Czechoslovakia, where a young *Dr. Zdenek Ceplecha* had established a small-camera network following the closure of the Harvard patrol camera project in 1951. The meteorite was recovered before the orbit could be calculated, but for a long time this was the only recovered fall.[12] Years passed and specialized networks were established for meteorite recoveries in Czechoslovakia by Ceplecha (Euorpean Fireball Network – EN: 1963–present), in the USA by *Fred Whipple* (Prairie Network – PN: 1963–1973) and in Canada by *Ian Halliday* (Meteorite Observation and Recovery Project – MORP: 1971–1985). The latter recovered only two more meteorites: *Lost City* in Oklahoma in 1970 and *Innisfree* in Canada in 1977.[13]

Lack of funding soon affected the other networks, but the European Network continued (Fig. 30.4) and was expanded with stations in Germany (with support of the DLR) and in the Netherlands (supported by the Dutch Meteor Society in an effort led by *Hans Betlem*). Then history turned its wheels, and in 1989 the iron curtain fell and shortly after Czechoslovakia became the Czech Republic and Slovakia. A now retired Ceplecha passed on running the network to a younger *Pavel Spurný* and both managed to continue this important work. Year after year, orbits of potential meteorite falls were calculated, but no meteorites were found.

Then one day Pavel calculated yet another fireball orbit and found the orbit to be the same (within error) as that fireball responsible for the Príbřam meteorite fall. Príbřam fell on April 7, 1959. The new fall had happened 43 years later at 20:20:18 UT on April 6, 2002 and the projected impact site was in Bavaria near the idyllic castle of *Neuschwanstein*. The rumbling sound of the fireball traveled as far as the Netherlands. Initial searches for this meteorite in the steep mountainous terrain were unsuccessful[14] and all this would have stayed a curiosity, had not, a few months later, on July 14 a 1.75 kg meteorite been recovered from the impact site.[15] A second 1.63 kg piece was found on May 27, 2003, more than a year after the fall. The original body was estimated to weigh 300 kg.

Pavel expressed expectations best in a message on the institute's web site: *From the perfect similarity of both heliocentric orbits we can predicate, that both bodies also had*

[11] I. A. Halliday, Detection of a meteorite "stream": observations of a second meteorite fall from the orbit of the Innisfree chondrite. *Icarus* **69** (1987), 550–556.

[12] Z. Ceplecha, Príbřam. 1. Double-station photographs of the fireball and their relations to the found meteorites. *Bull. Astron. Inst. Czechoslov.* **12** (1961), 21–47.

[13] R. E. McCrosky, A. Posen, G. Schwartz and C.-Y. Shao, Lost City meteorite – its recovery and a comparison with other fireballs. *J. Geophys. Res.* **76** (1971), 4090–4108; I. Halliday, A. A. Griffin and A. T. Blackwell, The Innisfree meteorite fall. *Meteoritics* **16** (1981), 153–170.

[14] D. Heinlein, Neuschwanstein: sensational meteorite fall in the Bavarian/Austrian Alps. *Meteorite* **8** (2002), 39–40.

[15] P. Spurný, J. Obsert and D. Heinlein, Photographic observations of Neuschwanstein, a second meteorite from the orbit of the Píbram chondrite. *Nature* **423** (2003), 151–153.

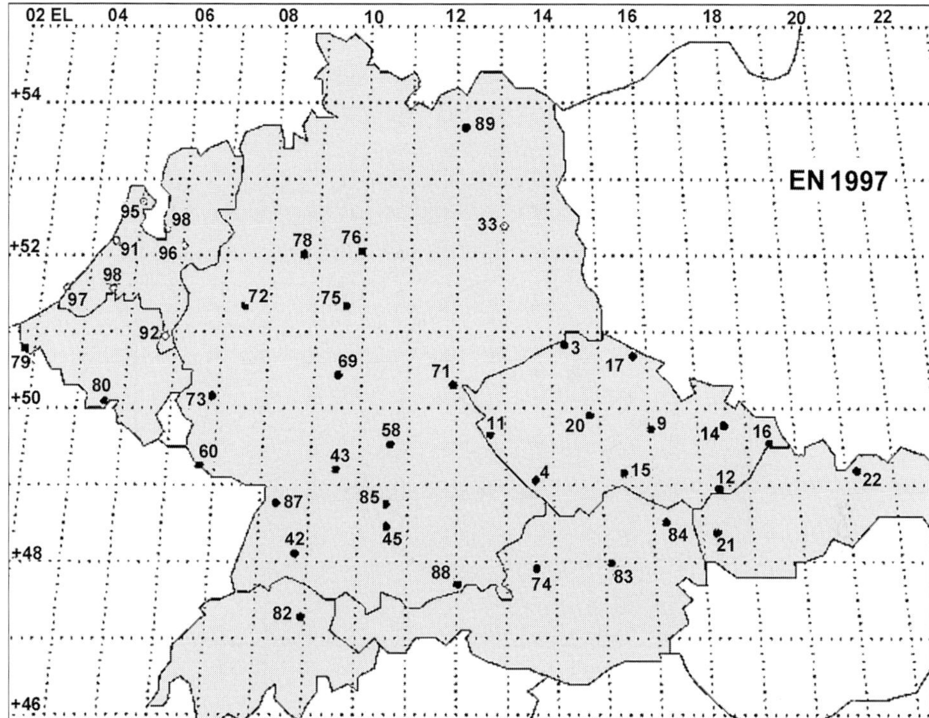

Fig. 30.4 Stations of the European Network in 1997.

the same composition and therefore we can expect that meteorites produced by the April 6 fireball are H5 ordinary chondrites.

The asteroid families (Chapter 10) have similar colors,[16] or taxonomic class, and are therefore chips of the same rock. One might expect that an asteroidal meteoroid stream would also have the same physical properties and any recovered meteorites would be of similar type.

Surprisingly enough, Neuschwanstein was found to be a rare (1.5% of all falls) enstatite chondrite (EL6).[17] EL6 meteorites, unlike others, contain small amounts of moderately volatile elements such as zinc, showing that this meteorite had a heating event, while in the parent body, up to 1070 K (800 °C), and changed considerably. Príbřam did not.

The agreement between the orbital elements, however, is very close, with a D-criterion of only $D = 0.025$. The chance of such matching orbital elements among the sample of 200 suspected meteorite falls is 1 in 100 000.[18] Hence, this pairing is not a coincidence, and the meteorites should be considered members of an asteroidal

[16] S. J. Bus, Compositional structure in the asteroid belt: results of a spectroscopic survey. Doctoral thesis, Massachusetts Institute of Technology (1999).

[17] A. Bischoff and J. Zipfel, Mineralogy of the Neuschwanstein (EL6) chondrite – first results. *Lunar Planetary Sci. Conf.* **34** (2003), 1212.

[18] Z. Ceplecha, Geometric, dynamical, orbital and photometric data on meteoroids from photographic fireball network. *Bull. Astron. Inst. Czechoslov.* **38** (1987), 222–234.

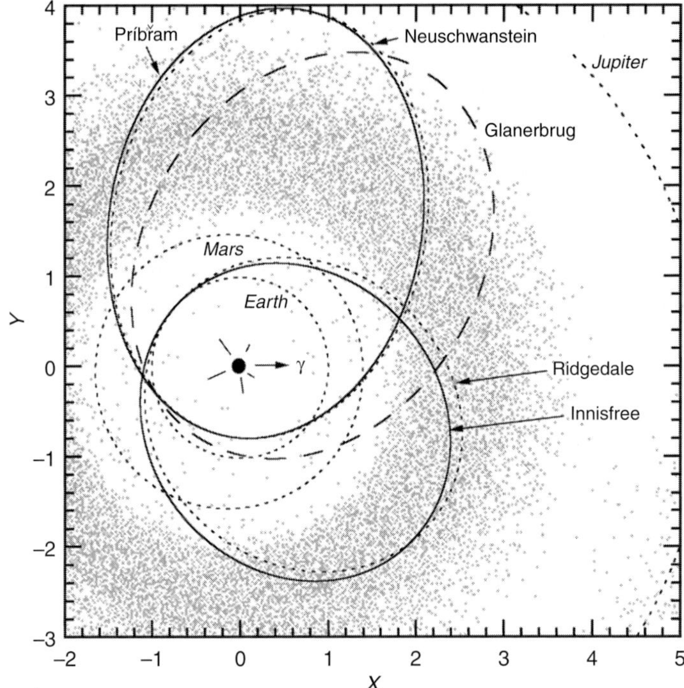

Fig. 30.5 A pre-atmospheric orbit of recovered meteorites, for the orbit pairs listed in Table 8.

meteoroid stream with a radiant near R.A. = 190°, Decl. = +21° (Table 9). Considering the rarity of a collision with Earth from this stream, it should contain at least 10^9 members, all of which would form an asteroid with a minimum diameter of 600 m.[19]

If different materials are permitted in one stream, other recovered meteorites may be part of this stream as well, perhaps including the LL5 chondrite breccia *Glanerbrug*.[20] Glanerbrug has a cosmic ray exposure age of 22.5 ± 3.0 million yr,[21] the same as Príbram (19 ± 2 million yr), but less than Neuschwanstein (48 million yr). The meteorite is a rare type of brecciated chondrite consisting of dark gray LL-type rock mixed in with light gray L/LL-type inclusions, different from both Príbram and Neuschwanstein. The initial mass of this meteorite was only 8–70 kg, measuring 20–40 cm in size. The node was identical to Príbřam, but the longitude of perihelion of the orbit derived from visual observations (even after finding the nearest solution in the large error range) is off (Fig. 30.5).

[19] J. Oberst, P. Spurný and D. Heinlein, Neuschwanstein and Príbřam: solitaire meteorites or members of a stream. *Geophys. Res. Abstracts* **5** (2003), 11 512.

[20] P. Jenniskens, J. Borovicka, H. Betlem *et al.*, Orbits of meteorite producing fireballs. The Glanerbrug – a case study. *Astron. Astrophys.* **255** (1992), 373–376; P. Jenniskens, J. Borovicka, H. Betlem *et al.*, The Glanerbrug meteorite fall. *Publ. Astron. Inst. Czechoslov. Acad. Sci.* **79** (1992), 1–18.

[21] T. Loeken, P. Scherer, H. W. Weber *et al.*, Noble gases in eighteen stone meteorites. *Chem. Erde* **52** (1992), 249–259; L. Lindner and K. C. Welten, The Glanerbrug revisited. *Radiant, J. DMS* **24** (2002), 114–118.

The Glanerbrug meteorite fell in the Netherlands on April 7, 1990 and was only the fourth recovered and preserved fall in the Netherlands, and the first since 1925. I had the privilege to be at Leiden Observatory at that time and responsible for answering the public's questions about anything seen in the sky. I first learned of this fall after being called by a journalist. The meteorite had fallen through the roof of a house in the Dutch town of *Glanerbrug*, near Enschede in the eastern part of the Netherlands just a few kilometers from the German border, and had shattered into thousands of fragments on impact, now strewn on the attic floor. The occupants of the house were not at home at the time of the impact, and the lady of the house discovered the hole in the roof only after finding some broken roof tiles on her door mat the following morning. I asked to have some fragments sent for closer inspection. The next day, I picked up a copy of a daily newspaper, which featured a brief report with a person holding in his hand several of the fragments. From that picture alone it was clear: this was a genuine meteorite. The fragments had a dark fusion crust covering a light-gray interior. When the fragments arrived later that morning, the find was confirmed by *Dr. Charley Arps* of the National Museum of Natural History and the news was announced at the police station in Glanerbrug at a joint press conference later that day. The police had confiscated the material, while investigating a possible case of vandalism. With members of the *Dutch Meteor Society*, we gathered eye witness accounts of the fireball, which occurred at 18:32:38 UT under daytime conditions in the early evening of April 7. We also visited eye witnesses, and searched for other meteorite fragments. None were found. From the visual observations of the fireball, the pre-atmospheric orbit for the Glanerbrug meteorite was determined (Fig. 30.5), which differs from Príbřam in inclination and argument of perihelion (Table 9).[22]

Asteroid streams could form very rapidly in a collision, but should then also disperse quickly. *Galina Ryabova* studied ejecta emitted from the S-type asteroid *(1620) Geographos*, which passes close to Earth's orbit at both nodes. It measured about 5 by 2 by 2 km, rotating about the axis of maximum momentum of inertia with a period of 5.22 h. If the meteoroids are released with low 1–2 m/s, it takes about 4000 yr to distribute the meteoroids around the orbit. If a collision is responsible and the ejection speed is of order 100 m/s, then it only takes about 55 yr. Ryabova found that speeds in excess of 500 m/s would permit encounters with Earth very quickly. Using the *D*-criterion to search among observed orbits, naturally some low inclination orbits were found that might be associated with this asteroid, but they do not stand out well from the sporadic background and no information points to an asteroidal origin of the meteoroids.

[22] M. Langbroek, Príbřam and Glanerbrug. *Radiant, J. DMS* **23** (2001), 76–78; P. Jenniskens, J. Borovicka, H. Betlem, *et al.*, Orbits of meteorite producing fireballs. The Glanerbrug, a case study. *Astron. Astrophys.* **255** (1992), 373–376.

Based on the similarity of their orbits, Terentjeva and Barabanov linked several asteroids and fireballs to Príbřam.[23] *Lubor Kresák* had earlier associated the Príbřam meteorite with the δ-Leonid shower (Table 7).[24] This shower has a high dispersion, from February to March, much greater than most other streams, but not big enough to agree with the high exposure age of Príbřam. Kresák suggested that the mass of Príbřam was near the surface of the original body and released more recently, but that, too, is excluded by cosmic ray exposure studies. The nearby larger body would betray its presence by inducing thermal neutron reactions in the meteorite at the surface.[25] Hence, the δ-Leonid stream associated with the formation event of Príbřam is excluded. Any similarity in orbits demands a recent separation, possibly linked to a bright comet on the sky in prehistoric times.[26]

The exposure ages can be used to search for evolved meteorite streams that have lost any similarity in orbit. Indeed, H-chondrites have exposure ages clustering narrowly around 7 million years, L-chondrites tend to be around 25 million years, while LL-chondrites tend to be about 15 million years (albeit with a wide dispersion).[27] An H-chondrite stream may exist among fallen meteorites.[28] This could hint at three collisions among asteroids in the main belt. Meteoroids from such a continued erosion of fragments in the main belt would form zodiacal bands (Chapter 31), rather than meteoroid streams.

An October shower. Could there be fireball showers from meteorite streams? For example, on the evening of October 4, 1902, *G. Percy Bailey* reported seeing 50 light tracks behind clouds from Stonyhurst College, Blackburn, UK, in the evening between 19:45 and 20:00 UT ($\lambda_\odot = 192.009°$)![29] Forty years later during the second World War in 1942, from *Garten eines Lazarettes hinter der Ostfront (Südabschnitt)*, Dr. Werner Sandner reported having seen a significant meteor shower in the evening of October 5 ($\lambda_\odot = 192.7 \pm 0.1°$), radiating from Cassiopeia, typically $+3^m$ and yellow in color.[30] He had to interrupt the observations for a visit to the out house. Then, on the evening of October 5, 1976 (between 02:55 and 04:37 UT, October 6, $\lambda_\odot = 193.31–193.38°$), *Mr. E. Root* at Pompano Beach,

[23] A. K. Terentjeva and S. I. Barabanov, Family of minor bodies connected with the Príbřam meteorite. *Solar Syst. Res.* **36** (2002), 431–439.

[24] L. Kresak, Multiple fall of Príbřam meteorites photographed V. The association of the Príbřam fall with the sigma Leonid stream. *Bull. Astron. Inst. Czech.* **14** (1963), 49–52.

[25] K. C. Welten, K. Nishiizumi, J. Masarik *et al.*, Cosmic-ray exposure history of two Frontier Mountain H-chondrite showers from spallation and neutron-capture products. *Meteorit. Planet. Sci.* **36** (2001), 301–317.

[26] W. T. Reach, On the origin of the interplanetary dust within recorded history. *Meteoritics* **2** (1992), 353–360.

[27] K. Marti and T. Graf, Cosmic-ray exposure history of ordinary chondrites. *Ann. Rev. Earth Planet. Sci.* **20** (1992), 221–243.

[28] R. T. Dodd, S. F. Wolf and M. E. Lipschutz, An H chondrite stream: identification and confirmation. *J. Geophys. Res.* **98** (1993), 15 105–15 118.

[29] G. P. Bailey, A possible meteor shower on Oct. 4. *Nature* **66** (1902), 577.

[30] W. Sandner, Sternschnuppenschwarm am 1942.X.5[d]. *Sterne* **23** (1943), 46; A. Teichgraeber, Bemerkung zu dem Sternschnuppenschwarm 1942 X 5[d]. *Sterne* **23** (1943), 172.

Fig. 30.6 **Meteors from the October Camelopardalids** captured with a low-light-level video camera by Jarmo Moilanen on October 5, 2005 (17:06–22:41 UT). Photo: Jarmo Moilanen.

Florida, saw 113 meteors moving from north to east. No further details are available.[31] In early October, 1996 one or two dozen fireballs were observed all over the world in a short period of time. It is not clear if those fireballs had the same radiant. Most recently, a meteorite fell at *Berthoud* near Boulder, Colorado, on October 5, 2004 at 20:30 UT ($\lambda_\odot = 192.877°$) in broad daylight. If all these events are linked to the same stream, it is remarkable that its node has not changed over nearly a century.

It is more likely that the shower in Cassiopeia is due to a long-period comet, with perhaps one other sighting in AD 1178 (Table 1). This shower may return in the near future.

Again, the future caught up with me faster than I could write this book. On October 5, 2005 such outburst was observed by video cameras, radiating from the border of Camelopardalis[32] and Draco (Fig. 30.6). Jarmo Moilanen (Finland) detected 12 meteors from a compact geocentric radiant at R.A. $= 164.1 \pm 2.0°$, Decl. $= +78.9 \pm 0.5°$. The differential mass distribution index was a low $s = 1.4 \pm 0.2$ (+0 to −6 magnitudes). The new shower was confirmed by Esko

[31] E. Root, Unusual displays. *Meteor News* **36** (1976), 13; R. A. MacKenzie, *Solar System Debris* (Dover: British Meteor Society, 1976), p. 42.

[32] P. Jenniskens, October Camelopardalids. CBET 309, *IAU* Central Bureau for Astronomical Telegrams, ed. D. W. E. Green, 2005; P. Jenniskens, J. Moilanen, E. Lyytinen, I. Yrjölä, J. Brower, 2005 October 5 outburst of October Camelopardalids. *WGN* **33** (2005), *WGN* **33** (2005), 145.

Lyytinen (two meteors, data only in the early period) and Illeka Yrjölä (four meteors) at nearby locations, and by Sirko Molav in Germany (seven meteors). One two-station meteor had $V_g = 47.3 \pm 0.5$ km/s. The orbit confirms that this outburst was caused by the *i*-revolution dust trail of a long-period comet, rather than a short period asteroid. This shower was subsequently moved from Table 9 to Table 7.

31

Sporadic meteors and the zodiacal cloud

The small 40–500 μm meteoroids that dominate the influx of matter to Earth's atmosphere are nearly all *sporadic*. In the 26 MHz AMOR radar orbit survey (limiting magnitude +14), for example, no obvious peak in flux occurs that is due to meteoroid streams throughout the year. Nonetheless, meteoroid streams are present in the data, but at smaller numbers relative to the background. This reflects the fact that the sporadic meteors have a steeper magnitude distribution index ($\chi = 4.3 \pm 0.2$). As said in Chapter 26, visual observers often include minor showers in their counts of sporadics, resulting in a lower $\chi = 3.4 \pm 0.2$[1] compared to annual streams ($\chi \sim 2.5$).

31.1 The zodiacal cloud

Most sporadic meteoroids do not impact on the planets, but ultimately end up being destroyed in collisions with other meteoroids, the small fragments of which are blown out of the Solar System by radiation pressure. Before they do, the particles spend a long time in a diffuse cloud of dust that surrounds the Sun and is concentrated towards the plane of the planets. This *zodiacal cloud* can be seen as a diffuse glow on new-Moon nights along the ecliptic plane, particularly bright just after sunset and just before dawn (Fig. 31.1).

The zodiacal cloud was perhaps first described in an account by *Joshua Childrey* in 1661[2] and first explained by *Jean Dominique Cassini* (1683), better known for the Cassini division in Saturn's rings. He explained the light of the zodiacal cloud as scattered light from a ring of dust particles around the Sun.[3] The zodiacal cloud is also a bright source of heat (infrared) emission. From IRAS (and COBE) satellite data, we now know that this ring is slightly tilted and warped with respect to the ecliptic plane,

[1] M. Kresáková, The magnitude distribution of meteors in meteor streams. *Contrib. Astr. Obs. Skalnaté Pleso* **3** (1966), 75–112; M. Weber, Meteor observations at Prerov during 1934–1943. *WGN* **22** (1994), 96–103; B.-A. Lindblad, Luminosity function of sporadic meteors and extrapolation of influx rate to micrometeorite region. *Smiths. Contrib. Astrophys.* **11** (1967), 171–180; P. Brown, J. Jones, A determination of the strengths of the sporadic radio-meteor sources. *Earth, Moon, Planets*, **68** (1995), 223–245.

[2] J. Childrey, *Britannia Baconica or the Natural Rarities of England, Scotland and Wales* (London, 1661) (Printed for the author).

[3] J. D. Cassini, *Découverte de la Lumière qui paroist dans le Zodiaque* (Paris: Académie Royale des Sciences, 1683).

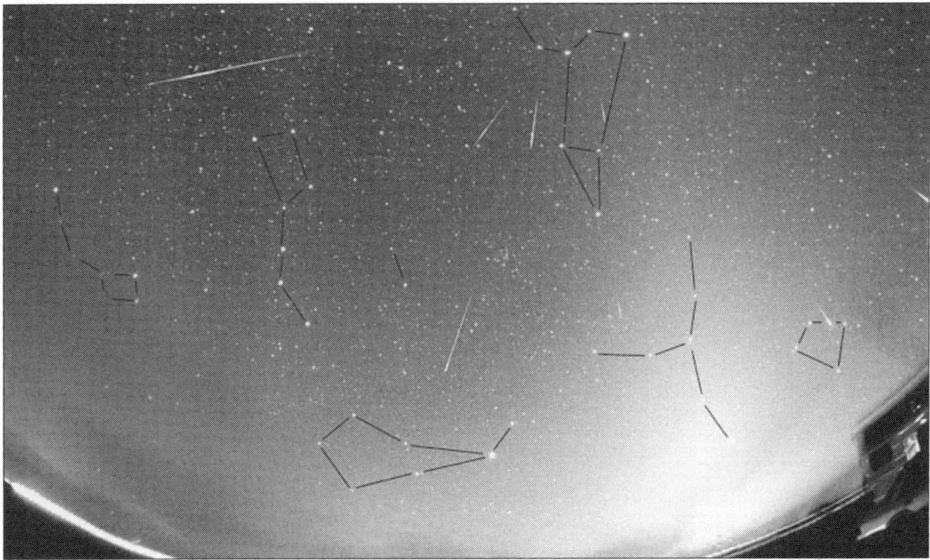

Fig. 31.1 The zodiacal light in the zodiac constellations of Virgo and Leo during the November 18, 2001 Leonid storm in a 30 min exposure by David Harvey near Nachita, New Mexico, USA.

with an ascending node at $95 \pm 20°$ and an inclination of $1.5 \pm 0.4°$ near Earth[4] and $i \sim 3°$ in the inner solar system.[5] The visible zodiacal cloud has a node at $77.1 \pm 0.4°$ and inclination $1.71 \pm 0.02°$ (and $2.03 \pm 0.02°$, respectively).[6] The infrared spectrum measured by the ISO satellite shows only a weak 10 μm silicate band with little structure from mostly larger than 10 μm grains amorphous in nature.[7] Smaller grains are blown away by radiation pressure. Indeed, the peak of the grain cross-section distribution is at about 100 μm, with most scattering from 10–500 μm grains ($\sim 10^{-8}$–10^{-5} g). The dust temperature is that expected for a rapidly rotating black-body, also meaning that the grains are large. The grain temperature falls off with the heliocentric distance r, in AU, as $T = 286\, r^{-0.44}$ K. The total amount of matter in this cloud is estimated to be about 3×10^{17} kg.[8] Near Earth, the dust density at any given time is about 1 particle >1 μm in a $1 \times 1 \times 1$ km^3 sized cube.

Up to 30% of particles, most with inclinations of less than 12° are thought to be of asteroidal origin (Chapter 10). The main sources for asteroidal dust are the asteroid

[4] R. Dumont and A. C. Levasseur-Regourd, Zodiacal light photopolarimetry IV, annual variations of brightness and the symmetry plane of the zodiacal cloud. *Astron. Astrophys.* **64** (1978), 9–16.

[5] C. Leinert, M. Hanner, I. Richter and E. Pitz, The plane of symmetry of interplanetary dust in the inner solar system. *Astron. Astrophys.* **82** (1980), 328–336.

[6] W. T. Reach, Zodiacal emission. I – Dust near the Earth's orbit. *Astrophys. J.* **335** (1988), 468–485; T. Kelsall, J. L. Weiland, B. A. Franz *et al.*, The COBE diffuse infrared background experiment search for the cosmic infrared background. II. Model of the interplanetary dust cloud *Astrophys. J.* **508** (1998), 44–73.

[7] T. Ootsubo, T. Onaka, I. Yamamura *et al.*, IRTS observation of the mid-infrared spectrum of the zodiacal emission. *Earth, Planets Space* **50** (1998), 507–511.

[8] D. W. Hughes, Meteoroids – an overview. In *Meteoroids and Their Parent Bodies*, ed. J. Stohl and I. P. Williams. (Bratislava: Astron. Inst. Slovak Acad. Sci., 1993), pp. 15–28.

disruptions of the Hirayama families, with 85% of asteroidal particles coming from the Veritas-family source that feeds the 9.35° band in the zodiacal dust, and 15% originating from two sources (including the Karin cluster) that feed the near-ecliptic bands.[9] These dust bands consist of relatively large dust particles (with $s = 1.4$).[10] The Veritas family was created in a collision about 50 million years ago. The bands formed on a time scale of 10–100 thousand years, one on each side of the ecliptic plane. The bands form because the particle orbits precess and nutate (Chapter 27) at different rates as a function of their semimajor axes with particles spending most time at aphelion. The result is a torus with peak particle number densities at latitudes near the mean proper orbital inclination of its particles, and near the loci of perihelia and aphelia of the particle orbits, forming a perihelion and aphelion pair that overlap along the line of sight from Earth. In light of the recent evidence of meteoroid streams still being in close connection with minor bodies, much of the zodiacal cloud (Chapter 21) and even the zodiacal dust bands could be products of the recent fragmentation of dormant comet nuclei. Especially if there are more than just a few such bands, as pointed out by *Mark Sykes*.[11] Moreover, there is no evidence that the asteroid families, which are very old on the timescale of recent fragmentations of comets, are still sources of abundant asteroidal dust, more so than the rest of the asteroid belt.

31.2 Meteoroid stream dispersion

Over 70% of zodiacal dust originates from the disintegrating, mostly Jupiter-family type, comets. Figure 31.2 shows the first stage of the formation of a diffuse zodiacal cloud from a narrow cometary meteoroid stream, the π-Puppids. The Puppids disperse rapidly into a diffuse cloud on a time scale of only 1000 years. The initial stages of meteoroid stream dispersion are linked to the planetary perturbations that can change the orbits significantly.

The distribution in semimajor axis widens only slightly, and most orbits continue to be of Apollo-type, with an aphelion near Jupiter and a perihelion near Earth's orbit. The inclination distribution broadens significantly and the eccentricity of the particles attains a wider distribution (Fig. 31.3). Differential precession will rotate the orbits and change the node as a function of the semimajor axis, thus creating a torus of dust that stretches along the comet orbit to high Z-distances, not unlike that expected for the dust evolving from the breakup of asteroids. These effects work on a time scale as brief as a few hundred up to a 1000 yr.[12]

[9] S. F. Dermott, T. J. J. Kehoe, D. D. Durda, K. Grogan and D. Nesvorny, Recent rubble-pile origin of asteroidal solar system dust bands and asteroidal interplanetary dust particles. *ESA SP* **500** (2002), 319–322.

[10] K. Grogan, S. F. Dermott and D. D. Durda, The size–frequency distribution of the zodiacal cloud: evidence from the solar system dust bands. *Icarus* **152** (2001), 251–267.

[11] M. Sykes, IRAS observations of extended zodiacal structures. *Astrophys. J.* **334** (1988) L55–L58.

[12] G. Cremonese, M. Fulle, F. Marzari and V. Vanzani, Orbital evolution of meteoroids from short period comets. *Astron. Astrophys.* **324** (1997), 770–777.

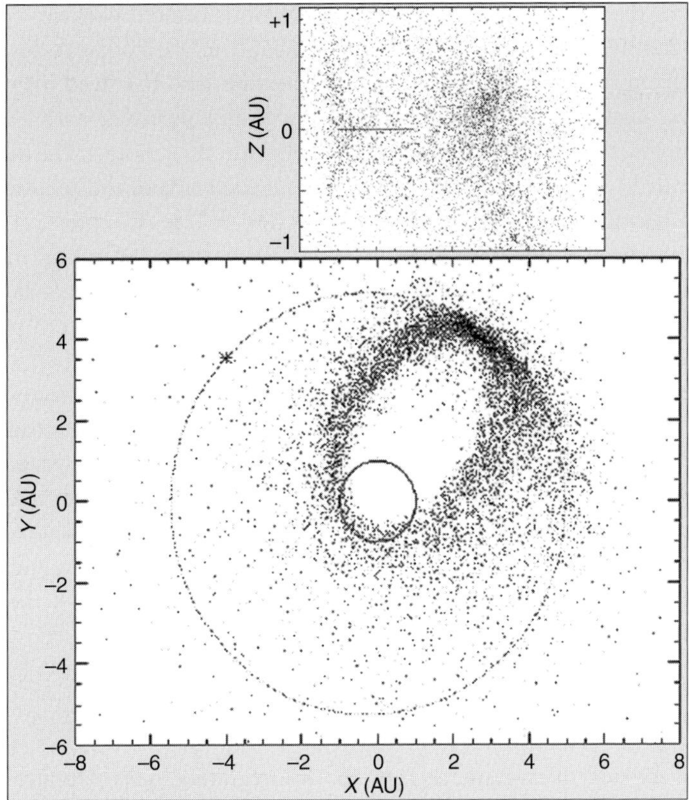

Fig. 31.2 Puppid shower ejected in 1848 as seen in 2774. Calculations by Jérémie Vaubaillon, IMCCE.

At this stage, meteoroids cease to be trapped in the orbital resonances. Indeed, Kortenkamp and Dermott calculated that meteoroids ejected from approximately 175 different Jupiter-family comets decayed into the Sun or were ejected from the Solar System after 30 000–150 000 yr (if surviving collisions that long), but only about 10% of those meteoroids were trapped at one point in mean-motion resonances and a higher percentage only when the meteoroids were released from an orbit close to one of the resonances. The remainder evolved into the inner Solar System without trapping.[13]

31.3 Nonradial effects of radiation forces

On longer time scales, >10 000 yr, the evolution in the zodiacal cloud is dominated by subtle effects of absorption and emission of light (and to a lesser extent solar wind particles), that can change the meteoroid orbit.

[13] S. J. Kortenkamp and S. F. Dermott, Accretion of interplanetary dust particles by the Earth. *Icarus* **135** (1998), 469–495.

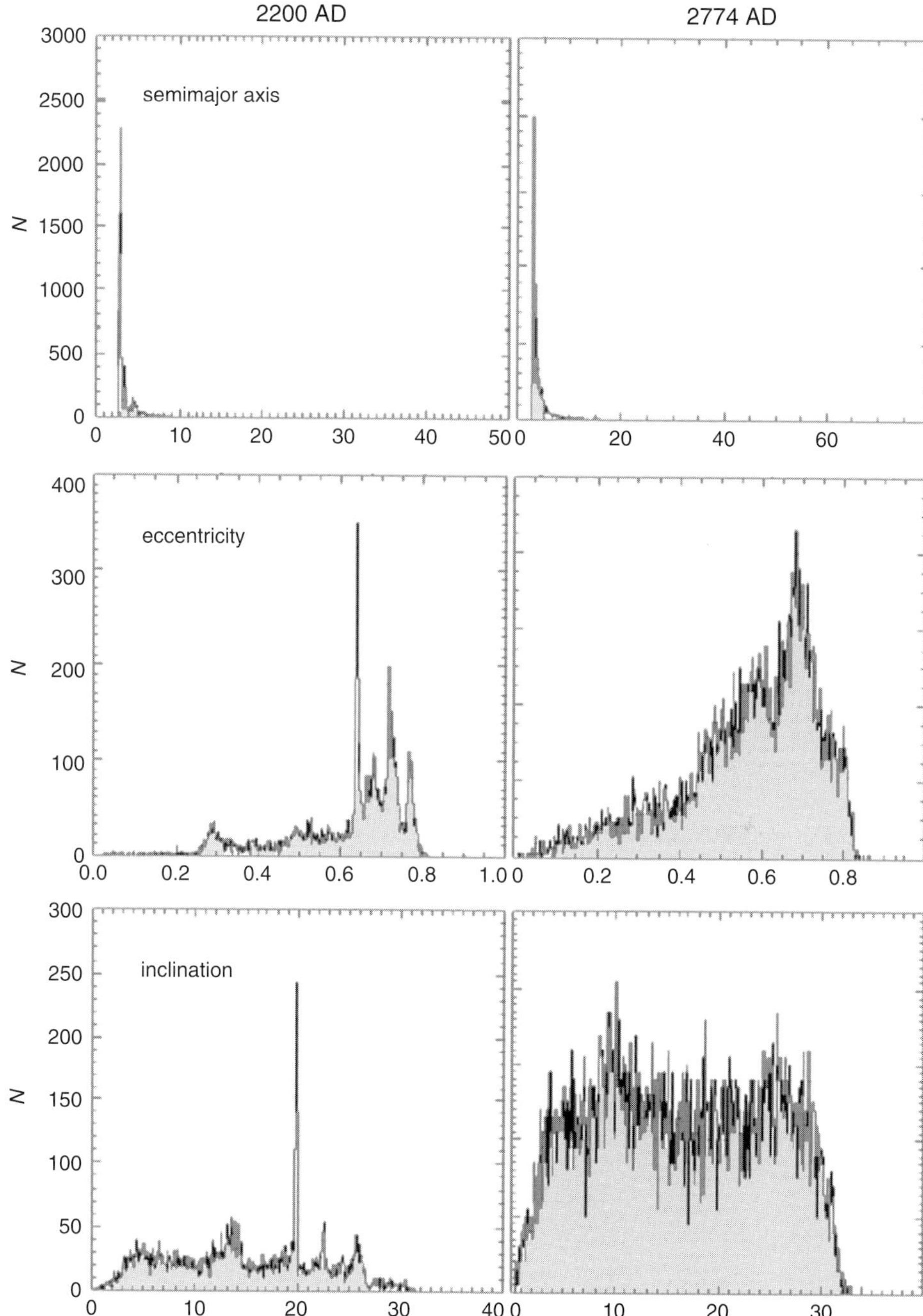

Fig. 31.3 Dispersion over time of orbital elements of the Puppid shower ejected in 1848. Calculations by Vaubaillon.

The most familiar force is called the *Poynting–Robertson (PR) drag*, and is due to the fact that absorbed light is re-emitted with a small Doppler shift in wavelength, different in different directions relative to the direction of motion of the meteoroid. The light emitted in the forward direction is slightly more energetic, causing a net reaction force on the particle. This slows the particle down over time, more so near perihelion where the Sun shines most brightly, which has the effect of gradually turning an elliptical orbit into a circular orbit and causing the particle to spiral inward towards the Sun on a time scale of about $10^7\ M^{1/3}$ yr, where M is the meteoroid mass in grams.[14] Solar wind particles cause a similar effect, about 30% of the Poynting–Robertson drag force. If this effect is important, then the dust density inside the source region (in the case of a steady state) should fall off with r^{-1}.

The Poynting–Robertson effect assumes *isotropic emission*. *Anisotropic emission* and *scattering* can also change the orbit of the particles.[15] If the particle is irregularly shaped and spins around an axis with a fixed orientation in space, then the grains can scatter light in a nonuniform manner, with reflections at an angle exerting an effective force on the particle (Fig. 31.4). The effect changes with the orientation of the grain relative to the Sun and can work in different ways depending on the surface properties and spin orientation of the grains. The net result is a dispersion of grains over time.

A third evolutionary effect is the *Yarkovsky–Radzievskii* effect:[16] nonisotropic thermal emission from an illuminated rotating object. This effect comes in several types. One is due to the fact that the particle absorbs the light on one hemisphere, then spins around somewhat before radiating the energy back into space as heat. That causes a reaction force in a direction different from that of the radial direction away from the Sun.

The effect depends on the size and direction of rotation of the particle and also comes in "seasonal" varieties, where a delay in re-emitting can cause the warmer summer hemisphere to absorb light from a different direction than that in which it is re-emitted (Fig. 31.5). The spin of meteoroids is determined by the effects of collisions, gravitational torques due to close encounters with the Sun and planets, emission of heat from irregular surface features, and internal dissipative effects, the latter tending to stop the grain from precessing and cause it to rotate preferentially around the principal axis.[17] In the end, grains tend to have arbitrary orientations of spin axes.

[14] The change over time is about (with $\eta = GM_\odot\beta/c$):

$$\Delta a/\Delta t = -\eta\,(2+3e^2)/[a(1-e^2)^{3/2}] \tag{31.1}$$

$$\Delta e/\Delta t = -5\eta e/[2a^2\sqrt{(1-e^2)}] \tag{31.2}$$

I. P. Williams, Meteoroid streams and showers. In *Dust in the Solar System and other Planetary Systems*, ed. S. F. Green, I. P. Williams, J. A. M. McDonnell and N. McBride. (Amsterdam: Pergamon, 2002), pp. 3–14.

[15] E. J. Lyytinen and T. Van Flandern, Predicting the strength of Leonid outbursts. *Earth, Moon Planets* **82/83** (2000), 149–166.

[16] V. V. Radzievski, A mechanism for the disintegration of asteroids and meteorites. *Astron. Zh.* **29** (1952), 162–170.

[17] D. Vokrouhlicky and D. Capek, YORP-induced long-term evolution of the spin state of small asteroids and meteoroids: Rubincam's approximation. *Icarus* **159** (2002), 449–467; D. Capek and D. Vokrouhlicky, The YORP effect with finite thermal conductivity. *Icarus* **172** (2004), 526–536.

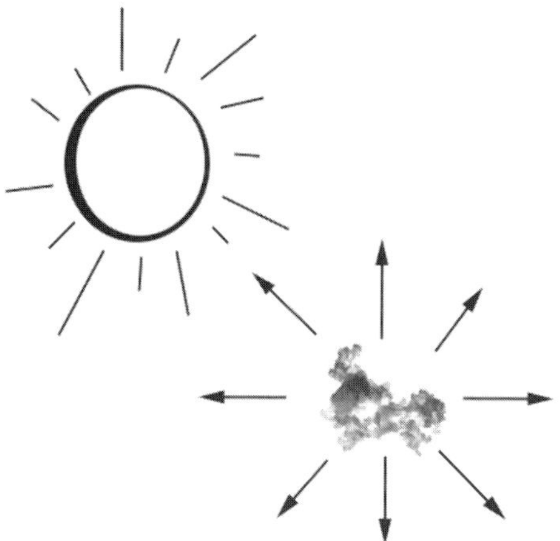

From perspective of dust grain

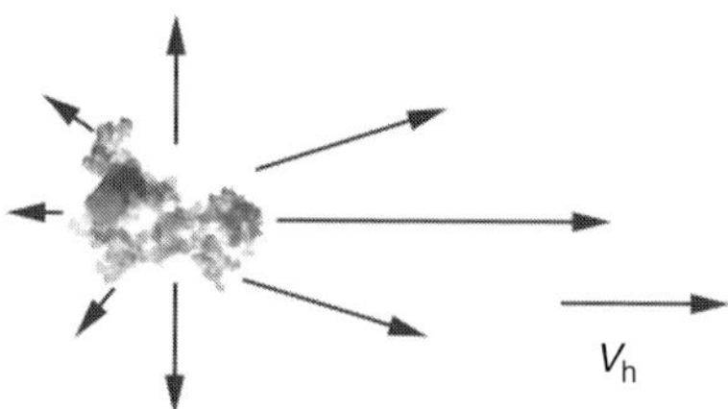

From perspective of the Sun

Fig. 31.4 The mechanism of Poynting–Robertson drag, whereby light absorbed by the grain and isotropically emitted from a moving dust grain does not appear to be isotropically emitted from the perspective of the Sun.

Because of this, all these effects add to the dispersion of the dust, rather than to systematic secular changes of their orbits.

The result is not very dramatic because the temperature difference between the day- and nightside of a meteoroid is not that great. That temperature difference is about:[18]

$$\Delta T_{\mathrm{d}} \sim (1 - A) d c_{\odot}(r) / 4\kappa \tag{31.3}$$

where A is the grain albedo, κ the thermal conductivity, d is the grain diameter, $c_{\odot}(r)$ is the solar flux at the heliocentric distance r ($= 1.36 \times 10^3$ J m^{-2} s^{-1} at 1 AU). For a silicate glass grain of 200 μm diameter, $\kappa = 1.4$ J/m/s/K and $A = 0.1$, so that the

[18] J. A. Burns, P. L. Lamy and S. Soter, Radiation forces on small particles in the solar system. *Icarus* **40** (1979), 1–48.

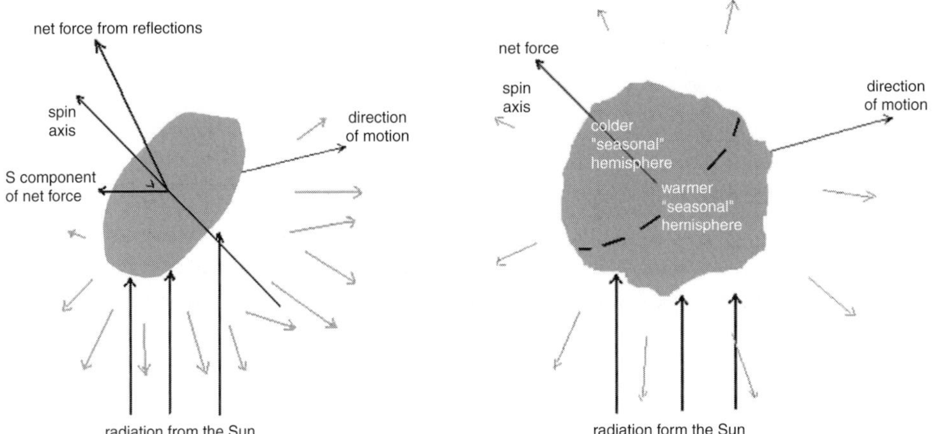

Fig. 31.5 The nonisotropic scattering (left) and the seasonal Yarkovsky–Radzievskii effect from light absorption and re-emission (right) can also influence the orbit of the particle. Both result from a particle spinning around an axis fixed in space.

day–night difference is 0.04 K. An iron grain has $\kappa = 76$ J/m/s/K, $A = 0.05$, so that $\Delta T_d \sim 0.0009$ K.[19]

In addition to radiation forces, there are magnetic forces from the charge carried by the grains. A 10 μm grain has been measured by Cassini dust detectors to carry a charge of 10^{-15} Coulomb on average, or have a 5 V surface potential.

None of these effects has been proven to leave a mark on meteor showers, despite many efforts to invoke the Yarkovsky–Radzievskii and Poynting–Robertson drag effects to understand the evolution of cometary dust trails and the asymmetry of the Geminid shower, for example. The effects are found to be important only on the longer time scales relevant to the evolution of the zodiacal dust. Nevertheless, most meteoroid stream formation models are easily adapted to, and often include, the Poynting–Robertson drag.

Even among zodiacal dust the Poynting–Robertson drag does not show up clearly in the orbits of large meteoroids. Most mm–cm sized sporadic meteoroids appear to have aphelion distances close to Jupiter and relatively high eccentricities, instead of the more circular orbits expected from the action of the Poynting–Robertson effect over a period of 1–10 million years. Fred Whipple[20] pointed out that the observed meteoroids with shorter semimajor axes do not tend to align their apside lines with that of Jupiter, as is the case for asteroids. This should occur over a timescale of >1 million years. Radar surveys of smaller meteoroids 100–500 μm in size show that these

[19] T. Mukai, J. Bum, A. M. Nakamura, R. E. Johnson and O. Havnes, Physical processes on interplanetary dust. In *Interplanetary Dust*, ed. E. Grün, B.Å.S. Gustafson, S. F. Dermott and H. Fechtig. (Berlin: Springer, 2001), pp. 445–507.

[20] F. L. Whipple, Meteoritic erosion in space. *Smithsonian Contrib. to Astrophys.* 7 (1963), 239–248.

also arrive from Jupiter-family comet orbits. This argues against an age of the zodiacal cloud longer than 100 000 years.

31.4 Erosion in space

There are a number of processes that can change the physical properties of the meteoroids and the dust size distribution. Meteoroid streams can fade away by the breaking and eroding of the meteoroids over time. The meteoroids can break catastrophically by mutual collisions, or more gradually by noncatastrophic collisions, heating and cooling, cosmic ray impacts, chemical attack, and other processes that are collectively called *space weathering*.[21]

The net effect of space weathering is a gradual increase of χ and particle mean density. The magnitude distribution index (χ) in showers increases by about 0.23 per 1000 yr when age/P/q is plotted versus χ (Table 8). This would imply an age of about 20 000 yr for the sporadic background.

Mutual collisions among meteoroids are important. Evidence for that are meteoroids with small semi-major axis in Halley-type streams. Several of the Leonid meteors observed by the Dutch Meteor Society during the 1997–2001 Leonid outbursts had a speed significantly less than $V_g = 70.6$ km/s and a semimajor axis much less than that of 55P/Tempel–Tuttle. Collisions with sporadic dust grains close to the ecliptic plane could be responsible. Six out of 562 orbits (1%) were perturbed in this manner, over a timescale of 100–1000 years.[22]

It is now estimated that the mm-sized grains responsible for visible meteors tend to survive the erosive effects of collisions with other meteoroids in the interplanetary environment for periods of only about 62 000 yr,[23] depending on many factors. Cm-sized grains can survive for up to 200 000 years. Hence, the effects of space weathering are important only on about the same time scale as the age of the meteoroid population. Not surprisingly, the mean density of ~1 g sporadic meteoroids is not much different from that of similar annual shower meteoroids,[24] consistent with a young age for the sporadic background as a whole.

Collisions with so-called "beta-meteoroids" are important. Groups of meteoroids resulting from the breakup of a larger particle are called *swarms*. In early dust detection experiments on the Earth's orbiting HEOS 2 satellite (February, 1970–August, 1972) 15 swarms of μm-sized particles were detected within 10 Earth radii, presumably because particles became charged, trapped in the Earth's magnetic field and then

[21] F. L. Whipple, Meteoritic erosion in space. *Astron. J.* **67** (1962), 285–286 (abstract).
[22] J. M. Trigo-Rodríguez, H. Betlem, E. Lyytinen, Leonid meteoroid orbits perturbed by collisions with interplanetary dust. *Astrophys. J.* **621**, 1146–1162.
[23] S. Nikolova and J. Jones, Lifetimes of meteoroids in interplanetary space: the effect of erosive collisions and planetary perturbations. *ESA SP* **495** (2001), 581–585.
[24] F. Verniani, Meteor masses and luminosity. *Smithsonian Contrib. Astrophys.* **10** (1967), 181–195.

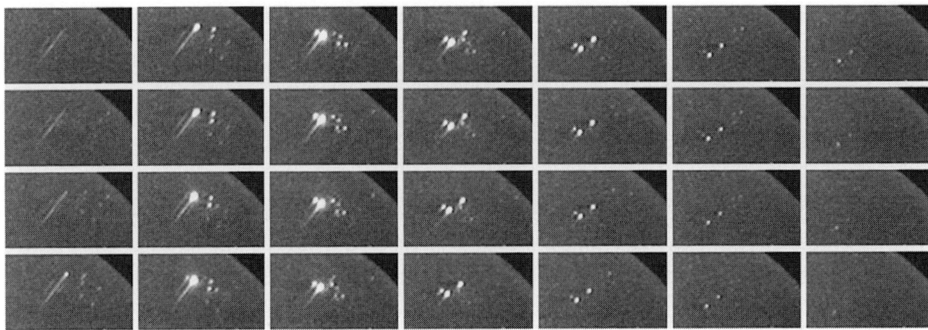

Fig. 31.6 Meteor cluster at 13:31:51 UT, November 17, 1997 by Masao Kinoshita of the Nippon Meteor Society. Each intensified video frame is 1/30 second apart.[25]

broken up by electrostatic forces.[26] Dust collisions in space are responsible for the *beta-meteoroids*, first discovered in the dust experiments on Pioneer 8 and 9.[27] Beta-meteoroids are the small submicron-sized fragments produced when dust grains collide or fall apart by partial evaporation near the Sun. Due to their small size, they have a very high surface-to-mass ratio ($\beta > 1$) and move on hyperbolic orbits away from the Sun. This produces a stream of very tiny particles with a radiant near the Sun.

That is not to say that no space erosion occurs by other processes. Some disruptions of the meteoroids of 55P/Tempel–Tuttle have been observed during the recent Leonid storms as *meteor clusters*, groups of meteors in a particular part of the sky that appear in a brief period of time. In the well documented November 17, 1997 Leonid of 13:31:51 UT, which had broken in 100–150 fragments, the larger fragments were found on one side of the cluster (Fig. 31.6).[28] The breakup appeared to have occurred when reaching perihelion, a week prior to impacting Earth, corresponding to a relative speed of the fragments of order 20 cm/s.[29]

Other weathering processes include exposure to energetic particles. While in the interplanetary medium, the grains are exposed to solar wind particles and cosmic rays.[30] The solar wind consists mostly of protons moving at ~400 km/s with a density of ~4.5 cm^{-3} at 1 AU, or a total flux of (all particles) 2×10^{12} $(r/\text{AU})^{-2}$ m^{-2} s^{-1}.

[25] M. Kinoshita, M. Maruyama, T. Sagayama, Preliminary activity of Leonid meteor storm observed with a video camera in 1997. *Geophys. Res. Lett.* **26** (1999) 41–44.

[26] H. J. Hoffmann, H. Fechtig, E. Grün and J. Kissel, Temporal fluctuations and anisotropy of the micrometeoroid flux in the Earth–Moon system measured by HEOS 2. *Planet. Space Sci.* **23** (1975), 985–991.

[27] O. E. Berg and E. Grün, Evidence of hyperbolic cosmic dust particles. In *Space Research XIII*, ed. M. J. Rycroft and S. K. Runcorn. (Berlin: Akademie-Verlag, 1973), pp. 1047–1055; A. E. Zook and O. E. Berg, A source for hyperbolic cosmic dust particles. *Planet. Space Sci.* **23** (1975), 183–203.

[28] M. Kinoshita, M. Maruyama, and T. Sagayama, Preliminary activity of Leonid meteor storm observed with a video camera in 1997. *Geophys. Res. Lett.* **26** (1999), 41–44.

[29] J.-I. Watanabe, I. Tabe, H. Hasegawa *et al.*, Meteoroid clusters – evidence of fragmentation in space. *ESA SP* **500** (2002), 277–279.

[30] T. Mukai, J. Bum, A. M. Nakamura, R. E. Johnson and O. Haynes, Physical processes on interplanetary dust. In *Interplanetary Dust*, ed. E. Grün, B.Å.S. Gustafson, S. F. Dermott and H. Fechtig. (Berlin: Springer, 2001), pp. 445–507.

These protons are light and easily stopped, with a penetration depth of $\sim 10^{-6}$ g/cm^2. More energetic particles are produced in solar flares and by galactic cosmic rays. At 1 AU, a typical solar flare would give 10^{14} particles/m^2 with energies >1 MeV. The galactic cosmic rays are mostly protons with energies in the GeV range with many penetration depths of $\sim 10^{-2}$ g/cm^2. These fast particles penetrate the minerals and deposit their energy, briefly heating the grain, turning crystalline minerals into glasses, and leaving *solar wind tracks*. They chemically change the organic matter into a darker more cross-linked material with larger aromatic structures. Such effects have been observed in collected IDPs.

The sintering and melting of grains occurs when they approach the Sun. The example of the Geminid shower was discussed earlier. Other evidence for this includes the fact that many meteoroids with small perihelion distances ($q < 0.25$ AU, I > 1200 K) and in prograde low to moderately inclined orbits ($i < 50°$) appear to have lost their more volatile minerals. Many have a conspicuous lack of sodium-containing minerals (or, alternatively, have a hard time ablating the sodium-containing minerals during the meteor phase).[31]

Closer in towards the Sun, the zodiacal cloud merges into what is called the *F-Corona* of the Sun: a bright glow seen between about 0.5° and 6° from the Sun, or 2–25 times the Sun's radius. Near the Sun, a dust-free zone (for meteoroids <1 μm) is suspected due to the heating and the sublimation of the grains. Such a zone has not been confirmed with certainty, however. Models suggest that this zone has a diffuse border between 3 and 9 solar radii (0.014–0.042 AU). Hence, meteoroids with perihelion distances $q < 0.042$ AU (I > 1200 K) will suffer from a significant loss of matter by heating. Compare this to the perihelion distance of the Sunskirter streams of Arietids and δ-Aquariids ($q = 0.09$, I ~ 860 K) and the shower of Geminids ($q = 0.14$), which come to within a factor of 2–4. No known meteor streams have $q < 0.07$ AU (I > 1000 K).

31.5 Distribution of orbital elements

It is mostly dynamic processes that appear to determine how long meteoroids stay in the inner solar system. The distribution of meteoroid orbits in the zodiacal cloud can be measured from meteor observations (Fig. 31.7). In order to do so, it has to be understood how efficiently the various techniques observe meteors of a certain mass and from a certain direction of motion. This aspect has led to a lot of confusion and only in recent years has the picture become clearer. *Andrew Taylor* showed that the velocity distribution derived from the *Harvard Radio Meteor Project* had a software bug and an incorrect velocity bias, and that, when corrected, the velocity distribution is close to that derived from photographic observations. The radar was less efficient than expected in picking up fast meteors due to the rapid dispersion of ionization at

[31] J. Borovicka, P. Spurný and P. Koten, Evidence for the existence of non-chondritic compact material on cometary orbits. *ESA SP* **500** (2002), 265–268.

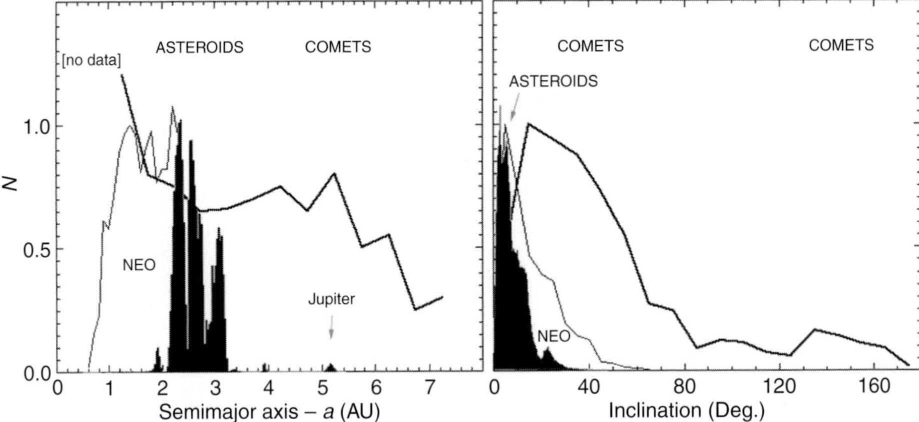

Fig. 31.7 The distribution of orbital elements of radar meteoroids after correction for observing selection effects by Taylor and Elford.[32] The distributions (solid line) are compared to that of main-belt asteroids (data from the Minor Planet Center) and near-Earth objects (mostly asteroids thin line) detected by LONEOS in early 2004.

high altitudes (called the *echo height ceiling effect*). He concluded that the meteoroids hit Earth not at 15 km/s on average, as was thought before, but at 19.1 km/s when averaged by number, or 29.4 km/s when averaged by the observed magnitude of the meteors.[33,34]

Fig. 31.8 shows the result. The dust density falls off exponentially, away from the Sun. At Earth's orbit, the dust density is about 1 particle >1 μm/km^3, as said. Other sources of dust, such as the Earth's dust ring, the Kuiper belt dust, and dust from the interstellar medium are only a small fraction of the total dust density. The Earth's ring does not contribute many meteors (called the *Cyclids*, when they have $e \sim 0$, $a \sim 1.0$ AU), because the grains tend to get trapped and librate around the Lagrange points, keeping the dust away from Earth.

Most large meteoroids appear to be derived from Jupiter-family comets, rather than from the main asteroid belt, because they have aphelia near Jupiter, rather than in the asteroid belt. Only about 15% of all meteoroids (>200 μm) arrive from retrograde orbits. The highest dust denstiy is found in a torus peaking at about 0.3 AU from the Sun (Fig. 31.9),[35] with lower dust densities (but not fully empty) very near the position of the Sun. For lower dust densities, that torus becomes broader and thicker and the contour center shifts outward. Dust from the asteroid families (the zodiacal dust

32 A. D. Taylor and W. G. Elford, Meteoroid orbital element distributions at 1 AU deduced from the Harvard Radio Meteor Project observations. *Earth, Moon Planets*, **50** (1998), 569–575.

33 *Ibid.*

34 D. P. Galligan and W. J. Baggaley, The orbital distribution of radar-detected meteoroids of the Solar System dust cloud. *MNRAS*, **353** (2004), 422–446.

35 T. Kelsall, J. L. Weiland, B. A. Franz *et al.*, The COBE diffuse infrared background experiment search for the cosmic infrared background. II. Model of the interplanetary dust cloud. *Astrophys. J.* **508** (1998), 44–73.

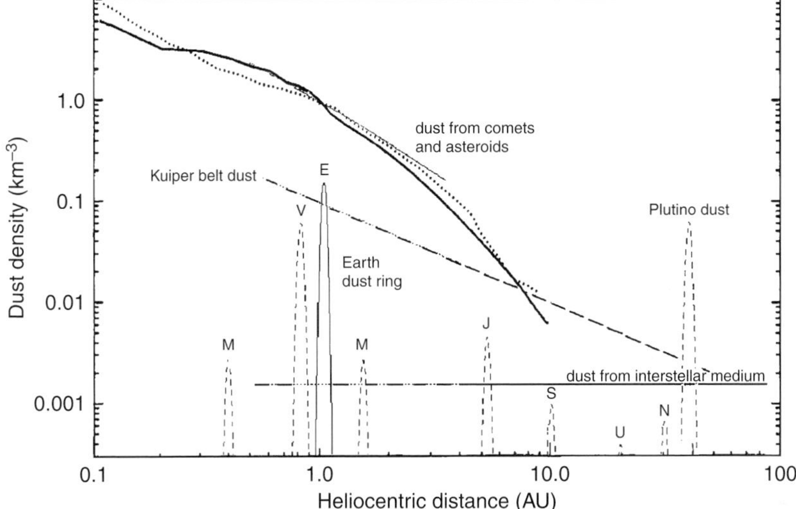

Fig. 31.8 The radial distribution of dust in the solar system, from a graph by William Reach (IPAC) and Eberhardt Grün (MPI), with superposed (scaled) radar measurements for the radial distribution at 0° ecliptic latitude by AMOR (solid line) and the Harvard Radio Survey of 1962–1965 (dotted line).[36]

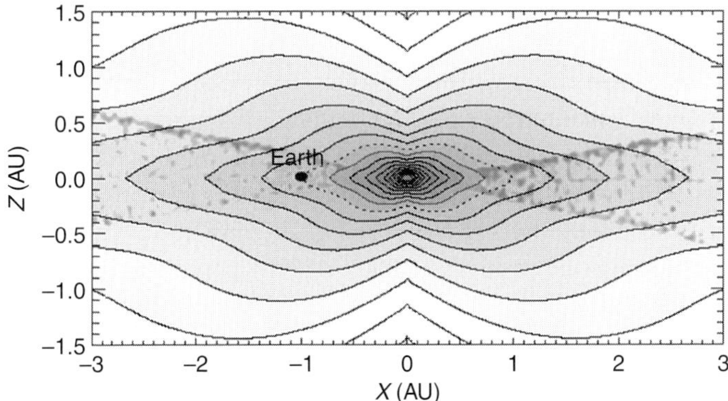

Fig. 31.9 Zodiacal dust distribution in and out of the plane of the ecliptic as derived from *Clementine* data, with the position of a model of the dust bands from the asteroid families superposed and the position of Earth for reference. The Sun is at the center.[37] Contours are factors of 1.5 in dust density.

[36] D. P. Galligan and W. J. Baggaley, Determination of the spatial distribution of the solar system meteoroid population using AMOR. *ESA SP* **500** (2002), 229–232.

[37] J. M. Hahn, H. A. Zook, B. Cooper and B. Sunkara, Clementine observations of the zodiacal light and the dust content of the inner solar system. *Bull. Am. Astron. Soc.* **33** (2001), 1132 (abstract); S. F. Dermott, D. D. Durda, B. Å. S. Gustafson *et al.*, Zodiacal dust bands. In *Asteroids, Comets, Meteors 1993*, ed. A. Milani *et al.* (Dordrecht: Kluwer, 1994) pp. 127–1142.

bands) is significant to just within 1 AU for the low-i component, and perhaps up to 1.6 AU for the 10° band.

This agrees with *in situ* observations. Pioneer 10 and 11 carried a "beer-can" dust detector consisting of 108 pressurized cells designed to measure the low flux of 10 μm and larger grains in the asteroid belt. Pioneer 10 traveled through the Asteroid belt in early 1973.[38] The rate of impacts of small grains did not significantly increase when passing the asteroid belt and remained constant at least to 18 AU.[39] Zodiacal light cameras onboard the spacecraft detected a steady decrease of meteoroids from 1 to 3.3 AU, with <30% contribution from asteroidal dust between 2.3 and 3.3 AU. The *Gegenschein*, from backward scattered light, was measured beyond the asteroid belt and found to be associated with 1 μm sized dust grains, again presumably of cometary origin. Finally, ULYSSES did not detect the expected asteroidal dust among other small <1 μm dust grains when it passed the asteroid belt in late 1990.[40]

Past Jupiter, the dust density of 1 μm grains declined approximately with the square of the distance. Dust ejected from prograde and retrograde moving Halley-type comets is expected to evolve into a distribution that falls off as $r^{-1.65 \pm 0.15}$, each grain spending a lot of time in mean-motion resonances.[41] This dust is released near perihelion in eccentric orbits, but spends much time at aphelion. Beyond Saturn, the Pioneer measured a constant dust flux out to 20 AU from a source in the Edgeworth–Kuiper belt. While this dust may initially spiral inward towards the Sun by Poynting–Robertson drag, its fate will depend on the giant planets it meets on the way in.

Over time, close encounters with the planets will eject the meteoroids into hyperbolic orbits or send them into the Sun. This demands a continuous replenishment. The current total of periodic comets (80 t/s from the Jupiter family and 300 t/s from Halley types) of sizes in the range 6 mm–1 m, does not appear to supply enough dust, but could account for the total mass in the zodiacal cloud in about 13 000 yr.

Comet Hale–Bopp alone produced 1000 t of dust per second when at perihelion. Hence, a single giant comet can easily offset the balance of input and loss. Indeed, it has been proposed that the current zodiacal cloud is not typical. ^{3}He isotope concentrations in deep sea sediments, presumably originating from micrometeorites, suggest that 37 and 50 million years ago the dust influx was much larger than today.

[38] M. Landgraf, J.-C. Liou, H. A. Zook and E. Grün, Origins of solar system dust beyond Jupiter. *Astron. J.* **123** (2002), 2857–2861.

[39] E. Grün, M. Baguhl, H. Svedhem and H. A. Zook, *In situ* measurements of cosmic dust. In *Interplanetary Dust*, ed. E. Grün, B. Å. S. Gustafson, S. F. Dermott and H. Fechtig. (Berlin: Springer, 2001), pp. 295–346.

[40] I. Mann, E. Grun and M. Wilck, The contribution of asteroid dust to the interplanetary dust cloud: the impact of ULYSSES results on the understanding of dust production in the asteroid belt and of the formation of the IRAS dust bands. *Icarus* **120** (1996), 399–407.

[41] J.-C. Liou, H. A. Zook and A. A. Jackson, Orbital evolution of retrograde interplanetary dust particles and their distribution in the solar system. *Icarus* **141** (1999), 13–28.

Part VI

Impact and relevance of meteor showers

32
Impact!

Meteor showers are a threat to satellites in orbit and an early warning of comet impacts that could one day threaten our very existence. Each year, Earth is hit by 20 000 tons of meteoroids, <1 kg in mass, 10 million meteors brighter than $+6.5^m$, at a rate of $\sim$1000 visible meteors per second, but by only one superbolide. And only once every 100 million years does a 10 km sized minor planet hit. Despite these disparate frequencies, each group brings in about the same amount of mass (Fig. 32.1).

32.1 Giant impacts

A 10 km sized minor planet would instantly blind any casual observer by its blazing -45^m fireball and create a 100 km sized crater from the explosion on impact, if not all of the cosmic speed is dissipated in the atmosphere. The last eye witness of such a giant impact was likely to have been a dinosaur 65 million years ago, when all its relatives but the birds died after a 10 km minor body hit the *Yucatan Peninsula* of Mexico in what was then a shallow sea. The 180–300 km diameter crater is now buried by 300–1000 m of limestone. At that time,[1] much of the asteroid or comet survived the atmosphere and when it hit the ground, so much energy was released that the rock (a limestone ocean bed rich in carbonates and sulfates) was vaporized out to a distance of 6–12 km from the impact point. The vapor was super hot at thousands of degrees Celsius, expanding rapidly into a plume rising from the ground into space along the fireball track and ultimately encircling the whole Earth. For thousands of miles around the crater there was a rain of molten droplets (*tektites*) and debris. Low-lying areas near the coast of the shallow sea were inundated by a tsunami. For weeks, the impact heat of the explosion and the heated atmosphere would have caused fires that would have released soot and carbon dioxide into the atmosphere. In the following years, clouds blocked out sunlight, causing a rapid drop of temperatures at the Earth's surface, which may have persisted for about 10 yr. The ozone layer was damaged. The vapor

[1] D. Morrison, The Spaceguard survey. In *Report of the NASA International Near-Earth-Object Detection Workshop*, David Morrison, chair. (Pasadena, CA: JPL, 1992); D. Steel, *Rogue Asteroids and Doomsday Comets* (New York: Wiley, 1995); D. Steel, Project Spaceguard: will humankind go the way of the dinosaurs? *Irish Astron. J.* **24** (1997), 19–30; D. Morrison, The Spaceguard survey – protecting the Earth from cosmic impacts. *Mercury* **21** (1992), 103–106.

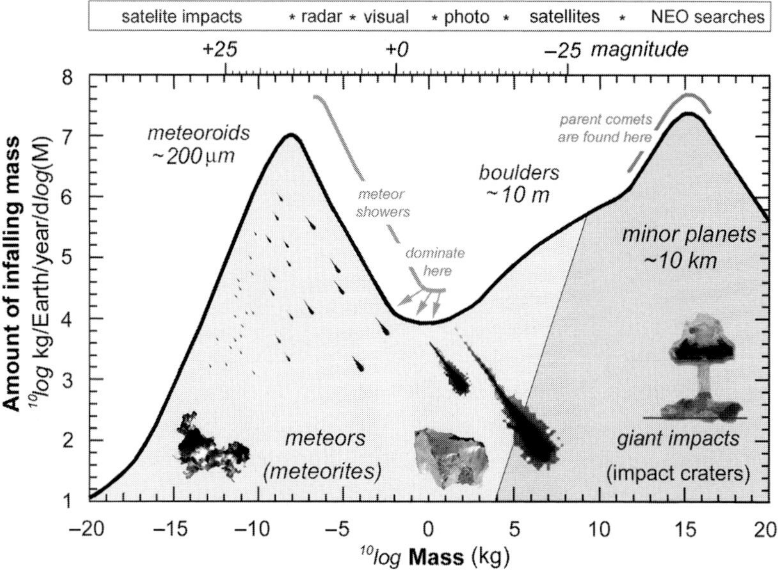

Fig. 32.1 The amount of infalling matter on the whole Earth per year (on a logarithmic scale) in different intervals of mass (bottom) or magnitude intervals (top). For masses smaller than 10^{-7} kg, the curve is based on the Grün model[2] (Eq. 32.1), scaled upward by a factor of 2.[3] Between masses of 10^{-7} and 1 kg, I used the LDEF calibration by Love and Brownlee, scaled down slightly (factor 0.7) within their error margin[4] to match the Grün curve and the flux calibrated radar data by Brown and Jones,[5] and I calculated a smooth curve for visual meteors based on the assumption that sporadic ($x = 4.3$), streams ($x = 2.5$), and fireballs ($x = 2.0$) have a continuous size distribution in this range. The curve for masses larger than 1 kg is based on the observed frequency of full-Moon fireballs and superbolides,[6] as well as the observed frequency of asteroids and comets in near-Earth's orbits,[7] as shown in Fig. 32.2. The light-gray zone is that of meteors, only a tiny fraction of which survives as (micro-)meteorites, the bulk of mass being deposited as meteoric debris, atoms, and molecules in the Earth's atmosphere.

from the explosion contained trillions of tons of sulfur dioxide gas, carbon dioxide gas, and water vapor. It was highly toxic and very efficient at trapping the heat through the greenhouse effect. As a result, the cold and dark period was followed by a warm period with acidic rains containing dissolved sulfur dioxide. Along with the greenhouse vapor there was also a large amount of dust and soot ejected, all of which ended up changing the climate of the Earth and raising the temperature at the surface a

[2] E. Grün, H. A. Zook, H. Fechtig and R. H. Giese, Collisional balance of the meteoritic complex. *Icarus* **62** (1985), 244–272.
[3] Z. Ceplecha, Influx of interplanetary bodies onto Earth. *Astron. Astrophys.* **263** (1992), 361–366.
[4] S. G. Love and D. E. Brownlee, A direct measurement of the terrestial mass accretion rate of cosmic dust. *Science* **262** (1993), 550–553.
[5] P. Brown and J. Jones, A determination of the strengths of the sporadic radio-meteor sources. *Earth, Moon, Planets* **68** (1995), 223–245.
[6] P. Brown, R. E. Spalding, D. O. ReVelle, E. Tagliaferri and S. P. Worden, The flux of small near-Earth objects colliding with the Earth. *Nature* **420** (2002), 294–296.
[7] J. S. Stuart and R. P. Binzel, Bias-corrected population, size distribution, and impact hazard for the near-Earth objects. *Icarus* **170** (2004), 295–311.

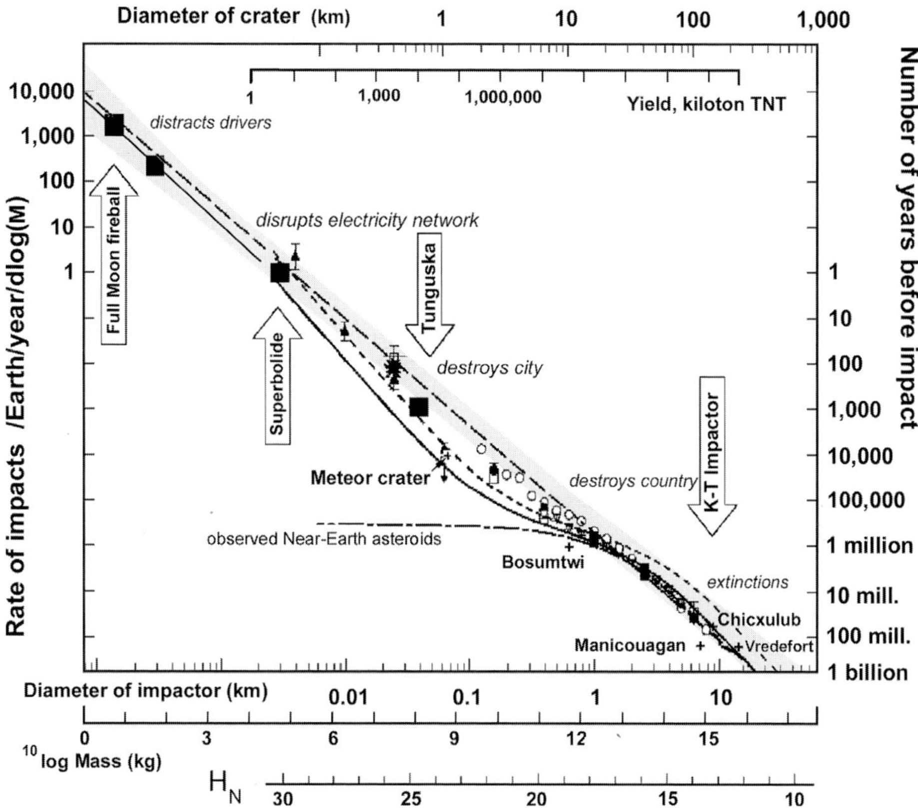

Fig. 32.2 The terrestial impact rate for minor planets and boulders of different sizes. The curve is based on the observed frequency of full-Moon fireballs and superbolides,[8] as well as the observed frequency of asteroids and comets in near-Earth's orbits.[9] The rate of impacts is shown as a function of diameter of the impactor, mass of the impactor, H_N is the magnitude of the minor planet for albedo $=0.14$, the yield of the explosion, and the diameter of the resulting crater. The different scales depend on impact speed, density of the impactor, and properties of the terrain and impact geometry, and are therefore only approximate.

few degrees Celsius. Changing currents caused ocean waters to become almost 10 °C warmer at places, killing off half of the living species: not only land animals and plants, but also ocean life. This lasted for thousands of years and it took 100 000 yr before species flourished again.

Smaller impacts are more frequent (Fig. 32.2). Objects as big as 1 km in diameter, weighing about a trillion kg hit once every 500 000–1000 000 yr. Such impacts are at the limit of causing global consequences. The most recent impact was about 700 000

[8] S. G. Love and D. E. Brownlee, A direct measurement of the terrestial mass accretion rate of cosmic dust. *Science* **262** (1993), 550–553.

[9] J. S. Stuart and R. P. Binzel, Bias-corrected population, size distribution, and impact hazard for the near-Earth objects. *Icarus* **170** (2004), 295–311.

years ago, when a minor body hit southeast Asia, possibly on the border of Thailand and Laos, or in Cambodia, and splattered molten rock over an area that stretched from southern China in the North to Australia in the South, including the Philippines and large parts of the Indian ocean. Tektites can now be found where river gravel from those times is exposed at the surface. The devastation may have changed the local ecosystem and helped the human migration into the Indonesian archipelago.

These giant impacts are the only natural mortal danger to mankind against which we can protect ourselves. The danger of dying in one of these impacts is comparable to other common hazards. For example, the odds of being killed in a vehicle accident is 1 in 100, while the odds of being murdered stand at about 1 in 300. Dying in a fire occurs to 1 in 800 people, while dying in a firearms accident occurs to 1 in 2500 in the United States.

In comparison, the odds of an asteroid or comet impact being your cause of death stands at 1 in 20 000, the same as an airplane crash. The nature of the danger, of course, is very different. The time between airplane crashes is short, but relatively few people are affected each time. Asteroid and comet impacts are much less frequent, but many more people die if they do. It is the biggest impacts that are the largest danger.[10]

32.2 Impact frequency

The first step towards protection, project *Space Guard*, is to learn the orbit of all potentially damaging objects approaching Earth now or in the near future. So far, about half of all objects larger than 1 km in diameter have been found. NASA hopes to have found 95% by 2008. This search has been a bonanza for the discovery of potential parent bodies of meteoroid streams, the results of which have been discussed in previous chapters.

Meteor showers can help to identify extinct comet nuclei among the population of asteroids and provide data on the main element composition of their parent bodies. There are now thought to be about 900 NEOs with a diameter $D > 1$ km ($H < 18$), with a differential size distribution $\alpha = 2.75 \pm 0.10$, a size distribution index $s = 1.58 \pm 0.03$, and a corresponding magnitude distribution index $\chi = 1.71 \pm 0.06$, if each object hit Earth at the same speed and produce light at the same luminous efficiency. Most are asteroids, some are Jupiter-family comets, and a few are Halley-type and Oort-cloud comets. Nurmi *et al.* gave the following relative contributions from various sources: 54% asteroids and 46% comets including 26% Jupiter-family comets, 11% Halley-type comets, and 9% long-period comets. Others have estimated the contribution from comets to be in the range 20–50%.[11]

[10] D. Morrison, The Spaceguard survey. In *Report of the NASA International Near-Earth-Object Detection Workshop*, David Morrison, chair. (Pasadena, CA: JPL, 1992); J. S. Stuart and R. P. Binzel, Bias-corrected population, size distribution, and impact hazard for the near-Earth objects. *Icarus* **170** (2004), 295–311.

[11] H. Rickman, J. Fernández, A. Tancredi and G. Licandro, The cometary contribution to planetary impact rates. In *Collisional Processes in the Solar System*, ed. M. Y. Marov and H. Rickman, *Astrophysics and Space Science Library*, vol. 261 (Dordrecht: Kluwer, 2001) pp. 131–142.

The impact crater record on the Moon and other satellites in the solar system shows that both comets and asteroids have size distributions determined by a collisional cascade, whereby an equal amount of cross-sectional area is contained in each bin of magnitude. Asteroids have a crater population size index $\alpha \sim 2.84$, or $N(>D) \sim D^{-1.84}$, while comet impacts on the Galilean satellites (where few asteroids are believed to hit) imply $\alpha \sim 3.2$.

Meteor showers are particularly good at finding the path of long-period comets such as Hale–Bopp. They can be very big and approach Earth in a retrograde orbit, and are therefore the suspected cause of the biggest impacts.

Long-period comets approach Earth very infrequently and tend to sneak up on us. Meteor showers can betray the presence of the intermediate long-period comets, that return every 200–10 000 yr, and tell us much about them, before they are a danger. They have 1 revolution dust trails that are dense enough to cause intense showers at Earth (Chapter 13). These dust trails are detected a century before the comet itself returns, but they wander into Earth's path only once or twice every 60 years.

The impact rate of objects larger than 1 km may have varied from as low as 0 to as high as 700 impacts over a given million year period.[12] Fig. 32.2 is only a global average. A strong perturbation of the Oort cloud can create a shower of comets, raising the impact rate on the planets for periods of 2–3 million years. Their aphelion directions should cluster in a region of the sky. In reality, the only clustering found (except for the fragments of Sungrazers) is along the galactic plane, suggesting most comets were perturbed by the tidal forces of the galaxy and no current comets are from a recent star encounter.[13]

Over the course of our history that was often not the case. The Solar System is not isolated from our environment of stars and molecular clouds. Over the past 4.55 billion years since formation, there have been many close encounters with stars and dense molecular clouds, and also the tidal pull of the galaxy itself has varied over time. During that time, the Sun has completed about 19 orbits around the galactic center at a not always safe distance between 8.4 and 9.7 kpc (1 pc = 206 265 AU) from the galactic center (being now near to and moving towards perigalacton). At the same time, the Sun has moved in a periodic motion in and out of the galactic plane with an amplitude of ~49–93 pc, having passed about 150 times through the plane, the last time about 2–3 million years ago in a northward direction. That created tidal forces, perturbing the outer Oort cloud comets in new orbits. In addition, the Sun has felt the gravitational pull of about 3–11 dense molecular clouds. 55 000 star systems passed within 1 pc from the Sun, the nearest approaching to perhaps only 900 AU (based on

[12] P. Nurmi, M. J. Valtonen and J. Q. Zheng, Periodic variation of Oort Cloud flux and cometary impacts on the Earth and Jupiter. *Mon. Not. R. Astron. Soc.* **327** (2001), 1367–1376.

[13] J. A. Fernández, Dynamics of comets: recent developments and new challenges. In *Asteroids, Comets, Meteors 1993*, ed. A. Milani *et al.*, (Dordrecht: Kluwer, 1994) pp. 223–240.

fragments from exploded satellites, and even needles that were once released in an ill-conceived science experiment.

At the altitude of the *International Space Station* where atmospheric drag is a powerful removal mechanism of small orbital debris, meteoroids dominate impacts at sizes below 0.1 mm, depending on the solar cycle. Natural meteoroids dominate the ~0.1 mm size regime at 800 km cruising altitude in *low Earth's orbit* (LEO). Below 0.01 mm and above 1 mm sizes, the orbital debris impacts are dominant. In *geo-stationary orbit* (GEO), the man-made debris is less and the Earth's gravity can significantly concentrate meteoroids by gravitational focussing, increasing the rate of slow sporadic meteoroids 4–40 fold.[24]

The satellite impact hazard was of particular concern during the recent Leonid storms. Not only because of the thousand fold increase in the meteoroid influx for a short period of time, but also because of the high impact speed (72 km/s), resulting in a relatively large plume of charged particles upon impact that can set up dangerous voltages and cause electrical failure.

Back in 1966, the satellite park was not extensive enough to be harmed by the Leonid storm that year. During the rich encounter with the Leonid Filament in 1965, the Pegasus II and III satellites had 200 m^2 of detector area in orbit to measure dust impacts in the 10–100 μm size range. In the 24 h following midnight November 17, four particles hit the spacecraft, a rate of two standard deviations above the mean influx for that time of year. *Martin Beech et al.* examined this case again and concluded that at least one of the particles could have been a Leonid.[25] They calculated that the whole park would have had three impacts of meteoroids $>10^{-7}$ g in 1965 and 26 in 1966. Both values are probably too high, because the mass distribution index was less than $s = 2.0$ in 1965, and the peak flux was less than ZHR = 150 000 during the 1966 storm. With $s = 1.6$, there would only have been ~0.1 impact in 1965, and with ZHR = 15 000 at the 1966 storm peak, there would only have been three impacts for the whole park in 1966. None caused mission-ending failure.

When ESA's Olympus satellite was hit during the 1993 Perseid outburst in geostationary orbit, causing a shortcut and failure of one of the gyros, and the space-craft was subsequently decommissioned from fear of becoming unmanageable,[26] concern rose again for the upcoming Leonid storms.[27] Then, on March 13, 1994, the tether of the SEDS-2 (Small Expendable Deployer System) satellite suffered a

[24] J. Jones, Gravitational focussing of meteoroids. *Presented at Meteoroids 2004 Conf.*, University of Western Ontario, London, Canada. *Earth, Moon* and *Planets* (2006) (in press).

[25] M. Beech, R. Jehn, P. Brown and J. Jones, Satellite impact probabilities: annual showers and the 1965 and 1966 Leonid storms. *Acta Astronautica* **44** (1999), 281–292.

[26] R. D. Caswell, N. McBride and A. D. Taylor, Olympus end of life anomaly – a Perseid meteoroid impact event? *Int. J. Impact Eng.* **17** (1995), 139–150.

[27] M. Beech and P. Brown, Space platform impact probabilities – the threat from the Leonds. *ESA J.* **18** (1994), 63–73; M. Beech, P. Brown and J. Jones, The potential danger to space platforms from meteor storm activity. *Q. J. R. Astron. Soc.* **36** (1995), 127–152.

cut 3.7 d after deployment, presumably due to a meteoroid impact, and the payload re-entered within hours.[28]

Since 1966, the total surface area of all satellites has dramatically increased from a low 4846 m^2 due to 1686 objects (with the Echo I and II satellites counting for 76% of the total surface area in LEO) to about 160 000 m^2 by some 8000 objects in 1998 (of which perhaps 6% are functioning satellites). Satellite insurers and investment companies were concerned. Shuttle launches for human space flight were postponed. Numerous satellite instruments were shut down during the storm. The Hubble Space Telescope was pointed away from the shower radiant, and many solar panels were re-oriented by a small angle so that they were positioned with the smallest cross section towards the stream. In part perhaps thanks to these mitigation efforts, no impact damage was recorded. The lack of failure relaxed much of the concern.

The 2001 Leonid storm was observed from the now manned International Space Station, which is shielded with *Whipple shields* (multiple layers of thin foil) that shatter and stop any impacting meteoroid for objects <2 cm in size. *Fred Whipple* had called this the "meteor bumper."[29] By the year 2002, it was again difficult to get information about possible meteoroid impacts because of insurance concerns, but the danger had not gone away. Operators of NASA's large Chandra X-ray Observatory satellite, had fewer qualms about reporting such minor calamities and presented data showing it was hit close to the peak of the Leonid storm by a meteoroid with sufficient momentum to torque the spacecraft and change its attitude.[30]

We now know that what mattered was the position of the youngest dust trails at the time of encounter. The dust trails are so narrow that the impact rate can vary dramatically even between Earth and geostationary orbit, and certainly at the Moon or the Earth's *L1 Lagrange point* (where the gravitational and centrifugal forces of Sun and Earth cancel each other out). Earth narrowly missed the young trails in 1998 when the danger was highest (Fig. 32.3). Only WIND received some of the brunt of the 1866 dust trail at around 01:37 UT, November 18. It was just back from the L1 point and was put into a petal orbit on November 17, 1998, sitting close to the ecliptic plane at the time. Operations of some instruments appear to have been halted for impact hazard mitigation from about 14 to 24 UT, November 17, but many were back in operation at the time of the dust trail crossing! Being a satellite with a relatively small surface area, it survived the encounter unharmed.

Interplanetary satellites can encounter dust trails that do not cause meteor storms on Earth. *Bill Cooke* at the Marshall Space Flight Center has pointed out that Mariner 4 crossed a stream of meteoroids for about 45 min after completing a flyby of Mars on September 15, 1967, and again on December 10 and 11. Of all of the NASA Mars spacecraft, Mariner 4 was the only one that carried a micrometeoroid dust

[28] S. T. Lai, E. Murad and W. J. McNeil, Spacecraft interactions with hypervelocity particulate environment. *38th Aerospace Sciences Meeting & Exhibit*, 10–13 January 2000, Reno, NV, *AIAA-2000-0103* (2000) 10pp.

[29] F. L. Whipple, Meteorites and space travel. *Astron. J.* **52** (1947), 131.

[30] Chandra Science Team, COSPAR 2004 presentation (2004).

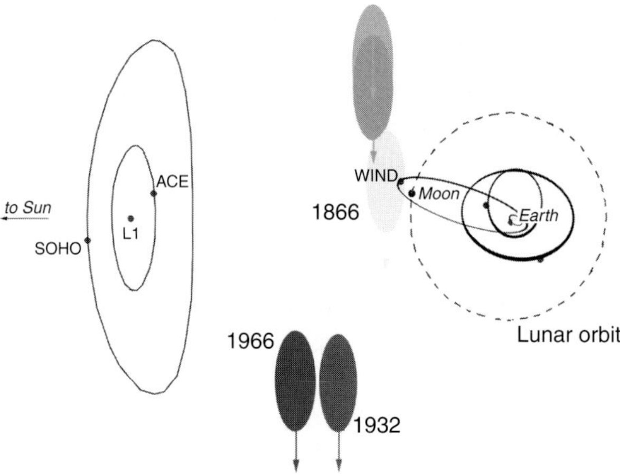

Fig. 32.3 Spacecraft positions at 01:37 UT, November 18, 1998, during the Leonid meteor shower.

detector. The impacts on September 15 ripped away bits of insulation and temporarily changed the orientation of the spacecraft in space.[31] During the encounter, the impact rate increased to 10 000 times above the normal rate of impacts in prior months, having 17 hits within 15 min, while at a distance of 1.273 AU from the Sun. There were a total of 5000 hits over the entire surface. The hits caused a torque about the roll axis. The event was dramatized in the movie "Red Mars."

Engineering meteoroid models

In the normal operation of satellites, with different faces turned to different directions over the cause of their life span, the natural meteoroid impact rate can be calculated from an analytical description known as the "Grün model," after the German astronomer *Eberhard Grün*. This model is the present standard reference model.[32] It is a magic recipe with many constants that gives the cumulative flux values $F(>M)$ averaged over a year and as a number of particles larger than a given mass (M, in gram) that impact a flat square meter surface area at 1 AU perpendicular to the ecliptic plane and spinning with the spin axis perpendicular to the ecliptic plane (no gravitational focussing included) during 1 s:

$$
\begin{aligned}
F(>M) = {} & 3.156 \times 10^7 (2200\, M^{0.306} + 15)^{-4.38} \\
& + 1.3 \times 10^{-9} (M + 10^{11} M^2 + 10^{27} M^4)^{-0.36} \\
& + 1.3 \times 10^{-16} (M + 10^6 * M^2)^{-0.85}
\end{aligned} \tag{32.1}
$$

[31] Original source: Anonymous, Mariner-Venus 1967: Final Report. *NASA Report SP* 190 (1968); Also: *Aviation Week and Space Technology* **88** (1968), 22; *Astronautics Aeronautics* **5** (1967), 270–271.

[32] E. Grün, H. A. Zook, H. Fechtig and R. H. Giese, Collisional balance of the meteoritic complex. *Icarus* **62** (1985), 244–272.

In this Grün model, the meteoroid environment is assumed to be isotropic, with no meteoroid streams. A three-dimensional extension of the Grün model, valid for the whole solar system, was devised by *Neil Divine*, known as the Divine model.[33] An analytical approximation of the, computing intensive, Divine model is given by *Rüdiger Jehn*.[34]

Meteor showers are important for meteoroids larger than 200 μm, when a satellite remains oriented in space or when a very strong shower hits at a discrete moment in time, such as during a meteor storm. *Neil McBride* of the Open University; Milton Keynes, UK, used all the activity measurements by the DMS and WAMS meteor observers, compiled and calculated by the author in preparation for meteor outburst research,[35] and translated the ZHR values into meteoroid influxes for ESA (Fig. 32.4).[36] The result of this unexpected forward use of the data by Neil has been called the *Jenniskens & McBride model*.[37] The model refers to the approach taken, with the sample of streams likely to evolve when the showers become better characterized over time.

For a given orientation of the satellite surface, the meteor showers cause large daily variations in the influx of meteoroids that are not considered in the Grün model. Fig. 32.4 shows results from the thesis work by *Torsten Bieler*.[38] These variations are highly directional, with very different effects on the solar panels of satellites positioned in different orientations.

Speed is important, in general terms, and this is valid for entry into Earth's atmosphere as well as for impacts on the satellites in orbit,[39] the penetration depth of meteoroids (as in the amount of matter along the trajectory needed to stop the particle) is proporational to $V^{0.81}$, the charge production of meteoroid impacts is proportional to $V^{3.48}$, while the plasma current created is proportional to $V^{4.48}$. Meteor showers have characteristic speeds that can be higher than the sporadic mean (Fig. 32.5).

The size distribution index is important as well. Most showers are richer in large meteoroids than the sporadic background. As a result, they are best recognized in the photographic and visual regime and mostly responsible for life-ending satellite impacts.

33 N. Divine, Five populations of interplanetary meteoroids. *J. Geophys. Res.* **98** (1993), 17 029–17 048; see also: R. Jehn, An analytical model to predict the particle flux on spacecraft in the solar system. *Planet. Space Sci.* **48** (2000), 1429–1435.

34 R. Jehn, An analytical model to predict the particle flux on spacecraft in the solar system. *Planetary Space Sci.*, **48** (2000), 1429–1435.

35 P. Jenniskens, Meteor stream activity, I. The annual streams. *Astron. Astrophys.* **287** (1994), 990–1013.

36 N. McBride, The importance of the annual meteoroid streams to spacecraft and their detectors. *Adv. Space Res.* **20** (1997), 1513–1516; T. McDonnell, N. McBride, S. F. Green *et al.*, Near Earth environment. In *Interplanetary Dust*, ed. E. Grün, B.Å.S. Gustafson, S. F. Dermott and H. Fechtig. (Berlin: Springer, 2001), pp. 163–231.

37 C. Lemcke, G. Scheifele and M.-C. Maag, *ESABASE/DEBRIS Release 2, Technical Description*, European Space Agency (1998); C. Lemcke, Particle fluxes on orbiting structures. *ESA SP* **393** (1997), 64.

38 T. Bieler, Modeling of impacts from meteoroid streams. Ph.D. thesis, Fachhochschule Aachen, Aachen, Germany (2000).

39 J. A. M. McDonnell, N. McBride and D. J. Gardner, The Leonid meteor stream: spacecraft interactions and effects. *ESA SP* **393** (1997), 391–396.

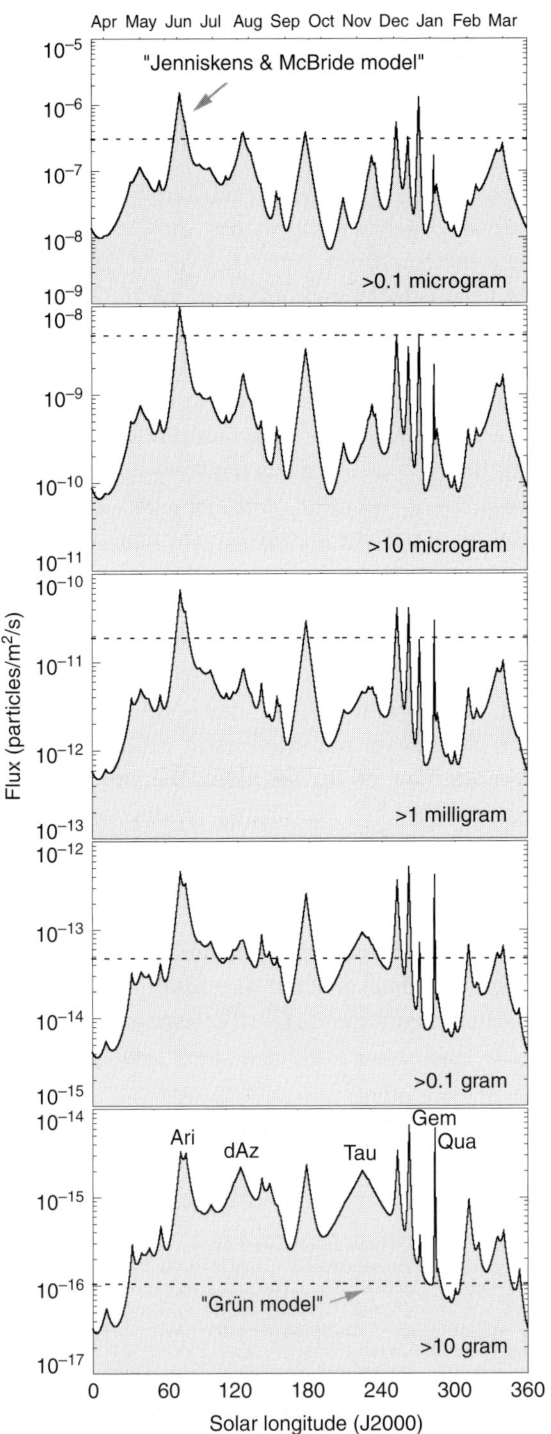

Fig. 32.4 The meteoroid influx rate for particles of different sizes in the Jenniskens–McBride model. Image from the Ph.D. thesis work by Torsten Bieler.

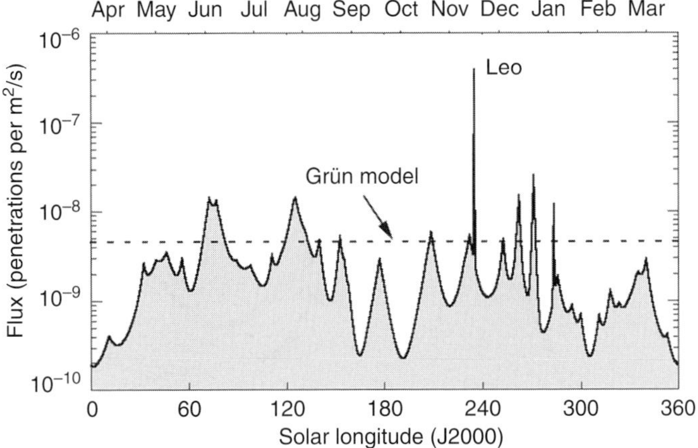

Fig. 32.5 The impact penetration flux of meteoroids for a 0.1 cm thick target in the presence of a Leonid storm of ZHR = 1000/h (3 h wide) that hits the target at an impact angle of 45°. Graph by Neil McBride.

The fast meteoroids are the most dangerous for creating unwanted discharges in unshielded electronics on satellites.[40] Fig. 32.6 shows the distribution of impact speeds of natural meteoroids.

32.5 Interstellar meteoroid streams

Among the smallest meteoroids hitting Earth's atmosphere are those from interstellar space. The spacecraft *ULYSSES* discovered what Edington could not believe, a continuous stream of very small meteoroids (~0.1 μm in size) that stream into the Solar System with a radiant in Hercules. A recent supernova in the nearby Scorpius–Centaurus association of massive stars created a bubble in the interstellar medium 120 pc across, at the edge of which the Sun is located. Within that bubble is a small warm interstellar cloud 2–5 pc across, known as the *local interstellar cloud*, which appears to be a fragment of the expanding shell of gas from the supernova explosion, mostly swept-up interstellar matter in the path of the blast wave. The Sun is moving through that cloud in the direction of Hercules towards R.A. = 270°, Decl. = +30°. The warm neutral hydrogen gas flows in from a direction R.A. = 254°, Decl. = +07°, and so are the dust particles embedded in the gas.

At Earth, *interstellar meteoroids* arrive with a heliocentric speed larger than the 41.9–42.6 km/s that would be caused by the Sun's gravitational pull alone. With the Earth's velocity and gravitational pull added in, apparent speeds are anywhere in the range 21–78 km/s. Only 1% of all detected >40 μm meteoroids by the AMOR

[40] J. A. M. McDonnell, N. McBride and D. J. Gardner, The Leonid meteoroid stream: spacecraft interactions and effects. *ESA SP* **393** (1997), 391–396.

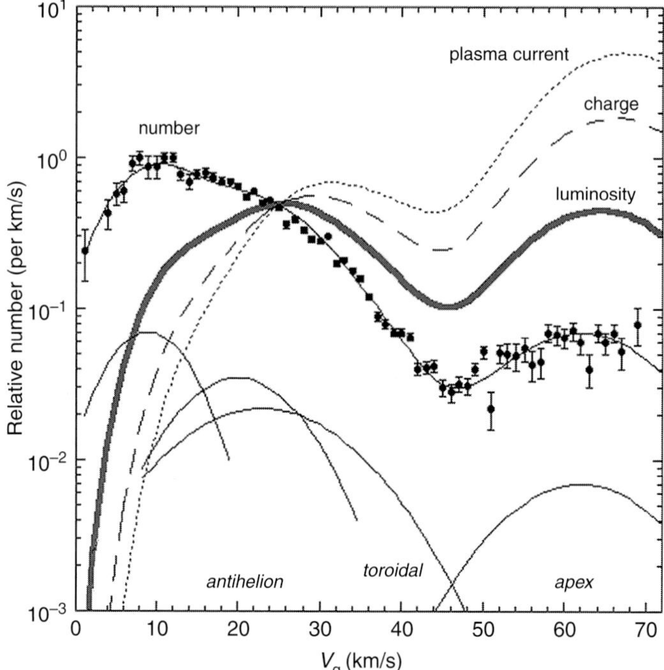

Fig. 32.6 The observed velocity distribution of sporadic meteoroids (~200 μm) at Earth according to the reanalyzed Harvard Meteor Project data by Andrew Taylor.[41] The two bumps are from prograde and retrograde moving material. Shown are the relative amount of luminosity, charge, and current produced during impacts of those meteoroids.

radar in Christchurch, New Zealand, appear to be of interstellar origin, and that fraction rapidly decreases with increasing mass. There are no photographed meteors that are with certainty interstellar.

On the southern hemisphere, interstellar meteoroids appear to arrive from a discrete source, possibly from the nearby young star with dust disk *Beta-Pictoris*, according to New Zealand meteor astronomer *Jack Baggaley* and coworkers.[42] If those grains originate from beta-Pictoris, this would imply a huge mass of dust being ejected at speeds of order 20 km/s. It is possible, however, that Earth experiences a brief episode of ejection along the plane of the dust disk some time ago. The event of those interstellar particles hitting Earth would also qualify as a meteor shower.

[41] A. D. Taylor, The Harvard Radio Meteor Project meteor velocity distribution reappraised. *Icarus* **116** (1995), 154–158. A. D. Taylor, N. McBride, A radiant-resolved meteoroid model. *ESA SP* **393** (1997), 375–380.

[42] A. D. Taylor, W. J. Baggaley and D. I. Steel, Discovery of interstellar dust entering the Earth's atmosphere. *Nature* **380** (1996), 323–325.

33
Meteor showers on other planets

Some day in the future, we will perhaps travel to Mars or Venus to watch a particularly nice meteor shower. Super fast meteoroids entering at 85 km/s, or meteors in slow-motion at 5 km/s. Until that time, remote observing and robotic missions may glimpse these meteors, or in the absence of an atmosphere, the impact flashes. They will probe the large-grain dust environment away from Earth's orbit and sample comets not encountered by Earth. And they probe the ablation process in atmospheres more like that of the early Earth. A call to include meteor surveys on space landers has been made.[1]

Meteoroid impacts may also be traced from the neutral atom debris layers (and their associated narrow ionospheric layers of electrons) that they leave behind high in the atmosphere. Theoretical models predict that such layers exist at Venus, Mars, Jupiter, Saturn, Neptune, and Saturn's moon Titan.[2] Even the sparse atmospheres of Triton and Pluto may on occasion be lit up by slow meteors.

33.1 Meteor showers on the Moon

Our Moon does not have a permanent atmosphere, but impact flashes have been observed. Meteoroids hit the surface continuously. They add meteoric metals to the Moon's surface and are responsible for the release of the sodium atoms that cause our Moon's sodium atmosphere.

In 1939, *Lincoln LaPaz*[3] estimated that 5 kg sized boulders hitting the Moon would produce visible flashes at a rate of 10 a year. Unconfirmed reports of impact flashes included that of *F. H. Thornton*,[4] who detected a point of light on the floor of Plato in October of 1945. In August of 1948, *A. J. Woodward*[5] saw a flash "like a bright sparkle of frost" that lasted three seconds. He described it as having "the appearance of an

[1] R. D. McGown, B. E. Walden, T. L. Billings *et al.*, Mars Meteor Survey. Concepts and approaches for Mars exploration. In *Workshop on Concepts and Approaches for Mars Exploration*, July 18–20, 2000, (Houston, Tx: Lunar and Planetary Institute, 2000), pp. 217–218.

[2] W. D. Pesnell and J. M. Grebowsky, Meteoric ions in planetary ionospheres. *Adv. Space Res.* **27** (2001), 1807–1814.

[3] L. LaPaz, The atmosphere of the Moon and lunar meteorite erosion. *Pop. Astron.* **46** (1938), 277–282.

[4] F. H. Thornton, *J. Brit. Astron. Assoc.* **57** (1947), 143.

[5] A. J. Woodward, An unusual observation of the Moon. *J. R. Astron. Soc. Canada* **42** (1948),194.

object striking the moon." Another case was that of *H. P. Wilkins*,[6] who observed a 1 s point of light in the crater Gassendi in May of 1951. Coordinated attempts by the *Association of Lunar and Planetary Observers* (ALPO) to observe such impacts, a project coordinated by *Robert M. Adams* from 1955 to 1965, never succeeded in obtaining simultaneous, independent observations of "Lunar transient phenomena."

Attention was again focused on this issue when *John E. Westfall* proposed the use of video cameras for recording impact flashes at an ALPO meeting in June 1997[7] and visible impact flashes were predicted by *Martin Beech* and *Simona Nikolova* in a paper published in 1998.[8] A mass of order 1 kg is needed to produce a flash of magnitude +4.6 (La Paz) or +6.2 (Gehring *et al.*[9]). In 1999, the Moon was closer to the center of the 1899 dust trail than was Earth and several efforts were made to watch for impact flashes. The Leonid shower was known to contain such kilogram sized meteoroids, which had produced many spectacular fireballs the year before.

Two teams were successful in recording the anticipated flashes (Fig. 33.1). *David W. Dunham*[10] at Mount Airy, MD, coordinated the effort in the United States and teamed up with *George Varros*, an amateur astronomer, later to participate in the 2002 Leonid MAC mission. Other observers that used similar PC-23 C low-light level video cameras were *Rick Frankenberger*, and *David M. Palmer*. *Brian Cudnik* provided visual observations. Seven impact flashes were recorded on video and confirmed by other observers, with magnitudes in the range +4.9 to +6.2. The flashes decayed rapidly, with only a faint $\sim +8^{m}$ afterglow after 1/60 s (one video frame). On one occasion, the flash was seen for 1/6 s. The brightest flash, at 3:49:40 UT November 18 (Fig. 33.1), was recorded 140 km SW of Rocca.

Luis R. Bellot Rubio and *José L. Ortiz*[11] worked with Spanish speaking amateur observers to set up a similar effort. While observing efforts in Europe were clouded out, observer *Pedro Valdes Sada* of the Universidad de Monterrey in Mexico was successful. By comparing the rate of impacts with those expected from the shower, it was found that 0.1% of kinetic energy was transferred into visible light.[12]

Each year, 83 boulders larger than 1 kg hit the face of the Moon, and about six larger than 5 kg. Good opportunities for viewing are summarized in Table 10c. The

[6] H. P. Wilkins, *Our Moon* (London: Frederick Muller, 1954), p. 128.

[7] J. E. Westfall, *Worthy of Resurrection: Two Past A.L.P.O. Lunar Projects*. Presented at June 1997 ALPO meeting (1997).

[8] M. Beech and N. Simona, Leonid flashers – meteoroid impacts on the Moon. *Nuovo Cimento, Note Brevi* **21C** (1998), 577–581.

[9] J. W. Gehring, A. C. Charters and R. L. Warnica, *Meteoroid Impact on the Lunar Surface* ed. J. W. Salisbury and P. F. Glaser, (New York: Academic, 1964), pp. 215–263.

[10] B. M. Cudnik, D. W. Dunham, D. M. Palmer *et al.*, Ground-based observations of high velocity impacts on the Moon's surface – the lunar Leonid phenomena of 1999 and 2001. *33rd Annual Lunar and Planetary Science Conf.*, March 11–15, 2002, (Houston, Tx: Lunar and Planetary Institute, 2002), abstract no. 1329.

[11] L. R. Bellot Rubio, J. L. Ortiz and P. V. Sada, Luminous efficiency in hypervelocity impacts from the 1999 lunar Leonids. *Astrophys. J.* **542** (2000), L65-L68; J. L. Ortiz, P. V. Sada, L. R. Bellot Rubio *et al.*, Optical detection of meteoroidal impacts on the Moon. *Nature* **405** (2000), 921–923.

[12] J. L. Ortiz, J. A. Quesada, J. Aceituno, F. J. Aceituno and L. R. Bellot Rubio, Observation and interpretation of Leonid impact flashes on the Moon in 2001. *Astrophys. J.* **576** (2002), 567–573; L. R. Bellot Rubio, J. L. Ortiz and P. V. Sada, Observation and interpretation of meteoroid impact flashes on the Moon. *Earth, Moon Planets* **82/83** (2002), 575–598.

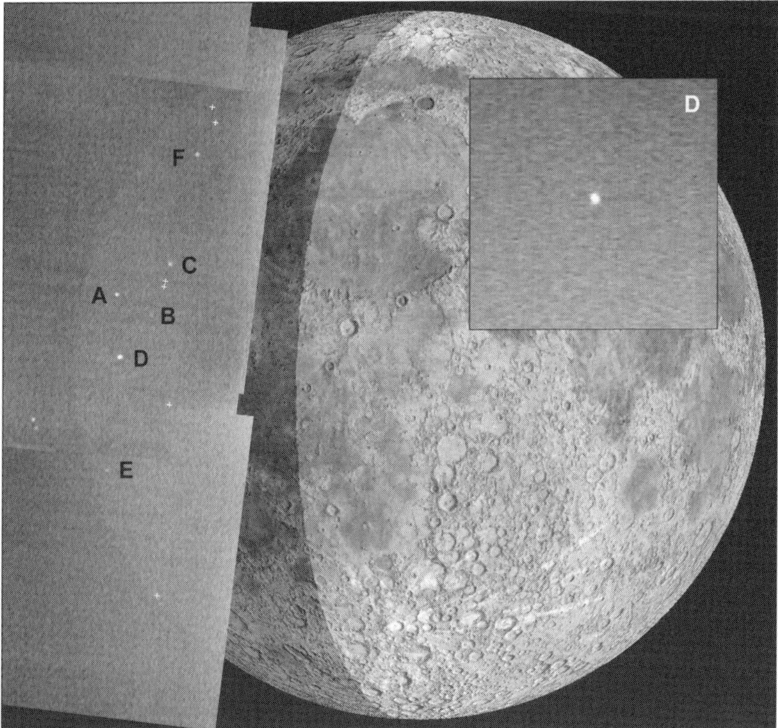

Fig. 33.1 Compilation of impact flashes detected during the November 18, 1999 Leonids, by David Dunham observing at Mount Airy, MD. Crosses mark faint flashes. Bright flash "D" at 03:39:40 UT was also recorded by David Palmer.

Moon is inside the Earth's gravity well, but meteoroids will only be accelerated by Earth to 1.40–1.49 km/s[13] (range because Moon can be at the apogee or the perigee of orbit) and 2.38 km/s from the Moon's gravity itself. This adds (quadratically) to the 41.8–42.5 km/s for meteoroids falling from afar (Earth at aphelion or perihelion of its slightly elliptical orbit). With a speed of the Earth–Moon system around the sun of 29.3–30.3 km/s and that of the Moon around Earth of 0.97–1.08 km/s, and <0.01 km/s from the Moon spinning, this adds up to a peak impact speed (for Leonid-like meteoroids) of 73.9 km/s, and a low, minimum impact speed of 2.8 km/s.

Meteoroid impact gardening of sodium containing surface minerals is not necessarily the main source of the sodium release from the soil. The sodium atmosphere is larger at full Moon (a total lunar eclipse), suggesting that photon sputtering[14] or thermal desorption, rather than solar wind or micrometeorite sputtering, is responsible for the release

[13] $V_{esc} = \sqrt{(2GM_{\odot}/R)}$, where G is the gravitational constant, $M_{\odot}$ is the planet mass and R is the distance of the meteor layer from the center of the planet.

[14] B. V. Yakshinskiy and T. E. Madey, Photon-stimulated desorption as a substantial source of sodium in the lunar atmosphere. *Nature* **400** (1999), 642–644.

of the sodium.[15] The sodium atmosphere is gradually blown away by light and energetic particles from the Sun. On November 19, 1998, during the new Moon, two days after the peak of the Leonid meteor shower, a surprising discovery was made by Wilson and coworkers.[16] All-sky images through a sodium filter showed a diffuse glow in the sky opposite the position of the Moon. It was shown that these were sodium atoms streaming away from the Moon's atmosphere like a comet tail. When Earth travels through the stream, the atoms are focused towards the tail center by the Earth's gravity.

33.2 Meteoroid impacts on Mercury

Impact flashes also occur on Mercury. The dust density at Mercury's orbit is higher than that at Earth. Deeper in the Sun's gravity well, the meteoroids tend to impact with a higher speed. The planet is smaller than Earth, but relatively heavy, as a result of which the meteoroids are accelerated to 4.25 km/s before hitting the surface by Mercury's gravity alone. The impact speed increases to 61.7–76.0 km/s as a result of Mercury being deeper in the Sun's gravity well, and 38.9–59.0 km/s from the planet's orbit around the Sun. This adds up to a very high, peak impact speed of 135.0 km/s, nearly twice that of Leonids on Earth. Hence, impact flashes may be seen in future satellite missions.

Only four Jupiter-family or Halley-type comets are known to have a perihelion inside Mercury's orbit, with comet 2P/Encke passing by Mercury at only 0.026 AU (Table 10a). *Frank Selsis* has pointed out that the Taurid meteoroids are significantly dispersed and impacts are likely. Mercury will cross the stream on January 24, April 22, July 18, . . ., 2009, when the planet is at the position shown in Fig. 33.2.[17] If any of these impacts could be seen from Earth, viewing would be best if Mercury passed by the orbit of 2P/Encke in the Earth-months of December–January.

33.3 Meteor showers on Venus

More than 11 short-period comets (and many other minor bodies) approach the orbit of Venus to within 0.1 AU.[18] The most promising candidate for meteor outbursts is comet *45P/Honda–Mrkos–Pajdusáková* (Table 10a), passing the Venusian orbit recently at only 0.0016 AU.[19] The comet arrived in the vicinity of its present orbit

[15] M. Mendillo and J. Baumgardner, Constraints on the origin of the Moon's atmosphere from observations during a lunar eclipse. *Nature* **377** (1995), 404–406.

[16] J. K. Wilson, S. M. Smith, J. Baumgardner and M. Mendillo, Modeling an enhancement of the lunar sodium tail during the Leonid meteor shower of 1998. *GLR SPA* **26** (1999), 313–314; S. M. Smith, J. K. Wilson, J. Baumgardner and M. Mendillo, Discovery of the distant lunar sodium tail and its enhancement following the Leonid meteor shower of 1998. *GLR SPA* **26** (1999), 384–353.

[17] F. Selsis, J. Brillet and M. Rapaport, Meteor showers of cometary origin in the Solar System: revised predictions. *Astron. Astrophys.* **416** (2004), 783–789.

[18] M. Beech, Venus-intercepting meteoroid streams. *Mon. Not. R. Astron. Soc.* **294** (1998), 259–264; F. Selsis, J. Brillet and M. Rapaport, Meteor showers of cometary origin in the Solar System: revised predictions. *Astron. Astrophys.* **416** (2004), 783–789.

[19] A. A. Christou, Prospects for meteor shower activity in the Venusian atmopshere. *Icarus* **168** (2004), 23–33.

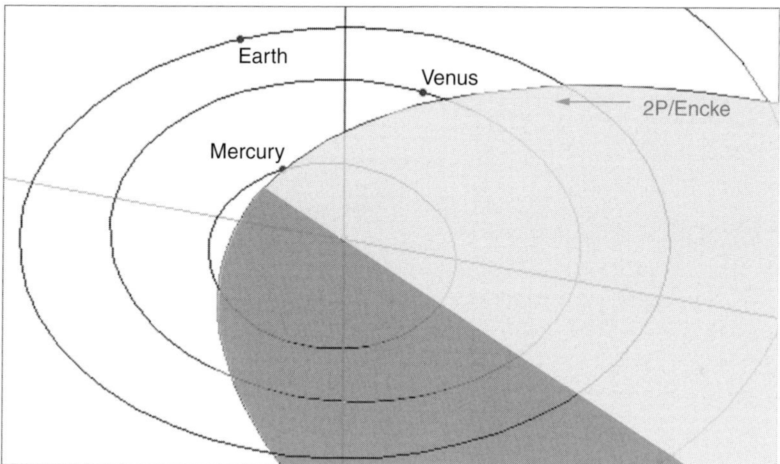

Fig. 33.2 Mercury during the closest approach of the orbit of 2P/Encke (on January 24, 2009).

only in 1983, after an encounter with Jupiter. It will encounter Venus itself on two instances in the coming 20 years, in June, 2006 (0.085 AU) and in January, 2017 (0.18 AU) respectively. In 2006, Venus will pass the critical longitude only five days after the comet, following the comet by 38 days on August 31, 2017.

Most striking about potential meteor activity at Venus is the coincidence in nodes of several of the most promising showers (Table 10a). There are four Halley-type comets, the biggest is 12P/Pons–Brooks, and one known stream (Geminids),[20] which are all encountered in the period from solar longitude 251° to 269° (Fig. 33.3). The Geminids are of particular interest, because observations on Venus could measure the distribution of the dust at a different true anomaly of the orbit (Fig. 33.4).[21]

In addition, sporadic rates are good too. The rate of large meteoroids impacting Venus is about four times that at Earth and those meteoroids will give rise to 1–2^{m} brighter meteors due to higher entry speeds (albeit that they will be slightly shorter in length).[22] Venus has 82% of the mass of the Earth and is closer to the Sun (0.72 AU), so that meteoroids approaching from afar are accelerated by 49.4–49.7 km/s in the Sun's gravitational well and by 10.3 km/s in the gravitational well of Venus. These numbers compare to 42.1 km/s and 11.1 km/s (at 100 km altitude), respectively, for Earth. The orbital speed is 34.8–35.3 km/s, adding to a peak impact speed of 85.6 km/s. This compares to a rare top speed of 74.1 km/s on Earth.

Venus is also exposed to the small β-meteoroids (<1 μm in size) that originate closer in towards the Sun and are blown outward by solar radiation pressure. These

[20] G. O. Ryabova, Mathematical model of the Geminid meteor stream formation. *ESA SP* **495** (2001), 77–81.

[21] G. O. Ryabova, The Geminid meteoroid stream: a new model. *Proc. Int. Meteor Conf.*, Bollmannsruh 2003. (Potsdam: International Meteor Organization, 2003), pp. 131–135.

[22] V. M. Kolmakov, The flux density of sporadic meteoroids on Venus, the Earth, and Mars. *Astronomicheskii Vestnik* **25** (1991), 217–224 (In Russian).

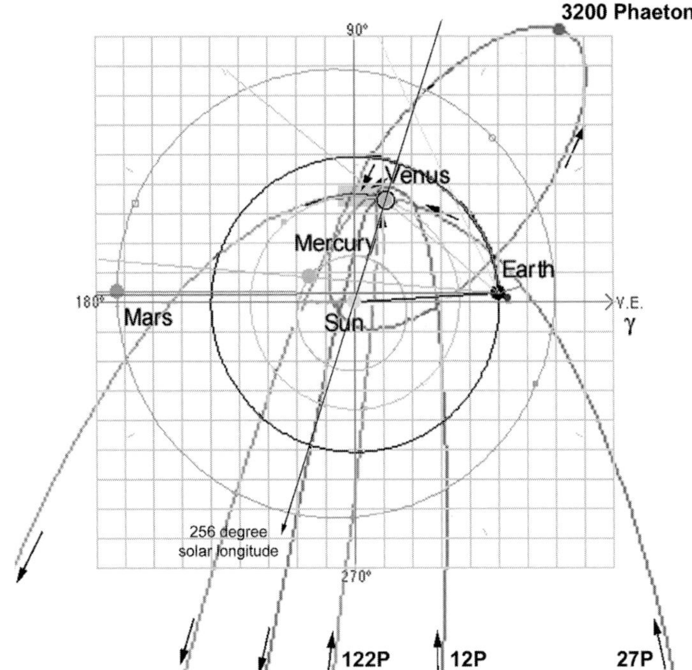

Fig. 33.3 A cluster of showers at solar longitude 251°–264°. The position of the terrestrial planets on September 25, 2004. "V.E." is vernal equinox.

β-meteoroids move in hyperbolic orbits and were measured by the Pioneer 10 and 11 spacecraft at Venus to impact at an amount of 10^{-14} g/cm^2/s.

Meteors will appear above the haze and clouds of sulfuric acid aerosols that obscure the surface of Venus. The meteors are expected to have their peak brightness at about 100–115 km altitude, where the atmospheric density is about 10^{-7} to 10^{-9} g/cm^3. At the surface, the atmosphere of Venus is about 90 times denser than at that of Earth, 96.5% of which is carbon dioxide and the remainder is mostly molecular nitrogen. The density falls off rapidly above the cloud deck. The mesopause is only 25 km higher than on Earth (Fig. 33.5).

Venus is close enough to Earth to see really bright meteors on the night-time hemisphere.[23] Best viewing opportunities are listed in Table 10b. None have yet been reported. The main hurdle is the large distance between Venus and Earth, even at conjunction, making meteors fainter by 28–30 magnitudes than when seen from a distance of 100 km. To see a -12^{m} (at 100 km) fireball, typically the brightest shower members, a telescope has to be able to detect brief $+16^{m}$ flashes in, say 1/30 s exposures, against the night-time sky of Venus. The most promising wavelength

[23] M. Beech and P. Brown, On the visibility of bright Venusian fireballs from Earth. *Earth, Moon Planets* **68** (1995), 171–179.

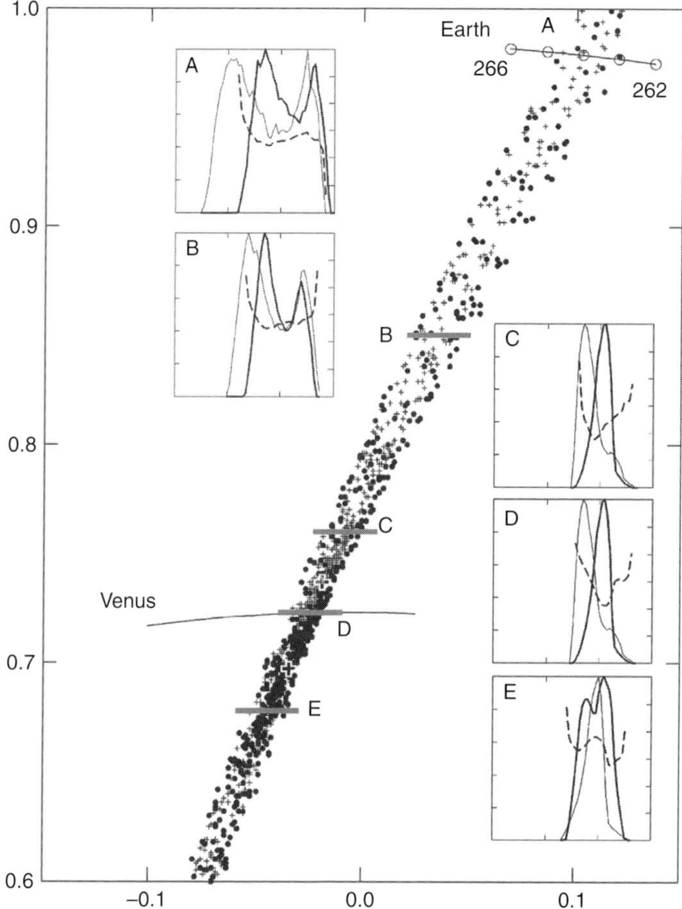

Fig. 33.4 A Geminid stream model for two masses 0.002 14 g (•, thin line) and 0.000 214 g (+), with expected activity and mass-distribution index profiles (- - -), by Galina O. Ryabova of Tomsk State University. Note that the highest dust density is at the orbit of Venus.

range would be a narrow band at 777 nm, where meteors may generate atomic oxygen emission from the dissociation of CO_2, and where Venus' surface features are not seen through the sulfuric acid haze.

The forbidden line at 577 nm may also be observed in the wake of the meteor. This line would be fainter, but persists for tens of seconds instead of $\sim 10^{-2}$ s. One atmosphere-grazing fireball track may have already been observed (some time after the meteor) by the Pioneer Venus Orbiter Ultraviolet Spectrometer instrument orbiting Venus in 1979. The satellite measured three correlated spectra during orbit 75 containing a strong NO (C-X) emission at 155–257 nm, which was confined in the Venus atmosphere in a narrow band at least 900 km in length and less than 5 km wide.[24]

[24] D. L. Huestis and T. G. Slanger, New perspectives on the Venus nightglow. *J. Geophys. Res.* **98** (1993), 10 839–10 847.

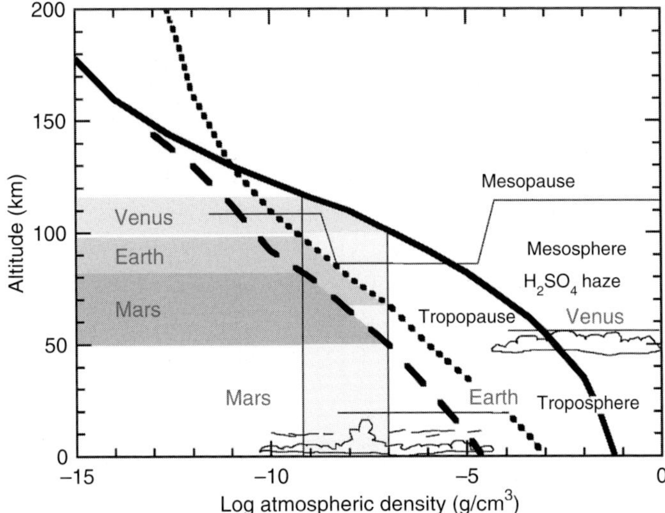

Fig. 33.5 Atmospheric density profiles of Venus (—), Earth (• • •), and Mars (- - -). Meteors have peak luminosity at the gray altitudes. Stratopause and mesopause are marked. Figure after Christou (2004).[25]

Venus is predicted to have up to a factor of 10 lower sodium content in the upper atmosphere than Earth.[26] No neutral atom debris layer has yet been observed. Some layers of electrons might be associated with meteoric metal ions.[27] However, other sources of ionization exist too. Venus does not have a strong magnetic field, allowing small charged particles to hit the atmosphere and ablate. Moreover, the solar wind penetrates deep into its upper atmosphere, as do cosmic ray particles.

33.4 Meteor showers on Mars

Why is Mars red? Since the Viking Landers, the Martian soil is understood to be a mixture of weathered basalt and weathered meteoric dust. The red color of Mars is at least partially due to oxidized meteoritic metal.[28] Initial efforts to explain the soil as aeolian dust from weathered volcanic glass (palagonite) and the alteration products of impact melt rocks failed to explain why iron atoms were found in disordered (glassy) sites in the dust. The *Mars Pathfinder* mission detected more magnesium and iron in

[25] A. A. Christou, Prospects for meteor shower activity in the Venusian atmopshere. *Icarus* **168** (2004), 23–33.

[26] V. A. Krasnopolskii, Threshold estimates of the content of some substances in the atmospheres of Mars and Venus from the results of spectrocopy of the twilight glow on the Mars 5, Venera 9, and Venera 10 satellites. *Cosmic Res.* **16** (1979), 713–718.

[27] M. Dubin and R. K. Soberman, Generation of Venus' nightside ionospheric structure by particle streams from cosmoids. *Adv. Space Res.* **15** (1995), 123–129.

[28] G. J. Flynn and D. S. Mckay, An assessment of the meteoritic contribution to the Martian soil. *J. Geophys. Res.* **95** (1990), 14 497–14 509; A. S. Yen, Meteoritic nickel at the surface of Mars? *American Astronomical Society, DPS meeting #35*, #09.01 (abstract) (2003).

the topsoil than in sampled rocks. The elemental compositions pointed towards olivines and pyroxenes being the dominant minerals, rather than clays, carbonates, and other low-temperature aqueous sediments that would be expected if liquid water had interacted extensively with the Mars surface at that location.

Meteoroids are an abundant source of meteoric metals in the atmosphere, which will quickly oxidize under UV light, while micrometeorites can bring metallic iron directly to the surface. With today's estimated meteoric influx rate, about 2 cm of exogenous matter (at 1 g/cm^3) should have settled on the surface of Mars every billion years. Unlike Earth, this cosmic dust is not buried by plate tectonics.

The Martian atmosphere is 95.3% carbon dioxide (CO_2), and the remainder is molecular nitrogen N_2 (2.7%), Argon (1.6%), and a trace of molecular and atomic oxygen (0.13%). This compares to Earth's 20% O_2 and the remainder N_2. The pressure at the Martian surface is 100 times less than that at Earth's (Fig. 33.5). On the other hand, small meteoroids are ablated at around the same altitude as on Earth, which is where they have met their own mass in air. At this altitude (80–140 km) there is a cold $T = 120$ K mesopause ($T = 150$ K at Earth) warmed from below by surface heat. The predicted distribution of meteoric metals in the upper atmosphere of Mars[29] is more spread out in altitude than in Earth's atmosphere and meteor tracks will be longer. *Lars Adolfsson* and colleagues simulated a meteor in the Martian atmosphere to find that high speed (>30 km/s) meteoroids will have the same magnitude as on Earth, but slow meteors are dimmer.[30] The meteoric metal layer has not been observed yet, but the layer is expected to be partially ionized, leading to the equivalent of sporadic-E layers.[31]

The meteor rate at Mars is 0.5 times higher than that on Earth, in terms of the influx of particles of a given size or mass (Fig. 31.6).[32] Fast meteoroids may arrive from comets 1P/Halley and 13P/Olbers, and perhaps from minor planet *5335 Damocles* (Table 10d).[33] About 80% more long-period comets cross the orbit of Mars than cross Earth's orbit.

A visual observer will see fewer meteors, because the meteoroids will tend to impact Mars at a lower entry speed, because of its smaller size and the fact that Mars, at 1.52 AU, is located less deeply in the gravitational well of the Sun. The meteoroids accelerate to 100 km altitude by only 5.0 km/s (compared to 11.1 km/s on Earth) from the gravity of Mars and up to 32.6–35.9 km/s from that of the Sun. The orbital speed

[29] G. J. Molina-Cuberos, O. Witasse, J.-P. Lebreton, R. Rodrigo, J. J. López-Moreno, Meteoric ions in the atmosphere of Mars. *Planet. Space Sci.* **51** (2003), 239–249; W.D. Pesnell and J. Grebowsky, Meteoric magnesium ions in the Martian atmosphere. *J. Geophys. Res.* **105** (2000), 1695–1708.

[30] L. G. Adolffson, B. Å. S. Gustafson and C. D. Murray, The Martian atmosphere as a meteoroid detector. *Icarus* **119** (1996), 144–152.

[31] W. D. Pesnell and J. M. Grebowsky, Meteoric ionization layers in the Martian atmosphere. *American Astronomical Society, 191st AAS Meeting*, #27.03 (abstract) (1997); *Bull. Am. Astron. Soc.* **29** (1997), 1254 (abstract).

[32] V. M. Kolmakov, The flux density of sporadic meteoroids on Venus, the Earth, and Mars. *Astronomicheskii Vestnik* **25** (1991), 217–224 (in Russian); L. G. Adolffson, B. Å. S. Gustafson and C. D. Murray, The Martian atmosphere as a meteoroid detector. *Icarus* **119** (1996), 144–152.

[33] A. Christou and K. Beurle, Meteoroid streams at Mars: possibilities and implications. *Planet. Space Sci.* **47** (1999), 1475–1485; A. H. Treiman and J. S. Treiman, Cometary dust streams at Mars: preliminary predictions from meteor streams at Earth and from periodic comets. *J. Geophys. Res.* (*Planets*) **105** (2000), 24 571–24 582.

of Mars oscillates between 22.0 and 26.5 km/s. This still adds up to a peak impact speed of 39.1 km/s in a head-on collision with a meteoroid in a long-period orbit. Showers and sporadic meteors on Mars have a similar distribution in origins to those on Earth and produce a similar velocity distribution. Prograde moving Jupiter-family comets would hit at speeds of order 10–30 km/s, while retrograde moving comets would hit at about 30–62 km/s.[34] With lower impact speed, large meteoroids survive in greater numbers. The first meteorite at the surface was detected by the Mars Exploration Rover *Opportunity*.

33.5 Meteor showers on Jupiter

The conditions for meteoroid streams near the giant planets are very different than those near the terrestrial planets. Each encounter will perturb the meteoroids into significantly different orbits, but they will return into the vicinity of the giant planet's orbit in later returns. Many Jupiter-family comets tend to have their aphelion near Jupiter, where meteoroids move slowly and dust densities are high (Table 10e).

On Jupiter (as well as Saturn) the meteors are expected to appear of order 100 km above the 1 bar layer. That is significantly above the haze of high altitude ammonia (NH_3) crystals and methane (CH_4) clouds, and the darker NH_3SH and sulfur clouds that drift about 25 km lower in the predominantly hydrogen (76%) and helium (24%) atmosphere. From Earth, we miss the advantage of being able to observe a large part of the night-time side of Jupiter. The global influx of dust on Jupiter is about 20 000 t/yr, and the peak of the mass ablation occurs at an altitude of about 350 km above the 1 bar level.[35]

The meteoric metal layer has been observed as an increase of the electron density in Jupiter's atmosphere, as derived from the attenuation of the *Voyager 2* radio transmission during an occultation exit. What appears to be an E-layer with an electron density of 10 000 cm^{-3} was detected at a height of 350–450 km above the 1 bar level.[36] Most ionization is thought to be carried by meteoric Mg^+ and Fe^+ and ions, which are present at a density of about $2\text{–}4 \times 10^4\,cm^{-3}$.

Meteors from meter-sized bodies can also be observed,[37] although none have been reported so far. It is interesting to note, however, that a fireball probably caused by a decimeter-sized meteoroid was observed in Jupiter's atmosphere by the Voyager 1 spacecraft during its March, 1979 encounter with the planet.[38] Jupiter's large mass will accelerate the meteors more than Earth's does (up to 60.2 km/s down to the meteor layer), so that even small meteoroids cause quite bright meteors. Falling from afar into

[34] Y. Ma, I. P. Williams and W. Chen, The velocity distribution of periodic comets and the meteor shower on Mars. *Astron. Astrophys.* **394** (2002), 311–316.
[35] Y. H. Kim, D. W. Pesnell, J. M. Grebowsky and J. L. Fox, Meteoric ions in the ionosphere of Jupiter. *Icarus* **150** (2001), 261–278.
[36] G. F. Lindall and V. R. Eshleman, The atmosphere of Jupiter: an analysis of Voyager radio occultation measurements. *Bull. Am. Astron. Soc.* **12** (1980), 683–684 (abstract).
[37] I. Nemtchinov and I. Kosarev, Meteor flashes in the atmospheres of giant gaseous planets. *ESA SP* **500** (2002), 293–295.
[38] A. F. Cook and T. C. Duxbury, A fireball in Jupiter's atmosphere. *J. Geophys. Res.* **86** (1981), 8815–8817.

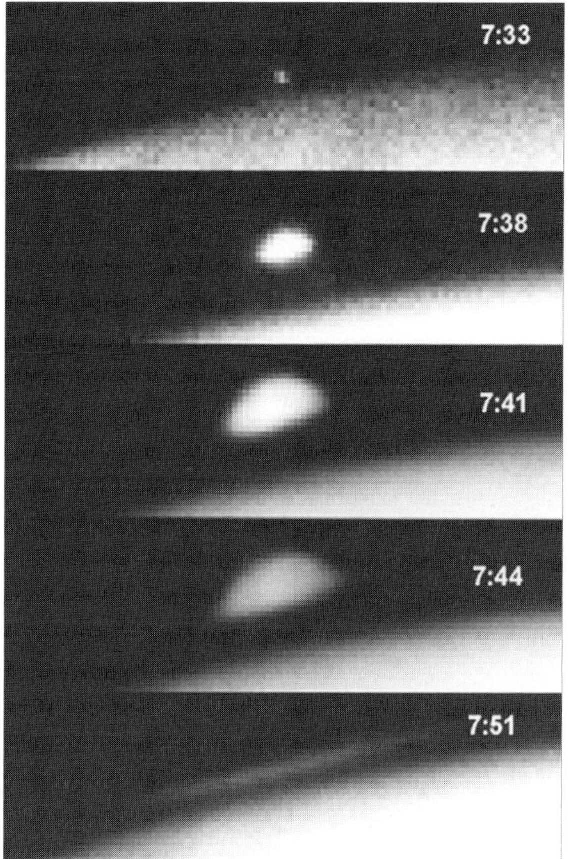

Fig. 33.6 Impact plume of fragment G, impacting Jupiter at about 07:32:00 UT July 18, 1994. The images show the evolution of the plume over the next minutes. Photo: NASA/HST.

the Sun's gravitational well, this adds between 18.0–18.9 km/s, Jupiter's motion around the Sun 12.4–13.7 km/s and its spin add another 13.7–18.9 km/s, for a peak impact speed of 79.2 km/s, with a lowest impact speed of 60.2 km/s, providing a very narrow range of impact speeds.

About 7.5 times more long-period comets cross the orbit of Jupiter than do so at Earth.[39] Even at these distances, they are expected to lose dust and can create meteoroid streams, especially when they first arrive inside Jupiter's orbit. In addition, Jupiter's large mass exerts the strongest tidal force on a comet, the difference in gravity force between the front and back side of the comet. This can cause a breakup, such as in comet *Shoemaker–Levy 9*, which was captured by Jupiter in an orbit around the planet. All the fragments hit the planet in a series of dramatic explosions (Fig. 33.6).

[39] D. Olsson-Steel, Collisions in the solar system – IV. Cometary impacts upon the planets. *Mon. Not. R. Astron. Soc.* **227** (1987), 501–524.

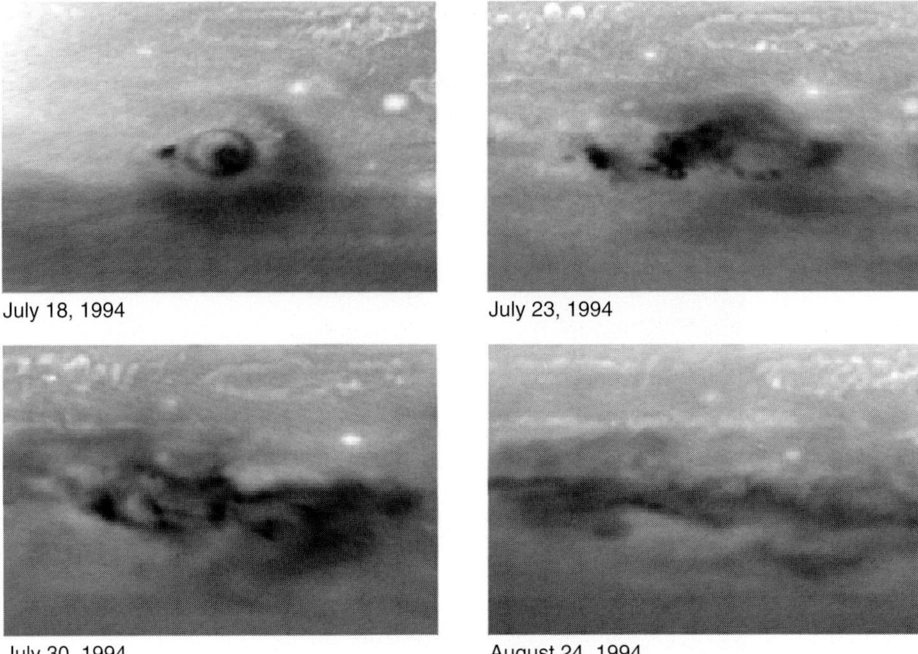

Fig. 33.7 Impact debris field in Jupiter's atmosphere surrounding the impact sites of fragments D (small spot), 17 July 1994, at 10:45 UT and G (large region, 18 July, 1994 at 7:28 UT) of D/Shoemaker–Levy 9. The central dark spot of impact G measures 1550 miles (2500 km) in diameter. The outermost dark ring has an inner edge about the size of Earth. Photo: NASA/Hubble Space Telescope.

The impact rate on Jupiter by 1.5 km diameter comets is currently $N(D > 1.5\,\text{km}) = 0.005/\text{yr}$.[40]

One established way to watch for such impacts is to look for the dark spots of meteoric dust spread around the impact site from the impact plume (Fig. 33.7). Those spots persist for a while before being dispersed by the upper atmosphere winds after several days.

33.6 Meteors on Jupiter's Moons

Jupiter's moons Europa, Ganymede, and Callisto are covered by impact craters that can be used to measure the frequency of impacts from comets of different sizes. These data show that even at Jupiter there is a marked lack of small comets <1 km in size. Impact flashes on Europa are particularly interesting, because of the water-rich surface ice layer, with relatively low binding energy. As a result, this could perhaps lead to

[40] K. Zahnle, P. Schenk, S. Sobieszczyk, L. Dones and H. Levison, Differential cratering of synchronously rotating satellites by ecliptic comets. *Icarus* **153** (2001), 111–129; K. Zahnle, P. Schenk, H. Levison and L. Dones, Cratering rates in the outer Solar System. *Icarus* **163** (2003), 263–289.

particularly large vapor plumes that may be observed in satellite images. The reflected light from the surface of Europa has absorption features from meteoric metal atoms (Fe, Mg).[41] A very exotic place to look for meteor showers is the sulfur dioxide (SO_2) atmosphere of Io. The atmosphere is thicker above the volcanic plumes. Above the plume of *Loki*, the density can increase sufficiently for a mesopause to develop at an altitude of ~30 km.

33.7 The outer planets

Conditions on Saturn (Table 10f) are somewhat similar to Jupiter, with meteors expected to occur above the main cloud layer. Saturn's atmosphere is also predominantly hydrogen (94%) and helium (6%). Saturn has a lower escape speed (36.1 km/s), a lower orbital speed (9.1–10.2 km/s) and a lower speed for long-period meteoroids (13.3–14.0 km/s), giving a total impact speed in the range 36–49 km/s. In addition, it has a lower local dust density. The atmospheric wind speeds are very high, up to 500 m/s near the equator, which will disperse any train rapidly.

More exciting are Saturn's moons and ring system, prominent dark spokes in Saturn's broad B Ring are filamentary markings about 12 000 km long, which rotate around the planet with the motion of the particles in the rings. It is believed that these features are dust levitated above the ring plane by electric fields. Dust impacts may play a role in creating the dust.

In Titan's relatively dense atmosphere, made up of mostly nitrogen (98%) and hydrocarbons such as methane ($1.6 \pm 0.2\%$), meteors would appear in the layer of haze at about 500 km altitude. The mesopause is at ~615 km. Though perhaps difficult to observe, they could provide important clues about the evolution of prebiotic organic compounds on the early Earth.

Observing meteors on Uranus and Neptune brings progressively lower benefit from the gravity of the planet and the local dust environment. At Uranus, the escape velocity is about 21.3 km/s and at Neptune it is 23.5 km/s. Meteoroids falling from long-period orbits will only add 9.4–9.9 and 7.6–7.7 km/s, respectively, adding to a contribution from the orbital speed of 6.5–7.1 and 5.4–5.5 km/s, respectively, and ~2.6 km/s from spin, for peak impact speeds of 28.9 and 28.4 km/s. Their atmospheres are 89% hydrogen, 11% helium, and traces of methane.

Triton and Pluto are considered to be part of the family of Kuiper Belt objects, Triton is now captured in an orbit around Neptune while Pluto is in a heliocentric orbit that is in resonance with Neptune. Both have an atmosphere rich in nitrogen and methane that can freeze out on the dark side of the body. Although the Tritonian atmosphere is very thin, with an atmospheric pressure 10^{-5} of that of the Earth at the surface, it is dense enough to make meteors visible, albeit at very low altitudes.[42] The

[41] L. M. Shulman, Meteor phenomena on the outer planets. *ESA SP* **500** (2002), 289–292.
[42] D. W. Pesnell, J. M. Grebowsky and A. L. Weissman, Watching meteors on Triton. *Icarus* **169** (2004), 472–481.

crater population of Triton shows more frequent impacts from small comets, presumably from the Kuiper Belt.[43] Small comets are five times more frequent at Triton than at Europa.[44]

Pluto's gravity would accelerate the meteoroids to about 1.3 km/s. Meteoroids would fall at speeds of less than 6.0–7.7 km/s and Pluto's orbital speed of 3.7–6.1 km/s adds to a maximum impact speed of about 13.9 km/s. A large meteoroid would be needed for a meteor to be seen. One-revolution dust trails of long-period comets may be compact enough to detect as meteor outbursts. Even though 40 times more long-period comets cross the orbit of Neptune, few will have a perihelion far enough inside Jupiter's orbit to generate dust with a node at Pluto's.

[43] K. Zahnle, P. Schenk, H. Levison and L. Dones, Cratering rates in the outer Solar System. *Icarus* **163** (2003), 263–289.
[44] C. B. Phillips, J. Moore and K. Zahnle, Small comet abundance and Solar System location. *Am. Geophys. Union, Fall Meeting 2001*, #P32A-0546 (abstract) (2001).

34
Meteors and the origin of life

One of the motivations for the study of meteor showers here and on other planets is to learn their impact on the atmosphere of the early Earth. Meteoroids brought to our young planet molecules made of carbon, nitrogen, and oxygen, called *organic*.[1] During the entry, the organic content of the meteoroids was certainly chemically altered, and perhaps into compounds that were needed to make life possible.[2] Moreover, the atmosphere in those days was more similar to the N_2/CO_2 atmospheres of Mars and Venus, but with significant amounts of methane and water vapor that kept the surface warm. In such atmospheres, the chemistry induced by the kinetic energy of the meteors can also produce compounds important for the start of life.

34.1 The cradle of life

The Solar System evolved from the collapse of a dense globule in an interstellar cloud, triggered perhaps by the shock wave of a nearby supernova.[3] As the cloud collapsed in the next 100 000 years, it heated up, and compressed into a protostar at the center. Conservation of angular momentum caused the weakly rotating cloud to rotate faster and flatten to form a disk (Fig. 34.1). Most matter falling onto the disk passed an accretion shock and continued to stream to the central star, leaving most angular momentum contained in the rotating disk. Some of the matter flowing into the star was ejected in bipolar outflows north and south, and part of that may have fallen back onto the disk.

The grains started to grow. Most infalling interstellar grains were small 0.01–1 μm in size. Once inside the disk, they were sufficiently close together to collide and stick

[1] C. F. Chyba, P. J. Thomas, L. Brookshaw and C. Sagan, Cometary delivery of organic molecules to the early Earth. *Science* **249** (1990), 366–373; G. J. Flynn, The delivery of organic matter from asteroids and comets to the early surface of Mars. *Earth, Moon Planets* **72** (1996), 469–474; M. Maurette, Carbonaceous micrometeorites and the origin of life. *Origins Life Evol. Biosphere* **28** (1998), 385–412.

[2] P. Jenniskens, M. A. Wilson, D. Packan *et al.*, Meteors: a delivery mechanism of organic matter to the early Earth. *Earth, Moon Planets* **82/83** (2000), 57–70; P. Jenniskens, Meteors as a delivery vehicle for organic matter to the early Earth. *ESA SP* **495** (2001), 247–254.

[3] See: P. J. Thomas, C. F. Chyba and C. P. McKay, (eds), *Comets and the Origin and Evolution of Life* (New York: Springer, 1997), 296pp.

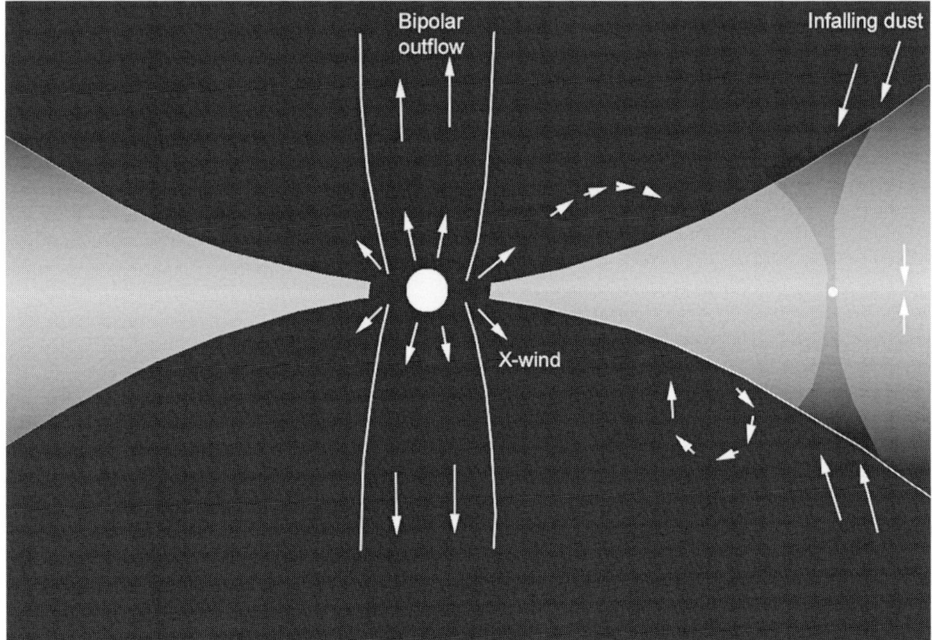

Fig. 34.1 Model of the disk of dust from which the planets formed.

together. Relative speeds had to be low, <1 m/s for μm-sized silicate spheres, otherwise they would break again. The growth process was a *hierarchical accretion*, meaning two small units made a bigger one, then two bigger ones made an even bigger one, etc., each step driven by different physical mechanisms of sticking and accumulation. Initially, the interstellar dust grains grew by sticking from Van der Waals forces into larger μm- and mm-sized porous aggregate particles, with a mass–size relation of $M \sim D^{1.91}$. When the grains got heavier, they were no longer dragged by the gas as much, and settled to the central plane of the nebula. Now the predominant growth was through very slow sticking collisions, with relative speeds of 1–10 cm/s, between dust clusters of similar sizes, because of sediment drift and gas turbulence resulting in *aggregates* with a mean *fractal dimension* of 1.85, with a FWHM of 0.36, where the particle diameter (d) related to the mass of the grain as: $M \sim D^{1.85}$, instead of $M \sim D^3$ for a solid sphere. Hence the density decreased with increasing size. In the subsequent collision of the aggregates, particles were compacted when the collision speeds were high enough, or even broken apart, to reform later, absorbing the impact energy and facilitating further sticking. The surviving accumulated aggregates were the size of about 6 mm near Neptune, 6 cm at Jupiter, and 20 cm at Earth. These are the meteoroids that we know as visible meteors on Earth. These *pebbles* accumulated more dust and other pebbles as they decoupled from the gas and started to spiral inward towards

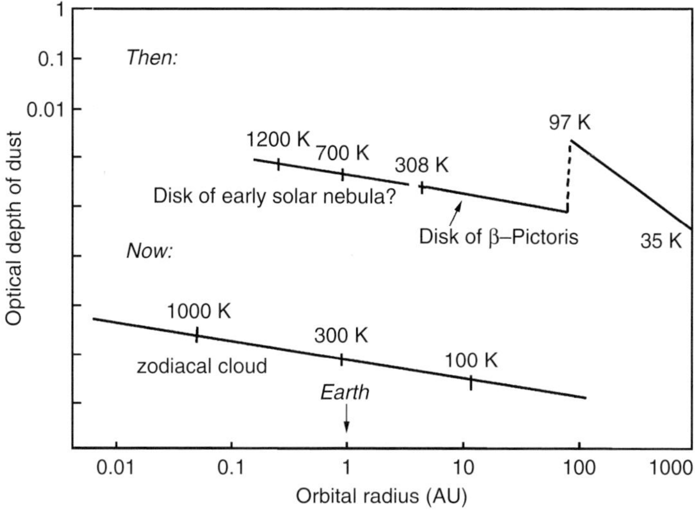

Fig. 34.2 Models of dust optical depth and temperature in the current solar system zodiacal dust cloud (from COBE data), compared to that of the β-Pictoris disk, as an example of the young solar system.

the Sun, accumulating into several 100 m sized to km sized *cometesimals*.[4] Gravity, finally, brought these cometesimals together into *comets*.

The region where Earth was formed ($r \sim 1$ AU) was dry and poor in organics, more so than the region where the asteroids were formed ($r \sim 3$–4 AU), while matter beyond the orbit of Jupiter retained much water ice. The grains were heated when they first fell into the disk and later by friction with the gas (thermal heating and shock waves), by mutual collisions, by lightning, and by stellar winds. Heating was most severe in the terrestrial region ($T \sim 700$ K at 1 AU, Fig. 34.2), where the gas was warm and dense, gravity was strong, the disk rotation speeds were highest, and occasionally shock waves evaporated matter altogether (brief flashes $T > 1300$ K).

Further out was a narrow zone where rocky grains were molten rather than evaporated in the wake of shocks: meteorites from the asteroid belt contain mm-sized molten droplets called "chondrules" and partially evaporated droplets, called *calcium–aluminium-rich inclusions* (CAI). Beyond that, grains were not intensely heated. Comets from the Uranus–Neptune region contain water ice in an interstellar ortho-para ratio of the hydrogen spins, which implies that the ice never warmed above 50 K.

Gravity continued to cause the growth of comets into protoplanets the size of Mars and Pluto. The protoplanets large enough to hold onto hydrogen gas continued to grow by accretion of gas from the disk and form the giant planets. The process of

[4] S. J. Weidenschilling and J. N. Cuzzi, *Protostars and Planets III*, ed. E. H. Levy and J. I. Lunine. (Tucson, AZ: University of Arizona Press, 1997), pp. 1031–1060; G. Wurm and J. Blum, Experiments on preplanetary dust aggregation. *Icarus* **132** (1998), 125–136; J. A. Nuth, F. Rietmeijer and H. G. M. Hill, Condensation processes in astrophysical environments: the composition and structure of cometary grains. *Meteoritics Planet. Sci.* **37** (2002), 1579–1590.

accumulation was stopped when the star started to illuminate the disk, periodically erupted brightly, and gradually blew the gas away in a strong solar wind. This was about 1 million years after the initial collapse. It stopped the further growth of the major planets, but also slowed down the process of planetesimal accretion by taking away gas drag.

The final phase of this evolution was marked by the large collisions of proto-planets, the effects of which were no longer erased by later collisions. After 30 million years, a proto-Earth collided with another object the size of Mars (10% of the mass of the proto-Earth) in an oblique impact to form our iron-poor Moon from the mantle material of both objects. Earth at that collision was already 95% of its present mass, and differentiated with a core and mantle. Earth came out at its final mass with a mantle containing oxidized minerals much like today. Our Moon set the stage for making our planet a habitable world by stabilizing the tilt of the Earth's spin axis.

In the next 500 million years, Jupiter and Saturn continued to clean out the Solar System and sent many comets our way.[5] The Moon was bombarded, as was Earth, adding a "late veneer" of volatile compounds (water and organics). Later, objects continued to arrive from the regions beyond Saturn. The impacting comets were scattered and many were on hyperbolic orbits, but in my opinion those impacting Earth were more likely (returning more often) on elliptical orbits, in which case they should have caused the same intense meteor showers as today and populate a zodiacal cloud. The amount of mass hitting Earth from rare impacting comets would have been about the same as that from the steady influx of $\sim$200 μm sized meteoroids. Most organic matter in the high-pressure, high-temperature comet impacts was burned. Hence, if reduced organic compounds survived the meteor phase, then meteors were their dominant pathway from space to Earth.[6]

34.2 The origin of life

When life on our planet first arose 4.0 ± 0.4 billion years ago, meteor rates were at least 100 times higher than today. The night sky would have looked much like the brief peak of recent Leonid storms. Life is traced in the geologic record back to at least 3500 million years ago, and possibly as far back as 3800 million years. Life on Earth is thought to have originated in the period before that, when giant impacts were still frequent and volcanism was rampant. From this period, called the *Hadean*, no sedimentary rocks survive. Only small fragments in later rock can be dated back to almost the time of the Earth's formation 4550 million years ago. It is therefore not easy to verify the conditions faced by early living organisms.[7]

What kind of life could that have been? Multicellular life as we know it dates back to only about 543 million years ago. Before that, there were only single cell organisms

[5] A. H. Delsemme, Cometary origin of the biosphere. *Icarus* **146** (2000), 313–325.

[6] C. F. Chyba and C. Sagan, Endogenous production, exogenous delivery, and impact-shock synthesis of organic molecules: an inventory for the origins of life. *Nature* **355** (1992), 125–132.

[7] See: S. J. Mojzsis, R. Krishnamurthy and G. Arrhenius, Before RNA and after: geophysical and geochemical constraints on molecular evolution. In *The RNA World*, 2nd edn. (Cold Spring Harbor, NY: Cold Spring Harbor Laboratory Press, 1993), pp. 1–47.

that invented photosynthesis to use sunlight as a source of energy, thereby generating waste oxygen molecules. Initially, this oxygen was dissolved in the oceans, rusting dissolved iron, which settled onto the ocean bottom. Only since about 2500 million years ago was there enough oxygen accumulated in our atmosphere to permit breathing. Before photosynthesis, bacteria used hydrogen sulfide (H_2S), methane (CH_4), and other small molecules to generate energy. Nowadays, such bacteria are found in hot springs and lagoons of shallow lakes and seas. It happens that sedimentation in such environments also left the best and oldest fossil records. That does not mean that life necessarily originated in such environments. It is likely, though, that the cradle of life was a wet and organic-rich place.

What was the nature of that organic material? Comet and asteroid impacts are an abundant source of organics and water.[8] It was long believed that the low ratio of deuterium over hydrogen in the water of our ocean argued against having comets play an important role in the delivery of water, because the D/H ratio of ocean water is more like that of asteroids than some Oort cloud comets. But now comets have been observed with the low ratio as well, presumably originating later in time, or further inward, for the water to have had time to exchange isotopes with the nebular gas.[9]

We know much about the organics present in asteroids and comets from collected meteorites and interplanetary dust particles, which provide convenient laboratory samples. These are typically highly polymerized carbon compounds with a low content of the functional groups needed to do interesting chemistry. Many studies have searched for useful compounds such as amino acids in the organics of meteorites, but they are always present in minute amounts, if at all. It is more likely that the terrestrial environment chemically changed the matter into useful compounds.

In my opinion, that started in the meteor phase, a potentially important step in the chemical evolution.[10] Some enrichment in oxygen, for example, could make the exogenous organic matter useful for prebiotic chemistry. In the Earth's early atmosphere, that oxygen came from CO_2, the dominant component of the atmosphere, or water and volcanic sulfuric acid in the Earth's environment. This is not a new idea. Even *Lord Kelvin* speculated at one point that life might have come to Earth via a meteor.[11] On the other hand, it is an idea that is very difficult to prove.

Unfortunately, products of this organic chemistry in today's atmosphere are now rapidly diluted in and contaminated by the Earth's biomass and we have not been able yet to sample meteoric debris, except perhaps for nm-sized meteoritic particles in stratsopheric aerosols.[12] Hence, much of our understanding of this organic

[8] J. Oró, Comets and the formation of biochemical compounds on the primitive Earth. *Nature* **190** (1961), 389–390.

[9] F. Robert, D. Gautier and B. Dubrulle, The solar system D/H ratio: observations and theories. *Space Sci. Rev.* **92** (2000), 201–224; A. Morbidelli, J. Chambers, J. I. Lunine *et al.*, Source regions and timescales for the delivery of water to the Earth. *Meteoritics Planet. Sci.* **35** (2000), 1309–1320.

[10] P. Jenniskens, Meteors as a vehicle for the delivery of organic matter to the early Earth. *ESA-SP* **495** (2001), 247–254.

[11] M. McCartney and A. Whitaker, *Physicists of Ireland: Passion and Precision* (Bristol: Institute of Physics Publishing, 2002), 298pp.

[12] D. M. Murphy, D. S. Thomson and M. J. Mahoney, *In situ* measurements of organics, meteoritic material, mercury, and other elements in aerosols at 5 to 19 kilometers. *Science* **282** (1998), 1664–1669.

Fig. 34.3 Mission patch and logo of the 1998 Leonid MAC mission.

chemistry will have to come from a better understanding of the physical conditions in meteors.

34.3 The Leonid MAC missions

The fate of organic matter in the meteors remains particularly hard to address because organic molecules are so difficult to detect by remote sensing. Atomic carbon has no strong optical emissions in the visible and near-IR. Small molecules such as CN, CH, and C_2 can be detected more readily at optical wavelengths, but no certain detections have been made to date. Any more complex molecules will be present at low abundance. Modern instruments are needed to probe the physical conditions in meteors and to detect the signatures of the organic matter in the meteoroids.

A great opportunity to study meteors in this manner was the occasion of the recent Leonid storms. We are still pouring over the results from the deployment of a wide range of spectrographs, imagers, and resonant lidar during the four missions of the *Leonid Multi-Instrument Aircraft Campaign*, in 1998, 1999, 2001, and 2002. The new instruments included optical, mid-IR, and near-IR spectrographs, a sub-mm spectrometer, and high-speed imagers. Spectacular showers were observed on all occasions.[13] These were NASA's first Astrobiology missions (Fig. 34.3).

[13] P. Jenniskens, S. Butow and M. Fonda, The 1999 Leonid Multi-Instrument Aircraft Campaign – an early review. *Earth, Moon Planets* **82/83** (2000), 1–26; P. Jenniskens and R. W. Russell, The 2001 Leonid Multi-Instrument Aircraft Campaign. *ISAS SP* **15** (2002), 3–15 (http://Leonid.Arc.nasa.gov).

34.4 The fate of organics in meteoric ablation

We searched in vain for CN emission and set a strong lower limit [CN/Fe] < 0.03.[14] If all nitrogen in the complex organic matter of comet Halley's dust would have come off as CN, the value would have been [N]/[Fe] = 0.79. This implies that nitrogen in the organic matter is not lost in the form of CN.

There was some evidence of the loss of functional groups from the organic matter, in the sense that emission from atomic hydrogen was detected.[15] The hydrogen does not originate in the atmosphere, but rather in the hot phase of the meteor emission, being excited during the cascade phase after having been evaporated into the vapor cloud. Meteor spectra also show the emission of OH radicals at 310 nm, which are readily formed when hydrogen atoms are present in an oxygen plasma. The hydrogen can originate from meteoroid organics or be due to the dissociation of water molecules present in the meteoroid as mineral water.[16]

The plasma temperature was measured in the wake of the meteor, signifying the environment to which ablated organic compounds are first exposed.[17] We found that volatile and non-volatile mineral components ablate at the same time and, therefore, also expect much of the organics to remain part of the grains while penetrating into the atmosphere. We measured a plasma excitation temperature of $T \sim 4400$ K and found it to be nearly independent of meteor mass or speed. This temperature is of interest for inducing chemistry in the ambient atmosphere because it is at a temperature where CO_2 does not only break up into CO + O, but CO also breaks up into C + O. Such dissociation can lead to interesting organic compounds. Carbon atom emission was not detected, perhaps only because it is difficult to observe.

The temperature in the meteor wake was measured as a function of time from the emissions in meteor afterglow and persistent trains (Fig. 4.11). We found that the extended wake of debris in large meteoroids caused a more gradual cooling of the plasma. For a -12^{m} fireball, the ablation vapor cooled from ~4400 to ~1200 K in a few seconds, while a -3^{m} Leonid's plasma cooled in about ~0.1 s. This compares to timescales of the cooling of the electrons created in collisional ionization of order 0.001 s at 80 km, to about 0.1 s at 115 km, faster for brighter meteors.[18] Even fast

[14] P. Jenniskens, E. L. Schaller, C. O. Laux *et al.*, Meteors do not break exogenous organic matter into di-atomic molecules. *Astrobiology* **4** (2003), 67–79.

[15] P. Jenniskens and A. M. Mandell, Hydrogen emission in meteors as a potential marker for the exogenous delivery of organics and water. *Astrobiology* **4** (2004), 123–134.

[16] S. Abe, H. Yano, N. Ebizuka *et al.*, First results of OH emission from meteor and afterglow: search for organics in cometary meteoroids. *ESA SP* **500** (2002), 213–216; P. Jenniskens, C. O. Laux and E. L. Schaller, Search for the OH ($X^2\Pi$) Meinel band emission in meteors as a tracer of mineral water in comets: detection of N_2^+ (A–X). *Astrobiology* **4** (2004), 109–121.

[17] P. Jenniskens, C. O. Laux, M. A. Wilson and E. L. Schaller, The mass and speed dependence of meteor air plasma temperatures. *Astrobiology* **4** (2004), 81–94; P. Jenniskens and H. C. Stenbaek-Nielsen, Meteor wake in high frame-rate images – implications for the chemistry of ablated organic compounds. *Astrobiology* **4** (2004), 95–108.

[18] W. J. Baggaley and T. H. Webb, The thermalization of meteoric ionization. *J. Atmosph. Terrest. Phys.* **39** (1977), 1399–1403.

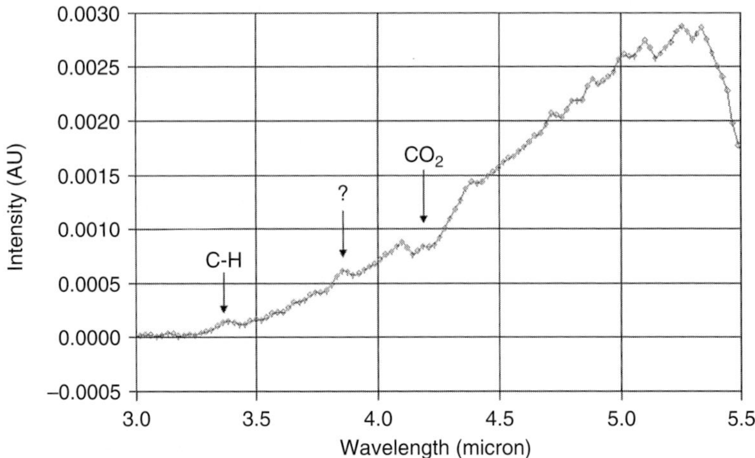

Fig. 34.4 Mid-infrared emission of a persistent train. Telluric CO_2 and water vapor account for the absorptions. Courtesy: George Rossano, Ray Russell, Aerospace Corporation.

(72 km/s) and large (100–1000 g) meteoroids appear to leave solid particles in their path, creating a red continuum in the meteoric afterglow emissions.[19]

A team from the Aerospace Corporation, led by *Ray W. Russell* and *George S. Rossano*, detected the 3.4 µm fingerprint of C-H bonds in organic molecules in the mid-infrared emission of persistent trains, even when the gas and dust had cooled to just 50 K over the ambient temperature a few minutes after the fireball (Fig. 34.4).[20] A continuum emission was also observed, which may originate from meteoric debris, adding to other evidence that even super fast Leonids leave solid debris in their wake.[21] The 3.4 µm band might be due to organic matter still intimately mixed with the solid silicate dust debris.

In this context, cometary dust shows a similar mid-infrared emission band (Fig. 34.5), a significant part of which is caused by methanol and other gaseous molecules containing C-H bonds.[22] However, a small but distinct feature at 3.29 µm in the spectrum of Perseid comet 109P/Swift–Tuttle is evidence for an aromatic ring structure in complex organic molecules and, according to *Michael A. DiSanti et al.*, that could mean that part of the 3.3–3.55 µm band also originated from complex nonaromatic organic molecules or dust.

[19] J. Borovicka and P. Jenniskens, Time resolved spectroscopy of a Leonid fireball afterglow. *Earth, Moon Planets* **82/83** (2000), 399–428.

[20] X. Chu, A. Z. Liu, G. Papen *et al.*, Lidar observations of elevated temperatures in bright chemniluminscent meteor trails during the 1998 Leonid shower. *Geophys. Res. Lett.* **27** (2000), 1815–1818.

[21] R. W. Russell, G. S. Rossano, M. A. Chatelain *et al.*, Mid-infrared spectroscopy of persistent Leonid trains. *Earth, Moon Planets* **82–83** (2000), 439–456.

[22] M. A. DiSanti, M. J. Mumma, T. R. Geballe and J. K. Davies, Systematic observations of methanol and other organics in comet P/Swift–Tuttle: discovery of new spectral structure at 3.42 µm. *Icarus* **116** (1995), 1–17.

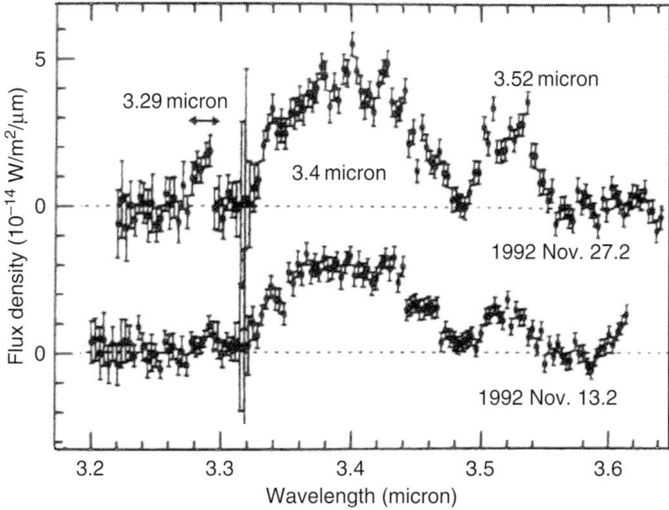

Fig. 34.5 3.4 μm feature in emission from Perseid comet 109P/Swift–Tuttle. Data by DiSanti, Mumma, Geballe and Davies (1995).

34.5 The future

This research is ongoing. New techniques are being developed to enable the measurement of the signatures of carbon in the light of meteors from remote sensing. Future observations may include spectroscopic observations from space, where the strongest carbon atom emission lines in the ultraviolet are no longer shielded by the atmosphere. Such observations could measure how much carbon is in the dust of particular comets and how much of that carbon survives space weathering.

Such future studies may also address what happened to the organic matter brought in by fireballs large enough to create a shockwave. We may learn this from the difficult remote sensing of rare natural fireballs or from the study of artificial fireballs created by the hypervelocity re-entry of carbon-coated sample return capsules from interplanetary space. For that reason, the Leonid MAC missions are being followed by similar airborne observing campaigns to monitor the return of the Genesis and Stardust sample return capsules. Upon their return from interplanetary space, they plunge into our atmosphere at the speed of natural asteroids, at 11.0 and 12.8 km/s, respectively, creating a bright light. The study of meteor outbursts as a tool for studying comets has just begun. The remainder of this book helps to shine a light on what is ahead in the next 50 years. Table 11 is an index to the information contained in the other tables and chapters throughout this book. Please bear in mind that all the predictions mentioned in this book are uncertain for a number of reasons, not in the least part because of the unknown distribution of dust in the trails and Filaments. The data listed in the tables serve only as a guideline in the hope that they may increase your chances of observing a meteor outburst. Be aware of this uncertainty and observe in good observing conditions for some time interval around the given peak time. None of the lists are complete and other encounters may be announced. Observers are encouraged to check the internet for updates closer to the date of the expected occurrences. If you do go out to observe, never put yourself in danger. When you need to travel, rest before driving. With so many parent comets just identified and new tools for predicting future showers, there is much to explore. Happy storm chasing! May all of your wishes come true.

Appendix

Appendix A

The Whipple ejection speed

Whipple's recipe[1] for calculating the ejection speed of a dust particle V_{ej} (m/s) has two terms, describing the speeding up of grains by the push from collisions with water vapor and the slowing down by the pull of the gravity of the comet

$$V_{ej} = \sqrt{(43.0\, D_c/(\rho\Lambda\ dr^{9/4}) - 0.559\rho_c D_c^2)} \tag{A.1}$$

This equation is based on the assumptions: that the amount of energy carried out by the outflow of water vapor is proportional to the energy input from sunlight, that the outflow is uniform over the nucleus, that the grains are spherical, that water vapor is the dominant driver, and that the outflowing gas has a Maxwellian distribution of velocities. The formula includes the size (d, in cm) and density (ρ, in g/cm^3) of the meteoroid, the size (D_c, in km) and density (ρ_c, in g/cm^3) of the comet nucleus, and the distance from the Sun (r, in AU), as well as the heat absorption efficiency (Λ), which is the fraction of all light absorbed.

Other assumptions have resulted in modified versions of this equation by later authors. *Jim Jones* and *Peter Brown*, for example, considered an adapted version of Whipple's ejection model, which corrected the assumption for a blackbody-limited nucleus temperature, instead of temperature limited, and the neglect of the adiabatic expansion of the gas[2]

$$V_{ej} = \sqrt{(32.3 D_c/(\rho\Lambda\ dr))} \tag{A.2}$$

This equation follows earlier work by Jones, except for ignoring an $r^{-1.038}$ -dependence in favor of an empirically derived variation of (small grain) dust ejection speed.

[1] F. L. Whipple, A comet model. II. Physical relations for comets and meteors. *Astrophys. J.* **113** (1951), 464–474.

[2] P. Brown and J. Jones, Simulation of the formation and evolution of the Perseid meteoroid stream. *Icarus* **133** (1998), 36–68.

The Crifo ejection formulas

One of the latest incarnations by J.-F. Crifo[3] and A. V. Rodinov (the *Crifo ejection model*) was developed for the interpretation of comet images and is used in the meteor storm prediction model of Vaubaillon. The Crifo model describes the dust ejection speed (V_d) as a multiplication of the rate of coupling between gas and dust (ϕ) and the gas ejection speed (V_g) as follows:

$$V_d = \phi\sqrt{((\gamma + 1)/(\gamma - 1))}V_g \tag{A.3}$$

$$V_g = \sqrt{(\gamma k_B T_g / M_{H_2O})} \tag{A.4}$$

Where $\gamma = 4/3$ is the ratio of the specific heats of water, k_B is the Boltzmann constant $= 1.38 \times 10^{-23} m^2$ kg s^{-2} K^{-1}, and M_{H_2O} the mass of a water molecule, which is equal to 18 atomic mass units or $M = 3.0 \times 10^{-26}$ kg. T_g is the kinetic temperature of the outflowing water vapor and is derived from the functions below, κ is a parameter depending on the incident angle of sunlight (z, angle from the subsolar point in degrees), and r is the heliocentric distance of ejection ($r < 3.0$ AU)

$$\kappa = (1-A)\cos(z)/r^2 \tag{A.5}$$

$$T_g(\mathrm{K}) = 275.062 + 0.040\,0845\,\kappa - 72.1812\,\kappa^2 \tag{A.6}$$

For some observations, the thermal dust and ice temperature are needed as well, given by

$$T_d(\mathrm{K}) = 398.944 + 0.249\,975\,\kappa - 0.033\,1083\,\kappa^2 \tag{A.7}$$

$$T_{ice}(\mathrm{K}) = 210.663 + 0.042\,7156\,\kappa - 47.8234\,\kappa^2 \tag{A.8}$$

"A" is the albedo of the nucleus ($\sim$0.04).[4] The coupling factor between dust and gas (the heart of the Crifo model) is assumed to be

$$\phi = 1/(1.2 + 0.72\sqrt{(a_d/a^*)}) \tag{A.9}$$

Where a_d is the radius of the particle and a^* is a critical radius below which the effectiveness of the gas–particle interaction decreases. If $a_d \ll a^*$, then the ejection speed becomes the same as the gas speed, with $\phi = 1.0$. In the case of meteoroids responsible for visible meteors, $a_d \gg a^*$ and $\phi = \sqrt{a^*} / (0.72\sqrt{a_d})$. The critical radius

$$a^* = M(1 - A)c_\odot/(\rho_d \Lambda L_s V_g) \times R_c f(r)\cos z/r^2 \tag{A.10}$$

[3] J.-F. Crifo and A. V. Rodinov, The dependence of the circumnuclear coma structure on the properties of the nucleus. *Icarus* **127** (1997), 319–353; A. V. Rodionov, J.-F. Crifo, K. Szege, J. Lagerros and M. Fulle, An advanced physical model of cometary activity. *Planet. Space Sci.* **50** (2002), 983–1024.

[4] Y. R. Fernández, D. C. Jewitt and S. Sheppard, Low albedos among extinct comet candidates. *Astrophys. J.* **553** (2001), L197–L200.

with M in atomic mass units and r in astronomical units, is about $a^* = 10\ \mu m$ at 1 AU for a comet such as 55P/Tempel–Tuttle. Here, $c_\odot$ is the energy flux at 1 AU called the solar constant ($c_\odot = 1353\ W/m^2$), ρ_d is the density of the meteoroid (~1000 kg/m³), Λ is the fraction of heat that goes into sublimation ($\Lambda = 1.0$), L_s is the latent heat of sublimation ($L_s = 2.833 \times 10^6$ J/kg), R_c is the radius of the nucleus (where D_c is the diameter), and $f(r)$ is the fraction of surface area that is active at a given heliocentric distance (usually assumed to be constant, and taken to be $f(r) = 0.24$ for our meteor storm prediction models of 55P/Tempel–Tuttle).

The gravity of the comet will pull the meteoroids back, decelerating. The ejection velocity is

$$V_{esc}^2 = 2\,GM_c/R_c = (8\pi/3)G\rho_c R_c^2 \quad \text{(A.11)}$$

$$V_{ej}^2 = V_d^2 - V_{esc}^2 \quad \text{(A.12)}$$

where G is the gravitational constant ($= 6.673 \times 10^{-11}\ m^3\ kg^{-1}\ s^{-2}$).

Using all of this for the Leonids $a_d \gg 10\ \mu m$ gives to a first approximation

$$V_{ej}(m/s) = \sqrt{(200 D_c f(r) \cos z/(\rho\Lambda\, dr^2) - 0.559\rho_c D_c^2)} \quad \text{(A.13)}$$

with D and D_c in km, r in AU, and ρ and ρ_c in g/cm³, almost identical to Whipple's equation for $f(r) = 0.21$. To calculate the amount of matter ejected as a function of heliocentric distance, the temperature of the surface needs to be evaluated from the thermodynamic balance between heating from sunlight and the warming of the gray-body nucleus added to the energy lost from outflow:

$$\varepsilon\sigma_b T_{surface}^4 + \alpha_s L_s F_A = (1-A)c_\odot \cos(z)/r^2 \quad \text{(A.14)}$$

where ε is the material emissivity ($\varepsilon = 0.9$ for carbonaceous materials) and σ_b = Stefan–Boltzmann constant $= 5.67 \times 10^{-8}\ W\ m^{-2}\ K^{-4}$. The total molecular flux over the whole surface at a distance of $r = 1$ AU is Q, while F_A is the flux per unit of surface

$$F_A = Q/f\pi R_c^2 \quad \text{(A.15)}$$

Whereby the total loss of water vapor can be derived from the absolute magnitude of the comet, using the fact that the comet brightness is proportional to the rate at which ices are lost[5]

$$\log Q_{H_2O} = 30.78 - 0.26 H_{10} \quad \text{(A.16)}$$

[5] L. Jorda, J. Crovisier and D. W. E. Green, The correlation between cometary water production rates and visual magnitudes. *Bull. Am. Astron. Soc.* **24** (1992), 1006–1007 (abstract).

By choosing subsolar points (z) and points equally spaced in time along the comet orbit, ejection speed distributions can be calculated, assuming an empirical distribution with probability[6]

$$P(V - V_{\mathrm{ej}}) = 1 - \left(\frac{V}{V_{\mathrm{ej}}} - 1\right)^2 \quad \text{(A.17)}$$

The largest size of a meteoroid lifted from the comet surface, according to Whipple, follows from Eq. (A.1) with $V_{\mathrm{ej}} = 0$ m/s

$$d_{\max}(\mathrm{cm}) = 76.9/\Lambda\rho(\mathrm{g/cm^3})\ \rho_{\mathrm{c}}(\mathrm{g/cm^3})D_{\mathrm{c}}(\mathrm{km})r(\mathrm{AU})^{9/4} \quad \text{(A.18)}$$

Appendix B

The change of orbital elements due to radiation pressure

The change of orbital elements upon release, without ejection speed, from a comet at a heliocentric distance r(AU) are (for grains scattering with efficiency Q_{pr})[7]

$$a_d = a + \Delta a, \text{with } \Delta a = a((1 - \beta)/(1 - 2a\beta/r) - 1) \quad \text{(B.1)}$$

$$e_{\mathrm{d}}^2 = 1 - (1 - e_{\mathrm{c}}^2)(1 - 2\,a\beta/r)/(1 - \beta)^2 \quad \text{(B.2)}$$

$$q_{\mathrm{d}} = a_{\mathrm{d}}(1 - e_{\mathrm{d}}) \quad \text{(B.3)}$$

$$\beta = 1.148 \times 10^{-4}\, Q_{\mathrm{pr}}/(\rho\, d) \quad \text{(B.4)}$$

where d is the particle diameter in cm, ρ is the particle density in g/cm^3, and β is the ratio of radiation over gravitational forces, so that the effective force of gravity on the meteoroid is decreased by a factor $(1 - \beta)$. The first equation follows from the conservation of energy

$$0.5\mathrm{M}V^2 + GM_{\odot}M(1 - \beta)/r = GM_{\odot}M(1 - \beta)/2a \quad \text{(B.5)}$$

where the last term is the orbital energy. The second equation follows from the conservation of specific angular momentum $h^2 = GM_{\odot}a\,(1 - e^2)$:

$$GM_{\odot}(a + \Delta a)(1 - e_{\mathrm{d}}^2)(1 - \beta) = GM_{\odot}a(1 - e^2) \quad \text{(B.6)}$$

[6] J. Jones, The ejection of meteoroids from comets. *Mon. Not. R. Astron. Soc.* **275** (1995), 773–780.

[7] Z. Sekanina, Meteoric storms and formation of meteor streams. In *Asteroids, Comets, Meteoric Matter*, ed. C. Cristesu, W. J. Klepczynski and B. Millet. (Bucharest: Academic Socialist Republic), pp. 239–267; L. Kresák, Orbital evolution of the dust streams released from comets. *Bull. Astron. Inst. Czech.* **27** (1976), 35–46.

Radiation pressure changes the radius vector of the node, according to Sekanina by

$$\Delta r_\Omega / r = \beta(1 \pm \cos(v + \omega)) r / (a(1 - e^2)) \tag{B.7}$$

$$\Delta\Omega = 0 \tag{B.8}$$

The change of orbital elements due to ejection speed

The change in semimajor axis (Δa) from ejection of a meteoroid follows from the change in orbital energy (E) by adding the extra kinetic energy of the meteoroid from ejection

$$\Delta E_{\mathrm{kin}} = 1/2M(V + v_{||})^2 - 1/2\,MV^2 \sim MVv_{||} \tag{B.9}$$

the last step for $v_{||} \ll V$. Because the total orbital energy $E = GM_\odot M/2a$:[8]

$$\begin{aligned} \Delta E/E \sim \Delta E_{\mathrm{kin}}/E = \Delta a/a = 2/3\Delta P/P = 2(2a/r - 1)v_{||}/V \\ \Delta a/a = 2(1 + e)/(1 - e) \times v_{||}/V \qquad \text{if } r = q = a(1 - e) \end{aligned} \tag{B.10}$$

where r is the heliocentric distance of the point of ejection, $v_{||}$ is the ejection speed in the direction of cometary motion, $v_{||} = v_{\mathrm{ej}} \cos(\phi) \sin(\theta)$, and V is the heliocentric velocity of the comet during ejection.

The change of the eccentricity of the orbit follows from the angular momentum $h = Mr \times V$

$$\Delta h/h = \Delta a/2a - e\,\Delta e/(1 - e^2) = v_{||}/V \tag{B.11}$$

$$\Delta e = (1 - e^2)(2 - 2a/r)ev_{||}/V \tag{B.12}$$

If ejection has a component in the transverse direction perpendicular to the direction of motion of the comet, then the angular momentum is changed accordingly, and the longitude of the node, following Williams, with $v_| = v_{\mathrm{ej}} \sin(\phi)$[9]

$$\begin{aligned} \Delta\Omega &= v_|\; r \sin(\omega + v)/h \;\sin(i) \quad \text{(in radians)} \\ &= \sqrt{((2r^2 - ra)/(1 - e^2))}/a \times \sin(\omega + v)/\sin(i) \times v_|/V \end{aligned} \tag{B.13}$$

Where v is the true anomaly of the point of ejection, $v_|$ is the component of the ejection velocity perpendicular to the orbital plane. Ejection at perihelion would have $v = 0$. The change in nodal distance (along the nodal line) follows from

$$r_\Omega = q(1 + e)/(1 \pm e\,\cos(\omega)) \tag{B.14}$$

8 I. P. Williams, The determination of the ejection velocity of meteoroids from cometary nuclei. *ESA SP* **495** (2001), 33–42.

9 I. P. Williams, The evolution of meteoroid streams. In *Meteors in the Earth's Atmosphere*, ed. E. Murad and I. P. Williams. (Cambridge: Cambridge University Press, 2002), pp. 13–32.

Where the $\pm$ sign gives the result for one node or the other. Both Δr_Ω and $\Delta\omega$ can be expressed analytically, but the math is unwieldy. To first order

$$\Delta r_\Omega / r_\Omega \sim 2(1+\cos(\omega))/(1 \pm e\ \cos(\omega)) \times v_{||}/V \tag{B.15}$$

The velocity vector from given orbital elements

The heliocentric velocity V can be decomposed into the component in the radial direction, normal to the ecliptic plane, and perpendicular to those two in the $\lambda_\odot$ direction (if a, r in AU, then $GM_\odot = 8.872 \times 10^8$ (m/s)^2 per AU)[10]

in the radial direction: $$V_R^2 = GM_\odot(1-\beta)(2/r - 1/a - a/r^2 \times (1-e^2)) \tag{B.16}$$

in the normal direction: $$V_N^2 = GM_\odot(1-\beta)a/r^2 \times (1-e^2)\sin^2(i) \tag{B.17}$$

in the perpendicular direction: $$V_T^2 = GM_\odot(1-\beta)\, a/r^2 \times (1-e^2)\cos^2(i) \tag{B.18}$$

Its geocentric velocity V_g, or the apparent speed as seen from Earth, of the meteoroids (before being accelerated due to falling into the Earth's gravity well), will be (at $r = 1$ AU, no Earth spin)

$$V_g^2 = GM_\odot(1-\beta)(3 - 1/a - 2\ \cos(i)\sqrt{[q(1+e)]}) \tag{B.19}$$

Correction for zenith attraction

Shiaparelli's equations for correcting the observed zenith angle (Z_o) and azimuth (Az_o) of the radiant and speed V_∞ of a meteor (strictly the apparent speed at altitude H) to the geocentric radiant (Z, Az) and speed (V_g)

$$V_g^2 = V_\infty^2 - 2GM_E/(R_E + H) = V_\infty^2 - 123.06 \text{ for } H = 100\,\text{km} \tag{B.20}$$

$$\begin{aligned} Z - Z_o &= \Delta Z = 2\arctan[(V_\infty - V_g)/(V_\infty + V_g) \times \tan(Z_o/2)] \\ Az - Az_o &= \Delta Az = 0 \end{aligned} \tag{B.21}$$

A better formalism for correcting the radiant of an individual meteor for zenith attraction, which depends on the zenith angle (Z_m) and azimuth (Az_m) of the position of the meteor as seen by the observer, was given by *Peter Gural* as follows[11]

[10] E. Grün, M. Baguhl, H. Svedhem and H. A. Zook, *In situ* measurements of cosmic dust. In *Interplanetar Dust*, ed. E. Grün, B. Å. S. Gustarson, S. F. Dermott and H. Fechtig. (Berlin: Springer 2001), pp. 295–346.

[11] P. Gural, Fully correcting for the spread in meteor radiant positions due to gravitational attraction. *WGN* **29** (2000), 134–138.

$$\gamma = Z_m - \arcsin[R_E/(R_E + H) \times \sin Z_m] \tag{B.22}$$

$$\cos Z^* = \cos(Z_o) \cos\gamma + \sin Z_o \sin\gamma \cos(Az_m - Az_o) \tag{B.23}$$

$$\Delta Z^* = 2 \arctan\left[((V_\infty + V_g)/(V_\infty + V_g)) \times \tan\frac{Z^*}{2}\right] \tag{B.24}$$

$$\cos(Z_o + \Delta Z) = (\cos Z_o \sin(Z^* + \Delta Z^*) - \sin(\Delta Z^*) \cos(\gamma))/\sin(Z^*) \tag{B.25}$$

$$\cos \Delta Az = \cos \Delta Z^* - \cos Z_o \cos(Z_o + \Delta Z))/\sin(Z_o) \sin(Z_o + \Delta Z) \tag{B.26}$$

where the Earth's radius, $R_E = 6378$ km, M_E is the Earth's mass, H is the elevation of the meteor, and with the correct sign of ΔAz after taking the inverse cosine is the sign of the quantity sin $(Az_m - Az_o)$ for positive azimuth defined East of North. The true radiant is at zenith distance $Z = Z_o + \Delta Z$ and azimuth $Az = Az_o + \Delta Az$.

Appendix C

The mass of a meteoroid causing a meteor

For the typical blue plate response of the photographic plates (370–500 nm) used in the Harvard Super-Schmidt cameras, the mass of a meteoroid (M) relates to the absolute photographic meteor magnitude (m_{ph}), in the range $-2 < m_{ph} < +3$ according to[12]

$$\log M(\text{g}) = 5.15 - 0.44m_{ph} - 3.89 \log V_\infty(\text{km/s}) - 0.67 \log(\sin(h_r)) \tag{C.1}$$

where the angle of incidence h_r = the radiant altitude (90° in the zenith) and V_∞ is the atmospheric velocity $\sim V_\infty = \sqrt{(V_g^2 + 11.2^2)}$, log is always $\log_{10}$.

This result rates among the most important in meteor physics, because it relates meteors to meteoroids. However, it is no more than a scaling of parameters with a normalization called "luminous efficiency" and depends critically on the assumptions made on how the luminous efficiency depends on speed, and how "magnitude" is defined for different colors of light.

The Greeks invented the magnitude scale by classifying the brightest star as being of magnitude 1 and the faintest visible star as magnitude 6. Later, the magnitude was defined more formally as five magnitudes corresponding to an increase in brightness by a factor of 100, with the star Vega (spectral type A0V) being magnitude 0 *at all wavelengths*

$$m_V = -2.5 \log(F/F_0) \tag{C.2}$$

[12] L. G. Jacchia, F. Verniani and R. E. Briggs, Analysis of the atmospheric trajectories of 413 precise photographic meteors. *Smithsonian Contrib. Astrophys.* **10** (1967), 1–139.

where F is the flux from a star in a certain wavelength range and F_0 the flux of Wega in the same wavelength range. Because of the way the magnitude is defined, the zero point for the magnitude scale varies with wavelength[13] where the different colors,

Color	Wavelength (nm)	FWHM (nm)	F_0 for $m = 0.0$ (W/m^{-2}/nm)
Blue (B)	444.3	83.1	6.40×10^{-11}
Green(G)	548.3	82.7	3.67×10^{-11}
Red(R)	685.5	174.2	1.92×10^{-11}
near-IR(I)	863.7	197.0	9.39×10^{-12}

central wavelength, and wavelength range refer to the Johnson photometric filter system (filter "V" is given here as "G"). When comparing photographic and visual magnitudes, for example, one has to apply a color index ($= m_\lambda - m_V$), which contains the factor ($-2.5 \log (F_{0,\lambda}/F_{0,V})$), but which should also take into account changes in the spectrum of the meteor. One normally compares the brightness of meteors as they would be at a distance of 100 km, this is called the *absolute magnitude*[14]

$$m_V^{abs} = m_V - 5 \log(D(\text{km})/100) \tag{C.3}$$

When meteors are further away than 100 km, they are fainter, while those overhead can be brighter. This relationship expresses the fact that apparent magnitudes are a measure of flux and decrease with increasing distance (D) according to $F \sim (D(\text{km})/100)^{-2}$.

The color of meteors is redish or greenish depending on whether the air plasma or the metal atoms radiate most brightly. In the 370–550 nm range, most emissions in the meteor spectrum originate from metal atom lines. The naked eye is sensitive to light in the range 380–675 nm. Corresponding photographic film is called panchromatic (360–675 nm). That wavelength range includes two broad bands of molecular nitrogen emission and the sodium D line. As a result, meteors appear brighter to the naked eye than on photographic plates when compared to the continuum (~blackbody) emission of stars. From this, the "color index" was measured by comparing visual magnitude estimates with those derived from photographic film and found to be -1.9^m. Indeed, I find that -1.3^m of that is due to the difference in spectral coverage and -0.6^m is due to the difference in the magnitude scale definition.

[13] L. Colina, R. Bohlin and F. Castelli, Absolute flux calibrated spectrum of Vega. *Space Telescope Science Institute, Instrument Science Report CAL/SCS-008* (1996).

[14] In spectroscopy, those absolute magnitudes, converted to flux in units of W/m^2/nm with Eq. (C.1), are then converted to intensity (W/m^2/str/nm) by correcting for the solid angle over which the light is received on a 1 m^2 surface area, the angle of incidence, and the size of the source region. If the meteor plasma is an emitting sphere with radius R_m (in meters), then the correction is

$$I_\lambda = F_\lambda \cos(\theta)\ 100\,000^2/4\pi R_m^2 \tag{C.4}$$

where θ is the angle between the direction of the camera and the direction of the meteor.

A different color index should be used to calculate the visual magnitude from magnitudes derived by other instruments. Many red-sensitive CCD cameras tend to be sensitive from about 400 to 900 nm in the near-IR. The camcorders and digital photographic cameras that use CCD detectors have filters that only transmit blue (B), green (V), and red (R) light, but with different response curves for each camera. Normal eyes are sensitive from about 450 to 700 nm.

No dependence on meteor velocity was found, which we now understand to be because the meteor spectrum does not change much with speed. Large differences occur only when the ratio of air-plasma (N_2) to metal atom line emissions change, resulting in a conspicuous color change from pink (more air plasma) to green (more metal atom lines). In addition, fast and bright meteors can have a significant "hot component" in the spectrum with strong H and K lines of Ca^+, which are more readily detected on photographic film.

Now the more difficult question: how does this relationship (Eq. C.1) relate to meteors of a broader range of meteor magnitudes than observed in the Super-Schmidt program, and how was it scaled in the first place? It is sufficient to point out that the empirical relationship (C.1) is derived simply from the assumption that the observed luminosity is merely a fraction (the luminosity efficiency, τ) of the kinetic energy of the meteor. Eq. (C.1) is derived by considering the integrated intensity of the meteor and assuming that to be proportional to the kinetic energy of the meteoroid with instantaneous speed V

$$\int F(t)\delta t = F_{\max}\ \Delta t = \tau 0.5 M V^2 \tag{C.5}$$

The fraction τ describes the amount of energy that is radiated over an effective band, in the optical, of $\Delta\lambda = 90.2$ nm wide (the effective width for the dark-adapted eye). The peak visual magnitude is defined as $m_v = -2.5 \log(F_{\max}/F_0)$, with $F_0 = 3.67 \times 10^{-11}$ W/m^{-2}/nm. Eq. (C.1) now follows from

$$m_v = -2.5\ \log(F/F_0 D_\lambda) = -2.5\ \log(0.5\tau\ MV^2/\Delta t\ F_0 D_\lambda) \tag{C.6}$$

That leaves two things to be determined: the effective duration of the meteor (Δt) and the luminous efficiency, τ. The duration can be derived from photographic and video data (taken as 0.7 times the measured duration), which should be proportional to $\sin\ (h_r)^{-1}$ and V_∞^{-1}. Empirically, from DMS photographic and video surveys ($-5 < m_v < +6$) we find from plotting log (Δt) versus $\log(V_\infty)$, etc.

$$\Delta t(\mathrm{s}) = 9.2 m_v - 0.07 \sin(\mathrm{h_r}) - 0.41 V_\infty(\mathrm{km/s}) - 0.92 \tag{C.7}$$

In comparison, the Harvard photographic data resulted in an effective duration

$$\Delta t(\mathrm{s}) \sim m_v - 0.04 \sin(\mathrm{h_r}) - 0.67 V_\infty(\mathrm{km/s}) - 0.89 \tag{C.8}$$

From available laboratory measurements at low speeds, Öpik[15] derived a linear velocity dependence of the luminosity for bright visual meteors

$$\tau = 8.5 \times 10^{-5} V(\mathrm{km/s}) \tag{C.9}$$

which was later used by Whipple and Verniani,[16] but this result is uncertain and has been regarded with skepticism. Most observations point towards τ being constant for $V > 20$ km/s. Including Eq. (C.7) and $V \sim V_\infty$ gives the steep velocity dependence of Eq. (C.1):

$$\log M(\mathrm{g}) = 5.86 - 0.47 m_\mathrm{V}^\mathrm{abs} - 3.92 \log\ V_\infty(\mathrm{km/s}) - 0.41 \log\ \sin(h_\mathrm{r}) \tag{C.10}$$

This relationship predicts a mass of 0.045 g for a zero magnitude Leonid and $\tau = 0.61\%$; Note that $\log M(\mathrm{g}) \sim -0.4\ m_\mathrm{v}$, if the light curve shape did not change with meteoroid mass. If time played no role, then the dependence on $\sin(h_\mathrm{r})$ would be $-1.00 \log \sin(h_\mathrm{r})$.

The real uncertainty here is in the luminous efficiency factor and this has been a topic of ongoing work. Most recently, *Doug ReVelle* and *Zdenek Ceplecha*[17] derived an average luminous efficiency in the panchromatic photographic passband (360–650 nm) of 5.57% for asteroidal fireballs (type I), 1.35% for carbonaceous chondrite-like fireballs (type II), and 0.242% for cometary type III fireballs. They found the following mass and speed dependence for the apparent *differential luminous efficiencies*[18] of bright cometary fireballs, expressed in percent (V in km/s, M in kg)

$$\begin{aligned}
\ln(\tau) &= -2.670 - 10.307(\ln V) + 9.781(\ln V)^2 - 3.0414(\ln V)^3 \\
&\quad + 0.3213(\ln V)^4 + 1.15 \tanh(0.38 \ln(M)) \qquad \text{if } V > 25.372\,\mathrm{km/s} \quad \text{(C.11)} \\
\ln(\tau) &= -4.674 + \ln V + 1.15 \tanh(0.38 \ln(M)) \qquad \text{if } V < 25.372\,\mathrm{km/s}
\end{aligned}$$

Note that the luminosity efficiency is still assumed to be proportional to speed. For $M < 1$ kg, the last part of the function[19] is nearly constant at -1.15, changing to $+1.15$ for $M \gg 1$ kg, where a shockwave is formed. When this equation is incorporated into the previous formulas, and ignoring changes in lightcurve shape, I have for $V_\infty > 25.4$ km/s and $M < 1$ kg

$$\log M(\mathrm{g}) = 6.31 - 0.40 m_\mathrm{V}^\mathrm{abs} - 3.92 \log\ V_\infty(\mathrm{km/s}) - 0.41\ \log(\sin(h_\mathrm{r})) \tag{C.12}$$

[15] E. Öpik, Researches on physical theory of meteor phenomena. III Basis of physical theory of meteors. *Publ. Obs. Astr. Tartu* **29** (1937), 3–69.

[16] F. Verniani, On the luminous efficiency of meteors. *Smithsonian Contrib. Astrophys.* **8** (1965), 141–172.

[17] D. O. Revelle and Z. Ceplecha, Bolide physical theory with application to PN and EN fireballs. *ESA SP* **495** (2001), 507–512.

[18] There is a difference between the apparent luminous efficiencies considered here and the intrinsic values derived when taking fragmentation into account. Intrinsic luminous efficiencies are larger, with $\log(\tau) = -0.534 - \ldots + 0.174 \tanh(0.38 \ln(m))$ for $V < 25.372$ km/s and $\ln(\tau) = -1.401 + \ln V + 0.174 \tanh(0.38 \ln(m))$. This is 0.91% for a 1 g Leonid. If we want to calculate the mass of a visual magnitude meteor, then the apparent luminous efficiency should be used.

[19] Hyperbolic tangent $\tanh(z) = (e^z - e^{-z})/(e^z + e^{-z})$.

Appendix D

Meteoroid in the atmosphere

In a time interval δt, a meteoroid with instantaneous speed V and mass M strikes a mass of air $A_E\,\rho V\,\delta t$, where A_E is the surface area of the particle, ρ is the air mass density, and V is the speed of the meteoroid. This provides a kinetic energy of $0.5A_E\rho V^3\,\delta t$ that is used to heat, melt, vaporize, and fragment the body, causing it to lose a mass δM. Hence[20]

$$\delta M = -0.5\tau_E A_E \rho V^3\,\delta t \tag{D.1}$$

where τ_E is now the fraction of kinetic energy that is used to lose mass. The transfer of impulse leads to a loss of speed

$$\delta V/\delta t = -0.5C_d A_E/M\rho V^2 \tag{D.2}$$

where C_d is the drag coefficient. For supersonic speeds, C_d has a nearly constant value close to 2 for free-molecule flow (0.92 in the case of continuum flow). Combining Eqs. (D.1) and (D.2) gives

$$\delta M/\delta V = \sigma MV \tag{D.3}$$

which defines the ablation coefficient $\sigma = \tau_E/C_d$. The ablation coefficient is expected to be nearly constant for bodies of similar composition that experience a gradual and regular ablation in the course of their atmospheric trajectory. However, if the body dissolves in a train of fragments, then sigma will change. This obscures the meaning of this parameter. The deceleration of the meteoroid usually proceeds more rapidly than expected due to fragmentation. The variation of log (observed $\delta V/\delta t$ divided by expected $\delta V/\delta t$) is approximately proportional to that of the quantity log ($M_{inf}/M - 1$). This proportionality factor is the progressive fragmentation index χ_f. For a single body with no fragmentation, $\chi_f = 0$. Higher values of χ_f indicate a progressively more serious fragmentation cascade.

Appendix E

Secular perturbation to form northern and southern branches

Secular perturbations of short-period meteoroid streams such as those of the Geminids, Taurids, and Quadrantids are described approximately by the integrals of motion, from Babadzhanov[21]

[20] D. W. R. McKinley, *Meteor Science and Engineering* (New York: McGraw-Hill, 1961), 309pp.

[21] P. B. Babadzhanov, Formation of twin meteor showers. In *Asteroids, Comets, Meteors III* (Uppsala: Astronomical Observatory, 1989), pp. 497–503.

$$(1 - e^2)\cos^2(i) = C_1 \tag{E.1}$$

$$e^2(0.4 - \sin^2 i \sin^2 \omega) = C_2 \tag{E.2}$$

$$\omega + \Omega = \Pi = C_3 \tag{E.3}$$

where C_1, C_2, and C_3 are constants derived from the observed orbital elements of a parent body or meteoroid stream. With different evolution rates for different semi-major axes, these perturbations will fill out the total volume in space constrained by the above Eqs. (E.1)–(E.3). For a given set of (a, e) there are four possible values for the argument of perihelion that results in an encounter with Earth[22]

$$\pm\cos(\omega) = [a(1 - e^2) - 1]/e \tag{E.4}$$

Appendix F

D-*criterion for stream association*

The similarity between two orbits can be expressed in terms of a criterion developed by Southworth and Hawkins[23] and called the *D-criterion* (a, e, i only) or D_{SH}-*criterion* (including the node and Π). It defines an ellipse in q, e, i, and Π-space, whereby the difference of angular orbital elements is measured by their chords (twice the sin of half the angle):[24]

$$D^2 = (e_1 - e_2)^2 + (q_1 - q_2)^2 + (2\sin[(i_1 - i_2)/2])^2 \tag{F.1}$$

$$\begin{aligned} D_{SH}^2 = {} & D^2 + \sin(i_1)\ \sin(i_2)(2\ \sin[(\Omega_1 - \Omega_2)/2])^2 \\ & + (0.5(e_1 + e_2) \times 2\ \sin[(\Pi_1 - \Pi_2)/2])^2 \end{aligned} \tag{F.2}$$

for small inclinations of less than about 20°. The variation of $\Pi_1 - \Pi_2$ is more complicated for higher inclinations. The orbit is more similar to the proposed mean of the stream, and the stream association is more likely, if the D-criterion is smaller. To validate stream association, one may adopt $D_{SH} < 0.2$ (but see Chapter 26). Alternative versions of this criterion have been developed, including that by Drummond (D_D), Valsecchi *et al.* (D_N), and Jopek (D_H). They all tend to retrieve about 70% of simulated stream members in a noisy dataset of well-chosen cut-off levels.[25]

[22] D. I. Steel, D. J. Asher and S. V. M. Clube, The structure and evolution of the Taurid complex. *Mon. Not. R. Astron. Soc.* **251** (1991), 632–648.

[23] R. B. Southworth and G. S. Hawkins, Statistics of meteor streams. *Smithsonian Contrib. Astrophys.* **7** (1963), 261–285.

[24] *Ibid.*

[25] D. P. Galligan, Performance of the D-criteria in recovery of meteoroid stream orbits in radar data set. *Mon. Not. R. Astron. Soc.* **327** (2003), 623–628.

A more objective criterion would be to use the Tisserand invariant with respect to Jupiter ("J"), best expressed in terms of q and e, rather than the semimajor axis (*T criterion*):

$$T_J = a_J(1 - e)/q + 2\cos i\, \sqrt{[q\ (1 + e)/a_J]} \tag{F.3}$$

whereby a_J refers to the perihelion distance of Jupiter. ΔT_J defines an error ellipse in Tisserand invariant space. Whipple used a variant of this, which he called the *K criterion*.

Appendix G

Zenith Hourly Rate of meteors and meteoroid Influx

The Zenith Hourly Rate[26]

$$\mathrm{ZHR} = (N/T_{\mathrm{eff}})\chi^{(L_{\mathrm{m}}-6.5)} C_{\mathrm{p}}\ \sin(h_{\mathrm{r}})^{-1} \tag{G.1}$$

where N is the number of shower meteors seen in a period T_{eff} (hr). L_{m} is the star limiting magnitude, the faintest star visible in averted vision. h_{r} is the radiant elevation. C_{p} is the personal correction coefficient = 1.0 for a standard observer having a probability of detecting meteors above elevations of 32° while watching the zenith: P(+6) = 0.001, P(+5) = 0.009, P(+4) = 0.09, P(+3) = 0.32, P(+2) = 0.48, P(+1) = 0.63, P(+0) = 0.70, P(−1) = 0.73, and P (−2) = 0.75.

The influx of particles of magnitude 0 per cm^2 per second is given by

$$F_{\mathrm{N}}(m = 0) = \mathrm{ZHR}_{\mathrm{max}}\ (0.4 + 0.6\chi)/(\Sigma_{\mathrm{m}} P(\mathrm{m})\chi^{\mathrm{m}})\ A_{\mathrm{e}}\ 3600 \tag{G.2}$$

A_{e}, the effective surface area visible for an observer above elevations of 32°, ranges from $3.84 \times 10^{14}\,\mathrm{cm}^2$ to $7.74 \times 10^{14}\,\mathrm{cm}^2$ for apparent speeds of 20 and 72 km/s, respectively. $\Sigma_{\mathrm{m}}\ P(m)\ \chi^{\mathrm{m}}$ equals 15 for $\chi = 2.5$ and 43 for $\chi = 3.5$. The total mass influx of a stream over a range of meteor magnitudes is

$$F_{\mathrm{M}} = F_{\mathrm{N}}(m = 0)\Sigma_{\mathrm{m}}\mathrm{M}(m)\chi^{\mathrm{m}} \tag{G.3}$$

[26] P. Jenniskens, Meteor Stream Activity. I. The annual streams. *Astron. Astrophys.* **287** (1994), 990–1013.

Table 1 *Historic reports of meteor showers*

Compilation of dated historic records of meteor showers from catalogues by Eduard Biot, Ishiro Hasegawa, Sang-Hyeon Ahn, Umberto Dall'Olmo, W. S. Rada, and others.[1] The shaded entries are possibly due to the same meteor shower. Dates are given according to the Julian calendar before AD 1582 and to the Gregorian calendar after AD 1583. Solar longitudes are in Equinox J2000. Asterisks mark identified outbursts.

$\lambda_\odot$	Date (UT)	Description	Seen from	Shower
5.0	571 Mar. 03.9	Planets shooting	Mecca	
6.0	1857 Mar. 24.7	Stars fell like rain	China	
21.9	581 Mar. 20.7	Stars fell like rain	Korea	**Shower 1**
22.2	36 Mar. (17.2)	>100 to NW, N and NE for two nights	China	(HT)
25.4	1008 Mar. 26.7	More than ten stars flew speedily and disappeared	China	
27.8	1891 Apr. 16.6	At 22 h, stars fell like rain for hours	China	
32.5	687 BC Mar. 23.7	Stars file like rain	China	**Lyrids**
32.2	15 BC Mar. (27.7)	During night, stars fell like rain until cock-crow	China[2]	(LP)
32.0	1803* Apr. 20.3	A great shower in morning	China	
34.6	464 Apr. 01.7	Stars fell	China	
33.4	1040 Apr. 04.1	Stars fell, early in the morning	Europe	**Shower 3**[3]
33.6	1094 Apr. 05.2	A shower till dawn in all directions	Europe	(JF)
32.2	1095 Apr. 04.0	Fall of stars, westward, all night	Europe	
33.0	1096 Apr. 04.1	Stars moved from cock-crow to dawn	Europe	
32.3	1122 Apr. 04.0	Innumerable stars fell	Europe	
32.2	1123 Apr. 04.2	Almost a rain of stars, before dawn	Europe	

[1] I. Hassegawa (1993) Historical records of meteor showers. In: *Meteoroids and their Parent Bodies*. ed. J. Stohl and I. P. Williams (Bratislava, Astron. Inst. Slovak Acad. Sci., 1993), pp. 209–223; U. Dall'Olmo, Meteors, meteor showers and meteorites in the middle ages: from European medieval sources. *JHA* **9** (1978), 123–134; M. Biot, Ed. (1841) *Catalog Général des Etoiles Filantes et des Autres Météores Observés en Chine pendant 24 Siècles*; M. Biot, Ed. Chinese meteors. *Mem. Acad. Sci. Inst. Nat. France* **10** (1848), 192; Z. Tian-shan, Ancient Chinese records of meteor showers. *Chinese Astronomy* **1** (1977), 197–220 (first published in 1966); S. Imoto, I. Hasegawa, Historical records of meteor showers in China, Korea and Japan. *Smiths. Contr. Astrophys.*, **2** (1958), 131 (updated in 1993); W. S. Rada, F. R. Stephenson, A catalogue, of meteor showers in Medieval Arab chronicles. *Q. J. R. Astron. Soc.* **33** (1992), 5–16; M. R. Kidger, Some comments on the identification of medieval meteor showers recorded by the Arabs. *Q. Jl. Roy. Astron. Soc.* **34** (1993), 331–334; S.-H. Ahn, Meteoric activity in the 11th century. *Mon. Not. R. Astron. Soc.*, **358** (2005), 1105–1115.

[2] Date uncertain. Could be March 24.8 ($\lambda_\odot = 29.3$), in which case this is not a Lyrid return.

[3] This is not likely to be the Lyrid shower. Although the period could be $P = 11.8$ yr, reflecting Jupiter's perturbation of a dust trail, the variations in the node are too large for a long-period comet dust trail. A shorter orbital period with $P \sim 5.9$ yr is more likely.

Table 1 (*cont.*)

$\lambda_{\odot}$	Date (UT)	Description	Seen from	Shower
32.4	1136 Apr. 03.7	Large number of stars flew from NE to SW	Korea	**Shower 3**
29.3	1204 Apr. 01.0	Sparking stars during an aurora	Europe	
41.3	74 BC Apr. (06.0)	At dawn, a number fell to W together	China	**EAQ**[4] (HT)
41.5	401 Apr. 08.7	Many red-hued stars flew to the west	China[5]	
42.0	443 Apr. 09.9	A countless number of meteors flew N till dawn	China	
41.0	466 Apr. 08.8	Countless number to the West, ceased at dawn	China	
41.5	530 Apr. 09.7	Bright meteors moved to NW, thousand in number	China	
40.2	839 Apr. 10.6	Nearly 200 meteors flew to W	China	
43.2	839 Apr. 13.7	More than 200 meteors flew to W	China	
43.3	905 Apr. 13.7	One large, many small meteors moving to S like rain	China	
42.7	927 Apr. 13.7	Many stars flew to NW	China	
42.9	934 Apr. 13.7	Many stars flew to NW	China	
42.8	935 Apr. 13.9	From 4th to 5th watch, >100 flew to the west	China	
45.0	1179 Apr. 17.7	More than a hundred stars flew from E to W	Korea	
52.7	461 Apr. 20.7	Several hundreds of long stars (large and small)	China[6]	
53.4	245 BC Apr. 16.7	Uncountable meteors went to W	China	
57.4	1212 Apr. 30.0	Large number of stars shooting	Cairo	
66.6	839 May 08	Very fast meteors E > W, many nights	Europe	
69.2	839 May 10.7	From 1st to 5th watch, >100 meteors crossed	China	
70.7	935 May 12.9	Stars shooting, never seen before	Baghdad	
83.1	466 May 22.7	>100 meteors went to SW	China	**Shower 5**[7]
84.1	1539 May 30.7	Rain	China	(HT)

[4] These outbursts have been associated with the η-Aquariids. The node shifts by +0.0039°/yr. The returns occur in a periodic pattern with $P = 3.815$ yr and $T_p = 919.84$ AD, one-third the orbital period of Jupiter, suggesting Halley's older dust trails were brought periodically in Earth's path.

[5] Zhuang Tian-shan derives the approximate radiant at R.A. = 326°, Decl. = −1°.

[6] *Possibly* associated with the comet C/461 – M. Kresáková, Associations between ancient comets and meteor showers. *Astron. Astrophys.* **187** (1987), 935–936.

[7] Periodic pattern suggests a Halley-type comet with $P = 178.6$ yr. Possibly associated with sightings of C/1539 H_1 – M. Kresáková, Associations between ancient comets and meteor showers. *Astron. Astrophys.* **187** (1987), 935–936; The

Table 1 (*cont.*)

$\lambda_\odot$	Date (UT)	Description	Seen from	Shower
83.1	1733 June 10.4	At sunset, rain	China	**Shower 5**
83.0	1883 June 12.7	Rain	China	
84.9	12 BC May (21.4)	Stars to SE at 3–5 p.m. in daytime, ceased at dusk	China	
88.9	735 May 30.7	During the night many stars flew chaotically	Japan	
92.9	1651 June 19.7	A meteor large as a cup followed by many small	China	
94.1	1861 June 23.7	In NE, meteors fell like rain	China	
96.6	1547 June 12.9	Stars fell like rain at 4 a.m.	China	π-Cetids?
98.9	1884 June 28.7	Stars fell like rain	China	
99.4	1414 June 14.7	In evening a fireball. Continued until midnight	Korea	
102.3	1644 June 28.7	Stars fell like rain	China	
102.8	1122 June 16.5	Evening, three nights, meteors N > S	China	
104.1	1814 July 04.2	Shower	Canada	**Shower 6**
104.2	1636 June 30.7	Stars fell like rain	China	(LP)
104.9	1645 July 01.7	At night, large and small	China	
107.2	1098 June 21.0	Several angels flying. A shower	Europe	
110.0	1643 July 07.6	During 1st and 2nd watch all stars in sky trembled	Korea	
111.8	841 June 23.9	During 5th watch, >50 meteors scattered to S	China	**Shower 7**[8] (HT/LP)
111.9	1511 June 28.7	Flaming stars at night	China	
111.4	1560 June 27.7	Stars fell like rain	China	
111.0	1862 July 11.7	Numerous stars fell to SW	China	
116.4	1563 July 03.7	Meteors fell like rain	China	
118.5	1519 July 05.7	Stars scattered	Korea	
119.2	1645 July 16.7	Meteors fell like rain all night	China	
119.8	1235 July 05.3	In daytime, meteors fell like rain	China	
125.5	1666 July 23.7	Stars fell like weaving	China	**Shower 8**[9]
125.6	1646 July 23.7	Stars fell like rain	China	(JF)
125.9	1645 July 23.7	Small stars fell like snow just before midnight	China	
123.2	1847 July 24.7	Many stars fell	China	

theoretical radiant of this comet is at R.A. = 170°, Decl. = +64°, V_g = 17 km/s on May 16 (node = 55.1), two weeks off. Associated with June Lyrids by G. Kronk.

[8] Periodic pattern with P = 343 yr, or fraction thereof. Periodic return with the orbit of Jupiter also possible.

[9] Strong change of node by −0.0129°/yr. Most likely P = 5.17 or 6.70 yr. On the other hand, reports could refer to δ-Aquariid shower.

Table 1 (*cont.*)

$\lambda_\odot$	Date (UT)	Description	Seen from	Shower
128.4	784 July 10.7	Stars fell by tens and ...	China	
132.6	708 July 14.6	Uncountable in all directions	China	**Shower 9**[10]
133.2	714 July 15.8	Uncountable stars flew to the NW till dawn	China	(JF)
139.8	36 July 17.9	At dawn, >100 meteors flew in four directions	China	**Perseids**[11] (HT)
140.4	466 July 21.7	Over 100 going SW		
139.0	830 July 22.6	From dusk till 5th watch, uncountable stars fell	China	
140.3	833 July 23.7	Over a hundred, all night		
137.8	835 July 21.6	From dusk till 3rd watch >20 meteors	China	
138.4	841 July 21.7	From 1st until 5th watch, >50 meteors scattered	China	
138.1	924 July 21.7	Many stars flew chaotically across one another	China	
140.0	924 July 23.7	Many stars flew crossing each other	China	
137.9	925 July 21.7	After first watch, more than 70 stars to SW	China	
138.8	925 July 22.7	Many stars flew at midnight	China	
139.8	925 July 23.7	Many small stars flew in SW	China	
138.6	926 July 22.7	Many small stars flew chaotically	China	
136.8	933 July 20.7	Many stars flew across one another	China	
141.7	933 July 25.7	Many stars flew across one another	China	
140.3	989 July 24.7	Several stars were scattered	Japan	
135.8	1007 July 20.6	Meteors flew to N	Japan	
139.7	1007 July 24.6	Many meteors flew till dawn	Japan	
140.6	1007 July 25.6	Meteors appeared again	Japan	
140.8	1042 July 25.7	A large number of stars flew and rolled	Korea	
139.2	1065 July 24.0	Trains of small stars	Baghdad	
140.2	1095 July 25.7	One large, a large number of small stars flew to S	Korea	

[10] Zhuang Tian-shan gives the radiant at R.A. = 303°, Decl. = +24° (AD 708) and R.A. = 356°, Decl. = +60° (AD 714).

[11] Least-squares fit: node shifts only +0.00038°/yr. There are also reports from AD 1846–1902 at $\lambda_\odot = 130.0°–135.9°$, which are also likely to be Perseids.

Table 1 (*cont.*)

$\lambda_{\odot}$	Date (UT)	Description	Seen from	Shower
142.3	1106 July 27.7	Large number of small stars flew to W	Korea	**Perseids**
136.7	1243 July 23.0	Stars fell	Europe	
138.6	1243 July 26.0	Running stars lasting whole night	Europe	
139.6	1243 July 26.0	Numerous stars fell (peak shower)	Europe	
140.1	1451 July 27.8	More than 80 large or small meteors appeared	China	
139.1	1556 July 26.7	More than 30 stars flew to S	China	
137.8	1565 July 25.7	Stars flew chaotically to the N or the S	Japan	
138.3	1581 Aug. 05.7	In SW, stars fell like rain	China	
139.4	1590 Aug. 06.7	Stars fell like rain	China	
137.5	1625 Aug. 04.7	Every night, meteors fell like weaving	China	
139.3	1645 Aug. 06.7	Stars flew to S like rain	China	
140.2	1645 Aug. 07.7	Stars flew crossing until dawn	China	
140.4	1851 Aug. 11.7	Stars fell like arrows to S for 2 hours	China	
142.3	1859 Aug. 13.7	Stars fell like weaving	China	
139.9	1861 Aug. 10.7	Meteors fell from N to S unceasingly	China	
143.7	1861 Aug. 14.7	Stars fell like rain	China	
138.7	1862 Aug. 09.67	At 22:00, uncountable number of stars NE > SW	China	
139.7	1862 Aug. 10.7	A countable number small + large fell like rain	China, Japan	
140.5	1862 Aug. 11.6	Stars fell crossing	China	
141.5	1862 Aug. 12.6	Stars went to SW	China	
143.5	1866 Aug. 14.7	Stars flew to S	China	
139.4	1871 Aug. 10.7	In NE, stars fell like rain	China	
141.3	1906 Aug. 13.7	Many stars fell from N to S, like rain	China	
144.4	551 July 26.7	Countless number flew during night to N or NW	China	
145.4	465 July 26.7	A large number of meteors flew to the SW	China	
146.9	268 July 26.7	Many stars flew to the west like rain	China	

Table 1 (*cont.*)

$\lambda_{\odot}$	Date (UT)	Description	Seen from	Shower
147.8	865 July 31.7	1st watch, fireball followed by group of small N > S	China[12]	
148.9	1645 Aug. 16.7	Stars fell like rain	China	
146.9	1860 Aug. 17.7	Stars fell like rain	China	**Shower 11**
148.4	1853 Aug. 19.5	At 20 h, in SW rain	China	(JF)
148.7	1861 Aug. 19.7	In evening, to SW	China	
149.7	1861 Aug. 20.7	Meteors flew to SW till dawn	China	
150.5	464 July 31.7	Many stars went to W or SW	China	**Shower 12**[13]
150.2	551 Aug. 01.7	Uncountable, small, crossed in all four directions	China	(HT/LP)
150.3	1581 Aug. 07.7	In SW, stars fell like rain	China	
153.8	1621 Aug. 21.6	At 11 p.m., large meteor followed by several 100s from NW	China	
152.0	1839 Aug. 23.6	At 11 p.m., like weaving for hours	China	**Shower 13**[14] (JF)
152.1	1858 Aug. 23.6	About 1000 stars fell NW > SE	China	
152.7	1860 Aug. 23.7	In W, stars fell like rain	China	
151.5	1888 Aug. 22.7	Stars fell like rain N to S	China	
154.3	1557 Aug. 11.7	In SE, stars fell, soon after in NW	China	
154.2	1877 Aug. 25.7	For three nights, stars fell like arrows	China	
156.7	1863 Aug. 28.7	To SE, until dawn	China	
157.9	1589 Aug. 25.7	Many flew in the full Moon	China	**Aurigids**[15]
159.0	1885 Aug. 30.7	Uncountable meteors	China	(LP)
160.6	1898 Sep. 01.7	Meteors fell like weaving	China	
161.7	1882 Sep. 02.7	At midnight, like rain for hours	China	
162.1	902 Aug. 15.9	Planets scattered all night	Baghdad	
163.6	1886 Sep. 04.7	Meteors fell like rain	China	
167.1	1037 Aug. 21.7	Several 100s flew to SW, many bright	China	**Shower 15**[16] (LP)
167.4	1063 Aug. 22.7	Hundreds of meteors crossed in W	China	

[12] Zhuang Tian-shan gives the approximate radiant at R.A. = 352°, Decl. = −34°.
[13] Perhaps periodic with $P = 86$ yr, or fraction thereof. Perhaps related to κ-Cygnids. Zhuang Tian-shan gives the radiant at R.A. = 201°, Decl. = +60°.
[14] Evening shower, NW → SE.
[15] Associated with comet C/1911 H_1 (Kiess).
[16] Periodic with 11.86 yr period of Jupiter. No change in node. If long period, then it is unlikely that this stream is associated with sightings of a comet in Korea on August 22, 1063 at same time of shower, as proposed by M. Kresáková, Associations between ancient comets and meteor showers. *Astron. Astrophys.* **187** (1987), 935–936.

Table 1 (*cont.*)

$\lambda_{\odot}$	Date (UT)	Description	Seen from	Shower
166.1	1548 Aug. 23.6	From first to second watch countless meteors	Korea	**Shower 15**
167.2	1548 Aug. 24.7	Meteors fell in W	Korea	
167.1	1560 Aug. 24.7	During 4th to 5th watch, like rain in all directions	Korea	
167.1	1560 Aug. 24.7	Meteors fell like rain	China	
167.9	1857 Sep. 08.7	Many, from NE > W for hours	China	
170.6	1698 Sep. 08.7	At that night, meteors flew like fabrics of silk	Japan	
173.4	1682 Sep. 11.5	At the 2nd watch, 100s of stars fell	Taiwan	
174.1	884 Aug. 27.6	From 20:00 to 24:00, meteor showers appeared	Japan	
176.1	884 Aug. 29.7	At midnight, uncountable fell in N	Japan	**Shower 16**[17] (LP)
175.9	1560 Sep. 02.7	Stars fell like rain	China	
176.0	1610 Sep. 13.7	In N, stars moved for hours	China	
177.3	532 Aug. 28.7	Stars fell like rain	China, Korea	
182.8	1103 Sep. 07.7	Three meteors appeared and hundred small stars	Korea	
184.4	1898 Sep. 26.1	From Triangulum (Mr. Hansay, from balloon)	Europe	
188.2	881 Sep. 11.2	Two nights: stars fell like rain	China	**Shower 17**[18] (HT)
188.6	881 Sep. 11.7	Stars fell again	China	
188.1	1012 Sep. 11.7	More than 20 stars went to N	China	
187.1	1683 Sep. 25.7	Meteors often appeared	Korea	
192.5	1178 Sep. 17.7	Seven meteors + countless stars in all directions	Korea	
199.4	1798 Oct. 09.7	Stars flew all around, next few nights too	China	
206.1	288 Sep. 25.3	Stars flew like rain in daytime	China	**Orionids** (HT)
207.5	288 Sep. 26.7	Stars fell like rain	China	
202.4	585 Sep. 23.7	Hundreds of meteors flew to the four directions	China	
203.2	903 Sep. 27.0	A shower lasted all night	Europe	
201.9	930 Sep. 25.7	Many stars flew chaotically, fell beyond horizon	China	

[17] Based on a constant node over a long time interval. 1610–1560 = 50 yr.

[18] Based on a two-day observing interval (makes LP unlikely). Zhuang Tian-shan gives an approximate radiant at R.A. = 31°, Decl. = +19°.

Table 1 (*cont.*)

$\lambda_{\odot}$	Date (UT)	Description	Seen from	Shower
206.9	1012 Sep. 30.7	More than 20 stars small and large flew to N	China	**Orionids**
207.5	1111 Oct. 02.7	A large number of small stars in all directions	Korea	
206.2	1436 Oct. (02.7)	From dusk till dawn, >100 large + small meteors	China	
208.4	1439 Oct. 05.7	About 260 meteors appeared	China	
206.8	1465 Oct. 03.7	Meteors flew from NE to SW	Japan	
206.1	1651 Oct. 14.7	Stars fell like rain	China	
207.1	1651 Oct. 15.7	Stars fell like rain	China	
204.5	1864 Oct. 15.7	Meteors crossed like weaving	China	
216.3	1178 Oct. 11.7	Uncountable stars flew to W	Korea	**Shower 19?**[19]
213.8	1535 Oct. 11.7	Many stars moved	China	
212.2	1623 Oct. 20.7	Stars fell like rain	China	
210.7	1902 Oct. 23.7	Stars fell like rain	China	
217.5	1622 Oct. 25.7	Stars fell like rain	China	
218.1	1683 Oct. 26.9	Meteors appeared in 3rd through 5th watch	Korea	
219.4*	902[20] Oct. 13.0	A shower seen in Taormina, Sicily	Egypt, Italy	**Leonids** (HT)
221.8	931 Oct. 15.8	More than 100 meteors in all directions	China	
221.9	931 Oct. 15.96	From 5th watch till dawn, >100 small meteors flew	China	
222.7	931 Oct. 16.7	Many stars fell crossing	China	
219.9*	934 Oct. 13.7	Stars flew like rain in the SW	China[21]	
220.2*	934 Oct. 14.0	A shower	Europe	
220.9	934 Oct. 14.7	Uncountable stars flew crossing	China	
220.9	934 Oct. 14.7	Uncountable stars flew crossing	China	
219.0	935 Oct. 13.1	Planets shooting, never seen before	Baghdad	
220.4*	967 Oct. 14.7	Stars scattered from NE to SW	Japan	
220.5*	1002 Oct. 14.7	Meteors scattered chaotically during the night	Japan	
220.5*	1002 Oct. 14.7	A few thousands of meteors Cnc > feet of Uma	China	
220.6*	1002 Oct. 14.8	Meteors fell all night	Japan	

[19] Could be HT with precession –0.0079°/yr, or sightings of Orionid shower.
[20] Rada and Stevenson give the date as October 14, 902.
[21] * = storm identified with dust trail encounter (Table 4).

Table 1 (*cont.*)

$\lambda_{\odot}$	Date (UT)	Description	Seen from	Shower
221.6	1002 Oct. 15.8	Meteors fell during 0 and 4 a.m. watch	Japan	**Leonids**
220.1	1035 Oct. 14.8	Disaster of meteors happened during the night	Japan	
220.5*	1037 Oct. 14.7	Disaster of meteors happened during the night	Japan	
222.4	1101 Oct. 17.0		France	
223.4*	1202 Oct. 18.9	Stars fell from the sky to E, W, and scattered like locusts	Egypt	
223.3*	1237 Oct. (18.8)	Meteors appeared in the morning	Japan	
223.0	1238 Oct. (18.7)	From 22:00 to 02:00 a countable number	Japan	
224.4	1366 Oct. 21.0	Sparkling of stars	China	
225.5	1466 Oct. 22.7	Meteors appeared	Japan	
227.5	1498 Oct. 24.9	Fireball, followed by tens small	China	
227.6*	1532 Oct. 24.7	During night seven meteors + stars fell like rain	Korea	
228.8	1532 Oct. 25.92	During fourth watch, stars fell like rain	China	
237.7	1532 Nov. 03.7	Stars fell like rain in all directions	China	
227.5	1533 Oct. 24.9	Meteors fell like rain in all directions	Korea	
227.6	1533 Oct. 24.95	From 4th watch till dawn, countless fell	China	
228.6	1533 Oct. 25.92	During 4th watch till dawn thousands of stars fell	China	
228.6	1533 Oct. 26.0	At dawn, stars fell like snow	Japan	
229.7	1533 Oct. 26.95	From 4th watch till dawn, countless fell (intense)	China	
231.9	1533 Oct. 29.25	In daytime, stars fell like a rain	China	
237.4	1533 Nov. 03.7	Stars fell like rain in three nights	China	
229.1	1538 Oct. 26.7	Meteors appeared in all directions	Korea	
226.9	1554 Oct. 24.7	Meteors appeared at intervals	Korea	
227.9	1566 Oct. 25.7	Stars fell like a rain, following two nights also	China	
228.9*	1566 Oct. 26.7	Uncountable small meteors in all directions	Korea	
229.9	1566 Oct. 27.7	Stars fell like rain	China	
228.7	1594 Nov. 05.7	Thousands of meteors scattered from all directions	China	
229.1	1601 Nov. 05.9	At dawn, stars fell like rain	China	

Table 1 (*cont.*)

$\lambda_{\odot}$	Date (UT)	Description	Seen from	Shower
228.6	1602 Nov. 05.7	Hundreds of large or small stars fell chaotically	China	**Leonids**
229.7*	1602 Nov. 06.7	Hundreds of large or small stars fell chaotically	China	
235.7	1602 Nov. 12.7	Rain and hail of stars in all directions	Korea	
227.8	1625 Nov. 04.7	Meteors appeared in all directions	Korea	
228.8*	1625 Nov. 05.7	Many stars fought in W	Korea	
229.8	1625 Nov. 06.7	More than 10 meteors in midair	Korea	
230.4	1666 Nov. 07.8	Large star, followed by many small, in NE	China	
231.1	1698 Nov. 08.8	Meteors fell like weaving	Japan	
226.4	1798 Nov. 05.7	Fell like rain	China	
232.8*	1799 Nov. 12.3	Leonids observed by Humboldt and others	Venezuela	
233.2*	1832 Nov. 13.2	Stars scattered late in night	Saudi Arabia	
233.1	1833 Nov. 13.0	An unusual number of meteors	USA	
231.7	1867 Nov. 12.7	Stars fell like rain in NW	China	
232.7	1867 Nov. 13.7	Stars flew frequently	Japan	
214.2	1623 Oct. 22.7	In Guyuan Xian, stars fell like rain	China	
215.1	1885 Oct. 26.7	Stars fell like rain	China	
217.5	1622 Oct. 25.7	Stars fell like rain in Shanxi province Guyuan Xian	China	
217.1	1885 Oct. 28.7	Stars fell like rain	China	
220.3	1841 Oct. 31.7	Stars fell like rain, at midnight	China	**Shower 21**
220.5	1856 Oct. 31.7	Many stars fell	China	(JF)
220.8	1858 Nov. 01.5	In daytime, meteors fell from W > E	China	
221.4	1872 Nov. 01.7	At midnight, stars fell like rain	China	
221.5	1731 Oct. 31.7	Stars fell like rain	China	
221.6	1885 Nov. 02.2	Two nights, like weaving, like rain	China	
223.4	855 Oct. 16.9	Westward meteors, whole night	Europe	
229.8	1625 Nov. 06.7	Tens of bright red meteors 3rd to 5th watch	Korea	**Shower 22** (Taurids?)
228.4	1798 Nov. 07.7	Weaving, lasted for two nights	China	
228.3	1818 Nov. 08.7	Fell like rain in S	China	
229.6	1883 Nov. 10.7	Rain	China	
230.0	1893 Nov. 10.7	Rain	China	

Table 1 (*cont.*)

$\lambda_\odot$	Date (UT)	Description	Seen from	Shower
234.2	1855 Nov. 14.8	Like rain, after midnight	China	**Shower 23**
234.6	1661 Nov. 11.7	Stars fell like rain from SE to NW	China	(LP)
237.8	1395 Nov. 03.7	Stars flew and fell	China	
238.6	1533 Nov. 04.8	After 3rd watch, stars fell like rain	China	
239.2	1885 Nov. 19.7	At midnight, stars fell like rain	China	**α-Monocerotids?**
240.7	643 Nov. 01.7	A large number of stars flew to W	Korea	**Shower 25**[22] (HT/LP)
240.8	970 Nov. 03.7	Numerous stars flew to N	China	
240.8	1665 Nov. 17.9	At 6 a.m., stars fell like rain	China	
241.5	1798 Nov. 20.7	N > S rain	China	
242.0	1235 Nov. 06.8	Stars were moving during the 5th watch	Japan	
242.9	1630 Nov. 20.0	White lighting at head of Orion in form of whisps	Germany[23]	
243.2	1799 Nov. 22.6	Stars fell like rain	China	
247.3	1534 Nov. 13.7	Stars fell like rain	China	
247.5	1798 Nov. 26.6	At twilight, stars fell like rain	China	**Shower 26**[24]
248.6	1798 Nov. 27.7	Stars crossing like weaving	China	
247.2	1799 Nov. 26.6	At evening twilight in SW, weaving	China	
248.2	1799 Nov. 27.6	At 18 h, stars flew crossing	China	
250.4	1798 Nov. 29.5	At 21 h, stars fell like rain in NE	China	
249.1	1531 Nov. 15.7	Rain	China	
252.1	1539 Nov. 18.7	At midnight, rain	China	
257.6	1741 Dec. 06.0	St. Petersburg	Russia	**Andromedids**
257.7	1798 Dec. 06.7	In NW, stars fell like rain	China	(JF)
257.8	1798 Dec. 06.8	400 in few hours Brandes	Germany	
257.8	1798 Dec. 06.8	In the night, a large number flew chaotically	Japan	
260.8	1798 Dec. 09.7	Small meteors fell like rain as before	Japan	
253.5	1825 Dec. 03.5	Two nights, moved like weaving NW > SE	China	
257.7	1830 Dec. 07.9	Andromedids	France	

[22] Based on constant node over long time interval.
[23] R. Gorelli, The meteor shower of November 19, 1630. *WGN* **19** (1991), 47–50.
[24] Possibly linked to Andromedids or annual Taurids.

Table 1 (*cont.*)

$\lambda_\odot$	Date (UT)	Description	Seen from	Shower
256.4	1838 Dec. 06.7	Stars fell in the whole sky	China	**Andromedids**
256.7	1838 Dec. 06.9	Andromedids	Europe and America	
247.7	1872* Nov. 27.8	At midnight, stars fell like rain	China and Europe	
254.9	1875 Dec. 05.7	Rain	China	
252.6	1884 Dec. 02.7	Rain	China	
247.4	1885* Nov. 27.8	Andromedids storm	Europe	
245.0	1886 Nov. 25.7	Stars fell like rain	China	
243.7	1887 Nov. 24.7	Stars fell crossing for hours	China	
248.8	1887 Nov. 29.7	Like rain, in NW	China	
252.9	1887 Dec. 03.8	Medium brightness, long yellow	Austria	
254.4	1534 Nov. 20.7	Rain	China	
258.7	1533 Nov. 24.7	Stars fell like rain	China	**Geminids**
259.2	1566 Nov. 25.7	259.2 meteors fell like weaving	China	(JF)
259.0	1797 Dec. 07.7	Stars fell like rain	China	
259.9	1856 Dec. 09.7	Stars fell like rain	China	
257.2	1882 Dec. 07.7	Many stars fell like rain from NE to SW	China[25]	
258.7	1884 Dec. 08.7	Many stars flew to W about 4 hours	China	
259.7	1884 Dec. 09.7	Stars fell like weaving for hours	China	
260.7	1884 Dec. 10.7	Stars fell like weaving for hours	China	
261.7	1884 Dec. 11.7	Stars fell like weaving	China	
263.8	1798 Dec. 12.7	Meteors fell like rain	China	
264.3	685 Nov. 24.7	Stars fell like a shower	Japan	
265.7	1869 Dec. 15.7	Rain NE > SW	China	
265.3	1886 Dec. 15.7	Stars fell like weaving, for hours	China	
266.3	1886 Dec. 16.7	At midnight, rain for hours	China	
267.8	1732 Dec. 15.7	Rain	China	
268.2	1797 Dec. 16.7	Rain	China	
269.2	1797 Dec. 17.7	At midnight, like rain for hours	China	
271.6	1433 Dec. 06.7	Brilliant objects flying in successive nights	Japan	**Ursids?** (HT)
271.2	1532 Dec. 06.8	In all directions till dawn	China	
273.0	1794 Dec. 21.7	Stars fell like rain for hours	China	
271.7	1795 Dec. 20.7	Stars fell like rain	Japan	
272.4	1882 Dec. 22.7	Meteors fell like fire	China	
269.6	1799 Dec. 18.6	Rain	China	
277.1	1798 Dec. 25.7	Rain	China	

[25] Zhuang Tian-shan gives direction SE to E.

Table 1 (*cont.*)

$\lambda_\odot$	Date (UT)	Description	Seen from	Shower
280.5	1824 Dec. 29.7	Like rain, in NW	China	**Shower 30**[26]
278.1	1872 Dec. 27.7	Many stars fell crossing	China	(JF)
276.7	1885 Dec. 26.7	N > SE, continually throughout night	China	
278.8	1885 Dec. 28.7	At 5 p.m., many in SE	China	
277.2	1891 Dec. 27.7	Rain, for hours	China	
284.0	1179 Dec. 17.7	Large or small stars were moving during night	Japan	
283.1	1798 Dec. 31.6	Stars fell like rain	China	**Quadrantids**
283.6	1886 Jan. 02.5	Stars fell like rain from NW to S	China	(JF)
290.0	1566 Dec. (25.9)	Large star, followed by group of hundreds like rain	China	
294.5	784 Dec. 24.7	From 10 p.m. to 4 a.m., all stars were trembled	Japan	
296.3	1858 Jan. 14.7	Rain	China	
297.1	1886 Jan. 15.7	Many to SW	China	
299.7	609 Dec. (28.9)	Devils were struck by meteors	Mecca	**Shower 32**[27]
302.3	685 Jan. 00.7	At dusk, seven stars flew together to NE	Japan	(JF/HT)
304.1	685 Jan. 02.54	In Xu ke (22:00) stars fell like rain	Japan	
302.7	743 Jan. 02.0	Jets of fire (*Ictus ignei*)	Europe	
302.2	745 Jan. 01.0	Jets of fire (*Ictus ignei*)	Europe	
301.2	764 Dec. 31.2	From 2nd watch till dawn (two nights)	China	
303.6	1643 Jan. 18.7	Stars fell like rain	China	
305.1	685 Jan. 03.5	At 8 p.m., stars fell like rain	Japan	
306.0	1431 Jan. 09.7	Brilliant objects appeared during recent nights	Japan	
308.9	765 Jan. (07.7)	Uncountable meteors appeared	Korea	
310.6	1546 Jan. 14.7	Stars fell like rain	China	
311.9	773 Jan. 10.7	Stars fell like rain	Japan	
312.5	1909 Jan. 31.7	Many in all directions	China	
315.3	1637 Jan. 29.7	Stars fought in SE	Korea	
316.8	1799 Feb. 02.7	Stars crossed like weaving	China	
320.9	1060 Jan. 22.2	Shooting stars, huge scale	Baghdad	
322.8	1842 Feb. 09.7	Weaving	China	

[26] If related, rapid decrease of node: −0.047°/yr.

[27] Short visibility period, large variations of node. Possibly associated with comet seen in Japan during December 12, 684–January 10, 685 – M. Kresáková, Associations between ancient comets and meteor showers. *Astron. Astrophys.* **187** (1987), 935–936.

Table 1 (*cont.*)

$\lambda_\odot$	Date (UT)	Description	Seen from	Shower
324.3	308 Jan. 20.7	Stars flew and scattered, trembling	China	**Shower 33** (LP)
324.3	308 Jan. 20.7	Stars flew and were scattered with sound	China	
324.8	1729 Feb. 09.7	Many stars trembled, group moved to E	China	
324.9	1869 Feb. 11.7	Rain	China	
326.4	1606 Feb. 09.7	Stars fell in W	China	
327.9	1216 Jan. 30.2	A swarm of stars, before dawn	Europe	
334.4	913 Feb. (02.9)	Stars scattered towards E	Baghdad	
335.5	1680 Feb. 18.7	Meteors appeared for days	Korea	
339.7	1578 Feb. 12.7	Many stars flew to W	China	
340.8	1437 Feb. 12.7	During Wei ke (14:00) many brilliant objects	Japan	
343.0	1106 Feb. 13.0	Falling stars, long lasting shower	Italy	**Shower 34**[28] (HT)
343.8	1410 Feb. 15.7	Many stars trembled	Korea	
345.4	913 Feb. 13.9	Stars were flying, until midnight	Europe	
346.1	1892 Mar. 04.7	Stars fell like shooting arrows	China	
356.7	1648 Mar. 11.7	Stars fell like rain in NE	China	
357.1	808 Feb. 24.7	Stars fell like rain	China	
358.1	1888 Mar. 16.7	Stars fell like rain for three days	China	
360.6	1768 Mar. 17.45	In evening, stars fell like rain	China	

[28] If due to the same shower, this would have a relatively long visibility and constant node, changing less than –0.0027°/yr.

Table 2 *Possible showers coexisting with extinct comet nuclei*

Candidate (>1% probability) extinct Jupiter-family comets (JFC) discovered before August, 2005. The likelihood of an NEO being a JFC was evaluated by Bill Bottke. In the calculation, the strength of each of five NEO source regions are assumed to be: real alpha = 0.063 932 095, 0.080 195 352, 0.230 062 30, 0.252 756 68, and 0.373 054 60, for Jupiter-family comets, outer belt (OB), 3:1 resonance, Mars crossers (MC), and ν_6 secular resonance, respectively.[1] These are average probabilities from all of the nine surrounding a, e, and i bins, with 1.00 = 100%. The position of the stream relative to Earth's orbit is given as "inside" = node towards Sun, "below" = nearest point is below ecliptic and the node some distance from Earth's orbit.

Name	T_p	(Nearest star) Date	Shower $\lambda_\odot$	R.A.	Decl.	Probable source region: JFC	OB V_g	3:1 Δr	MC Ω	ν_6 $H_N{}^2$
2003 EH$_1$		(κ-Bootis)	**Quadrantids**			**0.437**	0.303	0.207	0.022	0.031
	2003-02-24.4	Jan. 04.0	282.9	229.9	+49.6		40.20	+0.214	Outside	**16.67**
2002 XO$_{14}$		(α-Arietis)	–.–			0.124	**0.703**	0.069	0.091	0.013
	2003-01-15.8	Jan. 09.9	289.0	33.0	+25.6		8.82	−0.006	Inside	21.84
2002 GZ$_8$		(φ-Pegasi)	(Daytime)			**0.264**	**0.543**	0.082	0.089	0.023
	2006-10-06.0	Jan. 30.3	309.7	358.8	+22.4		9.78	+0.002	Outside	18.15
2001 ME$_1$*			**Daytime Capricornid/Sagittarids**			**0.547**	0.236	0.146	0.049	0.021
	2005-12-06.4	Feb. 08.2	318.7	311.0	−21.2		29.55	−0.079	Outside	**16.81**
2005 FH			–.–			0.078	0.082	**0.619**	0.149	0.073
	2004-12-12.5	Feb. 12.1	322.7	19.3	−68.1		23.19	−0.044	Inside	17.45
2003 YM$_{137}$*		(δ-Cancri)	–.–			0.187	0.478	0.181	0.116	0.039
	2004-03-31.6	Feb. 19.0	329.6	131.5	+16.0		15.44	+0.045	Above	18.72
2002 MT$_3$		(σ-Hydri)	–.–			**0.335**	**0.580**	0.030	0.046	0.009
	2002-04-26.5	Feb. 20.73	31.3	121.0	+01.9		14.02	−0.037	Inside	19.68
1999 RD$_{32}$*		(θ-Leonis)	N.δ-Leonids?			**0.464**	**0.293**	0.134	0.077	0.032
	2003-09-29.8	Mar. 05.7	345.4	167.6	+13.5		22.8	+0.068	Above	**16.32**
2005 EM$_{69}$			Daytime q-Pegasids?			**0.388**	**0.481**	0.055	0.061	0.015
	2005-01-28.1	Mar. 07.9	347.6	351.6	+18.9		18.37	−0.002	Inside	24.67
2003 XM		(π-Persei)	–.–			**0.586**	0.402	0.007	0.003	0.001
	2004-02-07.2	Mar. 15.8	355.5	38.9	+37.0		10.55	+0.067	Outside	19.12

2003 SK_{84}*		(φ-Pegasi)	(Daytime)			0.009	**0.206**	**0.416**	**0.261**	0.108
	2003-11-24.9	Mar. 22.2	0.8	358.3	+16.6		20.62	+0.053	Below	21.12
2001 HA_4*		(α-Pegasi) $i = 17°$	–.–			**0.314**	**0.216**	**0.264**	0.132	0.074
	2005-07-19.8	Mar. 20.7	0.4	348.7	+17.5		24.95	−0.010	Inside	**17.48**
2004 GA		(α-Persei)	–.–			0.176	**0.719**	0.032	0.063	0.010
	2004-03-16.4	Apr. 02.0	12.6	49.3	+57.0		11.32	+0.057	Outside	20.04
2002 JZ_8		(χ-Leporis)	(Daytime)			0.106	**0.645**	0.078	0.149	0.022
	2002-03-31.2	Apr. 10.5	20.9	78.7	−12.6		10.99	−0.001	Inside	20.75
1998 SH_2*		(γ-Virginis)	**α-Virginids**			**0.330**	**0.420**	0.139	0.077	0.034
	1998-07-18.2	Apr. 16.8	27.1	195.3	−01.7		18.50	+0.010	Below	20.81
2005 GK_{141}			α-Bootids?			0.238	0.491	0.107	0.130	0.033
	2005-05-20.0	Apr. 26.3	36.4	200.5	+32.4		14.58	0.013	Outside	22.16
2000 DN_1		(α-Aurigae)	–.–			**0.335**	**0.580**	0.030	0.046	0.009
	2005-02-18.0	Apr. 29.7	39.7	80.6	+48.2		11.8	+0.008	Inside	19.72
2002 FC		(γ-Comae Berenices)	**N. γ-Virginids**			**0.335**	**0.580**	0.030	0.046	0.009
	2002-05-28.9	May 01.4	41.3	183.2	+24.6		11.18	+0.023	Inside	18.82
2004 BZ_{74}		(ε-Scorpii)	**α-Scorpiids**			**0.909**	0.070	0.016	0.002	0.003
	2003-12-11.5	May 19.9	59.2	249.1	−34.0		32.26	+0.030	Above	18.39
2002 KG_4		(α-Ursae Minoris) $i = 28°$	ι-Cassiopeiids?			**0.535**	**0.354**	0.075	0.019	0.018
	2002-05-02.9	May 24.0	63.2	97.2	+88.2		19.00	−0.034	Inside	20.85
1999 LT_1		(ε-Draconis) $i = 43°$	–.–			**0.674**	0.134	0.148	0.019	0.025
	2004-07-15.4	May 28.2	67.2	299.5	+73.7		26.0	+0.035	Inside	**17.23**
1986 JK (Hypnos)*		(μ-Librae)	–.–			**0.427**	**0.512**	0.021	0.035	0.005
	2005-05-01.1	June 4.1	73.8	221.5	−10.3		13.2	+0.008	Below	18.30
2004 HW*		(α-Virginis)	**Corvids**			0.124	**0.703**	0.069	0.091	0.013
	2004-05-25.2	June 8.0	77.5	205.1	−12.7		10.23	+0.009	Above	**17.12**
2003 YM_{137}*		(ω-Geminorum)	(Daytime)			0.187	0.478	0.181	0.116	0.039
	2004-03-31.6	June 13.1	82.4	103.3	+28.8		15.55	−0.004	−H	18.72
2005 OW*			**July Centaurids**			0.059	**0.503**	0.214	0.188	0.036
	2005-08-22.6	June 23.1	92.0	253.7	−26.7		13.43	−0.060	Inside	20.81

Table 2 (*cont.*)

Name	T_p	(Nearest star) Date	Shower $\lambda_\odot$	R.A.	Decl.	Probable source region: JFC	OB V_g	3:1 Δr	MC Ω	ν_6 $H_N{}^2$
2004 MP_7			–.–			0.263	0.433	0.152	0.107	0.045
	2004-05-12.8	June 24.6	93.4	109.1	+51.6		19.63	−0.030	Inside	21.67
2001 ME_1*		(ξ-Sagittarii)	**N. σ-Sagittariids**			**0.547**	0.236	0.146	0.049	0.021
	2005-12-06.4	June 24.5	93.2	284.7	−18.1		29.65	+0.011	–	**16.81**
1983 LC*		(ξ-Ophiuchi)	–.–			**0.330**	**0.420**	0.139	0.077	0.034
	2004-10-05.4	June 29.2	97.8	263.6	−21.9		17.52	+0.023	Inside	19.33
2005 NG_{56}			ε-Ursae Majorids?			0.120	**0.738**	0.052	0.074	0.015
	2005-06-22.2	July 14.9	112.7	182.2	+53.1		14.29	−0.017	Inside	22.51
2003 KU_2*		(ζ-Scuti)	–.–			0.187	0.413	0.197	0.156	0.047
	2003-10-18.7	July 24.6	122.0	276.7	−10.4		14.32	+0.027	Below	**17.65**
1997 QK_1		(ε-Centauri)	July Centaurids?			0.124	**0.703**	0.069	0.091	0.013
	2002-04-17.2	July 22.4	119.9	222.7	−28.5		9.20	+0.009	Inside	20.10
2004 BG_{121}		(β-Cygni) $i = 20°$, short a	–.–			**0.621**	0.236	0.101	0.021	0.021
	2003-04-22.7	July 30.5	127.6	296.7	+28.1		23.69	−0.034	Outside	19.02
2004 HW*		(ξ-Virginis)	(Daytime)			0.124	**0.703**	0.069	0.091	0.013
	2004-05-25.2	July 30.7	127.8	177.8	+0.8		10.20	+0.015	Above	**17.12**
2002 PD_{43}		$i = 26°$, $q = 0.11$	Daytime ξ-Orionids?			0.010	0.102	**0.562**	0.229	0.098
	2002-06-15.2	Aug. 7.6	135.4	107.5	+14.0		38.49	+0.017	Outside	19.01
2003 BK_{47}		(γ-Sagittae) $i = 21°$	–.–			**0.203**	**0.373**	**0.229**	0.127	0.068
	2002-07-25.6	Aug. 11.0	138.6	298.5	+21.1		19.29	−0.017	Inside	18.51
1999 JM_8		(γ-Bootis) $i = 14°$	–.–			0.102	**0.553**	0.105	0.209	0.031
	2004-01-14.7	Aug. 12.1	139.7	211.4	+37.0		12.54	−0.049	Inside	15.04
1992 UY_4*		(λ-Scorpii)	–.–			0.059	**0.503**	0.214	0.188	0.036
	2005-09-01.3	Aug. 19.7	147.0	266.4	−33.8		9.25	+0.017	Above	17.55
2000 QS_7*		(α-Capricorni)	η-Serpentids?			**0.187**	**0.478**	**0.181**	0.116	0.039
	2005-02-21.1	Aug. 22.7	149.8	297.8	−11.6		12.6	+0.004	Outside	19.55

1999 RD_{32}*		(β-Sextantids)	Daytime γ-Leonids?			**0.464**	**0.293**	0.134	0.077	0.032
	2003-09-29.8	Aug. 28.7	155.6	156.3	+01.1		22.8	−0.051	+	**16.32**
1986 JK (Hypnos)*		(η-Virginis)	(Daytime)			**0.427**	**0.512**	0.021	0.035	0.005
	2005-05-01.1	Aug. 30.1	157.1	187.9	−04.1		13.1	+0.036	Below	18.30
1998 SH_2*		(φ-Leonis)	–.–			**0.330**	**0.420**	0.139	0.077	0.034
	1998-07-18.2	Sep. 4.2	161.9	173.2	−01.3		18.50	+0.023	Below	20.81
1979 VA (Wilson–Harrington)*			**Sep. Sagittariids**			0.059	**0.503**	0.214	0.188	0.036
	2005-07-10.8	Sep. 06.5	195.1	258.6	−21.1		8.87	+0.048	Above	15.99
2003 RS_1		(β-Librae)	(Daytime)			0.270	0.676	0.017	0.034	0.004
	2003-08-30.4	Sep. 16.71	74.1	231.3	−08.7		9.52	+0.055	Outside	21.73
1992 UY_4*		(π-Scorpii)	(Daytime)			0.059	**0.503**	0.214	0.188	0.036
	2005-09-01.3	Sep. 21.2	178.5	241.2	−28.7		8.87	+0.037	Above	17.55
2001 HA_4*		(β-Ceti) $i = 17°$	ω-Piscids?			**0.314**	**0.216**	**0.264**	0.132	0.074
	2005-07-19.8	Sep. 21.61	78.9	10.6	−17.6		24.93	+0.002	Outside	**17.48**
2001 WS_1		(ζ-Pavonis) $i = 13°$	–.–			0.055	**0.320**	**0.278**	**0.293**	0.054
	2006-01-05.0	Sep. 24.3	181.6	284.7	−73.9		11.14	+0.014	Outside	16.79
2004 TL_{10}			–.–			0.187	0.413	0.197	0.156	0.047
	2004-09-03.4	Oct. 03.8	190.9	212.6	−41.2		13.10	−0.004	Inside	21.29
1998 US_{18}		(φ-Virginis)	(Daytime)			**0.187**	**0.413**	**0.197**	**0.156**	0.047
	1998-09-10.3	Oct. 10.3	197.3	220.5	+04.7		16.2	+0.034	Outside	20.68
2003 SK_{84}*		(θ-Ceti)	–.–			0.009	**0.206**	**0.416**	**0.261**	0.108
	2003-11-24.9	Oct. 15.6	202.5	20.6	−3.2		20.45	+0.097	Above	21.12
6344 P-L*		(θ-Pegasi)	γ-Piscids?			0.059	**0.503**	0.214	0.188	0.036
	1960-11-09.3	Oct. 17.0	203.9	339.0	+6.0		11.06	+0.029	Below	21.36
1998 SY_{14}		(b-Aquarii)	(Daytime)			**0.427**	**0.512**	0.021	0.035	0.005
	2003-09-16.8	Oct. 30.5	217.4	351.9	−18.0		10.2	+0.003	Outside	20.62
2003 KU_2*		(π-Scorpii)	(Daytime)			0.187	0.413	0.197	0.156	0.047
	2003-10-18.7	Oct. 31.5	218.3	241.6	−26.4		14.02	+0.087	Below	17.65
2004 TN_1			μ-Pegasids?			0.264	0.543	0.082	0.089	0.023
	2004-12-10.9	Nov. 01.4	219.3	3.9	+25.0		14.48	+0.010	Above	21.73

Table 2 (*cont.*)

Name	T_p	(Nearest star)	Shower	Probable source region:			JFC	OB	3:1	MC	ν_6
		Date		$\lambda_\odot$	R.A.	Decl.		V_g	Δr	Ω	H_N[2]
1983 LC*		(β-Scorpii)	(Daytime)				**0.330**	**0.420**	0.139	0.077	0.034
	2004-10-05.4	Nov. 9.8		228.0	240.2	−19.5		17.50	+0.025	Inside	19.33
2000 QS_7*		(ξ-Ophiuchi)	(Daytime)				**0.187**	**0.478**	**0.181**	0.116	0.039
	2005-02-21.1	Nov. 12.5		230.4	263.5	−21.1		12.4	+0.054	Below	19.55
2003 WC_{158}		(θ-Serpentis Cauda)	-.-				0.059	0.405	0.231	0.258	0.046
	2003-11-10.2	Nov. 23.8		241.8	286.0	+8.2		10.53	+0.050	Outside	20.39
2002 JB_9		(β-Centauri) $i = 47°$	-.-				**0.563**	0.034	0.280	0.056	0.067
	2002-08-18.4	Dec. 2.6		250.7	206.8	−58.5		33.37	−0.026	Inside	**15.67**
2003 WY_{25}		(χ-Sculptoris)	**Phoenicids**				**0.825**	0.152	0.013	0.008	0.002
	2003-12-11.6	Dec. 1.3		249.4	5.3	−25.0		9.49	+0.019	Outside	20.88
2004 UV_1			-.-				0.043	0.195	**0.345**	**0.329**	0.089
	2005-01-08.7	Dec. 08.5		256.7	54.2	−45.5		15.37	+0.044	Outside	**17.85**
6344 P-L*		(τ-Capricorni)	(Daytime)				0.059	**0.503**	0.214	0.188	0.036
	1960-11-09.3	Dec. 17.0		265.3	312.0	−14.9		10.41	+0.080	Below	21.36
2001 XP_1		(π-Ursae Majoris) $i = 39°$	-.-				**0.684**	0.079	0.162	0.047	0.027
	2002-03-06.7	Dec. 19.7		268.1	112.3	+72.2		28.51	+0.010	Outside	**17.55**
2000 YG_{29}		(ζ-Horlogii) $i = 19°$	-.-				**0.684**	0.284	0.020	0.006	0.004
	2006-08-05.7	Dec. 24.0		272.5	35.4	−54.9		14.36	−0.022	Inside	18.63
1983 TB (Phaethon)		(α-Gemini) $i = 22°$	**Geminids**				0.000	0.000	0.200	0.263	**0.537**
	2003-09-26.0	Dec. 17.1		265.4	115.6	+31.7		34.55	−0.014	Inside	14.60

Notes:

[1] W. F. Bottke, A. Morbidelli, R. Jedicke *et al.*, Debiased orbital and absolute magnitude distribution of the near-Earth objects. *Icarus* **156** (2002), 399–433.

[2] $\Delta r = r_{comet} - r_{Earth}$ and is negative if Earth passes outside of the minor planet orbit..

* Uncertain encounter conditions due to low inclination.

Table 3 *Predicted meteor outbursts from long-period comet one-revolution dust trails*

A working list of potential meteor outbursts from known long-period comets (name given) and potential returns of meteor outbursts seen in the past (codes in Table 7). Results from Lyytinen and Jenniskens.[1] Bold entries are the more certain predictions. Times can be uncertain by at least 1 h (solar longitudes > 0.05°, in J2000) if the orbit of the parent body is uncertain. Miss-distance Δr_{D-E} is positive if the dust trail is outside Earth's orbit.

Date (yr:mo:day)	Time (UT)	Name	$\Delta r(D-E)$ (AU)	R.A. (°)	Decl. (°)	$\lambda_\odot$ (°)	Comments
2005 08 07	20:11	bPe	<0.002	52	+40	135.332	Uncertain
2006 01 02	**06:21**	**C/1969 T_1**	**−0.000 28**	**231**	**−57**	**281.633**	**Barely**
2006 07 16	23:23	kPa	<0.0007	275	−67	114.189	Uncertain
2006 08 08	02:50	bPe	+0.000 19	52	+40	135.357	Steep/faint
2007 04 28	17:28	aBo	+0.000 27	219	+19	037.954	Faint
2007 09 01	**11:37**	**Aur**	**−0.000 42**	**90**	**+39**	**158.561**	**!!**
2007 12 21	**03:40**	**aLy**	**−0.000 11**	**138**	**+44**	**268.769**	
2008 12 20	08:10	aLy	+0.000 40	138	44	268.691	R.A.+
2011 01 02	20:47	C/1969 T_1	+0.000 58	231	−57	281.952	Too far?
2011 06 05	**05:44**	**aCi**	**−0.000 01**	**218**	**−70**	**074.094**	
2012 06 04	10:43	aCi	+0.000 27	218	−70	074.041	R.A.+
2012 06 11	04:24	gDe	+0.000 36	312	+17	080.484	Barely
2012 11 26	09:47	NoO	+0.0011	85	+04	244.368	If R.A.+
2013 06 11	**08:28**	**gDe**	**+0.000 21**	**312**	**+17**	**080.402**	
2015 02 08	11:28	aCe	+0.000 21	210	−58	319.161	Rapid, R.A.+
2016 12 20	14:45	aLy	+0.000 22	138	+44	268.922	If R.A.+
2018 03 31	11:47	C/1907 G_1	−0.000 48	309	−60	010.463	Steep, far
2019 03 31	17:26	C/1907 G_1	+0.000 17	309	−60	010.444	Steep
2019 08 17	03:19	bHy	−0.000 46	23	−76	143.693	Too far?
2019 11 22	04:52	aMo	−0.000 36	117	+01	239.306	Far
2020 08 16	**14:18**	**bHy**	**+0.000 27**	**023**	**−76**	**143.886**	
2021 03 01	09:32	C/1976 D_1	−0.000 35	13	−64	340.729	Too far?
2021 08 12	**04:22**	**C/1852 K_1**	**−0.000 10**	**43**	**−13**	**139.402**	
2022 08 12	11:59	C/1852 K_1	−0.000 40	43	−13	139.465	
2027 06 11	**21:55**	**gDe**	**+0.000 10**	**312**	**+17**	**080.378**	
2029 08 07	23:33	bPe	<0.0003	52	+40	135.319	Uncertain
2030 07 17	03:33	kPa	<0.001	275	−67	114.211	Uncertain
2031 07 17	19:30	kPa	+0.000 46	275	−67	114.600	R.A.+
2032 03 01	06:35	C/1976 D_1	−0.000 25	13	−64	340.789	
2033 06 05	**00:32**	**aCi**	**+0.000 04**	**218**	**−70**	**74.229**	
2034 06 11	22:21	gDe	+0.000 22	312	+17	80.591	Too far?

[1] E. Lyytinen and P. Jenniskens, Meteor outbursts from long-period comet dust trails. *Icarus* **162** (2003), 443–452.

Table 3 (*cont.*)

Date (yr:mo:day)	Time (UT)	Name	$\Delta r(D-E)$ (AU)	R.A. (°)	Decl. (°)	$\lambda_{\odot}$ (°)	Comments
2035 02 15	22:22	C/1854 R_1	+0.000 69	305	+38	326.574	Far, steep
2035 11 27	02:32	NoO	+0.0004	85	+04	244.160	Steep, far?
2037 06 04	19:18	aCi	−0.000 31	218	−70	73.988	Too far?
2038 03 01	21:46	C/1976 D_1	−0.000 11	13	−64	340.875	
2038 03 06	18:52	aPx	+0.000 55	135	−35	345.767	Too far?
2038 09 11	**12:43**	**eEr**	**+0.000 31**	**56**	**−14**	**168.364**	**(If long period)**
2039 02 08	**15:28**	**aCe**	**+0.000 36**	**210**	**−58**	**319.174**	**Too far?**
2039 03 07	**08:24**	**aPx**	**+0.000 09**	**135**	**−35**	**346.074**	
2039 09 11	**16:27**	**eEr**	**−0.000 54**	**56**	**−14**	**168.261**	**(If long period)**
2040 04 22	**00:25**	**Lyrids**	**−0.000 22**	**271**	**+34**	**31.942**	**Faint meteors**
2040 09 10	**20:25**	**eEr**	**−0.000 09**	**56**	**−14**	**168.172**	**(If long period)**
2041 04 22	**08:40**	**Lyrids**	**−0.000 25**	**271**	**+34**	**32.022**	**Faint meteors**
2042 03 31	15:58	C/1907 G_1	−0.000 26	309	−60	10.484	Steep
2043 03 31	**21:36**	**C/1907 G_1**	**−0.000 05**	**309**	**−60**	**10.464**	
2043 11 22	**10:58**	**aMo**	**−0.000 08**	**117**	**+01**	**239.409**	
2044 03 31	03:16	C/1907 G_1	+0.000 16	309	−60	10.446	Steep
2045 03 31	09:13	C/1907 G_1	+0.000 13	309	−60	10.436	Steep
2048 03 31	15:45	C/1907 G_1	−0.000 39	309	−60	10.448	Steep
2048 06 11	09:48	gDe	+0.000 13	312	+17	80.484	
2049 06 11	14:02	gDe	+0.000 23	312	+17	80.407	
2050 03 01	22:52	C/1976 D_1	−0.000 11	13	−64	340.846	

Note: "far", large calculated distance from Earth's orbit in a nominal case; "barely", calculated encounter is far from the trail center; "R.A.+", only if the radiant position is off by up to 10°; "steep", a rapid change of the trail position with time; "uncertain", rapid changes with radiant position.

Table 4a *Leonid showers from 55P/Tempel–Tuttle dust trail encounters*

Past and future encounters with the trails of comet 55P/Tempel–Tuttle. All calculations of peak time are made for the Earth's center. Moon cycle is the Moon's synodic period fraction. Dates are in Julian Calendar prior to 1582, and in Gregorian/Western Calendar at later times. (Post-)predictions of peak activity are from: $\text{ZHR} = 27\,000\, f_{\text{m}}\, 10^{(-1050\,|\Delta r - 0.000\,77|)}\, f(\Delta a)$, with

$\Delta a > +0.11$ AU: $f(\Delta a) = (0.12/2)^2 / ((0.35\,\Delta a - 0.04)^2 + (0.12/2)^2)$

$\Delta a < +0.11$ AU: $f(\Delta a) = (0.12/2)^2 / ((\Delta a - 0.11)^2 + (0.12/2)^2)$

The magnitude distribution index χ is derived from an assumed Δa-dependence only, and is uncertain:

$\chi = 3.5 - 2.2\,(0.4/2)^2 / ((\Delta a + 0.05)^2 + (0.4/2)^2)$

The trail width is calculated from: $W(°) = 0.024 \times 10^{346\,|\Delta r\,(\text{AU}) - 0.000\,77|}$

Year Trail	Δr $(D-E)$ (AU)	Δa_0 (AU)	f_m	$\lambda_\odot$ (°)	Time (UT)	χ	W (°)	ZHR (/h)	src
902:	*Moon in Cap. (63%)*			***219.4 ± 0.2***	*Shower in Egypt and Italy*				
802	+0.000 99	+0.0478	0.323	219.247	Oct. 12 20:54	1.72	0.029	2500	V
904:	*Moon in Vir. (0%)*			*No record – over Americas (End of Maya Classic Era ~899 AD)*					
802	+0.000 92	+0.2064	0.417	218.161	Oct. 11 07:11	2.67	0.027	6000	V
934:	*Moon in Oph. (8%)*			***219.9 ± 0.2***	*In S.W., stars fell like rain in China*				
834	−0.000 12	−0.0082	0.114	220.078	Oct. 13 21:42	1.39	0.049	70	V
802	+0.000 03	−0.0049	0.074	220.086	Oct. 13 21:54	1.41	0.043	70	V
769	+0.000 04	−0.0039	0.060	220.090	Oct. 13 22:00	1.41	0.043	60	V
736	+0.000 03	−0.0034	0.062	220.094	Oct. 13 22:05	1.41	0.043	60	V
702	−0.000 04	−0.0037	0.097	220.092	Oct. 13 22:02	1.41	0.046	80	V
967:	*Moon in Cap. (63%)*			***220.4 ± 0.2***	*Stars scattered from NE to SW in Japan*				
901	+0.000 07	−0.0509	0.720	220.098	Oct. 14 09:16	1.30	0.042	440	V
834	+0.000 76	−0.0090	0.130	220.168	Oct. 14 10:56	1.39	0.024	690	V
802	+0.000 93	−0.0054	0.063	220.177	Oct. 14 11:09	1.40	0.027	240	V
769	+0.000 96	−0.0043	0.040	220.182	Oct. 14 11:17	1.41	0.028	150	V

Table 4a (*cont.*)

Year Trail	Δr $(D-E)$ (AU)	Δa_0 (AU)	f_m	$\lambda_\odot$ (°)	Time (UT)	χ	W (°)	ZHR (/h)	src
736	+0.000 94	−0.0038	0.062	220.185	Oct. 14 11:21	1.41	0.027	240	V
702	+0.000 88	−0.0041	0.070	220.183	Oct. 14 11:18	1.41	0.026	240	V
669	+0.000 51	−0.0050	0.112	220.202	Oct. 14 11:45	1.41	0.030	350	V
1002:	*Moon in Sgr. (35%)*			***220.5 ± 0.1***	*A few thousand China, meteors fell all night Japan*				
935	+0.000 85	+0.1418	0.615	220.244	Oct. 14 12:06	2.35	0.026	13 300	V
1037:	*Moon in Oph. (12%)*			***220.5 ± 0.1***	*Appeared at midnight, disaster of meteors Japan*				
935	+0.000 91	+0.2235	0.408	220.328	Oct. 14 13:27	2.73	0.027	5600	V
1101:	*Moon in Cnc. (54%)*			***222.6 ± 0.2***	*Stars fell from the sky in France*[1]				
968?	+0.001 86	−0.0550	0.308	222.188	Oct. 16 19:48	1.30	0.057	70	V
935?	−0.000 31	−0.0507	0.254	222.336	Oct. 16 23:20	1.30	0.057	60	V
1202:	*Moon in Lib. (1%)*			***223.4 ± 0.2***	*Stars flew like the flight of locust*[2]				
1167	+0.001 16	+0.3436	0.732	223.262	Oct. 18 18:59	3.05	0.033	2800	V
1237:	*Moon in Vir. (3%)*			***223.3 ± 0.1***	*Meteors appeared in morning Japan*				
736?	+0.001 93	+0.1044	0.091	223.504	Oct. 19 00:07	2.12	0.060	150	V
1366:	*Moon in Tau. (98%)*			***224.4 ± 0.1***	*Sparkling of stars – possibly Filament*[3]				
1135?	+0.000 40	+0.0037	2.468	225.440	Oct. 22 00:10	1.45	0.18	10	V
1069?	+0.000 57	+0.0008	2.283	225.434	Oct. 22 00:01	1.43	0.15	30	V

[1] Mentioned by A. S. Hershell, *Silliman's Journal (II)* **36** (1863), 377. The 12th century French text *Chronicon Sancti Maxentii* says that on October 17, "stars were seen to fall from the sky."

[2] Several Arabic sources in 13th and 14th centuries talk about a shower in mid-October Al-Dhahabi (1274–1348) wrote in *Siyar a'lam al-nubala'*, "the stars were disquieted and flew like the flight of locust. This continued until the dawn and the people were terrified and they made haste with prayers."

[3] Gary Kronk (online Meteor showers, a descriptive catalogue) quotes the Portuguese text of *Duarte Munes de Leão* (1600). *Cronicas dos reis de Portugal* (revised version published in 1975 by Porto) for on the morning of October 23 ($\lambda_\odot = 226.4$): "There was in the heavens a movement of stars, such as men never before saw or heard of. From midnight onward, all the stars moved from the east to the west; and after being together, they began to move, some in one direction, and others in another. And afterward they fell from the sky in such numbers, and so thickly together, that as they descended low in the air, they seemed large and fiery, and the sky and the air seemed to be in flames, and even the earth appeared as if ready to take fire. Those who saw it were filled with such great fear and dismay, that they were astounded, imagining they were all dead men, and that the end of the world had come." Dall'Olmo (1978) gives the date as October 21.

1532:	*Moon in Vir. (16%)*			***227.7 ± 0.1***	*Stars flew like a shower in Korea*				
1201	+0.000 52	+0.0491	0.397	227.808	Oct. 24 22:08	1.73	0.029	2900	V
1533:	*Moon in Cap. (50%)*			***227.5 ± 0.1***	*Countless fell in China and Korea*				
1366?	+0.001 85	+0.0482	0.222	227.471	Oct. 24 20:15	1.73	0.057	210	V
1400?	−0.000 24	+0.0417	0.056	227.370	Oct. 24 17:51	1.68	0.054	60	V
1538:	*Moon in Sgr. (13%)*			***229.1 ± 0.2***	*Meteors appeared Korea – probably Filament*				
1035?	+0.002 44	+0.1795	0.072	229.101	Oct. 26 17:54	2.55	0.091	30	V
1538:	*Moon in Sgr. (16%)*			*Not observed (over Americas)*					
1167	+0.000 84	+0.0979	0.097	229.729	Oct. 27 08:51	2.08	0.025	2200	V
1566:	*Moon in Psc. (97%)*			***228.9 ± 0.2***	*Stars fell like a shower in Korea (Munhon-Piko)*				
1167	+0.001 20	+0.0303	0.790	228.776	Oct. 26 14:17	1.61	0.034	2800	V
1594:	*Moon in Leo (49%)*			***228.7 ± 0.2***	*Thousands of meteors – probably Filament*				
1366?	−0.004 28	+0.8442	0.123	228.633	Nov. 05 15:24	3.40	–.–	0	V
1601:	*Moon in Psc. (76%)*			***228.9 ± 0.1***	*At dawn, stars fell like rain – Filament (1 day off?)*				
1466?	+0.001 59	+0.0495	0.109	229.859	Nov. 06 15:32	1.74	0.046	200	V
1499?	+0.002 02	+0.0636	0.158	229.937	Nov. 06 17:24	1.84	0.065	193	V
1602:	*Moon in Cnc. (50%)*			***229.7 ± 0.2***	*Hundreds fell in China*				
1433	+0.000 11	+0.0729	0.184	229.690	Nov. 06 17:47	1.90	0.041	950	V
1623:	*Moon in Psc. (92%)*			*Not observed (over Americas)*					
1333	−0.000 20	+0.4136	0.382	226.804	Nov. 04 06:03	3.15	0.052	250	V
1625:	*Moon in Sgr. (32%)*			***228.8 ± 0.2***	*Many stars fought Korea*				
1366	+0.001 07	+0.4889	1.540	228.825	Nov. 05 18:33	3.23	0.030	3500	V
1625:	*Moon in Cap. (39%)*			***229.8 ± 0.2***	*More than 10 meteors in midair*				
1234?	−0.000 05	−0.0547	0.389	229.626	Nov. 06 13:39	1.30	0.046	170	V
1698:	*Moon in Sgr. (20%)*			***231.1 ± 0.2***	*Meteors fell like weaving in Japan*				
1300?	−0.000 69	−0.0335	0.161	231.071	Nov. 08 17:23	1.63	0.077	50	V
1268?	+0.001 27	+0.0258	0.130	230.801	Nov. 08 10:57	1.58	0.036	350	V

Table 4a (*cont.*)

Year Trail	Δr $(D-E)$ (AU)	Δa_0 (AU)	f_m	$\lambda_\odot$ (°)	Time (UT)	χ	W (°)	ZHR (/h)	src
1977:	*Moon in Aqr. (62%)*			*Not observed (over Asia, E. Europe)?*					
1234	+0.000 19	+0.1377	0.150	236.856	Nov. 18 23:55	2.33	0.038	980?	V (!)
1998:	*Moon in Lib. (2%)*			***235.30 ± 0.05***	*1.91 ± 0.09*	*0.130 ± 0.005*		*65 ± 5*	
1899?	−0.004 28	+0.0572	0.008	235.281	Nov. 17 19:45	–.–	–.–	0	V
1899?	−0.0044	+0.050	–.–	235.270	Nov. 17 19:35	–.–	–.–	0	A
1932?	−0.005 27	+0.0767	0.573	235.277	Nov. 17 18:43	–.–	–.–	0	V
1998:	*Moon in Vir. (4%)*			*Filament (Tab. 4b)*					
1333?	−0.000 71	−0.0228	0.464	234.529	Nov. 17 01:55	1.34	0.078	60	V
1333?	+0.000 15	−0.023	–.–	234.460	Nov. 17 00:18	–.–	–.–	–.–	R.A.
1333?	<0.001	−0.024	–.–	234.5	Nov. 17 02 h	–.–	–.–	–.–	ABE
1999:	*Moon in Aqr. (67%)*			***235.282 ± 0.001***	*2.38 ± 0.04*	*0.031 ± 0.005*		*4200 ± 100*	
1899	+0.001 22	+0.1355	0.448	235.292	Nov. 18 02:10	2.32	0.035	4050	V
1899	+0.000 67	+0.136	–.–	235.300	Nov. 18 02:23	153.6	+21.7	70.7	IS
1899	+0.000 67	+0.138	0.398	235.291	Nov. 18 02:10	–.–	–.–	500	A
1899	+0.000 66	+0.14	0.38	235.290	Nov. 18 02:08	–.–	0.032	1200	MA
1899	–.–	+0.14	0.38	235.291	Nov. 18 02:10	–.–	–.–	4000	L
1899	+0.000 70	+0.137	–.–	235.291	Nov. 18 02:10	–.–	–.–	–.–	KR
1932	−0.001 10	+0.1872	1.055	235.280	Nov. 18 01:53	2.59	0.106	260	V
1932	−0.0019	+0.191	0.54	235.274	Nov. 18 01:45	153.73	+21.98	70.76	MS
1932	−0.0019	+0.19	0.53	235.27	–.–	–.–	–.–	–.–	MA
1600	+0.000 61	+0.026	–.–	235.176	Nov. 17 23:25	153.5	+21.9	70.7	IS

1999:	*Moon in Aqr. (74%)*			***235.910 ± 0.010***	*1.96 ± 0.09*	*0.12 ± 0.04*		*50 ± 10*	
1866	−0.001 35	+0.0829	0.203	236.045	Nov. 18 20:05	1.97	0.130	30	V
1866	−0.001 58	+0.083	–.–	235.996	Nov. 18 18:57	154.1	+21.5	70.7	IS
1866	−0.0016	+0.084	0.19	236.038	Nov. 18 19:54	154.19	+21.70	70.74	MS
1866	–.–	–.–	–.–	236.040	Nov. 18 20:00	–.–	–.–	–.–	L
1866	−0.001 60	+0.080	0.17	236.039	Nov. 18 19:59				MA
2000:	*Moon in Cnc. (64%)*			***235.276 ± 0.001***	*>2.0 ± 0.1*	*0.070 ± 0.005*		*350 ± 20*	
1932	+0.002 42	+0.3163	1.790	235.273	Nov. 17 08:00	2.99	0.090	370	V
1932	+0.001 18	+0.295	–.–	235.349	Nov. 17 09:45	153.6	+21.7	70.7	IS
1932	+0.001 19	+0.302	0.574	235.270	Nov. 17 07:55		0.063	0	A
1932	+0.0012	+0.30	0.53	235.27	Nov. 17 07:53			0	MA
1932	+0.001 20	+0.300	0.553	235.270	Nov. 17 07:55			215	L
2000:	*Moon in Leo (53%)*			***236.276 ± 0.001***	*2.21 ± 0.10*	*0.070 ± 0.005*		*250 ± 20*	
1866	−0.000 36	+0.1121	0.158	236.286	Nov. 18 08:05	2.17	0.059	280	V
1866	−0.000 78	+0.111	–.–	236.246	Nov. 18 07:09	154.2	+21.4	70.7	IS
1866	−0.000 78	+0.114	0.138	236.276	Nov. 18 07:52	–.–	0.063	120?	A
1866	−0.000 80	+0.116	0.135	236.279	Nov. 18 07:56	–.–	0.081	700	L
1866	−0.000 77	+0.11	0.13	236.28	Nov. 18 07:51	–.–	–.–	80–5000	MA
1866	−0.000 70	+0.114	–.–	236.273	Nov. 18 07:48	–.–	–.–	–.–	KR
2000:	*Moon in Leo (55%)*			***236.090 ± 0.010***	*2.11 ± 0.10*	*0.080 ± 0.005*		*170 ± 20*	
1733	−0.000 63	+0.0635	0.334	236.107	Nov. 18 03:50	1.84	0.073	280	V
1733	−0.000 78	+0.061	–.–	236.071	Nov. 18 02:59	153.7	+21.6	70.6	IS
1733	−0.000 69	+0.066	0.138	236.096	Nov. 18 03:35	–.–	0.063	120	A
1733	−0.000 77	+0.064	0.292	236.102	Nov. 18 03:44	–.–	0.063	170	A
1733	−0.000 86	+0.060	0.049	236.115	Nov. 18 04:02	–.–	0.063	18	A
1733	−0.000 80	+0.065	0.250	236.103	Nov. 18 03:45	–.–	–.–	700	L
1733	−0.000 77	+0.06	~0.27	236.102	Nov. 18 03:44	–.–	0.12	90–5000	MA

Table 4a (*cont.*)

Year Trail	Δr ($D-E$) (AU)	Δa_0 (AU)	f_m	$\lambda_\odot$ (°)	Time (UT)	χ	W (°)	ZHR (/h)	src
2001:	*Moon in Sgr. (14%)*			***236.451*** ± ***0.003***	*2.29 ± 0.11*	*0.040 ± 0.010*		*1400 ± 100*	
1866	+0.000 20	+0.1392	0.155	236.471	Nov. 18 18:31	2.34	0.041	1000	V
1866	−0.000 24	+0.137	–.–	236.446	Nov. 18 17:55	154.5	+21.3	70.8	IS
1866	−0.0002	+0.142	0.14	236.463	Nov. 18 18:20	154.45	+21.68	70.76	MS
1866	−0.000 23	+0.142	0.139	236.464	Nov. 18 18:21	–.–	0.036	460	A
1866	−0.000 22	+0.142	0.13	236.463	Nov. 18 18:19	–.–		13 000−35 000	MA
1866	−0.000 25	+0.144	0.135	236.467	Nov. 18 18:27	–.–	0.035	6100	L
1866	−0.000 20	+0.141	–.–	236.469		153.6	+21.6		KR
2001:	*Moon in Sgr. (11%)*			***236.140*** ± ***0.003***	*1.80 ± 0.10*	*0.055 ± 0.002*		*600 ± 50*	
1767	+0.000 69	+0.0789	0.185	236.113	Nov. 18 09:59	1.95	0.026	4000	V
1767	+0.000 44	+0.081	0.157	236.114	Nov. 18 10:02	–.–	0.053	600	MA
1767	+0.000 05	+0.068	0.005	236.160	Nov. 18 11:08	–.–	0.069	33	MA
1767	+0.000 42	+0.079	–.–	236.112	Nov. 18 09:59	154.1	+21.6	70.7	IS
1767	+0.0004	+0.082	0.16	236.113	Nov. 18 10:00	154.30	+21.88	70.71	MS
1767	+0.000 43	+0.081	0.140	236.115	Nov. 18 10:03	–.–	0.04	2000	L
1767	+0.000 43	+0.081	∼0.14	236.114	Nov. 18 10:01	–.–	–.–	2500	MA
2001:	*Moon in Sgr. (14%)*			***236.433*** ± ***0.009***	*1.80 ± 0.10*	*0.090 ± 0.020*		*680 ± 100*	
1699	−0.000 06	+0.0426	0.269	236.451	Nov. 18 18:02	1.69	0.042	580	V
1699	−0.000 14	+0.044	–.–	236.409	Nov. 18 17:03	154.6	+21.4	70.9	IS
1699	−0.00 02	+0.042	0.40	236.427	Nov. 18 17:27	154.49	+21.76	70.74	MS
1699	−0.000 15	+0.041	0.395	236.428	Nov. 18 17:30	–.–	0.069	920	A
1699	−0.000 15	+0.041	0.43	236.429	Nov. 18 17:31	–.–	–.–	9000	A

1699	−0.000 10	+0.043	0.260	236.433	Nov. 18 17:37	–.–	0.095	1750	L
1666	−0.000 54	+0.0285	0.125	236.409	Nov. 18 17:02	–.–	0.068	50	V
1666	−0.000 08	+0.031	0.010	236.483	Nov. 18 18:47	154.50	+21.69	70.74	MS
1666	−0.000 10	+0.030	0.160	236.423	Nov. 18 17:22	–.–	–.–	600	L
2001:	*Moon in Sgr. (14%)*			*Blended with 1699 and 1866 trail*					
1633	−0.000 25	+0.0261	0.021	236.445	Nov. 18 17:54	1.58	0.054	16?	V
1633	−0.0002	+0.026	0.021	236.435	Nov. 18 17:40	154.48	+21.71	70.74	MS
1633	−0.000 40	+0.026	0.160	236.425	Nov. 18 17:25	–.–	–.–	260?	L
2001:	*Moon in Sgr. (8%)*			*Unobserved?*					
1433	+0.000 40	+0.0200	0.009	235.702	Nov. 18 00:12	1.54	0.032	30?	V
2001:	*Moon in Sgr. (10%)*			*Unobserved?*					
1366	+0.000 62	+0.0237	0.015	235.924	Nov. 18 05:29	1.56	0.027	90?	V
2002:	*Moon in Ari. (10%)*			***236.898*** ± ***0.005***	*3.05 ± 0.09*	*0.024 ± 0.001*		*2200 ± 100*	
1866	+0.000 69	+0.1688	0.168	236.898	Nov. 19 10:51	2.50	0.026	3400	V
1866	+0.000 04	+0.167	–.–	236.876	Nov. 19 10:19	154.2	+21.5	71.5	IS
1866	−0.000 05	+0.172	0.151	236.890	Nov. 19 10:40	–.–	0.041	2900−6000	A
1866	+0.000 04	+0.174	0.148	236.894	Nov. 19 10:44	–.–	0.028	7400	L
1866	−0.000 05	+0.17	0.15	236.888	Nov. 19 10:36			25 000	MA
1866	−0.000 10	+0.170	–.–	236.885	Nov. 19 10:32	152.9	+21.9		KR
2002:	*Moon in Ari. (99%)*			***236.616*** ± ***0.001***	*2.58 ± 0.07*	*0.022 ± 0.001*		*2000 ± 100*	
1767	+0.000 54	+0.1113	0.163	236.614	Nov. 19 04:06	2.17	0.029	2500	V
1767	+0.000 14	+0.110	–.–	236.601	Nov. 19 03:48	154.3	+21.5	70.9	IS
1767	+0.000 15	+0.113	0.13	236.610	Nov. 19 04:00	–.–	–.–	15 000	A
1767	+0.000 13	+0.114	0.130	236.612	Nov. 19 04:02	–.–	0.033	4500	L
1767	+0.000 15	+0.113	0.132	236.610	Nov. 19.04:00	–.–	0.057	25−60	MA

Table 4a (*cont.*)

Year Trail	Δr $(D-E)$ (AU)	Δa_0 (AU)	f_m	$\lambda_\odot$ (°)	Time (UT)	χ	W (°)	ZHR (/h)	src
2003:	*Moon in Gem. (82%)*			***(230.9)***	–.–	–.–		*20 ± 5*	
1499	+0.003 10	+0.2467	0.926	230.905	Nov. 13 18:21	2.81	0.154	55?	V
1499	–.–	+0.28	~1.6	230.834	Nov. 13 16:40		0.5	100	L
1499	–.–	+0.26	0.8	230.904	Nov. 13 18:20				A
2003:	*Moon in Vir. (21%)*			–.–	–.–	–.–		*<20*	
1333	+0.001 73	+0.1151	0.036	237.240	Nov. 20 01:13	2.33	0.052	95?	V
2004:	*Moon in Aqr. (50%)*			–.–	–.–	–.–		*<20*	
1333	+0.002 10	+0.1375	0.040	237.223	Nov. 19 06:47	2.33	0.069	42?	V
1333	+0.0018	+0.14	0.022	237.218	Nov. 19 06:39	154.95	+21.28	70.64	MS
—									
2005:	*Moon in Cnc. (76%)*								
1167	−0.001 04	+0.1291	0.032	238.747	Nov. 21 01:16	2.28	0.101	10?	V
2006:	*Moon in Vir. (3%)*								
1932	+0.000 88	+1.7643	0.906	236.624	Nov. 19 04:58	3.47	0.026	200	V
1932	+0.0002	+0.94	0.470	236.618	Nov. 19 04:50	–.–	–.–	50	L
1932	+0.000 09	+0.961	0.53	236.615	Nov. 19 04:46	–.–	–.–	100	MA
1932	+0.000 10	+0.883	–.–	236.507	Nov. 19 02:12	154.1	+21.2	71.2	IS
1932	+0.000 10	+0.963	0.62	236.615	Nov. 19 04:46	154.32	+21.09	70.80	MS
2007:	*Moon in Aqr. (61%)*								
1932	+0.000 98	+1.9822	0.562	236.118	Nov. 18 23:03	3.48	0.028	80	V
1932	+0.0004	+1.06	0.560	236.107	Nov. 18 22:36	–.–	–.–	30	L
1932	−0.000 37	+0.981	–.–	236.108	Nov. 18 22:37	154.0	+21.4	70.9	IS
2008:	*Moon in Cnc. (61%)*								
1932	−0.001 03	+2.2035	0.294	236.803	Nov. 18 21:38	1.96	0.100	3	V
1466	−0.002 72	+0.0805	0.517	234.949	Nov. 17 01:32	3.48	>0.1	1?	V

2009:	*Moon in Sco. (1%)*		*Caution: distorted old trail, similar situation in 1974 did not result in storm.*						
1433	+0.002 29	+0.1099	0.090	235.408	Nov. 17 18:29	2.16	0.081	60?	V
1466	−0.000 65	+0.0997	0.151	235.549	Nov. 17 21:49	2.09	0.074	130?	V
1533	+0.000 69	+0.1308	0.221	235.555	Nov. 17 21:58	2.29	0.025	5000?	V (!)
1102	−0.000 53	+0.0810	0.051	235.775	Nov. 18 03:12	1.96	0.068	60	V
2033:	*Moon in Vir. (15%)*								
1899	−0.001 06	+0.1753	0.421	235.403	Nov. 17 22:07	2.53	0.103	120	V
1899	−0.001 61	+0.18	0.35	235.39	Nov. 17 21:42	–.–	–.–	0	MA
1899	–.–	+0.18	–.–	235.388	Nov. 17 21:45	–.–	–.–	430	L
2034:	*Moon in Aqr. (39%)*								
1932	+0.000 19	+0.2961	1.698	235.385	Nov. 18 03:45	2.95	0.038	5300	V (sic!)
1932	−0.000 98	+0.29	0.44	235.37	Nov. 18 03:20	–.–	–.–	0–1000	MA
1932	–.–	+0.30	–.–	235.342	Nov. 18 02:42	–.–	–.–	470	L
2034:	*Moon in Cap. (50%)*								
1866	−0.000 70	+0.1292	0.149	236.481	Nov. 19 05:51	2.28	0.077	115	V
1866	−0.001 19	+0.13	0.13	236.46	Nov. 19 05:19	–.–	–.–	0–100	MA
1866	–.–	+0.11	–.–	236.463	Nov. 19 05:25	–.–	–.–	280	L
2034:	*Moon in Cap. (48%)*								
1767	−0.000 72	+0.0713	0.233	236.181	Nov. 18 22:42	1.89	0.079	160	V
1767	–.–	+0.07	–.–	236.164	Nov. 18 22:17	–.–	–.–	200	L
1733	+0.002 49	+0.0850	0.275	236.094	Nov. 18 20:37	1.99	0.094	110	V
2034:	*Moon in Cap. (50%)*								
1699	−0.000 22	+0.0407	0.140	236.481	Nov. 19 05:50	1.68	0.053	150	V
1699	–.–	+0.04	–.–	236.478	Nov. 19 05:46	–.–	–.–	100–200	L
1666	−0.001 08	+0.0261	0.242	236.431	Nov. 19 04:39	1.58	0.105	25?	V
2035:	*Moon in Aqr. (78%)*								
1633	+0.000 27	+0.0366	0.0731	237.241	Nov. 11 06:12	1.65	0.036	240?	V
1633	–.–	+0.02	–.–	237.210	Nov. 11 05:33	–.–	–.–	420	L

Table 4a (*cont.*)

Year Trail	Δr $(D-E)$ (AU)	Δa_0 (AU)	f_m	$\lambda_\odot$ (°)	Time (UT)	χ	W (°)	ZHR (/h)	src
2035:	*Moon in Gem. (78%)*								
1366	+0.001 13	+0.0280	0.017	236.710	Nov. 19 17:35	1.59	0.032	70?	V
2035:	*Moon in Cnc. (70%)*								
1666	+0.000 16	+0.0431	0.016	237.318	Nov. 20 08:02	1.69	0.039	45	V
1699	+0.001 63	+0.0632	0.167	237.327	Nov. 20 08:15	1.83	0.048	520	V
2036:	*Moon in Lib. (0%)*								
1965	−0.000 61	+0.5872	0.225	235.428	Nov. 17 17:06	3.30	0.072	25	V
1965	−0.0013	+0.60	0.52	235.41	Nov. 17 16:40	–.–	–.–	–.–	MA
2037:	*Moon in Psc. (79%)*								
1965	+0.001 20	+0.6947	0.132	235.389	Nov. 17 22:18	3.35	0.034	100	V
1965	+0.0007	+0.71	0.52	235.38	Nov. 17 22:05	–.–	–.–	–.–	MA
2037:	*Moon in Psc. (91%)*								
1800	+0.000 96	+0.1263	0.079	237.307	Nov. 19 19:57	2.26	0.028	1300	V
1800	+0.0006	+0.130	0.064	237.290	Nov. 19 19:33	154.40	+19.87	70.37	MS
1800	–.–	+0.13	–.–	237.293	Nov. 19 19:37	–.–	–.–	810	L
2037:	*Moon in Psc. (92%)*								
1833	+0.001 26	+0.1401	0.079	237.526	Nov. 20 01:09	2.34	0.035	640	V
1833	+0.0007	+0.143	0.069	237.504	Nov. 20 00:38	154.53	+19.84	70.37	MS
1833	–.–	+0.14	–.–	237.509	Nov. 20 00:45	–.–	–.–	570	L
2038:	*Moon in Leo (45%)*								
1767	+0.002 27	+0.1592	0.199	237.547	Nov. 20 07:53	2.45	0.080	130	V
2038:	*Moon in Leo (33%)*								
1866	+0.001 45	+0.2096	0.059	238.666	Nov. 21 10:29	2.68	0.041	230	V
1866	–.–	+0.22	–.–	238.651	Nov. 21 10:07	–.–	–.–	180	L

2040:	*Moon in Tau. (100%)*								
1366	+0.002 77	+0.0929	0.326	236.615	Nov. 18 21:56	2.04	0.118	70?	V
2043:	*Moon in Tau. (99%)*								
1366	−0.000 19	+0.4694	0.335	234.603	Nov. 17 16:38	3.22	0.051	170?	V
2043:	*Moon in Tau. (93%)*								
1533	+0.000 38	+0.1441	0.083	235.915	Nov. 18 23:53	2.37	0.033	850?	V
2043:	*Moon in Gem. (89%)*								
1400	−0.000 71	+0.1551	0.067	236.653	Nov. 19 17:26	2.43	0.078	50?	V
2047:	*Moon in Sgr. (24%)*								
1567	+0.000 96	+0.1505	0.043	238.692	Nov. 21 18:26	2.40	0.028	700?	V
2048:	*Moon in Tau. (98%)*								
1567	+0.000 79	+0.1541	0.031	239.586	Nov. 21 21:56	2.42	0.024	760?	V
2049:	*Moon in Sex. (46%)*								
1833	+0.000 26	+0.5870	0.192	235.437	Nov. 18 01:22	3.30	0.036	180	V
2052:	*Moon in Vir. (12%)*								
1866	−0.000 19	+0.7125	0.121	235.242	Nov. 17 15:05	3.36	0.052	24	V
2059:	*Moon in Sgr. (31%)*								
1135	+0.000 62	+0.4848	0.225	227.938	Nov. 11 04:05	3.23	0.027	750?	V
2063:	*Moon in Lib. (0%)*								
1234	+0.000 19	+0.0467	0.119	236.886	Nov. 20 01:57	1.72	0.038	370?	V
2068:	*Moon in Leo (30%)*								
1767	+0.000 07	+0.0939	0.248	236.463	Nov. 18 22:42	2.05	0.042	1200	V
1767	–.–	+0.10	–.–	236.419	Nov. 18 21:38	–.–	–.–	600–1200	L
2069:	*Moon in Sgr. (16%)*								
1932	+0.001 10	+0.3400	0.297	235.467	Nov. 18 05:05	3.04	0.031	1300	V
2069:	*Moon in Aqr. (35%)*								
1699	+0.000 00	+0.0707	0.090	237.725	Nov. 20 10:48	1.89	0.044	350	V
1699	–.–	+0.073	–.–	237.66	Nov. 20 09:13	–.–	–.–	400	L
2071:	*Moon in Vir. (10%)*								
1965	+0.001 82	+0.5360	0.451	235.582	Nov. 18 20:11	3.27	0.055	140	V

Table 4a (*cont.*)

Year Trail	Δr $(D-E)$ (AU)	Δa_0 (AU)	f_m	$\lambda_\odot$ (°)	Time (UT)	χ	W (°)	ZHR (/h)	src
2074:	*Moon in Cap. (25%)*								
1800	+0.001 57	+0.1557	0.037	240.228	Nov. 23 05:01	2.43	0.045	140	V
2075:	*Moon in Tau. (100%)*								
1333	−0.000 07	+0.1695	0.052	239.947	Nov. 23 04:34	2.50	0.047	170?	V
2078:	*Moon in Tau. (94%)*								
1533	−0.000 17	+0.1911	0.095	238.557	Nov. 21 14:06	2.60	0.051	220?	V
2081:	*Moon in Tau. (100%)*								
1366	+0.001 21	+0.5295	0.075	234.080	Nov. 16 21:55	3.27	0.034	100?	V
2082:	*Moon in Cnc. (63%)*								
1300	−0.000 19	+0.4121	0.301	229.134	Nov. 12 06:12	3.15	0.052	200?	V
2085:	*Moon in Cap. (33%)*								
1102	+0.000 23	+0.1645	0.049	240.079	Nov. 22 21:11	2.48	0.037	330?	V
2095:	*Moon in Leo (32%)*								
1699	+0.000 90	+0.1554	0.020	237.371	Nov. 20 18:29	2.43	0.027	370	V
2096:	*Moon in Sgr. (21%)*								
1102	+0.001 21	+0.0551	0.066	237.436	Nov. 20 02:07	1.78	0.034	335?	V

Sources: KR = E. D. Kondrat'eva and E. A. Reznikov, Periodic comet Tempel–Tuttle and Leonid meteor shower. *Astron. Herald* **2** (1985), 144–151 (in Russian); E. D. Kondrat'eva, I. N. Murav'eva and E. A. Reznikov, On the forthcoming return of the Leonid meteoric swarm. *Solar Syst. Res.* **31** (1997), 546–549.

L = E. Lyytinen, Leonid predictions for the years 1999–2007 with the satellite model of comets. *Meta Res. Bull.* **8** (1999), 33–40; E. Lyytinen and T. van Flandern, Predicting the strength of Leonid outbursts. *Earth, Moon, Planets* **82/83** (2000), 149–166.

ABE = D. J. Asher, M. E. Bailey and V. V. Emel'yanenko, Resonant meteoroids from comet Tempel Tuttle in 1333: the cause of the unexpected Leonid outburst in 1998. *Mon. Not. R. Astron. Soc.* **304** (1999), L53–L56.

MA = R. H. McNaught and D. J. Asher, Leonid dust trails and meteor storms. *WGN* **27** (1999), 85–102; with later updates in various papers.

A = Alternate updates by David Asher (Armagh Observatory website).

JB = P. Jenniskens and H. Betlem, Massive remnant of evolved cometary dust trail detected in the orbit of Halley-type comet 55P/Tempel–Tuttle. *Astrophys. J.* **531** (2002), 1161–1167.

J = Calculations by P. Jenniskens (this work).

MS = Sato M. (2004), personal website.

IS = Sato I. (2004), personal website.

V = Model by J. Vaubaillon, calculations by J.Vaubaillon and P. Jenniskens (this work).

Table 4b *Leonid showers from 55P/Tempel–Tuttle filament*

Observed values in bold. Predictions below are based on the assumption that these older trails move in and out of Earth's orbit much like the younger trails. Results are not based on modeling of the old trails and are therefore much less reliable, perhaps to ±22 h.

Year Trail	Δr (*D–E*) (AU)	Δa_0 (AU)	f_m	$\lambda_\odot$ (°)	Time (UT)	χ	*W* (°)	ZHR (/h)	src
1960:	*Moon in Vir. (6%)*			–.–	–.–	–.–	–.–	*<10*	*JB*
Fil.	–.–	−0.865	1.0	–.–	–.–	2.1	0.8	<20	J
1961:	*Moon in Aqr. (67%)*			***234.6 ± 0.2***	Nov. 16 16 h	2.3	–.–	*~200*	*JB*
Fil.	–.–	−0.669	1.0	–.–	–.–	2.1	0.8	<20	J
1962:	*Moon in Cnc. (70%)*			–.–	–.–	–.–	–.–	*<40*	*JB*
Fil.	–.–	−0.474	1.0	–.–	–.–	2.0	0.8	<20	J
1963:	*Moon in Lib. (0%)*			–.–	–.–	–.–	–.–	*~20*	*JB*
Fil.	–.–	−0.279	1.0	–.–	–.–	2.0	0.8	<20	J
1964:	*Moon in Cet. (90%)*			***235.1 ± 0.1***	Nov. 16 22 h	–.–	–.–	*120 ± 50*	*JB*
Fil.	–.–	−0.083	1.0	–.–	–.–	1.8	0.8	<20	J
1965:	*Moon in Leo (43%)*			***234.7 ± 0.2***	Nov. 16 19 h	*1.7*	*~0.8*	*300 ± 50*	*JB*
Fil.	–.–	+0.113	1.0	234.77	Nov. 16 21 h	1.5	0.8	220	J
1966:	*Moon in Sgr. (24%)*			***235.0 ± 0.3***	Nov. 17 08 h	–.–	–.–	*<150*	*JB*
Fil.	–.–	+0.148	1.0	235.14	Nov. 17 12 h	1.5	0.8	40	J
1967:	*Moon in Tau. (100%)*			***235.3 ± 0.2***	Nov. 17 22 h	–.–	–.–	*100 ± 50*	*JB*
Fil.	–.–	+0.503	1.0	237.52	Nov. 17 02 h	2.0	0.8	30	J
1968:	*Moon in Vir. (12%)*			***235.9 ± 0.1***	Nov. 17 18 h	–.–	–.–	*100 ± 50*	*JB*
Fil.	–.–	+0.699	1.0	235.47	Nov. 17 08 h	2.0	0.8	20	J
1969:	*Moon in Aqr. (60%)*			–.–	–.–	–.–	–.–	*<10*	*JB*
Fil.	–.–	+0.894	1.0	235.41	Nov. 17 13 h	2.1	0.8	20	J

1993:	*Moon in Sgr. (21%)*			–.–	–.–	–.–	–.–	*<10*	*JB*
Fil.	–.–	−0.719	1.0	–.–	–.–	2.1	0.8	<20	J
1994:	*Moon in Tau. (100%)*			***235.90 ± 0.04***	Nov. 18 10 h	*2.1 ± 0.3*	*0.8 ± 0.1*	*75 ± 15*	*JB*
Fil.	–.–	−0.524	1.0	–.–	–.–	2.0	0.8	<20	J
1995:	*Moon in Vir. (25%)*			***235.33 ± 0.10***	Nov. 18 03 h	*2.0 ± 0.3*	~0.8	*30 ± 10*	*JB*
Fil.	–.–	−0.328	1.0	–.–	–.–	2.0	0.8	<20	J
1996:	*Moon in Aqr. (41%)*			***235.26 ± 0.04***	Nov. 17 07 h	*1.9 ± 0.9*	*0.7 ± 0.2*	*70 ± 10*	*JB*
Fil.	–.–	−0.132	1.0	–.–	–.–	1.9	0.8	<20	J
1997:	*Moon in Ori. (89%)*			***235.3 ± 0.1***	Nov. 17 14 h	–.–	*0.6 ± 0.1*	*120 ± 20*	*JB*
Fil.	–.–	−0.047	1.0	235.22	Nov. 17 12 h	1.7	0.8	<20	J
1998:	*Moon in Vir. (4%)*			***234.52 ± 0.01***	Nov. 17 02 h	*1.5 ± 0.1*	*0.8 ± 0.1*	*206 ± 20*	*JB*
Fil.	–.–	+0.148	1.0	234.56	Nov. 17 03 h	1.5	0.8	200	J
1999:	*Moon in Aqr. (69%)*			***235.55 ± 0.1***	Nov. 18 08 h	–.–	*~0.8*	*75 ± 10*	*J*
Fil.	–.–	+0.344	1.0	236.15	Nov. 18 23 h	1.8	0.8	80	J
2000:	*Moon in Leo (57%)*			***235.97 ± 0.1***	Nov. 18 01 h	–.–	*(0.8)*	*40 ± 20*	*J*
Fil.	–.–	+0.539	1.0	236.21	Nov. 18 06 h	2.0	0.8	120	J
2001:	*Moon in Sgr. (14%)*			***236.51 ± 0.1***	Nov. 18 20 h	–.–	*0.8 ± 0.2*	*70 ± 20*	*J*
Fil.	–.–	+0.735	1.0	236.15	Nov. 18 11 h	2.0	0.8	100	J
2002:	*Moon in Ari. (99%)*			***236.41 ± 0.1***	Nov. 18 23 h	*2.0 ± 0.1*	*0.8 ± 0.1*	*30 ± 10*	*J*
Fil.	–.–	+0.930	1.0	236.63	Nov. 19 05 h	2.1	0.8	70	J
2003:	*Moon in Vir. (24%)*			***236.85 ± 0.1***	Nov. 19 16 h	–.–	*(0.8)*	*60 ± 10*	*J*
Fil.	–.–	+1.126	1.0	237.22	Nov. 20 01 h	2.1	0.8	30	J
2004:	*Moon in Aqr. (55%)*			***~237.6***	Nov. 19 16 h	–.–	*(0.8)*	*~30*	*J*
Fil.	–.–	+1.320	1.0	237.54	Nov. 19 15 h	2.1	0.8	30	J
2005:	*Moon in Gem. (81%)*								
Fil.	–.–	+1.516	1.0	238.00	Nov. 20 07:31	2.1	0.8	20	J

Table 4b (*cont.*)

Year Trail	Δr (*D–E*) (AU)	Δa_0 (AU)	f_m	$\lambda_\odot$ (°)	Time (UT)	χ	*W* (°)	ZHR (/h)	src
2006:	*Moon in Vir. (3%)*								
Fil.	–.–	+1.711	1.0	236.68	Nov. 19 06:20	2.1	0.8	20	J
2007:	*Moon in Aqr. (62%)*								
Fil.	–.–	+1.906	1.0	236.17	Nov. 19 00:19	2.1	0.8	40	J
2008:	*Moon in Cnc. (63%)*								
Fil.	–.–	+2.101	1.0	235.69	Nov. 18 18:59	2.1	0.8	20	J
2009:	*Moon in Sco. (1%)*								
Fil.	–.–	+2.297	1.0	235.37	Nov. 17 17:35	2.1	0.8	50	J
2028:	–.–	−0.675	1.0	–.–	–.–	2.1	0.8	20	J
2029:	–.–	−0.480	1.0	–.–	–.–	2.0	0.8	20	J
2030:	–.–	−0.284	1.0	–.–	–.–	2.0	0.8	20	J
2031:	–.–	−0.089	1.0	–.–	–.–	1.8	0.8	20	J
2032:	–.–	+0.106	1.0	–.–	–.–	1.5	0.8	20	J
2033:	*Moon in Vir. (15%)*								
Fil.	–.–	+0.302	1.0	235.41	Nov. 17 22:18	1.8	0.8	20	J
2034:	*Moon in Cap. (49%)*								
Fil.	–.–	+0.497	1.0	236.28	Nov. 19 01:05	2.0	0.8	200	J
2035:	*Moon in Cnc. (73%)*								
Fil.	–.–	+0.692	1.0	237.09	Nov. 20 02:38	2.0	0.8	70	J
2036:	*Moon in Oph. (3%)*								
Fil.	–.–	+0.888	1.0	237.19	Nov. 19 11:02	2.1	0.8	45	J

2037:	*Moon in Psc. (91%)*								
Fil.	–.–	+1.083	1.0	237.30	Nov. 19 19:35	2.1	0.8	30	J
2038:	*Moon in Leo (39%)*								
Fil.	–.–	+1.279	1.0	238.19	Nov. 20 23:13	2.1	0.8	40	J
2039:	*Moon in Cap. (31%)*								
Fil.	–.–	+1.474	1.0	(238.12)	Nov. 21 03:33	2.1	0.8	20	J
2040:	*Moon in Tau. (97%)*								
Fil.	–.–	+1.669	1.0	(238.13)	Nov. 20 10:01	2.1	0.8	<20	J
2041:	*Moon in Vir. (16%)*								
Fil.	–.–	+1.865	1.0	(237.04)	Nov. 19 14:13	2.1	0.8	<20	J
2042:	*Moon in Aqr. (54%)*								
Fil.	–.–	+2.060	1.0	236.05	Nov. 18 20:49	2.1	0.8	30	J
2043:	*Moon in Aur. (91%)*								
Fil.	–.–	+2.255	1.0	236.20	Nov. 19 06:41	2.1	0.8	50	J
2044:	*Moon in Lib. (1%)*								
Fil.	–.–	+2.451	1.0	236.64	Nov. 18 23:15	2.1	0.8	20	J
2045:	*Moon in Psc. (80%)*								
Fil.	–.–	+2.646	1.0	(237.40)	Nov. 19 23:29	2.1	0.8	<20	J
2046:	*Moon in Leo (54%)*								
Fil.	–.–	+2.842	1.0	(238.10)	Nov. 20 22:21	2.1	0.8	<20	J
2047:	*Moon in Sgr. (24%)*								
Fil.	–.–	+3.037	1.0	(238.69)	Nov. 21 18:23	2.1	0.8	<20	J
2048:	*Moon in Tau. (98%)*								
Fil.	–.–	+3.232	1.0	(239.59)	Nov. 21 22:04	2.1	0.8	<20	J
2049:	*Moon in Sex. (46%)*								
Fil.	–.–	+3.428	1.0	235.44	Nov. 18 01:27	2.1	0.8	20	J
2050:	*Moon in Sgr. (14%)*								
Fil.	–.–	+3.623	1.0	235.43	Nov. 18 07:17	2.1	0.8	30	J

Table 4b (*cont.*)

Year Trail	Δr (D–E) (AU)	Δa_0 (AU)	f_m	$\lambda_\odot$ (°)	Time (UT)	χ	W (°)	ZHR (/h)	src
2051:	*Moon in Tau. (100%)*								
Fil.	–.–	+3.818	1.0	235.50	Nov. 18 15:10	2.1	0.8	<20	J
2052:	*Moon in Vir. (11%)*								
Fil.	–.–	+4.014	1.0	235.59	Nov. 17 23:23	2.1	0.8	20	J

Table 5a *Ursid showers from 8P/Tuttle dust trails*

Year Trail	Δr (*D–E*) (AU)	Δa_0 (AU)	f_m	$\lambda_\odot$ (°)	Time (UT)	χ	W (°)	ZHR (/h)	src
1945:	*Moon in Leo (84%)*			***>271.324***		–.–	***0.05 ± 0.01***	***>120***	
1392	+0.0005	+0.048	0.024	271.326	Dec. 22 18:16	2.8	0.05	160	JL
1959:				*(No data)*					
	No dust trail encounters								JL
1972:				*(No data)*					
	No dust trail encounters								JL
1986:	*Moon in Leo (65%)*			***270.935 ± 0.005***		***2.8***	***0.05 ± 0.01***	***110***	
1378a:	+0.0030	+0.038	0.028	270.934	Dec. 22 21:36	2.8	0.05	1	JL
1378b:	+0.0004	–.–	–.–	270.816	Dec. 22 18:49	–.–	–.–	130	JL
2000:	*Moon in Lib. (10%)*			***270.780 ± 0.005***		***3.2 ± 0.5***	***0.09 ± 0.01***	***90***	
1405	−0.001 24	+0.068	–.–	270.754	Dec. 22 07:29	2.8	0.05	(4)	JL
1392	+0.0038	–.–	–.–	270.808	Dec. 22 08:45	2.8	0.05	0	JL
1378a:	+0.0069	+0.044	–.–	270.82	Dec. 22 09:00	–.–	–.–	0	JL
1378b:	+0.0250	+0.044	–.–	–.–	–.–	–.–	–.–	0	JL
2002:	*Moon in Cnc. (89%)*			*(No data)*					
1302?	+0.0010	+0.0002	–.–	270.919	Dec. 22 23:40	2.8	0.05	100	L
1090a	−0.006	+0.026	–.–	270.726	Dec. 22 19:07	–.–	–.–	0	L
1090b	<0.03	+0.040	–.–	270.79	Dec. 22 20:40	–.–	–.–	0	L
2004:	*Moon in Ari. (82%)*			*(No data)*					
1048	−0.003	+0.092	0.045	270.49	Dec. 22 01:43	–.–	–.–	(1)	L
2006:	*Moon in Cap. (6%)*								
966	+0.000	+0.039	–.–	270.72	Dec. 22 19–21	2.8	0.10	35	L
2014:	*Moon in Sgr. (1%)*								
1405	−0.0007	+0.066	0.02	270.838	Dec. 22 23:38	2.8	0.05	10	L
2016:	*Moon in Vir. (37%)*								
1076	<0.003	+0.025	–.–	270.76	Dec. 22 10:05	2.8	0.05	30	L
2020:	*Moon in Psc. (52%)*								
829	<0.001	+0.092	0.038	270.57	Dec. 22 06:10	2.8	0.05	490	L
815	–.–	+0.078	–.–	270.44–270.92	Dec. 22 03–22	2.8	0.05	420	L
2028:	*Moon in Psc. (38%)*								
1378	<0.005	+0.034	–.–	270.875	Dec. 22 14:29	2.8	0.06	3	L
1392	+0.0007	+0.038	0.02	270.880	Dec. 22 14:36	2.8	0.05	190	L
2030:	*Moon in Sco. (5%)*								
1090	<0.001	+0.021	0.04	270.612	Dec. 22 20:46	2.8	0.05	180	L
1062	+0.005	+0.038	0.044	270.702	Dec. 22 22:54	2.8	0.05	0	L

Table 5a (*cont.*)

Year Trail	Δr (*D–E*) (AU)	Δa_0 (AU)	f_m	$\lambda_\odot$ (°)	Time (UT)	χ	W (°)	ZHR (/h)	src
1062	−0.003	+0.046	–.–	270.740	Dec. 22 23:47	2.8	0.05	0	L
1076	+0.0055	+0.031	0.019	270.562	Dec. 22 19:35	2.8	0.05	0	L
2032:	*Moon in Sex. (71%)*								
952	<0.003	+0.057	–.–	270.73	Dec. 22 12:	2.8	0.05	50	L
883	+0.004	+0.016	–.–	270.73	Dec. 22 12:	2.8	–.–	0	L
856	–.–	+0.017	–.–	270.72	Dec. 22 11:45	2.8	–.–	0	L
843	−0.005	+0.019	0.06	270.73	Dec. 22 12:00	2.8	–.–	0	L
2033:	*Moon in Sgr. (1%)*								
925	+0.0035	+0.054	0.003	270.448	Dec. 22 11:24	2.8	0.05	0	L
2034:	*Moon in Tau. (89%)*								
829	−0.003	+0.068	–.–	270.44	Dec. 22 17:20	2.8	0.16	0	L
815	+0.000	+0.078	–.–	270.59	Dec. 22 20:53	2.8	–.–	65	L
2042:	*Moon in Ari. (83%)*								
1378	−0.0004	+0.028	0.04	270.737	Dec. 23 01:28	2.8	0.08	20	L
1378	+0.004	+0.023	0.15	270.706	Dec. 23 00:44	2.8	–.–	0	L
1419	−0.001	+0.049	0.083	270.663	Dec. 22 23:44	2.8	–.–	15	L
2043:	*Moon in Vir. (49%)*								
1117	<0.0004	+0.025	–.–	270.65	Dec. 23 06:10	2.8	0.06	70	L
1158	−0.0005	+0.033	0.038	270.678	Dec. 23 06:49	2.8	–.–	18	L
1144	−0.0015	+0.031	0.021	270.665	Dec. 23 06:31	2.8	–.–	0	L
1131	+0.007	+0.031	–.–	270.63	Dec. 23 05:40	2.8	–.–	0	L
2044:	*Moon in Cap. (8%)*								
1090	<0.0003	+0.055	–.–	270.58	Dec. 22 10:05	2.8	0.09	100	L
1076	+0.006	+0.0491	–.–	270.55	Dec. 22 09:25	2.8	–.–	0	L
1117	+0.007	+0.068	–.–	270.5	Dec. 22 08:15	2.8	–.–	0	L
2046:	*Moon in Vir. (29%)*								
1048	−0.004	+0.088	–.–	270.076	Dec. 22 10:45	2.8	0.05	0	L
2047:	*Moon in Psc. (39%)*								
925	<0.002	+0.068	–.–	270.33	Dec. 22 22:45	2.8	0.05	310	L

Sources: AR = R. Arlt and J. Rendtel, First analysis of the 1997 Perseids. *WGN* **25** (1997), 207–209.

BR = P. Brown and J. Rendtel, The Perseid meteoroid stream: characterization of recent activity from visual observations. *Icarus* **124** (1996), 414–428.

IS = I. Sato, Personal website (2004).

J = Calculations by P. Jenniskens, this work.

JB = P. Jenniskens, H. Betlem, M. de Lignie *et al.*, On the unusual activity of the Perseid meteor shower (1989–96) and the dust trail of comet 109P/Swift–Tuttle. *Mon. Not. R. Astron. Soc.* **301** (1998), 941–954.

JL = P. Jenniskens and E. Lyytinen, Dust trails of 8P/Tuttle and the unusual outbursts of the Ursid shower. *Icarus* **159** (2002), 197–209.

L = Model by Esko Lyytinen, calculations by Esko Lyytinen and Markku Nissinen, this work.

MS = J. W. Mason and I. D. Sharp, *J. Brit. Astron. Assoc.* **91** (1981), 4.

R = P. Roggemans, The ZHR profile of the 1985 Perseids, recomputed. *WGN* **15** (1987), 181–187; P. Roggemans, On the Perseid meteor stream 1986 (II). Zenithal hourly rate data. *WGN* **16** (1988), 12–31.

R.A. = J. Rendtel and R. Arlt, Perseids 1995 and 1996 – an analysis of global data. *WGN* **24** (1996), 141–147.

V = Model by Jérémie Vaubaillon, calculations by J. Vaubaillon and P. Jenniskens (this work).

Table 5b *Ursid showers from 8P/Tuttle Filament*

I assumed that the peak time is proportional to the position of old (AD 815–1158) trails shifted by +0.1°. A mean minimum distance (Δr, dust–Earth) was calculated from those trail positions. The peak rate then is $\mathrm{ZHR} = 125 \times 10^{-7.72|\Delta r|}$, with Δr in AU. I assume that the dust is evenly distributed along the orbit. Rates may be higher upto a factor 3 around the return of the comet, because of dust trapped in the same resonance as the comet. Without knowing the number of revolutions of the contributing dust (likely to be a wide range), Δa_0 is undefined. Here, Δa_0 is the value that would result in the observed lag at perihelion after one orbit.

Year Trail	Δr (D–E) (AU)	Δa_0 (AU)	f_m	$\lambda_\odot$ (°)	Time (UT)	χ	W (°)	ZHR (/h)	src
	(Comet back at perihelion on Dec. 15, 1980)								
1981:	*Moon in Lib. (12%)*			**270.82**		–.–	–.–	**55 ± 25**	**O**[1]
Fil.	+0.076	+0.3264	1.0	270.82	Dec. 22 12:00	2.6	0.35	32	J = this
1982:	*Moon in Aqr. (41%)*			**<271.04**		–.–	–.–	**>35**	**JL**
Fil.	+0.068	+0.6007	1.0	270.73	Dec. 22 15:59	2.6	0.35	37	J
1993:	*Moon in Psc. (64%)*			**270.77 ± 0.04**		**2.5**	**0.35 ± 0.07**	**100 ± 10**	**BL**[2]
Fil.	+0.068	−0.1118	1.0	270.72	Dec. 22 11:23	2.6	0.35	37	J
	(Comet back at perihelion on June 25, 1994)								
1994:	*Moon in Leo (81%)*			**270.56 ± 0.04**		**2.6**	**0.35 ± 0.07**	**50 ± 6**	**JL**
Fil.	+0.044	+0.1760	1.0	270.67	Dec. 22 16:28	2.6	0.35	57	J
1995:	*Moon in Sgr. (1%)*			(No data)					
Fil.	+0.089	+0.4549	1.0	270.53	Dec. 22 19:22	2.6	0.35	26	J
1996:	*Moon in Tau. (92%)*			**270.6 ± 0.1**		–.–	**(0.35)**	**25 ± 5**	**JL**
Fil.	+0.100	+0.7254	1.0	270.54	Dec. 22 01:43	2.6	0.35	21	J
1997:	*Moon in Vir. (47%)*			**270.61 ± 0.05**		–.–	**0.30 ± 0.08**	**16 ± 4**	**JL**
Fil.	+0.094	+0.9881	1.0	270.41	Dec. 22 04:58	2.6	0.35	24	J
1998:	*Moon in Cap. (12%)*			**270.48 ± 0.05**		–.–	**0.35 ± 0.10**	**13 ± 3**	**JL**
Fil.	+0.113	+1.2435	1.0	270.42	Dec. 22 11:19	2.6	0.35	17	J

[1] K. Ohtsuka, Enhanced activity of the 1994 Ursids from Japan. *WGN* **23** (1995), 69–72.
[2] B. Lunsford, *Meteor News* **103** (1994).

Table 5b (*cont.*)

Year Trail	Δr (*D–E*) (AU)	Δa_0 (AU)	f_m	$\lambda_\odot$ (°)	Time (UT)	χ	*W* (°)	ZHR (/h)	src
1999:	*Moon in Ori. (100%)*			(No data)					
Fil.	+0.100	+1.4922	1.0	270.48	Dec. 22 18:49	2.6	0.35	21	J
2000:	*Moon in Lib. (10%)*			**270.85 ± 0.05**		–.–	**(0.35)**	**22 ± 5**	
Fil.	+0.115	+1.7348	1.0	(270.84)	Dec. 22 09:30	2.6	0.35	16	J
2001:	*Moon in Aqr. (45%)*			**270.60 ± 0.05**		–.–	**0.30 ± 0.08**	**17 ± 5**	
Fil.	+0.103	+1.9518	1.0	270.67	Dec. 22 10:23	2.6	0.35	20	J
2002:	*Moon in Cnc. (90%)*			(No data)					
Fil.	+0.096	+2.1842	1.0	(270.83)	Dec. 22 21:33	2.6	0.35	23	J
2003:	*Moon in Sgr. (0%)*			(No data)					
Fil.	+0.066	+2.4120	1.0	270.82	Dec. 23 03:29	2.6	0.35	39	J
2004:	*Moon in Ari. (85%)*			**270.96 ± 0.02**		–.–	**0.35 ± 0.08**	**80 ± 10**	
Fil.	−0.018	+2.6360	1.0	270.97	Dec. 22 13:01	2.6	0.35	91	J
2005:	*Moon in Leo (62%)*								
Fil.	+0.060	−0.6159	1.0	270.72	Dec. 22 13:23	2.6	0.35	43	J
2006:	*Moon in Sgr. (6%)*								
Fil.	+0.065	−0.3117	1.0	(270.64)	Dec. 22 17:38	2.6	0.35	39	J
2007:	*Moon in Tau. (98%)*								
Fil.	+0.082	−0.0175	1.0	270.48	Dec. 22 20:00	2.6	0.35	29	J
	(Comet back at perihelion on Jan. 27, 2008)								
2008:	*Moon in Vir. (24%)*								
Fil.	+0.047	+0.2674	1.0	270.54	Dec. 22 03:42	2.6	0.35	54	J
2009:	*Moon in Aqr. (27%)*								
Fil.	+0.096	+0.5435	1.0	270.38	Dec. 22 06:00	2.6	0.35	23	J
2010:	*Moon in Gem. (99%)*								
Fil.	+0.097	+0.8114	1.0	270.28	Dec. 22 09:51	2.6	0.35	22	J
2011:	*Moon in Sco. (6%)*								
Fil.	+0.108	+1.0717	1.0	270.24	Dec. 22 15:05	2.6	0.35	18	J

Table 5b (*cont.*)

Year Trail	Δr (*D–E*) (AU)	Δa_0 (AU)	f_m	$\lambda_\odot$ (°)	Time (UT)	χ	*W* (°)	ZHR (/h)	src
2012:	*Moon in Ari. (76%)*								
Fil.	+0.110	+1.3249	1.0	270.33	Dec. 21 23:13	2.6	0.35	18	J
2013:	*Moon in Leo (79%)*								
Fil.	+0.097	+1.5715	1.0	270.49	Dec. 22 09:17	2.6	0.35	22	J
2014:	*Moon in Sgr. (1%)*								
Fil.	+0.110	+1.8343	1.0	270.56	Dec. 22 17:05	2.6	0.35	18	J
2015:	*Moon in Tau. (93%)*								
Fil.	+0.104	+2.0691	1.0	270.70	Dec. 23 02:25	2.6	0.35	20	J
2016:	*Moon in Vir. (36%)*								
Fil.	+0.023	+2.2991	1.0	270.87	Dec. 22 12:40	2.6	0.35	83	J
2017:	*Moon in Cap. (17%)*								
Fil.	+0.013	+2.5250	1.0	270.81	Dec. 22 17:17	2.6	0.35	99	J
2018:	*Moon in Ori. (100%)*								
Fil.	−0.018	+2.7472	1.0	(270.64)	Dec. 22 19:29	2.6	0.35	91	J
2019:	*Moon in Lib. (13%)*								
Fil.	+0.080	−0.4618	1.0	270.49	Dec. 22 21:39	2.6	0.35	30	J
2020:	*Moon in Psc. (52%)*								
Fil.	+0.073	−0.1627	1.0	(270.54)	Dec. 22 05:27	2.6	0.35	34	J
	(Comet back at perihelion Aug. 28, 2021)								
2021:	*Moon in Cnc. (91%)*								
Fil.	+0.087	+0.1268	1.0	270.33	Dec. 22 06:47	2.6	0.35	27	J
2022:	*Moon in Oph. (1%)*								
Fil.	+0.085	+0.4072	1.0	270.22	Dec. 22 10:21	2.6	0.35	28	J
2023:	*Moon in Ari. (79%)*								
Fil.	+0.095	+0.6791	1.0	270.14	Dec. 22 14:29	2.6	0.35	23	J
2024:	*Moon in Leo (59%)*								
Fil.	+0.100	+0.9431	1.0	270.27	Dec. 21 23:49	2.6	0.35	21	J

Table 5b (*cont.*)

Year Trail	Δr (*D–E*) (AU)	Δa_0 (AU)	f_m	$\lambda_\odot$ (°)	Time (UT)	χ	*W* (°)	ZHR (/h)	src
2025:	*Moon in Sgr. (4%)*								
Fil.	+0.090	+1.1997	1.0	270.26	Dec. 22 05:39	2.6	0.35	25	J
2026:	*Moon in Tau. (96%)*								
Fil.	+0.107	+1.4495	1.0	270.34	Dec. 22 13:41	2.6	0.35	19	J
2027:	*Moon in Vir. (23%)*								
Fil.	+0.095	+1.6931	1.0	270.51	Dec. 22 23:54	2.6	0.35	23	J
2028:	*Moon in Aqr. (34%)*								
Fil.	+0.090	+1.9195	1.0	270.49	Dec. 22 05:25	2.6	0.35	25	J
2029:	*Moon in Gem. (97%)*								
Fil.	+0.108	+2.1525	1.0	270.71	Dec. 22 16:50	2.6	0.35	18	J
2030:	*Moon in Sco. (5%)*								
Fil.	+0.007	+2.3810	1.0	270.63	Dec. 22 21:11	2.6	0.35	110	J
2031:	*Moon in Psc. (72%)*								
Fil.	(+0.047)	+2.6054	1.0	270.56	Dec. 23 01:37	2.6	0.35	54	J
2032:	*Moon in Leo (74%)*								
Fil.	+0.028	+2.8264	1.0	270.42	Dec. 22 04:38	2.6	0.35	76	J
2033:	*Moon in Sgr. (1%)*								
Fil.	+0.073	−0.3529	1.0	270.29	Dec. 22 07:41	2.6	0.35	34	J
2034:	*Moon in Ari. (88%)*								
Fil.	−0.003	−0.0573	1.0	(270.31)	Dec. 22 14:15	2.6	0.35	119	J
	(Comet back at perihelion Apr. 18, 2035)								
2035:	*Moon in Vir. (39%)*								
Fil.	+0.090	+0.2288	1.0	270.20	Dec. 22 17:54	2.6	0.35	25	J
2036:	*Moon in Cap. (17%)*								
Fil.	+0.071	+0.5061	1.0	270.26	Dec. 22 01:19	2.6	0.35	35	J
2037:	*Moon in Tau. (100%)*								
Fil.	+0.097	+0.7751	1.0	270.28	Dec. 22 08:01	2.6	0.35	22	J

Table 5b (*cont.*)

Year Trail	Δr (*D–E*) (AU)	Δa_0 (AU)	f_m	$\lambda_\odot$ (°)	Time (UT)	χ	*W* (°)	ZHR (/h)	src
2038:	*Moon in Lib. (16%)*								
Fil.	+0.092	+1.0364	1.0	270.35	Dec. 22 15:53	2.6	0.35	24	J
2039:	*Moon in Cet. (56%)*								
Fil.	+0.082	+1.2905	1.0	270.36	Dec. 22 22:06	2.6	0.35	29	J
2040:	*Moon in Cnc. (87%)*								
Fil.	+0.082	+1.5380	1.0	270.41	Dec. 22 05:31	2.6	0.35	29	J
2041:	*Moon in Oph. (1%)*								
Fil.	+0.066	+1.7795	1.0	270.53	Dec. 22 14:26	2.6	0.35	39	J
2042:	*Moon in Ari. (79%)*								
Fil.	+0.084	+1.9558	1.0	270.36	Dec. 22 16:34	2.6	0.35	28	J
2043:	*Moon in Vir. (51%)*								
Fil.	+0.105	+2.1881	1.0	(270.49)	Dec. 23 01:57	2.6	0.35	19	J
2044:	*Moon in Cap. (7%)*								
Fil.	+0.045	+2.4159	1.0	270.39	Dec. 22 05:40	2.6	0.35	56	J
2045:	*Moon in Tau. (97%)*								
Fil.	+0.047	+2.6397	1.0	270.47	Dec. 22 13:46	2.6	0.35	54	J
2046:	*Moon in Vir. (28%)*								
Fil.	+0.057	+2.8603	1.0	270.14	Dec. 22 12:14	2.6	0.35	45	J
2047:	*Moon in Psc. (38%)*								
Fil.	+0.050	−0.3067	1.0	270.14	Dec. 22 18:13	2.6	0.35	51	J
2048:	*Moon in Gem. (97%)*								
Fil.	+0.055	−0.0125	1.0	270.28	Dec. 22 03:46	2.6	0.35	47	J
	(Comet back at perihelion Jan. 21, 2049)								
2049:	*Moon in Lib. (7%)*								
Fil.	+0.041	+0.2722	1.0	270.24	Dec. 22 08:58	2.6	0.35	60	J

Table 5c *109P/Swift–Tuttle – parent of the Perseid shower: dust trail encounters*

The activity is estimated from the Leonid shower model, scaled by factor 2.6 for ejection speed ($\sim\sqrt{R_c}$) and factor 1.6 for longer orbital period: for $\Delta a < 0.11 \times 2.6 \times 1.6 = 0.46$ AU: $\text{ZHR} = 33.3/135 \times 27\,000 \times 10^{(-1050/2.6|\Delta r - 0.000\,77|)}\,(0.12 \times 2.6/2)^2/((\Delta a - 0.11 \times 2.6 \times 1.6)^2 + (0.12 \times 2.6/2)^2)$, while for $\Delta a > 0.46$ AU different in a similar way as in Table 4. Only those encounters are listed with ZHR > 200 (~twice the normal peak rate). The width of each dust trail encounter is derived from scaled Leonid trail:
$W(\text{AU}) = 2\sqrt{((0.33 \times 2.6 \times 0.000\,32/2)^2 + (\Delta r/2.6 - 0.000\,77)^2)}$
The magnitude distribution index is from:
$\chi = 3.5 - 2.2\,(0.4 \times 2.6 \times 1.3/2)^2/((\Delta a/1.33 + 0.05)^2 + (0.4 \times 2.6 \times 1.3/2)^2)$

Year Trail	Δr (*D–E*) (AU)	Δa_0 (AU)	f_m	$\lambda_\odot$ (°)	Time (UT)	χ	*W* (°)	ZHR (/h)	src
		(comet at perihelion Sep. 6.4, 698)							
734:	*Moon in Aqr. (97%)*				(Not reported, over Americas)				
59	+0.002 02	+0.6595	0.599	139.541	July 22 12:57	3.3	0.02	1200	V
772:	*Moon in Aqr. (93%)*				(Not reported, over Americas)				
188	+0.000 49	−1.4107	3.746	139.488	July 22 05:16	3.5	0.07	320	V
		(comet at perihelion Apr. 19.5, 826)							
830:	*Moon in Cnc. (1%)*			**139.2**	From dusk till 5th watch uncountable China				
698?	+0.001 23	+0.6019	0.089	139.432	July 23 00:40	3.3	0.04	290	V
59?	−0.002 62	+0.0729	0.580	139.407	July 23 00:01	1.9	0.20	50	V
860:	*Moon in Leo (1%)*				(Not reported, Europe, not that strong)				
59	−0.000 89	0.1171	0.635	139.570	July 22 20:26	2.2	0.13	310	V
865:	*Moon in Gem. (22%)*				(Not reported, Europe??)				
59	+0.002 24	+0.7371	0.957	139.384	July 22 22:42	3.4	0.02	1400	V
		(comet at perihelion Apr. 19.5, 950)							
948:	*Moon in Cap. (100%)*				(Not reported, Europe??)				
569	+0.004 56	−0.1853	11.201	139.467	July 23 00:07	2.0	0.11	290	V
188	+0.002 58	+0.0636	0.403	139.467	July 23 00:07	1.8	0.03	140	V
59	−0.001 89	−0.1296	2.050	139.297	July 23 00:09	1.6	0.17	170	V

Table 5c (*cont.*)

Year Trail	Δr (D–E) (AU)	Δa_0 (AU)	f_m	$\lambda_\odot$ (°)	Time (UT)	χ	W (°)	ZHR (/h)	src
948:	*Moon in Cap. (100%)*				(Not reported, over Americas)				
698	+0.002 09	−0.0782	1.030	139.416	July 23 05:48	1.3	0.02	350	V
1066:	*Moon in Cnc. (1%)*				(Not reported, over Americas)				
569	+0.000 32	−0.4419	1.007	139.368	July 24 10:15	3.2	0.08	300	V
		(comet at perihelion Sep. 17.6, 1079)							
1085:	*Moon in Cnc. (0%)*				(Not reported, over Americas)				
59	+0.001 96	−0.1346	0.914	139.419	July 24 08:22	1.6	0.02	300	V
1089:	*Moon in Cap. (99%)*				(Not reported, Europe??)				
59	+0.001 14	0.0872	0.303	139.622	July 24 00:14	2.0	0.04	440	V
1102:	*Moon in Lib. (50%)*				(Not reported, E. Europe??)				
188	−0.001 16	0.2823	0.666	139.519	July 24 19:24	2.9	0.14	490	V
1141:	*Moon in Psc. (80%)*				(Not reported, over Americas)				
59	−0.000 52	0.6603	1.377	139.891	July 25 04:35	3.3	0.11	2600	V (!)
1196:	*Moon in Gem. (7%)*				(Not reported, Europe??)				
698	+0.002 46	−0.0684	7.116	139.654	July 25 00:01	1.3 (!)	0.03	1800	V
1197:	*Moon in Oph. (69%)*				(Not reported, over Americas)				
569	+0.001 76	−0.2764	3.400	139.629	July 25 06:21	2.5	0.02	900	V
		(comet at perihelion Nov. 6.1, 1212)							
1243:	*Moon in Lib. (50%)*			**139.6**	(Not reported in W. Europe)				
569	+0.000 06	+0.5373	0.042	139.680	July 26 02:36	3.3	0.09	140	V
698	−0.003 87	+1.2987	0.443	139.681	July 26 02:38	3.5	0.26	17	V
1269:	*Moon in Tau. (22%)*				(Not reported, Asia??)				
569	+0.003 40	+1.2755	1.607	139.593	July 25 16:20	3.5	0.06	410	V
1316:	*Moon in Vir. (29%)*				(Not reported, Asia, too faint?)				
569	−0.000 76	+2.0428	1.429	139.753	July 25 21:26	3.5	0.12	390	V

1318:	*Moon in Gem. (7%)*				(Not reported, over Americas)				
188	+0.002 57	+0.0318	5.752	139.541	July 26 04:26	1.6	0.03	1800	V
1323:	*Moon in Tau. (41%)*				(Not reported, over Americas)				
569	+0.000 73	−0.2323	0.445	139.627	July 26 13:20	2.3	0.06	320	V
1340:	*Moon in Leo (0%)*				(Not reported, Europe, too faint)				
1212	+0.002 37	−0.9542	5.762	139.658	July 25 22:41	3.4	0.02	260	V
188	+0.002 54	+0.0057	0.620	139.644	July 25 22:21	1.5	0.03	180	V
		(comet at perihelion May 2.3, 1348)							
1362:	*Moon in Vir. (23%)*				(Not reported, Pacific)				
569	+0.001 62	+0.1240	0.231	139.575	July 26 11:48	2.2	0.02	250	V
1365:	*Moon in Lib. (53%)*				(Not reported, over Americas)				
698	+0.001 38	0.5785	1.278	139.659	July 26 08:20	3.3	0.03	4700	V
1366:	*Moon in Cet. (84%)*				(Not reported, Asia??)				
698	+0.002 49	0.5718	0.590	139.664	July 26 14:44	3.3	0.03	790	V
1367:	*Moon in Cnc. (0%)*				(Not reported, Europe??)				
698	+0.002 48	0.6044	1.208	139.678	July 26 21:15	3.3	0.03	1600	V
1369:	*Moon in Ari. (53%)*				(Not reported, over Americas)				
569	−0.000 10	+0.1102	0.683	139.576	July 26 06:54	2.2	0.09	670	V
1387:	*Moon in Oph. (79%)*				(Not reported, Europe??)				
188	+0.001 69	+0.5056	0.384	139.647	July 26 23:23	3.2	0.02	1100	V
1420:	*Moon in Aqr. (99%)*				(Not reported, over Americas)				
569	+0.000 69	−0.6086	3.187	139.376	July 26 03:30	3.3	0.06	980	V
1423:	*Moon in Cet. (85%)*				(Not reported, Europe??)				
569	−0.001 01	+1.7795	1.088	139.497	July 27 01:09	3.5	0.13	310	V
1479:	*Moon in Oph. (66%)*				(Not reported, Pacific)				
1079	+0.003 98	+0.0642	1.976	139.570	July 27 11:17	1.8	0.09	190	V
698	+0.005 04	+0.0458	1.130	139.570	July 27 10:51	1.7	0.13	40	V
		(comet at perihelion Oct. 18.3, 1479)							
1506:	*Moon in Lib. (52%)*				(Not reported, Americas)				
698	−0.001 60	0.6187	0.342	139.568	July 27 09:18	3.3	0.16	240	V

Table 5c (*cont.*)

Year Trail	Δr (*D–E*) (AU)	Δa_0 (AU)	f_m	$\lambda_\odot$ (°)	Time (UT)	χ	*W* (°)	ZHR (/h)	src
1508:	*Moon in Cnc. (1%)*				(Not reported, Europe??)				
698	−0.002 02	+0.5853	1.708	139.628	July 26 23:06	3.3	0.18	850	V
1514:	*Moon in Lib. (39%)*				(Not reported, Pacific)				
698	+0.001 14	1.3301	0.383	139.692	July 27 13:31	3.5	0.04	720	V
1565:	*Moon in Leo (0%)*			**139.1**	China: >(Not reported, Pacific)				
569	+0.001 02	−0.2266	4.510	139.577	July 27 12:20	2.3	0.05	2700	V
1567:	*Moon in Ari. (60%)*				(Not reported, Americas)				
569	+0.001 77	−0.1436	2.305	139.725	July 28 04:29	1.7	0.02	870	V
1589:	*Moon in Ori. (18%)*				(Not reported, Asia??)				
1212	+0.000 48	−1.0440	2.178	139.694	Aug. 6 18:58	3.4	0.07	290	V
1590:	*Moon in Vir. (35%)*			**139.4**	Stars fell like rain in China – unidentified				
826??	+0.015 88	−0.1598	(0.1)	139.535	Aug. 06 21:05	1.8	0.61	0	V
1601:	*Moon in Sco. (60%)*				(Not reported, E. Europe, just before dawn in China??)				
569	+0.002 13	+0.0706	3.159	139.603	Aug. 06 18:23	1.9	0.02	1700	V
		(comet at perihelion Feb. 6.7, 1610)							
1645:	*Moon in Cap. (100%)*			**139.3**	Stars flew to S. like rain China				
569?	+0.001 03	+0.1017	0.391	139.664	Aug. 07 02:28	2.1	0.05	650	V
1646:	*Moon in Ori. (18%)*				(Not reported, Americas)				
569	+0.001 78	0.0963	0.559	139.708	Aug. 07 09:46	2.1	0.02	460	V
1647:	*Moon in Lib. (53%)*				(Not reported, Asia??)				
698	−0.002 31	+1.2078	4.643	139.706	Aug. 07 15:49	3.4	0.19	860	V
1648:	*Moon in Psc. (89%)*				(Not reported, Europe??)				
698	+0.001 59	+0.7564	1.386	139.667	Aug. 06 21:01	3.4	0.02	3700	V
1649:	*Moon in Cnc. (2%)*				(Not reported, Europe??)				
698	+0.003 86	+1.1339	1.885	139.619	Aug. 07 00:02	3.4	0.08	390	V

1704:	*Moon in Lib. (44%)*				(Not reported, Americas?)				
698	−0.000 06	−0.0343	1.748	139.639	Aug. 08 04:34	1.3	0.09	1070	V
1705:	*Moon in Cet. (84%)*				(Not reported, Pacific)				
698	+0.000 57	−0.0357	0.936	139.615	Aug. 08 10:17	1.3	0.07	1020	V
1706:	*Moon in Leo (0%)*				(Not reported, Asia??)				
698	−0.000 10	−0.0273	1.892	139.637	Aug. 08 17:00	1.3(!)	0.09	1100	V
1711:	*Moon in Tau. (27%)*				(Not reported, Europe??)				
698	+0.001 19	−0.0398	0.724	139.665	Aug. 09 00:28	1.3	0.04	640	V
1733:	*Moon in Cnc. (2%)*				(Not reported, Asia??, too usual?)				
698	−0.000 64	+0.0842	0.549	139.662	Aug. 08 15:46	2.0	0.12	300	V
		(comet at perihelion June 15.9, 1737)							
1766:	*Moon in Leo (5%)*				(Not reported, Europe ??)				
569	+0.001 22	−0.5844	2.673	139.402	Aug. 08 00:20	3.3	0.04	620	V
1767:	*Moon in Cap. (99%)*				(Not reported, over America's)				
569	+0.000 68	+0.1125	1.566	139.624	Aug. 09 07:48	2.2	0.06	3200	V
1780:	*Moon in Oph. (66%)*				(Not reported, E. Europe??)				
1079	+0.002 86	+0.7925	0.316	139.735	Aug. 08 18:34	3.4	0.04	250	V
1807:	*Moon in Lib. (54%)*				(Not reported, Europe??) "St. Lawrence Tears"?				
1079	+0.000 15	+1.1380	0.251	139.873	Aug. 10 19:58	3.4	0.08	500	V
1814:	*Moon in Tau. (31%)*				(Not reported, Asia??)				
698	+0.001 93	−0.7493	2.892	139.939	Aug. 10 16:46	3.3	0.02	260	V
1851:	*Moon in Cap. (99%)*			**140.5**	Stars fell like arrows to S. for 2 hrs China – unidentified				
1212??	+0.001 92	−0.9623	0.584	139.846	Aug. 11 01:57	3.4	0.02	40	V
1861:	*Moon in Vir. (21%)*			**139.9**	Meteors fell from N to S unceasingly China – probably Filament				
1479	−0.000 76	−0.8751	2.406	139.753	Aug. 10 13:11	3.4	0.12	130	V
1862:	*Moon in Cap. (100%)*			**139.6**	Countable number fell like rain China – probably Filament				
1737?	+0.002 69	−0.0293	0.701	139.378	Aug. 10 09:53	1.3	0.03	160	V
1479	−0.004 79	−0.4182	17.454	139.578	Aug. 10 14:52	3.0	0.30	50	V
		(comet at perihelion Aug. 23.4, 1862)							

Table 5c (*cont.*)

Year Trail	Δr (D–E) (AU)	Δa_0 (AU)	f_m	$\lambda_\odot$ (°)	Time (UT)	χ	W (°)	ZHR (/h)	src
1873:	*Moon in Aqr. (94%)*				(Not reported, Asia??)				
121	+0.002 55	+0.1216	0.574	139.770	Aug. 10 15:27	2.2	0.03	260	V
1886:	*Moon in Sgr. (86%)*				(Not reported, E. Europe??)				
1079	+0.001 11	+0.1640	0.378	139.608	Aug. 10 19:13	2.5	0.04	750	V
1891:	*Moon in Vir. (33%)*				(Not reported, Europe/E. USA??)				
1079	+0.000 05	+0.4513	0.260	139.644	Aug. 11 02:49	3.2	0.09	890	V
1897:	*Moon in Sgr. (96%)*				(Not reported, E. Europe??)				
698	+0.001 50	+0.5308	0.415	139.790	Aug. 10 19:28	3.3	0.03	1400	V
1956:	*Moon in Vir. (38%)*				(Not reported, Americas??)				
1212	+0.000 41	+1.9508	0.542	140.047	Aug. 12 04:43	3.5	0.07	470	V
1969:	*Moon in Cnc. (1%)*				(Not reported, Americas??)				
1212	+0.001 81	+1.9515	1.019	139.817	Aug. 12 06:59	3.5	0.02	480	V
1979:	*Moon in Psc. (74%)*				(Not reported, W. USA, bad weather)				
1862	−0.000 71	−0.2699	(0.5)	139.522	Aug. 12 13:05	2.9	0.12	800	V
1862	−0.000 55	−0.3491	–.–	139.497	Aug. 12 12:28	46.1	+57.5	59.6	IS
826	+0.004 65	1.0744	0.244	139.736	Aug. 12 18:26	3.4	0.12	26	V
1980:	*Moon in Leo (1%)*				(Observed from Switzerland?)				
1479	−0.000 33	−0.9687	2.037	139.611	Aug. 11 21:30	3.4	0.10	140	V
1479	−0.000 54	−0.598	–.–	139.605	Aug. 11 21:35	46.1	+57.7	59.5	IS
1348	+0.000 91	−0.598	–.–	139.772	Aug. 12 01:45	46.2	+57.8	59.4	IS
1348	+0.001 01	−0.604	–.–	139.781	Aug. 12 01:58	46.2	+57.8	59.4	IS
1981:	*Moon in Sgr. (86%)*				(Not reported, Europe, not strong enough)				
1862	−0.002 01	−1.4978	4.988	139.416	Aug. 11 22:49	3.5	0.18	40	V
1862	+0.000 52	+0.0417	–.–	139.407	Aug. 11 22:49	45.7	+57.8	59.4	IS

1988:	*Moon in Cnc. (1%)*			**139.59 + 0.2, −0.02**		−.−	−.−	**21 ± 4**	
1737?	+0.002 47	−0.0790	0.097	139.519	Aug. 11 20:19	1.3	0.03	23	V
1989:	*Moon in Oph. (74%)*				(Not reported, Europe, too faint)				
1479	−0.002 91	−0.8673	0.546	139.766	Aug. 12 08:38	3.4	0.04	17	V
1479	+0.000 76	−1.059	−.−	139.748	Aug. 12 08:15	46.1	+57.6	59.5	IS
1990:	*Moon in Tau. (64%)*			**139.57 (2)**		**−.−**	**∼0.04**	**150 ± 75**	**J**
				139.55 (5)		**−.−**	**−.−**	**100 ± 10**	**BR**
1610	+0.001 91	−0.3041	(0.1)	139.600	Aug. 12 10:36	2.7	0.02	100	V
1610	+0.001 23	−0.252	−.−	139.610	Aug. 12 10:53	46.0	+57.6	59.5	IS
1991:	*Moon in Leo (10%)*			**139.564 ± 0.005**		**−.−**	**0.034 ± 0.06**	**350 ± 75**	**J**
				139.55		**−.−**	**−.−**	**219 ± 63**	**BR**
				139.580 ± 0.005		**−.−**	**0.06**	**350 ± 20**	**R.A.**
1737	+0.002 23	−0.0422	0.114	139.504	Aug. 12 14:20	1.3	0.02	38	V
1610	+0.002 63	−0.3345	4.479	139.607	Aug. 12 16:56	2.8	0.03	470	V
1610	−0.000 20	−.−	−.−	139.576	Aug. 12 16:11	46.4	+57.6	59.7	IS
1992:	*Moon in Cap. (98%)*			**139.453**		**−.−**	**0.054**	**550 ± 150**	**J**
				139.50 ± 0.04		**2.05 ± 0.05**	**< 0.06**	**350**	**R.A.**
1862	−0.001 14	+1.1475	0.083	139.502	Aug. 11 20:33	3.4	0.14	50	V
1862	−0.001 73	+1.56	−.−	139.483	Aug. 11	45.9	+57.7	59.5	IS
1610	−0.002 49	−0.2110	3.544	139.541	Aug. 11 21:32	2.2	0.20	139	V
826	+0.002 39	+0.2040	0.350	139.574	Aug. 11 22:21	2.7	0.02	253	V
		(comet at perihelion Dec. 12.3, 1992)							
***1993*:**	*Moon in Tau. (37%)*			**139.445**	***“Nodal blanket”***	**1.72 ± 0.02**	**0.109 ± 0.015**	**80 ± 20**	**J**
1862	+0.000 95	+0.1129	0.069	139.429	Aug. 12 00:53	2.2	0.05	130	V
1862	−.−	−.−	−.−	139.433	Aug. 12 00:59	−.−	−.−	−.−	L
1862	+0.003 42	+1.606	−.−	139.449	Aug. 12 01:26	45.9	+57.7	59.3	IS
1862	+0.003 42	+1.606	−.−	139.449	Aug. 12 01:26	45.9	+57.7	59.3	IS

Table 5c (*cont.*)

Year Trail	Δr (D–E) (AU)	Δa_0 (AU)	f_m	$\lambda_\odot$ (°)	Time (UT)	χ	W (°)	ZHR (/h)	src
1994:	*Moon in Vir. (26%)*				(Not reported, Europe, too faint)				
1079	+0.004 88	−0.1412	0.243	139.513	Aug. 12 00:09	2.2	0.15	13	V
1997:	*Moon in Lib. (58%)*			**139.72**		**1.78**	**–.–**	**137 ± 7**	**AR**
1079	+0.003 83	+0.0780	0.665	139.675	Aug. 12 07:36	1.9	0.08	80	V
1998:	*Moon in Psc. (75%)*				(Not reported, too faint)				
1479	+0.005 20	−0.4659	1.363	139.702	Aug. 12 14:23	3.1	0.14	10	V
1479	+0.001 79	+0.461	–.–	139.721	Aug. 12 15:03	46.2	+57.9	59.3	IS
2003:	*Moon in Aqr. (99%)*				No outburst				
513	−0.0003	to + 0.0008	–.–	139.82	Aug. 13 00:07	–.–	–.–	(100)	L
2004:	*Moon in Gem. (16%)*			**139.44 ± 0.01**		**–.–**	**0.05 ± 0.01**	**130**	**J**
1862	−0.000 00	+1.5371	0.101	139.439	Aug. 11 20:49	3.5	0.09	100	V
1862	−0.0012			139.441	Aug. 11 20:54	–.–	–.–	(400)	L
1862	−0.001 30	+2.9034	–.–	139.424	Aug. 11 20:41	45.9	+57.7	59.5	IS
2004:	*Moon in Gem. (13%)*				(Not observed?)				
698	+0.002 11	+0.4162	0.093	139.823	Aug. 12 06:25	3.2	0.02	170	V
2005:	*Moon in Vir. (40%)*								
1862	+0.002 54	+1.6673	0.100	139.407	Aug. 12 02:08	3.5	0.03	34	V
1862	+0.0048	+1.583	1.0	139.414	Aug. 12 02:19	45.95	+57.73	59.32	MS
2005:	*Moon in Lib. (42%)*								
1479	+0.000 83	+0.4109	0.040	139.654	Aug. 12 08:18	3.2	0.05	240	V
1479	−0.0008	–.–	–.–	139.68	Aug. 12 08:58	–.–	–.–	–.–	L
1479	−0.000 83	+0.301	–.–	139.670	Aug. 12 08:53	46.0	+57.7	59.5	IS
1479	−0.000 87	+0.322	0.006	139.681	Aug. 12 08:59	45.87	+57.67	59.49	MS

2005:	*Moon in Lib. (46%)*								
1212	+0.003 35	+2.0838	0.302	139.975	Aug. 12 16:20	3.5	0.06	30	V
1212	+0.0017	+0.078	0.001	139.762	Aug. 12 11:09	45.89	+57.72	59.41	MS
2006:	*Moon in Psc. (86%)*								
1479	+0.0014	–.–	–.–	139.71	Aug. 12 15:33	–.–	–.–	–.–	L
1479	+0.001 41	+0.301	–.–	139.712	Aug. 12 15:58	45.9	+57.7	59.5	IS
513	−0.0005–+0.0001	–.–	–.–	139.77 (.78)	Aug. 12 17:18	–.–	–.–	–.–	L
2007:	*Moon in Leo (0%)*								
1479	+0.002 36	+0.3169	0.017	139.740	Aug. 12 22:42	3.0	0.02	20	V
513	+0.0003–+0.0010	–.–	–.–	139.81	Aug. 13 00:27	–.–	–.–	–.–	L
1479	+0.001 85	+0.301	–.–	139.745	Aug. 12 22:55	46.0	+57.6	59.5	IS
2008:	*Moon in Sgr. (80%)*								
1479	−0.0001	–.–	–.–	139.765	Aug. 12 05:30	–.–	–.–	(670)	L
1479	−0.000 02	+0.301	–.–	139.760	Aug. 12 05:27	46.4	+57.6	60.5	IS
1479	−0.000 41	+0.314	–.–	139.660	Aug. 12 02:57	45.9	+57.6	59.5	IS
513	−0.0001–−0.0007	–.–	–.–	139.83	Aug. 12 07:07	–.–	–.–	–.–	L
2009:	*Moon in Ari. (65%)*								
1610	+0.001 94	+0.7631	0.016	139.590	Aug. 12 07:16	3.4	0.02	≫14	V
1610	+0.000 86	+0.732	–.–	139.618	Aug. 12 08:01	45.8	+57.6	59.5	IS
1610	+0.0009			139.615	Aug. 12	–.–	–.–	(170)	L
2010:	*Moon in Vir. (11%)*								
1479	+0.001 93	+0.2920	0.041	139.722	Aug. 12 16:39	2.9	0.02	65	V
1479	+0.000 56	+0.314	–.–	139.618	Aug. 12 14:06	45.7	+57.7	59.4	IS
2013:	*Moon in Vir. (32%)*								
1079	+0.001 21	−0.3616	0.221	139.903	Aug. 12 15:43	2.9	0.04	80	V
1479	+0.003 18	+0.3048	0.009	139.533	Aug. 12 06:28	3.0	0.05	4	V
2015:	*Moon in Snc. (4%)*								
1862	+0.002 63	+2.9076	0.115	139.483	Aug. 12 17:32	3.5	0.03	11	V
2016:	*Moon in Sco. (64%)*								
1079	+0.000 23	+0.1523	0.365	139.683	Aug. 12 04:43	2.4	0.08	580	V
1862	−0.003 27	+3.0340	0.110	139.438	Aug. 11 22:36	3.5	0.23	1	V

Table 5c (*cont.*)

Year Trail	Δr (D–E) (AU)	Δa_0 (AU)	f_m	$\lambda_\odot$ (°)	Time (UT)	χ	W (°)	ZHR (/h)	src
2027:	*Moon in Sgr. (85%)*								
1862	+0.003 06	+4.3401	0.835	139.484	Aug. 12 19:18	3.5	0.05	21	V
1079	+0.002 30	−0.2821	1.771	139.663	Aug. 12 23:46	2.6	0.02	280	V
2028:	*Moon in Ari. (62%)*								
1862	+0.000 73	+4.4601	0.117	139.402	Aug. 11 23:32	3.5	0.06	22	V
1479	+0.000 98	+0.6257	0.070	139.633	Aug. 12 05:20	3.3	0.05	370	V
1479	+0.0004	–.–	–.–	139.641	Aug. 12 05:31	–.–	–.–	(3000)	L
2038:	*Moon in Sgr. (94%)*								
1862	−0.000 23	+5.6482	0.119	139.493	Aug. 12 15:05	3.5	0.10	$\gg$6	V
569	−0.000 53	+0.090 20	<0.1	139.754	Aug. 12 21:36	2.0	0.11	$<$30	V
2039:	*Moon in Tau. (39%)*								
1479	+0.000 54	+1.3230	0.039	139.612	Aug. 13 00:21	3.4	0.36	85	V
2067:	*Moon in Leo (10%)*								
1212	+0.004 20	+0.6257	0.379	139.810	Aug. 13 09:25	3.3	0.10	100	V
2075:	*Moon in Leo (2%)*								
1479	+0.002 61	+2.1020	0.322	139.712	Aug. 13 08:14	3.5	0.03	60	V
2076:	*Moon in Sgr. (92%)*								
1212	+0.002 08	+1.7118	0.445	139.641	Aug. 12 12:39	3.5	0.02	210	V
		(comet at perihelion July 12.4, 2126)							

Table 5d *109P/Swift–Tuttle – Perseid shower Filament outbursts*

Peak time and activity are not calculated from a rigorous model, but estimated from a combination of the movement of AD 188–698 trails and the Sun's reflex motion (which do not always agree). Where $f_m < 1$, I have assumed that Jupiter has cleared part of the Filament. An upper limit for the Filament activity is obtained by taking $f_m = 1$. Peak rate: $ZHR = 260\, f_m\, 10^{(-100|\Delta r - 0.01|)}$

Year Trail	Δr (*D–E*) (AU)	Δa_0 (1-rev) (AU)	f_m	$\lambda_\odot$ (°)	Time (UT)	χ	*W* (°)	ZHR (1/h)	src
1862:	–.–	–.–	–.–	**139.61**	Aug. 10 15:38	–.–	**>0.3**	–.–	**JB**
1863:	–.–	–.–	–.–	**139.64**	Aug. 10 22:34	–.–	**0.5**	–.–	**JB**
1950–1979:	–.–	–.–	–.–	**139.67 ± 0.05**	–.–	–.–	–.–	**<6**	**Lindblad**
1979:	−0.0009	+1.646	1.0	(139.59)	Aug. 12 15:00	1.9	0.043	21	J
1980:	*Moon in Leo (2%)*			**139.8 ± 0.1**	Aug. 12 02:00	–.–	–.–	**~40**	**MS**
Fil.	+0.0013	+1.646	1.0	(139.41)	Aug. 11 16:00	1.9	0.043	35	J
1981:	*Moon in Sgr. (85%)*								
Fil.	+0.0043	+1.646	1.0	(139.26)	Aug. 11 19:00	1.9	0.043	69	J
1982:	*Moon in Ari. (56%)*								
Fil.	+0.0074	+1.646	1.0	(139.18)	Aug. 11 23:00	1.9	0.043	143	J
1983:	*Moon in Vib. (16%)*								
Fil.	+0.0099	+1.646	1.0	(139.18)	Aug. 12 05:00	1.9	0.043	256	J
1984:	*Moon in Cap. (100%)*								
Fil.	+0.0113	+1.646	1.0	(139.26)	Aug. 11 13:00	1.9	0.043	191	J
1985:	*Moon in Tau. (18%)*			**139.78 ± 0.1**	Aug. 12 08:00	–.–	–.–	**~30**	**R**
Fil.	+0.0114	+1.646	(0.1)	(139.40)	Aug. 11 23:00	1.9	0.043	19	J
1986:	*Moon in Lib. (43%)*			**(139.7)**	Aug. 12 12:00	–.–	–.–	**<10**	**R**
Fil.	+0.0102	+1.646	(0.1)	(139.55)	Aug. 12 09:00	1.9	0.043	25	J
1987:	*Moon in Psc. (85%)*			**(139.7)**	Aug. 12 19:00	–.–	–.–	**<10**	**R**
Fil.	+0.0083	+1.646	(0.1)	(139.68)	Aug. 12 18:00	1.9	0.043	18	J
1988:	*Moon in Leo (0%)*			**139.59 + 0.2, −0.02**	Aug. 11 22:00	–.–	–.–	**<20**	**JB**
				139.78 ± 0.03	Aug. 12 03:00	–.–	–.–	**21 ± 4**	**BR**
Fil.	+0.0063	−0.564	0.2	139.53 (±0.2°)	Aug. 11 22:00	1.9	0.043	22	J

Table 5d (*cont.*)

Year Trail	Δr (D–E) (AU)	Δa_0 (1-rev) (AU)	f_m	$\lambda_\odot$ (°)	Time (UT)	χ	W (°)	ZHR (1/h)	src
1989:	*Moon in Oph. (73%)*			**139.56 + 0.05, −0.01**	Aug. 12 04:00	–.–	**~15**	**70 ± 15**	**JB**
				139.56 ± 0.03	Aug. 12 04:00	–.–	–.–	**37 ± 10**	**BR**
Fil.	+0.0049	−0.434	1.0	139.50 (±0.2°)	Aug. 12 02:00	1.9	0.043	80	J
1990:	*Moon in Ari. (64%)*			**139.57 ± 0.02**	Aug. 12 10:00	–.–	**~15**	**150 ± 75**	**JB**
				139.55 ± 0.05	Aug. 12 09:00	–.–	–.–	**10 ± 10**	**BR**
Fil.	+0.0046	−0.304	1.0	139.50 (±0.2°)	Aug. 12 08:00	1.9	0.043	75	J
1991:	*Moon in Leo (10%)*			**139.564 ± 0.005**	Aug. 12 15:51	**1.86**	**0.034(6)**	**350 ± 75**	**JB**
				139.402 ± 0.005	Aug. 12 11:48	–.–	–.–	**50 ± 10**	**R.A.**
				139.66 ± 0.01	Aug. 12 18:15			**40 ± 10**	**R.A.**
Fil.	+0.0055	−0.174	1.0	139.51 (±0.2°)	Aug. 12 15:00	1.9	0.043	92	J
1992:	*Moon in Cap. (90%)*			**139.453 ± 0.005**	Aug. 11 19:20	**2.12 ± 0.15**	**0.054**	**550 ± 150**	**JB**
				139.48 ± 0.02	Aug. 11 20:00	–.–	–.–	**155 ± 22**	**BR**
Fil.	+0.0074	−0.044	1.0	139.48 (±0.2°)	Aug. 11 20:00	1.9	0.043	143	J
1993:	*Moon in Tau. (36%)*			**139.508 ± 0.01**	Aug. 12 02:52	**2.02 ± 0.04**	**~0.04**	**250 ± 35**	**JB**
				139.53 ± 0.01	Aug. 12 03:25	–.–	–.–	**199 ± 17**	**BR**
Fil.	+0.0099	+0.087	1.0	139.46 (±0.2°)	Aug. 12 02:00	1.9	0.043	252	J
1994:	*Moon in Vir. (30%)*			**139.586 ± 0.010**	Aug. 12 10:54	**1.9 ± 0.2**	**0.043± 0.007**	**190 ± 30**	**JB**
				139.59 ± 0.01	Aug. 12 11:00	**1.82 ± 0.05**	–.–	**173 ± 17**	**BR**
Fil.	+0.0121	+0.216	1.0	139.55 (±0.2°)	Aug. 12 10:00	1.9	0.043	159	J
1995:	*Moon in Psc. (95%)*			**139.66 ± 0.02**	Aug. 12 18:56	–.–	**0.041 ± 0.005**	**120 ± 25**	**JB**
				139.62 ± 0.05	Aug. 12 17:56	–.–	–.–	**106 ± 30**	**BR**
Fil.	+0.0136	+0.346	1.0	139.63 (±0.2°)	Aug. 12 18:00	1.9	0.043	114	J

1996:	*Moon in Gem. (5%)*			**139.64 ± 0.03**	Aug. 12 00:37	–.–	**0.059 ± 13**	**100 ± 40**	**JB**
				139.67 ± 0.03	Aug. 12 02:08	–.–	–.–	**56 ± 17**	**BR**
Fil.	+0.0137	+0.476	1.0	139.66 (±0.2°)	Aug. 12 01:00	1.9	0.043	110	J
1997:	*Moon in Lib. (50%)*			**139.72 ± 0.03**	Aug. 12 08:45	–.–	–.–	**137 ± 7**	**J**
Fil.	+0.0124	+0.606	(0.1)	139.67 (±0.2°)	Aug. 12 07:00	1.9	0.043	15	J
1998:	*Moon in Psc. (75%)*				(No data)				
Fil.	+0.0100	+0.736	(0.1)	(139.71)	Aug. 12 15:00	1.9	0.043	26	J
1999:	*Moon in Leo (3%)*			**139.793 ± 0.004**	Aug. 12 22:48	**<2.1**	**0.043 ± 15**	**20 ± 3**	
Fil.	+0.0069	+0.866	(0.1)	(139.83)	Aug. 13 00:00	1.9	0.043	13	J
2000:	*Moon in Sgr. (93%)*				(No data)				
Fil.	+0.0039	+0.996	(0.2)	139.97 (±0.2°)	Aug. 12 09:00	1.9	0.043	12	J
2001:	*Moon in Tau. (47%)*				(No data)				
Fil.	+0.0016	+1.126	1.0	140.04 (±0.2°)	Aug. 12 17:00	1.9	0.043	38	J
2002:	*Moon in Vir. (22%)*			–.–		–.–	–.–	**<15**	
Fil.	+0.0007	+1.256	1.0	139.85 (±0.2°)	Aug. 12 19:00	1.9	0.043	31	J
2003:	*Moon in Aqr. (99%)*				(No data)				
Fil.	+0.0012	+1.386	1.0	139.87 (±0.2°)	Aug. 13 01:00	1.9	0.043	34	J
2004:	*Moon in Gem. (15%)*			**139.63 ± 0.01**		**–.–**	**0.15 ± 0.02**	**60 ± 10**	**J**
Fil.	+0.0029	+1.516	1.0	139.67 (±0.2°)	Aug. 12 02:00	1.9	0.043	50	J
2005:	*Moon in Lib. (44%)*								
Fil.	+0.0052	+1.646	1.0	(≤139.86)	Aug. 12 13:00	1.9	0.043	≤86	J
2006:	*Moon in Psc. (84%)*								
Fil.	+0.0075	+1.776	1.0	(≤139.94)	Aug. 12 22:00	1.9	0.043	≤147	J
2007:	*Moon in Leo (0%)*								
Fil.	+0.0092	+1.906	1.0	(≤139.97)	Aug. 13 04:00	1.9	0.043	≤214	J
2008:	*Moon in Oph. (79%)*								
Fil.	+0.0096	+2.036	1.0	(139.57)	Aug. 12 01:00	1.9	0.043	≤239	J
2009:	*Moon in Ari. (62%)*								
Fil.	+0.0089	+2.166	(0.1)	(≤139.96)	Aug. 12 17:00	1.9	0.043	20	J

Table 5d (*cont.*)

Year Trail	Δr (D–E) (AU)	Δa_0 (1-rev) (AU)	f_m	$\lambda_\odot$ (°)	Time (UT)	χ	W (°)	ZHR (1/h)	src
2010:	*Moon in Vir. (11%)*								
Fil.	+0.0072	+2.296	(0.1)	(139.81)	Aug. 12 19:00	1.9	0.043	14	J
2011:	*Moon in Aqr. (100%)*								
Fil.	+0.0050	+2.426	(0.1)	(139.89)	Aug. 13 03:00	1.9	0.043	8	J
2012:	*Moon in Tau. (25%)*								
Fil.	+0.0031	+2.556	(0.2)	139.95 (±0.2°)	Aug. 12 11:00	1.9	0.043	10	J
2013:	*Moon in Vir. (33%)*								
Fil.	+0.0020	+2.686	1.0	140.01 (±0.2°)	Aug. 12 18:00	1.9	0.043	(41)	J
2014:	*Moon in Psc. (95%)*								
Fil.	+0.0023	+2.816	1.0	139.73 (±0.2°)	Aug. 12 17:00	1.9	0.043	44	J
2015:	*Moon in Cnc. (3%)*								
Fil.	+0.0038	+2.946	1.0	139.72 (±0.2°)	Aug. 12 23:00	1.9	0.043	63	J
2016:	*Moon in Sco. (64%)*								
Fil.	+0.0064	+3.076	1.0	(≤139.72)	Aug. 12 06:00	1.9	0.043	113	J
2017:	*Moon in Cet. (77%)*								
Fil.	+0.0094	+3.206	1.0	(≤139.78)	Aug. 12 13:00	1.9	0.043	≤225	J
2018:	*Moon in Leo (3%)*								
Fil.	+0.0120	+3.336	1.0	(≤139.79)	Aug. 12 20:00	1.9	0.043	≤164	J
2019:	*Moon in Sgr. (95%)*								
Fil.	+0.0136	+3.466	1.0	(≤139.78)	Aug. 13 02:00	1.9	0.043	≤112	J
2020:	*Moon in Tau. (43%)*								
Fil.	+0.0139	+3.596	1.0	(≤139.89)	Aug. 12 10:00	1.9	0.043	≤106	J
2021:	*Moon in Vir. (20%)*								
Fil.	+0.0127	+3.726	(0.1)	139.85 (±0.2°)	Aug. 12 16:00	1.9	0.043	14	J

2022:	*Moon in Aqr. (99%)*								
Fil.	+0.0103	+3.856	(0.1)	139.85 (±0.2°)	Aug. 12 22:00	1.9	0.043	24	J
2023:	*Moon in Gem. (10%)*								
Fil.	+0.0074	+3.986	(0.1)	139.83 (±0.2°)	Aug. 13 03:00	1.9	0.043	14	J
2024:	*Moon in Lib. (47%)*								
Fil.	+0.0047	+4.116	(0.2)	(≥139.81)	Aug. 12 09:00	1.9	0.043	16	J
2025:	*Moon in Psc. (87%)*								
Fil.	+0.0028	+4.246	1.0	(≥139.83)	Aug. 12 16:00	1.9	0.043	≤50	J
2026:	*Moon in Leo (0%)*								
Fil.	+0.0022	+4.376	1.0	139.97 (±0.2°)	Aug. 13 01:00	1.9	0.043	≤43	J
2027:	*Moon in Sgr. (86%)*								
Fil.	+0.0028	+4.506	1.0	139.96 (±0.2°)	Aug. 13 07:00	1.9	0.043	49	J
2028:	*Moon in Ari. (60%)*								
Fil.	+0.0043	+4.636	1.0	139.87 (±0.2°)	Aug. 12 11:00	1.9	0.043	70	J
2029:	*Moon in Vir. (10%)*								
Fil.	+0.0062	+4.766	1.0	139.91 (±0.2°)	Aug. 12 18:00	1.9	0.043	109	J
2030:	*Moon in Aqr. (100%)*								
Fil.	+0.0078	+4.896	1.0	139.89 (±0.2°)	Aug. 13 00:00	1.9	0.043	158	J
2031:	*Moon in Tau. (21%)*								
Fil.	+0.0085	+5.026	1.0	139.89 (±0.2°)	Aug. 13 06:00	1.9	0.043	≤185	J
2032:	*Moon in Vir. (33%)*								
Fil.	+0.0080	+5.155	1.0	139.93 (±0.2°)	Aug. 12 13:00	1.9	0.043	≤163	J
2033:	*Moon in Psc. (94%)*								
Fil.	+0.0062	+5.285	(0.1)	139.94 (±0.2°)	Aug. 12 20:00	1.9	0.043	11	J
2034:	*Moon in Cnc. (2%)*								
Fil.	+0.0036	+5.415	(0.1)	139.94 (±0.2°)	Aug. 13 02:00	1.9	0.043	6	J
2035:	*Moon in Oph. (72%)*								
Fil.	+0.0008	+5.545	(0.1)	139.92 (±0.2°)	Aug. 13 07:00	1.9	0.043	3	J

Table 5d (*cont.*)

Year Trail	Δr (*D–E*) (AU)	Δa_0 (1-rev) (AU)	f_m	$\lambda_\odot$ (°)	Time (UT)	χ	*W* (°)	ZHR (1/h)	src
2036:	*Moon in Psc. (75%)*								
Fil.	−0.0015	+5.675	(0.2)	139.90 (±0.2°)	Aug. 12 13:00	1.9	0.043	4	J
2037:	*Moon in Leo (3%)*								
Fil.	+0.0298	+5.805	1.0	139.87 (±0.2°)	Aug. 12 19:00	1.9	0.043	14	J
2038:	*Moon in Sgr. (96%)*								
Fil.	+0.0298	+5.935	1.0	(139.85)	Aug. 13 00:00	1.9	0.043	15	J
2039:	*Moon in Tau. (38%)*								
Fil.	+0.0282	+6.065	1.0	(139.67)	Aug. 13 02:00	1.9	0.043	22	J
2040:	*Moon in Vir. (16%)*								
Fil.	+0.0256	+6.195	1.0	(139.51)	Aug. 12 04:00	1.9	0.043	40	J
2041:	*Moon in Cap. (100%)*								
Fil.	+0.0226	+6.325	1.0	(139.40)	Aug. 12 07:00	1.9	0.043	81	J
2042:	*Moon in Gem. (11%)*								
Fil.	+0.0198	+6.455	1.0	(139.37)	Aug. 12 13:00	1.9	0.043	≤150	J
2043:	*Moon in Lib. (51%)*								
Fil.	+0.0180	+6.585	1.0	(139.43)	Aug. 12 20:00	1.9	0.043	≤219	J
2044:	*Moon in Psc. (89%)*								
Fil.	+0.0175	+6.715	1.0	(139.55)	Aug. 12 06:00	1.9	0.043	≤236	J
2045:	*Moon in Leo (0%)*								
Fil.	+0.0182	+6.845	(0.1)	(139.70)	Aug. 12 16:00	1.9	0.043	19	J
2046:	*Moon in Sgr. (88%)*								
Fil.	+0.0199	+6.975	(0.1)	(139.83)	Aug. 13 01:00	1.9	0.043	12	J
2047:	*Moon in Ari. (53%)*								
Fil.	+0.0218	+7.105	(0.1)	(139.92)	Aug. 13 09:00	1.9	0.043	7	J
2048:	*Moon in Leo (10%)*								
Fil.	+0.0233	+7.235	(0.2)	(139.94)	Aug. 12 16:00	1.9	0.043	10	J

2049:	*Moon in Aqr. (96%)*								
Fil.	+0.0238	+7.365	1.0	(139.89)	Aug. 12 21:00	1.9	0.043	39	J
2050:	*Moon in Ori. (19%)*								
Fil.	+0.0230	+7.495	1.0	(139.79)	Aug. 13 00:00	1.9	0.043	43	J
2051:	*Moon in Vir. (35%)*								
Fil.	+0.0210	+7.625	1.0	(139.67)	Aug. 13 04:00	1.9	0.043	61	J
2052:	*Moon in Aqr. (96%)*								
Fil.	+0.0183	+7.755	1.0	(139.57)	Aug. 12 08:00	1.9	0.043	102	J

Table 5e *1P/Halley dust trails and the η-Aquariids*

1P/Halley is the parent of the η-Aquariids: historic dust trail encounters. Uncertain. Calculated with scaled Leonid trail model and $f_m = 1/N$.

Year Trail	Δr (*D–E*) (AU)	Δa_0 (AU)	f_m	$\lambda_\odot$ (°)	Time (UT)	χ	*W* (°)	ZHR (/h)	src
511:	*Moon in Aqr. (20%)*			***Not observed***					
374	+0.000 98	−1.4411	–.–	41.594	Apr. 09 23:39	–.–	–.–	60	V
531:	*Moon in Cnc. (54%)*			***Not observed***					
218	+0.001 09	+0.0472	–.–	41.935	Apr. 10 11:10	–.–	–.–	900	V (!)
539:	*Moon in Cnc. (46%)*			***Not observed***					
141	+0.000 90	+0.4818	–.–	42.388	Apr. 10 23:27	–.–	–.–	1200	V (!)
543:	*Moon in Sgr. (64%)*			***Not observed***					
66	+0.002 59	+0.5216	–.–	42.263	Apr. 10 21:12	–.–	–.–	20	V
550:	*Moon in Cnc. (44%)*			***Not observed***					
451	+0.000 66	+2.8680	–.–	41.342	Apr. 09 17:06	–.–	–.–	150	V
601:	*Moon in Tau. (3%)*			***Not observed***					
451	+0.001 22	−0.6420	–.–	41.713	Apr. 10 04:12	–.–	–.–	470	V
619:	*Moon in Sgr. (73%)*			***Not observed***					
374	+0.000 34	+0.6629	–.–	41.518	Apr. 10 14:13	–.–	–.–	190	V
641:	*Moon in Aqr. (27%)*			***Not observed***					
374	+0.000 43	+1.6986	–.–	41.827	Apr. 10 13:07	–.–	–.–	60	V
647:	*Moon in Ari. (0%)*			***Not observed***					
451	+0.001 13	+2.7166	–.–	41.377	Apr. 10 14:53	–.–	–.–	370	V
650:	*Moon in Tau. (10%)*			***Not observed***					
66	+0.002 24	+0.8581	–.–	41.139	Apr. 10 03:28	–.–	–.–	50	V
672:	*Moon in Cnc. (52%)*			***Not observed***					
451	+0.000 82	−0.8771	–.–	41.843	Apr. 10 12:17	–.–	–.–	40	V
692:	*Moon in Sgr. (67%)*			***Not observed***					
530	−0.000 24	+0.8601	–.–	43.604	Apr. 12 10:51	–.–	–.–	20	V
713:	*Moon in Vir. (91%)*			***Not observed***					
141	+0.001 00	−0.8347	–.–	43.469	Apr. 12 16:43	–.–	–.–	40	V
719:	*Moon in Lib. (99%)*			***Not observed***					
218	+0.000 68	+0.4707	–.–	40.263	Apr. 09 22:17	–.–	–.–	410	V
796:	*Moon in Psc. (3%)*			***Not observed***					
218	+0.000 84	+1.2751	–.–	41.636	Apr. 11 01:54	–.–	–.–	250	V
964:	*Moon in Cet. (9%)*			***Not observed***					
218	+0.001 52	+0.4755	–.–	41.973	Apr. 12 11:44	–.–	–.–	1100	V (!)

Table 6a *Andromedids from Jupiter-family comet 3D/Biela dust trails*

Year Trail	Δr (D–E) (AU)	Δa_0 (AU)	f_m	$\lambda_\odot$ (°)	Time (UT)	χ	W (°)	ZHR (/h)	src
1741:	*Moon in Lib. (5%)*			***257.6***	***Dec. 6.0***	***Krafft (St. Petersburg)***		Large number	
	(Not modeled)								
1798:	*Moon in Sco. (1%)*			***257.8***	***Dec. 06.8***	***Brandes (Bremen)***		***HR ~ 100***[1]	
	(No trails from after 1751)								
1798:	*Moon in Oph. (0%)*			***260.7***	***Dec. 09.7***	***Small meteors fell like rain as before Japan***			
	(No trails from after 1751)								
1825:	*Moon in Vir. (35%)*			***254.5***		***Two nights, moved like weaving NW > SE in China***			
	(No dust trail encounter)								
1829:	*Moon in Cet. (88%)*					*No report (bad Moon)*			
1758	−0.000 93	+0.1329	0.046	257.637	Dec. 07 13:33	–.–	–.–	–.–	V
1830:	*Moon in Vir. (42%)*			***257.8***	***Dec. 07.9***	***Abbé Raillard (France)***[2]		***"Many"***	
	(No dust trail encounter)								
1833:	*Moon in Cnc. (76%)*				*No report*				
1799	−0.000 52	+0.0807	0.063	251.917	Dec. 01 22:58	–.–	–.–	–.–	V
1838:	*Moon in Leo (69%)*			***256.7***	***Dec. 07.0***	***Europe + US***		***~150***[3]	
	(No dust trail encounter)								
1839:	*Moon in Vir. (21%)*				*No report*				
1832	+0.0020	+0.1461	0.109	250.547	Dec. 01 03:29	–.–	–.–	–.–	V

[1] J. F. Benzenberg and H. W. Brandes, *Sternschnuppen* (Hamburg: Friedrich Perthes, 1800), p. 80; Chinese observers reported activity for three nights Dec. 4.7–6.9, 1798.

[2] A. Raillard, Sur une apparition extraordinaire de ces météores observeé le 7 decémbre 1830 *Comptes Rendus* des Séances de l'Académie des Sciences. Feb. 4 issue (1839), 177.

[3] Andromedid shower was active from Dec. 6 to 15 in 1838, strongest on Dec. 6 and 7, and was seen from New Haven (E. C. Herrick, C. P. Bush, A. B. Haile, J. D. Whitney, and B. Silliman, Jr.); E. C. Herrick, Note sur le nombre moyen des étoiles filantes pour un temps ordinaire et pour quatre observateurs. – Observation d'une pluie d'étoiles filantes dans la nuit du 7 au 8 decémbre 1839 *Comptes Rendus*. Jan. 21 and Feb. 04 issues (1839), 86. P. Flaugergues, *Bull. Ac. Roy. de Belgique* Vol. 6 (June 1839).

Table 6a (*cont.*)

Year Trail	Δr (*D–E*) (AU)	Δa_0 (AU)	f_m	$\lambda_\odot$ (°)	Time (UT)	χ	*W* (°)	ZHR (/h)	src
1847:	*Moon in Sco. (1%)*			***256.3***	***Dec. 06.8***	***R.A. = 21°, Decl. = +54.8°***		***HR ~ 150***[4]	
	(No dust trail encounter)								
1950:	*Moon in Vir. (16%)*				***Nov. 29***	***R.A. = 17.3°, Decl. = +62.8°***		***HR ~ 5?***[5]	
	(No dust trail encounter)								
1852:	*Moon in Ari. (98%)*				*No report*				
1758	−0.003 25	+0.0012	0.015	245.506	Nov. 25 12:05	–.–	–.–	–.–	V
1854:									
1792	−0.001 07	+0.0910	0.046	251.478	Dec. 01 21:52	–.–	–.–	–.–	V
1859:	*Moon in Sco. (0%)*				*No report*				
1799	−0.004 42	+0.0220	0.136	243.582	Nov. 24 09:40	–.–	–.–	–.–	V
1859:	*Moon in Sco. (0%)*				*No report*				
1806	−0.006 99	+0.0242	0.088	243.977	Nov. 24 19:02	–.–	–.–	–.–	V
1867:	*Moon in Cap. (19%)*			***250.0***	***Nov. 30.8***	***R.A. = 19.2°, Decl. = +48.8°***		***HR ~ 7?***[6]	
	(No dust trail encounter)								
1872:	*Moon in Vir. (10%)*			***247.713***	***Nov. 27 20:40***	***R.A. = 27.3°, Decl. = +43.8°***		***7400*** ($W = 0.062 \pm 0.005°$)	
1832	−0.005 63	+0.0190	0.242	247.839	Nov. 27 22:32	–.–	–.–	–.–	V
1839	−0.001 64	+0.0199	0.184	247.909	Nov. 28 00:12	–.–	–.–	–.–	V
1839	−0.0016	+0.021	–.–	247.851	Nov. 27 22:46	–.–	–.–	–.–	R
1846	+0.001 19	+0.0222	0.249	247.749	Nov. 27 20:24	–.–	–.–	–.–	V
1846	+0.0011	+0.023	–.–	247.689	Nov. 27 18:56	–.–	–.–	–.–	R
1852	+0.004 04	+0.0268	0.2872	247.701	Nov. 27 19:16	–.–	–.–	–.–	V
1872:	*Moon in Vir. (13%)*				No report				
1826	−0.006 05	+0.0175	0.2644	247.323	Nov. 27 10:18	–.–	–.–	–.–	V

[4] E. Heis, *Die Periodische Sternschnuppen* (Köln, 1849).
[5] E. Heis (1877) *Resultate* (Münster) p. 31, 159, and 175 (as reported in Hawkins *et al.*, 1959).
[6] J. V. Schiaparelli (1871) *Sternschnuppen* (Stettin, German edition by G. von Boguslavski), p. 92 and 100; J. V. Schiaparelli (1871) Preliminary results of shooting-star observations made by Zezioli, Part II. *Effemerides Astronomische del Osservatorio di Milano.*

1875:	*Moon in Aqr. (47%)*			***254.9***	***Dec. 05.7***	***Stars fell like rain in China***			
	(No dust trail encounter)								
1879:	*Moon in Ari. (98%)*				*No report*				
1852	−0.005 14	+0.0505	0.157	246.279	Nov. 27 04:38	−.−	−.−	−.−	V
1884:	*Moon in Tau. (100%)*			***252.6***	***Dec. 02.7***	***Stars fell like rain in China***			
	(No dust trail encounter)								
1885:	*Moon in Leo (65%)*			***247.336***	***Nov. 27 19:49***[7]	***R.A. = 26.2°, Decl. = +45.3°***		***6400*** ($W = 0.062 \pm 0.010°$)[8]	
1839	+0.002 06	−0.0071	0.696	247.311	Nov. 27 17:59	−.−	−.−	−.−	V
1846	+0.000 32	−0.0060	0.285	247.385	Nov. 27 19:44	−.−	−.−	−.−	V
1846	+0.0007	−0.0053	−.−	247.323	Nov. 27 18:15	−.−	−.−	−.−	R
1852	−0.000 70	−0.0057	0.211	247.399	Nov. 27 20:04	−.−	−.−	−.−	V
1852	−0.0003	−0.0052	−.−	247.333	Nov. 27 18:29				R
1886:	*Moon in Sco. (0%)*			***245.0***	***Nov. 25.7***	***Stars fell like rain in China***[9]			
1852	−0.004 25	+0.069	0.091	244.782	Nov. 25 12:08	−.−	−.−	−.−	V
1887:	*Moon in Psc. (71%)*			***243.7***	***Nov. 24.7***	***Stars fell crossing for hours in China***			
	(No dust trail encounter)								
1887:	*Moon in Tau. (100%)*			***248.8***	***Nov. 29.7***	***Like rain in NW (China)***			
	(No dust trail encounter)								
1887:	*Moon in Gem. (91%)*			***252.9***	***Dec. 03.8***	***Medium brightness, long yellow, Austria***			
	(No dust trail encounter)								
1887:	*Moon in Cet. (94%)*			***247.1***	***Nov. 28.1***	***Many seen from Mexico***[10]			
	(No dust trail encounter)								
1892:	*Moon in Sgr. (18%)*			***243.8***	***Nov. 24 02:00***	***R.A. = 27.2°, Decl. = +40.8°***			HR ~ 300
1832	−0.009 06	+0.0229	0.202	243.905	Nov. 24 04:19				V
1852	+0.004 28	+0.0211	0.214	243.880	Nov. 24 03:43				V

[7] H. A. Newton, *Am. J. Sci.* **31** (1886), 409.
[8] N. Nogami, Local records of the 1862 Perseids and the 1885 Andromedids. *Earth, Moon Planets* **68** (1995), 435–441; Chinese observers reported activity Nov. 26–29.
[9] Also: F. Schwab, Beobachtung der Biela-Meteore 1886 zu Klausenburg. *Astron. Nachr.* **116** (1887), 119.
[10] H. A. Newton, *Am. J. Sci.* **45** (1892), 61.

Table 6a (*cont.*)

Year Trail	Δr (D–E) (AU)	Δa_0 (AU)	f_m	$\lambda_\odot$ (°)	Time (UT)	χ	W (°)	ZHR (/h)	src
1899:	*Moon in Leo (55%)*			***243.7***	***Nov. 24 18:00***	***R.A.* = *25.2°*, *Decl.* = *+43.0°***			***HR* ~ *150***[11]
1846	−0.008 53	+0.0386	0.090	243.843	Nov. 24 22:00				V
1852	−0.002 31	+0.0394	0.189	243.983	Nov. 25 01:18				V
1904:	*Moon in Ari. (99%)*			***240.954***	***Nov. 22 08:10***	***R.A.* = *27.5°*, *Decl.* = *+43.7°***			***HR* ~ *20***[12]
	(No dust trail encounter)								
1906:	*Moon in Aqr. (56%)*				*No report*				
1839	−0.004 66	+0.0582	0.131	241.721	Nov. 23 14:39	–.–	–.–	–.–	V
1846	+0.001 98	+0.0546	0.102	242.422	Nov. 24 07:17	–.–	–.–	–.–	V
1913:	*Moon in Leo (38%)*				*No report*				
1846	−0.001 52	+0.0688	0.059	240.638	Nov. 22 08:03	–.–	–.–	–.–	V
1928:	*Moon in Vir. (2%)*			***229.446***	***Nov. 11 02:00***	**0:30–3:30 UT, 10 telescopic meteors**[13]			
	(No dust trail encounter)								
1940:	*Moon in Tau. (99%)*			***234.2***	***Nov. 15 peak of faint meteors***				***HR* ~ *30/h***[14]
1852	−0.000 29	+0.0722	0.043	234.174	Nov. 15 20:30	–.–	–.–	–.–	V

[11] C. P. Olivier, *Meteors* (Baltimore: Williams & Wilkins, 1925), pp. 69–70. Seen by Olivier himself: 75 meteors in all during a 2 hour watch. Meteors were faint with short paths. Diffuse radiant.

[12] W. F. Denning, The meteors from Biela's comet. *Mon. Not. R. Astron. Soc.* **65** (1905), 851–855; K. Bohlin, *Astronomiska Iakttagelser och Undersökningar* (Stockholm) 8, No. 2 (1905); Rev. W. F. A. Ellison of Enniscorthy (UK) observed 8 Andromedids in 15 seconds on Nov. 21 at 8:25 UT, and 24 between 7h25 m and 8h25 m UT. Denning predicted a brilliant return for Nov. 17 or 18, 1905, which did not materialize. Activity continued until Nov. 28.

[13] C. P. Olivier, Telescopic Biela meteors. *Astron. Nachr*. **236** (1929), 15.

[14] Experienced meteor observer R. M. Dole (Cape Elizabeth, Maine) observed an outburst of faint meteors on Nov. 15, while J. P. M. Prentice (UK) counted 5/hr during Nov. 27–Dec. 04 (J. P. M. Prentice, *Brit. Astron. Assoc. Handbook*, (London: British Astronomy Association, 1947), p. 42); A. C. B. Lovell, *Meteor Astronomy*, (Oxford: Clarendon Press, 1954), pp. 349–354.

Table 6b *7P/Pons–Winnecke Parent of the June Bootids: dust trail encounters*

Year Trail	Δr $(D–E)$ (AU)	Δa_0 (AU)	f_m	$\lambda_\odot$ (°)	Time (UT)	χ	W (°)	ZHR (/h)	src
1873:	*Moon in Oph. (96%)*				*No report*				
1796	+0.008 44	+0.0077	0.010	108.189	July 08 14:31	–.–	–.–	–.–	V
1891:	*Moon in Gem. (0%)*				*No report*				
1796	+0.013 71	+0.0046	0.004	105.555	July 06 10:43	–.–	–.–	–.–	V
1910:	*Moon in Cet. (51%)*				*No report*				
1808	−0.021 73	+0.0132	0.027	98.713	June 30 03:25	–.–	–.–	–.–	V
1813	+0.004 04	+0.0039	0.011	98.230	June 29 15:16	–.–	–.–	–.–	V
1819	+0.005 51	+0.0042	0.070	98.080	June 29 11:30	–.–	–.–	–.–	V
1825	+0.008 97	+0.0037	0.035	97.939	June 29 07:57	–.–	–.–	–.–	V
1916:	*Moon in Tau. (2%)*			***98.1 ± 0.1***	***June 29 00:00***	***~1.7***	***~0.25***	***200 ± 50***	
1813	−0.003 11	+0.0044	0.270	98.209	June 29 03:35	–.–	–.–	–.–	V
1819	+0.000 97	+0.0042	0.052	98.015	June 28 22:42	–.–	–.–	–.–	V
1819	–.–	–.–	–.–	98.250	June 29 04:37	218.9	+52.0	14.7	MS
1819	−0.000 413	+0.0018	–.–	98.034	June 28 23:11	217.4	+53.3	15.1	IS
1819	−0.0005	+0.0018	–.–	98.039	June 28 23:17	–.–	–.–	–.–	R
1825	+0.003 70	+0.0035	0.058	97.907	June 28 19:59	–.–	–.–	–.–	V
1830	+0.005 94	+0.0030	0.032	97.800	June 28 17:18	–.–	–.–	–.–	V
1836	+0.007 43	+0.0031	0.064	97.756	June 28 16:10	–.–	–.–	–.–	V
1917:	*Moon in Sgr. (99%)*				*No report*				
1847	+0.001 72	+0.0095	0.037	102.712	July 04 03:10	–.–	–.–	–.–	V
1852	+0.008 07	+0.0080	0.147	102.147	July 03 12:57	–.–	–.–	–.–	V
1921:	*Moon in Ari. (30%)*					**–.–**	**–.–**	***~7 ± 5***	
1802	+0.017 37	+0.0017	3.000	99.261	June 30 13:01				V
1758	+0.002 09	+0.0039	–.–	104.212	July 05 17:41	158.7	−07.0	10.7	IS
1758	+0.002 99	+0.0074	–.–	97.413	June 28 14:34	193.9	+47.6	12.6	IS

Table 6b (*cont.*)

Year Trail	Δr $(D–E)$ (AU)	Δa_0 (AU)	f_m	$\lambda_\odot$ (°)	Time (UT)	χ	W (°)	ZHR (/h)	src
1923:	*Moon in Aqr. (67%)*				*No report*				
1808	−0.001 49	+0.0156	0.025	102.280	July 04 05:10	–.–	–.–	–.–	V
1813	−0.002 79	+0.0110	0.107	103.734	July 05 17:45	–.–	–.–	–.–	V
1927:	*Moon in Gem. (0%)*				*June 27 21:00*	–.–	–.–	***~350***[1]	
						R.A. = 198.8°, Decl. = +53.6°			
1784	−0.000 63	+0.0030	–.–	97.745	June 29 11:44	203.8	+9.9	10.2	IS
1790	−0.001 34	+0.0038	–.–	101.428	July 03 08:25	199.9	+5.7	9.7	IS
1802	+0.026 04	−0.000 13	5.514	99.179	June 30 23:34	–.–	–.–	–.–	V
1936:	*Moon in Cap. (98%)*				*No report*				
1852	−0.002 24	+0.0098	0.134	104.682	July 06 01:37	–.–	–.–	–.–	V
						R.A. = 206.5°, Decl. = +53.7°			
1942:	*Moon in Psc. (42%)*					***Low rates (telescopic)***			
1825	+0.008 51	+0.0104	0.304	104.266	July 06 04:00	–.–	–.–	–.–	V
1830	−0.005 96	+0.0114	0.102	105.654	July 07 14:56	–.–	–.–	–.–	V
1954:	*Moon in Cnc. (3%)*				*No report*				
1808	+0.005 74	+0.0349	0.014	100.137	July 01 21:55	–.–	–.–	–.–	V
1967:	*Moon in Gem. (1%)*				*No report*				
1813	−0.001 28	+0.0132	0.014	104.341	July 06 15:41	–.–	–.–	–.–	V
1808	+0.004 48	+0.0256	0.033	101.559	July 03 17:40	–.–	–.–	–.–	V
1973:	*Moon in Cnc. (7%)*				*No report*				
1808	−0.007 44	+0.0409	0.036	100.928	July 02 14:38	–.–	–.–	–.–	V

[1] W. F. Denning, Pons–Winnecke's comet and meteors. *Observatory* **50** (1927), 189–190; J. Rendtel, R. Arlt and V. Velkov, Surprising activity of the 1998 June Bootids. *WGN* **26** (1998), 165–172.

1986:	*Moon in Ari. (26%)*				*No report*				
1836	−0.006 09	+0.0067	0.091	99.716	July 01 16:07	–.–	–.–	–.–	V
1819	−0.014 70	+0.0072	0.012	100.326	July 02 07:29	–.–	–.–	–.–	V
1992:	*Moon in Ari. (27%)*				*No report*				
1813	+0.001 13	+0.0034	0.001	94.575	June 25 19:40	–.–	–.–	–.–	V
1993:	*Moon in Vir. (75%)*				*No report*				
1796	−0.012 45	+0.0046	0.001	97.612	June 29 06:12	–.–	–.–	–.–	V
1998:	*Moon in Cnc. (13%)*			***95.69 ± 0.01***	June 27 12:27	***2.1***	***0.25 ± 0.03***	***250 ± 550***	
						R.A. = 223°, Decl. = +48°			
1852	+0.001 43	+0.0049	0.106	95.627	June 27 10:51	–.–	–.–	–.–	V
1847	−0.005 94	+0.0049	0.076	95.725	June 27 13:18	–.–	–.–	–.–	V
1841	−0.004 49	+0.0045	0.227	95.807	June 27 15:22	–.–	–.–	–.–	V
1836	−0.005 31	+0.0044	0.156	95.912	June 27 18:01	–.–	–.–	–.–	V
1830	−0.006 91	+0.0044	0.188	96.023	June 27 20:49	–.–	–.–	–.–	V
1825	−0.013 99	+0.0050	0.503	96.189	June 28 00:59	–.–	–.–	–.–	V
1825	+0.001 41	+0.0025	–.–	96.036	June 27 21:09	228.6	+45.4	14.2	IS
1819	+0.001 77	+0.0049	0.092	96.203	June 28 01:20	–.–	–.–	–.–	V
1813	−0.004 52	+0.0051	0.011	96.296	June 28 03:40	–.–	–.–	–.–	V
1808	−0.020 69	+0.0100	0.126	96.754	June 28 15:12	–.–	–.–	–.–	V
2004:	*Moon in Leo (26%)*			***92.21 ± 0.01***	June 23 09:55	**2.3 ± 0.2**	**0.5 ± 0.1**	**18 ± 2**	
						R.A. = 227.3°, Decl. = +48.1°			
1852	−0.032 00	+0.0041	0.015	92.125	June 23 07:46	–.–	–.–	–.–	V
1847	−0.033 16	+0.0040	0.030	92.166	June 23 08:47	–.–	–.–	–.–	V
1841	−0.034 03	+0.0037	0.023	92.201	June 23 09:41	–.–	–.–	–.–	V
1836	−0.034 43	+0.0037	0.030	92.227	June 23 10:19	–.–	–.–	–.–	V
1830	−0.035 18	+0.0039	0.081	92.270	June 23 11:24	–.–	–.–	–.–	V
1830	+0.0033	+0.0018	–.–	92.276	June 23 13:45	223.0	+47.1	14.14	MS
1825	−0.036 87	+0.0041	0.042	92.368	June 23 13:53	–.–	–.–	–.–	V
1825	+0.0031	+0.0020	–.–	92.403	June 23 14:46	223.0	+47.0	14.13	MS
1819	−0.040 17	+0.0048	0.179	92.474	June 23 16:32	–.–	–.–	–.–	V

Table 6b (*cont.*)

Year Trail	Δr (D–E) (AU)	Δa_0 (AU)	f_m	$\lambda_\odot$ (°)	Time (UT)	χ	W (°)	ZHR (/h)	src
1819	+0.0062	+0.0023	–.–	29.473	June 23 16:32	223.4	+47.2	14.10	MS
1813	−0.042 44	+0.0053	0.021	92.631	June 23 20:29	–.–	–.–	–.–	V
1813	+0.0034	+0.0029	–.–	92.581	June 23 19:15	223.1	+47.0	14.10	MS
2010:	*Moon in Sco. (94%)*								
1836	−0.038 34	+0.0037	0.008	92.207	June 23 22:40	–.–	–.–	–.–	V
1830	−0.039 39	+0.0037	0.043	92.264	June 24 00:07	–.–	–.–	–.–	V
1825	−0.041 23	+0.0043	0.009	92.314	June 24 01:22	–.–	–.–	–.–	V
1819	−0.044 06	+0.0047	0.023	92.414	June 24 03:53	–.–	–.–	–.–	V
2028:	*Moon in Psc. (28%)*								
1796	−0.020 06	+0.0025	0.010	86.505	June 17 14:10	–.–	–.–	–.–	V

Table 6c *Encounters with dust trails of 15P/Finlay*

Calculations by Serjev Shanov and Serjev Dubrovski. Because of the very small inclination, these showers could be very broad (>1 d).

Year Trail	Δr (D–E) (AU)	Δa_0 (AU)	f_m	$\lambda_\odot$ (°)	Time (UT)	χ	W (°)	ZHR (/h)	src
2021:	*Moon in Tau. (67%)*								
1988	+0.0004	+0.0077	–.–	184.067	Sep. 27 06:14	264.1	−60.5	11.00	SD
2021:	*Moon in Gem. (54%)*								
1995	+0.0001	+0.0071	–.–	185.569	Sep. 28 18:58	261.6	−57.7	10.82	SD
2027:	*Moon in Cnc. (51%)*								
1960	+0.0009	+0.0077	–.–	208.689	Oct. 22 17:17	255.4	−38.4	11.58	SD
2027:	*Moon in Cnc. (43%)*								
1953	+0.0003	+0.0092	–.–	209.226	Oct. 23 06:14	256.1	−37.9	11.60	SD
2034:	*Moon in Ori. (57%)*								
1995	+0.0011	+0.0028	–.–	190.485	Oct. 04 02:53	257.5	−51.6	10.65	SD
2041:	*Moon in Vir. (2%)*								
1995	+0.0013	+0.0085	–.–	183.654	Sep. 26 23:02	264.1	−60.0	10.96	SD

Table 6d *21P/Giacobini–Zinner – parent of the Draconids: dust trail encounters*

Year Trail	Δr $(D–E)$ (AU)	Δa_0 (AU)	f_m	$\lambda_\odot$ (°)	Time (UT)	χ	W (°)	ZHR (/h)	src
1893:	*Moon in Vir. (0%)*				*No report*				
1805	+0.000 37	−0.0075	0.044	198.126	Oct. 09 17:21	–.–	–.–	–.–	V
1926:	*Moon in Lib. (9%)*			**~196.9**	**Oct. 09 ~22h**	**R.A. = 263°**	**Decl. = +54°**		**JP**[1]
1920	+0.000 627	−0.1018	–.–	196.910	Oct. 09 22:58	262.9	+54.4	20.6	IS
1920	+0.0007	−0.1015	–.–	196.933	Oct. 09 23:15	–.–	–.–	–.–	R
1933:	*Moon in Tau. (68%)*			**196.999**	**Oct. 09 20:03**	**3.6**	**0.025 ± 0.003**	**10 000**	
1907	+0.000 05	+0.0217	0.036	197.001	Oct. 09 20:07	–.–	–.–	–.–	V
1907	−0.000 172	+0.0309	–.–	196.859	Oct. 09 16:44	259.2	+52.9	21.2	IS
1907	−0.0002	+0.0309	–.–	196.996	Oct. 09 19:58	–.–	–.–	–.–	R
1900	+0.000 59	+0.0156	0.146	197.008	Oct. 09 20:17	–.–	–.–	–.–	V
1900	−0.0001	+0.0229	–.–	197.002	Oct. 09 19:58	–.–	–.–	–.–	R
1900	−0.000 01	+0.0229	–.–	196.989	Oct. 09 19:53	260.1	+53.2	23.3	IS
1894	+0.001 33	+0.0055	–.–	196.834	Oct. 09 16:07	262.7	+53.3	20.2	IS
1880	+0.000 88	+0.0108	0.029	196.812	Oct. 09 15:32	–.–	–.–	–.–	V
1935:	*Moon in Aqr. (91%)*				*No report*				
1831	−0.000 40	+0.0531	0.0145	196.446	Oct. 09 18:54	–.–	–.–	–.–	V
1940:	*Moon in Sgr. (62%)*				*No report*				
1894	+0.000 27	+0.0325	0.066	196.814	Oct. 09 10:33	–.–	–.–	–.–	V
1946:	*Moon in Cet. (99%)*			**196.992**	**Oct. 10 03:45**	**3.2**	**0.035 ± 0.003**	**12 000**	
1940	+0.001 79	+0.0209	0.299	196.993	Oct. 10 03:47	–.–	–.–	–.–	V
1940	+0.0012	+0.0208	–.–	196.988	Oct. 10 03:40	–.–	–.–	–.–	R
1933	+0.001 40	+0.0105	0.257	196.993	Oct. 10 03:47	–.–	–.–	–.–	V
1933	+0.001 07	+0.0104	–.–	196.963	Oct. 10 03:21	262.7	+54.1	20.6	IS

[1] J. P. M. Prentice, The great meteor shower of comet Giacobini-Zinner. *J. Brit. Astron. Assoc.* **44** (1934), 108–111: 1916 Draconids seen by J.P.M. Prentice between 20:20–23:20 UT.

1933	+0.0010	+0.0102	–.–	196.988	Oct. 10 03:40	–.–	–.–	–.–	R
1926	+0.001 26	+0.0070	0.885	196.993	Oct. 10 03:46	–.–	–.–	–.–	V
1926	+0.000 99	+0.0066	–.–	196.963	Oct. 10 03:21	262.8	+54.2	20.6	IS
1926	+0.0010	+0.0066	–.–	196.988	Oct. 10 03:40	–.–	–.–	–.–	R
1920	−0.000 42	+0.0053	1.274	196.894	Oct. 10 03:34	–.–	–.–	–.–	V
1920	+0.000 87	+0.0050	–.–	196.966	Oct. 10 03:25	262.9	+54.2	20.6	IS
1920	+0.0008	+0.0051	–.–	196.991	Oct. 10 03:45	–.–	–.–	–.–	R
1913	+0.000 88	+0.0042	1.717	196.997	Oct. 10 03:54	–.–	–.–	–.–	V
1913	+0.000 63	+0.0042	–.–	196.972	Oct. 10 03:34	263.1	+54.2	20.6	IS
1913	+0.000 60	+0.0042	–.–	196.994	Oct. 10 03:49	–.–	–.–	–.–	R
1907	−0.000 07	+0.0032	1.398	197.004	Oct. 10 04:03	–.–	–.–	–.–	V
1907	+0.000 04	+0.0040	–.–	196.984	Oct. 10 03:51	271.8	+56.1	20.4	IS
1907	+0.0001	+0.0040	–.–	196.998	Oct. 10 03:55	–.–	–.–	–.–	R
1900	+0.000 21	+0.0021	0.955	197.005	Oct. 10 04:05	–.–	–.–	–.–	V
1900	−0.000 05	+0.0032	–.–	196.989	Oct. 10 03:58	252.9	+50.4	20.1	IS
1900	−0.000 10	+0.0032	–.–	197.004	Oct. 10 04:34	–.–	–.–	–.–	R
1873	+0.000 76	+0.0019	2.110	197.001	Oct. 10 03:59	–.–	–.–	–.–	V
1866	−0.000 21	+0.0180	1.562	196.990	Oct. 10 03:42	–.–	–.–	–.–	V
1952:	*Moon in Gem. (62%)*			**196.934**	**Oct. 09 15:25**	**–.–**	**0.027 ± 0.005**	**∼250**	J
1805	+0.000 54	−0.0081	0.044	197.187	Oct. 09 21:34	–.–	–.–	–.–	V
1920	+0.000 040	−0.0656	–.–	196.933	Oct. 09 15:29	266.0	+55.1	20.4	IS
1920	+0.000 10	−0.0620	–.–	196.938	Oct. 09 15:32	–.–	–.–	–.–	R
1985:	*Moon in Cnc. (37%)*			**195.256**	**Oct. 08 09:36**	**3.4**	**0.045 ± 0.005**	**550 ± 50**	
1894	−0.001 28	+0.0020	0.151	195.240	Oct. 08 09:12	–.–	–.–	–.–	V
1887	+0.002 81	+0.0019	0.141	195.249	Oct. 08 09:25	–.–	–.–	–.–	V
1946	+0.011 14	+0.0197	–.–	195.253	Oct. 08 09:45	262.3	+56.2	20.8	IS
1985:	*Moon in Cnc. (38%)*			**195.174 ± 0.01**	**Oct. 08 07:36**	**–.–**	**0.017 ± 0.005**	**150 ± 50**	
1933	−0.011 40	+0.0035	0.068	195.174	Oct. 08 07:36	–.–	–.–	–.–	V
1933	−0.016 54	+0.0110	–.–	195.283	Oct. 08 10:30	264.0	+54.9	21.1	IS

Table 6d (*cont.*)

Year Trail	Δr (D–E) (AU)	Δa_0 (AU)	f_m	$\lambda_\odot$ (°)	Time (UT)	χ	W (°)	ZHR (/h)	src
1998:	*Moon in Tau. (89%)*			**195.10 ± 0.02**	**Oct. 08 13:46**	–.–	**0.090 ± 0.007**	**300 ± 40**	
	Older trails?								
1998:	*Moon in Tau. (89%)*			**195.078 ± 0.002**	**Oct. 08 13:14**	–.–	**0.017 ± 0.002**	**500 ± 100**	
1926	+0.000 33	−0.0043	1.025	195.076	Oct. 08 13:10	–.–	–.–	–.–	V
1926	−0.000 53	−0.0187	–.–	195.077	Oct. 08 13:12	–.–	–.–	–.–	R
1926	−0.000 24	−0.0186	–.–	195.075	Oct. 08 13:22	261.1	+55.1	21.4	IS
1926	−0.0004	−0.0181	–.–	195.091	Oct. 08 22:33	–.–	–.–	–.–	MS
1926	−0.0004	−0.0181	–.–	195.090	Oct. 08 23:25	–.–	–.–	–.–	MS
2011:	*Moon in Aqr. (91%)*								
1900	−0.000 76	+0.0085	0.041	195.034	Oct. 08 20:07	–.–	–.–	–.–	V
1900	−0.0014	+0.009	0.051	195.034	Oct. 08 20:07	–.–	–.–	–.–	MS
1894	+0.0011	+0.002	0.012	194.951	Oct. 08 18:07	–.–	–.–	–.–	MS
1894	+0.001 05	+0.0026	–.–	194.931	Oct. 08 17:41	263.7	+55.6	20.9	IS
1887	−0.0009	+0.002	0.007	194.909	Oct. 08 17:05	–.–	–.–	–.–	MS
1880	−0.0019	+0.001	0.005	194.888	Oct. 08 16:35	–.–	–.–	–.–	MS
1805	−0.000 72	+0.0740	0.121	195.062	Oct. 08 20:49	–.–	–.–	–.–	V
1798	−0.000 59	+0.0716	0.089	195.056	Oct. 08 20:40	–.–	–.–	–.–	V
1791	−0.000 18	+0.0694	0.064	195.069	Oct. 08 20:59	–.–	–.–	–.–	V
1785	−0.000 53	+0.0673	0.046	195.063	Oct. 08 20:49	–.–	–.–	–.–	V
1778	−0.001 07	+0.0653	0.042	195.067	Oct. 08 20:55	–.–	–.–	–.–	V
1765	+0.000 61	+0.0617	0.028	195.069	Oct. 08 20:59	–.–	–.–	–.–	V
2012:	*Moon in Gem. (47%)*								
1959	+0.0014	+0.0607	–.–	195.629	Oct. 08 16:48	–.–	–.–	–.–	R

2020:	*Moon in Tau. (78%)*								
1704	+0.000 11	+0.1203	2.111	193.960	Oct. 07 01:25	–.–	–.–	–.–	V
1711	+0.000 19	+0.1231	1.293	193.981	Oct. 07 01:57	–.–	–.–	–.–	V
2024:	*Moon in Sco. (25%)*								
1859	+0.000 42	+0.3107	–.–	195.149	Oct. 08 06:53	–.–	–.–	–.–	V
1852	–0.000 73	+0.2971	–.–	195.138	Oct. 08 06:36	–.–	–.–	–.–	V
2025:	*Moon in Ari. (97%)*								
2012	+0.001 48	+0.0654	0.0693	195.269	Oct. 08 15:18	–.–	–.–	–.–	V
2030:	*Moon in Psc. (97%)*								
1628	+0.001 15	0.1573	3.027	196.003	Oct. 09 16:34	–.–	–.–	–.–	V
2035:	*Moon in Lib. (17%)*								
1887	+0.000 16	+0.5964	–.–	191.675	Oct. 05 14:02	–.–	–.–	–.–	V
2037:	*Moon in Vir. (1%)*								
1615	+0.000 42	+0.1941	–.–	196.539	Oct. 10 00:36	–.–	–.–	–.–	V
1609	+0.000 40	+0.1909	–.–	196.526	Oct. 10 00:17	–.–	–.–	–.–	V
1596	+0.000 27	+0.1847	–.–	196.541	Oct. 10 00:39	–.–	–.–	–.–	V
2042:	*Moon in Cnc. (31%)*								
1655	–0.000 35	+0.2496	–.–	194.660	Oct. 08 00:47	–.–	–.–	–.–	V
2044:	*Moon in Psc. (99%)*								
1900	+0.009 47	+0.0035	0.015	192.955	Oct. 06 04:00	–.–	–.–	–.–	V

Table 6e *26P/Grigg–Skjellerup – parent of the π-Puppids: dust trail encounters*

Year Trail	Δr $(D–E)$ (AU)	Δa_0 (AU)	f_m	$\lambda_\odot$ (°)	Time (UT)	χ	W (°)	ZHR (/h)	src
1972:	*Moon in Leo (75%)*			–.–	**Activity > 4 d**	–.–	–.–	~2	
1967	+0.001 27	+0.0217	0.372	33.346	Apr. 23 00:24	–.–	–.–	–.–	V
1961	–0.017 07	+0.0136	0.775	33.315	Apr. 22 22:38	–.–	–.–	–.–	V
1977:	*Moon in Ori. (23%)*			**>33.666**	**>Apr. 23 16:00**	**>1.6**	**0.047 ± 0.005**	**>180 ± 60**	
1848	–0.003 87	–0.0001	0.094	33.653	Apr. 23 14:56	–.–	–.–	–.–	V
1917	–0.004 10	–0.0053	0.744	33.355	Apr. 23 07:36	–.–	–.–	–.–	V
1858	–0.006 72	+0.0001	0.106	33.649	Apr. 23 14:50	–.–	–.–	–.–	V
1863	–0.007 75	+0.0003	0.323	33.649	Apr. 23 14:50	–.–	–.–	–.–	V
1868	–0.010 23	+0.0006	0.339	33.612	Apr. 23 13:55	–.–	–.–	–.–	V
1902	–0.010 38	–0.0023	0.710	33.387	Apr. 23 08:23	–.–	–.–	–.–	V
1883	–0.010 59	–0.0017	0.164	33.555	Apr. 23 12:32	–.–	–.–	–.–	V
1912	–0.010 70	–0.0028	1.668	33.426	Apr. 23 09:20	–.–	–.–	–.–	V
1888	–0.010 70	–0.0018	0.155	33.514	Apr. 23 11:31	–.–	–.–	–.–	V
1897	–0.010 91	–0.0021	0.388	33.433	Apr. 23 09:30	–.–	–.–	–.–	V
1892	–0.010 96	–0.0019	0.251	33.474	Apr. 23 10:32	–.–	–.–	–.–	V
1878	–0.011 49	–0.0037	1.988	33.462	Apr. 23 10:14	–.–	–.–	–.–	V
1907	–0.011 92	–0.0025	1.401	33.424	Apr. 23 09:18	–.–	–.–	–.–	V
1873	–0.013 91	+0.0019	0.062	33.510	Apr. 23 11:24	–.–	–.–	–.–	V
1982:	*Moon in Psc. (0%)*			**<33.25**	**<Apr. 23 11:45**	**1.9**	**0.056 ± 0.012**	**>20**	
1848	+0.000 86	+0.0001	0.008	33.527	Apr. 23 18:33	–.–	–.–	–.–	V
1892	–0.007 42	+0.0021	0.037	32.894	Apr. 23 02:57	–.–	–.–	–.–	V
1888	–0.008 30	+0.0021	0.018	32.830	Apr. 23 01:23	–.–	–.–	–.–	V
1897	–0.008 88	+0.0023	0.051	32.998	Apr. 23 05:31	–.–	–.–	–.–	V
1902	–0.011 58	+0.0024	0.130	33.133	Apr. 23 08:51	–.–	–.–	–.–	V
1907	–0.016 30	+0.0052	0.048	33.282	Apr. 23 12:31	–.–	–.–	–.–	V

1912	−0.023 36	+0.0062	0.037	33.292	Apr. 23 12:46	–.–	–.–	–.–	V
1917	−0.026 28	+0.0066	0.034	33.194	Apr. 23 10:21	–.–	–.–	–.–	V
1984:	*Moon in Cap. (45%)*					**–.–**	**–.–**	**3.3 ± 1.7**	
	(No dust trail encounters)								
1987:	*Moon in Aqr. (24%)*					–.–	–.–	**(< 1)**	
1897	+0.005 80	+0.0042	0.001	32.846	Apr. 23 08:26	–.–	–.–	–.–	V
1988:	*Moon in Gem. (37%)*				**Apr. 22.6**			**∼4/h**	
	(No dust trail encounters)								
1990:	*Moon in Psc. (10%)*				**Apr. 22.6**	**–.–**	**–.–**	**∼1**	
	(No dust trail encounters)								
1992:	*Moon in Sgr. (72%)*				**Apr. 23.6**	**–.–**	**–.–**	**2.3 ± 1.2**	
	(No dust trail encounters)								
2003:	*Moon in Cap. (50%)*				**Apr. 23 17:10**			**Low activity**	
1961	–.–	–.–	–.–	33.07	Apr. 23 16:30	Faint	0.4	–.–	V
1961	+0.0060	+47.2	–.–	32.935	Apr. 23 13:05	110.2	–45.0	15.0	MS
1957	−0.0166	+45.9	–.–	33.192	Apr. 23 15:55	112.0	–42.0	14.9	MS
2006:	*Moon in Aqr. (22%)*								
1848	+0.047 82	+0.0026	0.0305	33.407	Apr. 23 19:16	–.–	–.–	–.–	V
1858	+0.049 45	+0.0040	0.0015	33.469	Apr. 23 20:48	–.–	–.–	–.–	V
2029:	*Moon in Cnc. (51%)*								
1897	+0.033 67	+0.0034	0.0028	31.582	Apr. 21 19:52	–.–	–.–	–.–	V
2034:	*Moon in Tau. (11%)*								
1972	−0.002 04	+0.0122	0.006	32.188	Apr. 22 17:30	–.–	–.–	–.–	V
1937	+0.009 55	+0.0136	0.005	31.970	Apr. 22 12:08	–.–	–.–	–.–	V
1967	+0.014 61	+0.0096	0.008	32.137	Apr. 22 16:15	–.–	–.–	–.–	V
1947	+0.022 39	+0.0095	0.011	31.956	Apr. 22 11:47	–.–	–.–	–.–	V
1957	+0.027 75	+0.0078	0.023	31.950	Apr. 22 11:38	–.–	–.–	–.–	V
1868	+0.039 43	+0.0018	0.065	32.785	Apr. 23 08:11	–.–	–.–	–.–	V
1892	+0.039 91	−0.0033	0.169	32.276	Apr. 22 19:39	–.–	–.–	–.–	V
1863	+0.040 64	+0.0008	0.315	32.772	Apr. 23 07:52	–.–	–.–	–.–	V

Table 6e (*cont.*)

Year Trail	Δr $(D–E)$ (AU)	Δa_0 (AU)	f_m	$\lambda_\odot$ (°)	Time (UT)	χ	W (°)	ZHR (/h)	src
1902	+0.040 65	−0.0052	0.421	32.369	Apr. 22 21:56	–.–	–.–	–.–	V
1888	+0.040 67	−0.0029	0.109	32.364	Apr. 22 21:50	–.–	–.–	–.–	V
1858	+0.040 96	+0.0005	0.353	32.769	Apr. 23 07:48	–.–	–.–	–.–	V
1853	+0.041 10	+0.0003	0.311	32.740	Apr. 23 07:04	–.–	–.–	–.–	V
1848	+0.041 12	+0.0002	0.133	32.749	Apr. 23 07:17	–.–	–.–	–.–	V
1897	+0.041 59	−0.0044	0.330	32.487	Apr. 23 00:50	–.–	–.–	–.–	V
1883	+0.041 60	−0.0028	0.094	32.543	Apr. 23 02:13	–.–	–.–	–.–	V

Table 6f *45P/Honda–Mrkos–Pajdusáková – associated with the August Capricornids*
Calculations by Serjev Shanov and Serjev Dubrovski.

Year Trail	Δr (*D–E*) (AU)	Δa_0 (AU)	f_m	$\lambda_\odot$ (°)	Time (UT)	R.A. (°)	Decl. (°)	V_g (Km/s)	src
2005:	*Moon in Lib. (44%)*								
1959	±0.001 66	−0.010	–.–	139.859	Aug. 12 13:26	325.1	−11.3	25.85	SD
1959	±0.0017	−0.010	0.026	139.938	Aug. 12 15:25	325.02	−11.11	25.84	MS
2015:	*Moon in Cnc. (0%)*								
1969	±0.001 55	−0.015	–.–	140.990	Aug. 14 07:12	325.8	−10.7	25.63	SD

Table 6g *73P/Schwasmann–Wachmann 3 – parent of the τ-Herculids: dust trail encounters*

Year Trail	Δr (D–E) (AU)	Δa_0 (AU)	f_m	$\lambda_\odot$ (°)	Time (UT)	χ	W (°)	ZHR (/h)	src
1914:	*Moon in Sco. (99%)*								
1903	+0.000 28	+0.0493	0.196	77.226	June 07 15:47	–.–	–.–	–.–	V
1925:	*Moon in Sgr. (98%)*								
1914	−0.000 51	+0.0765	0.071	77.711	June 07 23:37	–.–	–.–	–.–	V
1908	−0.002 35	+0.0517	0.110	77.696	June 07 23:14	–.–	–.–	–.–	V
1903	−0.013 01	+0.0463	0.222	77.171	June 07 10:03	–.–	–.–	–.–	V
1930:	*Moon in Lib. (91%)*			**~78.00**	**June 08 13:40**	**>4?**	**–.–**	**50??**	
1930	+0.005 91	+0.0071	24.616	77.727	June 08 06:49	–.–	–.–	–.–	V
1925	+0.005	+0.012	–.–	77.73	June 08 07:00	220	+45	13.7	MS
1925	+0.005	+0.012	–.–	78.02	June 08 14:00	220	+45	13.7	MS
1936:	*Moon in Sgr. (94%)*				*No report*				
1908	+0.0003	+0.051	–.–	77.69	June 07 18:45	221.5	+44.7	13.9	LAJ
1908	−0.004 02	+0.0463	0.140	77.718	June 07 19:25	–.–	–.–	–.–	V
1914	−0.001 20	+0.0557	0.119	77.719	June 07 19:27	–.–	–.–	–.–	V
1919	+0.000 66	+0.0722	0.104	77.736	June 07 19:52	–.–	–.–	–.–	V
1925	+0.001 58	+0.1084	0.048	77.752	June 07 20:16	–.–	–.–	–.–	V
1941:	*Moon in Lib. (97%)*				*No report*				
1930	+0.009 63	+0.0348	0.316	77.708	June 08 01:56	–.–	–.–	–.–	V
1930	+0.008	+0.012	–.–	78.15	June 08 13:00	220	+45	13.7	MS
1925	+0.007	+0.008	–.–	78.11	June 08 12:00	220	+45	13.8	MS
1903	+0.004 69	+0.0115	0.176	77.592	June 07 23:02	–.–	–.–	–.–	V
1942:	*Moon in Psc. (26%)*								
1914	−0.000 93	+0.0898	0.045	77.714	June 08 08:15	–.–	–.–	–.–	V

1947:	*Moon in Cap. (81%)*				*No report*				
1908	+0.001 37	+0.0442	0.129	77.692	June 08 14:25	–.–	–.–	–.–	V
1914	+0.005 19	+0.0482	0.080	77.701	June 08 14:38	–.–	–.–	–.–	V
1952:	*Moon in Oph. (100%)*				*No report*				
1930	+0.015	+0.012	–.–	78.40	June 08 15:00	220	+45	13.6	MS
1925	+0.014	+0.010	–.–	78.36	June 08 14:00	220	+45	13.6	MS
1953:	*Moon in Ari. (16%)*				*No report*				
1908	+0.000 19	+0.068	0.044	77.646	June 08 02:16	–.–	–.–	–.–	V
1959:	*Moon in Gem. (5%)*				*No report*				
1914	+0.000 54	+0.091	0.073	77.693	June 08 16:13	–.–	–.–	–.–	V
1964:	*Moon in Psc. (16%)*				*No report*				
1903	−0.011 55	+0.069	0.0481	76.730	June 06 22:51	–.–	–.–	–.–	V
1970:	*Moon in Psc. (12%)*				*No report*				
1925	+0.009	−0.0029	–.–	70.31	May 31 19:00	208	+29	12.1	MS
1930	+0.007	−0.0033	–.–	70.31	May 31 19:00	208	+29	12.1	MS
1935	+0.001	−0.0041	–.–	70.17	May 31 14:00	209	+30	12.1	MS
1970:	*Moon in Psc. (21%)*			**69.27**	May 30 16:24	**R.A. = 213 ± 1°, Decl. = +22 ± 2°**		**Low rates** By Ogasawara/Japan	
1930	+0.006	−0.0033	–.–	69.37	May 30 20:00	209	+29	12.0	MS
1925	+0.007	−0.0029	–.–	69.09	May 30 22:00	208	+29	11.9	MS
1980:	*Moon in Oph. (89%)*				*No report*				
1925	+0.001	−0.0074	–.–	43.72	May 03 18:00	199	+01	14.5	MS
1980:	*Moon in Sgr. (65%)*								
1930	+0.002	−0.0090	–.–	46.39	May 06 12:00	200	+04	14.3	MS
1985:	*Moon in Leo (54%)*								
1941	−0.007	−0.0139	–.–	66.67	May 27 20:00	209	+26	12.2	MS
2001:	*Moon in Leo (53%)*			**68.62–69.09**	**May 30 00:11**	**1.2 ± 0.8**	**–.–**	**–.–**	**R01**[1]
1941	+0.0026	−0.027	–.–	69.04	May 30 09:50	212.2	+28.4	12.5	LAJ

[1] R01 = R. Arlt, Results of the Schwassmann–Wachmann-3 meteors. *WGN* **29** (2001), 93–95.

Table 6g (*cont.*)

Year Trail	Δr $(D–E)$ (AU)	Δa_0 (AU)	f_m	$\lambda_\odot$ (°)	Time (UT)	χ	W (°)	ZHR (/h)	src
1941	−0.019	−0.0086	–.–	65.53	May 26 18:00	210	+27	13.0	MS
1925	−0.012	−0.0070	–.–	45.29	May 05 18:00	199	+03	14.6	MS
2006:	*Moon in Tau. (0%)*								
1952	−0.033	−0.0062	–.–	65.66	May 27 04:00	210	+29	12.9	MS
2011:	*Moon in Tau. (0%)*								
1952	+0.0011	−0.022	–.–	71.22	June 02 05:45	214.2	+33.5	12.9	LAJ
2017:	*Moon in Leo (35%)*								
1941	+0.0013	−0.012	–.–	69.64	May 31 03:16	212.6	+29.7	12.4	LAJ
1941	−0.002 81	−0.0129	0.116	69.564	May 31 01:18	–.–	–.–	–.–	V
2022:	*Moon in Tau. (0%)*				**(High rates expected from 1995 breakup)**				
1995*	+0.0004	−0.022	–.–	69.44	May 31 04:55	205.4	+29.2	12.1	LAJ
1995*	−0.002 14	−0.0214	0.118	69.459	May 31 05:17	–.–	–.–	–.–	V
1990	+0.003 23	−0.0204	0.146	69.439	May 31 04:47	–.–	–.–	–.–	V
1985	+0.008 27	−0.0195	0.182	69.392	May 31 03:36	–.–	–.–	–.–	V
1946	−0.011 35	+0.0149	0.039	68.010	May 29 17:01	–.–	–.–	–.–	V
1941	−0.006 43	+0.0123	0.018	68.507	May 30 05:28	–.–	–.–	–.–	V
1935	−0.002 65	+0.0107	0.006	68.740	May 30 11:18	–.–	–.–	–.–	V
1930	+0.000 15	+0.0094	0.052	68.887	May 30 14:57	–.–	–.–	–.–	V
1930	+0.000 15	+0.0094	0.052	68.887	May 30 14:57	–.–	–.–	–.–	V
2033:	*Moon in Cnc. (49%)*								
1914	+0.002 99	−0.0046	0.033	45.479	May 06 03:32	–.–	–.–	–.–	V
2049:	*Moon in Psc. (8%)*								
1990	−0.011 90	+0.0095	0.057	41.212	May 01 20:33	–.–	–.–	–.–	V

* comet breakup occurred in this return.

1985	−0.006 20	+0.0096	0.072	40.336	Apr. 30 22:53	–.–	–.–	–.–	V
1979	−0.007 74	+0.0097	0.029	40.841	May 01 11:21	–.–	–.–	–.–	V
1974	+0.002 31	+0.0088	0.021	38.848	Apr. 29 10:08	–.–	–.–	–.–	V
1968	+0.000 91	+0.0101	0.059	38.816	Apr. 29 09:20	–.–	–.–	–.–	V
1963	+0.003 64	+0.0111	0.007	38.462	Apr. 29 00:37	–.–	–.–	–.–	V
1935	+0.003 52	+0.0072	0.028	45.692	May 06 11:24	–.–	–.–	–.–	V
1930	−0.007 05	+0.0067	0.035	47.866	May 08 17:19	–.–	–.–	–.–	V
1925	−0.001 83	+0.0012	0.037	42.749	May 03 10:33	–.–	–.–	–.–	V
1919	−0.003 33	+0.0005	0.008	42.848	May 03 12:59	–.–	–.–	–.–	V
2065:	*Moon in Gem. (14%)*								
1930	−0.004 79	+0.0063	0.039	48.436	May 09 09:50	–.–	–.–	–.–	V
1935	−0.001 78	+0.0075	0.050	47.319	May 08 06:08	–.–	–.–	–.–	V
2006	+0.004 22	+0.0200	0.076	39.536	Apr. 30 05:32	–.–	–.–	–.–	V
2086:	*Moon in Oph. (97%)*								
1919	−0.012 58	+0.0086	0.003	93.295	June 25 13:27	–.–	–.–	–.–	V
2098:	*Moon in Psc. (64%)*								
1990	+0.023 86	−0.0006	0.050	116.661	July 20 03:24	–.–	–.–	–.–	V
1995	+0.029 27	−0.0027	0.055	115.934	July 19 09:06	–.–	–.–	–.–	V
2001	+0.019 30	+0.0063	0.122	114.634	July 18 00:24	–.–	–.–	–.–	V

Table 6h *76P/West–Kohoutek–Ikemura and its dust trail encounters*

Calculations by Esko Lyytinen. Lubor Kresák had pointed out that meteor activity may occur from this comet in the future. From 2069 to 2135, this comet passes close to Earth's orbit. From the current projected orbit of the comet. Trails from the returns of 1975 (after the first drop in perihelion distance) and from 2045 (1.5 revolutions before the next drop in q in 2055) were considered, as well as the 2057 trail. The meteoroids of the 2057 trail stay together as a dense clump and that is encountered just before Earth's passage by the node in 2135. Jupiter's encounter of 2137 will move the dust away from Earth's orbit.

Year Trail	Δr (*D–E*) (AU)	Δa_0 (AU)	f_m	$\lambda_\odot$ (°)	Time (UT)	χ	*W* (°)	ZHR (/h)	src
2135:	*Moon in Aqr. (50%)*								
2057	−0.0006	−0.0000	<14	259.085	Dec. 12 10:58	–.–	–.–	–.–	L

Table 6i *103P/Hartley 2 and its dust trail encounters*

Calculations by Esko Lyytinen. Only three trails have been considered so far, others may behave similarly.

Year Trail	Δr (*D–E*) (AU)	Δa_0 (AU)	f_m	$\lambda_\odot$ (°)	Time (UT)	χ	*W* (°)	ZHR (/h)	src
2062:	*Moon in Tau. (79%)*								
1973:	−0.011 68	+0.0105	0.036	208.570	Oct. 22 13:44	301.7	+28.9	12.07	L
1979:	−0.004 76	+0.0086	0.036	208.216	Oct. 22 05:11	–.–	–.–	–.–	L
1985	+0.006 81	+0.0052	0.030	208.187	Oct. 22 04:29	–.–	–.–	–.–	L
2068:	*Moon in Cnc. (35%)*								
1973:	−0.017 90	+0.0103	0.006	206.066	Oct.19 14:13	–.–	–.–	–.–	L
1979:	−0.003 74	+0.0082	0.005	205.155	Oct.18 16:12	–.–	–.–	–.–	L
1985	+0.018 00	+0.0048	0.003	204.305	Oct.17 19:38	–.–	–.–	–.–	L

Table 6j *P/2004 CB linear – potential future omicron Ursid showers from its dust trail encounters*

Calculations by Esko Lyytinen.

Year Trail	Δr (*D-E*) (AU)	Δa_0 (AU)	f_m	$\lambda_\odot$ (°)	Time (UT)	χ	*W* (°)	ZHR (/h)	src
2014: *Moon in Psc. (21%)*									
1818	+0.001 10	+0.0001	0.01	62.830	May 24 06:33	–.–	–.–	–.–	L
1853	+0.001 10	+0.0001	0.02	62.830	May 24 06:33	–.–	–.–	–.–	L
1903	+0.000 15	+0.0008	0.11	62.847	May 24 06:59	125	+78	15.86	L
1909	−0.000 30	+0.0012	0.14	62.858	May 24 07:15	–.–	–.–	–.–	L
1914	−0.001 60	+0.0021	0.24	62.880	May 24 07:49	–.–	–.–	–.–	L
1979	+0.003 00	−0.0006	0.17	62.810	May 24 06:04	–.–	–.–	–.–	L
2014: *Moon in Psc. (23%)*									
1929	+0.004 80	–0.001 01	0.35	62.770	May 24 03:19	–.–	–.–	–.–	L

Table 6k *D/Blanpain – parent of the Phoenicids: dust trail encounters*

Calculations by Esko Lyytinen.

Year Trail	Δr (*D–E*) (AU)	Δa_0 (AU)	f_m	$\lambda_\odot$ (°)	Time (UT)	χ	*W* (°)	ZHR (/h)	src
1887:	*Moon in Gem. (91%)* (No dust trail encounter)			**~252.6**	**Dec. 03 12:00**	-.-	-.-	**~50**	
1938:	*Moon in Ari. (96%)* (No dust trail encounter)			**~253.6?**	**Dec. 05 12:00**	-.-	-.-	-.-	
1956:	*Moon in Sgr. (11%)*			**~254.14**	Dec. 05 16:23	**2.9**	**0.46 ± 0.12**	**50 ± 30**	**J95**
1819	−0.001 95	+0.0043	0.066	254.135	Dec. 05 16:20	-.-	-.-	-.-	L
1754	+0.000 92	+0.0005	0.010	254.116	Dec. 05 15:50	3.63°	−42.03°	10.47	W
1760–1808	+0.000 65	+0.0010	0.213	254.148	Dec. 05 16:34	3.53°	−41.87°	10.46	W
1814–1830	−0.0014	+0.0043	0.478	254.146	Dec. 05 16:34	3.23°	−41.63°	10.48	W
1972:	*Moon in Sco. (1%)* (No dust trail encounter)			**~252.9**	**Dec. 04 14:00**	-.-	-.-	**~20**	

Sources:

V = Model by Jérémie Vaubaillon, calculations by J. Vaubaillon and P. Jenniskens (this work); Also: J. Vaubaillon, Dynamic of meteoroids in the solar system. Application to the prediction of meteoritic showers in general, and Leonids in particular. Ph.D. thesis, IMCCE, l'Observatoire de Paris, Paris, France (2003), 263pp.

IS = Isao Sato (2003) personal website.

L = Model by Esko Lyytinen, calculated by E. Lyytinen (this work).

LAJ = H. Lüthen, R. Arlt and M. Jäger, The disintegrating comet 73P/Schwassmann–Wachmann 3 and its meteors. *WGN* **29** (2001), 15–28.

MS = Mikiya Sato (2003) personal website.

R = E. A. Reznikov, The Giacobini–Zinner comet and Giacobinid meteor stream. *Workers Kazan' City Astronomical Observatory (Trudy Kazan. Gor. Astron. Obs.)* **53** (1993), 80–101 (in Russian).

SD = S. Shanov and S. Dubrovsky, personal website, (2004) Ulianovsk, Russia.

W = J.-I. Watanabe, *et al.*, Phoenicids in 1956 revisited. *ACM Conf.* (2005) Rio de Janeiro, Brazil (poster).

Table 7 *Working list of cometary meteor showers*

Recently active meteor showers and their parent comets. Radiant and orbital elements in equinox J2000. Dates given are for AD 2000. Shower activity is expressed with peak solar longitude $\lambda_{\odot}^{\max}$, full-width-half-maximum (W, in degrees, $\lambda_{\odot}$) and peak rate $\mathrm{ZHR}^{\max}$ (per hour), whereby: $\mathrm{ZHR} = \mathrm{ZHR}^{\max}\,(W/2)^2/((\lambda_{\odot} - \lambda_{\odot}^{\max})^2 + (W/2)^2)$. The magnitude distribution index is $\chi = N(m+1)/N(m)$. Sequence of mean orbit for each stream has most reliable values on top. In some cases, the original orbit (given from a mean of individual orbital elements) was made to intersect Earth's orbit. Parent body orbits are for the best corresponding theoretical orbit of a meteoroid intersecting the orbit of Earth, not to be confused with the orbit of the parent itself. Sources are given at end of Table.

Table legend:

IAU#	Code	Name			Dates (2000)	Peak	$\lambda_{\odot}^{\max}$ (°)	χ	W (°)	$\mathrm{ZHR}^{\max}$	src
	#Meteors	a (AU)	q (AU)	i (°)	ω (°) Node (°)	R.A. (°)	Decl. (°)	ΔR.A./$\Delta\lambda_{\odot}$	ΔDecl./$\Delta\lambda_{\odot}$	V_g (km/s)	src
	Outburst	Notes			Dates (2000)	Peak	$\lambda_{\odot}^{\max}$	χ	W (°)	$\mathrm{ZHR}^{\max}$	src
	(Epoch)	a (AU)	q (AU)	i (°)	ω (°) Node (°)	R.A. (°)	Decl. (°)	-.-	-.-	V_g (km/s)	src
	Parent	*Name*	*(Epoch)*	$\lambda_{\odot}^{max}$	*Tisserand parameter* T_J*, absolute magnitude* H_0 *(n) or* H_N *if asteroidal in appearance*						
	(Epoch)	a (AU)	q (AU)	i	ω (°) Node (°)	R.A. (°)	Decl. (°)	ΔR.A./$\Delta\lambda_{\odot}$	ΔDecl./$\Delta\lambda_{\odot}$	V_g (km/s)	Δ_{C-E} (AU)

	Code	Name			Dates (2000)		Peak	$\lambda_{\odot}^{\max}$	χ	W	$\mathrm{ZHR}^{\max}$	Notes
		a	q	i	ω	Node	R.A.	Decl.	ΔR.A./$\Delta\lambda$	ΔDecl./$\Delta\lambda$	V_g	Δ_{C-E}
	January:											
												Apex
#89	PVI	January π-Virginids			**01/10–01/21**		**Jan. 16**	**295**	~2	(4)	-.-	
	$N=7$	12.6	0.396	160.6	283.0	295.2	179	+09	+0.9	+0.1	64.4	SASY
	$N=22$	-.-	-.-	-.-	-.-	-.-	167	+10	-.-	-.-	-.-	RRR
#90	**JCO**	**Jan. Comae Berenicids**			**01/19–01/23**		**Jan. 22**	**301**	-.-	(1.6)	2	
	$N=3$	Inf.	0.512	137.3	267.8	300.5	188.9	+16.8	+1.3	−0.3	63.9	This work
	$N=6$	Inf.	0.548	136.8	263.1	297.4	175	+25	-.-	-.-	65	MP
	Parent?	*C/Lowe (1913 I) – assumed to be Halley type.*						$\lambda_{\odot}=304.4$	$T_J=-0.23$, $H_{10}=$-.-			*V13*
	(1913)	*(30)*	*0.405*	*120.5*	*280.7*	*304.4*	*187.9*	*+21.9*	-.-	-.-	*59.4*	*0.0045*

	Code	Name			Dates (2000)		Peak	$\lambda_{\odot}^{max}$	χ	W	ZHR^{max}	Notes
		a	q	i	ω	Node	R.A.	Decl.	ΔR.A./$\Delta\lambda$	ΔDecl./$\Delta\lambda$	V_g	Δ_{C-E}
												Antihelion
#91	JZA	Jan. (ζ-)Aurigids			**01/04–01/21**		**Jan. 13**	**292**	–.–	(6)	–.–	
	$N=2$	2.653	0.643	11.6	209.3	299.1	70.3	+60.1	–.–	–.–	12.1	This work
	$N=$–.–	2.341	0.923	8.7	213.4	292.3	69.9	+49.1	–.–	–.–	12.0	T89
	$N=22$	1.851	0.836	11.1	221.0	293.9	83.9	+58.3	–.–	–.–	11.9	ZS
#92	UER	υ-Eridanids			**01/14–01/30**		**Jan. 16**	**295.6**	–.–	(6)	–.–	
	$N=6$	1.89	0.971	12.3	17.4	115.6	69.3	−29.0	+2.12	+0.13	10.1	This work
	Parent?	*2004 TB$_{18}$*		*(Epoch 2006-03-06)*				$\lambda_{\odot}=301.1$	$T_J=+3.89$, $H_N=+17.7$			
	(2006)	*1.775*	*0.976*	*13.203*	*12.919*	*121.081*	*70.4*	*−34.8*	–.–	–.–	*9.89*	*+0.0211*
#93	**VEL**	**Puppids-Velids II**			**11/08–02/24**		**Jan. 17**	**296**	3.0	–.–	2.4 ± 0.4	PJ
		= Columbids = τ-Puppids = Jan. π-Puppids = γ-Velids = α-Pyxidids = δ-Velids?										
	Visual					$\lambda_{\odot}=296$	123	−48.3	+0.64	−0.19	33.1 ± 1.3	Mean
	$N=9$	2.4	0.93	62.0	33	145.7	152.4	−65.3	–.–	–.–	35.3	GE (2.13)
	$N=3$	7.1	0.86	49.9	43	143.7	133.4	−50.2	–.–	–.–	33.2	GE (2.09)
	Visual				01/10–01/15		115.0	−30.2				M (#62)
	Visual				12/01–12/06	73.0	101.7	−46.6	–.–	–.–	–.–	M (55)
#94	**RGE**	***ρ-Geminids***			**01/15–01/23**		**Jan. 17**	**296.3**	–.–	(4)	–.–	
	Outburst	7 slow meteors					**1993, Jan. 17**	**297.01**	1.4	>0.05	>14	
	(1993)	(2.96)	0.596	3	263	296.3	115.8	+24.9	–.–	–.–	23	MV
	$N=$–.–	1.822	0.658	3.3	262.0	302.6	123.7	+25.5	–.–	–.–	18.5	T89 (8a)
	$N=13$	2.229	0.594	6.4	266.3	287.7	109.8	+31.3	–.–	–.–	21.8	ZS73
	$N=2$	1.830	0.595	9.5	268.5	300.7	127.8	+33.8	–.–	–.–	20.1	L71B
	$N=6$	2.66	0.771	3.5	243.4	302.1	110.8	+28.9	1.1	−0.2	–.–	L71A
	$N=4$	2.78	0.780	5.2	242.7	301.2	110.2	+32.2	–.–	–.–	20.8	**SH**
	Fireballs	AD 1049–1216:			01/19–01/26		116	+27	–.–	–.–	–.–	AT (2)

RGE – after "MV:" from radiant meteor outburst, adopted speed.

#95	*(DCA)*	*δ-Cancrids*			**01/01–01/24**		**Jan. 17**	**296.3**	3.0	10	~3.2	
		= ecliptic antihelion source										
#96	NCC	N. δ-Cancrids			**01/01–01/24**		**Jan. 17**	**296.3**	3.0	10	~1.6	PJ
	Visual	= ecliptic antihelion source				$\lambda_\odot = 297$	130	+20	+0.75	−0.29	26	IMO
	$N = 37$	1.829	0.397	1.5	291.3	297.1	130.7	+19.7	-.-	-.-	26.4	ZS
	$N = 27$	1.901	0.425	1.2	287.9	292.9	124.8	+20.8	-.-	-.-	25.8	ZS73
	$N = 7$	2.273	0.448	0.3	282.9	297.1	126.6	+19.5	+0.9	−0.2	26.2	L71A
	$N = 4$	2.38	0.567	6.3	268.8	296.1	121.2	+27.7	-.-	-.-	25.7	**SH**
	$N = 6$	1.61	0.37	4.9	116.7	120.2	134.2	+14.1	+1.1	-.-	26.7	NL (61.1.1)
	Fireballs	AD 1005–1098:			01/30–02/05		125	+25	-.-	-.-	-.-	AT (3)
	Parent?	*1991 AQ*		*(Epoch 2005-01-30)*				$\lambda_\odot = 303.69$	$T_J = +3.16$, $H_N = +17.05$			
	(2005)	*2.214*	*0.486*	*3.192*	*241.006*	*341.504*	*131.8*	*+20.9*	-.-	-.-	*24.98*	*0.0336*
#97	SCC	S. δ-Cancrids			**01/01–01/24**		**Jan. 17**	**296.3**	3.0	10	~1.6	
		= ecliptic antihelion source, twin of DXL										
	$N =$ -.-	2.114	0.475	6.3	100.7	126.9	134.1	+10.1	-.-	-.-	25.2	T89 (6a)
	$N = 3$	3.03	0.58	9.9	84.7	120.6	118.2	+07.7	-.-	-.-	24.0	NL (61.1.2)
	Parent?	*2001 YB_5*		*(Epoch 2006-03-06)*				*286.97*	$T_J = +2.89$, $H_N = +20.62$			*MZ*
	(2006)	*2.348*	*0.323*	*5.484*	*116.742*	*106.841*	*121.2*	*+16.0*	-.-	-.-	*30.48*	*0.004*
#98	ECO	ε-Columbids			**01/14–01/30**		**Jan. 28**	**307.1**	-.-	(6)	-.-	
	$N = 3$	2.96	0.945	21.6	26.0	114.1	82.4	−34.7	+0.76	+0.02	16.6	This work
	$N =$ -.-	3.008	0.952	21.8	22.9	118.4	80.4	−36.0	-.-	-.-	16.6	T89 (7a)
	$N =$ -.-	2.430	0.846	14.3	50.2	124.7	102.6	−09.1	-.-	-.-	16.3	T89 (7b)
												Helion
#99	JSC	Daytime Scutids			**12/30–01/04**		**Jan. 02**	**280.4**	-.-	(2.4)	-.-	
		= ecliptic helion source										
	$N = 7$	2.406	0.551	12.4	89.4	280.4	278.2	−07.8	-.-	-.-	24.1	ZS73
#10	**QUA**	**Quadrantis Muralids = Quadrantids** (= BOO, Bootids)										
		= part of Machholz complex			**12/31–01/06**		**Jan. 04**	**283.28**	(2.2)	0.35 ± 0.05	130 ± 24	PJ
	$N = 85$	3.14	0.979	72.0	172.0	283.3	230.0	+49.5	+0.4	−0.2	41.36	This work
	$N = 6$	2.383	0.977	68.6	169.3	286.0	233.1	+49.1	-.-	-.-	38.9	PG
	$N = 14$	2.424	0.941	73.1	152.6	295.8	242.1	+41.0	-.-	-.-	41.2	ZS
	$N = 17$	2.612	0.974	72.4	170.5	283.5	229.4	+48.8	-.-	-.-	40.5	L71B

ECO – "this:" Diffuse radiant from Decl. = −40° to −18°, low V_g.

QUA – The proper name is *Quadrantis Muralids* after the full name of the constellation, but the Quadrantids concatenation is used. The activity profile has a main component and a background of fainter meteors. For more information, see Chapter 20.

	Code	Name a	q	i	Dates (2000) ω	Node	Peak R.A.	$\lambda_{\odot}^{max}$ Decl.	χ ΔR.A./$\Delta\lambda$	W ΔDecl./$\Delta\lambda$	ZHR^{max} V_g	Notes Δ_{C-E}
	Parent	*2003 EH$_1$*		*(Epoch 2005-01-30)*				$\lambda_{\odot}=282.9$	$T_J=+2.06$, $H_N=+16.67$			
	(2005)	*3.126*	*0.979*	*70.782*	*171.369*	*282.952*	*231.0*	*+50.0*	*+0.5*	*−0.3*	*40.77*	*0.2145*
#100	XSA	Daytime ξ-Sagittariids			**01/17–02/01**		**Jan. 23**	**304.9**	–.–	(6)	–.–	
	$N=15$	1.080	0.285	1.1	46.9	304.9	283.2	−21.9	–.–	–.–	24.4	ZS
	$N=14$	1.744	0.383	4.3	66.6	296.0	284.8	−18.6	–.–	–.–	26.3	ZS
February:												
												Apex
#101	PIH	π-Hydrids			**02/01–02/13**		**Feb. 06**	**317.1**	–.–	(5)	–.–	
	$N=8$	32.8	0.892	162.2	36.2	137.1	210.3	−23.0	+1.3	−0.4	70.7	This work
	$N=2$	2.3	0.96	178.8	159	145.7	236.7	−20.2	–.–	–.–	67	GE (2.54)
#102	**ACE**	**α-Centaurids**			**02/02–02/19**		**Feb. 08**	**319.4**	2.0	3.4 ± 0.6	7.3 ± 1.5	PJ
	Outburst	in 1980 and possibly 1974					**1980, Feb. 08**	**319.2**	2.2	0.011 ± 0.004	>230	PJ
	(1980)					$\lambda_{\odot}=319.2$	210.9	−58.2	–.–	–.–	–.–	JW
	Visual	(14)	0.977	107.0	348.9	138.9	212.1	−59.4	+1.9	−0.5	58.2	PJ
	$N=$–.–	Inf.	0.986	108.3	357.3	140.2	209.8	−58.4	+1.9	−0.5	59.3	K88
	$N=$–.–	2.5	0.973	105.0	344	(146.7)	223.6	−61.3	–.–	–.–	54.2	GE
#103	**TCE**	**Centaurids II**			**01/23–03/12**		**Feb. 09**	**321**	2.6	(19)	~3	PJ
		= θ-Centaurids = μ-Velids = β-Hydrids										
						$\lambda_{\odot}=321.0$	203.8	−39.6	+0.81	+0.01	60.2 ± 2.3	Mean
	visual	(7.2)	0.922	131.8	31.3	150.7	217.8	−43.7	+1.1	−0.4	64.6	PJ
	$N=2$	5.0	0.87	114.7	43	146.7	202.8	−48.3	–.–	–.–	59	GE (2.45)
	$N=2$	1.7	0.92	145.4	39	144.7	218.8	−35.2	–.–	–.–	63	GE (2.49)
	visual				02/10–02/13		207.5	−43.5				M (#112)

PIH – "this:" Diffuse radiant. q versus Π dependence.

#104	GBO	γ-Bootids			**02/06–02/13**		**Feb. 09**	**321.0**	-.-	-.-	-.-	
	$N=2$	Inf.	0.849	85.5	221.0	321.0	222.4	+39.1	+1.06	−0.42	50.3	This work
#34	DSE	δ-Serpentids			**02/12–02/22**		**Feb. 12**	**323**	-.-	(4)	-.-	
	$N=5$	2.3	0.98	118.0	171	323.7	241.6	+14.9	-.-	-.-	57.9	GE (2.16)
	$N=1$	9.229	0.986	130.5	184.7	324.1	237.0	+09.6	-.-	-.-	64.97	HV6429
	$N=1$	11.089	0.933	125.0	151.7	333.2	255.5	+08.8	-.-	-.-	63.22	HV6546
	Parent?	*C/1947 F_2 (Becvár)*		*(Epoch 1947-05-04)*				*$\lambda_\odot=323.07$*	*$T_J=-0.59$, $H_{10}=+11.2$ ($n=4$)*			
	(1974)	*(30)*	*0.987*	*129.156*	*182.129*	*323.071*	*237.1*	*+10.7*	*-.-*	*-.-*	*65.43*	*0.0249*
#105	OCN	**Centaurids I**			**01/06–02/19**		**Feb. 12**	**323.4 ± 0.4**	2.8	(18)	2.2 ± 0.3	PJ
		= o-Centaurids = α-Crucids = η-Craterids										
						$\lambda_\odot=311.0$	165.6	−60.3	+0.74	+0.41	43.7 ± 5.6	Mean
	visual	Inf.	0.876	88.3	39.3	142.9	174.6	−55.4	+0.9	−0.4	51.4	PJ
	visual	(2.4)	0.844	82.5	50.7	142.9	174.3	−55.3			45.5	PJ
	$N=3$	2.94	0.98	74.3	367	122.7	160.5	−63.3	-.-	-.-	41.5	GE (1.01)
	$N=3$	2.38	0.98	70	360	119.7	156.4	−65.3	-.-	-.-	38.4	N62
	visual						185.4	−64.6				M (#92)
#106	API	α-Pictorids			**02/11–02/15**		**Feb. 14**	**325**	-.-	(2.8)	-.-	
	$N=$-.-	(Inf.)	0.97	37	7	141	132	−56	-.-	-.-	(35)	PJ
	$N=3$	9.1	0.98	47.2	12	145.7	110.1	−65.1	-.-	-.-	30.0	GE (2.10)
	$N=5$	2.9	0.67	48.4	354	145.7	98.5	−76.0	-.-	-.-	28.9	GE (2.11)
	$N=2$	10.9	0.97	47.8	14	144	112.1	−64.1	-.-	-.-	31.0	GE (2.30)
#107	DCH	δ-Chamaleontids			**02/12–02/16**		**Feb. 14**	**325**	-.-	(1.6)	-.-	
	$N=47$	1.8	0.93	61.9	340	145.7	179.6	−83.3	-.-	-.-	34.2	GE (2.14)
	$N=4$	13.6	0.95	70.2	330	144.7	254.4	−86.1	-.-	-.-	42.6	GE (2.15)
#108	BTU	β-Tucanids			**02/27–03/02**		**Feb. 29**	**340.4**	-.-	(1.6)	-.-	
	?$N=11$	2.13	0.98	55.3	347	178.7	50.3	−80.8	-.-	-.-	32.1	GE (3.04)
	?$N=10$	10.0	0.98	58.3	346	178	49.6	−77.8	-.-	-.-	36.3	GE (3.05)
	?visual				02/25–02/27		66.3	−62.8				M (#42)
	Parent	*C/1976 D1 (Bradfield)*		*(Epoch 1976-03-03)*				*$\lambda_\odot=340.4$*	*$T_J=+0.82$, $H_N=+11.6$ (3.9)*			
	(1976)	*136.99*	*0.834*	*46.834*	*313.007*	*160.790*	*12.8*	*−63.5*	*-.-*	*-.-*	*32.86*	*0.0143*
												Antihelion
#109	ACN	α-Carinids			**01/24–02/09**		**Feb. 01**	**311.2**	2.5	3.9 ± 0.7	2.3 ± 0.6	PJ
		(2.5)	0.967	34.2	17.7	131.2	100.7	−54.8	+0.04	−0.60	21.7	PJ
	$N=1$	1.665	0.890	38.7	45.4	101.0	107.12	−45.28	-.-	-.-	23.5	HV9880

	Code	Name a	q	i	Dates (2000) ω	Node	Peak R.A.	$\lambda_{\odot}^{max}$ Decl.	χ ΔR.A./Δλ	W ΔDecl./Δλ	ZHR^{max} V_g	Notes Δ_{C-E}
#110	AAN	α-Antilids			**01/15–02/10**		**Feb. 02**	**313.1**	3.4	1.6	<2	GB
		= *α-Hydrids*				$\lambda_{\odot}=304.7$	140.0	−10.0	–.–	–.–	–.–	IMO
	$N=327$	1.788	0.142	64.3	141.9	133.1	161.9	−13.5	+0.68	−0.16	42.6	GB
	σ	–.–	±0.032	±7.8	±5.3	±2.4	±2.3	±3.1	±0.03	±0.01	±2.6	
	$N=6$						135.9	−10.0	+0.98	−0.57	42.6	GK
	Visual						178.3	−17.6				M (#85)
#111	FCM	Feb. Canis Majorids			**02/17–02/20**		**Feb. 19**	**330.6**	–.–	(1.6)	–.–	
	Outburst						1985, Feb. 19	330.60	(1.8)	–.–	>70	
	(1985)					$\lambda_{\odot}=330.60$	104.3	−25.6	–.–	–.–	Fast	Ch. 13
		Inf.	0.949	21.3	23.2	150.6	101.9	−28.1	–.–	–.–	19.5	This work
#29	*(DLE)*	*δ-Leonids*			**01/21–03/12**		**Feb. 23**	**334.7**	3.0	14 ± 6	1.1 ± 0.3	PJ
		= *ecliptic antihelion source*										
#112	NDL	N. δ-Leonids			**01/21–03/12**		**Feb. 23**	**334.7**	3.0	14 ± 6	0.6 ± 0.3	PJ
		= *ecliptic antihelion source*				$\lambda_{\odot}=336$	168.0	+16.0	+0.80	−0.30	–.–	IMO
	$N=8$	1.954	0.612	4.8	266.4	331.4	155.1	+17.9	–.–	–.–	20.6	ZS73
	$N=24$	2.618	0.643	6.2	259.0	338.8	159.7	+18.8	–.–	–.–	20.1	L72B
	Fireballs	AD 1043–1073:			02/19–02/23		115	+23	–.–	–.–	–.–	AT (4)
	Parent?	*1999 RD_{32}*		*(Epoch 2005-01-30)*				$\lambda_{\odot}=346.00$	$T_J=+2.87$, $H_N=+16.32$			
	(2005)	*2.640*	*0.605*	*5.927*	*264.128*	*345.995*	*168.2*	*+13.8*	*+0.53*	*−0.14*	*22.77*	*+0.0679*
#113	SDL	S. δ-Leonids			**01/21–03/12**		**Feb. 23**	**334.7**	3.0	14 ± 6	0.5 ± 0.3	PJ
		= *ecliptic antihelion source*										
	$N=$ –.–	1.950	0.804	9.8	237.2	334.7	148.6	+18.6	–.–	–.–	17.3	This work
	$N=$ –.–	2.182	0.729	4.3	69.0	146.4	137.7	+17.8	–.–	–.–	17.4	T89 (6c)
	$N=37$	2.473	0.702	6.4	91.3	134.5	136.1	+07.2	–.–	–.–	20.9	ZS
												Helion
#114	DXC	Daytime χ-Capricornids			**01/17–02/12**		**Feb. 01**	**311.3**	–.–	(10)	–.–	
	$N=5$	3.103	0.760	10.5	118.2	311.3	321.4	+06.0	–.–	–.–	18.5	ZS73
	$N=16$	2.473	0.702	1.5	108.0	314.2	324.7	−11.2	–.–	–.–	18.5	ZS

AAN – Stream detected in AMOR radar survey with limiting magnitude +14.

#115	DCS	Daytime Capricornids–Sagittariids										
					01/13–02/28		**Feb. 02**	**312.5**	–.–	(18)	~7	
	$N=15$	1.684	0.355	6.8	242.5	145.1	315.0	−23.3	–.–	–.–	26.8	ZS73
	$N=3$	2.08	0.36	4.5	246	144.7	316.7	−20.8	–.–	–.–	28.9	GE (2.01)
	$N=1$	1.5	0.28	2	234	143.7	308.7	−20.8	–.–	–.–	28.9	HV18043
	$N=29$	*1.712*	*0.415*	*6.2*	*69.8*	*309.1*	*299.6*	*−14.2*	–.–	–.–	*25.1*	*ZS*
	$N=26$	*1.991*	*0.314*	*6.8*	*60.0*	*314.0*	*299.8*	*−15.3*	–.–	–.–	*29.4*	*ZS73*
	Parent?	*2001 ME_1*		*(Epoch 2005-01-30)*				*$\lambda_\odot=318.80$*	*$T_J=+2.67$, $H_N=+16.81$*			
	(2005)	*2.652*	*0.356*	*3.558*	*247.831*	*138.795*	*311.0*	*−21.2*	–.–	–.–	*29.59*	*0.0782*
#116	DEQ	Daytime ε-Aquariids			**01/15–02/13**		**Feb. 04**	**315.8**	–.–	(12)	–.–	
		= ecliptic helion source										
	$N=17$	2.004	0.529	8.8	84.8	315.1	310.2	−6.8	–.–	–.–	23.1	ZS
#117	DCQ	Daytime *c*-Aquariids			**01/29–02/28**		**Feb. 13**	**325**	–.–	–.–	–.–	
		(Twin of Sep. κ-Aquariids)										
	$N=11$	2.3	0.82	2.1	299	144.7	346.7	−23.7	–.–	–.–	14.1	GE (2.18)
	Parent?	*2004 NL_8*		*(Epoch 2006-03-06)*				*$\lambda_\odot=314.7$*	*$T_J=+3.00$, $H_N=+17.12$*			
	(2006)	*2.584*	*0.736*	*4.412*	*293.079*	*135.129*	*332.4*	*−20.4*	–.–	–.–	*17.75*	*+0.0379*

March:

												Apex
#118	**GNO**	**γ-Normids**			**02/25–03/22**		**Mar. 13**	**353**	2.4	3.3 ± 0.5	5.8 ± 1.0	PJ
	Visual	(Inf.)	0.976	130.9	15.5	173.1	251.6	−51.3	+1.3	−0.1	(65.7)	PJ
	$N=6$	4.6	0.98	121.6	13	178.7	263.1	−56.0	–.–	–.–	58.9	GE (3.14)
	$N=3$	1.18	0.66	137.4	95.9	(179.7)	250.2	−42.7	–.–	–.–	56.8	GE (3.15)
	$N=2$	3.12	0.85	145.4	49	(179.7)	254.0	−41.2	–.–	–.–	64.0	GE
	radar					176	251.0	−50.1				WE55
	visual				03/06–03/11		242.0	−53.1				M (#149)
#119	LCE	λ-Centaurids			**03/17–03/21**		**Mar. 19**	**359**	–.–	(1.6)	–.–	
	$N=3$	7.74	0.774	59.3	58.0	178.7	173.7	−59.5	–.–	–.–	38.4	GE (3.09)
	?Visual				03/05–03/17		177.2	−56.6	–.–	–.–	–.–	M (84)

GNO – Also called Coronae Australids after a transcription error in the 1935 shower list by R. MacIntosh.

Table 7 (*cont.*)

	Code	Name a	q	i	Dates (2000) ω	Node	Peak R.A.	$\lambda_{\odot}^{max}$ Decl.	χ ΔR.A./$\Delta\lambda$	W ΔDecl./$\Delta\lambda$	ZHR^{max} V_g	Notes Δ_{C-E}
#43	ZSE	ζ-Serpentids			**03/20–04/01**		**Mar. 25**	**365**	–.–	(4.4)	–.–	
		Possibly related to δ-Aquilids										
	$N=1$	4.771	0.985	150.0	193.1	360.1	266.3	−06.3	–.–	–.–	67.42	HV5688
	$N=1$	6.358	0.946	150.3	207.9	11.5	273.7	−06.6	–.–	–.–	67.58	HV3024
#120	DPA	δ-Pavonids			**03/11–04/16**		**Mar. 31**	**11.1**	2.6	8.4 ± 1.6	5 ± 1	PJ
		= β-Pavonids = ϕ-Pavonids										
	Visual	(164)	0.959	108	337	191.1	309.1	−62.8	1.4	+0.2	58	PJ
	Visual				04/13–04/15		305.2	−66.3	–.–	–.–	–.–	M (#244)
	Parent	*C/1907 G_1 (Grigg-Mellish)*		*(Epoch 1907-03-28)*				*$\lambda_{\odot}=10.42$*	*$T_J=-0.23$, $H_{10}=+9.7$ ($n=8.4$)*			
	(1907)	*(29.96)*	*0.928*	*110.057*	*328.756*	*190.417*	*309.0*	*−60.4*	*–.–*	*–.–*	*58.72*	*0.0039*
	Parent?	*C/1742 C_1*		*(Epoch 1742-02-09)*				*$\lambda_{\odot}=7.71$*	*$T_J=-0.25$, $H_{10}=$–.–*			
	(1742)	*(29.96)*	*0.907*	*112.596*	*324.587*	*187.710*	*305.8*	*−58.9*	*–.–*	*–.–*	*59.50*	*0.1636*
												Antihelion
#121	NHY	ν-Hydrids			**03/01–03/06**		**Mar. 02**	**343.1**	–.–	(2.0)	–.–	
	$N=2$	1.622	0.626	12.5	88.4	163.1	159.1	−12.2	–.–	–.–	19.7	This work
	$N=$–.–	2.521	0.718	9.6	70.8	184.6	158.6	−11.3	–.–	–.–	19.3	T89 (18)
	$N=29$	2.138	0.618	0.5	84.7	163.0	162.0	+06.8	–.–	–.–	20.9	ZS
	$N=3$	12.5	0.75	7.3	62	178.7	159.6	−05.3	–.–	–.–	20.1	GE (3.02)
	Parent?	*1999 RM_{45}*		*(Epoch 2005-01-30)*				*$\lambda_{\odot}=340.88$*	*$T_J=+3.95$, $H_N=+19.33$*			
	(2005)	*1.681*	*0.598*	*10.881*	*90.716*	*160.888*	*158.4*	*−08.1*	*–.–*	*–.–*	*20.57*	*0.0071*
#122	APX	α-Pyxidids			**03/06–03/06**		**Mar. 05**	**345.9**	–.–	–.–	–.–	
	Outburst	Tim Cooper						<345.91	2.2	<0.015	>50	PJ
	(1979)	(Inf.)	0.852	30.1	44.1	166.1	135.5	−35.2	–.–	–.–	25.9	Ch. 13
	?Visual				03/07–03/14		175.2	−42.6	–.–	–.–	–.–	M (#82)
#50	*(VIR)*	*March Virginids*			**02/18–04/25**		**Mar. 14**	**354**	3.0	~22	4.5 ± 0.7	IMO/PJ

NHY – "this:" two video meteors from DMS database (+3.5 and +5.0^{m}). Intrinsically faint comet. Association uncertain.
VIR – This includes the March η-Virginids.

#123	**NVI**	N. March Virginids			**02/18–04/25**		**Mar. 14**	**354**	3.0	~22	2.2 ± 0.7	IMO/PJ
		= antihelion source										
	$N = -.-$	1.955	0.728	3.7	252.7	353.5	174.3	+08.7	+0.9	+0.4	(23.0)	GK
	$N = 18$	1.691	0.496	3.7	282.4	358.0	185.7	+02.3	-.-	-.-	23.0	ZS73
	Parent?	*1998 SJ_{70}*		*(Epoch 2005-01-30)*				$\lambda_\odot = 368.09$	$T_J = +3.25$, $H_N = +18.33$			
	(2005)	*2.234*	*0.657*	*7.157*	*260.045*	*8.232*	*188.0*	*+08.6*	-.-	-.-	*20.51*	*0.0341*
#124	**SVI**	*S. March Virginids*			**02/18–04/25**		**Mar. 14**	**354**	3.0	~22	2.3 ± 0.7	IMO/PJ
		= antihelion source = southern Virginids										
	$N = -.-$	3.607	0.606	3.8	81.8	160.4	158.3	+05.0	-.-	-.-	(22.9)	GK
	$N = 13$	2.160	0.565	6.1	91.2	182.0	179.7	−08.5	-.-	-.-	22.9	ZS73
	$N = 5$	1.95	0.64	0.1	83	175.7	172.6	+02.7	-.-	-.-	20	KL (1)
#11	**EVI**	**η-Virginids**			**03/03–03/23**		**Mar. 14**	**354**	3.0	(8)	<1.5	PJ
		= Southern Virginids										
	Outburst?			2.0			1953 Mar. 12–13		$\lambda_\odot = 352.4$			MP
	(1953)	2.19	0.387	0.8	113.8	172.4	181.6	−00.3	+1.00	−0.37	26.6	Ch. 28
	$N = 7$	2.562	0.382	3.5	349.1	280.5	182.1	+02.6	+1.00	−0.37	29.2	This work
	$N = 3$	4.336	0.424	4.0	282.4	352.4	180.7	+03.6	-.-	-.-	30.0	PG
	$N = -.-$	2.324	0.394	5.6	289.6	335.9	168.7	+10.7	-.-	-.-	27.6	T89 (14N)
	$N = 4$	2.637	0.234	3.5	308.0	334.5	174.3	+04.7	+0.90	+0.31	34.2	L71B
	$N = 3$	2.027	0.431	0.3	287	357.1	185.6	−02.3	-.-	-.-	28	L71B
	$N = 9$	1.94	0.36	5.7	296.9	356.2	188.8	+00.9	-.-	-.-	28.4	KL (2)
	$N = 3$	2.38	0.26	2.9	304.3	355.2	188.8	−04.1	-.-	-.-	32.8	NL (61.3.3)
	$N = -.-$	2.52	0.42	1.0	291.0	352.8	181.6	−00.3	-.-	-.-	25.6	MP
	$N = 63$	1.196	0.288	6.4	310.6	356.0	196.6	−01.5	-.-	-.-	26.2	ZS
	$N = 9$	1.869	0.325	9.8	299.7	345.5	183.3	+06.8	-.-	-.-	29.4	ZS73
	$N = 24$	1.329	0.222	8	315.7	360.2	202.6	−03.8	-.-	-.-	30.0	ZS73
	$N = 23$	1.815	0.312	0.0	121.8	184.4	198.1	−07.7	-.-	-.-	29.2	ZS73
	$N = 3$	2.82	0.403	5.2	285.8	355.1	183.6	+03.7	-.-	-.-	28.9	W57
	$N = -.-$	4.00	0.42	6	284	354.7	182.6	+03.7	-.-	-.-	32	W54
	$N = -.-$	2.578	0.384	0.3	109.4	160.8	170.6	+03.7	-.-	-.-	28.5	T89 (14Q)
	$N = -.-$	4.312	0.498	9.1	93.4	161.5	162.3	−02.3	-.-	-.-	27.5	T89 (14S)
	$N = -.-$	2.197	0.359	3.4	114.4	178.0	188.3	−05.3	-.-	-.-	28.2	T89 (19Q)
	$N = -.-$	4.126	0.397	12.3	105.4	175.2	179.3	−11.3	-.-	-.-	30.5	T89 (19S)

	Code	Name			Dates (2000)		Peak	$\lambda_\odot^{max}$	χ	W	ZHR^{max}	Notes
		a	q	i	ω	Node	R.A.	Decl.	ΔR.A./$\Delta\lambda$	ΔDecl./$\Delta\lambda$	V_g	Δ_{C-E}
	$N=2$	2.93	0.41	1	105	177.7	183.6	−03.3	–.–	–.–	27.8	GE (3.17)
	$N=$–.–	2.70	0.45	2	102	170.7	176.6	−00.3	–.–	–.–	29	JW
	Parent?	*D/1766 G$_1$ Helfenzrieder*		*(Epoch 2005-07-14)*				$\lambda_\odot = 333.77$	$T_J = +2.70$, $H_{10} =$ –.–			
	(1766)	*2.665*	*0.406*	*2.419*	*106.503*	*148.300*	*157.4*	*+7.1*	–.–	–.–	*28.61*	*0.1290*
	(2004)	*2.658*	*0.362*	*0.187*	*111.648*	*153.774*	*165.3*	*+6.1*	–.–	–.–	*30.01*	*0.0192*
	(2047)	*2.618*	*0.327*	*0.076*	*115.777*	*150.814*	*164.1*	*+6.1*	–.–	–.–	*31.07*	*0.0051*
#38	CUR	ξ-Ursae Majorids			**03/18–04/03**		**Mar. 19**	**358.0**	–.–	(6)	–.–	
	$N=3$	2.899	0.803	12.6	238.5	358.0	175.4	+30.1	–.–	–.–	18.9	This work
	$N=1$	3.780	0.810	12.6	234.8	358.0	172.4	+30.3	–.–	–.–	18.88	HV6915
	$N=1$	2.908	0.783	11.5	240.4	358.4	175.4	+26.1	–.–	–.–	18.59	HV6971
	$N=1$	2.553	0.804	19.3	238.4	360.2	187.4	+36.9	–.–	–.–	19.42	HV7040
#290	*(ALL)*	*a-Leonids*										
#39	NAL	N. α-Leonids			**02/25–03/25**		**Mar. 07**	**349**	–.–	(11)	–.–	
	$N=3$	2.016	0.907	6.9	43.3	173.9	158.7	+31.0	+0.4	+0.0	11.1	This work
	$N=23$	1.978	0.878	5.8	226.0	343.9	146.6	+31.3	–.–	–.–	12.3	ZS
	$N=10$	1.691	0.814	8.5	241.8	350.2	167.2	+28.0	–.–	–.–	13.9	ZS
	$N=$–.–	2.721	0.793	9.4	240.0	2.7	176.3	+22.7	–.–	–.–	–.–	GK
	$N=1$	2.638	0.912	7.0	218.2	2.5	160.2	+30.2	–.–	–.–	13.00	HV3076
	$N=1$	2.835	0.971	5.7	201.9	12.1	151.7	+33.6	–.–	–.–	10.68	HV10394
	$N=19$	1.119	0.853	2.4	256.4	2.6	181.3	+10.4	–.–	–.–	7.3	ZS
	Fireballs	AD 1071–1188:			03/22–03/30		111	+27	–.–	–.–	–.–	AT (9)
#125	SAL	S. α-Leonids			**02/25–04/24**		**Mar. 19**	**359**	–.–	(11)	–.–	
	$N=5$	2.016	0.907	6.9	221.8	358.8	141.8	−07.0	+1.1	+0.0	11.9	This work
	$N=$–.–	2.313	0.904	10.5	40.3	168.0	131.6	−12,2	–.–	–.–	13.2	T89 (16)
	Visual				03/13–03/23		148.2	−05.4	–.–	–.–	–.–	M (#71)
#126	SGE	March δ-Geminids			**03/22–04/15**		**Mar. 21**	**361**	–.–	(10)	–.–	
	$N=2$	1.66	0.94	2.1	32.8	152.9	114.5	+12.2	–.–	–.–	8.0	This work
	$N=1$	2.054	0.996	0.7	182.9	1.4	104.4	+28.1	–.–	–.–	4.9	IAU#332F1
	Fireballs	AD 1062–1188:					112.7	+20.9	–.–	–.–	–.–	AT (1)

#45	PDR	φ-Draconids			**03/11–04/17**		**Mar. 28**	**368**	–.–	(15)	–.–	
	$N=4$	3.020	0.995	36.3	1793	8.3	267.1	+69.4	–.–	–.–	22.9	This work
	$N=$–.–	2.770	0.996	37.5	171.1	14.4	281.0	+68.1	–.–	–.–	26.7	CL
	$N=$–.–	2.248	0.978	26.5	164.7	356.1	318.1	+72.2	–.–	–.–	17.3	T89 (15)
	$N=40$	2.156	0.988	33.0	169.0	06.5	285.4	+69.8	–.–	–.–	20.2	ZS
	$N=14$	1.173	0.972	30.9	166.3	12.2	291.5	+71.4	–.–	–.–	19.1	ZS73
												Helion
#127	MCA	Mar. Cassiopeiids			**02/11–04/10**		**Mar. 08**	**350**	–.–	(23)	–.–	
	$N=5$	2.40	0.931	14.8	146.5	345.2	352.8	+52.3	+1.16	−0.10	14.5	This work
	$N=4$	2.322	0.930	15.2	146.3	350.5	360.2	+50.9	–.–	–.–	13.4	PG
	$N=$–.–	2.567	0.934	14.3	148.8	349.7	4.7	+48.3	–.–	–.–	13.6	T89 (17)
#128	MKA	Daytime κ-Aquariids			**03/12–03/16**		**Mar. 14**	**354**	–.–	(2)	–.–	
	$N=7$	1.7	0.18	1.8	42	359.7	338.7	−7.7	–.–	–.–	33.2	GE (3.01)
	$N=3$	2.13	0.30	2.5	59.7	354.4	340.2	−7.3	–.–	–.–	29.8	NL (61.3.2)
#129	QPE	Daytime q-Pegasids			**03/11–03/16**		**Mar. 14**	**354**	–.–	(2)	–.–	
	$N=5$	2.94	0.88	9.7	101.3	354.5	352.3	+12.9	–.–	–.–	21.7	NL (61.3.1)
	Parent	*2005 EM_{169}*		*(Epoch 2005-03-13)*				*$\lambda_\odot=31.48$*	*$T_J=+347.54$, $T_J=+2.81$, $H_N=+24.67$*			
	(2005)	*2.845*	*0.755*	*10.925*	*115.757*	*347.537*	*351.4*	*+19.0*	*–.–*	*–.–*	*18.37*	*−0.0022*
#130	DME	δ-Mensids			**03/13–03/21**		**Mar. 17**	**356.7**	–.–	(3.2)	–.–	
	Visual					$\lambda_\odot=356.7$	58.0	−80.0	–.–	–.–	–.–	IMO
		(3.2)	0.982	56.1	345.6	177.1	58	−80	–.–	–.–	33	This work
	April:											
												Apex
#131	DAL	δ-Aquilids			**04/10–04/16**		**Apr. 13**	**23**				
	$N=2$	75	0.722	146.6	116.1	23.0	310.6	−00.2	+1.5	−0.2	66.2	This work
	visual				04/13–04/22		269.7	−10.0	B1900			M (185)
	Parent?	*C/1984 S_1 (Meier)*		*(Epoch 1984-10-14)*				*$\lambda_\odot=14.20$*	*$T_J=<-0.77$, $H_{10}=$–.–, P unknown*			
	(1984)	*Inf.*	*0.828*	*146.308*	*130.984*	*14.198*	*298.0*	*−2.2*	*–.–*	*–.–*	*67.40*	*0.0459*
#6	**LYR**	**April Lyrids**			**04/16–04/25**		**Apr. 22**	**32.4**	2.9	2.74 ± 0.13	12.8 ± 0.7	PJ
	(1982)							32.085	2.9	0.018 ± 0.006	250	
	(1945)							32.053	1.4	–.–	>97	
	(1922)							≤32.006	>2.7	~0.017	≥800	

Table 7 (*cont.*)

	Code	Name a	q	i	Dates (2000) ω	Node	Peak R.A.	$\lambda_{\odot}^{max}$ Decl.	χ ΔR.A./Δλ	W ΔDecl./Δλ	ZHR^{max} V_g	Notes Δ_{C-E}
	(1803)							31.950	–.–	–.–	~860	
	N = –.–	45.7	0.921	79.6	214.3	31.8	272.0	+33.3	–.–	–.–	46.6	DMS
	N = –.–	31.61	0.918	79.61	214.53	32.439	271.6	+33.3	–.–	–.–	46.5	AW
	dispesion (σ)	–.–	±0.013	±1.1	±2.6	±1.0	–.–	–.–	–.–	–.–	–.–	–.–
	N = 4	Inf.	0.928	80.3	212.0	32.6	272.7	+33.7	–.–	–.–	47.6	PG
	N = 7	56.0	0.918	79.5	214.5	32.2	271.9	+33.3	+1.23	+0.17	47.07	KP
	N = 5	28	0.919	79.0	214.3	32.5	271.9	+33.6	+1.1	+0.0	47.6	AC
	N = 5	25.812	0.879	78.6	217.2	32.3	271.5	+34.0	–.–	–.–	45.6	JW
	Parent	*C/1861 G_1 (Thatcher)*		*(Epoch 1861-05-25)*				*$\lambda_{\odot}$ = 31.48*	*T_J = +0.30, H_{10} = +5.5 (n = 4.0), P = 415.5 yr*			
	(1861)	*55.682*	*0.9225*	*79.776*	*213.484*	*31.860*	*271.9*	*+33.5*	*–.–*	*–.–*	*47.08*	*0.002*
#132	BPA	β-Pavonids			**03/11–04/16**		**Apr. 07**	**17.2**	–.–	(14)	–.–	
	Visual					$\lambda_{\odot}$ = 17.2	308.0	−63.0	+1.40	+0.20	59	IMO
		Inf.	0.992	108.8	349.2	197.7	308.0	−36.0	–.–	–.–	59	This work **Antihelion**
#46	BCR	β-Craterids			**04/01–04/06**		**Apr. 03**	**13.5**	–.–	(2)	–.–	
	N = 1	2.716	0.678	17.6	75.6	191.7	176.8	−26.9	–.–	–.–	22.6	HV10365
	N = 1	2.804	0.734	16.5	68.1	196.6	176.5	−28.3	–.–	–.–	21.0	HV10478
#27	**KSE**	**κ-Serpentids**			**04/01–04/07**		**Apr. 05**	**15.7**	–.–	(3)	<4	
	N = –.–	Inf.	0.45	65	275	15.7	230.6	+17.8	–.–	–.–	45	AC
	N = 1	41.7	0.417	63.0	279.9	16.5	232.6	+15.4	–.–	–.–	45.01	HV10098
#49	LVI	*λ-Virginids*			**03/03–04/24**		**Apr. 10**	**20**	–.–	(21)	–.–	
		= extension of η-Virginids?										
	N = –.–	2.188	0.257	8.8	306.1	24.5	218.7	−16.2	–.–	–.–	32.2	T89 (23)
	N = –.–	2.630	0.343	2.0	295.0	20.2	210.7	−10.2	–.–	–.–	26.8	L71B
	N = –.–	2.374	0.426	4.6	106.0	201.7	205.7	−16.3	–.–	–.–	27.0	T89 (24)
	N = 12	1.731	0.344	14.1	298.7	17.7	214.0	−01.3	–.–	–.–	28.6	Z73

	$N=63$	1.295	0.278	4.8	310.3	12.6	211.1	−08.7	-.-	-.-	27.4	ZS
	$N=7$	2.32	0.322	2.8	119.4	197.5	208.7	−14.2	-.-	-.-	30.0	L71A
#133	PUM	Apr. (ψ-)Ursae Majorids			**04/10–04/13**		**Apr. 12**	**23**	-.-	(1.2)	-.-	SH
	$N=$-.-	2.290	1.002	6.8	174.6	27.1	117.0	+52.9	-.-	-.-	8.4	T89 (25)
	$N=$-.-	1.805	0.984	14.0	203.0	22.2	188.6	+58.7	-.-	-.-	10	L71B
	$N=21$	1.865	0.993	9.4	183.5	28.9	153.2	+58.6	-.-	-.-	8.6	ZS
	$N=2$	2.67	0.909	11.0	187.1	27.3	157.8	+55.7	-.-	-.-	10.6	W57
	Outburst					1970	Apr. 02	11.535	-.-	-.-	20	
	(1970)					$\lambda_\odot=11.535$	160	+55	-.-	-.-	-.-	RM
#291	*(GVR)*	*γ-Virginids*										
#134	NGV	Northern γ-Virginids			**04/01–04/26**		**Apr. 13**	**24.3**	-.-	(10)	1.0	
	$N=$-.-	2.250	0.908	5.2	221.7	24.3	180.6	+17.7	-.-	-.-	11.7	T89 (22N)
	?Fireballs	AD 1030–1099:			04/02–04/10		151	+17	-.-	-.-	-.-	AT (11)
	Parent?	*2002 FC*		*(Epoch 2005-01-30)*				*$\lambda_\odot=37.30$*	*$T_J=+2.94$, $H_N=+18.82$*			
	(2005)	*2.831*	*0.960*	*6.630*	*208.091*	*37.300*	*183.2*	*+22.5*	-.-	-.-	*11.66*	*0.0234*
#135	SGV	Southern γ-Virginids			**03/27–04/28**		**Apr. 12**	**22.7**	-.-	(13)	-.-	
		$=\gamma$-Corvids = antihelion source meteors										
	$N=5$	2.262	0.867	5.0	50.0	212.7	183.2	−15.5	-.-	-.-	13.9	PG
	$N=$-.-	2.006	0.912	8.1	42.3	206.7	167.6	−22.3	-.-	-.-	11.6	T89 (22S)
	$N=$-.-	2.092	0.770	4.0	66.9	201.4	184.6	−11.3	-.-	-.-	15.5	T89 (20)
	Outburst!				**1841 Apr. 19**		**31.14 ± 0.06**		Faint	-.-	>34	
	(1841)	55 in 2.25 hr, Caleb G. Forshey, LA				$\lambda_\odot=31.14$	200	+07°	-.-	-.-	Slow	
	Parent?	*2003 BD_{44}*		*(Epoch 2005-01-30)*				*$\lambda_\odot=24.07$*	*$T_J=+3.62$, $H_N=+16.66$*			
	(2005)	*1.967*	*0.777*	*2.67*	*88.49*	*181.901*	*188.4*	*−09.3*	-.-	-.-	*15.7*	*+0.0116*
#136	SLE	σ-Leonids			**03/21–05/17**		**Apr. 17**	**27.7**	3.0	(23)	-.-	
	$N=$-.-	2.141	0.561	6.2	271.9	9.4	193.3	+03.1	-.-	-.-	23.0	PG
	$N=$-.-	2.278	0.605	2.2	266.3	14.5	192.6	−02.3	-.-	-.-	21.2	T89 (21)
	Visual	1.53	0.48	1.9	286	13.7	200.7	−06.3	-.-	-.-	-.-	H48
	Visual				04/04–04/30		209.9	−10.4				M (#116)
	Parent?	*2002 GM_5*		*(Epoch 2002-04-13)*				*$\lambda_\odot=25.65$*	*$T_J=+3.36$, $H_N=+21.44$*			
	(2002)	*2.126*	*0.647*	*7.355*	*262.328*	*25.794*	*205.0*	*+01.6*	-.-	-.-	*20.52*	*0.0255*
	Parent?	*1995 EK_1*		*(Epoch 2005-08-18)*				*$\lambda_\odot=14.41$*	*$T_J=+3.12$, $H_N=+17.79$*			
	(2005)	*2.266*	*0.509*	*5.903*	*277.490*	*14.414*	*199.5*	*−1.0*	-.-	-.-	*25.01*	*0.0510*

	Code	Name a	q	i	Dates (2000) ω	Node	Peak R.A.	$\lambda_{\odot}^{max}$ Decl.	χ ΔR.A./Δλ	W ΔDecl./Δλ	ZHR^{max} V_g	Notes Δ_{C-E}
#21	**AVB**	**α-Virginids**			**03/22–04/26**		**Apr. 18**	**28**	–.–	~8.6	~5	RK
		= ecliptic antihelion source meteors										
	N = –.–	2.58	0.688	4.2	240.4	21.7	185.4	+09.6	–.–	–.–	16.8	SH
	N = 19	2.349	0.753	0.7	247.5	28.9	195.6	−05.3	–.–	–.–	16.6	L71B
	N = 21	2.56	0.812	3.1	239.7	31.8	195.6	+00.7	–.–	–.–	–.–	L71A
	Parent?	*1998 SH_2*		*(Epoch 1998-09-21)*				*$\lambda_{\odot}$ = 28.95*	*T_J = +2.93, H_N = +20.81*			*PW04*
	(1998)	*2.703*	*0.760*	*2.401*	*245.257*	*28.946*	*195.9*	*−01.9*	*–.–*	*–.–*	*18.05*	*0.0110*
#137	**PPU**	**π-Puppids**			**04/15–04/28**		**Apr. 22**	**33.6**	2.0	(5)	<1	PJ
	Outbursts (see Tab. 6e)											
	Visual	2.97	1.00	21	359	33.640	110.4	−45.1	+0.40	−0.10	15	PJ
	Parent	*26P/Grigg–Skjellerup*		*(Epoch 1977-04-07)*				*$\lambda_{\odot}$ = 33.354*	*T_J = +2.81, H_{10} = +11.7*			
	(1977)	*2.999*	*1.006*	*21.098*	*359.311*	*213.354*	*109.8*	*−45.0*	*+0.46*	*+0.12*	*15.10*	*+0.0122*
	(2002)	*2.736*	*1.005*	*22.347*	*1.624*	*211.740*	*110.8*	*−48.5*	*–.–*	*–.–*	*15.46*	*+0.1127*
	(2044)	*2.779*	*1.005*	*22.397*	*2.243*	*211.470*	*111.5*	*−48.5*	*–.–*	*–.–*	*15.54*	*+0.0881*
#138	ABO	α-Bootids			**04/14–05/12**		**Apr. 28**	**36.7**	–.–	–.–	–.–	
	Outburst	Telescopic				**1984**	**Apr. 28**	**38.168**	≫4	0.016 ±0.002	High rate	
	(1984)	(155)	0.775	20.2	237.4	38.168	214.4	+19.3	–.–	–.–	(24.65)	Ch. 13
	N = 5	2.680	0.839	11.9	232.5	22.1	193.1	+22.9	–.–	–.–	16.7	PG
	N = 8	2.647	0.753	18.0	246.9	36.9	218.6	+18.8	+0.7	+0.2	20	L72B
	N = 2	2.02	0.73	18.5	2527	23.3	210.7	+21.0	–.–	–.–	22.9	SH
	?Fireballs	AD 1023–1099:			04/10–04/16		234	+30	–.–	–.–	–.–	AT (13)
#47	DLI	μ-Virginids			**04/01–05/12**		**Apr. 28**	**39 ± 2**	3.0	14 ± 3	1.1 ± 0.5	PJ
	N = 3	2.531	0.418	9.1	286.5	38.3	226.8	−08.7	–.–	–.–	28.3	PG
	N = 7	3.116	0.477	9.9	280.0	35.7	221.7	−05.2	+0.50	−0.19	26.8	L71B
#139	GLI	γ-Librids			**03/24–05/07**		**Apr. 28**	**39 ± 2**	3.0	14 ± 3	1.1 ± 0.5	PJ
	N = 38	0.926	0.191	5.8	326.7	15.9	223.9	−12.8	–.–	–.–	26.0	ZS
	N = 17	1.570	0.410	0.3	113.7	225.5	234.1	−19.7	–.–	–.–	24.7	ZS73

#140	XLI	Apr. χ-Librids			**03/24–05/07**		**Apr. 28**	**39 ± 2**	–.–	(18)	–.–	
	$N = 20$	1.269	0.101	2.7	332.0	28.6	236.3	−18.9	–.–	–.–	34.2	ZS
	$N = 23$	1.408	0.159	5.6	324.1	27.1	232.7	−16.3	–.–	–.–	32.2	ZS73
	Visual				04/17–04/27		227.4	−22.4	–.–	–.–	–.–	M (#128)
												Helion
#141	**DCP**	**Daytime χ-Piscids**			**03/28–04/21**		**Apr. 09**	**19.2**	–.–	(10)	–.–	
		= ecliptic helion source										
	$N = 12$	2.121	0.696	6.6	103.9	19.1	19.6	+21.0	–.–	–.–	18.0	ZS73
#40	ZCY	ζ-Cygnids			**04/08–04/11**		**Apr. 10**	**20.0**	–.–	(1.2)	–.–	
	$N = 30$	3.863	0.898	66.4	139.8	19.2	303.8	+44.8	–.–	–.–	39.0	ZS
	$N = 1$	Inf.	0.692	79.9	112.5	19.9	317.6	+32.8	–.–	–.–	47.59	HV7161
	$N = 1$	167.7	0.677	80.4	110.5	20.0	318.9	+32.4	–.–	–.–	47.77	HV10094
#142	MDR	μ-Draconids			**04/16–04/21**		**Apr. 18**	**29.2**	–.–	(2.0)	–.–	
	$N = 15$	2.359	0.997	48.1	172.6	19.9	281.0	+57.9	–.–	–.–	28.2	ZS
	$N = 2$	10.02	1.00	45.4	189.6	29.2	260.2	+60.4	–.–	–.–	31.0	**SH**
#143	LPE	Daytime λ-Pegasids			**04/15–05/12**		**Apr. 19**	**29.7**	–.–	(11)	–.–	
	$N = 16$	1.42	0.21	61	44	29.7	343.5	+19.9	–.–	–.–	38.4	KL (5)
	$N = 16$	1.412	0.108	42.0	25.6	47.1	362.2	+18.0	–.–	–.–	33.0	ZS
#144	**APS**	**Daytime Apr. Piscids**			**04/08–04/29**		**Apr. 20**	**30.3**	–.–	(8)	–.–	
		= extension of Taurid complex? Twin of NPI										
	$N = 34$	1.32	0.22	0.5	45	30.7	7.6	+03.3	–.–	–.–	28.9	KL (4)
	$N = 16$	1.55	0.34	6	59	30.7	12.7	+10.3	–.–	–.–	26.8	KL (6)
	$N = 3$	1.18	0.28	5.8	48.6	29.0	7.1	+04.6	–.–	–.–	25.1	NL (61.4.2)
	$N = 4$	2.44	0.34	12.8	64.9	28.3	9.1	+15.2	–.–	–.–	29.8	NL (61.4.1)
	Parent?	*2005 NZ*$_6$		*(Epoch 2006-03-06)*				$\lambda_\odot = 36.17$	$T_J = +3.43$, $H_N = +17.40$			
	(2006)	*1.835*	*0.250*	*8.512*	*51.562*	*36.166*	*14.3*	*+12.2*	*–.–*	*–.–*	*30.76*	*0.0096*

May:

												Apex
#31	**ETA**	**η-Aquariids**			**04/19–05/28**		**May 06**	**46.9**	2.7	11.0 ± 0.7	28 ± 4	PJ
	Outbursts (see Tab. 5e)											
	$N = 23$	16.16	0.581	163.9	97.9	44.44	336.90	−01.47	+0.760	+0.422	65.95	LOS
	$N = 942$	11.60	0.545	165.1	91.9	45.6	339.0	−01.4	+0.73	+0.31	65.0	GB
	σ	–.–	±0.080	±2.1	±13.5	±3.6	±1.8	±00.8	±0.02	±0.01	±3.5	

ETA – The activity profile has at least two components. See Chapter 18.

	Code	Name			Dates (2000)		Peak	$\lambda_{\odot}^{max}$	χ	W	ZHR^{max}	Notes
		a	q	i	ω	Node	R.A.	Decl.	ΔR.A./$\Delta\lambda$	ΔDecl./$\Delta\lambda$	V_g	Δ_{C-E}
	$N=11$	29.8	0.612	165.5	101.5	45.79	338.02	−01.75	+0.888	+0.456	–.–	L
	$N=9$	5.536	0.587	164.6	96.7	47.8	339.0	−00.9	–.–	–.–	65.0	ZS73
	$N=51$	3.34	0.54	161	90	47.7	338.6	+00.8	–.–	–.–	63	KL (17)
	$N=1$	13.0	0.560	163.5	95.2	43.1	336.8	−01.3	–.–	–.–	65.5	JW
	$N=$–.–	6.140	0.541	165.4	91.4	45.5	339.1	−01.5	–.–	–.–	64.8	G
	Parent	*1P/Halley*		*(Epoch 1986-02-19)*				$\lambda_{\odot}=46.21$	$T_J=-0.59$, $H_0=+2.10$ $(n=7.5)$			
	(1986)	*17.940*	*0.587*	*162.687*	*98.605*	*46.211*	*338.2*	*−0.3*	*–.–*	*–.–*	*65.93*	*0.0658*
#145	**ELY**	**η-Lyrids**			**05/07–05/12**		**May 10**	**49.1**	–.–	(2)	3 ± 1	
	$N=12$	6.03	0.995	79.4	190.0	45.7	292.5	+39.7	–.–	–.–	3 ± 1	
	$N=5$	Inf.	0.998	75.3	193.0	50.3	290.8	+42.7	–.–	–.–	45.3	KO
	σ	±0.042	±0.002	±1.4	±1.0	±1.0	±1.0	±0.8	–.–	–.–	±0.8	
	?Fireballs	AD 1049–1070:			05/08–05/12		292.3	+56.1				AT (15)
	Parent	*C/1983 H_1 (IRAS-Araki-Alcock)*			*(Epoch 1983-05-12)*			$\lambda_{\odot}=49.10$	$T_J=+0.41$, $H_{10}=+9.8$ $(n=0.9)$, $P=959$ *yr*			
	(1983)	*98.098*	*0.997*	*73.252*	*192.852*	*49.102*	*288.0*	*+44.0*	*–.–*	*–.–*	*43.77*	*0.0058*
#146	CAU	β-Coronae Australids			**04/23–05/30**		**May 15**	**54.7**	–.–	(11)	<3	PJ
	Visual	(12.1)	0.221	121.7	125.3	234.7	285.1	−39.9	+1.10	+0.00	56.4	IMO
	Visual				04/28–05/05		285.7	−36.9				M (#208)
#147	PAQ	φ-Aquariids			**05/19–05/23**		**May 21**	**60**	–.–	(2)	–.–	
	$N=$–.–	2.78	0.56	174.1	270.4	239.7	350.9	−3.5	–.–	–.–	64	NL (5.12) **Antihelion**
#148	MLV	May (λ-)Virgnids			**05/05–05/06**		**May 06**	**45**	–.–	(2)	–.–	
	$N=$–.–	3.232	0.547	13.3	90.5	223.4	217.7	−31.2	–.–	–.–	25.6	T89 (28S)
	$N=$–.–	6.705	0.686	3.5	72.0	225.2	210.7	−18.2	–.–	–.–	25	L71B
	$N=$–.–	2.157	0.482	1.3	101.4	224.3	227.7	−19.2	–.–	–.–	24.6	T89 (28Q)
#55	**ASC**	**α-Scorpiids**			**04/21–05/26**		**May 16**	**55.2 ± 0.9**	2.5	4.9 ± 1.1	3.2 ± 0.4	PJ
	$N=3$	2.640	0.330	9.8	116.5	236.8	247.0	−28.8	+0.73	−0.07	31.0	PG
	$N=$–.–	2.469	0.324	8.9	117.7	233.1	243.8	−28.1	–.–	–.–	30.6	T89 (31)

	$N=2$	2.235	0.212	3.5	132.0	230.2	247.8	−24.1	–.–	–.–	33	L71B
	$N=22$	1.502	0.264	3.5	130.0	237.8	255.1	−25.2	–.–	–.–	29.0	ZS
	$N=29$	1.458	0.243	0.6	132.9	232.5	251.1	−22.7	–.–	–.–	29.4	ZS73
	$N=3$	2.097	0.189	2.3	136.7	217.0	238.7	−21.1	–.–	–.–	32.1	L71B
	$N=5$	2.15	0.21	3	134	222.7	240.7	−22.1	+0.50	−0.19	35	AC
	$N=30$	1.464	0.283	13.5	308.3	58.4	257.1	−13.2	–.–	–.–	28.7	ZS
	$N=27$	1.707	0.282	7.4	306.2	56.9	253.2	−17.3	–.–	–.–	29.7	ZS73
	$N=4$	2.33	0.26	6.0	305.9	61.4	255.9	−19.2	–.–	–.–	32.5	NL (61.5.10)
	$N=3$	2.170	0.133	10.0	322.0	44.7	247.7	−18.1	–.–	–.–	36.3	L71B
	Fireballs	AD 988–1068:			05/25–06/01		240.7	−20.1	–.–	–.–	–.–	AT (18)
	Parent?	*2004 BZ_{74}*		*(Epoch 2006-03-06)*				*$\lambda_\odot=59.36$*	*$T_J=+2.37$, $H_N=+18.39$*			
	(2006)	*3.048*	*0.330*	*16.506*	*115.415*	*239.509*	*249.3*	*−34.0*	*–.–*	*–.–*	*32.19*	*0.0289*
#292	*(OPH)*	*May Ophiuchids*										
#149	NOP	N. May Ophiuchids			**04/25–05/31**		**May 13**	**49.7**	–.–	(14)	–.–	
		= ecliptic antihelion source										
						$\lambda_\odot=49.7$	249.0	−14.0	+0.90	−0.10	27.8	IMO
#150	SOP	S. May Ophiuchids			**04/25–05/31**		**May 20**	**56.7**	–.–	(14)	–.–	
		= ecliptic antihelion source										
						$\lambda_\odot=56.7$	258.0	−24.0	+1.00	+0.00	27.8	IMO
#151	EAU	ε-Aquilids			**05/09–05/22**		**May 20**	**59**	–.–	(5)	–.–	
	$N=17$	0.873	0.354	59.6	318.3	59.5	284.9	+15.6	–.–	–.–	30.8	ZS
	$N=30$	0.89	0.65	67	309	48.7	291.5	+29.1	–.–	–.–	32.1	KL (9)
	$N=16$	0.78	0.41	68	327	55.7	293.5	+22.1	–.–	–.–	31	KL (15)
												Helion
#293	*(DCE)*	*Daytime ω-Cetids*										
#152	**NOC**	**N. Daytime ω-Cetids**			**04/24–05/27**		**May 07**	**46.7**	–.–	(12)	<2	
		= ε-Arietids										
	$N=86$	1.27	0.10	27	28	32.7	356	+08.3	–.–	–.–	35.3	KL (3)
	$N=16$	0.967	0.108	42.0	25.6	47.8	2.3	+17.8	–.–	–.–	33.0	ZS
	$N=18$	1.412	0.147	29.1	35.9	48.0	12.8	+19.3	–.–	–.–	34.1	ZS
	$N=17$	1.64	0.11	30	32	52.7	17.7	+19.3	–.–	–.–	37.4	KL (13)
	$N=3$	1.01	0.08	15.5	22.8	62.1	24.4	+11.8	–.–	–.–	32.8	NL (61.5.13)
	$N=10$	2.44	0.17	10.2	42.6	64.4	37.7	+20.0	1.7	–.–	35.8	NL (61.5.3)

	Code	Name			Dates (2000)		Peak	$\lambda_{\odot}^{max}$	χ	W	ZHR^{max}	Notes
		a	q	i	ω	Node	R.A.	Decl.	ΔR.A./$\Delta\lambda$	ΔDecl./$\Delta\lambda$	V_g	Δ_{C-E}
	$N=11$	3.11	0.68	6	257	74.7	247.7	−13.1	+0.9	+0.5	21	AC
	$N=11$	3.112	0.679	6.0	256.7	74.6	246.7	−12.1	–.–	–.–	20.1	L71B
	$N=4$	2.63	0.755	3.3	249.1	70.5	238.7	−14.1	–.–	–.–	21	L71A
	Fireballs	AD 1037–1102:			06/07–06/19		260.7	−11.0				AT (21)
	Parent?	*1996 JG*	*(Epoch 2005-01-30)*					$\lambda_{\odot}=63.37$	$T_J=+3.77$, $H_N=+19.13$			
	(2005)	*1.802*	*0.611*	*5.200*	*269.733*	*63.370*	*242.5*	*−13.2*	*–.–*	*–.–*	*20.0*	*0.0164*
#161	SSC	S. ω-Scorpiids			**05/23–06/15**		**June 01**	**70**	–.–	(9)	~5	GK
		= *θ-Librids*			06/01–06/10		251.5	−22.2				M (163)
	$N=$–.–	2.852	0.693	1.7	74.7	250.0	243.7	−22.1	+0.90	−0.10	23.0	GK
	Fireballs	AD 995–1076:			06/14–06/19		246.8	−26.1				AT (18)
#61	**TAH**	**τ-Herculids**			**05/19–06/14**		**June 03**	**72**	–.–	(11)	<2	
	Outbursts (see Tab. 6g)											
	$N=14$	2.695	0.970	18.6	204.2	72.6	228.5	+39.8	+0.9	−0.1	15	L71B
	$N=8$	2.90	0.986	20.7	203.6	80.8	236.4	+40.8	–.–	–.–		L71A
	Parent	*73P/Schwassmann–Wachmann 3*			*(Epoch 1930-07-03)*			$\lambda_{\odot}=78.02$	$T_J=+2.78$, $H_{10}=+12.0$			
	1930 VI	*3.081*	*1.006*	*17.384*	*192.021*	*78.024*	*219.7*	*+44.6*	*–.–*	*–.–*	*13.79*	*0.0056*
	(2022)	*3.092*	*0.983*	*11.296*	*202.176*	*67.023*	*210.0*	*+27.4*	*–.–*	*–.–*	*12.36*	*0.0139*
	(2027)	*3.060*	*0.921*	*6.276*	*218.183*	*48.681*	*200.4*	*+10.2*	*–.–*	*–.–*	*13.50*	*0.0068*
	(2049)	*3.039*	*0.899*	*6.329*	*222.453*	*44.645*	*200.0*	*+09.3*	*–.–*	*–.–*	*14.25*	*0.0134*
#162	ACI	α-Circinids			**06/04–06/04**		**June 04**	**73.9**	–.–	–.–	–.–	
	Outburst	Belinda Bridge					**1977 June 04**	**74.02**	–.–	0.008	>100	PJ
	(1977)	(100)	0.855	33.4	46.8	253.92	218.6	−70.2	–.–	–.–	(27.1)	Ch. 13
#163	*(SAG)*	*μ-Sagittariids*			**04/21–07/12**		**June 28**	**97**	2.8	~16	(2.3)	PJ
		= Ophiuchids, ecliptic antihelion source										
#67	NSA	N. μ-Sagittariids			**05/26–07/10**		**June 09**	**78.0**	2.8	12	1.5	RK
		= θ-Ophiuchids, ecliptic antihelion source										
	$N=$–.–	2.390	0.566	4.5	271.2	91.7	271.9	−17.3	+0.73	+0.07	22.9	PG

	N = 28	2.224	0.503	0.3	279.3	86.2	270.3	−23.1	−.−	−.−	24.3	ZS
	N = −.−	2.415	0.544	3.2	274.8	87.5	270.7	−20.0	−.−	−.−	23.4	T89 (33)
	N = 30	1.858	0.665	1.5	263.3	99.4	275.5	−20.9	−.−	−.−	18.3	ZS
	Parent??	*D/1770 L_1 (Lexell)*		*(Epoch 1770-08-14)*				*$\lambda_\odot$ = 101.1*	*T_J = +2.61, H_{10} = −.−*			
	(1770)	*3.15*	*0.674*	*1.325*	*256.400*	*103.095*	*275.6*	*−21.3*	−.−	−.−	*20.94*	*0.0143*
#69	SSG	S. μ-Sagittariids			**05/26–07/12**		**June 09**	**78.0**	2.8	12	1.5	RK
		= θ-Ophiuchids, ecliptic antihelion source										
	N = −.−	2.142	0.650	2.6	84.3	279.4	276.4	−27.5	+0.73	+0.07	19.6	PG
	N = 4	2.33	0.52	1.0	97	260.7	262.8	−25.0	−.−	−.−	23.5	GE (6.01)
	N = 4	2.90	0.460	4.2	101.4	263.0	267.8	−28.0	−.−	−.−	26.7	CL
	N = 31	1.908	0.384	2.5	113.8	271.1	283.3	−25.1	−.−	−.−	26.8	ZS
	N = 11	1.981	0.430	3.9	108.4	280.2	290.8	−26.0	−.−	−.−	25.6	ZS
	N = 6	2.797	0.405	4.7	108.0	258.7	266.8	−28.0	−.−	−.−	28	L71B
	N = 3	2.08	0.52	5.0	97.0	266.0	268.0	−27.7	−.−	−.−	23.3	NL (61.6.9)
	Fireballs	AD 1080–1098:			06/22–07/17		283.8	−22.9				AT (23)
	Visual				06/28–07/05		286.5	−24.9				M (#211)
	Visual				06/08–06/15		277.6	−27.9				M (#195)
#295	*(JAQ)*	*June Aquilids*										
#164	NZC	N. June Aquilids			**06/09–07/02**		**June 17**	**86**	−.−	(9)	3	
	N = 19	1.364	0.150	45.0	325.0	81.7	292.7	−04.9	−.−	−.−	37.0	JVF
	N = 35	1.348	0.114	39.3	329.5	86.5	298.3	−07.1	−.−	−.−	36.3	ZS
	N = 11	1.143	0.152	43.1	326.7	99.7	311.1	+0.7	−.−	−.−	33.8	ZS
	N = 13	1.5	0.15	39.5	324.0	80.7	290.4	−05.8	−.−	−.−	36.3	GE (6.09)
	N = 2	1.8	0.18	37.9	318	78.7	285.7	−05.9	−.−	−.−	36.3	GE (6.33)
	N = 4	1.61	0.11	40.1	328.9	85.0	294.6	−08.3	−.−	−.−	38.5	NL (61.6.4)
	Visual				06/10–06/19		292.4	−11.8				M (#201)
#165	SZC	S. June Aquilids			**06/09–06/12**		**June 11**	**80**	−.−	(1.2)	1.43	GK
		= alpha Microscopiids										
	N = 4	1.15	0.11	33.5	152	259.7	297.8	−33.9	−.−	−.−	33.2	GE (6.08)
#166	JLY	June Lyrids			**06/11–06/21**		**June 15**	**85.167**	2.7	23	3.1	MK
		= *ξ-Draconids*										
	Outburst					**1996**	**June 15**	**85.167**	2.7	0.017	17	M96
	(1966)					$\lambda_\odot \sim$ 85.12	278.5	+30.0	−.−	−.−	−.−	Ch. 13
	(1996)	(100)	0.994	52.2	196.9	85.167	280.3	+55.0	−.−	−.−	(33.4)	Ch. 13

	Code	Name			Dates (2000)		Peak	$\lambda_\odot^{max}$	χ	W	ZHR^{max}	Notes
		a	q	i	ω	Node	R.A.	Decl.	ΔR.A./Δλ	ΔDecl./Δλ	V_g	Δ_{C-E}
	(1966)	(10)	0.84	50	231	84.5	278.4	+35.0	–.–	–.–	(31)	AC
	$N=11$	2.054	0.912	45.3	224.1	86.2	281.9	+43.8	–.–	–.–	27.1	ZS
	$N=$–.–	3.58	0.98	50	201	64	276.3	+52.0	–.–	–.–	32.4	T68
	$N=$–.–	Inf.	0.98	56	202	76	275.3	+50.0	–.–	–.–	39.3	T68
#296	*(SIS)*	*σ-Sagittariids*										
#167	NSS	N. σ-Sagittariids			**06/01–07/28**		**June 24**	**92.3**	–.–	(23)	–.–	
		= ρ-Sagittariids										
	$N=45$	1.133	0.332	8.2	309.4	92.3	293.1	−14.0	–.–	–.–	23.2	ZS
	$N=4$	1.72	0.38	4.5	296.6	86.6	278.4	−20.0	1.3	–.–	26.5	NL (61.6.6)
	$N=18$	1.970	0.386	3.9	293.3	73.1	263.5	−19.8	–.–	–.–	27.1	ZS73
	$N=4$	2.811	0.430	13.0	284.2	86.5	275.4	−11.0	–.–	–.–	28	L71B
	Parent?	*2001 ME$_1$*		*(Epoch 2005-01-30)*				*$\lambda_\odot=93.13$*	*$T_J=+2.67$, $H_N=+16.81$*			
	(2005)	*2.651*	*0.356*	*5.748*	*293.601*	*93.132*	*284.6*	*−18.1*	*–.–*	*–.–*	*29.70*	*+0.011*
#168	SSS	S. σ-Sagittariids			**06/01–07/28**		**June 18**	**87**	–.–	(23)	–.–	
		= γ-Sagittariids										
	$N=29$	2.594	0.361	2.8	113.6	267.4	278.6	−25.3	–.–	–.–	29.3	ZS73
	$N=2$	2.2	0.26	5.8	127	260.7	278.8	−27.0	–.–	–.–	31	GE (6.22)
	$N=3$	3.33	0.33	4.1	113.7	265.7	276.0	−24.5	–.–	–.–	31.1	NL (61.6.10)
	Fireballs	AD 1038–1078:			05/25–05/31		273.7	−28.0				AT (17)
#169	SCU	Scutids			**06/07–07/22**		**June 27**	**94.9**	–.–	(18)	–.–	
	$N=32$	1.361	0.599	13.7	278.8	95.6	281.7	+00.9	–.–	–.–	18.9	ZS
	$N=9$	1.367	0.615	10.3	276.8	93.4	277.4	−06.4	–.–	–.–	17.5	ZS73
	$N=2$	2.126	0.606	15.5	268.5	97.7	279.3	−02.0	–.–	–.–	18.9	L71B
#63	**COR**	**Corvids**			**06/04–06/30**		**June 26**	**94.9**	–.–	–.–	–.–	
	Outburst	seen by C. Hoffmeister				1937,	June	$\lambda_\odot=94.9$	1.9	(10)	13	H48
	(1937)	(3.0)	1.013	3.1	7.7	274.9	192.6	−19.4	–.–	–.–	–.–	H48
	$N=$–.–	3.00	1.014	2.9	7.1	275.0	191.4	−18.2	–.–	–.–	9.10	KP

	visual	2.5–10	1.012	3.5	7.7	274.9	192.6	−19.4	–.–	–.–	10 ± 2	AC
	(1953)	2.90	0.99	0.0	201	82.7	206	−09	–.–	–.–	10.0	HV7862
	?Fireballs	AD 1079–1111:			06/20–06/21		214	−14				AT (22)
	Parent?	*2004 HW*		*(Epoch 2006-03-06)*				$\lambda_{\odot}=77.41$	$T_J=+3.04,\ H_N=+17.13$			
	(2006)	*2.689*	*0.977*	*0.666*	*25.406*	*257.290*	*205.0*	*−13.0*	–.–	–.–	*10.23*	*0.0088*
#170	**JBO**	**June Bootids**			**06/26–06/30**		**June 28**	**96.3**	–.–	(1.6)	<1	
	Outbursts	(see Tab. 6b)										
	(1998)	3.3	1.016	18.4	183.65	96.046	222.88	+47.90	–.–	–.–	14.1	EN270698
	Error	±0.3	±0.000	±0.4	±0.07	±0.000	±0.16	±0.06	–.–	–.–	±0.4	
	(1998)	–.–	–.–	–.–	–.–	96.30	219.0	+49.0	+0.60	−0.40	13.9	TOM
	(1995)	3.157	1.016	26.7	182.1	92.633	237.6	+59.5	–.–	–.–	17.8	JN950624
	$N=$–.–	2.140	1.004	18.4	194.2	75.8	229.4	+49.8	–.–	–.–	12.8	T89 (34)
	$N=54$	2.479	1.000	21.7	168.0	91.2	208.9	+65.7	–.–	–.–	15.1	ZS
	(1952)	2.89	1.01	21	189	83.7	227.4	+50.8	–.–	–.–	14.8	HV4106
	Parent	*7P/Pons–Winnecke*		*(Epoch 1915-09-20)*				$\lambda_{\odot}=100.52$	$T_J=+2.67,\ H_{10}=+7.2$ *(var.)*			
	(1915)	*3.261*	*1.013*	*18.305*	*172.414*	*100.518*	*208.9*	*+53.0*	–.–	–.–	*14.24*	*−0.0419*
	(1995)	*3.435*	*1.013*	*22.301*	*172.648*	*93.098*	*214.0*	*+61.1*	–.–	–.–	*16.05*	*0.2422*
	(2039)	*3.359*	*1.016*	*17.194*	*177.484*	*89.230*	*207.6*	*+51.8*	–.–	–.–	*13.62*	*0.0339*
												Helion
#171	**ARI**	**Daytime Arietids**			**05/22–07/02**		**June 07**	**76.7**	2.7	6.6 ± 2.0	54 ± 12	PJ
	$N=1$	2.581	0.064	36.963	25.376	(80.411)	44.9	+25.7	–.–	–.–	42.1	U
	$N=48$	1.376	0.085	25.0	25.9	77.6	40.2	+23.8	–.–	–.–	35.7	ZS
	$N=55$	1.750	0.094	27.9	29.5	78.7	43.7	+25.9	–.–	–.–	37.6	ZS73
	$N=32$	2.0	0.08	17.4	28	81.7	49.7	+23.2	–.–	–.–	39.4	GE (6.05)
	$N=380$	1.67	0.10	18.7	29.9	77.3	44.0	+23.1	–.–	–.–	36.9	KL (14)
	$N=7$	2.27	0.05	38.9	20.3	85.3	47.3	+25.2	+1.5	–.–	42.8	NL (61.6.1)
	$N=8$	1.49	0.06	33.4	23.0	85.5	46.8	+26.3	–.–	–.–	38.8	NL (61.6.2)
	(1951)	1.61	0.090	21	29	77.5	44.1	+24.1	+0.70	+0.60	37	L54
	(1950)	1.49	0.10	18	29	77.7	44.5	+22.1	–.–	–.–	36	L54
	Parent	*Marsden-group of Sun skirters (Epoch 2004)*						$\lambda_{\odot}=81.5$	$T_J\sim+1.8,\ H_{10}>+18$			
	(2004)	*Inf.*	*0.0480*	*26.800*	*23.240*	*81.460*	*49.3*	*+23.2*	–.–	–.–	*45.68*	*0.006*
	(2004)	*3.33*	*0.0483*	*26.800*	*23.240*	*81.460*	*47.8*	*+23.5*	–.–	–.–	*43.03*	*0.000*

Table 7 (*cont.*)

	Code	Name			Dates (2000)		Peak	$\lambda_\odot^{max}$	χ	W	ZHR^{max}	Notes
		a	q	i	ω	Node	R.A.	Decl.	ΔR.A./Δλ	ΔDecl./Δλ	V_g	Δ_{C-E}
#172	**ZPE**	**Daytime ζ-Perseids**			**05/20–07/05**		**June 09**	**78.6**	2.7	(18)	~20	
		= *part of Taurid complex*, twin of STA										
	$N=1$	1.283	0.2860	5.506	50.012	(80.414)	58.3	+24.7	–.–	–.–	26.2	U
	$N=56$	1.492	0.365	6.5	60.5	81.5	64.5	+27.5	–.–	–.–	25.1	ZS
	$N=73$	1.918	0.319	5.3	59.2	78.3	61.1	+24.9	–.–	–.–	29.3	ZS73
	$N=6$	1.72	0.30	7.1	69	81.7	65.8	+27.1	–.–	–.–	27.8	GE (D.01)
	$N=27$	1.72	0.31	4.8	56	71.7	51.7	+22.2	–.–	–.–	27.8	NL
	$N=57$	1.74	0.33	3.2	59	73.7	55.7	+21.2	–.–	–.–	26.8	BF66
	$N=60$	1.55	0.31	5.7	57.0	71.3	52.2	+23.1	–.–	–.–	27.6	KL (12)
	$N=9$	1.67	0.30	5.7	55.4	84.5	65.0	+25.5	–.–	–.–	28.4	NL (61.6.3)
	$N=$–.–	2.33	0.35	8.0	61	77.0	63.8	+27.1	+1.1	+0.4	28.8	L54
	Parent?	*2P/Encke*		*(Epoch 1974-04-23)*				$\lambda_\odot = 98.6$	$T_J = +3.03$, $H_{10} = +11.5$ *(var.)*			
	(1974)	*2.217*	*0.338*	*5.793*	*62.907*	*98.539*	*85.6*	*+28.0*	–.–	–.–	*29.18*	*0.1778*
#173	**BTA**	**Daytime β-Taurids**			**06/05–07/17**		**June 28**	**96.7**	–.–	(17)	~10	
		= *part of Taurid complex*, twin of NTA										
	$N=41$	1.653	0.274	0.3	52.3	102.7	84.9	+23.5	–.–	–.–	29.0	ZS
	$N=57$	1.853	0.325	2.2	239.2	275.2	80.2	+21.4	–.–	–.–	28.2	ZS73
	$N=$–.–	2.2	0.34	6	246	276.4	86.7	+19.0	+0.8	+0.4	30	AC
	$N=2$	1.7	0.31	4.8	237	261.7	65.7	+18.1	–.–	–.–	27.8	GE (6.21)
	$N=7$	2.17	0.46	3.7	255.1	264.9	76.2	+20.4	+1.0	–.–	25.5	NL (61.6.5)
	$N=$–.–	2.2	0.34	6	224	278.1	87.3	+19.3	+0.8	+0.4	31.4	L54
	Parent	*2004 TG_{10}*		*(Epoch 2006-03-06)*				$\lambda_\odot = 101.88$	$T_J = +3.56$, $H_N = +19.46$			
	(2006)	*2.242*	*0.315*	*1.291*	*240.428*	*281.864*	*88.4*	*+22.5*	–.–	–.–	*29.90*	*0.0614*
#174	TAS	Daytime Aurigids			**06/09–07/25**		**June 27**	**96.0**	–.–	(18)	–.–	
		= *ecliptic helion source*										
	$N=$–.–	1.317	0.631	4.4	83.6	96.0	93.3	+31.3	–.–	–.–	15.8	ZS

July:

												Apex
#175	JPE	July Pegasids			**07/07–07/13**		**July 09**	**107.5**	3.0	3.5	2.7	IMO
	Visual	(44)	0.536	131.6	267.2	107.5	340.0	+15.0	+0.6	+0.2	61.3	PJ
	$N=1$	Inf.	0.64	131	243	113.7	339	+17	–.–	–.–	75.3	HV8047
	Parent??	*C/1979 Y_1 (Bradfield)*		*(Epoch 1980-01-24)*				*$\lambda_{\odot}=+108.61$*	*$T_J=\sim0.66$, $H_0=+11.8$ ($n=4.9$), $P=291$ yr*			
	(1980)	*45.017*	*0.565*	*146.367*	*263.932*	*108.611*	*346.5*	*+11.2*	*–.–*	*–.–*	*63.99*	*0.0673*
#176	PHE	July (γ-)Phoenicids			**06/24–07/18**		**July 12**	**110.3**	3.0	2.5 ± 0.5	4.0 ± 1.9	PJ
	$N=$–.–	2.5-Inf.	0.96	82–87	31–24	290.3	31.6	−47.7	+1.04	+0.53	47 ± 3	WE60
#177	BCA	β-Cassiopeids			**07/03–08/19**		**July 30**	**126.6**	3.43	5.0	10	D00
						$\lambda_{\odot}=125$	14	+63	–.–	–.–	60	
	$N=6$	18.194	0.995	89.1	192.1	125.5	353.5	+59.1	+1.04	+0.20	50.4	PG
	$N=12$					125.2	335.5	+53.3	–.–	–.–		T65
												Antihelion
#178	JCE	July Centaurids			**07/01–07/07**		**July 05**	**103.3**	–.–	(2.4)	–.–	
	Outburst	07:45–09:00 UT Sydney, 50 short swift, none next hr						$\lambda_{\odot}=103.282$	–.–	<0.030	>40	
	(1896)	(Inf.)	0.967	6.3	25.4	283.6	~225	−35	–.–	–.–	15.3	PJ
	$N=$–.–	4.170	0.944	4.9	32.8	276.9	233.7	−10.2	–.–	–.–	12.8	T89 (35)
	Parent??	*1997 QK_1*	*(Epoch 2005-01-30)*					*$\lambda_{\odot}=116.44$*	*$T_J=+2.98$, $H_N=+20.10$*			
	(2005)	*2.797*	*1.006*	*2.830*	*13.255*	*296.436*	*221.9*	*−28.5*	*–.–*	*–.–*	*9.11*	*0.0079*
#179	SCA	σ-Capricornids			**06/18–07/30**		**July 12**	**110**	–.–	(17)	–.–	
	$N=$–.–	1.726	0.491	3.9	284.5	112.9	302.7	−15.9	–.–	–.–	22.5	T89
	$N=40$	1.310	0.272	4.5	311.2	110.2	311.1	−14.5	–.–	–.–	26.9	ZS
	$N=35$	1.782	0.431	2.1	290.3	107.6	298.7	−18.7	–.–	–.–	24.6	ZS73
	$N=5$	2.86	0.37	3.9	289.8	117.8	307.6	−15.2	1.0	–.–	28.9	NL (61.7.8)
	$N=3$	1.56	0.23	6.9	312.5	126.0	326.9	−12.1	–.–	–.–	30.0	NL (61.7.11)
#180	MSE	μ-Serpentids			**07/03–07/31**		**July 16**	**114**	–.–	(11)	–.–	
	$N=$–.–	1.879	0.994	4.1	–.–	–.–	240.0	+11.0	–.–	–.–	7.7	J92
	$N=$–.–	2.447	0.992	3.3	197.3	115.0	232.7	−4.2	–.–	–.–	8.6	T89 (37)
#181	KPA	κ-Pavonids			**07/17–07/17**		**July 17**	**114.8**	–.–	–.–	–.–	
	Outburst						1986:	114.827	2.2	0.027	~60	PJ
	(1986)	(34.2)	0.865	25.6	45.7	294.83	282.1	−68.0	–.–	–.–	23.5	PJ
	(1986)	(3.2)	0.874	21.3	48.1	294.80	282.3	−66.9	–.–	–.–	18.9	PJ

	Code	Name			Dates (2000)		Peak	$\lambda_\odot^{max}$	χ	W	ZHR^{max}	Notes
		a	q	i	ω	Node	R.A.	Decl.	ΔR.A./$\Delta\lambda$	ΔDecl./$\Delta\lambda$	V_g	Δ_{C-E}
#88	ODR	*o*-Draconids			**07/06–07/24**		**July 17**	**115.5**	–.–	(7)	–.–	
	$N=3$	3.52	1.01	43	190	113.7	279.7	+61.8	–.–	–.–	28.6	CL
	$N=14$	4.329	1.006	46.2	192.2	115.5	285.0	+61.3	–.–	–.–	28.5	ZS
#182	OCY	*o*-Cygnids			**07/17–07/22**		**July 19**	**117.2 ± 0.5**	2.7	4.9 ± 1.2	2.5 ± 0.8	PJ
	Visual	= ψ-Cygnids ≠ α-Cygnids				$\lambda_\odot = 117.2$	304	+48	–.–	–.–	–.–	PJ
	Photo					–.–	304.5	+49.7	–.–	–.–	39.4	BK65
	$N=1$	4.57	0.97	60	206	112.7	306	+55	–.–	–.–	35.5	HV8018
#1	**CAP**	**α-Capricornids**			**07/19–08/18**		**July 30**	**127 ± 1**	2.2 ± 0.3	15.1 ± 2.6	2.0	PJ
								128 ± 1	3.2 ± 0.2	15.4	2.5	IMO
	$N=36$	2.618	0.602	7.68	266.67	128.9	306.6	−08.2	+0.54	+0.25	22.2	This work
	σ	±0.44	±0.037	±0.6	±4.4	±4.4	±2.6	±1.5	±0.02	±0.02	±1.5	
	$N=269$	2.155	0.550	7.7	273.3	122.3	306.7	−09.3	+0.91	+0.25	23.4	GB
	σ	±0.40	±0.036	±1.2	±5.3	±5.2	±2.7	±1.3	±0.02	±0.01	±1.7	
	$N=$–.–	2.540	0.594	7.2	267.6	123.8	303.4	−10.6	+0.75	+0.28	22.2	H02
	$N=15$	2.283	0.626	4.9	266.2	138.5	315.9	−08.7	–.–	–.–	20.6	PG
	$N=$–.–	2.038	0.544	7.0	275.9	123.5	306.4	−09.9	–.–	–.–	22.5	G
	$N=$–.–	2.636	0.580	6.0	268.0	134.7	314.7	−08.8	–.–	–.–	23.0	JVF
	$N=44$	1.920	0.620	6.1	267.9	136.6	315.9	−07.1	–.–	–.–	19.7	ZS
	$N=28$	1.850	0.630	0.9	267.2	147.5	327.1	−11.7	–.–	–.–	18.8	ZS73
	$N=18$	2.524	0.592	7.1	267.9	126.1	305.4	−09.6	–.–	–.–	25.0	L71B
	$N=21$	2.565	0.590	7.0	269.0	127.7	308.4	−09.6	+0.9	+0.3	22.8	AC
	Parent	*169 P/NEAT*	*(= 2002 EX_{12})*		*(Epoch 2005-09-18)*			$\lambda_\odot = 128.78$	$T_J = +2.89$, $H_N = +26.49$			
	(2005)	*2.603*	*0.605*	*7.622*	*265.952*	*128.784*	*306.6*	*−08.4*	*–.–*	*–.–*	*22.21*	*0.1434*
#183	PAU	Piscis Austrinids			**07/09–08/17**		**July 27**	**123.7 ± 0.7**	3.2	2.4 ± 0.5	2.9 ± 0.8	PJ
	$N=32$	4.31	0.17	45	114	303.7	340.7	−25.7	+0.90	+0.40	40.5	KL (32)
	radar					304	331.7	−17.8	+1.5	−0.3	–.–	WE60
	radar						328.7	−26.8	–.–	–.–	–.–	ER55
	Visual				07/26–08/08		338.4	−32.5				M (#290)
	Visual				07/14–07/22		331.9	−29.5				M (#274)

#184	GDR	γ-Draconids			**07/21–07/29**		**July 26**	**124.4**	–.–	(3.2)	–.–	
	$N=3$	6.4	0.972	38.7	203.1	124.4	278.8	+48.8	–.–	–.–	25.1	B63
	$N=1$	Inf.	0.973	43.6	203.7	117.3	281.0	+50.7	–.–	–.–	29.5	HV8089
#297	*(DAQ)*	*δ-Aquariids*										
#26	**NDA**	**Northern δ-Aquariids**			**07/15–08/25**		**July 26**	**123.4 ± 1.0**	3.3	11.8 ± 2.3	1.0 ± 0.2	PJ
		Part of Machholz complex					**Aug. 08**	**136.0**	3.4	20	3.5 ± 0.5	IMO
	$N=$–.–	2.536	0.071	23.0	332.6	139.0	344.7	+00.4	–.–	–.–	40.5	DMS
	$N=8$	2.80	0.096	19.8	327.4	140.7	345.3	+00.5	+0.75	+0.21	39.78	KP
	$N=5$	3.246	0.089	18.9	328.4	136.5	342.0	−01.6	–.–	–.–	40.5	PG
	$N=50$	2.16	0.06	18.0	338.8	128.5	337.4	−04.6	–.–	–.–	40.3	KL (26)
	$N=9$	2.102	0.085	20.7	330.8	141.2	347.6	+01.3	–.–	–.–	38.4	L71B
	$N=3$	3.57	0.18	21.8	311.2	122.1	320.3	−04.0	–.–	–.–	36.9	NL (61.7.9)
	$N=3$	2.43	0.126	21.2	324.4	143.9	347.3	+02.9	–.–	–.–	39.8	SH
	$N=9$	2.62	0.070	20.4	332.6	139.6	339.6	−04.7	+1.0	+0.2	42.3	J
#5	**SDA**	**Southern δ-Aquariids**			**07/08–08/19**		**July 29**	**125.6 ± 0.3**	3.3	13 ± 2	18 ± 4	PJ
		= part of Machholz complex										
	$N=$–.–	3.107	0.087	26.4	148.9	312.2	342.1	−15.4	–.–	–.–	40.5	DMS
	$N=2413$	1.971	0.067	30.8	154.5	307.2	340.4	−16.3	+0.73	+0.26	40.2	GB
	σ	–.–	±0.021	±9.8	±4.1	±5.8	±2.9	±1.5	±0.02	±0.01	±2.7	
	$N=99$	2.667	0.080	27.0	151.0	308.7	340.7	−15.7	–.–	–.–	41.0	JVF
	$N=10$	3.228	0.097	25.8	147.1	314.7	345.7	−14.6	–.–	–.–	40.5	PG
	$N=70$	1.630	0.069	28.2	155.4	306.4	343.0	−15.6	–.–	–.–	38.2	ZS
	$N=13$	2.643	0.074	28.4	151.6	307.8	341.4	−15.4	–.–	–.–	41.5	L71B
	$N=28$	2.80	0.079	25.5	150.6	306.7	339.6	−16.1	+0.75	+0.21	41.01	KP
	$N=151$	2.04	0.08	28.4	151.1	307.5	341.9	−16.1	–.–	–.–	39.6	KL (30)
	$N=$–.–	1.612	0.079	27.5	153.5	305.8	341.6	−15.9	–.–	–.–	38.5	G
	$N=13$	2.875	0.069	27.2	152.8	305.7	334.5	−16.1	+0.80	+0.18	41.4	AC
	$N=48$	2.33	0.07	32.5	152.4	306.5	340.1	−17.0	+0.9	+0.2	40.8	NL (61.7.1)
	$N=3$	2.63	0.05	50.5	155.0	302.7	340.1	−19.1	–.–	–.–	43.5	NL (61.7.2)
	radar	(2.87)	0.058	94.5	154.8	303.1	348.7	−17.7	+1.4	+0.1	(48.3)	WE60
	–.–	2.83	0.074	26.6	151.6	308.6	333.2	−17.6	–.–	–.–	42.7	J
	Parent	*Related to: Marsden Sungrazers (Epoch 2004)*						$\lambda_{\odot} = 123.3$	$T_J \sim +1.84$, $H_N > +20$			
	(2004)	*(3.265)*	*0.0480*	*26.94*	*156.881*	*303.268*	*337.0*	*−15.5*	–.–	–.–	*43.00*	*0.3347*

	Code	Name / a	q	i	Dates (2000) / ω	Node	Peak / R.A.	$\lambda_{\odot}^{max}$ / Decl.	χ / ΔR.A./$\Delta\lambda$	W / ΔDecl./$\Delta\lambda$	ZHR^{max} / V_g	Notes / Δ_{C-E}
												Helion
#185	DBA	D. (β-)Andromedids			**07/01–07/03**		**July 02**	**100.5**	–.–	(2)	–.–	
	$N=13$	0.599	0.139	59.4	8.9	100.5	15.0	+37.5	–.–	–.–	26.0	ZS
	Parent?	*96P/Machholz (Epoch 2002-07-25)*						$\lambda_{\odot}=94.61$	$T_J=+1.94$, $H_{10}=+12.5$ (*var.*)			
	(2002)	*0.731*	*0.057*	*60.181*	*14.584*	*94.609*	*37.7*	*+30.3*	–.–	–.–	*33.25*	*0.3434*
#186	EUM	ε-Ursae Majorids			**07/1–07/15**		**July 08**	**106**	–.–	(6)	–.–	
	$N=$–.–	2.994	0.980	20.0	156.3	105.8	192.5	+61.7	–.–	–.–	15.2	T89 (36)
#187	PCA	ψ-Casseiopeids			**07/15–07/19**		**July 17**	**114.8**	–.–	(2)	–.–	
	$N=25$	2.418	0.821	72.1	121.2	114.4	389.4	+71.5	–.–	–.–	40.3	ZS
	$N=23$	3.65	0.87	71	133	116.7	381.0	+75.3	–.–	–.–	40.5	KL (25)
	$N=41$	1.72	0.87	77	126	115.7	373.8	+66.3	–.–	–.–	40.5	KL (23)
	$N=23$	1.09	0.90	83	107	114.7	360.6	+56.3	–.–	–.–	39.4	KL (31)
	Parent	*1973 NA*			*(Epoch 2006-03-06)*			$\lambda_{\odot}=101.09$	$T_J=+2.53$, $H_N=+15.30$			
	(2006)	*2.434*	*0.825*	*67.674*	*121.797*	*100.275*	*12.2*	*+69.0*	–.–	–.–	*38.40*	*0.0854*
#188	XRI	Daytime ξ-Orionids			**07/04–07/29**		**July 20**	**117.7**				
	$N=3$	8.33	0.08	32.8	211.6	301.9	94.4	+15.0	–.–	–.–	44.0	NL (61.7.5)
	$N=23$	5.18	0.18	16	228	297.7	98.7	+16.0	–.–	(10)	38.4	KL (22)
#189	DMC	Daytime μ-Cancrids			**07/03–08/11**		**July 30**	**126.0**	–.–	(16)	–.–	
		= helion source										
	$N=43$	1.720	0.443	2.1	71.0	128.0	122.7	+22.4	–.–	–.–	24.3	ZS
	$N=5$	2.50	0.25	5.1	53.2	124.1	108.4	+24.8	+1.0	–.–	31.1	NL (61.7.6)
	August:											
												Apex
#190	BPE	β-Perseids			**08/07–08/08**		**Aug. 08**	**135.4**	–.–	–.–	–.–	
	Outburst	High rate of faint meteors					1935 Aug. 08	135.358	>4	0.011 ±0.004	High	PJ
	(1935)	(35.6)	0.863	142.7	134.3	135.4	52.8	+40.2	–.–	–.–	(66.2)	Ch. 13
	$N=1$	4.96	1.002	168.86	143.23	134.47	46.5	+44.8	–.–	–.–	63.8	DMS

#191	ERI	η-Eridanids			**08/03–08/14**		**Aug. 10**	**137.5**	–.–	(4)	<6	
	N = –.–	20.26	0.961	130.4	26.6	317.490	45.0	−12.9	–.–	–.–	64.0	O
	Parent?	*C/1852 K_1 (Chacornac)*		*(Epoch 1852-04-20)*				$\lambda_\odot$ = *139.38*	T_J = *~−0.74*, H_{10} = *+9.8 (n = 4.0), P unknown*			
	(1852)	*Inf.*	*0.910*	*131.097*	*37.367*	*319.380*	*42.9*	*−12.6*	*–.–*	*–.–*	*64.28*	*0.0052*
#192	TRI	Aug. Triangulids			**08/05–08/14**		**Aug. 12**	**139.8**	–.–	(4)	–.–	
	N = 7	1.062	0.886	150.0	264.7	139.8	37.5	+30.7	–.–	–.–	57.3	ZS
	N = 18	1.45	0.67	152	92	132.7	53.8	+33.2	–.–	–.–	59	KL (33)
	N = 15	1.52	0.98	141	173	137.7	38.8	+37.2	–.–	–.–	60	KL (35)
	Possibly related to #190 (BPE)											
#7	**PER**	**Perseids**			**07/17–08/24**		**Aug. 13**	**140.19**	2.5	3.01 ± 0.16	84 ± 5	PJ
	Outbursts (see Tab. 5c)											
	N = 87	71.4	0.953	113.22	151.3	140.19	48.33	+57.96	–.–	–.–	59.38	DMS
	σ (intrinsic)	>50	±0.009	±1.5	±2.3	±3.3	–.–	–.–	–.–	–.–	–.–	
	N = –.–	24.0	0.949	113.0	150.4	139.7	46.8	+57.7	+1.38	+0.18	59.49	KP
	N = 193	62.50	0.948	113.1	151.3	138.0	45.3	+57.7	–.–	–.–	59.5	PG
	N = –.–	25.366	0.963	112.8	154.5	135.6	38.9	+56.2	–.–	–.–	59.3	T89 (41)
	N = 8	8.04	0.960	110.2	152.5	140.5	46.2	+59.4	–.–	–.–	57.5	ZS
	N = 10	11.0	0.95	113.1	150.9	137.2	41.9	+56.8	–.–	–.–	58.8	KL (34)
	N = 20	22.527	0.934	113.2	147.9	139.5	46.9	+57.2	–.–	–.–	59	L71A
	N = –.–	22.53	0.934	113.2	147.9	139.4	47.0	+58.2	–.–	–.–	59	JW
	Parent	*109P/Swift–Tuttle*		*(Epoch 1995-10-10)*				$\lambda_\odot$ = *138.38*	T_J = *−0.274*, H_{10} = *+4.0 (n = 4.0)*			
	(1995)	*26.092*	*0.959*	*113.454*	*152.989*	*139.384*	*45.8*	*+57.7*	*–.–*	*–.–*	*59.41*	*0.0004*
#193	ZAR	ζ-Arietids			**08/13–08/25**		**Aug. 19**	**146**	–.–	–.–	–.–	
	N = 2	17.903	0.973	172.6	19.5	326.7	49.7	+14.2	–.–	–.–	70.1	L71B
	Visual					$\lambda_\odot$ = 146.9	04.6	+09.3	–.–	–.–	–.–	RM
	Fireballs	AD 1063–1077:			08/07–08/15		45	+15				AT (28)
	Parent?	*C/1862 N_1 (Schmidt–Tempel)*		*(Epoch 1862-07-04)*				$\lambda_\odot$ = *148.44*	T_J *~−1.19*, H_{10} = *+9.4 (n = 4.0), P unknown*			
	(1862)	*Inf.*	*0.956*	*172.109*	*27.166*	*328.437*	*49.5*	*+13.6*	*–.–*	*–.–*	*70.57*	*0.0136*
#194	UCE	υ-Cetids			**08/15–08/22**		**Aug. 19**	**146**	–.–	–.–	–.–	
	N = –.–	2.560	0.640	144.0	82.0	325.7	38.6	−2.8	–.–	–.–	61.0	JVF

Table 7 (*cont.*)

	Code	Name a	q	i	Dates (2000) ω	Node	Peak R.A.	$\lambda_\odot^{max}$ Decl.	χ ΔR.A./$\Delta\lambda$	W ΔDecl./$\Delta\lambda$	ZHR^{max} V_g	Notes Δ_{C-E}
#12	**KCG**	**κ-Cygnids**			**08/03–08/31**		**Aug. 20**	**145.2 ± 0.8**	2.2	8.8 ± 0.7	2.3 ± 0.4	PJ
					$\lambda_\odot = 145.0$		286	+59	+0.30	+0.10	24.8	IMO
	1993:	5.12	0.984	35.9	201.4	139.4	284.0	+52.7	-.-	-.-	24.0	DMS
	$N = 4$	±1.22	±0.005	±1.2	±1.5	±0.8	±0.8	±1.7			±1.1	
	1993:	4.10	0.957	34.7	206.2	140.4	287.1	+49.5	-.-	-.-	23.4	DMS
	$N = 5$	±0.57	±0.003	±1.3	±0.9	±0.5	±1.6	±0.8			±0.9	
	$N = 32$	3.19	0.995	32.6	197.4	140.9	276.9	+53.6	-.-	-.-	21.4	DMS
	σ	±0.91	±0.016	±3.7	±7.6	±5.9	±6.9	±6.8			±2.0	
	$N = 13$	3.513	0.991	33.0	197.7	142.0	278.4	+52.3	-.-	-.-	21.7	PG
	$N = 8$	2.583	0.979	42.9	203.1	1523.2	299.1	+63.2	-.-	-.-	25.9	ZS73
	$N = 3$	3.437	0.958	29.7	207.7	135.4	282.4	+42.1	-.-	-.-	20.1	L71B
	$N = 8$	3.09	0.99	38	194	145	286.2	+59.1	+0.0	+0.0	24.8	AC
	Fireballs	AD 1059–1098:			07/25–09/13		306.5	+40.2				AT (26 +38)
#200	ESE	η-Serpentids			**07/06–10/19**		**Aug. 24**	**151**	-.-	(39)	-.-	
		= *λ-Aquilids*										
	$N = 22$	1.795	0.956	4.5	213.2	149.5	287.7	−03.7	-.-	-.-	8.6	PG
	$N = $ -.-	1.792	0.916	5.2	209.5	151.7	277.7	−04.0	-.-	-.-	9.8	T89 (38)
	$N = $ -.-	2.294	0.898	2.8	39.7	356.2	322.4	−12.6	-.-	-.-	11.8	T89 (44Q)
	Parent?	*2000 QS$_7$*	*(Epoch 2005-01-30)*					*$\lambda_\odot = 149.69$ $T_J = +3.01$, $H_N = +19.55$*				
	(2005)	*2.682*	*0.905*	*3.187*	*222.591*	*149.690*	*297.8*	*−11.7*	-.-	-.-	*12.60*	*0.00437*
#201	GDO	γ-Doradids			**08/27–09/03**		**Aug. 29**	**155.7 ± 0.5**	**2.8**	3.7 ± 1.3	4.8 ± 1.6	PJ
	Visual	(28.8)	0.970	65.6	23.0	335.7	61.2	−50.2	+0.5	+0.2	(40.1)	PJ
												Helion
#202	ZCA	Daytime ζ-Cancrids			**08/07–08/22**		**Aug. 20**	**147**	-.-	-.-	-.-	
	$N = 3$	5.00	0.05	21.1	206.5	326.9	119.7	+19.0	-.-	-.-	43.8 (NL61.8.5)	
#203	GLE	Daytime γ-Leonids			**08/18–08/24**		**Aug. 22**	**148.7**	-.-	(3)	-.-	
		= *ecliptic helion source*										
	$N = 40$	1.527	0.569	2.1	262.0	321.8	139.9	+12.4	-.-	-.-	19.6	ZS

	$N=46$	0.968	0.860	6.4	59.0	149.2	156.6	+19.7	−.−	−.−	22.0	ZS
	$N=4$	2.38	0.60	7.5	90.6	148.7	153.4	+20.8	2.1	−.−	22.9	(NL61.8.4)
	Parent??	*1999 RD_{32}*		*(Epoch 2005-01-30)*				*$\lambda_\odot=155.3$*	*$T_J=+2.87$, $H_N=+16.32$*			
	(2005)	*2.640*	*0.605*	*6.147*	*274.453*	*335.309*	*156.1*	*+01.1*	*−.−*	*−.−*	*22.77*	*−0.0501*
#204	DXL	Daytime χ-Leonids = twin of SCC			**08/26–08/30**		**Aug. 27**	**154**	−.−	(3)	−.−	
	$N=25$	1.598	0.330	2.5	238.4	334.9	142.1	+12.8	−.−	−.−	27.4	ZS
	Parent?	*2001 YB_5*		*(Epoch 2005-01-30)*				*$\lambda_\odot=161.33$*	*$T_J=+2.89$, $H_N=+20.62$*			
	(2005)	*2.349*	*0.323*	*3.381*	*62.113*	*161.330*	*151.7*	*+14.3*	*−.−*	*−.−*	*30.46*	*0.0759*

September:

												Apex
#205	XAU	ξ-Aurigids			**08/30–09/03**		**Sep. 01**	**158**	−.−	(3)	−.−	
	$N=1$	28.24	0.971	106.4	128.0	157.5	93.5	+62.6	−.−	−.−	56.9	PK#620125
	$N=1$	11.49	0.793	117.5	123.8	159.4	92.6	+56.4	−.−	−.−	59.7	PK#621134
	$N=1$	2.50	0.785	111.5	117.1	160.4	92.3	+58.5	−.−	−.−	55.0	PK#621204
#206	**AUR**	**(α-) Aurigids** = θ-Aurigids			**08/25–09/05**		**Sep. 01**	**158.7**	2.6	2.7	4 ± 1	IMO
	Outburst				1994			158.700 ±0.020	1.7	0.019 ± 0.002	~400	PJ
	Outburst				1986		90.5	158.518 +34.6	1.3	0.018 ± 0.006	250 ± 30	PJ
	Outburst				1935		86.3	≥158.664 +41.0	2.2	0.021 ± 0.009	≥100	Ch. 13
	$N=3$	Inf.	0.683	148.5	110.2	158.7	89.8	+38.7	+1.35	+0.25	65.7	This work
	$N=$−.−	Inf.	0.802	146.4	121.5	158.6	85.5	+42.0	−.−	−.−	66.3	AC
	Parent	*C/1911 N_1 (Kiess)*		*(Epoch 1911-06-13)*				*$\lambda_\odot=158.98$*	*$T_J=-0.84$, $H_{10}=+7.9$*			
	(1911)	*184.6*	*0.684*	*148.421*	*110.378*	*158.978*	*91.6*	*+39.3*	*−.−*	*−.−*	*66.17*	*0.0027*
#207	SCS	Sep. (β-)Cassiopeids			**09/05–09/30**		**Sep. 16**	**173**	−.−	(10)	−.−	
	$N=8$	15.8	0.968	162.4	176.1	172.8	33.0	+68.9	−.−	−.−	69.1	This work
#208	SPE	Sep. (ε-)Perseids			**09/07–09/23**		**Sep. 12**	**170**	−.−	(6)	−.−	
	$N=8$	31.1	0.742	138.9	241.9	171.3	50.2	+39.4	−.−	−.−	64.5	This work

	Code	Name			Dates (2000)		Peak	$\lambda_{\odot}^{max}$	χ	W	ZHR^{max}	Notes
		a	q	i	ω	Node	R.A.	Decl.	ΔR.A./$\Delta\lambda$	ΔDecl./$\Delta\lambda$	V_g	Δ_{C-E}
	$N=10$	15.0	0.750	142.8	241.3	168.0	51.5	+39.5	–.–	–.–	65.6	R92
	σ	–.–	±0.033	±3.9	±3.6						±1.7	
	$N=3$	Inf.	0.733	140.5	242.6	166.9	47.2	+38.9	–.–	–.–	65.4	PG
	$N=$–.–	34.087	0.764	149.5	238.0	170.1	54.8	+36.2	–.–	–.–	67.1	T89 (46)
	Fireballs		AD 1046–1069:		09/03–09/09		33.8	+38.2				AT (40)
#209	EER	ε-Eridanids			**08/20–09/16**		**Sep. 12**	**170**	2.8	~12	1.5	PJ
		= π-Eridanids				$\lambda_{\odot}=170.0$	56.6	−13.8	+0.70	+0.20	59.0	IMO
	$N=1$	11.08	0.576	83.5	81.5	359.3	59.3	−17.4	–.–	–.–	52.6	HV83606
	Outburst				1981 Sep. 10.6			>168.12	–.–	0.06	>170	PJ
	Visual	(11.)	0.61	164	80	168.1	56.6	−13.8	–.–	–.–	~57	PJ
	Parent?	*C/1854 L_1 (Klinkerfues = 1854 III)*			*(Epoch 1854-06-23)*			*$\lambda_{\odot}=167.49$*	*$T_J=-0.11$, $H_{10}=+6.4$ ($n=4.0$)*			
	(1854)	*(25.27)*	*0.6481*	*108.110*	*73.871*	*347.487*	*53.5*	*−16.2*	*–.–*	*–.–*	*56.52*	*0.0076*
#210	BAU	β-Aurigids			**09/19–09/25**		**Sep. 21**	**179.3**	–.–	–.–	–.–	
	Outburst	19 meteors					1968 Sep. 21	179.258	2.3	–.–	–.–	
	(1968)					179.256	86	+43	–.–	–.–	Medium	Ch. 3
	$N=1$	4.57	0.98	148	159	176.7	93	+42	–.–	–.–	66.5	HV4453
	$N=1$	23.08	1.00	145	170	176.7	90	+45	–.–	–.–	67.9	HV4454
	$N=1$	Inf.	1.03	146	168	176.7	91	+44	–.–	–.–	68.8	HV4460
	$N=1$	8.04	0.97	147	157	182.7	102	+42	–.–	–.–	67.3	HV4554
	Parent?	*1790 A_1 (Hershel)*			*(Epoch 1790-01-16)*			*$\lambda_{\odot}=179.13$*	*$T_J<-0.70$, $H_{10}=$–.–*			
	(1790)	*Inf.*	*0.722*	*148.068*	*115.951*	*179.125*	*114.5*	*+38.2*	*–.–*	*–.–*	*66.76*	*0.0333*
#81	SLY	Sep. Lyncids			**09/26–09/29**		**Sep. 28**	**185.0**	–.–	(2)	–.–	
	$N=1$	9.05	0.880	138.0	221.9	187.0	81.2	+46.6	–.–	–.–	65.2	PK#572513
	$N=2$	76.970	0.770	136.5	152.5	185.9	110.9	+47.9	–.–	–.–	65.0	L71B
#211	AOR	Sep. α-Orionids			**09/21–09/28**		**Sep. 26**	**183**	–.–	(3)	–.–	
	$N=16$	1.52	0.91	152	47	2.7	86.7	+09.0	–.–	–.–	62	KL (37)
	$N=1$	1.2	0.68	157	91	3.7	80.7	+12.0	–.–	–.–	60.8	HV4597
	$N=1$	1.63	0.80	157	66	3.7	82.7	+11.0	–.–	–.–	63.4	HV4609

#212	**KLE**	***k*-Leonids**			**09/21–09/29**		**Sep. 24**	**181**	–.–	(3)	–.–	
		= twin shower of Dec. Monocerotids										
	$N=21$	48.0	0.11	26	39	180.7	162.7	+15.7	–.–	–.–	43.6	KL (39)
	$N=9$	17	0.17	24.8	31.9	184.2	162.2	+14.0	–.–	−0.2	44.8	NL (61.9.4)
	$N=3$	1.6	0.13	7.0	35.6	184.1	187.0	+5.8	–.–	–.–	35.4	NL (61.9.12)
	Parent	*C/1917 F_1 (Mellish)*		*(Epoch 1917-04-15)*				*$\lambda_\odot=168.9$*	*$T_J=+0.64$, $H_0=+7.4$ ($n=3.9$)*			
	(1917)	*27.65*	*0.190*	*36.339*	*51.122*	*169.558*	*155.9*	*+26.0*	–.–	–.–	*42.10*	*0.4955*
												Antihelion
#213	BRC	β-Gruids			**08/23–09/15**		**Sep. 03**	**160.3**	–.–	(9)	–.–	IMO
	Visual	(3.16)	0.852	16.9	50.8	340.3	337	−47	–.–	–.–	21.0	This work
#214	BCP	β-Capricornids			**09/05–09/15**		**Sep. 10**	**167.7**	–.–	(4)	–.–	
	$N=$–.–	2.429	0.170	13.0	43.0	327.7	305.7	−12.8	–.–	–.–	37.0	JFV
	Fireballs	AD 1045–1068:			09/24–09/30		304.7	−13.8	AT (52)			
#30	*(PSC)*	*Piscids*										
		= antihelion source meteors										
#215	NPI	Northern (δ-)Piscids			**09/12–09/31**		**Sep. 27**	**184**	3.0	<23	0.5	JW
		= extension of Taurid complex?										
	$N=7$	4.83	0.256	5.2	306.6	174.4	9.2	+07.7	+1.07	+0.12	31.2	This work
	$N=3$	1.977	0.272	5.2	305.2	173.3	7.0	+07.0	–.–	–.–	30.4	PG
	$N=93$	1.346	0.311	3.5	306.3	173.5	9.1	+07.2	–.–	–.–	25.6	ZS
	$N=42$	1.868	0.344	3.8	298.5	168.3	0.8	+03.9	–.–	–.–	27.4	ZS73
	$N=9$	2.062	0.399	3.4	290.8	199.8	26.7	+14.3	–.–	–.–	26.7	L71B
#216	SPI	Southern (δ-)Piscids			**09/12–09/23**		**Sep. 17**	**174**	3.0	<23	~1.5	IMO
		= extension of Taurid complex?										
	$N=5$	1.546	0.247	6.3	130.6	364.0	23.6	+05.1	–.–	–.–	26.5	This work
	$N=8$	1.55	0.284	5.3	126.6	359.7	18.0	+03.1	–.–	–.–	27.8	SS02
	σ	–.–	±0.021	±0.9	±3.0	(±0.1)	±1.82	±0.62	–.–	–.–	±1.3	
	$N=19$	1.64	0.28	5.4	125.7	363.0	18.7	+05.2	+0.7	+0.9	28.6	NL (61.9.1)
	$N=4$	2.5	0.28	8.4	123.9	362.9	20.3	+02.7	–.–	−0.7	31.9	NL (61.9.6)
	$N=16$	1.792	0.493	14.6	102.4	362.9	17.0	−11.0	–.–	–.–	23.6	ZS73
	Parent??	*2003 QC_{10}*		*(Epoch 2003-08-27)*				*$\lambda_\odot=181.02$*	*$T_J=+4.48$, $H_N=+17.83$*			
	(2003)	*1.376*	*0.369*	*5.035*	*119.760*	*361.015*	*16.3*	*+01.6*	–.–	–.–	*24.06*	*0.00010*

Table 7 (*cont.*)

	Code	Name			Dates (2000)		Peak	$\lambda_\odot^{max}$	χ	W	ZHR^{max}	Notes
		a	q	i	ω	Node	R.A.	Decl.	ΔR.A./$\Delta\lambda$	ΔDecl./$\Delta\lambda$	V_g	Δ_{C-E}
#217	OPC	ω-Piscids			**09/12–09/23**		**Sep. 17**	**174**	–.–	(4)	–.–	
		= ecliptic antihelion source										
	$N = 33$	2.864	0.525	1.5	273.6	190.8	10.6	+06.3	–.–	–.–	24.6	L71B
	$N = 6$	3.165	0.642	5.7	79.4	3.0	0.5	−08.8	–.–	–.–	21.4	This work
	$N = 7$	2.987	0.529	3.9	92.4	14.6	16.0	+01.7	–.–	–.–	24.5	PG
	$N = 3$	3.00	0.65	5.6	79.3	3.0	1.5	−08.6	–.–	–.–	24.6	B63 (II)
	Parent?	*2001 HA*$_4$		*(Epoch 2005-01-30)*				$\lambda_\odot = 179.00$	$T_J = +2.77$, $H_N = +17.48$			
	(2005)	*2.687*	*0.552*	*17.079*	*90.915*	*359.002*	*10.7*	*−17.7*	–.–	–.–	*24.89*	*0.0033*
#218	GSA	Sep. (γ-)Sagittariids			**09/11–09/15**		**Sep. 13**	**170.1**	–.–	(32)	–.–	
	$N =$ –.–	2.008	1.003	1.0	8.0	350.8	270.8	−31.0	–.–	–.–	6.2	T89 (48)
	Fireballs	AD 1071–1163:			09/23–10/12		283.8	−25.9				AT (51)
	Parent?	*107P/Wilson-Harrington*		*(Epoch 1936-11-08)*				$\lambda_\odot = 167.15$	$T_J = +5.66$, $H_N = +15.99$			
	(1936)	*2.643*	*0.996*	*0.856*	*13.940*	*347.153*	*279.1*	*−26.9*	–.–	–.–	*8.61*	*0.0467*
	(2001)	*2.643*	*1.001*	*0.377*	*8.948*	*353.036*	*276.1*	*−25.0*	–.–	–.–	*8.31*	*0.0481*
	(2048)	*2.645*	*0.999*	*0.043*	*10.475*	*352.143*	*277.9*	*−23.4*	–.–	–.–	*8.38*	*0.0487*
#76	**KAQ**	**κ-Aquariids**			**09/08–10/12**		**Sep. 22**	**179**	–.–	(14)	–.–	
	$N = 3$	2.564	0.884	1.0	45.2	359.8	334.0	−13.8	–.–	–.–	12.8	PG
	$N = 35$	2.606	0.741	7.6	247.8	179.5	343.0	+08.3	–.–	–.–	18.0	ZS
	$N = 6$	2.572	0.725	4.4	250.6	168.5	336.2	−01.2	–.–	–.–	18.1	ZS73
	$N = 9$	2.366	0.705	3.9	253.7	181.3	350.2	+03.2	–.–	–.–	18.2	ZS73
	$N = 5$	3.180	0.814	1.8	235.6	178.7	339.2	−04.4	–.–	–.–	16.5	L71B
	$N = 4$	2.940	0.867	2.1	229.2	186.6	342.3	−02.5	+0.80	+0.40	15.3	L71A
	$N = 3$	3.115	0.810	2.0	236.0	178.7	339.2	−04.4	–.–	–.–	16.0	AC
	visual				09/14–09/23		346.3	+00.5				M (#299)
	Fireballs	AD 1053–1163:		09/19–09/27		$\lambda_\odot \sim 166$	333.6	+00.2	–.–	–.–	–.–	AT (45)
	Parent?	*2006 AR*$_3$		*(Epoch 2006-03-06)*				$\lambda_\odot = 186.4$	$T_J = +3.17$, $H_N = +20.39$			
	(2006)	*2.458*	*0.878*	*4.743*	*46.724*	*6.389*	*345.4*	*−20.6*	–.–	–.–	*12.90*	*0.0095*

#219	SAR	Sep. μ-Arietids			**09/21–09/28**		**Sep. 22**	**179**	–.–	(3)	–.–	
	$N=17$	1.283	0.068	22.9	336.8	179.5	28.5	+18.6	–.–	–.–	36.3	ZS
	$N=22$	1.174	0.146	17.1	326.6	180.2	24.3	+19.0	–.–	–.–	31.2	ZS
	$N=46$	1.238	0.091	3.6	333.6	178.3	27.1	+12.5	–.–	–.–	33.7	ZS
	$N=83$	2.98	0.07	16	332	198.7	42.7	+20.2	–.–	–.–	41.5	KL (36)
	$N=5$	3.6	0.18	14.8	312.9	183.3	19.3	+15.6	1.4	–.–	36.7	NL (61.9.3)
												Helion
#220	NDR	ν-Draconids			**09/01–09/22**		**Sep. 13**	**170.3**	–.–	(8)	–.–	
	$N=$–.–	2.679	1.001	28.3	174.5	171.0	260.3	+54.0	–.–	–.–	18.4	T89 (45)
	$N=49$	2.565	1.004	32.3	181.4	162.6	265.4	+59.8	–.–	–.–	20.3	ZS
#221	**DSX**	**Daytime Sextantids**			**09/26–10/03**		**Sep. 30**	**188.35**	–.–	2.0 ± 0.2	20	O97
		= part of Phaethon complex										
	$N=410$	1.041	0.151	23.1	212.5	6.1	154.5	−01.5	+0.84	−0.43	31.2	GB
	σ		±0.025	±5.0	±3.5	±3.5	±2.8	±1.5	±0.02	±0.01	±2.2	
	$N=$–.–	1.25	0.16	22	213	4.3	153.2	−00.2	–.–	–.–	32.2	AC
	$N=14$	1.0	0.130	19.0	211.0	3.7	153.2	+02.6	–.–	–.–	32.0	JVF
	$N=9$	0.934	0.172	31.1	212.3	15.8	157.4	−08.6	–.–	–.–	29.7	ZS
	$N=$–.–	1.00	0.160	22.2	212.2	5.9	153.2	−01.2	–.–	–.–	30.1	G
	$N=9$	1.12	0.146	21.8	213.2	4.3	152.3	−00.3	–.–	–.–	32.2	NL (61.9.2)
	σ	±0.05	±0.01	±2.3	±2.1		±0.9	±1.5			±0.6	N64
	radar						155.6	−00.3				W60
	Parent	*2005 UD*		*(Epoch 2006-03-06)*				$\lambda_{\odot}=189.2$	$T_J=+4.51, H_N=+17.52$			
	(2006)	*1.275*	*0.163*	*21.560*	*217.151*	*9.181*	*159.9*	*−3.0*	–.–	–.–	*32.88*	*0.0790*
#222	DDL	Daytime δ-Leonids			**09/23–09/26**		**Sep. 25**	**183**	–.–	(2)	–.–	
	$N=3$	1.89	0.32	23.2	59.9	182.8	172.7	+21.2	–.–	–.–	31.1	NL (61.9.11)
#223	GVI	Daytime γ-Virginids			**09/23–09/29**		**Sep. 27**	**184**	–.–	(2)	–.–	
		= ecliptic helion source										
	$N=30$	2.037	0.616	0.9	273.2	345.8	168.4	+03.6	–.–	–.–	23.3	ZS
	$N=4$	2.04	0.61	5.8	92.9	184.0	187.0	+05.8	–.–	–.–	21.2	NL (61.9.7)
	$N=3$	2.6	0.59	6.3	272.6	3.4	179.6	−09.2	–.–	–.–	23.2	NL (61.9.8)

	Code	Name a	q	i	Dates (2000) ω	Node	Peak R.A.	$\lambda_{\odot}^{max}$ Decl.	χ ΔR.A./$\Delta\lambda$	W ΔDecl./$\Delta\lambda$	ZHR^{max} V_g	Notes Δ_{C-E}
	October:											
												Apex
#224	**DAU**	**Oct. δ-Aurigids**			**09/29–10/18**		**Oct. 08**	**191.0**	2	(8)	3.0	
	$N=14$	24.1	0.845	130.2	226.7	191.0	83.5	+50.4	+1.10	+0.10	64.9	R92
	σ	–.–	±0.085	±4.5	±12.2				–.–	–.–	±1.9	
#225	SOR	σ-Orionids			**09/26–10/26**		**Oct. 05**	**191.7**	3.0	(12)	–.–	
						$\lambda_{\odot}=191.7$	86.0	−03.0	+0.80	+0.00	65.00	IMO
	$N=1$	14.56	0.92	138	34	10	89	+00	–.–	–.–	65.9	HV8870
	$N=1$	57.18	0.66	145	71	28	95	+05	–.–	–.–	65.8	HV4974
#281	**OCT**	**October Camelopardalids**			**10/03–10/07**		**Oct. 06**	**193.0**	–.–	–.–	–.–	This work
	(1902)	(radiant unknown)				$\lambda_{\odot}=192.006$		50 light tracks behind clouds, U.K.				
	(1942)					$\lambda_{\odot}=192.8$	Cassiopeia		Meteor shower			
	(1976)					$\lambda_{\odot}=193.34$		113 meteors N > E, Florida				
	(2005)	368	0.993	78.6	170.6	192.57	166.0	+79.1	–.–	–.–	46.6	This work
#226	ZTA	ζ-Taurids			**10/07–10/24**		**Oct. 09**	**196**	–.–	(7)	–.–	
	$N=3$	21.3	0.715	162.4	70.9	16.5	86.1	+14.7	–.–	–.–	67.2	This work
	$N=6$	1.632	0.231	163.1	311.8	193.5	71.5	+28.2	–.–	–.–	56.5	ZS
	$N=17$	1.48	0.38	152	119	23.7	88.7	+12.0	–.–	–.–	57	KL (43)
#23	**EGE**	**ε-Geminids**			**10/14–10/27**		**Oct. 19**	**206.0**	1.6	15	3.0	IMO
	$N=3$	10.0	0.731	172.9	241.7	209.0	101.6	+26.7	–.–	–.–	68.8	This work
	$N=$–.–	14.90	0.770	173.0	236.7	208.5	102.0	+27.0	–.–	–.–	69	JW
	$N=7$	26.77	0.77	173	237	209.7	104.8	+26.9	+0.7	+0.0	69.4	AC
	$N=13$	3.58	0.88	175	223	203.7	104.8	+24.9	–.–	–.–	68	KL (41)
	$N=4$	14.895	0.770	173.0	236.7	208.2	102.8	+26.9	–.–	–.–	69.1	L71B
#227	OMO	Oct. Monocerotids			**10/18–10/22**		**Oct. 19**	**206**	–.–	(2)	–.–	
	$N=3$	5.2	0.865	135.0	46.4	25.7	101.9	−01.4	–.–	–.–	63.5	This work
	$N=2$	5.4	0.97	136	341	25.7	117.6	−05.1	–.–	–.–	65.0	GE (10.16)

	Parent?	*C/1723 T_1 (Keggler–Crossat–Saunderson)*		*(Epoch 1723-09-28)*				*$\lambda_{\odot} = 198.92$*	*$T_J \sim -0.01$, $H_{10} = +5.5$ ($n = 4$), P unknown*			
	(1723)	*Inf.*	*0.943*	*130.192*	*332.715*	*18.919*	*115.5*	*−07.8*			*65.18*	*0.0622*
#228	OLY	Oct. Lyncids			**10/10–10/23**		**Oct. 19**	**206**	−.−	(5)	−.−	
	$N = 6$	9.3	0.926	133.3	211.7	205.8	111.3	+48.8	−.−	−.−	64.8	This work
#229	NAU	ν-Aurigids			**10/20–10/22**		**Oct. 21**	**207.3**	−.−	(2)	−.−	
	$N = 7$	1.298	0.267	134.3	311.0	208.0	87.9	+39.6	−.−	−.−	53.1	ZS
#8	**ORI**	**Orionids**			**10/02–11/07**		**Oct. 22**	**208.6**	2.9	5.4 ± 0.6	23 ± 4	PJ
	Outburst	(1993)			1993 Oct. 7.8			204.5	2.0	1.0 ± 0.2	25.5 ± 5	PJ
	(1993)	616.2	0.613	163.5	76.5	24.069	92.0	+15.4	−.−	−.−	67.5	DMS93010
	$N = $ −.−	9.68	0.571	164.2	82.8	208.6	95.4	+15.9	+0.70	+0.11	66.2	This work
	$N = 30$	18.0	0.578	164.3	81.5	28.7	94.7	+15.9	+0.65	+0.11	66.53	KP
	$N = 12$	26.753	0.581	164.8	80.3	27.2	93.6	+16.2	−.−	−.−	67.3	PG
	$N = $ −.−	7.60	0.570	165.0	83.0	29.7	96.7	+16.0	−.−	−.−	66.0	JVF
	$N = 23$	16.72	0.570	163.9	83.4	29.9	96.4	+16.0	−.−	−.−	66.0	L71B
	$N = 17$	3.850	0.562	164.4	87.0	27.8	95.6	+16.1	−.−	−.−	64.6	ZS
	$N = 61$	4.76	0.57	164.2	86.1	26.1	93.8	+16.0	−.−	−.−	64.9	KL (49)
	Parent	*1P/Halley*		*(Epoch 1986-02-19)*				*$\lambda_{\odot} = 208.67$*	*$T_J = -0.59$, $H_0 = +2.1$ ($n = 7.5$)*			
	(1986)	*17.940*	*0.587*	*164.715*	*80.446*	*28.671*	*95.8*	*+16.0*	*−.−*	*−.−*	*66.79*	*0.1545*
#22	**LMI**	**Leonis Minorids**			**10/21–10/25**		**Oct. 22**	**209.0 ± 0.7**	1.9 ± 0.7	4.3 ± 1.2	2.7	PJ
	$N = 10$	286	0.616	125.32	102.73	208.36	159.5	+36.7	+1.42	−0.36	61.9	This work
	$N = 4$	33.6	0.641	124.5	106.3	209.9	160.7	+37.2	+0.96	+0.08	61.8	MLB
	$N = $ −.−	58.6	0.65	124	106	211.7	162.7	+36.7	−.−	−.−	61.8	AC
	Parent	*C/1739 K_1 (Zanotti)*		*(Epoch 1739-06-18)*				*$\lambda_{\odot} = 211.04$*	*$T_J \sim -0.01$, $H_{10} = +3.3$($n = 4$), P = unknown*			
	(1739)	*Inf.*	*0.674*	*124.260*	*110.665*	*211.044*	*161.0*	*+37.9*	*−.−*	*−.−*	*62.19*	*0.0568*
#230	ICS	October ι-Cassiopeiids			**10/15–10/28**		**Oct. 22**	**209**	−.−	(5)	−.−	
	$N = 6$	59.0	0.644	128.5	109.8	208.7	36.7	+66.0	−.−	−.−	66.3	This work
	Fireballs	AD 1032–1357:			10/22–11/06		8.7	+60.3	−.−	−.−	−.−	AT (61)
#231	ACM	Daytime α-Canis Majorids			**10/12–10/22**		**Oct. 17**	**204**	−.−	(4)	−.−	
	Visual				10/12–10/22		92.1	−14.0	−.−	−.−	−.−	M (#51)
		Inf.	0.812	110.5	50.9	24.8	92.1	−14.0	−.−	−.−	58.8	This work

	Code	Name a	q	i	Dates (2000) ω	Node	Peak R.A.	$\lambda_\odot^{max}$ Decl.	χ ΔR.A./$\Delta\lambda$	W ΔDecl./$\Delta\lambda$	ZHR^{max} V_g	Notes Δ_{C-E}
#232	BCN	Daytime β-Cancrids			**10/18–10/30**		**Oct. 26**	**213**	–.–	(5)	–.–	
	$N=2$	2.8	0.93	152	325	24.7	121.7	+04.9	–.–	–.–	66.1	GE (10.17)
	$N=2$	2.6	1.00	153	359	25.7	114.7	+05.9	–.–	–.–	66.1	GE (10.18)
	$N=3$	4.3	0.95	156	334	32.7	127.7	+05.8	–.–	–.–	68.1	N62 (26.1)
	$N=3$	6.1	0.98	148	10	35.7	120.6	+00.9	–.–	–.–	66.1	N62 (26.3)
												Antihelion
#233	**OCC**	**Oct. Capricornids**			**09/20–10/24**		**Oct. 03**	**189.7**	–.–	–.–	2.3 ± 1.5	PJ
	Outburst				1978		Oct. 03	189.70	–.–	>0.15	–.–	
						189.7	303.0	−10.0	+0.90	+0.20	10.0	IMO
	$N=$–.–	4.264	0.987	0.8	190.8	203.8	315	−14	–.–	–.–	10.4	T89 (53)
	$N=$–.–	(3.65)	0.99	2.8	193.2	189.3	301.5	−08.7	–.–	–.–	15 ± 3	W88
	Parent	*D/1978 R_1 (Haneda–Campos)*			*(Epoch 1979-01-07)*			*$\lambda_\odot = 183.49$*	*$T_J = +0.93$, $H_{10} = +12.5$ (var.)*			
	(1979)	*2.070*	*0.997*	*3.407*	*190.543*	*183.491*	*287.6*	*−05.1*	*–.–*	*–.–*	*7.34*	*0.1381*
#234	EPC	Oct. (ε-)Piscids			**10/02–10/12**		**Oct. 08**	**195**	–.–	(4)	–.–	
	$N=5$	2.919	0.616	2.8	262.5	195.0	8.7	+8.1	–.–	–.–	21.8	J92
	$N=$–.–	1.780	0.479	1.6	283.8	197.3	20.7	+11.3	–.–	–.–	23.2	T89 (52N)
	$N=45$	1.843	0.566	0.7	93.6	11.6	12.8	+4.4	–.–	–.–	20.5	ZS
	$N=6$	2.19	0.550	0.7	274.0	195.5	15.6	+7.3	–.–	–.–	23.5	LB71A
#235	LCY	λ-Cygnids			**10/20–11/07**		**Oct. 12**	**199**	–.–	(7)	–.–	
	$N=3$	2.768	0.972	13.9	199.4	213.2	319.1	+34.9	–.–	–.–	12.3	PG
	$N=$–.–	2.585	0.828	17.8	234.4	189.6	338.6	+31.3	–.–	–.–	18.0	T89 (49a)
	$N=$–.–	2.352	0.950	22.0	204.6	233.5	334.5	+57.3	–.–	–.–	15.8	T89 (49c)
	Parent?	*2005 CA*		*(Epoch 2006-03-06)*				*$\lambda_\odot = 202.8$*	*$T_J = +3.03$, $H_N = +15.34$*			
	(2006)	*2.359*	*0.967*	*17.006*	*203.186*	*202.759*	*313.3*	*+41.2*	*–.–*	*–.–*	*13.20*	*+0.131*
#236	GPS	γ-Piscids			**10/10–11/25**		**Oct. 13**	**200**	–.–	(18)	–.–	
	$N=$–.–	2.478	0.945	2.7	200.0	225.6	347.6	+0.3	–.–	–.–	10.4	T89 (54)

	N = –.–	2.866	0.871	1.4	224.9	229.8	377.7	+9.3	–.–	–.–	13.4	T89 (60)
	Outburst?	Radiant "SE Peg/NE Psc"				2003	Oct. 17	203.36	–.–	–.–	~20	
	Parent?	*6344 P-L*		*(Epoch 1960-09-26)*				$\lambda_\odot = 203.9$	$T_J = +3.06$, $H_N = +21.36$			
	(1960)	*2.645*	*0.931*	*4.058*	*213.516*	*203.905*	*339.0*	*+06.0*	–.–	–.–	*11.06*	*0.0280*
#237	SSA	σ-Arietids			**10/10–10/22**		**Oct. 15**	**202**	–.–	(5)	–.–	
		= part of Phaethon complex										
	N = 28	4.46	0.11	7	145	22.7	44.7	+14.2	–.–	–.–	40.5	KL (46)
	N = 24	1.74	0.05	17	158	21.7	52.7	+15.2	–.–	–.–	40.5	KL (47)
#86	OGC	Oct. γ-Cetids			**10/19–10/24**		**Oct. 20**	**206.4**	–.–	(2)	–.–	Ch. 13
	Outburst	Coast of Maine, USA: 25/hr 01 UT, 100/hr 03 UT						206.36	~2	~0.09	~100	RM
	(1935)			1935 Oct. 20		$\lambda_\odot = 206.36$	40.6	−04.8	–.–	–.–	–.–	RM
	N = 2	0.950	0.832	2.7	119.3	26.7	50.4	−06.9	–.–	–.–	3.3	This work
	N = –.–	2.442	0.791	11.6	58.4	28.1	18.6	−18.7	–.–	–.–	17.1	T89 (50)
	N = 2	1.760	0.783	8.5	67.0	27.7	22.6	−10.7	–.–	–.–	15.3	L71B
	Visual				10/15–10/23		48.8	−02.6	–.–	–.–	–.–	M (#35)
	Fireballs?	AD 1052–1093:			10/30–10/03		50	−10	–.–	–.–	–.–	AT (63)
#83	OCG	Oct. Cygnids			**10/04–11/07**		**Oct. 19**	**206**	–.–	(14)	–.–	
	N = –.–	3.428	0.904	25.0	218.1	190.7	322.5	+44	–.–	–.–	19.6	T89 (49b)
	N = 7	2.764	0.976	25.0	198.6	195.9	317.8	+52.6	–.–	–.–	17.2	ZS73
	N = 2	4.66	0.960	26.4	203.8	193.6	307.4	+48.2	–.–	–.–	22.2	W57
#299	*(OAR)*	*October Arietids*										
#25	**NOA**	**Northern Oct. (δ-)Arietids**			**10/01–10/24**		**(Nov. 03)**		–.–	>9	<3	
		= extension of NTA										
	N = –.–	2.168	0.334	5.2	117.1	42.4	52.7	+18.2	–.–	–.–	29.1	T89 (59)
	N = 15	4.05	0.22	12	307	201.7	34.7	+20.2	–.–	–.–	36.3	KL (45)
	Parent?	*2005 UR*		*(Epoch 2005-10-26)*				$\lambda_\odot = 216.2$	$T_J = +2.92$, $H_N = +21.6$			
	(2005)	*2.259*	*0.266*	*6.675*	*124.247*	*36.156*	*51.3*	*+14.1*	–.–	–.–	*31.65*	*0.0335*
#28	**SOA**	**Southern Oct. (δ-)Arietids**			**10/01–10/24**		**(Nov. 03)**		–.–	>9	<3	
		= extension of STA										
	N = –.–	2.883	0.381	6.6	108.8	33.0	41.7	+10.2	–.–	–.–	29.8	T89 (57)
	N = –.–	2.540	0.482	11.2	98.7	26.4	33.6	+01.2	–.–	–.–	26.2	T89 (52S)

Table 7 (*cont.*)

	Code	Name			Dates (2000)		Peak	$\lambda_{\odot}^{max}$	χ	W	ZHR^{max}	Notes
		a	q	i	ω	Node	R.A.	Decl.	ΔR.A./$\Delta\lambda$	ΔDecl./$\Delta\lambda$	V_g	Δ_{C-E}
	$N=83$	1.435	0.333	2.9	122.5	18.5	33.1	+10.6	–.–	–.–	25.6	ZS
	$N=58$	1.723	0.273	1.4	126.9	08.5	24.6	+09.1	–.–	–.–	24.2	ZS73
	$N=18$	1.74	0.24	1.2	130.7	23.2	40.3	+14.9	–.–	–.–	30.7	KL (44)
	$N=13$	1.75	0.30	5.8	122.5	24.9	39.5	+10.7	–.–	–.–	27.8	MP
	Parent?	*2P/Encke*		*(Epoch 2003-12-27)*				$\lambda_{\odot}=224.88$	$T_J=+3.03$, $H_{10}=+11.5$ *(var.)*			
	(2003)	*2.217*	*0.339*	*3.943*	*115.819*	*44.884*	*56.0*	*+16.5*	–.–	–.–	*29.06*	*0.1915*
#238	DOR	α-Doradids			**10/08–10/31**		**Oct. 21**	**208**	–.–	(9)	<2	
	Radar					$\lambda_{\odot}=207.7$	69.3	−55.9	–.–	–.–	–.–	WE60
												Helion
#9	**DRA**	**Oct. Draconids**			**10/02–10/16**		**Oct. 08**	**195.1**	2.6	–.–	<1	
	Outbursts (see Tab. 6d)											
	(1998)	3.572	0.9966	31.8	173.6	195.0188	263.40	+55.76	–.–	–.–	21.0	T
	(1998)	3.512	0.9964	31.8	173.4	195.081	263.16	+55.75	–.–	–.–	20.9	T
	(1998)	3.33	0.996	31.09	173.36	195.08	263.20	+55.42	–.–	–.–	20.5	SAYS
	σ $N=20$	±0.32	±0.001	±0.70	±0.83	–.–	±1.39	±0.70	–.–	–.–	0.6	
	$N=5$	3.02	0.996	31.4	172.9	196.4	264.1	+57.6	+1.9	+0.3	20.4	This work
	$N=7$	2.392	0.995	25.5	178.2	203.9	274.7	+52.4	–.–	–.–	16.7	PG
	$N=$–.–	2.120	0.992	27.6	171.4	204.6	267.2	+55.0	–.–	–.–	17.4	T89 (51b)
	$N=$–.–	2.855	0.994	29.8	183.8	202.6	283.2	+57.1	–.–	–.–	22.4	T89 (56a)
	$N=2$	3.51	0.996	30.7	171.8	197.0	262.4	+54.1	–.–	–.–	20.43	AC
	$N=2$	3.330	0.999	25.0	177.0	196.7	276.3	+49.0	–.–	–.–	17.8	L71B
	Parent	*21P/Giacobini–Zinner*		*(Epoch 1926-12-13)*				$\lambda_{\odot}=196.95$	$T_J=+0.73$, $H_{10}=+8.9$			
	(1926)	*3.513*	*0.994*	*30.73*	*171.748*	*196.946*	*261.9*	*+54.0*	–.–	–.–	*20.45*	*0.0005*
	(1998)	*3.391*	*0.996*	*31.859*	*172.543*	*195.398*	*261.9*	*+55.9*	–.–	–.–	*20.92*	*+0.0381*
	(2045)	*3.235*	*0.995*	*31.790*	*171.351*	*194.088*	*259.9*	*+55.9*	–.–	–.–	*20.80*	*+0.1040*
#239	GPU	γ-Puppids			**09/28–10/30**		**Oct. 16**	**202.7**	3.0	(1.3)	–.–	
						$\lambda_{\odot}=202.7$	109.0	−44.0	+0.60	−0.20	43.0	IMO
	$N=4$	17.8	0.980	71.6	16.1	58.0	110.1	−44.0	+1.74	+0.03	39.2	This work

#240	DFV	Daytime ψ-Virginids *= ecliptic helion source*			**09/28–10/24**		**Oct. 15**	**202**	–.–	(10)	–.–	
	$N=22$	1.513	0.525	2.6	258.1	22.3	193.7	−09.6	–.–	–.–	21.1	ZS
#241	OUI	Oct. Ursae Minorids			**10/10–10/27**		**Oct. 21**	**208**	–.–	(7)	–.–	
	$N=4$	3.98	0.995	51.5	179.1	209.1	246.6	+74.3	+1.4	−0.3	30.9	This work
	$N=$–.–	4.0	0.994	50.0	179.8	208.2	245.5	+73.8	–.–	–.–	29.6	This work
	$N=43$	2.294	0.992	40.0	183.0	210.6	267.0	+69.2	–.–	–.–	24.1	ZS#82
	$N=$–.–	2.851	0.984	48.7	191.1	208.0	276.5	+76.0	–.–	–.–	29.1	T89 (56b)
#242	XDR	ξ-Draconids			**10/05–11/05**		**Oct. 24**	**210.8**	–.–	(12)	–.–	
	$N=$–.–	2.886	0.984	70.6	183.0	220.8	192.4	+74.7	–.–	–.–	40.1	T89 (61)
	$N=38$	1.279	0.988	69.0	175.3	210.8	170.3	+73.3	–.–	–.–	35.8	ZS
	$N=8$	2.052	0.927	71.0	143.1	194.1	176.6	+70.9	–.–	–.–	39.3	ZS73
	$N=22$	2.543	0.992	67.1	169.7	196.3	190.1	+78.9	–.–	–.–	380	ZS73
	November:											
												Apex
#243	ZCN	ζ-Cancerids			**11/06–11/20**		**Nov. 07**	**225**	–.–	(6)	–.–	
	$N=7$	2.98	0.443	166.1	100.7	55.40	120.5	+14.3	–.–	–.–	63.4	This work
	Fireballs	AD 1026–1098:			11/02–11/18		101.7	+19.9				AT (68)
#244	PAR	ψ-Aurigids			**10/20–11/16**		**Nov. 09**	**226.98**	–.–	(11)	–.–	
	Outburst	15–20.5 UT, 43 meteors 19?					Nov. 09	226.98	–.–	–.–	~60	BMS#759
						$\lambda_\odot=226.98$	90.9	+40.0	–.–	–.–	–.–	RM
	Outburst	23:15–00:15 UT (Jan Janssens, Belgium)				1989 Nov. 04		222.687	–.–	–.–	>50	A90
						$\lambda_\odot=222.687$	29	+64	–.–	–.–	–.–	A90
	$N=$–.–	6.634	0.439	113.0	277.8	219.3	94.0	+50.0	–.–	–.–	56.7	T89 (62)
	Fireballs	AD 1024–1096:			10/30–11/03		80.9	+38.0				AT (64)
#245	NHD	Nov. Hydrids			**11/16–11/23**		**Nov. 17**	**235**	–.–		–.–	
	$N=2$	6.0	0.91	137.6	34.4	54.8	130.3	−06.3	0.92	–.–	65.5	This work
	$N=2$	5.4	0.97	136	34.1	25.7	117.6	−05.1	–.–	–.–	65.0	GE (10.16)
	Parent?	*C/1943 W_1 (van Gent–Peltier–Daimaca)*			*(Epoch 1994-01-12)*			*$\lambda_\odot=239.50$*	*$T_J\sim-0.03$, $H_0=+10.2$ ($n=3.5$),*			*P unknown*
	(1994)	*(200)*	*0.902*	*136.007*	*34.249*	*59.495*	*132.9*	*−08.3*	*–.–*	*–.–*	*66.69*	*0.0336*
#13	**LEO**	**Leonids**			**10/31–11/23**		**Nov. 17**	**235.1**	2.5	3.0 ± 0.6	13 ± 3	PJ
	Outbursts (see Tab. 4a)											
	(2001)	10.1	0.9853	162.36	173.50	236.15	154.24	+21.60	+0.659	−0.325	70.66	HB

	Code	Name			Dates (2000)		Peak	$\lambda_{\odot}^{max}$	χ	W	ZHR^{max}	Notes
		a	q	i	ω	Node	R.A.	Decl.	ΔR.A./Δλ	ΔDecl./Δλ	V_g	Δ_{C-E}
	(2000)	9.4	0.9853	162.42	173.47	236.10	154.16	+21.76	-.-	-.-	70.54	HB
	(1999)	9.9	0.9839	162.44	172.51	235.31	153.66	+21.76	-.-	-.-	70.57	HB
	(1998)	9.8	0.9839	162.05	171.89	234.68	153.29	+22.12	-.-	-.-	70.54	HB
	(1997)	8.5	0.9844	162.23	172.37	235.20	153.58	+21.90	-.-	-.-	70.38	HB
	(1995)	14.8	0.9827	161.55	171.85	235.24	154.08	+22.18	-.-	-.-	70.83	HB
	$N=5$	13.972	0.985	162.4	173.1	235.6	153.2	+22.0	-.-	-.-	70.9	PG
	$N=$-.-	11.5	0.985	162.6	172.5	235.2	153.0	+22.0	+0.70	−0.42	70.7	AC
	$N=9$	10.3	0.984	162.1	172.4	235.0	153.6	+22.1	+0.60	−0.45	70.26	KP
	$N=29$	15.2	0.984	162.53	172.36	235.7	153.9	+21.6	+0.944	−0.603	70.92	BPS
	σ	-.-	±0.003	±1.26	±3.06	±1.10	±1.0	±0.8	-.-	-.-	±0.92	
	Parent	*55P/Tempel–Tuttle*		*(Epoch 1998-08-15)*				$\lambda_{\odot}=235.02$	$T_J=-0.29$, $H_0=+8.5$ $(n=10.7)$			
	(1998)	*10.338*	*0.984*	*162.482*	*172.229*	*235.021*	*153.4*	*+21.8*	-.-	-.-	*70.63*	*0.0081*
#246	**AMO**	**α-Monocerotids**			**11/15–11/25**		**Nov. 21**	**239.3**	2.7	3.8	3.5	IMO
	(1995)	Outburst						239.322	1.80	0.008	500	PJ
	$N=10$	(500)	0.488	134.13	90.66	59.322	117.10	+00.83	-.-	-.-	63.0	DMS
	σ	(a > 28)	±0.005	±0.34	±0.78	±0.4	±0.13	±0.16	-.-	-.-	±0.2	
	(1985)				R. A. = 109.6°, Decl. = −7.1°			239.316	(2.7)	0.003	~600	Ch. 13
	(1935)				R. A. = 110.8°, Decl. = −5.1°			239.344	(~3)	~0.009	~1200	Ch. 13
	(1925)					-.-	-.-	239.384	-.-	~0.005	~2300	Ch. 13
												Antihelion
#247	*(TAU)*	*Taurids*			**09/16–12/29**		**Nov. 05**	**224 ± 1**	2.3	22 ± 3	7.3	PJ
#17	**NTA**	**Northern Taurids**			**09/16–12/29**		**Nov. 05**	**224 ± 1**	2.3	22 ± 3	4.0	PJ
								228 ± 3	2.3	46	3.5 ± 0.5	IMO
	$N=80$	2.12	0.350	3.1	294.9	226.2	58.6	+21.6	+0.80	+0.16	28.3	PK
	σ	±0.25	±0.053	±1.1	±6.5	±10.2	-.-	-.-	-.-	-.-	±1.9	
					$\lambda_{\odot}=234.673$		62.0	+23.2	-.-	-.-	27.1	U

	$N=12$	2.178	0.383	4.7	291.0	223.0	51.1	+22.8	–.–	–.–	28.2	PG
	$N=25$	2.20	0.284	2.9	302.3	212.7	44.0	+18.9	+0.82	+0.22	30.69	KP
	$N=$–.–	2.443	0.418	2.9	286.0	242.5	67.7	+22.1	–.–	–.–	27.3	T89 (69)
	$N=$–.–	2.454	0.557	3.8	91.3	60.6	58.7	+20.1	–.–	–.–	22.5	T89 (65)
	$N=13$	2.19	0.36	5.5	294.6	206.1	34.2	+18.4	–.–	–.–	28.4	KL (50)
	$N=45$	2.59	0.359	2.4	292.3	230.7	59.0	+22.4	+0.76	+0.10	29.2	AC
	Fireballs	AD 1062–1095:			11/21–11/25		55	+25	–.–	–.–	–.–	AT (74)
	Parent	*2004 TG_{10}*		*(Epoch 2006-03-06)*				*$\lambda_{\odot}=223.83$*	*$T_J=+2.99$, $H_N=+19.46$*			
	(2006)	*2.242*	*0.315*	*3.622*	*298.443*	*223.865*	*54.7*	*+22.3*	*–.–*	*–.–*	*29.89*	*+0.0128*
#2	**STA**	**Southern Taurids**			**09/25–12/19**		**Nov. 05**	**224 ± 1**	2.3	22 ± 3	3.3	PJ
								225 ± 2	2.3	50	4.0 ± 0.3	IMO
	$N=144$	2.07	0.352	5.4	115.4	37.3	49.4	+13.0	+0.73	+0.18	28.0	PK
	σ	±0.32	±0.058	±1.1	±7.2	±11.1	–.–	–.–	–.–	–.–	±2.1	
	$N=19$	2.096	0.357	5.6	114.4	38.7	49.8	+12.8	–.–	–.–	28.3	PG
	$N=49$	1.67	0.30	7.1	129	24.7	44.7	+11.2	–.–	–.–	29	GE (D.01)
	$N=46$	2.00	0.340	6.3	116.8	32.7	44.7	+11.5	+0.82	+0.22	28.35	KP
	$N=46$	1.93	0.375	5.2	113.2	40.7	51.2	+13.8	+0.79	+0.15	27.0	AC
	$N=73$	2.08	0.33	2.2	118.2	15.5	27.9	+08.8	–.–	–.–	28.8	KL (37)
	$N=17$	2.08	0.50	4.2	99.0	56.8	59.7	+16.7	1.0	–.–	23.8	NL (61.11.1)
	$N=8$	2.4	0.38	5.5	112	43.7	53.7	+14.2	–.–	–.–	27.8	**W54**
	Fireballs	AD 1052–1170:			11/19–12/07		43.7	+04.2	–.–	–.–	–.–	AT (73)
	Parent	*2P/Encke*		*(Epoch 2003-12-27)*				*$\lambda_{\odot}=224.88$*	*$T_J=+3.03$, $H_{10}=+11.5$ (var.)*			
	(2003)	*2.217*	*0.339*	*3.943*	*115.819*	*44.884*	*56.0*	*+16.5*	*–.–*	*–.–*	*29.06*	*0.1915*
#24	**PEG**	**μ-Pegasids**			**09/29–11/18**		**Nov. 12**	**230.4**	–.–	(5)	<2	
	Outburst?				1952, Nov. 12			$\lambda_{\odot}=230.41$	–.–	–.–	>50	Ch. 21
	HV3570	3.856	0.9735	8.10	195.80	230.412	335.53	+21.78	–.–	–.–	11.21	This work
	(error)	±0.16	±0.0003	±0.09	±0.10	–.–	±0.15	±0.12	–.–	–.–	±0.14	
	HV3570	3.860	0.974	8.1	195.66	230.412	335.42	+21.68	–.–	–.–	11.20	JW
	$N=2$	3.099	0.948	7.18	206.34	(235.04)	355.4	+26.1	–.–	–.–	11.1	U
	$N=6$	3.86	0.97	8	196	230.7	335.6	+21.3	–.–	–.–	11.2	AC

PEG – "U:" From two video meteors during the 1998 and 1999 Leonid campaigns.

	Code	Name a	q	i	Dates (2000) ω	Node	Peak R.A.	$\lambda_{\odot}^{max}$ Decl.	χ ΔR.A./$\Delta\lambda$	W ΔDecl./$\Delta\lambda$	ZHR^{max} V_g	Notes Δ_{C-E}
#18	**AND**	**Andromedids**			**10/08–11/22**		**Nov. 14**	**232**	-.-	(18)	-.-	
	Outbursts (see Tab. 6a)											
	$N=18$	2.76	0.789	10.0	238.9	231.0	24.2	+32.5	+0.63	+0.33	17.2	This work
	$N=5$	1.749	0.691	12.0	-.-	221	27.2	+34.9	-.-	-.-	17.6	J92
	$N=3$	2.375	0.760	14.3	245.2	207.2	3.3	+31.8	-.-	-.-	18.1	PG
	$N=$-.-	2.441	0.738	12.4	248.6	201.9	2.6	+26.3	-.-	-.-	18.7	T89 (55)
	$N=$-.-	1.824	0.854	13.8	232.4	234.8	17.7	+46.3	-.-	-.-	14.1	T89 (63)
	$N=23$	2.90	0.777	7.5	242.7	225.5	23.7	+09.3	-.-	-.-	18.9	SHS59
	$N=$-.-	2.7	0.74	6.8	247	226	27.7	+25.2	-.-	-.-	18.0	J
	Parent	*3D/Biela*		*(Epoch 1852-09-29)*				$\lambda_{\odot}=246.56$	$T_J=+0.78$, $H_{10}=+7.10$			
	(1852)	*3.524*	*0.864*	*12.654*	*224.712*	*246.556*	*5.8*	*+43.4*	-.-	-.-	*16.15*	*−0.0066*
	(2004)	*3.491*	*0.798*	*7.501*	*236.175*	*213.790*	*24.6*	*+27.2*	-.-	-.-	*17.21*	*0.0410*
	(2050)	*3.510*	*0.806*	*5.950*	*234.913*	*231.498*	*24.2*	*+23.9*	-.-	-.-	*16.76*	*0.0891*
#248	IAR	Nov. ι-Aurigids			**11/01–11/23**		**Nov. 16**	**233.637**	1.88 ± 0.12	(9)	8.2 ± 2.8	HM
				1998 Nov. 18		$\lambda_{\odot}=235.0$	77.0	+35.0	-.-	-.-	-.-	KZ99
	$N=25$	1.076	0.085	30.5	336.3	222.3	76.3	+33.3	-.-	-.-	34.0	ZS
	$N=32$	1.467	0.132	19.6	326.1	222.1	68.1	+30.2	-.-	-.-	34.2	ZS
	Fireballs	AD 1032–1083:		10/30–11/03			80.9	+38.0	-.-	-.-	-.-	AT (64)
#249	NAR	Nov. ν-Arietids			**11/17–11/29**		**Nov. 23**	**241**	-.-	(5)	-.-	
	$N=3$	1.54	0.790	2.8	246.4	236.18	39.0	+23.2	-.-	-.-	12.3	NMS
	$N=$-.-	2.608	0.794	5.3	238.2	241.3	38.7	+21.2	-.-	-.-	16.3	T89 (67a)
	$N=4$	3.257	0.784	9.7	238.0	228.2	22.7	+30.3	-.-	-.-	17.8	L71B
#250	**NOO**	**Nov. (ω-)Orionids**			**11/16–11/29**		**Nov. 27**	**245**	-.-	(5)	-.-	
	Outburst					1964	Nov. 25/26	244.12	1.8	0.020	140	(Outburst)
	(1964)	Inf.	0.23	43	123	244.12	85.7	+04.0	-.-	-.-	~45	Ch. 13
	$N=16$	12.7	0.088	26.9	145.8	(60.0)	85.2	+15.6	+0.71	−0.03	43.3	This work

	N = 8	7.32	0.089	29.1	146.6	57.5	83.8	+15.3	–.–	–.–	43.5	MB_{97}
	N = 7	12.26	0.045	41.5	161.4	41.9	79.4	+14.6	–.–	–.–	38.8	ZS
	Fireballs?	AD 1053–1082:			11/08–11/13		83.6	+00.0				AT (69)
												Helion
#251	**IVI**	**Daytime ι-Virginids**			**11/05–11/07**		**Nov. 05**	**223**	–.–	(2)	–.–	
	N = 8	1.217	0.985	10.1	60.7	224.3	210.4	−03.8	–.–	–.–	29.0	ZS
	December:											
												Apex
#19	**MON**	**(Dec.) Monocerotids**			**11/27–11/17**		**Dec. 13**	**260.9 ± 0.6**	3.0	4 ± 3	2.0 ± 0.4	PJ
	N = 11	50.7	0.193	35.2	128.1	80.2	101.8	+08.1	+0.83	−0.05	42.0	This work
	N = 15	19.9	0.188	34.9	128.9	80.2	102.0	+08.3	+0.95	−0.03	41.6	O89
	σ	–.–	±0.012	±3.1	±2.1	±2.2	±1.1	±1.2			±1.8	
	N = 12	27.6	0.187	34.9	128.9	81.1	102.2	+08.3	–.–	–.–	41.8	LO90
	N = 30	6.199	0.153	22.3	135.8	72.5	95.1	+14.5	–.–	–.–	40.0	ZS
	N = 52	7.114	0.119	24.7	141.2	68.0	92.1	+15.0	–.–	–.–	41.6	ZS73
	N = –.–	52.24	0.175	31.5	131.0	82.5	100.7	+08.0	+0.80	+0.00	42	LB
	N = 3	7.69	0.19	39.9	130	82.7	106.7	+05.9	–.–	–.–	40.5	GE (12.09)
	N = 6	5.56	0.11	39.0	138.9	76.9	102.3	+09.5	–.–	−0.5	42.2	NL (61.12.7)
	N = 4	11	0.11	22.6	135.3	73.9	95.5	+14.5	+1.0	–.–	41.3	NL (61.12.2)
	N = –.–	3.464	0.121	22.3	141.9	89.0	113.7	+13.9	–.–	–.–	41.6	T89 (75S)
	N = 4	20.00	0.20	18.7	131.5	77.3	96.8	+15.1	–.–	–.–	40.6	NL (60.12.9)
	N = 3	42	0.14	24.8	135.8	77.6	100.5	+14.0	–.–	–.–	42.4	J
	N = 2	Inf.	0.186	35.2	128.2	81.6	103.7	+07.9	–.–	–.–	42.4	W57
	Parent	*C/1917 F_1 (Mellish)*		*(Epoch 1917-04-15)*				*$\lambda_{\odot} = 260.27$*	*$T_J = +0.04$, $H_0 = +7.4$ ($n = 3.9$)*			
	(1917)	*27.65*	*0.190*	*35.981*	*128.263*	*80.267*	*102.0*	*+07.8*	*–.–*	*–.–*	*41.94*	*0.0615*
#32	**DLM**	**Dec. Leonis Minorids**			**12/12–12/17**		**Dec. 14**	**262.4**	–.–	(2)	–.–	
	N = 6	11.9	0.554	133.8	265.6	262.2	156.1	+32.7	–.–	–.–	62.3	This work
	N = 3	4.70	0.549	133.97	265.98	262.41	156.5	+32.0	–.–	–.–	62.1	DMS
	N = –.–	inf.	0.612	132.3	255.8	261.6	156.1	+34.6	–.–	–.–	63.7	CL
	N = 9	3.76	0.81	138	249	261.7	163.7	+39.7	–.–	–.–	64	KL (193r)
	N = 3	57.4	0.712	135.7	264.5	259.1	151.7	+32.8	–.–	–.–	63.4	W57
	?Outburst	44 meteors					**1921 Dec. 05**	**254.26**	9.1?	–.–	>80	
	(1921)					$\lambda_{\odot} = 254.26$	160	+37	–.–	–.–	–.–	Ch. 13

	Code	Name			Dates (2000)		Peak	$\lambda_{\odot}^{max}$	χ	W	ZHR^{max}	Notes
		a	q	i	ω	Node	R.A.	Decl.	ΔR.A./Δλ	ΔDecl./Δλ	V_g	Δ_{C-E}
	Parent?	*C/1798 X_1 (Bouvard)*		*(Epoch 1799-01-01)*				*$\lambda_{\odot}=266.65$*	*$> T_J=0.01$, $H_{10}=+11.0$ $(n=4)$, P unknown*			
	(1799)	*(200)*	*0.775*	*139.354*	*234.967*	*266.648*	*169.0*	*+28.3*	*-.-*	*-.-*	*66.62*	*0.1165*
#16	**HYD**	**σ-Hydrids**			**12/03–12/18**		**Dec. 17**	**265.5 ± 0.8**	3.0	6.6 ± 2.0	2.5 ± 0.5	PJ
	$N=18$	12.3	0.224	124.9	124.0	84.8	131.9	+00.2	+0.72	−0.21	58.0	This work
	$N=4$	37.046	0.294	132.5	114.4	73.2	124.2	+03.1	-.-	-.-	60.2	PG
	$N=$-.-	113.8	0.253	127.9	119.3	72.5	121.7	+03.9	-.-	-.-	58.5	T89 (71)
	$N=2$	11.525	0.230	125.0	124.0	82.7	129.6	+00.8	-.-	-.-	57.9	L71B
	$N=$-.-	11.53	0.230	125.0	124.0	82.7	127.6	+01.8	+0.80	−0.20	58	JW
	$N=8$	30.0	0.244	125.5	120.7	79.7	127.2	+01.4	+0.7	−0.2	58.4	J
#252	ALY	α-Lyncids			**12/10–01/03**		**Dec. 21**	**268.9**	-.-	-.-	-.-	
	Outburst	(M. Currie)					**1971 Dec. 20**	**<268.78**	-.-	60	>350	Table III
	(1971)	(25.4)	0.281	84.4	295.9	268.8	138.8	+43.8	-.-	-.-	50.4	PJ
	?$N=5$	1.193	0.206	153.8	318.4	264.6	149.8	+21.5	-.-	-.-	54.4	ZS
#15	**URS**	**Ursae Minorids (=Ursids)**			**12/17–12/26**		**Dec. 23**	**271.0 ± 0.3**	3.4	(0.7)	12 ± 3	PJ
	Outbursts (see Tab. 5b)							270.6 ± 0.1	3.0	1.6	12 ± 1	IMO
	$N=64$	4.62	0.944	51.5	204.9	270.74	219.35	+75.34	-.-	-.-	33.0	This work
	σ	±1.3	±0.007	±1.0	±1.9	±0.07	±4.0	±0.7	-.-	-.-	±0.9	
	$N=$-.-	2.62	0.89	52	224	270.7	190.5	+74.7	-.-	-.-	32	KL (195)
	Parent	*8P/Tuttle*		*(Epoch 2008-01-15)*			*270.34*	*$T_J=+0.31$, $H_0=+8.0$ $(n=6.0)$*				
	(2008)	*5.094*	*0.934*	*54.983*	*207.506*	*270.341*	*213.4*	*+74.4*	*-.-*	*-.-*	*34.27*	*0.0952*
	(2049)	*5.106*	*0.934*	*54.611*	*207.438*	*270.018*	*213.9*	*+74.8*	*-.-*	*-.-*	*34.08*	*0.1030*
#20	COM	Dec. Comae Berenicids			**12/12–01/15**		**Dec. 26**	**274**	3.0	26	3.2 ± 0.7	IMO
		= apex source										
	$N=$-.-	14.6	0.560	136.0	262.2	283.1	175.7	+24.7	-.-	-.-	64	L87
	$N=$-.-	11.1	0.86	149.4	137	260.7	190.6	+13.7	-.-	-.-	64	GE (12.22)
	$N=4$	14.4	0.541	139.4	265.0	283.3	175.2	+22.2	-.-	-.-	63.7	This work

HYD – "this:" Strong variation of q versus i. Values are for a cluster of orbits near the center of the range.
URS – The proper name is "Ursae Minorids." The activity curve consists of two components with $B=0.9\pm0.4$ (ZHR = 10) and $B+=0.08\pm0.03/B-=0.5\pm0.4$ (ZHR = 2.0), where B is the $\log_{10}$ exponent of ZHR versus solar longitude.

	$N=11$	Inf.	0.580	134.0	259.6	282.0	175.8	+25.4	+0.88	−0.45	64.0	AC
	$N=1$	18.92	0.556	136.4	263.2	284.3	176.4	+23.5	–.–	–.–	63.75	JW
												Antihelion
#253	CMI	Dec. Canis Minorids			**12/02–12/05**		**Dec. 03**	**252.4**	–.–	–.–	–.–	
	Outburst						**1988 Dec. 03**	**252.4**	–.–	–.–	~54	Ch. 13
	(1988)	93 meteors plotted				$\lambda_\odot=253.3$	112.9	+07.7	–.–	–.–	Swift	Ch. 13
	$N=1$	1.38	0.08	39	154	77	111	+11	–.–	–.–	38.8	HV5548
	$N=1$	3.79	0.08	36	149	80	109	+13	–.–	–.–	43.2	HV9544
#254	**PHO**	**Phoenicids**		(1956)	**12/04–12/06**		**Dec. 05**	**253**	2.9	2.8 ± 1.5	<3	PJ
	Outbursts (see Tab. 6k)											
	Visual	2.96	0.99	13	359	74	15.6	−44.7	–.–	–.–	11.7	AC
	Parent	*D/1819 W_1 (Blanpain)*		*(Epoch 1819-11-22)*				*$\lambda_\odot=249.80$*	*$T_J=+0.51$, $H_{10}=+8.5$*			
	(1819)	*2.957*	*0.986*	*8.970*	*360.164*	*69.795*	*357.4*	*−42.0*	–.–	–.–	*10.06*	*0.0971*
	Parent	*2003 WY_{25}*		*(Epoch 2005-01-30)*				*$\lambda_\odot=252.25$*	*$T_J=+0.54$, $H_N=+20.88$*			
	(2005)	*2.847*	*0.983*	*5.911*	*6.250*	*72.244*	*3.8*	*−26.6*	–.–	–.–	*9.15*	*0.0163*
	(1819)	*2.993*	*0.889*	*9.23*	*349.65*	*80.020*	*347.9*	*−44.0*	–.–	–.–	*10.85*	*0.089*
#255	**PUV**	**Puppids-Velids I**			**11/02–01/22**		**Dec. 06**	**254.0**	2.9	18 ± 3	4.5 ± 0.7	PJ
		= core of Puppid-Velid Complex										
	Visual					$\lambda_\odot=254.0$	128.0	−45.0	+0.50	−0.20	36.8 ± 2.2	Mean
	$N=6$	1.3	0.98	70	360	82.7	141.5	−43.2	–.–	–.–	35.3	GE (12.07)
	$N=3$	1.1	0.98	75	361	82.7	145.5	−45.2	–.–	–.–	37.4	GE (12.08)
	Visual				12/03–12/12		113.4	−45.2	–.–	–.–	–.–	M (61)
	Visual				12/03–12/04		118.4	−40.8	–.–	–.–	–.–	M (65)
#14	*(XOR)*	*χ-Orionids*			**12/01–01/5**		**Dec. 11**	**≤259**	–.–	(22)	(0.6)	
#256	**ORN**	**Northern χ-Orionids**			**12/01–01/5**		**Dec. 11**	**≤259**	–.–	(22)	(0.6)	
		= Extension of NTA into Dec.										
	$N=12$	2.22	0.449	2.5	283.9	257.3	83.9	+25.5	+0.54	+0.00	24.9	This work
	$N=7$	2.271	0.417	3.3	286.6	254.2	82.1	+26.4	–.–	–.–	26.4	PG
	$N=$–.–	2.185	0.523	4.0	275.0	261.7	82.8	+23.0	–.–	–.–	23.6	T89 (72)
	$N=15$	1.729	0.376	0.1	294.4	251.6	83.4	+23.4	–.–	–.–	26.2	ZS
	$N=49$	1.475	0.265	0.0	309.2	256.9	97.0	+23.3	–.–	–.–	28.7	ZS73
	$N=4$	2.216	0.472	2.5	281.0	259.0	83.8	+26.0	–.–	–.–	25.7	L71B
	$N=4$	2.22	0.47	2	281	258	84.8	+26.0	–.–	–.–	25.2	AC

PUV – Puppids–Velids – Name consists of two constellation names, both "-id".

	Code	Name a	q	i	Dates (2000) ω	Node	Peak R.A.	$\lambda_{\odot}^{max}$ Decl.	χ ΔR.A./$\Delta\lambda$	W ΔDecl./$\Delta\lambda$	ZHR^{max} V_g	Notes Δ_{C-E}
	Parent?	*2002 XM$_{35}$*		*(Epoch 2002-12-02)*				*$\lambda_{\odot}$ = 251.77*	*T_J = +2.96, H_N = +22.96*			
	(2002)	*2.331*	*0.376*	*2.854*	*290.836*	*251.773*	*81.3*	*+25.7*	*–.–*	*–.–*	*28.31*	*0.0196*
#257	**ORS**	**Southern χ-Orionids**			**12/01–01/16**		**Dec. 12**	**260**	–.–	(22)	(0.6)	
		= Extension of STA into Dec.										
	N = 12	2.23	0.594	5.2	86.4	80.1	78.7	+15.7	+0.88	−0.03	21.5	This work
	N = 6	2.193	0.528	4.4	94.3	78.5	81.1	+17.6	–.–	–.–	23.3	PG
	N = 8	2.224	0.471	6.9	100.6	79.8	85.7	+16.0	–.–	–.–	23.1	L71B
	N = 4	1.85	0.56	4.6	93.7	79.3	80.7	+16.8	+1.3	–.–	21.5	NL (60.12.2)
	N = 32	1.790	0.420	2.6	109.0	78.2	87.8	+20.6	–.–	–.–	25.2	ZS73
	N = 8	2.1	0.47	7	101	79	85.7	+16.0	–.–	–.–	25.5	AC
	N = 4	1.64	0.38	15.7	114.3	78.3	91.6	+08.9	–.–	–.–	26.9	NL (60.12.3)
#4	**GEM**	**Geminids**			**11/27–12/18**		**Dec. 14**	**262.08**	(2.0)	0.90 ± 0.10	120 ± 10	Ch. 22
		Main peak:										
	N = 221	1.372	0.1410	24.02	324.42	261.49	113.2	+32.5	+1.02	−0.15	34.58	DMS
	σ	±0.033	–.–	±1.29	±0.91	±0.30	±0.69	±0.24	–.–	–.–	±0.61	
	Mean	1.357	0.1400	24.27	324.63	261.433	At peak in year 2000					IAU
	$\Delta/\Delta\lambda_{\odot}$	−1.9e −3	+1.0e −3	−0.13	−0.41	+1.00						
	Δ/yr	−7.0e −4	+2.1e −6	+9.7e −3	+7.8e −3	+6.8e −4						
	N = 19	1.386	0.137	24.4	324.7	260.2	112.0	+32.6	–.–	–.–	35.0	PG
		Background component (radar):										
	N = 48	1.376	0.141	23.9	324.9	259.2	110.5	+32.9	+0.92	+0.07	34.63	This work
	σ	±0.12	±0.009	±2.1	±1.2	±1.7	±0.76	±0.36	–.–	–.–	±0.80	
	N = 118	1.306	0.139	23.2	325.2	262.1	114.2	+32.1	–.–	–.–	34.0	ZS
	N = 20	1.30	0.13	18.2	327	261.7	112.8	+29.9	–.–	–.–	34.2	GE
	N = 11	1.19	0.14	16.4	325.7	260.1	110.2	+29.9	+1.1	–.–	32.5	NL (60.12.1)
	N = 401	1.31	0.14	24	326	260.7	111.8	+32.9	–.–	–.–	34.2	KL (51)

	Parent	*3200 Phaethon*		*(Epoch 2005-08-18)*				*$\lambda_\odot = 262.50$*	*$T_J = +4.51$, $H_N = +14.6$*			
	(2005)	*1.271*	*0.140*	*24.186*	*325.246*	*262.495*	*115.0*	*+32.4*	–.–	–.–	*33.92*	*0.0208*
#258	DAR	Dec. α-Aurigids			**12/14–12/31**		**Dec. 14**	**262.2**	–.–	(7)	–.–	
	Outburst?				*1996 Dec. 13*			*262.218*	*1.8*	–.–	*~110*	
	(1996)					*262.218*	*79.7*	*+43.1*	–.–	–.–	–.–	*T89*
	$N=5$	2.279	0.668	7.2	257.7	270.0	84.9	+35.5	–.–	–.–	19.5	PG
	$N=$–.–	2.365	0.694	11.2	253.6	274.7	85.9	+42.0	–.–	–.–	19.5	T89 (77)
	$N=$–.–	2.096	0.760	21.8	245.4	276.5	91.2	+62.0	–.–	–.–	19.9	T89 (78)
	$N=22$	1.851	0.901	11.1	221.0	293.9	85.0	+58.3	–.–	–.–	11.9	ZS
	$N=6$	2.053	0.816	6.7	235.9	279.6	78.4	+39.1	–.–	–.–	14.2	ZS73
#259	**CAR**	**Carinids**			**12/04–02/04**		**Dec. 16**	**264**	3.0	(25)	–.–	
	Visual					$\lambda_\odot = 263.7$	129.0	−58.0	–.–	–.–	38.9 ± 2.5	Mean
	$N=4$	2.94	0.91	66.6	323.7	76.9	155.3	−60.9	–.–	–.–	38.9	NL (60.12.7)
	$N=16$	1.940	0.970	70.0	347.0	79.7	142.4	−54.2	–.–	–.–	39.0	JVF
	$N=3$	1.89	0.98	69.6	340.5	77.2	143.6	−54.3	–.–	–.–	38.5	NL (61.12.6)
	$N=5$	2.08	0.98	70.1	353.4	79.0	138.8	−53.1	–.–	–.–	39.1	NL (60.12.8)
	$N=7$	2.205	0.970	57.3	344	82.7	135.6	−63.4	–.–	–.–	33.2	GE (12.06)
#260	GTI	γ-Triangulids			**12/15–01/13**		**Dec. 24**	**272**	–.–	(12)	–.–	
		= Dec. β-Perseids										
	$N=3$	2.545	0.926	7.1	212.0	273.0	43.3	+41.6	–.–	–.–	11.3	PG
	$N=$–.–	2.292	0.932	4.7	209.3	272.3	40.8	+32.2	–.–	–.–	10.5	T89 (76)
	$N=14$	2.420	0.857	2.1	228.0	262.6	54.7	+25.2	–.–	–.–	12.8	L71B
												Helion
#261	DDC	Daytime δ-Scorpiids				**12/05–12/07**		**Dec. 06**	**254**	–.–	(2)	–.–
		= ecliptic helion source										
	$N=4$	2.38	0.50	4.0	262.6	74.4	246.6	−26.3	–.–	–.–	25.3	NL (61.12.3)
	$N=3$	2.27	0.43	5.6	255.4	79.5	248.8	−24.3	–.–	–.–	26.9	NL (60.12.5)
	Parent?	*2004 YD_5*		*(Epoch 2006-03-06)*				*$\lambda_\odot = 268.24$*	*$T_J = +0.39$, $H_N = +29.26$*			
	(2006)	*2.274*	*0.494*	*3.610*	*262.331*	*88.195*	*263.1*	*−27.4*	–.–	–.–	*24.89*	*0.0002*
#262	KLI	Daytime (κ-)Librids				**12/08 –12/12**		**Dec. 11**	**259**	–.–	(2)	–.–
	$N=3$	1.35	0.19	8.4	221.1	78.8	231.3	−20.8	–.–	–.–	31.9	NL (60.12.6)
	radar						235.7	−19.2				ER55

Notes:

A90 – P. Aneca, New evidence for a Cassiopeid meteor shower? *WGN* **18** (1990), 68.

AC – A. F. Cook, A working list of meteor streams. In *Evolutionary and Physical Properties of Meteoroids. Proc. IAU Colloq.* 13, held in Albany, NY, 14–17 June 1971, ed. C. L. Hemenway, P. M. Millman and A. F. Cook. NASA SP-319 (Washington, DC, 1973), pp.183–191.

AT – I. S. Astapovic and A. K. Terent'eva, Fireball radiants of the 1st–15th centuries. In *Physics and Dynamics of Meteors. Symp. no. 33*, held at Tatranska Lomnica, Czechoslovakia, 4–9 September 1967, ed. L. Kresák and P. Mackenzie Millman, *Int. Astronomical Union Symp.* no. 33. (Dordrecht: Reidel, 1986), p. 308.

AW – T. R. Arter and I. P. Williams, The mean orbit of the April Lyrids. *Mon. Not. R. Astron. Soc.* **289** (1997), 721–728.

B03 – H. Betlem, High precision photographic orbits from the 1995–2001 Leonid showers. Presented at: *Leonid MAC Workshop, NASA Ames Research Center (2003)*. Note: 1995–1997: Filament; 1999: 1866 trail. 2000: 1733 trail, 2001: 1767 trail.

B04 – B. A. Lindblad, L. Neslusan, V. Porubčan and J. Svoren, IAU meteor database of photographic orbits – version 2003. *Earth, Moon Planets* **93** (2003) 249–260.

B63 – P. Babadzhanov, Orbital elements of photographic meteors. *Smitsonian Contrib. Astrophys.* **7** (1963), 287–291.

B92 – P. B. Babadzhanov and Yu.V. Obrubov, Dynamics and relationships between interplanetary bodies. I. Comet Machholz and its meteor showers. In *Meteoroids and Their Parent Bodies*, ed. J. Stohl and I. P. Williams. (Bratislava: Acad. Inst. Slovak Acad. Sci., 1992), pp. 49–52.

BE88 – H. Betlem, C. R. Ter Kuile, M. de Lignie *et al.*, Precision meteor orbits obtained by the Dutch Meteor Society – photographic meteor survey (1981–1993). *Astron. Astrophys. Suppl. Ser.* **128** (1988), 179–185.

BF66 – K. Baker and G. Forti, *Harvard Radio Meteor Project Res. Rep.* No. 14 (1996).

BK65 – P. B. Babadzhanov and E. N. Kramer, Orbits of bright photographic meteors. *Smithsonian Contrib. Astrophys.* **11** (1965), 67–79.

BPS – B. A. Lindlbad, V. Porubčan and J. Stohl, The orbit and mean radiant motion of the Leonid meteor stream. In *Meteoroids and Their Parent Bodies*, ed. J. Stohl and I. P. Williams. (Bratislava: Astron. Inst. Slovak Acad. Sci., 1992), pp. 177–180.

CL – A. F. Cook, B.-A. Lindblad, B. G. Marsden, R. E. McCrosky and A. Posen, Yet another stream search among 2401 photographic meteors. *Smithsonian Contrib. Astrophys.* **15** (1972), 1–5.

D00 – A. Dubietis, Obervations of the Cassiopeid meteor shower. *WGN* **28** (2000), 108–113.

D82 – J. D. Drummond, A note on the Delta Aurigid meteor stream. *Icarus*, **51** (1982), 655–659.

DMS – H. Betlem *et al.*, *Dutch Meteor Society Photographic Orbit Database* (Dutch Meteor Society, 2001).

ER55 – C. D. Ellyett and K. W. Roth, The radar determination of meteor showers in the southern hemisphere. *Austr. J. Phys.* **8** (1955), 390–401.

G – D. P. Galligan, Radar meteoroid orbit stream searches using cluster analysis. *Mon. Not. R. Astron. Soc.* **340** (2003), 899–907.

G03 – D. P. Galligan, A direct search for significant meteoroid stream presence within an orbital data set. *Mon. Not. R. Astron. Soc.* **340** (2003), 893–898.

GB – D. P. Galligan and W. J. Baggaley, Wavelet enhancement for detecting shower structure in radar meteoroid data – II. Application to the AMOR data set. In *Dust in the Solar System and Other Planetary Systems*, ed. S. F. Green, I. P. Williams, J. A. M. McDonnell and N. McBride. (New York: Pergamon, 2002), pp. 48–60.

GE – G. Gartrell and W. G. Elford, Southern Hemisphere meteor stream determination. *Austr. J. Phys.* **8** (1975), 591–620.

GK – G. W. Kronk, *Meteor Showers. A Descriptive Catalog*, (Hillside, NJ: Enslow, 1988), 331pp.

H02 – I. Hasegawa, Parent objects of alpha-Capricornid meteor stream. *ESA SP* **495** (2001), 55–62.

H48 – C. Hoffmeister, *Meteorströme*, (Leipzig: Barth, 1948), 286pp.

HA – G. S. Hawkins and M. Almond, Radio echo observation of the major night-time meteor streams. *Mon. Not. R. Astron. Soc.* **112** (1952), 219–233.

HH70 – K. B. Hindley and M. A. Houlden, *Nature* **225** (1970), 1232.

HV# – These refer to individual meteors with Harvard numbers used in the publications by JW and MP.

IMO – J. Rendtel, R. Arlt, R. Koschack *et al.*. Chapter 6: Meteor shower descriptions. In *Handbook for Visual Meteor Observers*. Monograph N° 2. (Potsdam: International Meteor Organization, 1995).

IAU# – Meteoroid orbit from IAU Meteor Orbit Database (2003).

J – L. G. Jacchia, Meteors, meteorites, and comets: interrelations. In *The Moon, Meteorites and Comets*, ed. B. M. Middlehurst and G. P. Kuiper. (Chicago, IL: University of Chicago Press, 1963), Ch. 22, pp. 774–798.

J92 – T. Jopek, TV meteor stream searching. In *Meteoroids and Their Parent Bodies*, ed. J. Stohl and I. P. Williams. (Bratislava: Astron. Inst. Slovak Acad. Sci., 1992), pp. 269–272.

JN# – Meteoroid orbit from Nippon Meteor Society photographic orbit surveys.

JVF – T. J. Jopek and Cl. Froeschlé, A stream search among 502 TV meteor orbits. An objective approach. *Astron. Astrophys.* **320** (1997), 631–641; T. J. Jopek, G. B. Valsecchi and Cl. Froeschlé, Meteoroid stream identification: a new approach – II. Application to 865 photographic meteor orbits. *Mon. Not. R. Astron. Soc.* **304** (1998), 751–758.

JW – L. G. Jacchia and F. L. Whipple, Precision orbits of 413 photographic meteors. *Smithsonian Contrib. Astrophys.* **4** (1961), 97–129.

K71 – P. A. Koning, reference in B. A. Mackenzie *Solar System Debris*, (1971), pp. 39–40.

KL – B. L. Kashcheyev and V. N. Lebedinets, Radar studies of meteors. *Smithsonian Contrib. Astrophys.* **7** (1963), 183–199; B. L. Kascheev, V. N. Lebedinets and M. F. Lagutin, *Meteoric Phenomena in the Earth's Atmosphere*, No. 2. (Moscow: Nauka, 1967), pp. 130–258.

KO – K. Ohtsuka, Eta Lyrid meteor stream associated with comet Iras–Araki–Alcock, 1983 VII. In *Origin and Evolution of Interplanetary Dust*, ed. A. C. Levasseur-Regourd and H. Hasegawa. (Dordrecht: Kluwer, 1991), pp. 315–318.

KP – L. Kresák and V. Porubčan, The dispersion of meteors in meteor streams. I. The size of the radiant areas. *Bull. Astron. Inst. Czechoslov.* **21** (1970), 153–170.

KZ99 – D. Koschny and J. Zender, Possible new radiant in Auriga on November 17, 1998. *WGN* **27** (1999), 51–52.

L – B. A. Lindblad, The orbit of the eta Aquariid meteor stream. In *Asteroids, Comets Meteors III.* (Uppsala: Astronomical Observatory, 1990), pp. 551–553.

L54 – A. C. B. Lovell, *Meteor Astronomy* (Oxford: Clarendon Press, 1954).

L71A – B.A. Lindblad, A stream search among 865 precise photographic meteor orbits. *Smithsonian Contrib. Astrophys.* **12** (1971), 1–13.

L71B – B.A. Lindblad, 2. A computerized stream search among 2401 photographic meteor orbits. *Smithsonian Contrib. Astrophys.* **12** (1971), 14–24.

L87 – B.A. Lindblad, Physics and orbits of meteoroids. In *The Evolution of the Small Bodies of the Solar System. Proc. Int. School of Physics "Enrico Fermi"*, held at Villa Monastero, Varenna on Lake Como, Italy, August 5–10, 1985, ed. M. Fulchignoni and L. Kresák. (Amsterdam: North-Holland, 1987), pp. 229–251.

LK – V. N. Lebedinets, V. N. Korpusov and A. K. Sosnova, *Trudy Inst. Experiment. Meteorolog.* **1** (34) (1972), 88.

LO90 – B. A. Lindblad and D. Olsson-Steel, The Monocerotid meteor stream and comet Mellish. *Bull. Astron. Inst. Czechoslov.* **41** (1990), 193–200.

LOS – B. A. Lindblad, K. Ohtsuka and K. Shirakawa, The orbit of the Eta Aquarid meteor stream. *Planet. Space Sci.* **42** (1994), 113–116.

M – R. A. McIntosh, An index to southern meteor showers. *Mon. Not. R. Astron. Soc.* **95** (1935), 709–718.

M02 – H. Meng, Activity of iota-Aurigids in 2001 and the possible orbit of the meteoroid stream. *WGN*, **30** (2002), 175–180; H. Meng, Determination and analysis of the new iota-Aurigid meteor shower from 1998, 1999, and 2000 plotting data. *WGN* **30** (2002), 32–37.

M96 – M. Langbroek, A small meteor outburst on June 15–16, 1996. *WGN* **24** (1996), 115–118.

MB97 – M. de Lignie, H. Betlem, Simultane video meteoren van de leoniden aktie 1995. *Radiant* **19** (1997), 68–71.

MC – M. J. Currie, A short-duration telescopic shower. *WGN* **23** (1995), 151–154.

MK – M. R. Kidger, On the existence of the June Lyrid meteor shower. *WGN* **28** (2000), 171–176.

MLB – M. C. de Lignie and H. Betlem, A double-station video look on the October meteor showers. *WGN* **27** (1999), 195–201.

MP – R. E. McCrosky and A. Posen, Oribtal elements of photographic meteors. *Smithsonian Contrib. Astrophys.* **4** (1961), 15–84.

MV – M. van Vliet, Meteorenzwerm aktief of 17 January! *Radiant, J. DMS* **15** (1993), 52–53.

MZ – H. Meng, J. Zhu, X. Gong *et al.*, A new asteroid-associated meteor shower and notes on comet-asteroid connection. *Icarus* **169** (2004), 385–389.

N62 – C. S. Nilsson, Ph.D. thesis, University of Adelaide (1962).

N64 – C. S. Nilsson, The Sextantid meteor stream. *Aust. J. Phys.* **17** (1964), 158–160.

NL – C. S. Nilsson, A southern hemisphere radio survey of meteor streams. *Austr. J.* **17** (1964), 205–256.

NMS# – Meteoroid orbit from the various Nippon Meteor Society orbit surveys.

O89 – K. Ohtsuka, The December Monocerotids and P/Mellish. *WGN* **17** (1989), 93–96.

O97 – K. Ohtsuka, C. Shimoda, M. Yoshikawa and J.-I. Watanabe, Activity profile of the Sextantid meteor shower. *Earth, Moon Planets* **77** (1997), 83–91.

OT – K. Ohtsuka, T. Tanigawa, H. Murayama and I. Hasegawa, The new meteor shower η–Eridanids. *ESA SP* **495** (2001), 109–112.

PG – V. Porubčan and M. Gavajdová, A search for fireball streams among photographic meteors. *Planet. Space Sci.* **42** (1994), 151–155; M. Gavajdova, Search for associations between fireball streams and asteroids. *Earth, Moon Planets* **68** (1995), 289–292.

PJ – P. Jenniskens, Meteor stream activity I. The annual streams. *Astron. Astrophys.* **287** (1994), 990–1013; P. Jenniskens, Meteor stream activity II. Meteor outbursts. *Astron. Astrophys.* **295** (1995), 206–235.

PK – V. Porubčan and L. Kornoš, The Taurid meteor shower. *ESA SP* **500** (2002), 177–180.

PW04 – P. Wiegert, *Proc. Meteoroids 2004 meeting*, Ontario, London (in press) 2004.

R81 – R. A. MacKenzie, *Solar System Debris*. BMS. (New York: Dover. pp. 22, 40. Very uncertain radiant of outburst "Cassiopeiids" at 358.5 + 61.7 on same day may refer to same shower.

R92 – J. Rendtel, Delta Aurigids and September Perseids. In *Meteoroids and Their Parent Bodies*, ed. J. Stohl and I. P. Williams. (Bratislava: Astron. Inst. Slovak Acad. Sci., 1992), pp. 185–188.

RK – L. J. Rogers and C. S. L. Keay, Observations of some southern hemisphere meteor showers. In *Meteoroids and Their Parent Comets*, ed. J. Stohl and I. P. Williams., (Bratislava: Astron. Inst. Slovak Acad. Sci., 1992), pp. 273–276.

RRR – M. S. Rao, P. V. S. Rama Rao and P. Ramesh, Minor meteor showers of November, December, and January. *Aust. J. Phys.* **22** (1969), 767–774.

SASY – K. Suzuki, T. Akebo, S. Suzuki and T. Yoshida, A new minor shower belonging to the Coma Berenicid Complex? *WGN* **22** (1994), 50–51.

SAYS – S. Suzuki, T. Akebo, T. Yoshida and K. Suzuki, TV observations of the 1998 Giacobinid outburst. *WGN* **27** (1999), 214–218.

SH – R. B. Southworth and G. S. Hawkins, Statistics of meteor streams. *Smithsonian Contrib. Astrophys.* **7** (1963), 261–285.

SS02 – Y. Shigeno and H. Shioi, Outburst of faint Piscids in 2001. *WGN* **30** (2002), 56–58.

T – M. Tomita, A. Murasawa, C. Shimoda *et al.*, On two double-station photographic 1998 Draconids. *WGN* **27** (1999), 118–119.

T65 – A. K. Terentjeva, Malye meteornye roi. In *Issledovaniya Meteorov*. (Moscow: Nauka, 1965), pp. 62–132.

T68 – A. K. Terentjeva, Investigation of minor meteor streams. In: *Physics and Dynamics of Meteors*, ed. L. Kresák and P. Millman. (Dordrecht: Reidel, 1968), pp. 408–427.

T89 – A. K. Terentjeva, Fireball streams. In *Asteroids Comets Meteors III*. (Uppsala: Astronomical Observatory, 1989), pp. 579–584 (also: *WGN* **17**, 242–245).

T98 – A. K. Terentjeva, Outburst of activity of the alpha-Aurigid meteor shower. *WGN* **26** (1998), 77–78.

This – This publication.

TOM – M. Tomita, K. Ohtsuka, T. Maruyama and Y. Shiba, A Pons–Winneckid fireball? Japan, June 24, 1995, 13h04m39s UT. *WGN* **26** (1998), 180–182.

U – M. Ueda, Y. Fujiwara, M. Sugimoto and M. Kinoshita, Results of double-station TV observations in 1998 and 1999. *ESA SP* **495** (2001), 325–330.

V13 – M. Viljev, comet 1912 d (Lowe). *Astron. Nachr.* **195** (1913), 415.

W54 – F. L. Whipple, Photographic meteor orbits and their distribution in space. *Astron. J.* **59** (1954), 201–217.

W55 – A. A. Weiss, Radio echo observations of meteors in the southern hemisphere. *Austr. J. Phys.* **8** (1955), 148–166.

W57 – F. L. Whipple, Some problems of meteor astronomy. In *Radio Astronomy. Proc. IAU Symp.* No. 4 (Jodrell Bank, UK, August 25–27, 1955 H. C. van de Hulst (ed.), (Cambridge: Cambridge Univ. Press, 1957).

W60 – A. A. Weiss, Radio-echo observations of southern hemisphere meteor shower activity from 1956 December to 1958 August. *Mon. Not. R. Astron. Soc.* **120** (1960), 387–403.

WE57 – A. A. Weiss, The distribution of the orbits of sporadic meteors. *Aust. J. Phys.* **10** (1957), 77–102; *Aust. J. Phys.* **10** (1957), 299.

WE60 – A. A. Weiss, Southern hemisphere meteor shower activity in July and August. *Aust. J. Phys.* **13** (1960), 522–531.

W79 – J. Wood, Pisces Austrinids. *WAMS Bull.* Waarnemingsresultaten Australie 1979. *Radiant, J. DMS* **1** (1979), 7.

W81 – J. Wood, Waarnemingen van de alpha-Cruciden zwerm. *WGN* **9** (1981), 6–10.

W87 – J. Wood, Meteor streams of the southern hemisphere. *WGN* **15** (1987), 129–130.

W88 – J. Wood, The October Capricornid meteor stream. *WGN* **16** (1988), 191–194.

ZS – Z. Sekanina, Statisical model of meteor streams IV. A study of radio streams from the synoptic year. *Icarus* **27** (1976), 265–321.

ZS73 – Z. Sekanina, Statistical model of meteor streams III. Stream search among 19303 radio meteors. *Icarus* **18** (1973), 253–284.

Table 8 *Meteoroid stream formation ages*

Summary of meteoroid stream formation epochs of annual showers. The formation date is given as the moment of breakup that generated the meteoroid stream, or the period over which meteoroids were ejected more gradually by the drag of water vapor at perihelion.

Shower	Type	Formation Date	Remnants	Mass × 10^9 kg	Chapter
Andromedids	breakup	AD 1843	3D/Biela	14 000	20
			Andromedids	33	
Phoenicids	breakup	~AD 1819	D/1819 V1 (Blanpain)	<5600	20
			2003 WY_{25}	30	
			Phoenicids	100	
Quadrantids	breakup	~AD 1490	2003 EH_1	16 000	21
			Quadrantids	10 000	
			C/1490 Y_1	50 000	
Daytime Arietids	breakup	≥AD 1059	Marsden-group sungrazers	~10 000	23
			Daytime Arietids	8000	
Geminids	breakup	~AD 1030	3200 Phaethon (1983 TB)	69 000	22
			Geminids	28 000	
Machholz complex	breakup	AD 170, 340, 530, or 710	96P/Machholz 1	75 000	23
			Marsden-group sungrazer progenitor	~10 000?	
			Kracht-group sungrazer progenitor	~10 000?	
			δ-Aquariids South progenitor	1600	
			2003 EH_1	16 000	
(part of) N. Taurids	breakup	~AD 600	2004 TG_{10}	360	25
			Northern Taurids	10 300	
Capricornids	breakup	~AD 10	2002 EX_{12}	17 000	24
			α-Capricornids	5200	
Taurid complex	breakup	~8000 BC	2P/Encke	51 000	25
			2004 TG_{10} progenitor	≥360	
			Southern Taurids progenitors	–.–	
			Northern Taurids progenitors	–.–	
Sporadics	–.–	~18 000 BC	Zodiacal cloud	300 million	31
Orionids	gas drag	16 000 BC	1P/Halley – parent	230 000	27
		~20 000–1400 BC	Orionids	21 000	
		~AD 20 000–700	η-Aquariids (twin)	(included)	
Perseids	gas drag	160 000 BC	109P/Swift–Tuttle – parent	4000 000	27
		3000 BC	Main peak	11 000	
		160 000–3000 BC	Background	330 000	

Table 9 *Working list of possible asteroidal meteor showers*

These "streams" consist of pairings and small groupings among photographic fireballs. The criteria for this list is low end height in the "asteroid domain" (Chapter 30). A good portion may be mere chance alignments of typical low-inclination prograde orbits.

	Code	Name a(AU)	q(AU)	i	Dates (2000) ω	Node	Peak R.A.	$\lambda_\odot^{max}$ Decl.	χ ΔR.A./Δλ	W ΔDecl./Δλ	ZHR^{max} V_g	Notes $\Delta_{C\text{–}E}$ (AU)
#263	NAN	ν-Andromedids			**01/03–01/11**		**Jan. 07**	**286.3**	–.–	–.–	–.–	
	$N=2$	1.904	0.980	6.2	188.6	286.3	20	+40	–.–	–.–	7.7	T89
	Parent?	*2002 XO_{14}*	*(Epoch 2005-01-30)*					$\lambda_\odot=287.66$	$T_J=+2.98$, $H_N=+21.84$			
	(2005)	*2.778*	*0.974*	*2.608*	*192.579*	*287.658*	*34.0*	*+25.6*	–.–	–.–	*8.91*	*0.0054*
#264	XCE	ξ-Cetids			**01/04–01/09**		**Jan. 07**	**286.3**	–.–	–.–	–.–	
	$N=$–.–	1.986	0.978	1.2	7.8	106.3	37	+08	–.–	–.–	7.7	T89
#265	JGD	January γ-Delphinids			**01/04–01/16**		**Jan. 10**	**289.4**	–.–	–.–	–.–	
	$N=$–.–	2.414	0.896	12.6	140.0	289.0	311	+18	–.–	–.–	14.0	T89
	$N=2$	2.41	0.892	12.6	140.0	289.7	309.9	+18.5			13.7	This work
#266	ACC	α-Cancrids			**01/13–02/07**		**Jan. 25**	**303.5**	–.–	–.–	–.–	
	$N=3$	1.167	0.484	7.2	112.6	124.2	136.1	+06.1	–.–	–.–	19.3	PG
#267	JNO	January ν-Orionids			**01/01–02/04**		**Jan. 28**	**307.9**	–.–	–.–	–.–	
	$N=$–.–	1.866	0.854	4.1	51.7	112.5	88	+12	–.–	–.–	12.0	T89
	$N=2$	2.684	0.907	3.4	36.7	127.9	92.9	+13	–.–	–.–	12.1	This work
	Tagish Lake: 2.0		0.885	2.0	223.9	297.901	89.9	+29.8	–.–	–.–	11.1	TB05
	Parent?	*2003 AC_{23}*	*(Epoch 2005-01-30)*					$\lambda_\odot=310.03$	$T_J=+3.45$, $H_N=+21.54$			
	(2005)	*2.165*	*0.899*	*0.400*	*220.101*	*310.032*	*99.5*	*+24.4*	–.–	–.–	*11.27*	*0.0354*
#268	BCD	β-Cancrids			**01/31–02/10**		**Feb. 05**	**316.2**	–.–	–.–	–.–	
		= mixed bag (most collections have cometary meteoroids mixed in)										
	$N=2$	2.25	0.826	3.9	54.3	139.1	119.4	+11.0	–.–	–.–	14.6	This work
	$N=3$	1.981	0.794	4.6	61.4	136.2	121.3	+09.5	–.–	–.–	14.8	PG
		1.998	0.799	3.8	61.3	135.3	118	+12	–.–	–.–	14.1	T89 (8S)
	Parent?	*2002 MT_3*	*(Epoch 2005-01-30)*					$\lambda_\odot=328.90$	$T_J=2.91$, $H_N=+19.68$			
	(2005)	*2.808*	*0.871*	*3.941*	*44.739*	*148.899*	*121.5*	*+09.7*	–.–	–.–	*14.04*	*0.0354*

#269	OCS	ω-Cassiopeiids			**02/06–02/06**		**Feb. 06**	**317.5**	–.–	–.–	–.–	
	Innisfree	1.871	0.986	12.2	177.9	317.5	7.4	+66.4	–.–	–.–	9.4	H
	Ridgedale	1.885	0.984	12.3	186.6	316.7	31.5	+72.2	–.–	–.–	9.6	H
#270	FAO	Feb. α-Orionids			**02/04–02/18**		**Feb. 07**	**318.0**	–.–	–.–	–.–	
	$N=5$	2.16	0.954	3.5	22.4	318.0	88.8	+09.4	+0.83	−0.11	9.4	This work
#271	MLY	March Lyncids			**02/24–03/07**		**Mar. 01**	**339.4**	–.–	–.–	–.–	
	$N=3$	1.965	0.962	7.4	204.1	339.4	123.5	+50.3	–.–	–.–	9.4	PG
	$N=$–.–	2.195	0.938	9.4	209.2	329.2	114	+52	–.–	–.–	11.4	T89 (11)
#272	ACO	April α-Comae Berenicids			**04/05–04/21**		**Apr. 07**	**17.8**	–.–	–.–	–.–	
	$N=5$	2.680	0.839	11.9	232.5	22.1	193.1	+22.9	–.–	–.–	16.7	PG
	Pribram	2.401	0.790	10.48	241.74	17.803	192.34	+17.46	–.–	–.–	17.43	SHO
	Neusch.	2.40	0.793	11.43	241.1	16.827	192.33	+19.58	–.–	–.–	17.51	SHO
	Glanerb.	(2.40)	0.919	21.0	218.3	17.817	202 ± 7	$+47 \pm 6$	–.–	–.–	20.3 ± 4	JF
#273	PBO	ϕ-Bootids			**04/21–05/15**		**May 02**	**42.1**	–.–	–.–	–.–	
	$N=3$	2.67	0.996	20.8	195.6	53.6	217.5	+53.7	–.–	–.–	15.1	This work
	$N=3$	2.205	0.963	19.1	207.6	39.4	212.0	+47.7	–.–	–.–	14.8	PG
	$N=$–.–	2.404	0.995	15.6	194.0	54.0	203	+51	–.–	–.–	12.3	T89 (26)
	$N=6$	1.248	0.949	19.3	225.8	41.2	240	+51	–.–	–.–	11.4	L72B
	$N=2$	1.53	0.94	20.2	223.3	47.8	231.7	+41.0	–.–	–.–	16.9	**SH**
#274	NUM	ν-Ursae Majorids			**05/04–05/26**		**May 18**	**54.4**	–.–	–.–	–.–	
	$N=2$	2.22	1.005	9.0	188.3	44.3	173.5	+45.2	–.–	–.–	9.2	This work
	$N=3$	1.976	1.006	7.2	189.6	53.7	180.6	+35.3	–.–	–.–	8.2	PG
	Parent?	*2003 QO_{104}*	*(Epoch 2005-08-18)*					$\lambda_{\odot}=58.32$	$T_J=+3.51$, $H_N=+15.87$			
	(2005)	*2.126*	1.011	*11.619*	*183.492*	*58.324*	187.1	*+52.0*	–.–	–.–	*9.76*	*0.0033*
#275	CLI*	χ-Librids			**06/06–06/24**		**Jun. 10**	**79.7**	–.–	–.–	–.–	
	$N=4$	3.484	0.898	1.3	46.2	255.2	223.2	−20.4	–.–	–.–	12.2	This work
#276	ADR	α-Draconids			**06/04–06/15**		**Jun. 11**	**80.5**	–.–	–.–	–.–	
	$N=3$	1.56	1.014	18.8	173.5	80.5	220.0	+66.4	–.–	–.–	11.6	This work

* CLI – This stream has a Jupiter-family comet, type $I_J = 2.59$.

Table 9 (*cont.*)

	Code	Name a(AU)	q(AU)	i	Dates (2000) ω	Node	Peak R.A.	$\lambda_{\odot}^{max}$ Decl.	χ ΔR.A./$\Delta\lambda$	W ΔDecl./$\Delta\lambda$	ZHR^{max} V_g	Notes Δ_{C-E} (AU)
#277	GCA	γ-Camelopardalids			**07/01–07/23**		**Jul. 12**	**109.9**	–.–	–.–	–.–	
	$N=2$	1.38	0.764	37.3	100.8	109.9	88.7	+79.0	–.–	–.–	22.7	This work
#278	MSR	July μ-Serpentids			**07/03–07/28**		**Jul. 15**	**112.2**	–.–	–.–	–.–	
	$N=3$	2.42	0.990	6.4	201.7	112.2	240.8	+04.7	–.–	–.–	9.8	This work
#279	ZED	July ζ-Draconids			**07/07–07/30**		**Jul. 18**	**115.7**	–.–	–.–	–.–	
	$N=6$	2.757	1.016	32.5	176.7	115.7	251.6	+66.5	–.–	–.–	20.6	This work
#280	ADL	δ-Librids			**07/31–08/29**		**Aug. 14**	**141.7**	–.–	–.–	–.–	
	$N=3$	1.946	1.009	2.8	182.8	141.7	234.5	−02.5	–.–	–.–	7.1	This work
	Parent?	*2003 MU*	*(Epoch 2005-01-30)*					*$\lambda_{\odot}=139.08$*	*$T_J=+3.62$, $H_N=+20.37$*			
	(2005)	*1.874*	*1.013*	*5.881*	*178.180*	*140.999*	*231.2*	*+13.0*	–.–	–.–	*7.20*	*0.0253*
#282	DCY	δ-Cygnids			**10/04–10/28**		**Oct. 23**	**200.8**	–.–	–.–	–.–	
	$N=5$	2.50	0.991	19.6	191.5	200.8	294.6	+46.5	–.–	–.–	14.0	This work
	$N=3$	2.768	0.972	13.9	199.4	212.5	319.1	+34.9	–.–	–.–	12.3	PG
	Parent?	*2004 BE_{68}*	*(Epoch 2006-03-06)*					*$\lambda_{\odot}=211.00$*	*$T_J=+3.96$, $H_N=+18.37$*			
	(2006)	*1.781*	*0.989*	15.750	191.061	*211.001*	*299.4*	*+45.4*	–.–	–.–	*10.84*	*0.0118*
#283	OPL	π-Leonids			**10/25–11/06**		**Oct. 31**	**218.1**	–.–	–.–	–.–	
	$N=$–.–	0.756	0.510	4.0	189.0	38.1	146	+08	–.–	–.–	5.8	T89 (58)
#284	OMA	o-Ursae Majorids			**11/12–12/8**		**Nov. 11**	**228.9**	–.–	–.–	–.–	
	$N=$–.–	0.794	0.566	16.1	337.0	229.6	115	+70	–.–	–.–	9.4	T89 (64N)
#285	GTA	γ-Taurids			**11/01–11/21**		**Nov. 15**	**232.8**	–.–	–.–	–.–	
	$N=2$	1.13	0.634	0.4	101.5	52.8	60.0	+17.5	–.–	–.–	14.1	This work
#286	FTA	ω-Taurids			**11/13–11/27**		**Nov. 23**	**240.2**	–.–	–.–	–.–	
	$N=2$	1.79	0.529	3.2	97.4	60.2	58.2	+16.8	–.–	–.–	21.7	This work

	Parent?	*2002 UK_{11}*	*(Epoch 2005-01-30)*					*$\lambda_{\odot} = 233.24$*	*$T_J = +21.64$, $H_N = +21.64$*			
		1.322	*0.562*	*5.154*	*100.537*	*53.238*	*59.0*	*+12.0*	*–.–*	*–.–*	*17.80*	*0.0256*
#287	NER	Nov. ε-Eridanids			**11/18–11/28**		**Nov. 24**	**240.8**	–.–	–.–	–.–	
		1.380	0.825	8.4	65.6	60.8	52	−07	–.–	–.–	11.1	T89 (68)
	Parent?	*2000 KA*	*(Epoch 2005-01-30)*					*$\lambda_{\odot} = 243.33$*	*$T_J = +4.79$, $H_N = +21.35$*			
	(2005)	*1.333*	*0.716*	*6.707*	*82.368*	*63.333*	*62.1*	*+05.2*	*–.–*	*–.–*	*13.53*	*0.0013*
#288	DSA	S. Dec. δ-Arietids			**12/01–12/28**		**Dec. 08**	**256.5**	–.–	–.–	–.–	
		(Mixed bag)										
	$N = 4$	1.901	0.758	2.8	67.0	78.9	66.6	+15.6	–.–	–.–	15.2	PG
	$N =$ –.–	2.125	0.743	4.4	68.2	84.1	71	+14	–.–	–.–	16.6	T89 (74)
	$N =$ –.–	1.982	0.788	5.8	62.8	75.4	61	+07	–.–	–.–	14.9	T89
	$N =$ –.–	2.440	0.786	1.2	59.7	75.0	55	+18	–.–	–.–	15.8	T89
#289	DNA	N. Dec. δ-Arietids			**12/01–12/28**		**Dec. 08**	**256.5**	–.–	–.–	–.–	
	$N =$ –.–	1.826	0.766	4.2	247.3	251.1	53	+29	–.–	–.–	14.6	T89
	$N = 14$	2.420	0.857	2.1	228.0	262.6	54	+25	–.–	–.–	13	BL
	$N = 7$	2.13	0.838	1.8	232.8	257.6	52	+22	–.–	–.–	13.2	AC
	Parent?	*1990 HA*	*(Epoch 2005-08-18)*					*$\lambda_{\odot} = 252.64$*	*$T_J = +3.03$, $H_{10} = +16.18$*			ML
	(2005)	*2.570*	*0.781*	*1.471*	*240.412*	*252.640*	*53.1*	*+22.4*	*–.–*	*–.–*	*16.10*	*0.0623*

Legend to notes:

H – I. Halliday, Detection of a meteorite "stream" – observations of a second meteorite fall from the orbit of the Innisfree chondrite. *Icarus* **69** (1987), 550–556.

PG – M. Gavajdova, Search for associations between fireball streams and asteroids. *Earth, Moon Planets* **68** (1995), 289–292.

JF – P. Jenniskens, J. Borovicka, H. Betlem, *et al.*, Orbits of meteorite producing fireballs. The Glanerbrug – a case study. *Astron. Astrophys.* **255** (1992), 373–376; L. Foschini, P. Farinella, Ch. Froeschlé *et al.*, Long-term dynamics of bright bolides. *Astron. Astrophys.* **353** (2000), 797–812.

ML – M. Langbroek, The November–December delta Arietids and asteroid 1990 HA: on the trail of a meteoroid stream with meteorite-sized members. *WGN* **31** (2003), 177–182.

SHO – P. Spurný, D. Heinlein and J. Oberst, Na 43 jaar een tweede Pribram!! *Radiant, J. DMS* **24** (2002), 51–53.

TB05 – A. Terentjeva and S. Barabanov, The fireball stream of the Tagish Lake meteorite. *WGN* **32** (2005), 60–62.

Table 10a *Meteor showers on Mercury and Venus, as predicted from cometary orbits*

Published results were recalculated using the methods of Neslusan *et al.* Solar longitude now applies to the position of Venus. CH are new calculations by Apostolos Christou.

	Epoch	$\lambda_{\odot}$ (°)	R.A. (°)	Decl. (°)	V_g (km/s)	Δr $(C\text{-}P)$[1] (AU)	Type	src[2]
Mercury:								
2P/Encke	2004-08-23	334.6	113.5	+55.2	24.3	−0.026	J	
2004 TG_{10}	2005-01-30	53.2	74.3	+08.3	30.3	−0.014	J	
C/1989 A_3 (Bradfield)	1988-12-05	28.1	273.4	+34.1	69.6	+0.002	H	
Venus:								
2201 Oljato	2005-01-30	41.1	66.2	+19.7	16.9	+0.0065	J	C
1P/Halley	1994-02-17	69.8	353.9	+4.8	80.4	+0.0496	H	C
χ Orionids	–.–	78	84	+15	25.0	−0.0009	J	C, CH
1998 SH_2 (low i)	1998-09-21	98.0	177.7	−09.2	10.2	+0.037	J	C
1999 RD_{32}	2005-01-30	122.4	141.0	−4.3	20.1	+0.018	J	C
2101 Adonis (low i)	2005-01-30	127.9	308.8	−24.0	25.8	+0.022	J	C
δ Aquariids (N)	–.–	150	351	+1	45.9	−0.0004	J	C, CH
45P/Honda–Mrkos–Pajd	2001-06-20	168.4	338.0	−12.0	23.5	+0.010	J	C
35P/Herschel–Rigollet	1939-08-05	176.0	65.8	−46.5	46.3	+0.082	H	B, CH, SBR
72P/Denning–Fujikawa	1979-01-07	197.3	286.7	−49.5	11.3	+0.055	J	B, C
141P-A/Machholz 2	2005-04-20	246.2	282.4	+18.2	18.3	+0.068	J	C
27P/Crommelin	1984-06-29	250.9	321.6	+66.0	27.2	−0.026	H	B, C, SBR
122P/de Vico	1996-07-16	259.6	136.0	−42.9	56.3	−0.053	H	B, C
12P/Pons–Brooks	1954-09-15	255.9	201.4	+67.3	53.9	+0.073	H	B, C
C/1917 F_1 (Mellish)	1917-04-15	268.7	107.3	+08.4	46.2	−0.008	H	B, C, CH
3200 Phaethon (Geminids)	2005-08-18	265.4	116.4	+35.2	39.5	+0.042	J	C
2101 Adonis (low i)	2005-01-30	296.9	301.0	−15.6	25.9	−0.0138	J	C
45P/Honda–Mrkos–Pajd	2001-06-20	300.7	318.0	−21.2	23.5	+0.0016	J	C, SBR
2201 Oljato	2005-01-30	305.7	99.9	+25.3	16.9	+0.0083	J	C

Notes:

[1] Data by Christou and Selsis are 3-D miss-distances (δ), rather than difference at the node.

[2] C = A. A. Christou, Prospects for meteor shower activity in the venusian atmosphere. *Icarus* **168** (2004), 23–33.

Table 10b *Best dates for viewing the $\lambda_\odot = 252°–263°$ cluster on Venus from Earth*

Date (Oh UT)	$\lambda_\odot$ (°)	Geo. Elong. (°)		Sol. Elong. (°)	Distance (AU)	M_V Magn.	Phase (°)	Ill. %	Diameter (″)
2005 Dec. 21	255.40	301.28	Aqu.	31.7E	0.3351	−4.64	133.6	15.5	49.5
2010 Nov. 23	255.60	208.05	Lib.	32.9W	0.347 77	−4.62	131.1	17.2	47.5
2013 Dec. 20	255.34	298.93	Cap.	30.3E	0.327 21	−4.61	137.4	13.2	51.5
2018 Nov. 22	256.04	205.93	Lib.	34.0W	0.356 31	−4.63	129.0	18.5	46.4
2021 Dec. 19	255.27	296.5	Cap.	28.9E	0.3197	−4.59	139.8	11.8	52.7
2026 Nov. 21	255.97	203.9	Lib.	35.1W	0.365 13	−4.64	127.1	19.8	45.2
2029 Dec. 18	255.20	293.99	Cap.	27.3E	0.312 58	−4.56	142.3	10.5	53.9
2034 Nov. 20	255.90	201.93	Lib.	36.1W	0.3742	−4.64	125.2	21.2	44.2
2037 Dec. 17	255.14	291.37	Cap.	25.6E	0.305 83	−4.53	144.9	09.1	55.1
2042 Nov. 19	255.84	200.03	Lib.	37.1W	0.383 45	−4.64	123.4	22.5	43.1

Table 10c *Good opportunities to watch for Moon impacts from Earth*

The table gives days past the new moon for the peak flux dates of the indicated meteor streams. It covers the years 2005–2061 with the 19 yr lunar cycle causing a repeat in the lunar phase to within one day accuracy. Observable periods are the week near the first quarter (3–10 d old) and the last quarter (19.5–26.5 d old). Favorable geometries are shown as bold and grey. All showers in the table have observable impact regions for the first and last quarter except the Leonids and the η-Aquariids (first quarter only). The calculations of impact geometry and moon phase were made for this publication by Peter S. Gural.

Year	Year	Year	QUA	ETA	CAP	PER	STA	LEO	GEM
2005	2024	2043	**22**	27	**24**	**7**	4	16	13
2006	2025	2044	**4**	**8**	**5**	19	15	27	**24**
2007	2026	2045	15	18	15	0	**26**	**7**	**4**
2008	2027	2046	**26**	1	27	11	**7**	19	16
2009	2028	2047	**6**	12	**9**	21	18	1	27
2010	2029	2048	18	22	19	**4**	29	11	**8**
2011	2030	2049	29	**3**	29	14	**10**	22	19
2012	2031	2050	**10**	14	11	**25**	**21**	**4**	1
2013	2032	2051	**21**	26	**23**	**6**	2	15	12
2014	2033	2052	**3**	**7**	**3**	18	14	25	**22**
2015	2034	2053	14	17	14	28	**24**	**6**	**3**
2016	2035	2054	**24**	28	**25**	**9**	**5**	18	14
2017	2036	2055	**5**	**10**	**7**	20	16	29	**26**
2018	2037	2056	17	21	17	2	28	**9**	**6**
2019	2038	2057	28	2	28	13	**9**	20	17
2020	2039	2058	**8**	13	**10**	**23**	19	2	29
2021	2040	2059	19	24	**21**	**5**	1	13	**10**
2022	2041	2060	1	**5**	2	16	12	23	**20**
2023	2042	2061	12	16	12	27	**23**	**5**	2

Meteor outbursts:

Draconids - 2020 Oct. 07 01:57 UT→ good – shown in figure:
Perseids - 2009, 2016, 2028, 2039–2040
→ 2016, 2039, and 2040 good
Leonids - 2033–2038 → years 2034 and 2037 are good
Aurigids - 2007 Sep. 01 11:37 UT → all impacts on sunlit side
Draconids - 2011 Oct. 08 20:49 UT → moon near full (12 days)
Draconids - 2037 Oct. 10 00:36 UT → moon near new (1 day old)
τ-Herculids - 2022 May 31 04:55 UT → moon new (1 day old)

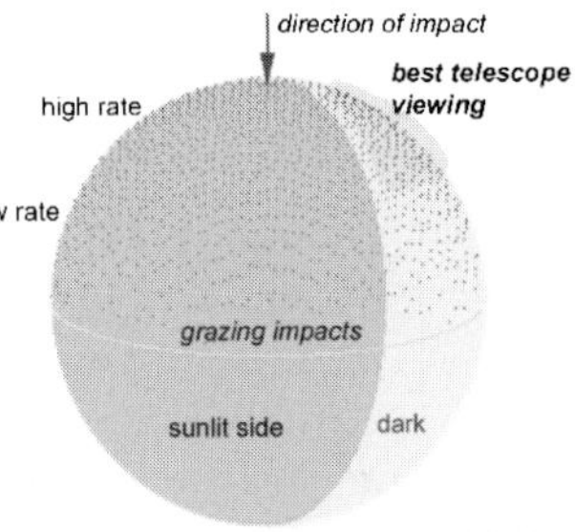

Oct. 7, 2020

Table 10c (*cont.*)

As the geometry changes are not very significant around the observable dates, one can generalize the focus regions on the non-illuminated face for telescopic video observations: QUA first quarter, days 3–10: northeast quadrant; QUA last quarter, days 17–25: northwest quadrant; ETA first quarter – central region; CAP first quarter – eastern half; CAP last quarter – western half; PER first quarter – northeast/central region; PER last quarter – northwest limb; STA first quarter – eastern half; STA last quarter – western half; LEO first quarter – eastern half; GEM first quarter – eastern half; GEM last quarter – western half.

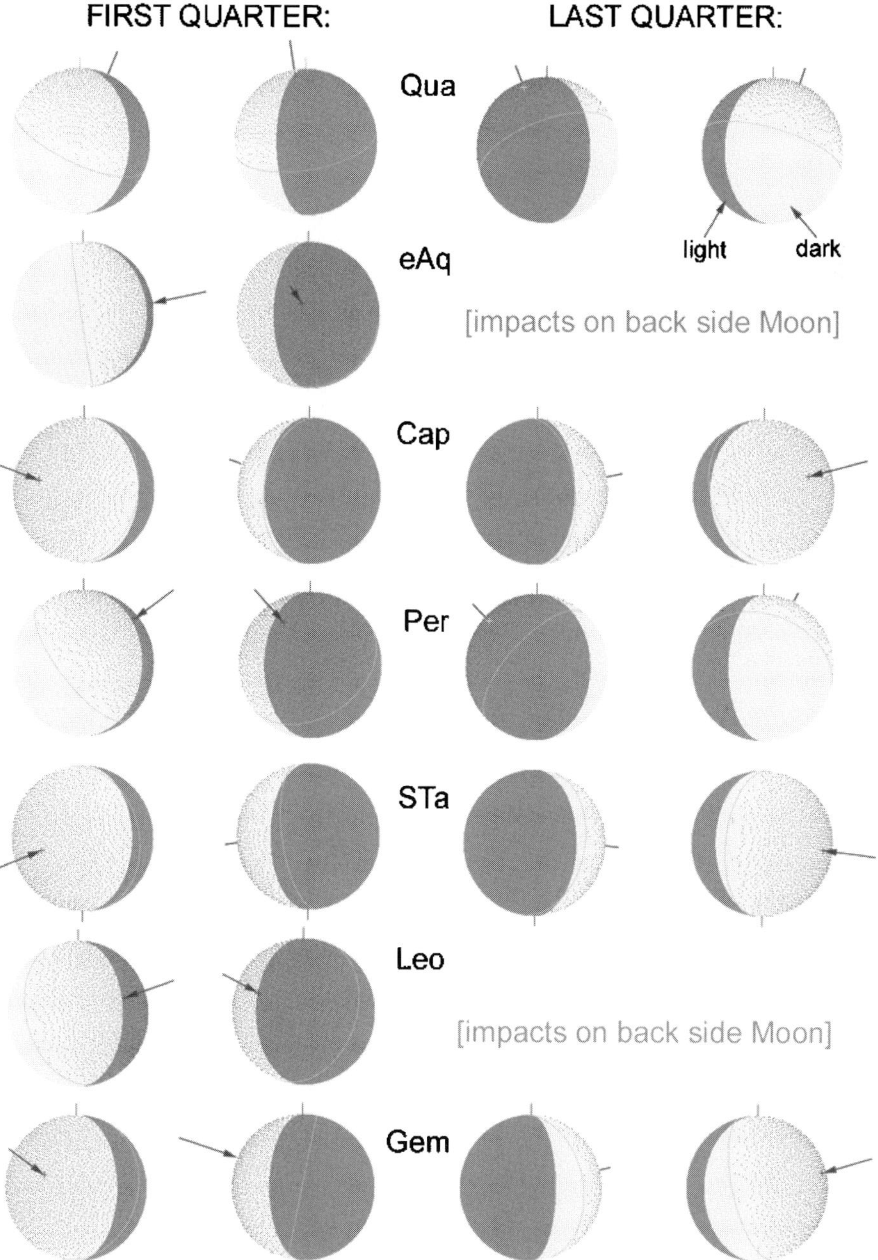

Table 10d *Meteor showers on Mars*[1]

Compilation of literature data and recalculated using Neslusan *et al.* Velocities and radiant are before adding the contribution from the acceleration due to the gravitational field of Mars.

	Epoch	$\lambda_\odot$ (°)	R.A. (°)	Decl. (°)	V_g (km/s)	Δr $(C\text{-}M)$ (AU)	Type	src
P/2001 J_1 (NEAT)	2001-05-07	19.6	25.2	−4.8	20.0	−0.038	J	
9P/Tempel 1	2005-03-11	71.9	151.5	+54.5	7.5	+0.007	J	T, CB, SBR, CH
10P/Tempel 2	2005-02-15	117.9	243.7	+33.2	7.6	−0.038	J	
(5660) 1974 MA	2005-08-18	121.3	346.7	−35.9	25.3	−0.024	J	CB, CH
45P/Honda–Mrkos–Apjdusakova	2001-06-20	123.2	319.0	−15.9	23.0	−0.013	J	T, SBR
58P/Jackson–Neujmin	1996-03-18	168.5	287.8	+28.1	10.1	+0.002	J	
P/2001 R_1 (LONEOS)	2002-01-06	210.4	5.8	−17.3	9.8	+0.0004	J	SBR
1P/Halley	1994-02-17	230.5	109.3	+16.1	53.9	+0.062	H	T, CB, CH, SBR
146P/Shoemaker–LINEAR	2001-02-20	237.4	246.1	−75.2	12.8	−0.008	J	SBR, CH
144P/Kushida	2000-08-04	256.8	359.8	+24.0	7.7	+0.084	J	
38P/Stephan–Oterma	1981-04-26	262.8	27.0	−46.5	11.8	+0.026	H	T, CHR
114P/Wiseman–Skiff	2000-05-16	267.8	328.1	+59.2	11.0	−0.006	J	CB, SBR, CH
13P/Olbers	1956-10-14	268.0	95.3	−21.5	26.9	+0.027	H	T, CB, CH
79P/du Toit–Hartley	2003-03-22	309.0	119.0	+27.5	13.1	+0.057	J	
(5335) Damocles	2005-08-18	312.9	239.8	+57.8	29.1	−0.056	J	CB, CH
81P/Wild 2	2004-08-23	329.0	98.3	+11.5	7.9	+0.035	J	
P/1999 D_1 (Hermann)	1999-05-22	349.4	15.5	+75.0	12.5	−0.009	J	T, SBR

Note:

[1]T = A. H. Treiman and J. S. Treiman, Cometary dust streams at Mars: preliminary predictions from meteor streams at Earth and from periodic comets. *J. Geophys. Res.* **105** (2000), 24 571–24 582.

M = Y. Ma, I. P. Williams, W. H. Ip and W. Chen, The velocity distribution of periodic comets and the meteor shower on Mars. *Astron. Astropys.* **394** (2002), 311–316.

A = L. G. Adolfsson, B. Å. S. Gustafson and C. D. Murray, The Martian atmosphere as a meteoroid detector. *Icarus* **119** (1996), 144–152.

CB = A. A. Christou and K. Beurle, Meteoroid streams at Mars: possibilities and implications. *Planet Space Sci.* **47** (1999), 1475–1485.

SBR = F. Selsis, J. Brillet and M. Rapaport, Meteor showers of cometary origin in the Solar System: revised predictions. *Astron. Astrophys.* **416** (2004), 783–789.

CH: A. Christou, private communication.

Table 10e *Meteor showers on Jupiter and its moons*[4]

Recalculated using methods by Neslusan *et al.*

	Epoch	$\lambda_\odot$ (°)	R.A. (°)	Decl. (°)	V_g (km/s)	Δr (*C-J*) (AU)	Type	src
132P/Helin–Roman–Alu 2	2005-08-18	9.8	340.9	−19.4	5.6	−0.027	J	LA, SBR
14P/Wolf	2004-02-05	23.5	261.7	−73.0	6.2	+0.039	J	LA, SBR
78P/Gehrels 2	2005-05-30	30.5	341.9	−26.2	3.9	−0.083	J	LA, SBR
P/1991 V_1 (Shoemaker–Levy 6)	1992-01-19	34.7	253.0	−9.4	9.3	−0.042	J	LA, SBR
18D/Perrine–Mrkos	1969-01-19	60.9	312.7	−43.9	6.3	+0.010	J	LA
104P/Kowal 2	2004-03-16	66.1	17.7	−18.8	6.1	−0.077	J	LA
43P/Wolf–Harrington	2005-05-30	74.6	20.6	−26.6	5.7	−0.016	J	LA, SBR
144P/Kushida	2000-08-04	74.9	40.0	+7.7	7.5	−0.007	J	LA, SBR
139P/Väisälä–Oterma(low *i*)	1999-04-12	82.7	311.6	−39.3	2.1	+0.040	J	LA, SBR
76P/West–Kohoutek–Ikemura	2001-05-11	84.1	343.2	+35.6	6.7	+0.057	J	LA, SBR
75P/Kohoutek	1994-06-17	89.7	6.8	−17.7	9.3	+0.023	J	LA, SBR
P/1998 W_1 (Spahr)	1999-01-02	102.0	339.2	+27.8	5.7	−0.026	J	LA, SBR
163P/NEAT	2005-03-11	103.7	345.7	+20.6	4.2	+0.063	J	
59P/Kearns–Kwee	2000-04-06	128.2	345.8	−28.4	6.1	−0.010	J	LA, SBR
124P/Mrkos	2002-05-06	181.4	94.0	−15.1	7.6	−0.023	J	LA, SBR
26P/Grigg–Skjellerup	1997-11-08	216.3	122.5	+44.4	7.2	+0.057	J	LA, SBR
135P/Shoemaker–Levy 8	1999-07-01	226.7	160.0	+40.5	2.8	−0.076	J	LA, SBR
83P/Russell 1	1985-06-24	233.9	151.4	+52.4	6.3	+0.058	J	LA, SBR
91P/Russell 3	2005-08-18	247.9	164.0	+56.1	4.2	+0.111	J	LA, SBR
86P/Wild 3	2001-06-20	249.0	142.6	−26.5	4.1	−0.094	J	LA, SBR
143P/Kowal–Mrkos	2004-11-11	249.2	94.3	+34.7	6.0	+0.067	J	LA
P/2004 R_1 (McNaught)	2004-09-15	296.0	203.3	+0.4	5.7	−0.018	J	
53P/Van Biesbroeck	2005-01-30	339.3	171.9	−3.7	7.8	+0.004	J	LA
P/1998 QP_{54} (LONEOS–Tucker)	1999-03-03	341.9	298.5	+7.5	6.4	+0.013	J	LA
P/1998 S_1 (LINEAR–Mueller)	2001-04-01	350.0	313.5	+7.5	4.6	+0.039	J	LA, SBR
168P/Hergenrother	2001-03-04	353.1	295.6	+6.7	6.7	+0.002	J	LA, SBR

Note:

[4] LA = L. Larson Shane, Determination of meteor showers on other planets using comet ephemerides. *Astron. J.* **121** (2001), 1722–1729.

Table 10f *Meteor showers on other outer planets and their moons*

Compilation of literature data. Recalculated.

	Epoch	$\lambda_{\odot}$ (°)	R.A. (°)	Decl. (°)	V_g (km/s)	Δr (C-P) (AU)	Type[1]	src
Saturn:								
C/1999 S_3 (LINEAR)	2000-04-06	11.2	327.1	+11.0	13.5	+0.097	L	SBR
155P/Shoemaker 3	2003-06-10	93.8	45.3	+19.4	5.5	+0.080	S	SBR
P/1999 V_1 (Catalina)	1999-12-23	114.3	51.2	−15.6	3.9	+0.023	S	SBR
P/1998 U_3 (Jäger)	1999-05-22	123.5	46.6	−13.3	4.5	−0.021	S	LA, SBR
140P/Bowell–Skiff (low i)	1999-05-22	140.2	93.9	+16.8	5.3	+0.072	S	SBR
134P/Kowal–Vavrova (low i)	1999-08-10	240.7	116.2	+28.2	4.4	−0.117	S	SBR
P/1994 J_3 (Shoemaker 4)	1994-10-15	272.9	188.8	−38.2	4.8	−2.95	S	SBR
126P/IRAS	1997-03-13	357.7	258.5	+10.6	7.3	−0.064	S	LA, SBR
Uranus:								
55P/Tempel–Tuttle	1998-08-15	55.3	311.3	−22.7	9.5	−0.235	U	SBR
122P/de Vico	1996-07-16	79.6	22.0	+21.1	8.7	+0.485	H	SBR
C/1999 XS_{87} (LINEAR)	2000-04-06	87.1	302.0	−28.8	6.7	+0.091	H	SBR
C/1999 E_1 (Li)	1999-09-10	128.3	339.2	+17.8	6.7	−0.035	H	LA, SBR
C/1999 G_1 (LINEAR)	1999-05-22	203.5	67.1	−3.4	9.3	−0.348	H	SBR
Neptune:								
C/1998 Y_1 (LINEAR)	1999-05-22	98.9	315.5	−10.3	6.5	−0.509	N	SBR
C/2000 J_1 (Ferris)	2000-06-25	208.4	68.4	+7.9	9.4	−0.895	L	SBR

Notes:

[1] S = "Saturn-type", with aphelion close to Saturn's orbit, orbital period ~15 yr.
U = "Uranus-type", with aphelion close to Uranus' orbit, orbital period ~33 yr.
N = "Neptune type", with aphelion close to Neptune's orbit, orbital period ~76 yr.

Table 11 *Calendar of exceptional meteor showers (2005–2052).* Chronological Index to Tables 3–6. Bold are the most promissing events. Included are also exceptional encounters with parent bodies.

	Date	Stream	Table
□	2005-08-07	bPe 1rev	3
□	2005-08-12	Per 1479	5c
□	2005-08-12	Per Filament	5d
□	2005-08-12	Per annual	Ch. 17
□	2005-08-12	Per impact Moon	10c
□	2005-08-12	45P 1959	6f
□	2005-11-21	Leo 1167	4a
□	2005-12-22	Urs Filament	5b
□	2006-01-02	C/1969 T_1 1rev	3
□	2006-01-03	Qua impact Moon	10c
□	2006-01-03	Qua annual	Ch. 20
□	2006-04-22	Lyrids annual	Ch. 26
□	**2006-04-23**	**Pup 1848**	**6e**
□	2006-05-12	Comet 73P at Earth	Ch. 8
□	2006-05-27	tHe 1952	6g
□	2006-07-16	kPa 1rev	3
□	2006-08-08	bPe 1rev	3
□	2006-08-12	Per 1479	5c
□	**2006-08-12**	**Per Filament**	**5d**
□	**2006-11-19**	**Leo 1932**	**4a**
□	2006-12-13	Gem annual	Ch. 21
□	2006-12-22	Urs 966	5a
□	2006-12-22	Urs Filament	5b
□	2007-04-28	aBo 1rev	3
□	2007-08-12	Per 1479	5c
□	**2007-08-12**	**Per Filament**	**5d**
□	2007-08-12	Per annual	Ch. 17
□	**2007-09-01**	**Aur 1-rev (!)**	**3**
□	**2007-11-18**	**Leo 1932**	**4**
□	2007-12-10	Phaethon at Earth	Ch. 9
□	**2007-12-21**	**aLy 1-rev**	**3**
□	2007-12-22	Urs Filament	5b
□	2008-01-03	Qua impact Moon	10c
□	2008-01-04	Qua annual	Ch. 20
□	2008-08-09	6P/d'Arrest at Earth	Ch. 19
□	**2008-08-12**	**Per 1479**	**5c**
□	**2008-08-12**	**Per Filament**	**5d**
□	2008-11-17	Leo 1932	4a
□	2008-12-20	aLy 1rev	3

Table 11 (*cont.*)

	Date	Stream	Table
☐	**2008-12-22**	**Urs Filament**	**5b**
☐	2009-01-03	Qua annual	Ch. 20
☐	2009-01-03	Qua impact Moon	10c
☐	2009-08-12	Per 1610	5c
☐	2009-11-17	Leo 1466	4a
☐	**2009-11-17**	**Leo 1533**	**4a**
☐	2009-11-17	Leo Filament	4b
☐	2009-12-13	Gem annual	Ch. 21
☐	**2010-06-24**	**JBo 1830**	**6b**
☐	2010-08-12	Per 1479	5c
☐	2010-08-12	Per annual	Ch. 17
☐	2010-08-12	Per impact Moon	10c
☐	2010-10-20	103P/Hartley 2 at Earth	Ch. 8
☐	2011-01-02	C/1969 T_1 1-rev	3
☐	2011-01-04	Qua annual	Ch. 20
☐	**2011-06-02**	**tHe 1952**	**6g**
☐	**2011-06-05**	**aCi 1-rev**	**3**
☐	2011-08-11	45P at Earth	Ch. 8
☐	**2011-10-08**	**Dra 1900 (!)**	**6d**
☐	2011-11-16	Leo 1800	4a
☐	2012-01-03	Qua impact Moon	10c
☐	2012-04-22	Lyrids annual	Ch. 26
☐	2012-06-04	aCi 1-rev	3
☐	2012-06-11	gDe 1-rev	3
☐	2012-08-12	Per impact Moon	10c
☐	2012-10-08	Dra 1959	6d
☐	2012-11-26	NoO 1-rev	3
☐	2012-12-13	Gem annual	Ch. 21
☐	2013-01-03	Qua impact Moon	10c
☐	**2013-06-11**	**gDe 1-rev**	**3**
☐	**2013-08-12**	**Per 1079**	**5c**
☐	2013-08-12	Per annual	Ch. 17
☐	2013-08-12	Per impact Moon	10c
☐	2014-01-03	Qua impact Moon	10c
☐	2014-01-03	Qua annual	Ch. 20
☐	**2014-05-24**	**2004 CB trails**	**6j**
☐	2014-05-29	2004 CB at Earth	Ch. 8
☐	2014-12-22	Urs 1405	5a
☐	2015-02-08	aCe 1-rev	3
☐	2015-08-12	Per Filament?	5d
☐	2015-08-14	45P 1969	6f

Table 11 (*cont.*)

	Date	Stream	Table
☐	2016-01-03	Qua annual	Ch. 20
☐	2016-01-03	Qua impact Moon	10c
☐	**2016-08-12**	**Per 1079 (!)**	**5c**
☐	2016-08-12	Per Filament	5d
☐	**2016-08-12**	**Per impact Moon**	**10c**
☐	2016-08-12	Per annual	Ch. 17
☐	2016-12-20	aLy 1-rev	3
☐	2016-12-22	Urs 1076	5a
☐	**2016-12-22**	**Urs Filament**	**5b**
☐	2017-01-03	Qua annual	Ch. 20
☐	2017-01-03	Qua impact Moon	10c
☐	2017-02-11	45P at Earth	Ch. 8
☐	2017-02-24	41P at Earth	Ch. 8
☐	2017-04-22	Lyrids annual	Ch. 26
☐	**2017-05-31**	**tHe 1941**	**6g**
☐	**2017-08-12**	**Per Filament**	**5d**
☐	2017-12-16	Phaethon flyby Earth	Ch. 9
☐	**2017-12-22**	**Urs Filament**	**5b**
☐	2018-03-31	C/1907 G_1 1rev	3
☐	**2018-08-12**	**Per Filament**	**5d**
☐	2018-08-12	Per annual	Ch. 17
☐	2018-12-16	46P/Wirtanen flyby	Ch. 8
☐	**2018-12-22**	**Urs Filament**	**5b**
☐	2019-01-03	Qua annual	Ch. 20
☐	2019-03-31	C/1907 G_1 1-rev	3
☐	2019-08-17	bHy 1-rev	3
☐	2019-08-13	Per Filament	5d
☐	2019-11-22	aMo 1-rev	3
☐	2019-12-22	Urs Filament	5b
☐	2020-01-03	Qua annual	Ch. 20
☐	2020-01-03	Qua impact Moon	10c
☐	2020-08-12	Per Filament	5d
☐	**2020-08-16**	**bHy 1-rev**	**3**
☐	2020-08-12	Per impact Moon	10c
☐	**2020-10-07**	**Dra 1704**	**6d**
☐	**2020-10-07**	**Dra impact Moon**	**10c**
☐	2020-12-13	Gem annual	Ch. 21
☐	**2020-12-22**	**Urs 829**	**5a**
☐	2020-12-22	Urs Filament	5b
☐	2021-01-03	Qua annual	Ch. 20
☐	2021-03-01	C/1976 D_1 1-rev	3
☐	2021-08-12	Per annual	Ch. 17

Table 11 (*cont.*)

	Date	Stream	Table
☐	**2021-08-12**	**C/1852 K_1 1-rev**	**3**
☐	2021-08-12	Per impact Moon	10c
☐	2021-09-27	15P/Finlay 1988	6c
☐	2021-09-28	15P/Finlay 1995	6c
☐	2022-01-03	Qua annual	Ch. 20
☐	**2022-05-31**	**tHe 1995 (!)**	**6g**
☐	2022-05-31	tHe impact Moon	10c
☐	2022-08-12	C/1852 K_1 1-rev	3
☐	2023-08-12	Per annual	Ch. 17
☐	2023-11-03	Tau impact Moon	10c
☐	2023-12-13	Gem annual	Ch. 21
☐	2024-01-03	Qua impact Moon	10c
☐	2024-08-12	Per impact Moon	10c
☐	2024-08-12	Per annual	Ch. 17
☐	**2024-10-08**	**Dra 1859**	**6d**
☐	2025-01-03	Qua annual	Ch. 20
☐	2025-01-03	Qua impact Moon	10c
☐	2025-08-12	Per Filament	5d
☐	2025-10-08	Dra 2012	6d
☐	2025-12-13	Gem annual	Ch. 21
☐	2026-08-12	Per annual	Ch. 17
☐	2026-08-13	Per Filament	5d
☐	2027-01-03	Qua impact Moon	10c
☐	**2027-06-11**	**gDe 1-rev**	**3**
☐	**2027-08-12**	**Per 1079 (!)**	**5c**
☐	2027-08-13	Per Filament	5d
☐	2027-10-22	15P/Finlay 1960	6c
☐	2027-10-23	15P/Finlay 1953	6c
☐	2028-01-03	Qua annual	Ch. 20
☐	2028-01-03	Qua impact Moon	10c
☐	**2028-06-17**	**JBo 1796**	**6b**
☐	**2028-08-12**	**Per 1479 (!)**	**5c**
☐	2028-08-12	Per Filament	5d
☐	2028-12-13	Gem annual	Ch. 21
☐	**2028-12-22**	**Urs 1392**	**5a**
☐	2029-04-13	2004 MN_4 ($+3^m$) flyby	Ch. 10
☐	2029-04-21	Pup 1897	6e
☐	2029 08-07	bPe 1-rev	3
☐	2029-08-12	Per Filament	5d
☐	2029-08-12	Per impact Moon	10c
☐	2030-01-03	Qua annual	Ch. 20
☐	2030-07-17	kPa 1-rev	3

Table 11 (*cont.*)

	Date	Stream	Table
☐	2030-08-13	Per Filament	5d
☐	2030-10-09	Dra 1628	6d
☐	**2030-12-22**	**Urs 1090**	**5a**
☐	**2030-12-22**	**Urs Filament**	**5b**
☐	2031-01-03	Qua impact Moon	10c
☐	2031-04-22	Lyrids annual	Ch. 26
☐	2031-07-17	kPa 1-rev	3
☐	2031-08-12	Per impact Moon	10c
☐	2031-08-13	Per Filament	5d
☐	2031-12-13	Gem annual	Ch. 21
☐	**2031-12-22**	**Urs Filament**	**5b**
☐	2032-01-03	Qua impact Moon	10c
☐	2032 03-01	C/1976 D_1 1rev	3
☐	2032-08-12	Per impact Moon	10c
☐	2032-08-12	Per Filament	5c
☐	2032-08-12	Per annual	Ch. 17
☐	2032-12-22	Urs 952	5a
☐	**2032-12-22**	**Urs Filament**	**5b**
☐	2033-01-03	Qua impact Moon	10c
☐	2033-01-03	Qua annual	Ch. 20
☐	**2033-05-06**	**tHe 1914**	**6g**
☐	**2033-06-05**	**aCi 1-rev**	**3**
☐	2033-11-17	Leo Filament	4b
☐	**2033-11-17**	**Leo 1899**	**4a**
☐	2033-12-22	Urs Filament	5b
☐	2034-04-22	Lyrids annual	Ch. 26
☐	**2034-04-22**	**Pup 1972**	**6e**
☐	2034-06-11	gDe 1-rev	3
☐	2034-10-04	15P/Finlay 1995	6c
☐	**2034-11-18**	**Leo 1932 (!)**	**4a**
☐	2034-11-18	Leo 1767	4a
☐	**2034-11-18**	**Leo impact Moon**	**10c**
☐	**2034-11-19**	**Leo Filament**	**4b**
☐	2034-11-19	Leo 1866	4a
☐	2034-11-19	Leo 1699	4a
☐	2034-12-22	Urs 815	5a
☐	**2034-12-22**	**Urs Filament**	**5b**
☐	2035-01-03	Qua impact Moon	10c
☐	2035-01-04	Qua annual	Ch. 20
☐	2035-02-15	C/1854 R_1 1rev	3
☐	2035-08-12	Per impact Moon	10c
☐	2035-08-12	Per annual	Ch. 17

Table 11 (*cont.*)

	Date	Stream	Table
☐	2035-10-05	Dra 1887	6d
☐	2035-11-11	Leo 1633	4a
☐	2035-11-19	Leo 1366	4a
☐	**2035-11-20**	**Leo 1699**	**4a**
☐	2035-11-20	Leo Filament	4b
☐	2035-11-27	NoO 1-rev	3
☐	2036-01-04	Qua annual	Ch. 20
☐	2036-01-03	Qua impact Moon	10c
☐	2036-11-17	Leo 1965	4a
☐	2036-11-19	Leo Filament	4b
☐	2036-12-22	Urs Filament	5b
☐	2037-06-04	aCi 1-rev	3
☐	2037-08-12	Per annual	Ch. 17
☐	2037-10-10	Dra 1609	6d
☐	2037-10-10	Dra impact Moon	10c
☐	**2037-11-17**	**Leo 1965**	**4a**
☐	**2037-11-19**	**Leo 1800**	**4a**
☐	**2037-11-19**	**Leo impact Moon**	**4a**
☐	2037-11-19	Leo Filament	4b
☐	**2037-11-20**	**Leo 1833**	**4a**
☐	**2037-11-20**	**Leo impact Moon**	**4a**
☐	2038-01-03	Qua annual	Ch. 20
☐	2038-03-01	C/1976 D_1 1rev	3
☐	2038-03-06	aPx 1-rev	3
☐	2038-08-12	Per 1862	5c
☐	**2038-09-11**	**eEr 1-rev**	**3**
☐	2038-11-20	Leo 1767	4a
☐	2038-11-20	Leo Filament	4b
☐	2038-11-21	Leo 1866	4a
☐	2039-01-03	Qua impact Moon	10c
☐	**2039-02-08**	**aCe 1-rev**	**3**
☐	**2039-03-07**	**aPx 1-rev**	**3**
☐	**2039-08-12**	**Per impact Moon**	**10c**
☐	**2039-08-13**	**Per 1479**	**5c**
☐	**2039-09-11**	**eEr 1-rev**	**3**
☐	2039-12-13	Gem annual	Ch. 21
☐	**2040-04-22**	**Lyr 1-rev**	**3**
☐	2040-08-12	Per Filament	5d
☐	**2040-08-12**	**Per impact Moon**	**10c**
☐	2040-08-12	Per annual	Ch. 20
☐	**2040-09-10**	**eEr 1-rev**	**3**

Table 11 (*cont.*)

	Date	Stream	Table
☐	2040-11-18	Leo 1366	4a
☐	2041-01-03	Qua annual	Ch. 20
☐	**2041-04-22**	**Lyr 1-rev**	**3**
☐	2041-08-12	Per Filament	5d
☐	2041-09-26	15P/Finlay 1995	6c
☐	2041-12-22	Urs Filament	5b
☐	2042-03-31	C/1907 G_1 1rev	3
☐	2042-08-12	Per Filament	5d
☐	2042-08-12	Per annual	Ch. 17
☐	2042-10-08	Dra 1655	6d
☐	2042-10-30	C/1991 L_3 return	Ch. 18
☐	2042-11-18	Leo Filament	4b
☐	2042-12-13	Gem annual	Ch. 21
☐	2042-12-23	Urs 1378	5a
☐	2043-01-03	Qua impact Moon	10c
☐	**2043-03-31**	**C/1907 G_1 1rev**	**3**
☐	2043-08-12	Per Filament	5d
☐	2043-08-12	Per impact Moon	10c
☐	2043-08-12	Per annual	Ch. 17
☐	2043-11-17	Leo 1366	4a
☐	2043-11-18	Leo 1533	4a
☐	2043-11-19	Leo 1400	4a
☐	2043-11-19	Leo Filament	4b
☐	**2043-11-22**	**aMo 1-rev**	**3**
☐	**2043-12-23**	**Urs 1117**	**5a**
☐	2044-01-03	Qua annual	Ch. 20
☐	2044-01-03	Qua impact Moon	10c
☐	2044-03-31	C/1907 G_1 1rev	3
☐	2044-08-12	Per Filament	5d
☐	2044-10-06	Dra 1900	6d
☐	2044-12-13	Gem annual	Ch. 21
☐	**2044-12-22**	**Urs 1090**	**5a**
☐	**2044-12-22**	**Urs Filament**	**5b**
☐	2045-03-31	C/1907 G_1 1rev	3
☐	2045-08-12	Per annual	Ch. 17
☐	**2045-12-22**	**Urs 843**	**5a**
☐	**2045-12-22**	**Urs Filament**	**5b**
☐	2046-01-03	Qua impact Moon	10c
☐	2046-12-22	Urs Filament	5b
☐	2047-01-03	Qua annual	Ch. 20
☐	2047-01-03	Qua impact Moon	10c

Table 11 (*cont.*)

	Date	Stream	Table
☐	2047-11-21	Leo 1567	4
☐	2047-12-13	Gem annual	Ch. 21
☐	**2047-12-22**	**Urs 925**	**5a**
☐	2047-12-22	Urs Filament	5b
☐	2048-03-31	C/1907 G_1 1rev	3
☐	2048-06-11	gDe 1-rev	3
☐	2048-08-12	Per impact Moon	10c
☐	2048-11-21	Leo 1567	4a
☐	2048-12-22	Urs Filament	5b
☐	2049-01-03	Qua annual	Ch. 20
☐	**2049-04-29**	**tHe 1968**	**6g**
☐	2049-06-11	gDe 1-rev	3
☐	2049-11-18	Leo 1833	4a
☐	**2049-12-22**	**Urs Filament**	**5b**
☐	2050-01-03	Qua impact Moon	10c
☐	2050 03-01	C/1976 D_1 1rev	3
☐	2050-08-12	Per impact Moon	10c
☐	2050-11-18	Leo Filament	4b
☐	2050-12-13	Gem annual	Ch. 21
☐	2051-01-03	Qua impact Moon	10c
☐	2051-08-12	Per annual	Ch. 21
☐	2051-08-12	Per impact Moon	10c
☐	2052-01-03	Qua annual	Ch. 20
☐	2052-01-03	Qua impact Moon	10c
☐	**2052-08-12**	**Per Filament**	**5d**
☐	2052-11-17	Leo 1866	4a

Index

Units and constants

$\pi = 3.141\,592\,653\,589\,793$
$e = 2.718\,281\,828\,459\,045$
$G = 6.672\,59 \times 10^{-11}$ m^3 kg^{-1} s^{-2}
1 mile = 1.609 3440 km
1 AU = $1.495\,978\,7066 \times 10^{11}$ m
1 Parsec = $3.08568025 \times 10^{16}$ m
1 mile/h = 0.447 04 m/s
1 ton of TNT = 0.907 184 748 99 tonne (metric) of TNT = $4.183\,999\,932 \times 10^9$ J

Earth, Sun and Jupiter (Epoch = J2000 = 2000 January 1.5)
Earth:
Earth Mass = 5.9742×10^{24} kg
Earth mean Radius = 6371.01 meter
Earth side real orbit period = 1.0000174 yr
Obliquity of the ecliptic = 23°.43929
General precession in longitude = 5029″.0966 per Julian century
Earth orbit eccentricity = 0.01671022
Earth orbit node = −11°.26064
Earth orbit angle of perihelion = 102°.94719
Earth orbit semimajor axis = 1.000 001 0178 AU
Sun:
Sun mass = 1.9891×10^{30} kg
Sun radius (1/2 diameter) = 695 000 000 m
Jupiter:
Jupiter Mass = 1898.86×10^{24} kg
Jupiter orbit semimajor axis = 5.20336301 AU
Jupiter orbit inclination = 1°.30530
Jupiter orbit eccentricity = 0.04839266
Jupiter orbit mean longitude of node = 100°.55615
Jupiter orbit angle of perihelion = 14°.75385
Jupiter sidereal orbit period = 11.862615 yr

Greek alphabet (used in the names of stars and meteor showers)

α	A	alpha	ι	I	iota	ρ	P	rho
β	B	beta	κ	K	kappa	σ	Σ	sigma
γ	Γ	gamma	λ	Λ	lambda	τ	T	tau
δ	Δ	delta	μ	M	mu	υ	Y	upsilon
ε	E	epsilon	ν	N	nu	ϕ	Φ	phi
ζ	Z	zeta	ξ	Ξ	xi	χ	X	chi
η	H	eta	o	O	omicron	ψ	Ψ	psi
θ	Θ	theta	$\pi(\varpi)$	Π	pi	ω	Ω	omega